Handbook of Combinatorial Optimization

VOLUME 1

Handbook of Combinatorial Optimization

Volume 1

Edited by

Ding-Zhu Du
*University of Minnesota,
Minneapolis, U.S.A.*

and

Panos M. Pardalos
*University of Florida,
Gainesville, U.S.A.*

KLUWER ACADEMIC PUBLISHERS
BOSTON / DORDRECHT / LONDON

A C.I.P. Catalogue record for this book is available from the Library of Congress.

ISBN 0-7923-5018-9 (HB)
ISBN 0-7923-5019-7 (Set)

Published by Kluwer Academic Publishers,
P.O. Box, 3300 AA Dordrecht, The Netherlands.

Sold and distributed in North, Central and South America
by Kluwer Academic Publishers,
101 Philip Drive, Norwell, MA 02061, U.S.A.

In all other countries, sold and distributed
by Kluwer Academic Publishers,
P.O. Box 322, 3300 AH Dordrecht, The Netherlands.

Printed on acid-free paper

All Rights Reserved
©1998 Kluwer Academic Publishers
No part of the material protected by this copyright notice may be reproduced or
utilized in any form or by any means, electronic or mechanical,
including photocopying, recording or by any information storage and
retrieval system, without written permission from the copyright owner

Printed in the Netherlands.

Contents

Preface vii

Mixed-Integer Nonlinear Optimization in Process Synthesis 1
    C. S. Adjiman, C. A. Schweiger, and C. A. Floudas

Approximate Algorithms and Heuristics for *MAX-SAT* 77
    R. Battiti and M. Protasi

**Connections between Nonlinear Programming
and Discrete Optimization** 149
    F. Giannessi and F. Tardella

Interior Point Methods for Combinatorial Optimization 189
    J.E. Mitchell, P.M. Pardalos, and M.G.C. Resende

Knapsack Problems 299
    D. Pisinger and P. Toth

Fractional Combinatorial Optimization 429
    T. Radzik

**Reformulation-Linearization Techniques
for Discrete Optimization Problems** 479
    H.D. Sherali and W.P. Adams

Gröbner Bases in Integer Programming 533
    R.R. Thomas

**Applications of Set Covering, Set Packing
and Set Partitioning Models: A Survey** 573
    R.R. Vemuganti

Author Index 747

Subject Index 773

Preface

Combinatorial (or discrete) optimization is one of the most active fields in the interface of operations research, computer science, and applied mathematics. Combinatorial optimization problems arise in various applications, including communications network design, VLSI design, machine vision, airline crew scheduling, corporate planning, computer-aided design and manufacturing, database query design, cellular telephone frequency assignment, constraint directed reasoning, and computational biology. Furthermore, combinatorial optimization problems occur in many diverse areas such as linear and integer programming, graph theory, artificial intelligence, and number theory. All these problems, when formulated mathematically as the minimization or maximization of a certain function defined on some domain, have a commonality of discreteness.

Historically, combinatorial optimization starts with linear programming. Linear programming has an entire range of important applications including production planning and distribution, personnel assignment, finance, allocation of economic resources, circuit simulation, and control systems. Leonid Kantorovich and Tjalling Koopmans received the Nobel Prize (1975) for their work on the optimal allocation of resources. Two important discoveries, the ellipsoid method (1979) and interior point approaches (1984) both provide polynomial time algorithms for linear programming. These algorithms have had a profound effect in combinatorial optimization. Many polynomial-time solvable combinatorial optimization problems are special cases of linear programming (e.g. matching and maximum flow). In addition, linear programming relaxations are often the basis for many approximation algorithms for solving NP-hard problems (e.g. dual heuristics).

Two other developments with a great effect on combinatorial optimization are the design of efficient integer programming software and the availability of parallel computers. In the last decade, the use of integer programming models has changed and increased dramatically. Two decades ago, only problems with up to 100 integer variables could be solved in a computer. Today we can solve problems to optimality with thousands of integer variables. Furthermore, we can compute provably good approximate solutions to problems with millions of integer variables. These advances have been made possible by developments in hardware, software and algorithm design.

The Handbooks of Combinatorial Optimization deal with several algorithmic approaches for discrete problems as well as with many combinatorial problems. We have tried to bring together almost every aspect of this enormous field with emphasis on recent developments. Each chapter in the Handbooks is essentially expository in nature, but of scholarly treatment.

The Handbooks of Combinatorial Optimization are addressed not only to researchers in discrete optimization, but to all scientists in various disciplines who use combinatorial optimization methods to model and solve problems. We are certain that experts in the field as well as nonspecialist readers will find the material of the Handbooks stimulating and helpful.

We would like to take this opportunity to thank the authors, the anonymous referees, and the publisher for helping us produce these volumes of the Handbooks of Combinatorial Optimization with state-of-the-art chapters. We would also like to thank Ms. Xiuzhen Cheng for making Author Index and Subject Index for this volume.

Ding-Zhu Du and Panos M. Pardalos

HANDBOOK OF COMBINATORIAL OPTIMIZATION (VOL. 1)
D.-Z. Du and P.M. Pardalos (Eds.) pp. 1-76
©1998 Kluwer Academic Publishers

Mixed-Integer Nonlinear Optimization in Process Synthesis

C. S. Adjiman, C. A. Schweiger, and C. A. Floudas[†]
Department of Chemical Engineering
Princeton University, Princeton, NJ 08544-5263
E-mail: claire@titan.princeton.edu, carl@titan.princeton.edu,
floudas@titan.princeton.edu

[†]Author to whom all correspondence should be addressed.

Contents

1	**Introduction**	**3**
2	**Optimization Approach in Process Synthesis**	**5**
3	**Algorithms for Convex MINLPs**	**8**
	3.1 Generalized Benders Decomposition	9
	3.1.1 Primal Problem	10
	3.1.2 Master Problem	11
	3.1.3 **GBD** Algorithm	13
	3.2 Outer Approximation	14
	3.2.1 Outer Approximation Primal Problem	15
	3.2.2 Outer Approximation Master Problem	15
	3.2.3 **OA** Algorithm	16
	3.3 Outer Approximation/Equality Relaxation	18
	3.3.1 **OA/ER** Algorithm	18
	3.4 Outer Approximation/Equality Relaxation/Augmented Penalty	20
	3.4.1 **OA/ER/AP** Algorithm	21
	3.5 Generalized Outer Approximation	23
	3.5.1 Primal Problem	23
	3.5.2 Master Problem	23
	3.5.3 **GOA** Algorithm	25
	3.6 Generalized Cross Decomposition	26

		3.6.1 **GCD** Algorithm	27
	3.7	Branch-and-Bound Algorithms	30
		3.7.1 Selection of the Branching Node	32
		3.7.2 Selection of the Branching Variable	32
		3.7.3 Generation of a Lower Bound	33
		3.7.4 Algorithmic Procedure	33
	3.8	Extended Cutting Plane (ECP)	34
		3.8.1 Algorithmic Procedure	36
	3.9	Feasibility Approach	36
	3.10	Logic Based Approach	37
		3.10.1 Logic-Based Outer Approximation	38
		3.10.2 Logic-Based Generalized Benders Decomposition	40
4	**Global Optimization for Nonconvex MINLPs**		**40**
	4.1	Branch-and-reduce algorithm	41
	4.2	Interval Analysis Based Algorithm	42
		4.2.1 Node Fathoming Tests for Interval Algorithm	42
		4.2.2 Branching Step	43
	4.3	Extended Cutting Plane for Nonconvex MINLPs	44
	4.4	Reformulation/Spatial Branch-and-Bound Algorithm	45
	4.5	The SMIN-αBB Algorithm	46
		4.5.1 Convex Underestimating MINLP Generation	47
		4.5.2 Branching Variable Selection	47
		4.5.3 Variable Bound Updates	48
		4.5.4 Algorithmic Procedure	49
	4.6	The GMIN-αBB algorithm	50
5	**Implementation: MINOPT**		**51**
6	**Computational Studies**		**56**
	6.1	Distillation Sequencing	56
	6.2	Heat Exchanger Network Synthesis	60
		6.2.1 Solution Strategy with the SMIN-αBB algorithm	66
		6.2.2 Specialized Algorithm for Heat Exchanger Network Problems	70
7	**Conclusions**		**70**
8	**Acknowledgments**		**71**
	References		

Mixed-Integer Nonlinear Optimization in Process Synthesis

Abstract

The use of networks allows the representation of a variety of important engineering problems. The treatment of a particular class of network applications, the *process synthesis problem*, is exposed in this paper. Process Synthesis seeks to develop systematically process flowsheets that convert raw materials into desired products. In recent years, the optimization approach to process synthesis has shown promise in tackling this challenge. It requires the development of a network of interconnected units, the process superstructure, that represents the alternative process flowsheets. The mathematical modeling of the superstructure has a mixed set of binary and continuous variables and results in a mixed-integer optimization model. Due to the nonlinearity of chemical models, these problems are generally classified as Mixed-Integer Nonlinear Programming (MINLP) problems.

A number of local optimization algorithms, developed for the solution of this class of problems, are presented in this paper: Generalized Benders Decomposition (GBD), Outer Approximation (OA), Generalized Cross Decomposition (GCD), Branch and Bound (BB), Extended Cutting Plane (ECP), and Feasibility Approach (FA). Some recent developments for the global optimization of nonconvex MINLPs are then introduced. In particular, two branch-and-bound approaches are discussed: the Special structure Mixed Integer Nonlinear αBB (SMIN-αBB), where the binary variables should participate linearly or in mixed-bilinear terms, and the General structure Mixed Integer Nonlinear αBB (GMIN-αBB), where the continuous relaxation of the binary variables must lead to a twice-differentiable problem. Both algorithms are based on the αBB global optimization algorithm for nonconvex continuous problems.

Once the theoretical issues behind local and global optimization algorithms for MINLPs have been exposed, attention is directed to their algorithmic development and implementation. The framework **MINOPT** is discussed as a computational tool for the solution of process synthesis problems. It is an implementation of a number of local optimization algorithms for the solution of MINLPs. The use of **MINOPT** is illustrated through the solution of a variety of process network problems. The synthesis problem for a heat exchanger network is then presented to demonstrate the global optimization SMIN-αBB algorithm.

1 Introduction

Network applications exist in many fields including engineering, applied mathematics, and operations research. These applications include problems such as facility location and allocation problems, design and scheduling of batch processes, facility planning and scheduling, topology of transportation networks, and process synthesis problems. These types of problems are typically characterized by both discrete and continuous decisions. Thus, the modeling aspects of these applications often lead to models involving both integer and continuous variables as well as nonlinear functions. This gives rise to problems classified as mixed-integer nonlinear optimization problems.

Major advances have been made in the development of mathematical programming approaches which address mixed-integer nonlinear optimization problems. The recent theoretical and algorithmic advances in mixed-integer nonlinear optimization have made the use of these techniques both feasible and practical. Because of this, optimization has become a standard computational approach for the solution of these networking problems.

Some of the major contributions to the development of mixed-integer nonlinear optimization techniques have come from the field of process synthesis. This is due to the natural formulation of the process synthesis problem as a mixed-integer nonlinear optimization problem. This has led to significant algorithmic developments and extensive computational experience in process synthesis applications. The research in this area has focused on the overall process synthesis problem as well as subsystem synthesis problems including heat exchanger network synthesis (HENS), reactor network synthesis, distillation sequencing, and mass exchange network synthesis, as well as total process flowsheets.

The process synthesis problem is stated as follows: given the specifications of the inputs (feed streams) and the specifications of the outputs, develop a process flowsheet which transforms the given inputs to the desired products while addressing the performance criteria of capital and operating costs, product quality, environmental issues, safety, and operability. Three key issues must be addressed in order to determine the process flowsheet: which process units should be in the flowsheet, how the process units should be interconnected, and what the operating conditions and sizes of the process units should be. The optimization approach to process synthesis has been developed to address these issues and has led to some of the major theoretical and algorithmic advances in mixed-integer nonlinear optimization.

The next section describes the optimization approach to process synthesis which leads to the formulation of a Mixed-Integer Nonlinear Program. In Section 3, the optimization algorithms developed for the solution of the posed optimization problem are presented. Although these methods have been developed for process synthesis, they are applicable to models that result in other network applications. Section 4 reports some recent developments for the global optimization of nonconvex MINLPs. Section 5 describes the algorithmic framework, **MINOPT**, which implements a number of the described algorithms. The final part of the paper describes the application of both global and local methods to a heat exchanger network synthesis problem.

2 Optimization Approach in Process Synthesis

A major advance in process synthesis has been the development of the optimization approach to the process synthesis problem. This approach leads to a mathematical programming problem classified as a Mixed Integer Nonlinear Program. Significant progress has been made in the development of algorithms capable of addressing this class of problems.

The optimization approach to process synthesis involves three steps: the representation of alternatives through a process superstructure, the mathematical modeling of the superstructure, and the development of an algorithm for the solution of the mathematical model. Each of these steps is crucial to the determination of the optimal process flowsheet.

The superstructure is a superset of all process design alternatives of interest. The representation of process alternatives is conceptually based on elementary graph theory ideas. Nodes are used to represent the inputs, outputs, and each unit in the superstructure. One-way arcs represent connections from inputs to process units, two-way arcs represent interconnections between process units, and one-way arcs represent connections to the outputs. The result is a bi-partite planar graph which represents the network of process units in the superstructure. This network represents all the options of the superstructure and includes cases where nodes in the graph may or may not be present. The idea of the process superstructure can be illustrated by a process which has one input, two outputs, and potentially three process units. The network representation of this is shown in Figure 1.

Since all the possible candidates for the optimal process flowsheet are embedded within this superstructure, the optimal process flowsheet that can be determined is only be as good as the postulated representation of alternatives. This superstructure must be rich enough to allow for a complete set of alternatives, but it must also be concise enough to eliminate undesirable structures.

Another example of a superstructure is illustrated by the two compo-

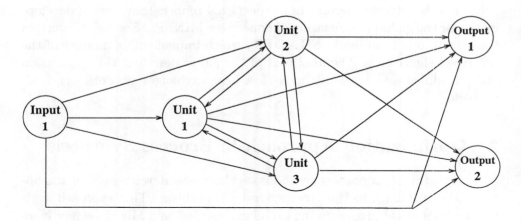

Figure 1: Network representation of superstructure

nent distillation scheme presented by [KG89]. This process consists of two feed streams of known composition and flowrate and two products streams with specified purities. The superstructure consists of a flash unit and a distillation unit and is shown in Figure 2.

Through the process synthesis, the structure flowsheet and the optimal values of the operating parameters are determined. The existence of process units leads to discrete decisions while the determination of operating parameters leads to continuous decisions. Thus, the process synthesis problem is mathematically classified as mixed discrete-continuous optimization.

The next step involves the mathematical modeling of the superstructure. Binary variables are used to indicate the existence of nodes within the network and continuous variables represent the levels of values along the arcs. The resulting formulation is a Mixed Integer Nonlinear Programming Problem (MINLP):

$$\begin{aligned} \min_{x,y} \quad & f(x,y) \\ \text{s.t.} \quad & h(x,y) = 0 \\ & g(x,y) \leq 0 \\ & x \in X \subseteq \mathbb{R}^n \\ & y \in Y \quad \text{integer} \end{aligned} \qquad (1)$$

where

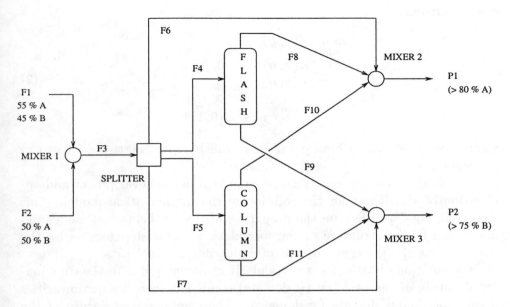

Figure 2: A Two-Column Distillation Sequence Superstructure

- x is a vector of n continuous variables representing flow rates, compositions, temperatures, and pressures of process streams and sizing of process units.

- y is a vector of integer variables representing process alternatives.

- $f(x, y)$ is the single objective function representing the performance criterion.

- $h(x, y) = 0$ are the m equality constraints that represent the mass and energy balances, and equilibrium expressions.

- $g(x, y) \leq 0$ are the p inequality constraints that represent design specifications, restrictions, and logical constraints.

This formulation is completely general and includes cases where nonlinearities occur in the x space, y space, and joint $x - y$ space.

The integer variables can be expressed as binary variables without loss of generality. Through an appropriate transformation, the general formulation can be written as

$$\begin{aligned}
\min_{x,y} \quad & f(x,y) \\
\text{s.t.} \quad & h(x,y) = 0 \\
& g(x,y) \leq 0 \\
& x \in X \subseteq \mathbb{R}^n \\
& y \in \{0,1\}^q
\end{aligned} \qquad (2)$$

where the y are the q binary variables which represent the existence of process units.

The final step of the optimization approach is the development and application of algorithms for the solution of the mathematical model. This step is highly dependent on the properties of the mathematical model and makes use of the structure of the formulation. This step focuses on the development of algorithms capable of addressing the MINLPs.

The solution of MINLPs is particularly challenging due to the combinatorial nature of the problem (y domain) combined with the nonlinearities in the continuous domain (x domain). The combinatorial nature of the problem becomes an issue as the number of y variables increases creating a large number of possible process structures. In the continuous domain, the models of chemical processes are generally nonlinear. The nonlinearities in the problem imply the possible existence of multiple solutions and lead to challenges in finding the global solution.

Despite the challenges involved in the solution of the MINLPs, there have been significant advances in the area of MINLPs on the theoretical, algorithmic and computational fronts. Many algorithms have been developed to address problems with the above form and a review of these developments is presented in the next section.

3 Algorithms for Convex MINLPs

A number of algorithms have been developed to address problems with the above form 2. Some deal with the formulation as stated, while others deal with a restricted class of the problem. The following is a chronological listing of these algorithms.

1. Generalized Benders Decomposition, **GBD** [Geo72, PF89, FAC89]

2. Branch and Bound, **BB** [Bea77, Gup80, OOM90, BM91]

3. Outer Approximation, **OA** [DG86]

4. Feasibility Approach, **FA** [MM86]

5. Outer Approximation with Equality Relaxation, **OA/ER** [KG87]

6. Outer Approximation with Equality Relaxation and Augmented Penalty, **OA/ER/AP** [VG90]

7. Generalized Outer Approximation, **GOA** [FL94]

8. Generalized Cross Decomposition, **GCD** [Hol90];

An overview of these MINLP algorithms and extensive theoretical, algorithmic, and applications-oriented descriptions of **GBD**, **OA**, **OA/ER**, **OA/ER/AP**, **GOA**, and **GCD** algorithms is found in [Flo95].

Some of these algorithms are applicable only to restricted classes of the general problem formulation. The general strategy of algorithms used to solve MINLPs is to formulate subproblems such that the subproblems are easier to solve than the original problem. This may involve fixing certain variable types, relaxing certain constraints, using duality, or using linearization. The algorithms iterate through solutions of the subproblems which provide upper and lower bounds on the optimal solution of the original problem. The nature of the subproblems and the quality of bounds provided by the subproblems are different for the various algorithms.

3.1 Generalized Benders Decomposition

The work of [Geo72] generalized the work of [Ben62] which exploits the structure of mathematical programming problems. The algorithm addresses problems with the form of problem 2. In fact, the algorithm is applicable to a broader class of problems for which the y variables may be continuous. The focus here is on MINLP models and thus the y variables will be treated as binary.

The basic idea behind **GBD** is the generation of upper and lower bounds on the solution of the MINLP model through the iterative solution subproblems formulated from the original problem. The upper bound is the result of the solution of the *primal* problem while the lower bound is the result of the solution of the *master* problem. The primal problem corresponds to the solution of the original problem 2 with the values of the y variables fixed. This problem is solved in the x space only and its solution provides information about the Lagrange multipliers for the constraints. The master problem is formulated by making use of the Lagrange multipliers and nonlinear duality theory. Its solution provides a lower bound as well as a new set of y variables. The algorithm iterates between the primal and master problems generating a sequence of upper and lower bounds which converge in a finite number of iterations.

3.1.1 Primal Problem

The primal problem results from fixing the values of the y variables. For values of y fixed to y^k where k is an iteration counter, the primal problem has the following formulation:

$$\begin{aligned}
\min_{x} \quad & f(x, y^k) \\
\text{s.t.} \quad & h(x, y^k) = 0 \\
& g(x, y^k) \leq 0 \\
& x \in X \subseteq R^n
\end{aligned} \quad (3)$$

The primal formulation is an NLP which can be solved by using existing algorithms. If the primal problem is feasible, then the optimal solution provides values for x^k, $f(x^k, y^k)$, and the Lagrange multipliers λ^k and μ^k for the equality and inequality constraints.

If the primal problem is found to be infeasible when applying a solution algorithm, a feasibility problem is formulated. This problem can be formulated by minimizing the ℓ_1 or ℓ_∞ sum of constraint violations. One possible formulation of the feasibility problem is the following:

$$\begin{aligned}
\min_{x,\alpha} \quad & \alpha_i + \alpha_e^+ + \alpha_e^- \\
\text{s.t.} \quad & g(x, y^k) - \alpha_i \leq 0 \\
& h(x, y^k) + \alpha_e^+ - \alpha_e^- = 0 \\
& x \in X \subseteq R^n \\
& \alpha_i, \alpha_e^+, \alpha_e^- \geq 0
\end{aligned} \quad (4)$$

Another possible form for the infeasible primal problem is the following where the equality constraints are not relaxed:

$$\begin{aligned}
\min_{x,\alpha} \quad & \alpha \\
\text{s.t.} \quad & g(x, y^k) \leq \alpha \\
& h(x, y^k) = 0 \\
& x \in X \subseteq R^n \\
& \alpha \geq 0
\end{aligned} \tag{5}$$

The solution of the feasibility problem provides values for $\bar{x}^k$ and the Lagrange multipliers $\bar{\lambda}^k$ and $\bar{\mu}^k$ for the equality and inequality constraints.

3.1.2 Master Problem

The formulation of the master problem for **GBD** makes use of nonlinear duality theory. The key aspects of the master problem formulation are the projection of the problem onto the y space and the dual representation.

For the projection of the problem onto the y space, problem 2 can be written as

$$\begin{aligned}
\min_{y} \inf_{x} \quad & f(x, y) \\
\text{s.t.} \quad & h(x, y) = 0 \\
& g(x, y) \leq 0 \\
& x \in X \subseteq R^n \\
& y \in Y \in \{0, 1\}^q
\end{aligned} \tag{6}$$

Let $v(y)$ and V be defined as follows:

$$\begin{aligned}
v(y) = \inf_{x} \quad & f(x, y) \\
\text{s.t.} \quad & h(x, y) = 0 \\
& g(x, y) \leq 0 \\
& x \in X \subseteq R^n
\end{aligned} \tag{7}$$

$$V = \{y : h(x, y) = 0, g(x, y) \leq 0 \text{ for some } x \in X \subseteq R^n\} \tag{8}$$

The projected problem can now be written as:

$$\begin{aligned}
\min_{y} \quad & v(y) \\
\text{s.t.} \quad & y \in Y \cap V
\end{aligned} \tag{9}$$

The difficulty in solving this problem is that V and $v(y)$ are known only implicitly. In order to overcome this, dual representations of V and $v(y)$ are used.

The dual representation of V is described in terms of a collection of regions that contain it. An element of Y also belongs to the set V if and only if it satisfies the system:

$$0 \geq \inf \bar{L}(x, y, \bar{\lambda}, \bar{\mu}), \quad \forall \bar{\lambda}, \bar{\mu} \in \Lambda$$
$$\text{where} \quad \Lambda = \left\{ \bar{\lambda} \in \mathrm{R}^m, \bar{\mu} \in \mathrm{R}^p : \bar{\mu} \geq 0, \sum_{i=1}^{p} \mu_i = 1 \right\} \quad (10)$$

This system corresponds to the set of constraints that have to be incorporated for the case of infeasible primal problems.

The dual representation of $v(y)$ is the pointwise infimum of a collection of functions that support it.

$$v(y) = \begin{bmatrix} \min_{x} f(x,y) \\ \text{s.t.} \quad h(x,y) = 0 \\ \quad\quad g(x,y) \leq 0 \\ \quad\quad x \in X \subseteq \mathrm{R}^n \end{bmatrix}$$
$$= \left[\sup_{\lambda, \mu \geq 0} \min_{x \in X} L(x, y, \lambda, \mu) \right] \quad \forall y \in Y \cap V \quad (11)$$

where $L(x, y, \lambda, \mu) = f(x, y) + \lambda^T h(x, y) + \mu^T g(x, y)$.

Now, the representation for V (10) and the representation for $v(y)$ (11) are substituted into problem 9 and the scalar μ_b is introduced to obtain the following master problem:

$$\min_{y \in Y, \mu_B} \mu_B$$
$$\text{s.t.} \quad \mu_B \geq \min_{x \in X} L(x, y, \lambda, \mu) \quad \forall \lambda, \forall \mu \geq 0 \quad (12)$$
$$\quad\quad 0 \geq \min_{x \in X} \bar{L}(x, y, \bar{\lambda}, \bar{\mu}) \quad \forall (\bar{\lambda}, \bar{\mu}) \in \Lambda$$

where $\quad L(x, y, \lambda, \mu) = f(x, y) + \lambda^T h(x, y) + \mu^T g(x, y)$
$\quad\quad\quad\quad \bar{L}(x, y, \bar{\lambda}, \bar{\mu}) = \bar{\lambda}^T h(x, y) + \bar{\mu}^T g(x, y)$ (13)

The key issue in the development of an algorithmic implementation of **GBD** is the solution of the master problem. The master problem consists of an outer optimization with respect to y whose constraints are two optimization problems with respect to x corresponding to the feasible and infeasible primal problems. These inner optimization problems need to be considered for all possible values of the Lagrange multipliers which implies that an infinite number of constraints need to be considered for the master problem.

One way to solve the master problem is to use relaxation of the problem where only a few of the constraints are considered. The inner optimization problems are considered only for fixed values of the multipliers which correspond to the multipliers from the solution of the primal problem. Furthermore, the inner optimization problems can be eliminated by evaluating the Lagrange function for fixed values of the x variables corresponding to the solution of the primal problem. This elimination assumes that the Lagrange function evaluated at the solution to the corresponding primal is a valid underestimator of the inner optimization problem. This is true when the projected problem $v(y)$ is convex in y.

3.1.3 GBD Algorithm

Step 1
   Obtain initial values: y^1
   Set the counter: $k = 1$
   Set the lower bound: $LBD = -\infty$
   Set the upper bound: $UBD = +\infty$
   Initialize the feasibility set: $\mathbf{F} = \emptyset$
   Initialize the infeasibility set: $\bar{\mathbf{F}} = \emptyset$
   Set the convergence tolerance: $\epsilon \geq 0$

Step 2
   Solve the primal problem for the fixed values of $y = y^k$:
   $$\min_{x} \quad f(x, y^k)$$
   $$\text{s.t.} \quad g(x, y^k) \leq 0$$
   $$\quad\quad h(x, y^k) = 0$$
   Obtain the optimal solution, optimal x^k and optimal Lagrange
      multipliers λ^k and μ^k
   If the primal is feasible
      Update the feasibility set: $\mathbf{F} = \mathbf{F} \cup k$
      If the optimal solution of the primal is less than UBD
         update the upper bound (UBD)
   Else

Solve the infeasible primal problem for the fixed values of $y = y^k$:
$$\min_{x,\alpha} \quad \alpha_i + \alpha_e^+ + \alpha_e^-$$
$$\begin{aligned} \text{s.t.} \quad g(x, y^k) - \alpha_i &\leq 0 \\ h(x, y^k) + \alpha_e^+ - \alpha_e^- &= 0 \\ \alpha_i, \alpha_e^+, \alpha_e^- &\geq 0 \end{aligned}$$
Obtain the optimal $\bar{x}^k$ and the Lagrange multipliers $\bar{\lambda}^k$ and $\bar{\mu}^k$
Update the infeasibility set: $\bar{\mathbf{F}} = \bar{\mathbf{F}} \cup k$

Step 3
Solve the relaxed master problem:
$$\min_{y, \mu_b} \quad \mu_b$$
$$\begin{aligned} \text{s.t.} \quad \mu_b &\geq f(x^l, y) + (\lambda^l)^T g(x^l, y) + (\mu^l)^T h(x^l, y) \quad l \in \mathbf{F} \\ 0 &\geq (\bar{\lambda}^l)^T g(\bar{x}^l, y) + (\bar{\mu}^l)^T h(\bar{x}^l, y) \quad l \in \bar{\mathbf{F}} \end{aligned}$$
Obtain optimal y^{k+1} and μ_b
Set the lower bound: $LBD = \mu_b$
If $UBD - LBD \leq \epsilon$
    Terminate
Else
    Update the counter: $k = k + 1$
    Go to step 2

This algorithm can be applied to general MINLP models, however it is only guaranteed to converge to the global solution for problems which meet specific conditions. First X must be a nonempty convex set, the functions f and g must be convex for each fixed $y \in Y$, and the function h must be linear in x for each $y \in Y$.

3.2 Outer Approximation

The basic ideas behind the Outer Approximation methods [DG86] is similar to those for **GBD**. At each iteration, upper and lower bounds on the solution to the MINLP are generated. The upper bound results from the solution of a primal problem which is formulated identically to the primal problem for **GBD**. The lower bound is determined by solving a master problem which is an outer linearization of the problem around the primal solution.

The outer approximation methods deal with a subclass of MINLP problems in which the functions $f(x, y)$, and $g(x, y)$ are linear in the y variables and separable in x and y. The set of y variables is also strictly binary variables. The formulation also does not allow for equality constraints. Thus, any equality constraints must be eliminated either algebraically or numerically in order to apply the **OA** algorithm. This class of MINLPs has the following formulation:

$$\begin{aligned}
\min_{x,y} \quad & f(x) + c^T y \\
\text{s.t.} \quad & g(x) + c^T y \leq 0 \\
& x \in X \subseteq R^n \\
& y \in \{0,1\}^q
\end{aligned} \quad (14)$$

The **OA** algorithm is similar to the **GBD** algorithm in that it iterates between upper and lower bounding primal and master subproblems. The difference is in the formulation of the master problem. The master problem for this method is formed by a projection onto the y space and an outer approximation of the objective function and feasible region.

3.2.1 Outer Approximation Primal Problem

As in **GBD**, the primal problem results from fixing the values of the y variables to y^k:

$$\begin{aligned}
\min_{x} \quad & f(x) + c^T y^k \\
\text{s.t.} \quad & g(x) + B y^k \leq 0 \\
& x \in X \subseteq R^n
\end{aligned} \quad (15)$$

If this problem is feasible, its solution provides an upper bound on the solution of the MINLP model. If the primal is infeasible, a feasibility problem similar to those used in **GBD** is formulated and solved.

3.2.2 Outer Approximation Master Problem

The master problem is formulated by projecting the problem onto the y space and using an outer approximation of the objective function and feasible region. The projected problem can be written as

$$\begin{aligned}
\min_{y} \quad & v(y) \\
\text{s.t.} \quad & y \in Y \cap V
\end{aligned} \quad (16)$$

where

$$\begin{aligned}
v(y) = c^T y + \inf_{x} \quad & f(x) \\
\text{s.t.} \quad & g(x) + B^T y \leq 0 \\
& x \in X \subseteq R^n
\end{aligned} \quad (17)$$

and

$$V = \{y : g(x) + By \leq 0, \text{ for some } x \in X \subseteq \mathrm{R}^n\} \tag{18}$$

The outer approximation of $v(y)$ is performed by linearizing $f(x)$ and $g(x)$ around the solution of the primal problem x^k. Provided that the functions $f(x)$ and $g(x)$ are convex, the linearizations represent valid support functions. Thus, replacing $v(y)$ with its outer approximation and replacing $y \in Y \cap V$ with $y \in V$ along with an integer cut constraint, the following master problem results:

$$\begin{aligned}
\min_{x,y,\mu_{OA}} \quad & c^T y + \mu_{OA} \\
\text{s.t.} \quad & \left. \begin{array}{l} \mu_{OA} \geq f(x^k) + \nabla_x f(x^k)(x - x^k) \\ 0 \geq g(x^k) + \nabla_x g(x^k)(x - x^k) + By \end{array} \right\} \forall k \in F \\
& x \in X \subseteq \mathrm{R}^n \\
& y \in Y \in \{0,1\} \\
& \sum_{i \in \mathrm{B}^k} y_i^k - \sum_{i \in \mathrm{NB}^k} y_i^k \leq |\mathrm{B}^k| - 1, \quad k \in F
\end{aligned} \tag{19}$$

where F is the set of all feasible solutions x^k to the primal problem.

Since the y variables participate linearly and are binary variables, this formulation is an MILP which can be solved by standard branch-and-bound algorithms. This formulation of the master problem requires that all of the feasible solutions to the primal problem be known which implies an exhaustive enumeration of the binary variables. In order to accommodate for this inefficiency, a relaxation is proposed where only linearizations around the currently known feasible points are included in the master problem. Additionally, to ensure that integer combinations which produce infeasible primal problems are also infeasible in the master problem, linearizations about the solution to the feasibility problem are also included in the master problem.

3.2.3 OA Algorithm

Step 1
  Obtain starting point: y^1
  Set the counter: $k = 1$
  Set the lower bound: $LBD = -\infty$
  Set the upper bound: $UBD = +\infty$
  Set the convergence tolerance: $\epsilon \geq 0$

Step 2
Solve the primal problem for the fixed values of $y = y^k$:
$$\min_{x} \quad f(x) + c^T y^k$$
$$\text{s.t.} \quad g(x) \leq -B y^k$$
Obtain the optimal x^k
If the primal is feasible
    If the optimal solution of the primal is less than UBD
        Update the upper bound (UBD)
Else
    Solve the infeasible primal problem for the fixed values of y^k:
$$\min_{x,\alpha} \quad \alpha_i + \alpha_e^+ + \alpha_e^-$$
$$\begin{aligned}\text{s.t.} \quad g(x) - \alpha_i &\leq -B y^k \\ \alpha_i, \alpha_e^+, \alpha_e^- &\geq 0\end{aligned}$$
Obtain the optimal $\bar{x}^k$

Step 3
Solve the relaxed master problem:
$$\min_{x,y,\mu_{OA}} \quad c^T y + \mu_{OA}$$
$$\text{s.t.} \quad \left.\begin{aligned} \mu_{OA} &\geq f(x^l) + \nabla_x f(x^l)(x - x^l) \\ 0 &\geq g(x^l) + \nabla_x g(x^l)(x - x^l) + By \end{aligned}\right\} \forall l \in \mathbf{F}$$
$$0 \geq g(\bar{x}^l) + \nabla_x g(\bar{x}^l)(x - \bar{x}^l) + By \ \} \ \forall l \in \bar{\mathbf{F}}$$
$$x \in X \subseteq \mathbb{R}^n$$
$$y \in Y \in \{0,1\}$$
$$\sum_{i \in B^l} y_i^l - \sum_{i \in NB^l} y_i^l \leq |\mathbf{B}^l| - 1, \quad \forall l \in \mathbf{F}$$
Obtain the solution and y^{k+1}
If the solution to the master is greater than the current lower bound
    Update the lower bound.
If $UBD - LBD \leq \epsilon$
    Terminate
Else
    Update the counter: $k = k + 1$
    Go to step 2

The stated algorithm may be applied to general problems whose formulation is of the form (14). However, it is not guaranteed to converge to the global solution unless some additional conditions are met. The functions f and g must be convex in x. If this is not the case, the linearizations employed in the master problem may not be valid and may possibly eliminate part of the feasible region.

3.3 Outer Approximation/Equality Relaxation

The **OA/ER** algorithm is a generalization of the Outer Approximation algorithm [KG87] to handle nonlinear equality constraints. The class of problems this algorithm can address is the following:

$$\begin{aligned}
\min_{x,y} \quad & f(x) + c^T y \\
\text{s.t.} \quad & g(x) + By \leq 0 \\
& h(x) + Cy = 0 \\
& x \in X \subseteq R^n \\
& y \in \{0,1\}^q
\end{aligned} \quad (20)$$

The basic idea behind this algorithm is to relax the equality constraints into inequalities and apply the **OA** algorithm. A square diagonal matrix T, whose diagonal elements are minus one, zero and one, is used for relaxing the equality constraints.

$$T^k (h(x) + Cy) \leq 0$$

The matrix T has the same number of rows as the number of equality constraints and the values of the diagonal elements depend on the signs of the corresponding multipliers obtained from the primal problem. The values of the elements are one for the multipliers which are positive, minus one for the negative multipliers, and zero for the zero valued multipliers.

$$T^k = \text{diag}(t_{ii}) \quad t_{ii} = \text{sign}(\mu_i^k)$$

With the equalities now relaxed to inequalities, the principles of the **OA** can be applied to the problem.

3.3.1 OA/ER Algorithm

Step 1
    Obtain starting point: y^1
    Set the counter: $k = 1$
    Set the lower bound: $LBD = -\infty$
    Set the upper bound: $UBD = +\infty$
    Set the convergence tolerance: $\epsilon \geq 0$

Step 2
Solve the primal problem for the fixed values of $y = y^k$:
$$\min_x \quad f(x) + c^T y^k$$
$$\text{s.t.} \quad g(x) \leq -By^k$$
$$\quad h(x) = -Cy^k$$
Obtain the optimal x^k and the Lagrange multipliers μ^k
If the primal is feasible
    If the optimal solution of the primal is less than UBD
        Update the upper bound (UBD)
Else
    Solve the infeasible primal problem for the fixed values of $y = y^k$:
$$\min_{x,\alpha} \quad \alpha_i + \alpha_e^+ + \alpha_e^-$$
$$\text{s.t.} \quad g(x) - \alpha_i \leq -By^k$$
$$\quad h(x) + \alpha_e^+ - \alpha_e^- = -Cy^k$$
$$\quad \alpha_i, \alpha_e^+, \alpha_e^- \geq 0$$
Obtain the optimal $\bar{x}^k$ and Lagrange multipliers $\bar{\mu}^k$

Step 3
Determine the matrix T:
$$T^k = \text{diag}(t_{ii}) \text{ where } t_{ii} = \text{sign}(\mu_i^k)$$
Solve the relaxed master problem:
$$\min_{x,y} \quad \mu + c^T y$$
$$\text{s.t.} \quad \begin{aligned} \mu &\geq f(x^l) + \nabla_x f(x^l)(x - x^l) \\ 0 &\geq g(x^l) + \nabla_x g(x^l)(x - x^l) + By \\ 0 &\geq T^l(h(x^l) + \nabla_x h(x^l)(x - x^l) + Cy) \end{aligned} \Bigg\} \forall l \in \mathbf{F}$$
$$\begin{aligned} 0 &\geq g(\bar{x}^l) + \nabla_x g(\bar{x}^l)(x - \bar{x}^l) + By \\ 0 &\geq T^l(h(\bar{x}^l) + \nabla_x h(\bar{x}^l)(x - \bar{x}^l) + Cy) \end{aligned} \Bigg\} \forall l \in \bar{\mathbf{F}}$$
$$x \in X \subseteq R^n$$
$$y \in Y \in \{0,1\}^q$$
$$\sum_{i \in B^l} y_i - \sum_{i \in N^l} y_i \leq |\mathbf{B}^l| - 1 \quad \forall l \in \mathbf{F}$$
Obtain the solution and y^{k+1}
If the solution to the master is greater than the current lower bound
    Update the lower bound.
If $UBD - LBD \leq \epsilon$
    Terminate
Else
Update the counter: $k = k + 1$
Go to step 2

This algorithm is not guaranteed to determine the global solution unless certain convexity conditions are met.

3.4 Outer Approximation/Equality Relaxation/Augmented Penalty

The **OA/ER/AP** algorithm [VG90] is a modification of the **OA/ER** algorithm. The objective of this algorithm is to avoid convexity assumptions necessary for finding the global solution using the **OA/ER** algorithm. This algorithm addresses the same class of problems as **OA/ER**:

$$\begin{aligned}
\min_{x,y} \quad & f(x) + c^T y \\
\text{s.t.} \quad & g(x) + By \leq 0 \\
& h(x) + Cy = 0 \\
& x \in X \subseteq \mathbb{R}^n \\
& y \in Y \in \{0,1\}^q
\end{aligned} \qquad (21)$$

This algorithm uses a relaxation of the linearizations of the master problem in order to expand the feasible region. Through this expansion of the feasible region, the probability of cutting part of the feasible region due to one of the linearizations is reduced. Note that this does not guarantee the possible elimination of part of the feasible region and thus the determination of the global solution cannot be guaranteed.

The linearizations in the master problem are relaxed by including slack variables in the constraints. The violations of the linearizations are penalized by including weighted sums of the slack variables in the objective function.

The difference between the **OA/ER** and **OA/ER/AP** algorithms is in the master problem formulations. The **OA/ER/AP** master problem has the following formulation:

$$\begin{aligned}
\min_{x,y,s,p,q} \quad & c^T y + \mu + \sum_l u_l s_l + \sum_l \sum_i v_{i,l} p_{i,l} + \sum_l \sum_i w_{i,l} q_{i,l} \\
\text{s.t.} \quad & \mu + s_l \geq f(x^l) + \nabla_x f(x^l)(x - x^l) \\
& \left. \begin{array}{rl} p_l \geq & g(x^l) + \nabla_x g(x^l)(x - x^l) + By \\ q_l \geq & T^l(h(x^l) + \nabla_x h(x^l)(x - x^l) + Cy) \end{array} \right\} \forall l \in \mathbf{F} \\
& \left. \begin{array}{rl} p_l \geq & g(\bar{x}^l) + \nabla_x g(\bar{x}^l)(x - \bar{x}^l) + By \\ q_l \geq & T^l(h(\bar{x}^l) + \nabla_x h(\bar{x}^l)(x - \bar{x}^l) + Cy) \end{array} \right\} \forall l \in \mathbf{\bar{F}} \\
& x \in X \subseteq R^n \\
& y \in Y \in \{0,1\}^q \\
& s_l, p_l, q_l \geq 0 \\
& \sum_{i \in \mathbf{B}^l} y_i - \sum_{i \in \mathbf{N}^l} y_i \leq |\mathbf{B}^l| - 1 \quad \forall l \in \mathbf{F}
\end{aligned} \tag{22}$$

where s^k is a slack scalar for iteration k and p_k and q_k are slack vectors at iteration k for the inequality and relaxed equality constraints respectively. The weights for the penalty terms, u_l, $v_{i,k}$, and $w_{i,k}$, are determined from the multipliers of the corresponding constraints from the solution of the primal problem. The correspondence between the constraints in the primal problem and the multipliers, μ_0^k, μ^k, and λ^k is as follows:

$$\begin{aligned}
\min_x \quad & b^T y^k + \mu_B \\
\text{s.t.} \quad & J(x) - \mu_B \leq 0 \leftarrow \mu_0^k \\
& g(x) - c^T y^k \leq 0 \leftarrow \mu^k \\
& h(x) - d^T y^k = 0 \leftarrow \lambda^k
\end{aligned}$$

The weights for the slack variables are assigned as follows ([VG90]):

$$\begin{aligned}
u_k &= 1000\mu_0^k \\
v_{i,k} &= 1000\mu^k \\
w_{j,k} &= 1000\lambda^k
\end{aligned}$$

3.4.1 OA/ER/AP Algorithm

Step 1

Obtain starting point: y^1
Set counter: $k = 1$
Set the lower bound: $LBD = -\infty$
Set the upper bound: $UBD = +\infty$

Set the convergence tolerance: $\epsilon \geq 0$

Step 2

Solve the primal problem for the fixed values of y^k:
$$\min_{x} \quad f(x) + c^T y^k$$
$$\text{s.t.} \quad \begin{aligned} g(x) &\leq By^k \\ h(x) &= Cy^k \end{aligned}$$

Obtain the optimal x^k and the Lagrange multipliers μ^k

If the primal is feasible

    If the optimal solution of the primal is less than UBD

        Update the upper bound (UBD)

Else

    Solve the infeasible primal problem for the fixed values of y^k:
$$\min_{x,\alpha} \quad \alpha_i + \alpha_e^+ + \alpha_e^-$$
$$\text{s.t.} \quad \begin{aligned} g(x) - \alpha_i &\leq c^T y^k \\ h(x) + \alpha_e^+ - \alpha_e^- &= By^k \\ \alpha_i, \alpha_e^+, \alpha_e^- &\geq 0 \end{aligned}$$

Obtain the optimal $\bar{x}^k$ and Lagrange multipliers $\bar{\mu}^k$

Step 3

Determine the matrix T as follows:
$$T^k = \text{diag}(t_{ii}) \text{ where } t_{ii} = \text{sign}(\mu_i^k)$$
Set the values for the penalty parameters, u, v, and w

Solve the relaxed master problem:
$$\min_{x,y,s,p,q} \quad c^T y + \mu + \sum_l u_l s_l + \sum_l \sum_i v_{i,l} p_{i,l} + \sum_l \sum_i w_{i,l} q_{i,l}$$
$$\text{s.t.} \quad \left.\begin{aligned} \mu + s_l &\geq f(x^l) + \nabla_x f(x^l)(x - x^l) \\ p_l &\geq g(x^l) + \nabla_x g(x^l)(x - x^l) + By \\ q_l &\geq T^l(h(x^l) + \nabla_x h(x^l)(x - x^l) + Cy) \end{aligned}\right\} \forall l \in \mathbf{F}$$
$$\left.\begin{aligned} p_l &\geq g(\bar{x}^l) + \nabla_x g(\bar{x}^l)(x - \bar{x}^l) + By \\ q_l &\geq T^l(h(\bar{x}^l) + \nabla_x h(\bar{x}^l)(x - \bar{x}^l) + Cy) \end{aligned}\right\} \forall l \in \bar{\mathbf{F}}$$
$$x \in X \subseteq \mathbb{R}^n$$
$$y \in Y \in \{0,1\}^q$$
$$s_l, p_l, q_l \geq 0$$
$$\sum_{i \in \mathbf{B}^l} y_i - \sum_{i \in \mathbf{N}^l} y_i \leq |\mathbf{B}^l| - 1 \quad \forall l \in \mathbf{F}$$

Obtain optimal y^{k+1}

If the solution to the master is greater than the current lower bound

    Update the lower bound

If $UBD - LBD \leq \epsilon$

    Terminate

Else
Update the counter: $k = k + 1$
Go to step 2

3.5 Generalized Outer Approximation

The Generalized Outer Approximation, **GOA**, algorithm [FL94] generalizes the **OA** approach to handle the following class of MINLP problems:

$$\begin{aligned} \min_{x} \quad & f(x,y) \\ \text{s.t.} \quad & g(x,y) \leq 0 \\ & x \in X \subseteq R^n \\ & y \in \{0,1\}^q \end{aligned} \qquad (23)$$

The difference between this formulation and that for the **OA** algorithm is that there is no restriction on the separability in the x and y variables and the linearity of the y variables.

The differences between the **GOA** algorithm and the **OA** algorithm are the treatment of infeasibilities, a new master problem formulation, and the unified treatment of exact penalty functions.

3.5.1 Primal Problem

The primal problem is formulated in the same manner as in **OA**. However, if the primal is infeasible, then the following feasibility problem is solved for $y = y^k$:

$$\begin{aligned} \min_{x} \quad & \sum_{i \in I'} w_i g_i^+(x, y^k) \\ \text{s.t.} \quad & g_i(x, y^k) \leq 0 \quad i \in I \\ & x \in X \subseteq R^n \end{aligned} \qquad (24)$$

where I is the set of feasible inequality constraints and I' is the set of infeasible inequality constraints. With this formulation of the feasibility problem, the linearizations of the nonlinear inequality constraints about the solution of the feasibility problem are violated.

3.5.2 Master Problem

The master problem for **GOA** is formulated based on the **OA** ideas of projection onto the y-space and the outer approximation of the objective function and feasible region. The difference in the master problem formulation is in the utilization of the infeasible primal information.

The projection onto the y-space is the same as for **OA**:

$$\min_{y} \; v(y) \quad \text{s.t.} \quad y \in Y \cap V \tag{25}$$

where

$$v(y) = c^T y + \inf_{x} \; f(x) \quad \text{s.t.} \; g(x) + B^T y \leq 0 \quad x \in X \subseteq \mathbb{R}^n \tag{26}$$

and

$$V = \{y : g(x) + By \leq 0, \text{ for some } x \in X\} \tag{27}$$

The formulation of the master problem follows from the outer approximation of the the problem $v(y)$ and a representation of the set V. For the outer approximation of $v(y)$, the linearizations of the objective function and constraints are used. The set V is replaced by linearizations of the constraints at y^k for which the primal is infeasible, and the feasibility problem has solution x^k. Thus, the master problem has the following formulation:

$$\begin{aligned}
\min_{x,y,\mu_{\text{GOA}}} \quad & c^T y + \mu_{\text{GOA}} \\
\text{s.t.} \quad & \\
& \left.\begin{array}{r} \mu_{\text{GOA}} \geq f(x^k, y^k) + \nabla f(x^k, y^k) \begin{pmatrix} x - x^k \\ y - y^k \end{pmatrix} \\ 0 \geq g(x^k, y^k) + \nabla g(x^k, y^k) \begin{pmatrix} x - x^k \\ y - y^k \end{pmatrix} \end{array}\right\} \forall k \in \mathbf{F} \\
& 0 \geq g(x^k, y^k) + \nabla g(x^k, y^k) \begin{pmatrix} x - x^k \\ y - y^k \end{pmatrix} \quad \forall k \in \bar{\mathbf{F}} \\
& x \in X \subseteq \mathbb{R}^n \\
& y \in Y \in \{0,1\}^q
\end{aligned} \tag{28}$$

where $\mathbf{F}$ is the set of all y^k such that the primal problem is feasible, and $\bar{\mathbf{F}}$ is the set of all y^k such that the primal problem is infeasible.

Mixed-Integer Nonlinear Optimization in Process Synthesis

As in **OA**, relaxation and an iterative procedure are used to solve the problem since the solution of the complete master problem (28) requires all feasible and infeasible solutions of the primal problem. For the solution of the relaxed master problem, the known feasible and infeasible solutions are used for the outer approximation.

3.5.3 GOA Algorithm

Step 1
  Obtain starting point: y^1
  Set the counter: $k = 1$
  Set the lower bound: $LBD = -\infty$
  Set the upper bound: $UBD = +\infty$
  Initialize the feasibility set: $\mathbf{F} = \emptyset$
  Initialize the infeasibility set: $\bar{\mathbf{F}} = \emptyset$
  Set the convergence tolerance: $\epsilon \geq 0$

Step 2
  Solve the primal problem for the fixed values of y^k:
  $$\min_{x} \quad f(x, y^k)$$
  $$\text{s.t.} \quad g(x, y^k) \leq 0$$
  If the primal is feasible
    Obtain the optimal x^k
    Update the feasibility set: $\mathbf{F} = \mathbf{F} \cup k$
    If the optimal solution of the primal is less than UBD
      Update the upper bound (UBD)
  Else
    Solve the infeasible primal problem for the fixed values of y^k:
    $$\min_{x} \quad \sum_{i \in I'} w_i g_i^+(x, y^k)$$
    $$\text{s.t.} \quad g_i(x, y^k) \leq 0 \quad i \in I$$
    $$\qquad\qquad x \in X \subseteq \mathrm{R}^n$$
    Obtain the optimal $\bar{x}^k$
    Update the feasibility set: $\bar{\mathbf{F}} = \bar{\mathbf{F}} \cup k$

Step 3
  Solve the relaxed master problem:

$$\min_{x,y,\mu_{\text{GOA}}} \quad c^T y + \mu_{\text{GOA}}$$

$$\text{s.t.} \quad \left. \begin{array}{rcl} \mu_{\text{GOA}} & \geq & f(x^l, y^l) + \nabla f(x^l, y^l) \begin{pmatrix} x - x^l \\ y - y^l \end{pmatrix} \\ \\ 0 & \geq & g(x^l, y^l) + \nabla g(x^l, y^l) \begin{pmatrix} x - x^l \\ y - y^l \end{pmatrix} \end{array} \right\} \forall l \in \mathbf{F}$$

$$0 \geq g(\bar{x}^l, y^l) + \nabla g(\bar{x}^l, y^l) \begin{pmatrix} x - \bar{x}^l \\ y - y^l \end{pmatrix} \quad \forall l \in \bar{\mathbf{F}}$$

$$x \in X$$
$$y \in Y$$

Obtain the solution and y^{k+1}
If the solution to the master is greater than the current lower bound
    Update the lower bound.
If $UBD - LBD \leq \epsilon$
    Terminate
Else
    Update the counter: $k = k+1$
    Go to step 2

3.6 Generalized Cross Decomposition

The Generalized Cross Decomposition, **GCD**, algorithm [Hol90] exploits the advantages of Dantzig-Wolfe Decomposition and **GBD** and simultaneously utilizes primal and dual information. This algorithm can address problems with the general MINLP formulation (2) where the constraints are partitioned into two sets:

$$\begin{aligned} \min_{x,y} \quad & f(x,y) \\ \text{s.t.} \quad & g_1(x,y) \leq 0 \\ & g_2(x,y) \leq 0 \\ & h_1(x,y) = 0 \\ & h_2(x,y) = 0 \\ & x \in X \subseteq \mathbb{R}^n \\ & y \in \{0,1\}^q \end{aligned} \qquad (29)$$

This algorithm consists of two phases and convergence tests. Phase I involves the solution of the primal and dual subproblems where the primal subproblem provides an upper bound on the solution along with Lagrange multipliers for the dual subproblem. The dual subproblem provides a lower bound on the solution of the problem and supplies new values of the y variables for the primal subproblem. Both the primal and dual subproblems provide cuts for the master problem (Phase II). In Phase II either a primal master problem or a Lagrange relaxation master problem is solved. The primal master problem is formulated using the same derivation as for the **GBD** master problem while the Lagrange relaxation master problem is formulated by using Lagrangian duality. The algorithm also uses several convergence tests to determine whether or not solutions of various subproblems can provide bound or cut improvement.

The algorithm for **GCD** is based on the idea that it is desirable to solve as few master problems as possible since these are generally more computationally intensive. The algorithm makes extensive use of the primal and dual subproblems to reduce the number of times the master problem must be solved to obtain the solution.

3.6.1 GCD Algorithm

Step 1

Obtain starting point: y^1
Set the counter: $k = 1$
Set the lower bound: $LBD = -\infty$
Set the upper bound: $UBD = +\infty$
Initialize the feasibility set: $\mathbf{F} = \emptyset$
Initialize the infeasibility set: $\bar{\mathbf{F}} = \emptyset$

Step 2

Solve the primal problem for the fixed values of y^k:
$$\begin{aligned}
\min_{x} \quad & f(x, y^k) \\
\text{s.t.} \quad & g_1(x, y^k) \leq 0 \\
& g_2(x, y^k) \leq 0 \\
& h_1(x, y^k) = 0 \\
& h_2(x, y^k) = 0 \\
& x \in X \subseteq \mathbb{R}^n
\end{aligned}$$
Obtain the optimal x^k and Lagrange multipliers λ_1^k, λ_2^k, μ_1^k, and μ_2^k
If the primal is feasible
    Update the feasibility set: $\mathbf{F} = \mathbf{F} \cup k$
    If the optimal solution of the primal is less than UBD
        update the upper bound (UBD)

Perform **CTD** test for λ_1^k and μ_1^k
Else
 Solve the infeasible primal problem for the fixed values of y^k:
$$\min_{x,\alpha} \quad \alpha_{i1} + \alpha_{i2} + \alpha_{e1}^+ + \alpha_{e1}^- + \alpha_{e2}^+ + \alpha_{e2}^-$$
$$\begin{aligned} \text{s.t.} \quad g_1(x, y^k) - \alpha_{i1} &\leq 0 \\ g_2(x, y^k) - \alpha_{i2} &\leq 0 \\ h_1(x, y^k) + \alpha_{e1}^+ - \alpha_{e1}^- &= 0 \\ h_2(x, y^k) + \alpha_{e2}^+ - \alpha_{e2}^- &= 0 \\ \alpha_{i1}, \alpha_{i2}, \alpha_{e1}^+, \alpha_{e1}^-, \alpha_{e2}^+, \alpha_{e2}^- &\geq 0 \end{aligned}$$
 Obtain the optimal $\bar{x}^k$ and the Lagrange multipliers $\bar\lambda_1^k, \bar\lambda_2^k, \bar\mu_1^k$ and $\bar\mu_2^k$
 Update the infeasibility set: $\bar{\mathbf{F}} = \bar{\mathbf{F}} \cup k$
 Perform **CTDU** test for $\bar\lambda_1^k$ and $\bar\mu_1^k$

Step 3
 If **CTD** or **CTDU** test from Step 2 is not passed
  Solve Relaxed Lagrange Relaxation Master problem:
$$\max_{\lambda_1,\mu_1,\mu_b} \quad \mu_b$$
$$\begin{aligned} \text{s.t.} \quad \mu_b &\leq f(x^l, y^l) + (\lambda_1)^T h_1(x^l, y^l) + (\mu_1)^T g_1(x^l, y^l) \quad l \in \mathbf{F} \\ 0 &\leq (\lambda_1)^T h_1(x^l, y^l) + (\mu_1)^T g_1(\bar{x}^l, y_l) \quad l \in \bar{\mathbf{F}} \\ \mu_1 &\geq 0 \end{aligned}$$
  Obtain optimal λ_1^k and μ_1^k

Step 4
 If $k \in \mathbf{F}$ or the **CTD** or **CTDU** test from Step 2 is not passed
  Solve the Dual Subproblem:
$$\min_{x,y} \quad f(x,y) + (\lambda_1^k)^T h_1(x,y) + (\mu_1^k)^T g_1(x,y)$$
$$\begin{aligned} \text{s.t.} \quad h_2(x,y) &= 0 \\ g_2(x,y) &\leq 0 \end{aligned}$$
  Obtain optimal y^{k+1}
  If the solution is greater than the lower bound
   Update the lower bound
 Else
  Solve the Dual Feasibility Subproblem:
$$\min_{x,y} \quad (\bar\lambda_1^k)^T h_1(x,y) + (\bar\mu_1^k)^T g_1(x,y)$$
$$\begin{aligned} \text{s.t.} \quad h_2(x,y) &= 0 \\ g_2(x,y) &\leq 0 \end{aligned}$$
  Obtain the optimal y^{k+1}

Step 5
 Perform the **CTP** test for y^{k+1}

If the **CTP** test fails
    Solve the Relaxed Primal Master Problem:
$$\min_{y,\mu_b} \mu_b$$
$$\text{s.t.} \quad \mu_b \geq f(x^l, y) + (\lambda^l)^T h(x^l, y) + (\mu^l)^T g(x^l, y) \quad l \in \mathbf{F}$$
$$0 \geq (\bar{\lambda}^l)^T h(\bar{x}^l, y) + (\bar{\mu}^l)^T g(\bar{x}^l, y) \quad l \in \bar{\mathbf{F}}$$
    Obtain the optimal y^{k+1}
    If the solution is greater than the lower bound
        Update the lower bound
If $UBD - LBD \leq \epsilon$
    Terminate
Else
    Update the counter: $k = k + 1$
    Go to step 2

The convergence tests are defined as follows:

CTP Test At the kth iteration, if the solution from the Dual Subproblem y^{k+1} satisfies

$$UBD \geq f(x^l, y^{k+1}) + (\lambda^l)^T h(x^l, y^{k+1}) + (\mu^l)^T g(x^l, y^{k+1}) \quad l \in \mathbf{F}$$
$$0 \geq (\bar{\lambda}^l)^T h(\bar{x}^l, y^{k+1}) + (\bar{\mu}^l)^T g(\bar{x}^l, y^{k+1}) \quad l \in \bar{\mathbf{F}}$$

then y^{k+1} will provide an upper bound or cut improvement. Otherwise, the Relaxed Primal Master is solved to obtain a new y^{k+1}.

CTD Test At the kth iteration, if the feasible solution of the Primal Problem satisfies:

$$LBD \leq f(x^l, y^l) + (\lambda_1^k)^T h(x^l, y^l) + (\mu_1^k)^T g(x^l, y^l) \quad l \in \mathbf{F}$$

then λ_1^k and μ_1^k will provide a lower bound or cut improvement. Otherwise, the Relaxed Lagrange Relaxation Master is solved to obtain new λ_1^k and μ_1^k.

CTDU Test At the kth iteration, if the Primal Problem is infeasible and the solution to the Infeasible Primal Problem satisfies

$$0 \leq (\bar{\lambda}_1^k)^T h(\bar{x}^l, y^l) + (\bar{\mu}_1^k)^T g(\bar{x}^l, y^l) \quad l \in \bar{\mathbf{F}}$$

then $\bar{\lambda}_1^k$ and $\bar{\mu}_1^k$ will provide cut improvement. Otherwise, the Relaxed Lagrange Relaxation Master is solved to obtain λ_1^k and μ_1^k.

The situations where the **GCD** algorithm reduces to Dantzig-Wolfe decomposition and **GBD** can be observed. First, when the tests **CTD** and **CTDU** are not passed at all iterations and only the relaxed primal master problem is used, **GCD** reduces to **GBD**. On the other hand, if the test **CTP** is not used at all iterations and only the relaxed primal Lagrange problem is used, then **GCD** reduces to Dantzig-Wolfe decomposition.

The algorithm for **GCD** is well-illustrated by the use of a flow diagram such as that in Figure 3.

3.7 Branch-and-Bound Algorithms

A number of branch-and-bound algorithms have been proposed to identify the global optimum solution of problems which are convex in the x-space and relaxed y-space. [Bea77, GR85, OOM90, BM91, QG92]. These algorithms can also be used for nonconvex problems of the form (2) but their convergence to the global optimum solution can only be guaranteed for convex problems. A basic principle common to all these algorithms is the generation of valid lower bounds on the original MINLP through its relaxation to a continuous problem. In most algorithms, the continuous problem is obtained by letting binary variables take on any value between 0 and 1. In most algorithms, this relaxation is an NLP problem. The only exception is the algorithm of [QG92], discussed in Section 3.7.3, in which an LP problem is obtained. If the NLP relaxation has an integer solution, this solution provides an upper bound on the global solution. The generation of lower and upper bounds in this manner is referred to as the *bounding step* of the algorithm. At first, all the binary variables are relaxed and the continuous problem corresponds to the first node of a branch-and-bound tree. At the second level, two new nodes are created by forcing one of the binary variables to take on a value of 0 or 1. This is the *branching step*. Nodes in the tree are pruned when their lower bound is greater than the best upper bound on the problem, or when the relaxation is infeasible. The algorithm terminates when the lowest lower bound is within a pre-specified tolerance of the best upper bound.

Although the size of the branch-and-bound tree is finite, and convergence is guaranteed, it is desirable to explore as few nodes of the tree as possible. The selection of the branching node and variable and the solution of the relaxed problem all affect the convergence characteristics of the algorithm. Several strategies have been suggested in the literature.

Mixed-Integer Nonlinear Optimization in Process Synthesis

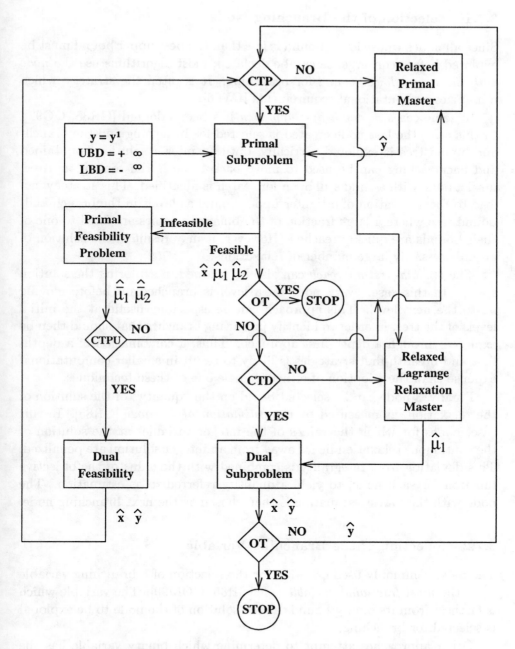

Figure 3: Generalized Cross Decomposition Flow Diagram

3.7.1 Selection of the Branching Node

Since all nodes whose lower bound is less than the best upper bound must be explored before convergence can be declared, most algorithms use the *node with the lowest lower bound* for branching. It is often the strategy which minimizes computational requirements [LW66].

In some cases, a *depth-first approach* has been adopted [GR85, QG92]. In this case, the last node created is selected for branching and the branch-and-bound tree is explored vertically, until an integer solution is obtained and backtracking can be used to move back towards the root of the tree, until a node with a child still open for search is identified. This strategy can lead to the generation of a tighter upper bound as levels in the branch-and-bound tree where a large fraction of the binary variables are fixed to one of their bounds are quickly reached. However, it may result in the solution of an unnecessarily large number of relaxations.

The *breadth-first approach* can also be followed in exploring the solution space. In this case, every node on a level is branched on before moving on to the next level. This approach can be especially useful at the initial levels of the tree, in order to identify promising branches that should then be explored through a depth-first approach. Thus, a combination of a depth-first and breadth-first strategies is likely to result in smaller computational expense than the adoption of either a single one of these techniques.

Finally, a node can be selected based on the "quality" of the solution of the relaxation, as measured by the *estimation of the node* [GR85]. In this case, nodes for which the values of the integer variables at the solution of the continuous relaxation lie far away from an integer solution are penalized. This deviation from integrality is combined with the value of the objective function for each node, to yield a quantity referred to as estimation. The node with the lowest estimation is then chosen as the next branching node.

3.7.2 Selection of the Branching Variable

The most commonly used criterion for the selection of a branching variable y^B is the *most fractional variable rule* [GR85, OOM90]. The variable which is farthest from its binary bounds at the solution of the node to be explored is selected for branching.

Other approaches attempt to determine which binary variable has the greatest effect on the lower bound of the problem. If the user has prior knowledge of the problem, *branching priorities* may be set to accelerate the

Mixed-Integer Nonlinear Optimization in Process Synthesis 33

increase of the best lower bound [GR85]. The quantitative analysis of the effect of each binary variable has also been proposed through the use of *pseudo-costs* [BGG+71, GR85]. The pseudo-cost of a variable y_j is defined by calculating the relative change in the objective function when y_j is fixed to 0 or 1. Thus, if the solution of the NLP with $0 \leq y_j \leq 1$ is $y_j = y_j^*$ and $f(x, y) = f^*$, and the optimum objective value with $y_j = 0$ is f_0, the lower pseudo-cost associated with y_j is $PC_j^L = (f_0 - f^*)/y_j^*$. Similarly, if f_1 is the optimum objective value for $y_j = 1$, the upper pseudo-cost is defined as $PC_j^U = (f_1 - f^*)/(1 - y_j^*)$. To avoid excessive computational requirements, pseudo-costs are only calculated once. At each node, the pseudo-costs of fractional variables are updated by computing the minimum of $PC_j^L y_j^*$ and $PC_j^U (1 - y_j^*)$, where y_j^* is the value of y_j at the solution of the NLP at the branching node. The variable with the maximum pseudo-cost is selected for branching as it is expected to lead to the greatest increase in the lower bound on the problem.

3.7.3 Generation of a Lower Bound

At each node, a relaxation of the MINLP problem must be solved in order to generate a lower bound. This procedure can be costly for large problems and [BM91] advocate the early detection of non-integer solutions in order to reduce the time spent solving NLPs. If the binary variables appear to be converging to values away from their bounds, the NLP solver should be interrupted before full convergence has been achieved and the current node should be branched on.

[QG92] suggest further relaxation of the problem to an LP in order to obtain lower bounds. When the lower bounding LP has an integer solution, this integer combination is used to formulate an NLP problem which yields an upper bound on the original problem. The LP problems are generated by linearizing the nonlinear functions at the solution of each NLP. In order to prevent the LPs from becoming excessively large, a reformulation which combines all nonlinearities into one inequality constraint is used. The first linearization is obtained by fixing the binary variables to an arbitrary combination and solving the resulting NLP.

3.7.4 Algorithmic Procedure

The general algorithmic statement for branch-and-bound approaches is as follows:

Step 1
    Set absolute tolerance ϵ; set $LBD^* = -\infty$ and $UBD^* = \infty$.
    Set node counter $k = 0$.
    Initialize set of nodes to be bounded, $J = \{0\}$;
    Set relaxed y-space, $Y_0 = [0,1]^q$.
    N, the list of nodes to be explored, is empty.

Step 2 *Bounding*
    Solve relaxed NLP problem, for $j \in J$:
$$\begin{aligned} LBD_j = \min\ & f(x,y) \\ \text{s.t.}\ & h(x,y) = 0 \\ & g(x,y) \leq 0 \\ & x \in X \subseteq \mathbb{R}^n \\ & y \in Y_j \end{aligned}$$
    If all y variables are integer at the solution:
        If $LBD_j \leq UBD^*$, $UBD^* = LBD_j$.
        Else, fathom the current node.
    If some y variables are fractional:
        If $LBD_j \leq UBD^*$, add the current node to N.
        Else, fathom the current node.

Step 3
    Set LBD^* to the lowest lower bound from the list N.
    If $UBD^* - LBD^* \leq \epsilon$, terminate with solution UBD^*.
    Otherwise, proceed to Step 4.

Step 4 *Branching*
    Select the next node to be explored, N_i ($i < k$), from list N. Its lower bound is LBD_i and the corresponding y-space is Y_i.
    Select a branching variable y^B.
    Create two new regions $Y_{k+1} = Y_i \cap \{y|y^B = 0\}$ and $Y_{k+2} = Y_i \cap \{y|y^B = 1\}$.
    Set $J = k+1, k+2$ and $k = k+2$. Go back to Step 2.

3.8 Extended Cutting Plane (ECP)

The cutting plane algorithm proposed by [Kel60] for NLP problems has been extended to MINLPs [WPG94, WP95]. This Extended Cutting Plane algorithm (ECP) can address problems of the form:

$$\begin{aligned} \min \quad & c_x^T x + c_y^T y \\ \text{s.t.} \quad & g(x,y) \leq 0 \\ & x \in X \subseteq R^n \\ & y \in Y \text{ integer} \end{aligned} \qquad (30)$$

where c_x and c_y are constant vectors.

Problems with a nonlinear objective function, $f(x,y)$, can be reformulated by introducing a new variable z such that $f(x,y) - z \leq 0$ and minimizing z.

The ECP algorithm relies on the linearization of one of the nonlinear constraints at each iteration and the solution of the increasingly tight MILP made up of these linearizations. The solution of the MILP problem provides a new point on which to base the choice of the constraint to be linearized for the next iteration of the algorithm. Unlike the Outer Approximation, described in Section 3.2, the ECP does not require the solution of any NLP problems for the generation of an upper bound. As a result, a large number of linearizations are required for the approximation of highly nonlinear problems and the algorithm does not perform well in such cases. Due to the use of linearizations, convergence to the global optimum solution is guaranteed only for problems involving inequality constraints which are convex in the x and relaxed y-space.

Given a point (x^0, y^0) in the feasible set, the function $g_i(x,y)$ can be underestimated by

$$g_i(x^0, y^0) + \left(\frac{\partial g_i}{\partial x}\right)_{x^0, y^0} (x - x^0) + \left(\frac{\partial g_i}{\partial y}\right)_{x^0, y^0} (y - y^0). \qquad (31)$$

The function $g_{j_0}(x,y)$, where $j_0 = \text{argmax}_i\, g_i(x^0, y^0)$, is then used to construct an underestimating MILP problem

$$\begin{aligned} \min \quad & c_x^T x + c_y^T y \\ \text{s.t.} \quad & l_0(x,y) \leq 0 \\ & x \in X \subseteq R^n \\ & y \in Y \text{ integer} \end{aligned} \qquad (32)$$

where $l_0(x,y) = g_{j_0}(x^0, y^0) + \left(\frac{\partial g_{j_0}}{\partial x}\right)_{x^0, y^0} (x - x^0) + \left(\frac{\partial g_{j_0}}{\partial y}\right)_{x^0, y^0} (y - y^0)$.

At any iteration k, a single constraint is chosen for linearization and the corresponding linearization $l_k(x, y)$ is added to the MILP. All linear constraints from the original MINLP problem are, of course, incorporated into the MILP from the start of the algorithm.

3.8.1 Algorithmic Procedure

The algorithmic procedure is:

Step 1
    Set absolute tolerance ϵ; set $LBD^* = -\infty$.
    Set iteration counter $k = 0$ and select starting point (x^0, y^0).

Step 2
    Solve kth MILP problem:
$$\begin{aligned} LBD^* = \min \quad & c_x^T x + c_y^T y \\ \text{s.t.} \quad & l_i(x, y) \leq 0, i = 0, \cdots, k-1 \\ & x \in X \subseteq \mathbb{R}^n \\ & y \in Y_j \end{aligned}$$
    The optimal solution of the MILP is at (x^k, y^k).
    Find $j_k = \text{argmax}_i \, g_i(x^k, y^k)$.

Step 3
    If $g_{j_k}(x^k, y^k) \leq \epsilon$, convergence has been reached. Terminate with solution LBD^*.
    Otherwise, proceed to Step 4.

Step 4
    Construct $l_k(x, y)$. Set $k = k + 1$. Return to Step 2.

3.9 Feasibility Approach

This algorithm was proposed by [MM85] as an extension to MINLP problems of the MINOS algorithm [MS93] for large-scale nonlinear problems.

The premise of their approach is that the problems to be treated are sufficiently large that techniques requiring the solution of several NLP relaxations, such as the branch-and-bound approach of Section 3.7, have prohibitively large costs. They therefore wish to account for the presence of the integer variables in the formulation and solve the mixed-integer problem directly. This is achieved by fixing most of the integer variables to one their bounds (the *nonbasic* variables) and allowing the remaining small subset (the *basic* variables) to take discrete values in order to identify feasible solutions. After each iteration, the reduced costs of the variables in the nonbasic set are computed to measure of their effect on the objective function. If a change causes the objective function to decrease, the appropriate variables are removed from the nonbasic set and allowed to vary for the next iteration. When no more improvement in the objective function is possible, the algorithm is terminated. This strategy leads to the identification of a local solution.

The basic and nonbasic sets are initialized through the solution of continuous relaxation of the NLP. The solution obtained must then be rounded to a feasible integer solution through a heuristic approach. The feasibility approach has been tested on two types of large-scale problems: quadratic assignment and gas pipeline network design. The second problem poses more difficulties as few variables are at either of their bounds when the continuous relaxation is solved. Therefore, the problem has relatively few nonbasic variables. This trend is preserved throughout the run, thus increasing the computational complexity of each iteration.

3.10 Logic Based Approach

An alternative to the direct solution of the MINLP problem was proposed by [TG96]. Their approach stems from the work of [KG89] on a modeling/decomposition strategy which avoids the zero-flows generated by the non-existence of a unit in a process network. The first stage of the algorithm is the reformulation of the MINLP into a generalized disjunctive program of the form:

$$\begin{aligned}
\min \quad & f(x) + \sum_i c_i \\
\text{s.t.} \quad & g(x) \leq 0 \\
& \begin{bmatrix} Y_i \\ h^i(x) \leq 0 \\ c_i = \gamma_i \end{bmatrix} \vee \begin{bmatrix} \neg Y_i \\ B^i x = 0 \\ c_i = 0 \end{bmatrix} \quad i \in D \\
& \Omega(Y) = \text{True} \\
& x \in X \subseteq \mathbb{R}^n \\
& c \geq 0 \\
& Y \in \{\text{True, False}\}^q
\end{aligned} \quad (33)$$

where c is the variable vector representing fixed charges, x is the vector representing all other continuous variables (flowrates, temperatures, ...) and

Y is the vector of Boolean variables, which indicate the status of a disjunction (True or False) and are associated with the units in the network. The set of disjunctions D allows the representation of different configurations, depending on the existence of the units. $\Omega(Y)$ is the set of logical relationships between the Boolean variables, representing the interactions between different network units. Instead of resorting to binary variables within a single model, the disjunctions are used to generate a different model for each different network structure. Since all continuous variables associated with the non-existing units are set to 0 ($B^i x = 0, c = 0$), this representation helps to reduce the size of the problems to be solved.

Two algorithms are suggested by [TG96] in order to solve problems such as (33). They are modifications of the Outer Approximation and Generalized Benders Decomposition presented in Sections 3.2 and 3.1 respectively.

3.10.1 Logic-Based Outer Approximation

In the case of disjunctive programming, the primal problem is obtained simply by fixing all the Boolean variables to a combination k of True and False, yielding an NLP problem of the form:

$$
\begin{aligned}
\min \quad & f(x) + \sum_i c_i \\
\text{s.t.} \quad & g(x) \leq 0 \\
& \begin{cases} h^i(x) \leq 0 \\ c_i = \gamma_i \end{cases} \quad \text{for } Y_i^k = \text{True} \\
& \begin{cases} B^i x = 0 \\ c_i = 0 \end{cases} \quad \text{for } Y_i^k = \text{False} \\
& x \in X \subseteq \mathrm{R}^n \\
& c \geq 0
\end{aligned}
\quad (34)
$$

The master problem is a disjunctive linear program. If K NLP problems have been solved, the nonlinear part of the objective and the nonlinear inequality constraints which are not part of disjunctions are linearized at the K solutions. The nonlinear constraints appearing in the ith disjunction are linearized at the solutions of the NLPs belonging to the subset $K_L^i = \{k | Y_i^k = \text{True}, k = 1, \ldots, K\}$. The master problem is then expressed as:

$$\begin{aligned}
\min \quad & \mu_{\text{OA}} + \sum_i c_i \\
\text{s.t.} \quad & f(x^k) + \nabla f(x^k)(x - x^k) \leq \mu_{\text{OA}}, \quad k = 1, \ldots, K \\
& g(x^k) + \nabla g(x^k)(x - x^k) \leq 0, \quad k = 1, \ldots, K \\
& \begin{bmatrix} Y_i \\ h^i(x^k) + \nabla h^i(x^k)(x - x^k) \leq 0 \\ c_i = \gamma_i \end{bmatrix} \vee \begin{bmatrix} \neg Y_i \\ B^i x = 0 \\ c_i = 0 \end{bmatrix} \quad i \in D \\
& \Omega(Y) = \text{True} \\
& x \in X \subseteq \mathbb{R}^n \\
& c \geq 0 \\
& Y \in \{\text{True, False}\}^q
\end{aligned}$$
(35)

This type of problem can be solved as a disjunctive problem [Bea90], or as an MILP [Bal85, RG94]. To ensure that all nonlinear disjunction constraints are present in the master problem at every iteration, several NLPs must be solved at the start of the algorithm. The structures to be optimized are chosen in such a way that each Boolean variable is True in at least one structure. A method for identifying the minimum number of combinations required to satisfy this condition has been developed [TG96].

The algorithmic procedure for the logic-based outer approximation algorithm is very similar to the original outer approximation algorithm:

Step 0
    Formulate the generalized disjunctive program of the form (33).

Step 1
    Set the counter: $k = 1$
    Set the lower bound: $LBD = -\infty$
    Set the upper bound: $UBD = +\infty$
    Set the convergence tolerance: $\epsilon \geq 0$
    Determine the minimum set of structures needed to obtain cuts for all constraints. Generate the corresponding NLPs.

Step 2
    Solve the NLP(s) for the fixed Boolean variables.
    Obtain the optimal x and c vectors.
    If the optimal solution of the NLP is less than the current upper bound, update the upper bound.

Step 3

Construct and solve the master problem.
Obtain its solution and Y^{k+1}
If the solution of the master is greater than the current lower bound
    Update the lower bound.
If $UBD - LBD \leq \epsilon$
    Terminate
Else
    Update the counter: $k = k + 1$
    Go to step 2

3.10.2 Logic-Based Generalized Benders Decomposition

The Generalized Benders Decomposition framework is not as readily adapted to disjunctive programming as the Outer Approximation. The master problem is generated according to the following scheme:

1. Construct a master problem, as was done for the logic-based OA algorithm (Problem (35)).

2. Transform the problem to an MILP.

3. Based on the values used for the Boolean variables in the previous NLPs, fix the binary variables: LPs are obtained.

4. Solve the LPs and obtain the values of the Lagrange multipliers for the constraints and the optimal continuous variables values.

5. Use these to construct an MILP problem: this is the master problem for the logic-based GBD.

Both algorithms identify the global solution of problems which are convex for all combinations of the Boolean variables.

4 Global Optimization for Nonconvex MINLPs

The algorithms discussed so far have a major limitation when dealing with nonconvex problems. While identification of the global solution for convex problems can be guaranteed, a local solution is often obtained for nonconvex problems. A number of algorithms that have been developed to address different types of nonconvex MINLPs are presented in this section.

4.1 Branch-and-reduce algorithm

[RS95] extended the scope of branch-and-bound algorithms to problems for which valid convex underestimating NLPs can be constructed for the nonconvex relaxations. The range of application of the proposed algorithm encompasses bilinear problems and separable problems involving functions for which convex underestimators can be built [McC76, AK90]. Because the nonconvex NLP must be underestimated at each node, convergence can only be achieved if the continuous variables are branched on. A number of tests are suggested to accelerate the reduction of the solution space. They are summarized here.

Optimality Based Range Reduction Tests For the first set of tests, an upper bound U on the nonconvex MINLP must be computed and a convex lower bounding NLP must be solved to obtain a lower bound L. If a bound constraint for variable x_i, with $x_i^L \leq x_i \leq x_i^U$, is active at the solution of the convex NLP and has multiplier $\lambda_i^* > 0$, the bounds on x_i can be updated as follows:

1. If $x_i - x_i^U = 0$ at the solution of the convex NLP and $\kappa_i = x_i^U - \frac{U-L}{\lambda_i^*}$ is such that $\kappa_i > x_i^L$, then $x_i^L = \kappa_i$.

2. If $x_i - x_i^L = 0$ at the solution of the convex NLP and $\kappa_i = x_i^L + \frac{U-L}{\lambda_i^*}$ is such that $\kappa_i < x_i^U$, then $x_i^U = \kappa_i$.

If neither bound constraint is active at the solution of the convex NLP for some variable x_j, the problem can be solved by setting $x_j = x_j^U$ or $x_j = x_j^L$. Tests similar to those presented above are then used to update the bounds on x_j.

Feasibility Based Range Reduction Tests In addition to ensuring that tight bounds are available for the variables, the constraint underestimators are used to generate new constraints for the problem. Consider the constraint $g_i(x, y) \leq 0$. If its underestimating function $\underline{g}_i(x, y) = 0$ at the solution of the convex NLP and its multiplier is $\mu_i^* > 0$, the constraint

$$\underline{g}_i(x, y) \geq -\frac{U-L}{\mu_i^*}$$

can be included in subsequent problems.

The branch-and-reduce algorithm has been tested on very small problems.

4.2 Interval Analysis Based Algorithm

An algorithm based on interval analysis was proposed by [VEH96] to solve to global optimality problems of the form (2) with a twice-differentiable objective function and once-differentiable constraints. Interval arithmetic allows the computation of guaranteed ranges for these functions [Moo79, RR88, Neu90]. Although the algorithm is not explicitly described as a branch-and-bound approach, it relies on the same basic concepts of successive partitioning of the solution space and bounding of the objective function within each domain. Branching is performed on the discrete and the continuous variables. The main difference with the branch-and-bound algorithms described in Section 3.7 is that bounds on the problem solution in a given domain are not obtained through optimization. Instead, they are based on the range of the objective function in the domain under consideration, as computed with interval arithmetic. As a consequence, these bounds may be quite loose and efficient fathoming techniques are required in order to enhance convergence. [VEH96] suggest a number of tests to determine whether the optimal solution lies in the current domain. In addition, they propose branching strategies based on local solutions to the problem. In order to avoid combinatorial problems, integrality requirements for the discrete variables are removed when performing interval calculations. Convergence is declared when best upper and lower bounds are within a pre-specified tolerance and when the width of the corresponding region is below a pre-specified tolerance.

4.2.1 Node Fathoming Tests for Interval Algorithm

The *upper-bound test* is a classical criterion used in all branch-and-bound schemes: if the lower bound for a node is greater than the best upper bound for the MINLP, the node can be fathomed.

The *infeasibility test* is also used by all branch-and-bound algorithms. However, the identification of infeasibility using interval arithmetic differs from its identification using optimization schemes. Here, an inequality constraint $g_i(x, y) \leq 0$ is declared infeasible if $G_i(X, Y)$, its inclusion over the current domain, is positive. As soon as a constraint is found to be infeasible, the current node is fathomed.

The *monotonicity test* is only used in interval-based approaches. If a region is feasible, the monotonicity properties of the objective function can be tested. For this purpose, the inclusions of the gradients of the objective with respect to each variable are evaluated. If all the gradients have a constant sign for the current region, the objective function is monotonic and only one point needs to be retained from the current node.

The *nonconvexity test* is used to test the existence of a solution (local or global) within a region. If such a point exists, the Hessian matrix of the objective function at this point must be positive semi-definite. A sufficient condition for this to occur is the non-negativity of at least one of the diagonal elements of its interval Hessian matrix. The interval Hessian matrix is the inclusion of a Hessian matrix computed for a given domain.

[VEH96] advocate two additional tests to accelerate the fathoming process. The first is the so-called *lower bound test*. It requires the computation of a valid lower bound on the objective function through a method other than interval arithmetic. If the upper bound at a node is less than this lower bound, the region can be eliminated. The generation of such an upper bound may occur in an interval-based approach as the constraints are not used when evaluating the objective. Thus, a region may be found feasible because of the overestimation inherent in interval calculations, and have an upper bound lower than the optimal solution. For general problems, the computation of a valid and tight lower bound on the objective function requires the use of rigorous convex lower bounding techniques such as those described in Section 4.5.

The second test, the *distrust-region method*, aims to help the algorithm identify infeasible regions so that they can be removed from consideration. Based on the knowledge of an infeasible point, interval arithmetic is used to identify an infeasible hypercube centered on that point.

4.2.2 Branching Step

The variable with the widest range is selected for branching. It can be a continuous or a discrete variable. In order to determine where to split the chosen variable, a relaxation of the MINLP is solved locally.

Continuous Branching Variable If the optimal value of the continuous branching variable, x^*, is equal to one of the variable bounds, branch at the midpoint of the interval. Otherwise, branch at $x^* - \beta$, where β is a very small scalar.

Discrete Branching Variable If the optimal value of the continuous branching variable, y^*, is equal to the upper bound on the variable, define a region with $y = y^*$ and one with $y^L \leq y \leq y^* - 1$, where y^L is the lower bound on y. Otherwise, create two regions $y^L \leq y \leq int(y^*)$ and $int(y^*) + 1 \leq y \leq y^U$, where y^U is the upper bound on y.

This algorithm has been tested on a small example problem and a molecular design problem [VEH96].

4.3 Extended Cutting Plane for Nonconvex MINLPs

The use of the ECP algorithm for nonconvex MINLP problems was suggested in [WPG94], using a slightly modified algorithmic procedure. The main changes occur in the generation of new constraints for the MILP at each iteration (Step 4). In addition to the construction of the linear function $l_k(x, y)$ at iteration k, the following steps are taken:

1. Remove all constraints for which $l_i(x^k, y^k) > g_{j_i}(x^k, y^k)$. These correspond to linearizations which did not underestimate the corresponding nonlinear constraint at all points due to the presence of nonconvexities.

2. Replace all constraints for which $l_i(x^k, y^k) = g_{j_i}(x^k, y^k) = 0$ by their linearization around (x^k, y^k).

3. If constraint i is such that $g_i(x^k, y^k) > 0$, add its linearization around (x^k, y^k).

The convergence criterion is also modified. In addition to the test used in Step 3, the following two conditions must be met:

1. $(x^k - x^{k-1})^T(x^k - x^{k-1}) \leq \delta$, a pre-specified tolerance.

2. $y^k - y^{k-1} = 0$.

The ECP algorithm has been used to address a nonlinear pump configuration problem [WPG94], where it was found to give good results for convex one-level problems, and to perform poorly for nonconvex problems. It has also been tested on a small convex MINLP from [DG86]. Finally, a comparative study between the Outer Approximation, the Generalized Benders Decomposition and the Extended Cutting Plane algorithm was presented in [SHW+96]. A parameter estimation problem from FTIR spectroscopy and a purely integer problem were addressed.

4.4 Reformulation/Spatial Branch-and-Bound Algorithm

A global optimization algorithm branch-and-bound algorithm has been proposed in [SP97]. It can be applied to problems in which the objective and constraints are functions involving any combination of binary arithmetic operations (addition, subtraction, multiplication and division) and functions that are either concave over the entire solution space (such as ln) or convex over this domain (such as exp).

The algorithm starts with an automatic reformulation of the original nonlinear problem into a problem that involves only linear, bilinear, linear fractional, simple exponentiation, univariate concave and univariate convex terms. This is achieved through the introduction of new constraints and variables. The reformulated problem is then solved to global optimality using a branch-and-bound approach. Its special structure allows the construction of a convex relaxation at each node of the tree. The integer variables can be handled in two ways during the generation of the convex lower bounding problem. The integrality condition on the variables can be relaxed to yield a convex NLP which can then be solved globally. Alternatively, the integer variables can be treated directly and the convex lower bounding MINLP can be solved using a branch-and-bound algorithm as described in Section 3.7. This second approach is more computationally intensive but is likely to result in tighter lower bounds on the global optimum solution.

In order to obtain an upper bound for the optimum solution, several methods have been suggested. A local MINLP algorithm as those described in Section 3 can be used. The MINLP can be transformed to an equivalent nonconvex NLP by relaxing the integer variables. For example, a variable $y \in \{0, 1\}$ can be replaced by a continuous variable $z \in [0, 1]$ by including the constraint $z - z \cdot z = 0$. The nonconvex NLP is then solved locally to provide an upper bound. Finally, the discrete variables could be fixed to some arbitrary value and the nonconvex NLP solved locally.

Branching variables in this algorithm can be either continuous or discrete variables. An "approximation error" is computed for each term in the problem as the distance between the original term and its convex relaxation. A variable that participates in the term with the largest such error is selected for branching. Finally, the authors perform bound updates on all variables in order to ensure tight underestimators are generated. This algorithm has been applied to several problems such as reactor selection, distillation column design, nuclear waste blending, heat exchanger network design and multilevel pump configuration.

4.5 The SMIN-αBB Algorithm

This algorithm, proposed in [AAF7a], is designed to address the following class of problems to global optimality:

$$\begin{aligned}
\min \quad & f(x) + x^T A_0 y + c_0^T y \\
\text{s.t.} \quad & h(x) + x^T A_1 y + c_1^T y = 0 \\
& g(x) + x^T A_2 y + c_2^T y \leq 0 \\
& x \in X \subseteq \mathbb{R}^n \\
& y \in Y \text{ integer}
\end{aligned} \qquad (36)$$

where c_0^T, c_1^T and c_2^T are constant vectors, A_0, A_1 and A_2 are constant matrices and $f(x)$, $h(x)$ and $g(x)$ are functions with continuous second-order derivatives.

The solution strategy for problems of type (36) is an extension of the αBB algorithm for twice-differentiable NLPs [AMF95, AF96, ADFN97]. It is based on the generation of two converging sequences of upper and lower bounds on the global optimum solution. A rigorous underestimation and convexification strategy for functions with continuous second-order derivatives allows the construction of a lower bounding MINLP problem with convex functions in the continuous variables. If no mixed-bilinear terms are present ($A_i = 0, \forall i$), the resulting MINLP can be solved to global optimality using the Outer Approximation algorithm (OA) described in Section 3.2. Otherwise, the Generalized Benders Decomposition (GBD) can be used, as discussed in Section 3.1, or the Glover transformations [Glo75] can be applied to remove these bilinearities and permit the use of the OA algorithm. This convex MINLP provides a valid lower bound on the original MINLP. An upper bound on the problem can be obtained by applying the OA algorithm or the GBD to problem (36) to find a local solution. This bound generation strategy is incorporated within a branch-and-bound scheme: a lower and upper bound on the global solution are first obtained for the entire solution space. Subsequently, the domain is subdivided by branching on a binary or a continuous variable, thus creating new nodes for which upper and lower bounds can be computed. At each iteration, the node with the lowest lower bound is selected for branching. If the lower bounding MINLP for a node is infeasible or if its lower bound is greater than the best upper bound, this node is fathomed. The algorithm is terminated when the best lower and upper bound are within a pre-specified tolerance of each other.

Before presenting the algorithmic procedure, an overview of the underestimation and convexification strategy is given, and some of the options available within the algorithm are discussed.

4.5.1 Convex Underestimating MINLP Generation

In order to transform an MINLP problem of the form (36) into a convex problem which can be solved to global optimality with the OA or GBD algorithm, the functions $f(x)$, $h(x)$ and $g(x)$ must be convexified. The underestimation and convexification strategy used in the αBB algorithm has previously been described in detail [AAMF96, AF96, ADFN97]. Its main features are exposed here.

In order to construct as tight an underestimator as possible, the nonconvex functions are decomposed into a sum of convex, bilinear, univariate concave and general nonconvex terms. The overall function underestimator can then be built by summing up the convex underestimators for all terms, according to their type. In particular, a new variable is introduced to replace each bilinear term, and is bounded by the convex envelope of the term [AKF83]. The univariate concave terms are linearized. For each nonconvex term $nt(x)$ with Hessian matrix $H_{nt}(x)$, a convex underestimator $L(x)$ is defined as

$$L(x) = nt(x) - \sum_i \alpha_i (x_i^U - x_i)(x_i - x_i^L) \tag{37}$$

where x_i^U and x_i^L are the upper and lower bounds on variable x_i respectively and the α parameters are nonnegative scalars such that $H_{nt}(x) + 2\text{diag}(\alpha_i)$ is positive semi-definite over the domain $[x^L, x^U]$. The rigorous computation of the α parameters using interval Hessian matrices is described in [AAMF96, AF96, ADFN97].

The underestimators are updated at each node of the branch-and-bound tree as their quality strongly depends on the bounds on the variables.

4.5.2 Branching Variable Selection

An unusual feature of the SMIN-αBB algorithm is the strategy used to select branching variables. It follows a hybrid approach where branching may occur both on the integer and the continuous variables in order to fully exploit the structure of the problem being solved. After the node with the lowest lower bound has been identified for branching, the type of branching variable must be determined according to one of the following two criteria:

1. Branch on the binary variables first.

2. Solve a continuous relaxation of the nonconvex MINLP locally. Branch on a binary variable with a low degree of fractionality at the solution. If there is no such variable, branch on a continuous variable.

The first criterion results in the creation of an integer tree for the first q levels of the branch-and-bound tree, where q is the number of binary variables. At the lowest level of this integer tree, each node corresponds to a nonconvex NLP and the lower and upper bounding problems at subsequent levels of the tree are NLP problems. The efficiency of this strategy lies in the minimization of the number of MINLPs that need to be solved. The combinatorial nature of the problem and its nonconvexities are handled sequentially. If branching occurs on a binary variable, the selection of that variable can be done randomly or by solving a relaxation of the nonconvex MINLP and choosing the most fractional variable at the solution.

The second criterion selects a binary variable for branching only if it appears that the two newly created nodes will have significantly different lower bounds. Thus, if a variable is close to integrality at the solution of the relaxed problem, forcing it to take on a fixed value may lead to the infeasibility of one of the nodes or the generation of a high value for a lower bound, and therefore the fathoming of a branch of the tree. If no binary variable is close to integrality, a continuous variable is selected for branching.

A number of rules have been developed for the selection of a continuous branching variable. Their aim is to determine which variable is responsible for the largest separation distances between the convex underestimating functions and the original nonconvex functions. These efficient rules are exposed in [AAF7b].

4.5.3 Variable Bound Updates

Variable bound updates performed before the generation of the convex MINLP have been found to greatly enhance the speed of convergence of the αBB algorithm for continuous problems [AAF7b]. For continuous variables, the variable bounds are updated by minimizing or maximizing the chosen variable subject to the convexified constraints being satisfied. In spite of its computational cost, this procedure often leads to significant improvements

Mixed-Integer Nonlinear Optimization in Process Synthesis

in the quality of the underestimators and hence a noticeable reduction in the number of iterations required.

In addition to the update of continuous variable bounds, the SMIN-αBB algorithm also relies on binary variable bound updates. Through simple computations, an entire branch of the branch-and-bound tree may be eliminated when a binary variable is found to be restricted to 0 or 1. The bound update procedure for a given binary variable is as follows:

1. Set the variable to be updated to one of its bounds $y = y_B$.

2. Perform interval evaluations of all the constraints in the nonconvex MINLP, using the bounds on the solution space for the current node.

3. If any of the constraints are found infeasible, fix the variable to $y = 1 - y_B$.

4. If both bounds have been tested, repeat this procedure for the next variable to be updated. Otherwise, try the second bound.

4.5.4 Algorithmic Procedure

The algorithmic procedure for the SMIN-αBB algorithm is formalized as follows:

Step 1
    Set absolute tolerance ϵ; set $LBD^* = -\infty$ and $UBD^* = \infty$.
    Set node counter $k = 0$.
    initialize set of nodes to be bounded, $J = \{0\}$;
    N, the list of nodes to be explored, is empty.

Step 2 *Bounding*
    For each node $N_j, j \in J$:
        Perform variable bound updates if desired.
        Generate a convex lower bounding MINLP.
        Solve convex MINLP using OA or GBD. Solution is LBD_j.
        If MINLP is infeasible, fathom the current node.
        If $LBD_j \leq UBD^*$, add the current node to N.
        Else, fathom the current node.

Step 3
    Set LBD^* to the lowest lower bound from the list N.
    If $UBD^* - LBD^* \leq \epsilon$, terminate with solution UBD^*.

Otherwise, proceed to Step 4.

Step 4 <u>Branching</u>

Select the node from the list N with the lowest lower bound for branching, N_i ($i < k$),

Its lower bound is LBD_i.

Select a branching variable y^B or x^B.

Create two new regions N_{k+1} and N_{k+2}.

Set $J = \{k+1, k+2\}$ and $k = k+2$. Go back to Step 2.

4.6 The GMIN-αBB algorithm

This algorithm operates within a classical branch-and-bound framework as described in Section 3.7. The main difference with the algorithms of [GR85], [OOM90] and [BM91] is its ability to identify the global optimum solution of a much larger class of problems of the form

$$\begin{aligned}
\min_{x,y} \quad & f(x,y) \\
\text{s.t.} \quad & h(x,y) = 0 \\
& g(x,y) \leq 0 \\
& x \in X \subseteq \mathbb{R}^n \\
& y \in \mathcal{N}^q
\end{aligned} \tag{38}$$

where $\mathcal{N}$ is the set of non-negative integers and the only condition imposed on the functions $f(x,y)$, $g(x,y)$ and $h(x,y)$ is that their continuous relaxations possess continuous second-order derivatives.

This increased applicability results from the use of the αBB global optimization algorithm for continuous twice-differentiable NLPs [AMF95, AF96, ADFN97]. The basic concepts behind the αBB algorithm were exposed in Section 4.5.

At each node of the branch-and-bound tree, the nonconvex MINLP is relaxed to give a nonconvex NLP, which is then solved with the αBB algorithm. This allows the identification of rigorously valid lower bounds and therefore ensures convergence to the global optimum. In general, it is not necessary to let the αBB algorithm run to completion as each one of its iterations generates a lower bound on global solution of the NLP being solved. A strategy of early termination leads to a reduction in the computational requirements of each node of the binary branch-and-bound tree and faster overall convergence.

The GMIN-αBB algorithm selects the node with the lowest lower bound for branching at every iteration. The branching variable selection strategy combines several approaches: branching priorities can be specified for some of the integer variables. When no variable has a priority greater than all other variables, the solution of the continuous relaxation is used to identify either the most fractional variable or the least fractional variable for branching.

Other strategies have been implemented to ensure a satisfactory convergence rate. In particular, bound updates on the integer variables can be performed at each level of the branch-and-bound tree. These can be carried out through the use of interval analysis. An integer variable, y^*, is fixed at its lower (or upper) bound and the range of the constraints is evaluated with interval arithmetic, using the bounds on all other variables. If the range of any constraint is such that this constraint is violated, the lower (or upper) bound on variable y^* can be increased (or decreased) by one. Another strategy for bound updates is to relax the integer variables, to convexify and underestimate the nonconvex constraints and to minimize (or maximize) a variable y^* in this convexified feasible region. The resulting lower (or upper) bound on relaxed variable y^* can then be rounded up (or down) to the nearest integer to provide an updated bound for y^*.

A number of small nonconvex MINLP test problems as well as the pump configuration problem of [WPG94] have been solved using this strategy.

5 Implementation: MINOPT

Although there are a number of algorithms available for the solution of MINLPs, there are relatively few implementations of these algorithms. The recent advances in the development of these algorithms has led to several automated implementations of these MINLP algorithms.

The earliest implementations make use of the modeling system **GAMS** [BKM92] which allows algebraic model representation and automatic interfacing with linear, nonlinear and mixed integer linear solvers. The algorithmic procedure, **APROS** [PF89], was developed for the automatic solution of mathematical programming problems involving decomposition techniques such as those used in the solution of MINLPs. **APROS** is an implementation of **GBD** and **OA** in **GAMS** where the modeling language is used to generate the NLP and MILP subproblems which are solved through the GAMS interface. **GAMS** also includes a direct interface to an implementation of **OA/ER** in the package **DICOPT++** [VG90]. The model can be written algebraically as an MINLP and the solver will perform the necessary decomposition.

More recently, a framework **MINOPT** [SF97b] has been developed for the solution of general mathematical programming problems. The primary motivation for the development of **MINOPT** (Mixed Integer Nonlinear OPTimizer) was to provide a user friendly interface for the solution of MINLPs. The development has expanded to include an interface for solving many classes of problems which include both algebraic and differential models. The next section describes this package in more detail and includes the results of its application to a number of example problems.

MINOPT has been developed as a framework for the solution of various classes of optimization problems. Its development has been brought about by the particular need for implementations of algorithms applicable to MINLPs. Further development has been done to address the solution of problems which involve dynamic as well as algebraic models. Extensive development of **MINOPT** has led to a highly developed computational tool.

MINOPT has a number of features including:

- Extensive implementation of optimization algorithms
- Front-end parser
- Extensive options
- Expandable platform
- Interface routines callable as a subroutine

MINOPT is capable of handling a wide variety of problems described by the variable and constraint types employed. **MINOPT** handles the following variable types:

- continuous time invariant
- continuous dynamic
- control
- integer

and recognizes the following constraint types:

- linear
- nonlinear

- dynamic
- dynamic path
- dynamic point

Different combinations of variable and constraint types lead to the following problem classifications:

- Linear Program (LP)
- Nonlinear Program (NLP)
- Mixed Integer Linear Program (MILP)
- Mixed Integer Nonlinear Program (MINLP)
- Nonlinear Program with Differential and Algebraic Constraints (NLP/DAE)
- Mixed Integer Nonlinear Program with Differential and Algebraic Constraints (MINLP/DAE)
- Optimal Control Program (OCP)
- Mixed Integer Optimal Control Program (MIOCP)

The **MINOPT** program has two phases: problem entry and problem solution. During the first phase **MINOPT** reads the input from a file, saves the problem information, and then determines the structure and consistency of the problem by analyzing the constraints and variables. After the problem has been entered, **MINOPT** proceeds to the second phase to solve the problem. Based on the problem structure determined by **MINOPT** and options supplied by the user, **MINOPT** employs the appropriate algorithm to solve the problem.

The entry phase of **MINOPT** features a parser which reads in the dynamic and/or algebraic problem formulation from an input file. The input file has a clear syntax and allows the user to enter the problem in a concise form without needing to specify the steps of the algorithm. The input file includes information such as variable names, variable partitioning (continuous, integer, dynamic), parameter definitions, and option specifications. The parser features index notation which allows for compact model representation. The parser allows for general constraint notation and has the ability to recognize and handle the various constraint types (i.e. linear, nonlinear, dynamic, point, path) and ultimately the overall structure of the problem. The **MINOPT** parser also determines the necessary analytical Jacobian information from the problem formulation.

The solution phase of **MINOPT** features extensive implementations of numerous optimization algorithms. Once the parser has determined the problem type, the solution phase applies the appropriate method to solve the problem. **MINOPT** utilizes available software packages for the solution of various subproblems. The solution algorithms implemented by **MINOPT** are listed in Table 1. The solution algorithms implemented by **MINOPT** are callable as subroutines from other programs.

Table 1: Solution algorithms implemented by **MINOPT**

Problem Type	Algorithm	Solver
LP	Simplex method	CPLEX MINOS LSSOL
MILP	Branch and Bound	CPLEX
NLP	Augmented Lagrangian/Reduced Gradient	MINOS
	Sequential Quadratic Programming	NPSOL
	Sequential Quadratic Programming	SNOPT
Dynamic	Integration (Backward Difference)	DASOLV
Optimal Control	Control Parameterization	DAEOPT
MINLP	Generalized Benders Decomposition	**MINOPT**
	Outer Approximation/Equality Relaxation	**MINOPT**
	Outer Approximation/Augmented Penalty	**MINOPT**
	Generalized Cross Decomposition	**MINOPT**

MINOPT has an extensive list of options which allows the user to fine tune the various algorithms.

- selection of different algorithms for a problem type
- selection of parameters for various algorithms
- solution of the relaxed MINLP

Mixed-Integer Nonlinear Optimization in Process Synthesis

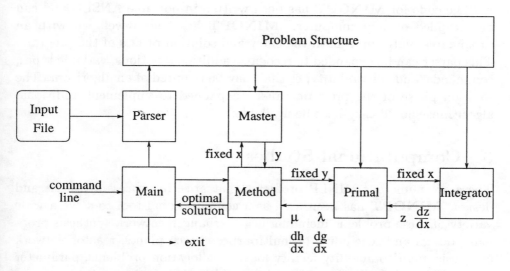

Figure 4: Program flow for **MINOPT**

- auto-initialization procedure—relaxed MINLP solved to determine the starting values for the y-variables.

- integer cuts for the **GBD** algorithm

- radial search technique for problems with discrete and continuous y variables (**GBD**)

- alternative feasibility formulation for infeasible primal

- solution of the **GBD** master problem in terms of both x and y rather than in y alone

- specification of parameters for external solvers

The flow of the program is described in Figure 4. The program is invoked from the command line and parses the input file and stores the information into a problem structure. The program then determines the appropriate method to solve the problem based on the problem type and options provided by the user. Based on the algorithm and parameters, **MINOPT** solves the problem by formulating and solving various subproblems. When needed, **MINOPT** draws necessary information from the problem structure.

The code for **MINOPT** has been written in portable ANSI C and can be compiled on any computer. **MINOPT** has been developed with an expandable platform in both the entry and solution phases of the program. This parser can be expanded to recognize additional options, variable types, commands, and constraint types that may be required of an algorithm. The solution phase of the program can be expanded to implement additional algorithms should they become available.

6 Computational Studies

There are numerous MINLP problems that arise in process synthesis and design. **MINOPT** has been used as a computational tool to solve a wide variety of these problems including heat exchanger network synthesis problems, design and scheduling of multipurpose batch plants, reactor network synthesis, multicommodity facility location-allocation problems, parameter estimation, and distillation sequencing problems. The computational results for these problems run on a Hewlett Packard 9000/780 (C-160) are shown in Table 2.

Two computational examples are selected from the area of process synthesis. Both of these examples illustrate the use of a superstructure, the mathematical modeling of the superstructure as well as the implementation of an appropriate algorithm to solve the problem. The first problem is a distillation sequencing problem and the second is a heat exchanger network synthesis problem.

6.1 Distillation Sequencing

The distillation sequencing problem is to determine the configuration of a separation system which will separate a given feed stream into desired products which meet desired specifications. The details for the problem description and model formulation can be found in [AF90].

The input flowrate and composition are given along with the desired product flowrates and compositions. The problem is to determine the flowrates and compositions of the streams and the interconnection of distillation units which minimizes the annualized cost. The superstructure postulating two distillation units for one input and two outputs is shown in Figure 5.

The continuous variables for the problem are the flowrates, F, the mole fractions, x, and the recoveries for the light key and heavy key, r^{lk} and r^{hk}. The binary variables represent the existence of a distillation column, y.

Table 2: Computational Results

Problem	X	Y	Z	L	N	T	ITER	CPU TIME
gbd_test	1	3	–	4	1	D	2/2/2	0.06/0.06/0.06
oaer_test	6	3	–	5	2	D	3/2/3	0.18/0.25
ap_test	2	1	–	3	1	D	2/–/2	0.07/–/0.07
minutil	206	–	–	87	–	A		0.09
minmatch	82	18	–	129	–	B		0.11
plan	22	–	–	14	–	A		0.05
schedule	12	16	–	27	–	B		0.13
batdes	10	9	–	18	2	D	5/2/2	0.28/0.12/0.13
complex	8	3	–	10	–	B		0.08
alky	10	–	–	3	5	C		1.8
cart	6	–	4	3	4	E		0.63
param	12	–	2	1	10	E		4.38
feedtray	93	9	–	34	62	D	2/2/2	1.43/1.74/5.38
procsel	7	3	–	6	2	D	6/4/3	0.31/0.27/0.28
alan	4	4	–	7	1	D	8/4/6	0.20/0.12/0.17
facility1	16	6	–	18	1	D	3/3/4	0.14/0.17/0.21
facility2	54	12	–	33	1	D	5/5/7	0.43/0.73/0.97
ciric1	5	1	–	6	1	C	15/–/–	0.29/–/–
duran86-1	3	3	–	4	3	D	4/4/4	0.27/0.28/0.24
duran86-2	6	5	–	11	4	D	8/4/3	1.16/0.50/0.40
duran86-3	9	8	–	19	5	D	14/4/7	3.24/0.66/1.30
duran86-4	5	25	–	6	26	D	69/2/11	42.1/0.4/38.50
kocis88-4a	16	6	–	60	2	D	16/–/–	3.74/–/–
kocis88-4b	22	24	–	72	2	D	31/4/3	20.28/1.86/0.89
kocis89-2a	27	2	–	29	6	D	3/3/2	0.54/0.57/0.50
kocis89-2b	27	8	–	34	1	D	3/–/–	0.55
meanv1	21	14	–	44	1	D	10/4/3	0.48/0.33/0.26
meanv2	28	14	–	51	1	D	7/2/3	0.35/0.21/0.28
nous1	22	28	–	49	1	D	12/–/–	2.38/–/–
vdv	22	14	4	27	8	F	2/–/–	16.02/–/–
bindis	72	60	114	67	4	F	4/3/–	2860.0/2100.0/–

X, Y, Z, L, N and T indicate the number of x, y, and z variables, number of linear and nonlinear constraints and the type of the problem (A–LP, B–MILP, C–NLP, D–MINLP, E–NLP/DAE, F–MINLP/DAE). ITER and CPU TIME indicate the number of iterations and cpu time for the MINLP problems (GBD/OAER/OAERAP).

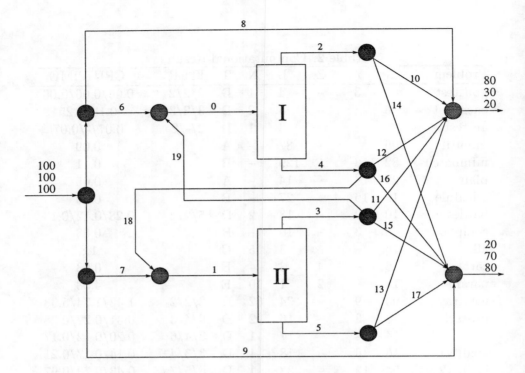

Figure 5: Superstructure for Nonsharp Separation System

The derivation of the mathematical model involves a number of indices and sets. The index set $I = \{i\}$ denotes components, $N = \{k\}$ denotes streams, $J = \{j\}$ denote columns, $S = \{s\}$ denote splitters, $M^c = \{m^c\}$ denote the mixers prior to each column, $M^f = \{m^f\}$ denote final mixers prior to each product, and $P = \{p\}$ denote the products. The splitter $s^0 \in S$ represents the initial splitting point of the feed stream. The following sets are defined for the connection of the sets of splitters and mixers with the streams in the superstructure.

$$
\begin{aligned}
S_s^{in} &\equiv \{l | l \in N \text{ is an inlet to splitter } s\} \; s \in S \\
S_s^{out} &\equiv \{l | l \in N \text{ is an outlet from splitter } s\} \; s \in S \\
M_{m^f}^{in} &\equiv \{l | l \in N \text{ is an inlet to mixer } m^f\} \; m^f \in M^f \\
M_{m^c}^{out} &\equiv \{l | l \in N \text{ is an outlet to mixer } m^c\} \; m^c \in M^c
\end{aligned}
$$

Mixed-Integer Nonlinear Optimization in Process Synthesis

The inlet and outlet streams of the column are given by the following:

$$SU_j \equiv \{n | n \in N \text{ is the inlet to column } j\} \; j \in J$$
$$SU_j^{top} \equiv \{p | p \in N \text{ is the top product of column } j\} \; j \in J$$
$$SU_j^{bot} \equiv \{q | q \in N \text{ is the bottom product of column } j\} \; j \in J$$

The key components for the column are given by the following:

$$LK_j \equiv \{i | i \in I \text{ is the light key for column } j\} \; j \in J$$
$$HK_j \equiv \{i | i \in I \text{ is the heavy key for column } j\} \; j \in J$$
$$LHK_j \equiv \{i | i \in I \text{ is lighter than the heavy key for column } j\} \; j \in J$$
$$HLK_j \equiv \{i | i \in I \text{ is heavier than the light key for column } j\} \; j \in J$$
$$ND_{ik} \equiv \{i | i \in I \text{ is not present in stream } k\} \; i \in I, k \in N$$

The problem formulation from [AF90] follows.

$$\min \sum_{j \in J} \left\{ a_{0j} + \left(a_{1j} + a_{2j} r_{ij}^{lk} + a_{3j} r_{i'j}^{hk} + \sum_{i \in I} b_{ij} x_{ik} \right) F_k \right\}$$

$$i \in LK_j, i' \in HK_j, k \in SU_j$$

s.t.
$$\sum_{k \in S_s^{in}} F_k - \sum_{k \in S_s^{out}} F_k = 0 \qquad s \in S$$

$$F_p x_{ik} - r_{ij}^{lk} f_{in} = 0 \qquad i \in LK_j, n \in SU_j, p \in SU_j^{top}, j \in J$$

$$F_q x_{ik} - r_{ij}^{hk} f_{in} = 0 \qquad i \in LK_j, n \in SU_j, q \in SU_j^{bot}, j \in J$$

$$f_{in} - F_n x_{in} = 0 \qquad i \in I, n \in SU_j$$

$$f_{in} - F_p x_{ip} - F_q x_{iq} = 0 \qquad i \in I, j \in J, n \in SU_j,$$
$$\qquad p \in SU_j^{top}, q \in SU_j^{bot}$$

$$\sum_{l \in M_{m^c}^{in}} F_l x_{il} - \sum_{l \in M_{m^c}^{out}} f_{il} = 0 \qquad i \in I, m^c \in M^c$$

$$x_{ik} = 0 \qquad (i,k) \in ND_{ik}$$

$$\sum_{l \in M_{mf}^{in}} -C_{ip} = 0 \qquad i \in I, p \in P$$

$$\sum_{i \in I} x_{ik} = 1 \qquad k \in S_s^{in}$$

$$F_k - U y_j \leq 0 \qquad j \in J, k \in SU_j$$

$$F_n - F_p - F_q = 0 \qquad n \in SU_j, p \in SU_j^{top}, q \in SU_j^{bot}$$

$$\sum_{l \in M_{m^c}^{in}} F_l - \sum_{l \in M_{m^c}^{out}} F_l = 0 \quad m^c \in M^c$$

$$\sum_{i \in M_{mf}^{in}} F_l - \sum_i C_{ip} \quad p \in P, i \in I$$

$$\sum_i f_{il} - F_l \quad l \in SU_j$$

$$F_l - \sum_{i \in LHK_j} f_{in} - \sum_{i \in HK_j} [1 - (r_{ij}^{hk})^L] f_{in} \leq 0$$

$$\quad n \in SU_j, l \in SU_j^{\text{top}}$$

$$F_l - \sum_{i \in HLK_j} f_{in} - \sum_{i \in LK_j} [1 - (r_{ij}^{lk})^L] f_{in} \leq 0$$

$$\quad n \in SU_j, l \in SU_j^{\text{bot}}$$

$$F_l \geq 0 \, f_{in} \geq 0 \quad i \in I, n \in N, n \in SU_j$$

$$x_{ik} \geq 0 \quad i \in I, k \in (SU_j \cup SU_j^{\text{top}} \cup SU_j^{\text{bot}})$$

$$(r^{lk})^L \leq r_{ij}^{lk} \leq (r^{lk})^U \quad i \in LK_j, j \in J$$

$$(r^{hk})^L \leq r_{ij}^{hk} \leq (r^{hk})^U \quad i \in HK_j, j \in J$$

The data for the problem are taken from Example 2 in [AF90].

The nonlinearities in the problem are due to bilinear terms. Some of the continuous variables in the problem are partitioned as **y** variables along with the binary variables such that the primal problem is linear and thus convex. Since the **y** variables consist of both continuous and binary variables, the **GBD** algorithm must be used.

The problem is solved using **MINOPT** which incorporates a radial search algorithm into the **GBD** algorithm. This option specifies that the full NLP problem with only the binary variables fixed is solved after each primal problem. The algorithm converges in 3 iterations and takes 1.36 seconds of CPU time on a Hewlett Packard 9000/780 (C-160). The optimal sequence utilizes a single column and the optimal flowsheet is shown in Figure 6.

6.2 Heat Exchanger Network Synthesis

The design of a heat exchanger network involving two hot streams, two cold streams, one hot and one cold utility is studied. The formulation of [YG91] is used. The annualized cost of the network is expressed as the summation of the utility costs, the fixed charges for the required heat-exchangers and an area-based cost for each each exchanger. The area is a highly nonlinear function of the heat duty and the temperature differences at both ends of

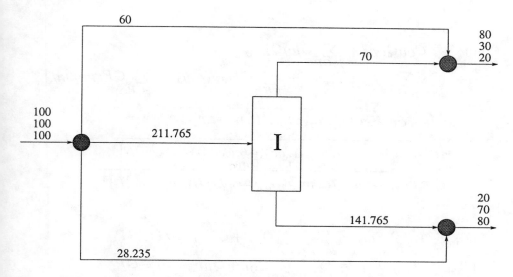

Figure 6: Optimal Flowsheet for the Nonsharp Separation Problem

the heat exchanger. The binary variables, which represent the existence of a given heat-exchanger, participate linearly in the problem. All the constraints are linear. This nonconvex MINLP therefore provides an opportunity to test the SMIN-αBB global optimization algorithm proposed in Section 4.5.

The stream data for the problem are summarized in Table 3. There are two temperature intervals. The steam utility costs $80/kW-yr and the cooling water costs $15/kW-yr. The fixed charges for the heat exchangers amount to $5500/yr. The cost coefficient for the area-dependent part of the heat exchanger costs is $300/yr. The overall heat transfer coefficients are 0.5 kW/m^{2}K for the hot stream-cold stream units, 0.83333 kW/m^{2}K for the cold stream-hot utility units and 0.5 kW/m^{2}K for the hot stream-cold utility units.

The superstructure for this problem is shown in Figure 7. There are 12 possible matches and therefore 12 binary variables. The global optimum configuration involves six heat exchangers and is shown in Figure 8. Given the set ST of K temperature locations, the set HP of hot process streams and the set CP of cold process streams, the general problem formulation is as follows:

$$\begin{aligned}
\min &\sum_{i \in HP} C_{CU} Q_{CU,i} + \sum_{j \in CP} C_{HU} Q_{HU,j} \\
&+ \sum_{i \in HP} \sum_{j \in CP} \sum_{k \in ST} CF_{ij} z_{ijk} + \sum_{i \in HP} CF_{i,CU} z_{CU,i} + \sum_{j \in CP} CF_{j,HU} z_{HU,j} \\
&+ \sum_{i \in HP} \sum_{j \in CP} \sum_{k \in ST} \frac{C_{ij} Q_{ijk}}{U_{ij}[\Delta T_{ijk} \Delta T_{ijk+1}(\Delta T_{ijk}+\Delta T_{ijk+1})/2]^{\frac{1}{3}}} \\
&+ \sum_{i \in HP} \frac{C_{i,CU} Q_{CU,i}}{U_{CU,i}[\Delta T_{CU,i}(T_{out,i}-T_{in,CU})(\Delta T_{CU,i}+T_{out,i}-T_{in,CU})/2]^{\frac{1}{3}}} \\
&+ \sum_{j \in CP} \frac{C_{j,HU} Q_{HU,j}}{U_{HU,j}[\Delta T_{HU,j}(T_{in,HU}-T_{out,j})(\Delta T_{HU,j}+T_{in,HU}-T_{out,j})/2]^{\frac{1}{3}}}
\end{aligned} \quad (39)$$

$$(T_{in,i} - T_{out,i}) Fcp_i = \sum_{k \in ST} \sum_{j \in CP} Q_{ijk} + Q_{CU,i} \quad \forall i \in HP$$

$$(T_{out,j} - T_{in,j}) Fcp_j = \sum_{k \in ST} \sum_{i \in HP} Q_{ijk} + Q_{HU,j} \quad \forall j \in CP$$

$$(T_{i,k} - T_{i,k+1}) Fcp_i = \sum_{j \in CP} Q_{ijk} \quad \forall k \in ST, \forall i \in HP$$

$$(T_{j,k} - T_{j,k+1}) F.cp_j = \sum_{i \in HP} Q_{ijk} \quad \forall k \in ST, \forall j \in CP$$

$$\begin{aligned}
T_{in,i} &= T_{i,1} & \forall i \in HP \\
T_{in,j} &= T_{j,K} & \forall j \in CP \\
T_{i,k} &\geq T_{i,k+1} & \forall k \in ST, \forall i \in HP \\
T_{j,k} &\geq T_{j,k+1} & \forall k \in ST, \forall j \in CP \\
T_{out,i} &\leq T_{i,K} & \forall i \in HP \\
T_{out,j} &\geq T_{j,1} & \forall j \in CP \\
(T_{i,K} - T_{out,i}) Fcp_i &= Q_{CU,i} & \forall i \in HP \\
(T_{out,j} - T_{j,1}) Fcp_j &= Q_{HU,j} & \forall j \in CP \\
Q_{ijk} - \Omega z_{ijk} &\leq 0 & \forall k \in ST, \forall i \in HP, \forall j \in CP \\
Q_{CU,i} - \Omega z_{CU,i} &\leq 0 & \forall i \in HP \\
Q_{HU,j} - \Omega z_{HU,j} &\leq 0 & \forall j \in CP \\
z_{ijk}, z_{CU,i}, z_{HU,j} &\in \{0,1\} & \forall k \in ST, \forall i \in HP, \forall j \in CP \\
T_{i,k} - T_{j,k} + \Gamma(1 - z_{ijk}) &\geq \Delta T_{ijk} & \forall k \in ST, \forall i \in HP, \forall j \in CP \\
T_{i,k+1} - T_{j,k+1} + \Gamma(1 - z_{ijk}) &\geq \Delta T_{ijk+1} & \forall k \in ST, \forall i \in HP, \forall j \in CP \\
T_{i,K} - T_{out,CU} + \Gamma(1 - z_{CU,i}) &\geq \Delta T_{CU,i} & \forall i \in HP \\
T_{out,HU} - T_{j,1} + \Gamma(1 - z_{HU,j}) &\geq \Delta T_{HU,j} & \forall j \in CP \\
\Delta T_{ijk} &\geq 10 & \forall k \in ST, \forall i \in HP, \forall j \in CP
\end{aligned}$$

Table 3: Stream data for heat exchanger network problem.

Stream	T_{in} (K)	T_{out} (K)	Fcp (kW/K)
Hot 1	650	370	10.0
Hot 2	590	370	20.0
Cold 1	410	650	15.0
Cold 2	350	500	13.0
Steam	680	680	—
Water	300	320	—

where the parameters are C_{CU}, the per unit cost of cold utility; C_{HU}, the per unit cost of hot utility; CF, the fixed charged for heat exchangers; C, the area cost coefficient; T_{in}, the inlet temperature of a stream; T_{out}, the outlet temperature; Fcp, the heat capacity flowrate of a stream; Ω, the upper bound on heat exchange; Γ, the upper on the temperature difference. The continuous variables are T_{ik}, the temperature of hot stream i at the hot end of stage k; T_{jk}, the temperature of cold stream j at the cold end of stage k, Q_{ijk}, the heat exchanged between hot stream i and cold stream j at temperature location k; $Q_{CU,i}$, the heat exchanged between hot stream i and the cold utility at temperature location k; $Q_{HU,j}$, the heat exchanged between cold stream j and the hot utility at temperature location k; ΔT_{ijk}, the temperature approach for the match of hot stream i and cold stream j at temperature location k; $\Delta T_{CU,i}$, the temperature approach for the match of hot stream i and the cold utility at temperature location k; $\Delta T_{HU,j}$, the temperature approach for the match of cold stream j and the hot utility at temperature location k. The binary variables are z_{ijk}, for the existence of a match between hot stream i and cold stream j at temperature location k; $z_{CU,i}$, for the existence of a match between hot stream i and the cold utility at temperature location k; $z_{HU,j}$, for the existence of a match between cold stream j and the hot utility at temperature location k.

Due to the linear participation of the binary variables, the problem can be solved locally using the Outer Approximation or Generalized Benders Decomposition algorithms described in Sections 3.2 and 3.1, and globally using the SMIN-αBB algorithm of Section 4.5.

This problem can be solved locally using **MINOPT**. For both **GBD** and **OAER** the problem is solved 30 times with random starting values for the binary variables. The starting values for the continuous variables are

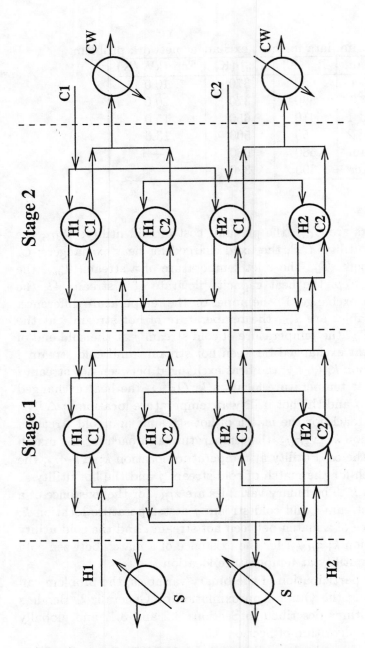

Figure 7: Superstructure for the heat exchanger network problem.

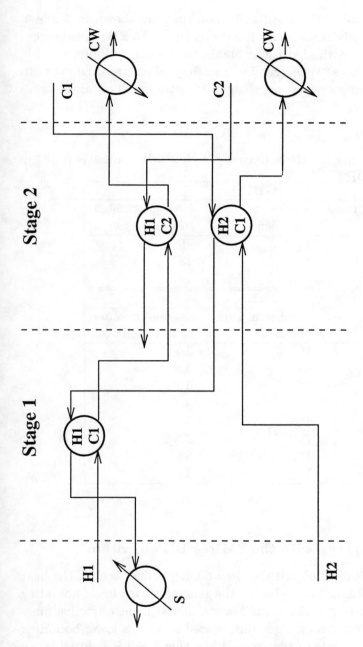

Figure 8: Optimum configuration for the heat exchanger network problem.

set to their lower bounds. The results of these runs are shown in Table 4. Whereas **GBD** generally takes more iterations than **OAER**, it converges to fewer local minima. Both algorithms obtain the global optimum roughly the same number of times. When random starting values are used for both the binary and continuous variables, the global optimum is obtained in all 30 runs.

Table 4: Local solutions for Heat Exchanger Network Synthesis problem obtained with **MINOPT**

GBD		
Local Solutions	number of times obtained	average number of iterations
154997	11	16
155510	18	18
161010	1	14

OAER		
Local Solutions	number of times obtained	average number of iterations
154997	10	3.1
155510	6	3.8
167602	3	6
180848	1	5
189521	1	5
197983	7	3.6
199196	1	5
212678	1	3

6.2.1 Solution Strategy with the SMIN-αBB algorithm

When using the SMIN-αBB algorithm, the area-dependent cost of the heat exchangers must be underestimated using the general convex lower bounding function (37), in order to generate valid lower bounds on the objective function. The Outer Approximation algorithm is used to solve a lower bounding convex MINLP at each node of the tree. When this MINLP is feasible, an upper bound on the objective function is obtained by solving the nonconvex MINLP locally in the same region. For the heat exchanger between hot stream i and cold stream j, the convex underestimator is expressed as

$$\frac{C_{ij}Q_{ijk}}{U_{ij}[\Delta T_{ijk}\Delta T_{ijk+1}(\Delta T_{ijk}+\Delta T_{ijk+1})/2]^{\frac{1}{3}}} \\ - \alpha_{ijk}^{Q}(Q_{ijk}^{U} - Q_{ijk})(Q_{ijk} - Q_{ijk}^{L}) \\ - \alpha_{ijk}^{\Delta T_k}(\Delta T_{ijk}^{U} - \Delta T_{ijk})(\Delta T_{ijk} - \Delta T_{ijk}^{L}) \\ - \alpha_{ijk}^{\Delta T_{k+1}}(\Delta T_{ijk+1}^{U} - \Delta T_{ijk+1})(\Delta T_{ijk+1} - \Delta T_{ijk+1}^{L}). \tag{40}$$

where α_{ijk}^{Q}, $\alpha_{ijk}^{\Delta T_k}$ and $\alpha_{ijk}^{\Delta T_{k+1}}$ are non-negative scalars obtained through one of the methods described by [AAF7a]. The convex underestimator for process stream-utility exchangers is similar, expect that one of the ΔT's is constant and only two α terms are therefore required. At the first level of the branch-and-bound tree, all binary variables can take on a value of either 0 or 1. As a result, every nonconvex term in the objective function must be underestimated to obtain a lower bound valid for the entire solution space. However, if branching occurs on the binary variables, the existence of some units is pre-determined for subsequent levels of the branch-and-bound tree. Thus, if some variable z_{ijk} is fixed to 0 at a node of the tree, proper updating of the variable bounds yields $Q_{ijk} = Q_{ijk}^{L} = Q_{ijk}^{U} = 0$. The bounds on ΔT_{ijk} and ΔT_{ijk+1} become $10 \leq \Delta T_{ijk} \leq T_{i,k} - T_{j,k} + \Gamma$ and $10 \leq \Delta T_{ijk+1} \leq T_{i,k+1} - T_{j,k+1} + \Gamma$. Since Γ is a large number, the convex terms corresponding the ΔT's do not naturally vanish from Equation (40). Even though unit the area of unit (ijk) is 0, its cost appears in the underestimating objective function as

$$-\alpha_{ijk}^{\Delta T_k}(\Delta T_{ijk}^{U} - \Delta T_{ijk})(\Delta T_{ijk} - \Delta T_{ijk}^{L}) \\ - \alpha_{ijk}^{\Delta T_{k+1}}(\Delta T_{ijk+1}^{U} - \Delta T_{ijk+1})(\Delta T_{ijk+1} - \Delta T_{ijk+1}^{L}). \tag{41}$$

In order to eliminate this redundant term, it is therefore necessary to introduce modified α parameters which account for the non-existence of a unit. These new parameters are defined as

$$\begin{aligned} \alpha_{ijk}^{*,\Delta T_k} &= \alpha_{ijk}^{\Delta T_k} z_{ijk}^{U} \\ \alpha_{ijk}^{*,\Delta T_{k+1}} &= \alpha_{ijk}^{\Delta T_{k+1}} z_{ijk}^{U}. \end{aligned} \tag{42}$$

where z_{ijk}^{U} is the current upper bound on variable z_{ijk}. According to Equation (42), if z_{ijk} is fixed to 0, its upper bound z_{ijk}^{U} is 0 and $\alpha_{ijk}^{*,\Delta T_k}$ and

$\alpha_{ijk}^{*,\Delta T_{k+1}}$ vanish. The convex underestimator for unit (ijk) no longer participates in the lower bounding objective function. On the contrary, if z_{ijk} is fixed to 1 or remains free to take on the value of 0 or 1, the convex underestimator is preserved.

This analysis of the objective function emphasizes the importance of the branching strategy in the generation of tight lower bounds on the objective function. Several branching strategies were used for this problem. First, the continuous variables were branched on exclusively (Run 1). Then, for Runs 2 and 3, the binary variables were branched on first, followed by the continuous variables. Finally, the "almost-integer" strategy described in Section 4.5.2 was used for Runs 4, 5 and 6. A binary variable was declared to have a low degree of fractionality if its value z^* at the solution of the relaxed MINLP was such that $\min\{z^*, 1-z^*\} \leq zdist$. For Run 4, $zdist = 0.1$ was used and for Runs 5 and 6, $zdist = 0.2$ was used.

A number of variable bound update strategies were also tested for this problem. In Runs 1 and 2, updates were performed only for the continuous variables. In all other runs, the bounds on the binary variables were also updated. In Run 6, the effect of updating the bounds on only a fraction of the continuous variables was studied.

The results are shown in Figure 9 and Table 5. Branching on the continuous variables only results in slow asymptotic convergence of algorithm to the global optimum solution (Run 1). The rate of convergence is greatly improved when the binary variables can be used for branching (Runs 2 to 6). Although the "almost-integer" branching strategy exhibits the best performance in terms of iterations (Runs 4 to 6), the lowest CPU requirements correspond to Run 3, which branches on all the binary variables before turning to the continuous variables. The average time spent on each iteration of the algorithm is therefore greater when the "almost-integer" strategy is applied. Two factors can account for this increase in the computational requirements. First, the selection of a binary branching variable requires the solution of a nonconvex MINLP. In addition, the generation of a lower bound on the solution at almost every node of the branch-and-bound tree for Runs 4 to 6 necessitates the solution of a convex MINLP. By comparison, only 58% of the nodes in the branch-and-bound tree for Run 3 involve the solution of a convex MINLP. Lower bounds at the remaining nodes are obtained by solving less expensive convex NLPs. Addressing the combinatorial aspects of the problem first by branching on the binary variables thus leads to the better performance of the SMIN-αBB algorithm.

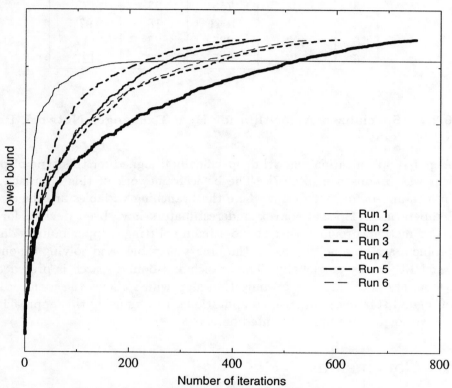
Figure 9: Progress of the lower bound for the heat exchanger network

Table 5: Global optimization of heat exchanger network – Note that Run 1 converges asymptotically.

Run	Iterations	CPU sec	Deepest level	Binary branches
1	800	2210	60	—
2	753	1116	26	343
3	604	755	23	173
4	451	1041	18	97
5	422	935	26	112
6	547	945	22	127

6.2.2 Specialized Algorithm for Heat Exchanger Network Problems

A global optimization algorithm specifically designed for this type of problem was proposed in [ZG97]. The basic framework of this approach is a branch-and-bound algorithm where the branching variables are the stream temperatures. Special convex underestimators have been devised for the cost function, provided there is no stream splitting. Upper bounds on the problem are obtained by fixing the binary variables and solving a nonconvex NLP locally or globally. The branch-and-bound search is preceded by a heuristic local MINLP optimization step which allows the identification of a good starting point. No computational results using this approach are known for the example presented here.

7 Conclusions

As was demonstrated in this paper, mathematical programming techniques are a valuable tool for the solution of process network applications. The optimization approach to process synthesis illustrates their use for an important industrial application. It was shown that this procedure generates Mixed-Integer Nonlinear Programming problems (MINLPs) and a number of algorithms capable of addressing such problems were presented, including decomposition-based methods, branch-and-bound and cutting plane techniques. Considerable progress has been made in handling both the combinatorial aspects of the problem as well as nonconvexity issues so that the global solution of increasingly complex problems can be identified. The development of the SMIN-αBB and GMIN-αBB algorithms has extended the class of problems that can rigorously be solved to global optimality.

The increasing capability of MINLP algorithms has permitted the development of automated frameworks such as **MINOPT**, in which general mathematical representations can be addressed. These developments have led researchers in numerous fields to employ mathematical modeling and numerical solution through MINLP optimization techniques in order to address their problems.

A number of issues must be resolved in order to develop algorithms that can handle more complex and realistic problems. Although computational power has increased, the ability for MINLP algorithms to solve large scale problems is still limited: a large number of integer variables leads to combinatorial problems, and a large number of continuous variables leads to the generation of large scale NLPs. In addition, rigorous models capable of accurately describing industrial operations usually involve complex mathematical expressions and result in problems which are difficult to solve using standard procedures. Finally, approaches to address important challenges such as the inclusion of dynamic models and optimal control problems into the MINLP framework are emerging [SF97a].

8 Acknowledgments

The authors gratefully acknowledge financial support from the National Science Foundation, the Air Force Office of Scientific Research, the National Institutes of Health, and Mobil Technology Company.

References

[AAF7a]   C. S. Adjiman, I. P. Androulakis, and C. A. Floudas, *Global optimization of MINLP problems in process synthesis and design*, Comput. Chem. Eng. **21** (1997a), S445–S450.

[AAF7b]   C. S. Adjiman, I. P. Androulakis, and C. A. Floudas, *A global optimization method, αBB, for general twice–differentiable NLPs – II. Implementation and computational results*, accepted for publication, 1997b.

[AAMF96] C. S. Adjiman, I. P. Androulakis, C. D. Maranas, and C. A. Floudas, *A global optimisation method, αBB, for process design*, Comput. Chem. Eng. Suppl. **20** (1996), S419–S424.

[ADFN97] C. S. Adjiman, S. Dallwig, C. A. Floudas, and A. Neumaier, *A global optimization method, αBB, for general twice-differentiable NLPs – I. Theoretical advances*, accepted for publication, 1997.

[AF90] A. Aggarwal and C.A. Floudas, *Synthesis of general distillation sequences—nonsharp separations*, Comput. Chem. Eng. **14** (1990), no. 6, 631–653.

[AF96] C. S. Adjiman and C. A. Floudas, *Rigorous convex underestimators for general twice-differentiable problems*, J. Glob. Opt. **9** (1996), 23–40.

[AK90] F. A. Al-Khayyal, *Jointly constrained bilinear programs and related problems : An overview*, Comput. Math. Applic. **19** (1990), no. 11, 53–62.

[AKF83] F. A. Al-Khayyal and J. E. Falk, *Jointly constrained biconvex programming*, Math. of Oper. Res. **8** (1983), 273–286.

[AMF95] I. P. Androulakis, C. D. Maranas, and C. A. Floudas, *αBB : A global optimization method for general constrained nonconvex problems*, J. Glob. Opt. **7** (1995), 337–363.

[Bal85] E. Balas, *Disjunctive programming and a hierarchy of relaxations for discrete optimization problems*, SIAM Journal on Algebraic and Discrete Methods **6** (1985), 466–486.

[Bea77] E. M. L. Beale, The State of the Art in Numerical Analysis, ch. Integer programming, pp. 409–448, Academic Press, 1977, pp. 409–448.

[Bea90] N. Beaumont, *An algorithm for disjunctive programs*, European Journal of Operations Research **48** (1990), no. 3, 362–371.

[Ben62] J. F. Benders, *Partitioning procedures for solving mixed-variables programming problems*, Numer. Math. **4** (1962), 238.

[BGG+71] M. Benichou, J. M. Gauthier, P. Girodet, G. Hentges, G. Ribiere, and O. Vincent, *Experiments in mixed-integer linear programming*, Math. Prog. **1** (1971), no. 1, 76–94.

[BKM92]   Anthony Brooke, David Kendrick, and Alexander Meeraus, *Gams: A user's guide*, Boyd & Fraser, Danvers, MA, 1992.

[BM91]   B. Borchers and J. E. Mitchell, *An improved branch and bound algorithm for mixed integer nonlinear programs*, Tech. Report RPI Math Report No. 200, Renssellaer Polytechnic Institute, 1991.

[DG86]   M. A. Duran and I. E. Grossmann, *An outer-approximation algorithm for a class of mixed-integer nonlinear programs*, Math. Prog. **36** (1986), 307–339.

[FAC89]   C. A. Floudas, A. Aggarwal, and A. R. Ciric, *Global optimal search for nonconvex NLP and MINLP problems*, Comput. Chem. Eng. **13** (1989), no. 10, 1117.

[FL94]   R. Fletcher and S. Leyffer, *Solving mixed integer nonlinear programs by outer approximation*, Math. Prog. **66** (1994), no. 3, 327.

[Flo95]   C. A. Floudas, *Nonlinear and mixed integer optimization: Fundamentals and applications*, Oxford University Press, 1995.

[Geo72]   A. M. Geoffrion, *Generalized benders decomposition*, J. Opt. Theory Applic. **10** (1972), no. 4, 237–260.

[Glo75]   F. Glover, *Improved linear integer programming formulations of nonlinear integer problems*, Management Sci. **22** (1975), no. 4, 445.

[GR85]   O. K. Gupta and R. Ravindran, *Branch and bound experiments in convex nonlinear integer programing*, Management Sci. **31** (1985), no. 12, 1533–1546.

[Gup80]   O. K. Gupta, *Branch and bound experiments in nonlinear integer programming*, Ph.D. thesis, Purdue University, 1980.

[Hol90]   K. Holmberg, *On the convergence of the cross decomposition*, Math. Prog. **47** (1990), 269.

[Kel60]   J. E. Kelley, *The cutting plane method for solving convex programs*, Journal of the SIAM **8** (1960), no. 4, 703–712.

[KG87]   G. R. Kocis and I. E. Grossmann, *Relaxation strategy for the structural optimization of process flow sheets*, Ind. Eng. Chem. Res. **26** (1987), no. 9, 1869.

[KG89]   G. R. Kocis and I. E. Grossmann, *A modelling and decomposition strategy for the MINLP optimization of process flowsheets*, Comput. Chem. Eng. **13** (1989), no. 7, 797–819.

[LW66]   E. L. Lawler and D. E. Wood, *Branching and bound methods: A survey*, Oper. Res. (1966), no. 14, 699–719.

[McC76] G. P. McCormick, *Computability of global solutions to factorable nonconvex programs : Part I – convex underestimating problems*, Math. Prog. **10** (1976), 147–175.

[MM85]   H. Mawengkang and B. A. Murtagh, *Solving nonlinear integer programs with large-scale optimization software*, Annals of Operations Research **5** (1985), no. 6, 425–437.

[MM86]   H. Mawengkang and B. A. Murtagh, *Solving nonlinear integer programs with large scale optimization software*, Ann. of Oper. Res. **5** (1986), 425.

[Moo79] R. E. Moore, *Interval analysis*, Prentice-Hall, Englewood Cliffs, NJ, 1979.

[MS93]   Bruce A. Murtagh and Michael A. Saunders, *Minos 5.4 user's guide*, Systems Optimization Laboratory, Department of Operations Research, Stanford University, 1993, Technical Report SOL 83-20R.

[Neu90] A. Neumaier, *Interval methods for systems of equations*, Encyclopedia of Mathematics and its Applications, Cambridge University Press, 1990.

[OOM90] G. M. Ostrovsky, M. G. Ostrovsky, and G. W. Mikhailow, *Discrete optimization of chemical processes*, Comput. Chem. Eng. **14** (1990), no. 1, 111.

[PF89]   G. E. Paules, IV and C. A. Floudas, *APROS: Algorithmic development methodology for discrete-continuous optimization problems*, Oper. Res. **37** (1989), no. 6, 902–915.

[QG92] I. Quesada and I. E. Grossmann, *An LP/NLP based branch and bound algorithm for convex MINLP optimization problems*, Comput. Chem. Eng. **16** (1992), no. 10/11, 937–947.

[RG94] R. Raman and I. E. Grossmann, *Modeling and computational techniques for logic based integer programming*, Comput. Chem. Eng. **18** (1994), 563–578.

[RR88] H. Ratschek and J. Rokne, *Computer methods for the range of functions*, Ellis Horwood Series in Mathematics and its Applications, Halsted Press, 1988.

[RS95] H. S. Ryoo and N. V. Sahinidis, *Global optimization of nonconvex NLPs and MINLPs with applications in process design*, Comput. Chem. Eng. **19** (1995), no. 5, 551–566.

[SF97a] C. A. Schweiger and C. A. Floudas, *Interaction of design and control: Optimization with dynamic models*, Optimal Control: Theory, Algorithms, and Applications (W. W. Hager and P. M. Pardalos, eds.), Kluwer Academic Publishers, 1997, accepted for publication.

[SF97b] C. A. Schweiger and C. A. Floudas, *MINOPT: A software package for mixed-integer nonlinear optimization*, Princeton University, Princeton, NJ 08544-5263, 1997, Version 2.0.

[SHW+96] H. Skrifvars, I. Harjunkoski, T. Westerlund, Z. Kravanja, and R. Pörn, *Comparison of different MINLP methods applied on certain chemical engineering problems*, Comput. Chem. Eng. Suppl. **20** (1996), S333–S338.

[SP97] E.M.B. Smith and C.C. Pantelides, *Global optimisation of nonconvex minlps*, Comput. Chem. Eng. **21** (1997), S791–S796.

[TG96] M. Türkay and I. E. Grossmann, *Logic-based MINLP algorithms for the optimal synthesis of process networks*, Comput. Chem. Eng. **20** (1996), no. 8, 959–978.

[VEH96] R. Vaidyanathan and M. El-Halwagi, *Global optimization of nonconvex MINLP's by interval analysis*, Global Optimization in Engineering Design (I. E. Grossmann, ed.), Kluwer Academic Publishers, 1996, pp. 175–193.

[VG90]   J. Viswanathan and I. E. Grossmann, *A combined penalty function and outer approximation method for MINLP optimization*, Comput. Chem. Eng. **14** (1990), no. 7, 769–782.

[WP95]   T. Westerlund and F. Pettersson, *An extended cutting plane method for solving convex MINLP problems*, Comput. Chem. Eng. Suppl. **19** (1995), 131–136.

[WPG94] T. Westerlund, F. Pettersson, and I. E. Grossmann, *Optimization of pump configuration problems as a MINLP probem*, Comput. Chem. Eng. **18** (1994), no. 9, 845–858.

[YG91]   T. F. Yee and I. E. Grossmann, *Simultaneous optimization model for heat exchanger network synthesis*, Chemical Engineering Optimization Models with GAMS (I. E. Grossmann, ed.), CACHE Design Case Studies Series, vol. 6, 1991.

[ZG97]   J.M. Zamora and I.E. Grossmann, *A comprehensive global optimization approach for the synthesis of heat exchanger networks with no stream splits*, Comput. Chem. Eng. **21** (1997), S65–S70.

Approximate Algorithms and Heuristics for MAX-SAT

Roberto Battiti
Dipartimento di Matematica, Università di Trento
38050 Povo (Trento), Italy
E-mail: battiti@science.unitn.it

Marco Protasi
Dipartimento di Matematica, Università di Roma "Tor Vergata"
Via della Ricerca Scientifica, 00133 Roma, Italy
E-mail: protasi@mat.utovrm.it

Contents

1 **Introduction** 78
  1.1 Notation and graphical representation 79

2 **Resolution and Linear Programming** 80
  2.1 Resolution and backtracking for *SAT* 80
  2.2 Integer programming approaches 83

3 **Continuous approaches** 85
  3.1 An interior point algorithm 85
  3.2 Continuous unconstrained optimization 86

4 **Approximate algorithms** 86
  4.1 Definitions and basic results 87
  4.2 Johnson's approximate algorithms 91
  4.3 Randomized algorithms for *MAX W-SAT* 99
      4.3.1 A randomized 1/2–approximate algorithm for *MAX W-SAT* 99
      4.3.2 A randomized 3/4–approximate algorithm for *MAX W-SAT*. 102
      4.3.3 A variant of the randomized rounding technique 107
  4.4 Another $\frac{3}{4}$–approximate algorithm by Yannakakis 110
  4.5 Approximate solution of *MAX W-SAT*: improvements 115
  4.6 Negative results about approximability 116

5	A different *MAX-SAT* problem and completeness results	117
6	**Local search**	**118**
	6.1 Quality of local optima	120
	6.2 Non-oblivious local optima	122
	6.2.1 An example of non-oblivious search	124
	6.3 Local search satisfies most *3-SAT* formulae	125
	6.4 Randomized search for *2-SAT* (Markov processes)	126
7	**Memory-less Local Search Heuristics**	**128**
	7.1 Simulated Annealing	128
	7.2 GSAT with "random noise" strategies	129
	7.3 Randomized Greedy and Local Search (GRASP)	130
8	**History-sensitive Heuristics**	**132**
	8.1 Prohibition-based Search: TS and SAMD	132
	8.2 HSAT and "clause weighting"	133
	8.3 Reactive Search	133
	8.3.1 The Hamming-Reactive Tabu Search (H-RTS) algorithm	135
9	**Experimental analysis and threshold effects**	**136**
	9.1 Models	137
	9.2 Hardness and threshold effects	138

References

1   Introduction

In the Maximum Satisfiability (*MAX-SAT*) problem one is given a Boolean formula in conjunctive normal form, i.e., as a conjunction of clauses, each clause being a disjunction. The task is to find an assignment of truth values to the variables that satisfies the maximum number of clauses.

In our work, n is the number of variables and m the number of clauses, so that a formula has the following form:

$$\bigwedge_{1 \leq i \leq m} \left(\bigvee_{1 \leq k \leq |C_i|} l_{ik} \right)$$

where $|C_i|$ is the number of literals in clause C_i and l_{ik} is a literal, i.e., a propositional variable u_j or its negation $\overline{u_j}$, for $1 \leq j \leq n$. The set of clauses in the formula is denoted by **C**. If one associates a *weight* w_i to each clause C_i one obtains the weighted *MAX-SAT* problem, denoted as

Approximate Algorithms for MAX-SAT

MAX W-SAT: one is to determine the assignment of truth values to the n variables that maximizes the sum of the weights of the satisfied clauses. Of course, *MAX-SAT* is contained in *MAX W-SAT* (all weights are equal to one). In the literature one often considers problems with different numbers k of literals per clause, defined as *MAX-k-SAT*, or *MAX W-k-SAT* in the weighted case. In some papers *MAX-k-SAT* instances contain *up to* k literals per clause, while in other papers they contain *exactly* k literals per clause. We consider the second option unless otherwise stated.

MAX-SAT is of considerable interest not only from the theoretical side but also from the practical one. On one hand, the decision version *SAT* was the first example of an $\mathcal{NP}$-complete problem [25], moreover *MAX-SAT* and related variants play an important role in the characterization of different approximation classes like $\mathcal{APX}$ and $\mathcal{PTAS}$ [8]. On the other hand, many issues in mathematical logic and artificial intelligence can be expressed in the form of satisfiability or some of its variants, like constraint satisfaction. Some exemplary problems are consistency in expert system knowledge bases [69], integrity constraints in databases [7, 35], approaches to inductive inference [49, 56], asynchronous circuit synthesis [46, 74].

The main purpose of this work is that of summarizing the basic approaches for the exact or approximated solution of the *MAX W-SAT* and *MAX-SAT* problem. The presentation of algorithms for the related *SAT* problem is therefore limited to a quick overview of some basic techniques and of methods that can be used also for *MAX-SAT*. Of course, given the impressive extension of the research in this area, we are not aiming at a comprehensive survey of the literature, and we have confined ourselves to citing the sources that we have used, some sources of historical significance, and some papers that are paradigmatic for the different approaches.

1.1 Notation and graphical representation

A clause will be represented either as $C = \overline{u} \vee v \vee z$ or as a set of literals, as in $C = \{\overline{u}vz\}$.

For the following discussion, it can be useful to help the intuition with a graphical representation of a formula in conjunctive normal form, as depicted in Fig. 1. In the figure, one has a case of *MAX 3-SAT*: all clauses have three literals and the formula is:

$$(u_1 \vee \overline{u_3} \vee u_5) \wedge (\overline{u_2} \vee \overline{u_4} \vee \overline{u_5}) \wedge (\overline{u_1} \vee u_3 \vee \overline{u_4})$$

Truth values to variables are assigned by placing a black triangle to the left if the variable is **true**, to the right if it is **false**. Each literal is depicted with a small circle, placed to the left if the corresponding variable is **true**, to the right in the other case. If a literal is *matched* by the current assignment (e.g., if the literal asks for a **true** value and the variable is set to **true**, or if is asks for **false** and the variable is **false**), it is shown with a gray shade. The *coverage* of a clause is the number of literals in the clause that are matched by the current assignment, and it is illustrated by placing a black square in the appropriate position of an array with indices ranging from 0 to the number of literals in each clause $|C|$.

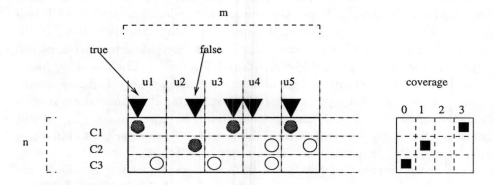

Figure 1: A formula in conjunctive normal form (CNF).

2 Resolution and Linear Programming

2.1 Resolution and backtracking for SAT

The basic method to solve SAT formulae is given by the recursive replacement of a formula by one or more formulae, the solution of which implies the solution of the original formula.

In *resolution* a variable is selected and a new clause, called the *resolvent* is added to the original formula. The process is repeated to exhaustion or until an empty clause is generated. The original formula is not satisfiable if and only if an empty clause is generated [77].

Let us now consider some details: A clause R is the *resolvent* of clauses C_1 and C_2 iff there is a literal $l \in C_1$ with $\bar{l} \in C_2$ such that $R = (C_1 \setminus \{l\}) \cup (C_2 \setminus \{\bar{l}\})$ and $u(l)$, the variable associated to the literal, is the only variable appearing both positively and negatively.

For the two clauses $C_1 = (l \vee a_1 \vee ... \vee a_A)$ and $C_2 = (\bar{l} \vee b_1 \vee ... \vee b_B)$ the resolvent is therefore the clause $R = (a_1 \vee ... \vee a_A \vee b_1 \vee ... \vee b_B)$. The resolvent

Approximate Algorithms for MAX-SAT

is a logical consequence of the logical *and* of the two clauses. Therefore, if the resolvent is added to the original set of clauses, the set of solutions does not change. It is immediate to check that, if both C_1 and C_2 are satisfied, i.e., have at least one matched literal, the resolvent must also be satisfied. In fact, if it is not, in the original clauses there are no matched literals apart from either $\bar{l}$ or l, but this implies that both clauses cannot be satisfied (see also Fig. 2 for a graphical illustration).

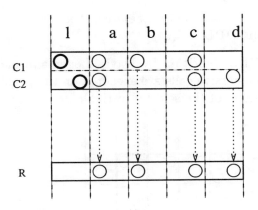

Figure 2: How to construct a resolvent, an example with variables l, a, b, c, d.

Davis and Putnam [29] started in 1960 the investigation of useful strategies for handling resolution. In addition to applying transformations that preserve the set of solutions they eliminate one variable at a time in a chosen order by using all possible resolvents on that variable. During resolution the lengths and the number of added clauses can easily increase and become extremely large.

Davis, Logemann and Loveland [28] avoid the memory explosion of the original DP algorithm by replacing the resolution rule with the *splitting rule* (Davis, Putnam, Logemann and Loveland, or DPLL algorithm for short). In splitting, a variable u in a formula is selected. Now, if there exist a satisfying truth assignment for the original formula then either u is **true** or $\bar{u}$ is **true** in the assignment. In the first case the formula obtained by eliminating all clauses containing u and by deleting all occurrences of $\bar{u}$ must be satisfied, see Fig 4. This derived formula is called $C(u)$ in Fig. 3. In the second case, the formula obtained by eliminating all clauses containing $\bar{u}$ and all occurrences of u must be satisfied. *Vice versa*, if both derived formulae cannot be satisfied, neither can the original problem.

A *tree* is therefore generated. At the root one has the original problem

DPLL(**C** : set of clauses)
Input: Boolean CNF formula $\mathbf{C} = \{C_1, C_2, \ldots, C_m\}$
Output: Yes or No (decision about satisfiability)
1   **if** C is empty **then return** Yes
2   **if** C contains an empty clause **then return** No
3   **if**   there is a pure literal l in C   **then return** DPLL(**C**(l))
4   **if**   there is a unit clause $\{l\} \in$ **C then return** DPLL(**C**(l))
5   Select a variable u in C
6   **if** DPLL(**C**(u)) = Yes **then return** Yes
7   **else return** DPLL(**C**($\overline{u}$))

Figure 3: The DPLL algorithm by Davis, Logemann and Loveland in recursive form. The recursive calls are executed on the problems derived after setting the truth value of the selected variable.

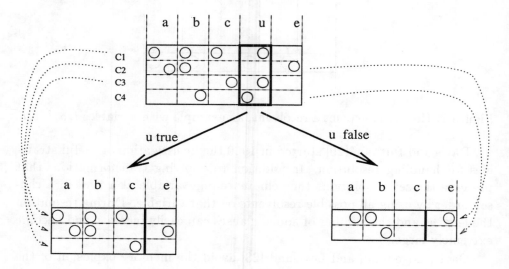

Figure 4: Example of splitting on a variable u.

and no variables are assigned values. At each node of the tree one generates two children by *selecting* one of the yet unassigned variables in the problem corresponding to the node and by generating the two problems derived by setting the variable to **true** or **false**. A trivial upper bound on the number of nodes in the tree is proportional to the number of possible assignments, i.e., $O(2^n)$. In fact, sophisticated techniques are available to reduce the number of nodes, that nonetheless remains exponential in the worst case.

Approximate Algorithms for MAX-SAT

The techniques include:

- avoiding the examination of a subtree when the fate of the current problem is decided (problems with an empty clause have no solutions, problems with no clauses have a solution). If the current problem cannot be solved, or if it is solved but one wants all possible solutions, one *backtracks* to the first unexplored branch of the tree. Note that, when splitting is combined with a depth-first search of the tree (as in the DPLL algorithm) one avoids the memory explosion because only one subproblem is active at a given time.

- *selecting* the next variable for the splitting based on appropriate criteria. For example, one can prefer variables that appear in clauses of length one (*unit clause rule*), or select a *pure literal* (such that it occurs only positive, or only negative), or select a literal occurring in the *smallest clause*.

A recent review of advanced techniques for resolution and splitting is presented in [45], and a summary of algorithms for deciding propositional tautologies is presented in [64].

It is worth noting that approaches based on the Davis-Putnam scheme tend to achieve the fastest speed when solving *SAT* problems. A recent state-of-the-art parallel implementation in given in [16]. Their previous sequential implementation turned out to be the fastest program in a *SAT* competition [18].

2.2 Integer programming approaches

The *MAX W–SAT* problem has a natural integer linear programming formulation (*ILP*). Let $y_j = 1$ if Boolean variable u_j is **true**, $y_j = 0$ if it is **false**, and let the Boolean variable $z_i = 1$ if clause C_i is satisfied, $z_i = 0$ otherwise. The integer linear program is:

$$\max \sum_{i=1}^{m} w_i z_i$$

subject to the following constraints:

$$\sum_{j \in U_i^+} y_j + \sum_{j \in U_i^-} (1 - y_j) \geq z_i, \ i = 1, \cdots, m$$

$$y_j \in \{0, 1\}, \ j = 1, \cdots, n$$
$$z_i \in \{0, 1\}, \ i = 1, \cdots, m$$

where U_i^+ and U_i^- denote the set of indices of variables that appear unnegated and negated in clause C_i, respectively.

Because the sum of the z_i w_i is maximized and because each z_i appears as the right-hand side of one constraint only, z_i will be equal to one if and only if clause C_i is satisfied.

If one neglects the objective function and sets all z_i variables to 1, one obtains an integer programming feasibility problem associated to the *SAT* problem [15].

The integer linear programming formulation of *MAX–SAT* suggests that this problem could be solved by a *branch-and-bound* method. A tree is generated, see also the DPLL method, where the root corresponds to the initial instance and two children are obtained by *branching*, i.e., by selecting one free variable and setting it **true** (left child) and **false** (right child). An *upper bound* on the number of satisfied clauses can be obtained by using a linear programming relaxation: the constraints $y_j \in \{0,1\}$ and $z_i \in \{0,1\}$ are replaced by $y_j \in [0,1]$ and $z_i \in [0,1]$. One obtains a Linear Programming (*LP*) problem that can be solved in polynomial time and, because the set of admissible solutions is enlarged with respect to the original problem, one obtains an upper bound.

Unfortunately this is not likely to work well in practice [48] because the solution $y_j = 1/2, j = 1, \cdots, n, z_i = 1, i = 1, \cdots, m$ is feasible for the *LP* relaxation unless there exist some constraint containing only one variable. The bounds so obtained would be very poor.

Better bounds can be obtained by using Chvátal cuts. In [49] it is shown that the resolvents in the propositional calculus correspond to certain cutting planes in the integer programming model of inference problems.

A general cutting plane algorithm for *ILP*, see for example [73], works as follows. One solves the *LP relaxation* of the problem: if the solution is integer the algorithm terminates, otherwise one adds linear constraints to the *ILP* that do not exclude integer feasible points. The constraints are added one at a time, until the solution to the *LP* relaxation is integer.

The application considered in [49] is to determine whether a formula in the propositional calculus implies another one. The propositional calculus is a formal logic involving propositions and logic connectives such as "not", "and", "or" and "implies". Any formula is equivalent to a conjunction of clauses (in particular, a rule in a knowledge base, such as "if A and B then C" can be written as a clause "not-A or not-B or C"). Thus, a set of clauses can be represented by a particular linear system $Ax \geq a$, also called *generalized set covering* problem. Finally, one can determine whether $Ax \geq a$ implies a formula F by expressing not-F as a system $Bx \geq b$ of clauses, see for example [65], and checking whether the combined system $Ax \geq a$, $Bx \geq b$ has a binary solution. If it does *not*, $Ax \geq a$ implies F. In the applications $Ax \geq a$ can represent a knowledge base known to be consistent.

Cutting planes are generated in [49] by finding *separating resolvents*, i.e., a resolvent (expressed as a linear inequality) that is violated by the current solution of the *LP* relaxation. When no separating resolvents are found, the current system is solved with branch-and-bound. The experimental results are that the cutting plane algorithm is orders of magnitude faster on problems in which the premises do not imply the given proposition (the majority of random problems), and moderately faster on other random problems [49].

LP relaxations of integer linear programming formulations of *MAX-SAT* have been used to obtained upper bounds in [47, 84, 39]. A linear programming and rounding approach for *MAX 2-SAT* is presented in [22]. Their cutting plane algorithm starts from the *LP* relaxation of *MAX 2-SAT*, and has separation routines for two families of cuts: cycle and wheel inequalities. Upper and lower bounds are found, the latter by using a *rounding* procedure to convert a fractional *LP* solution to a $\{0, 1\}$ solution. A method for strengthening the Generalized Set Covering formulation is presented in [70], where Lagrangian multipliers guide the generation of cutting planes.

3 Continuous approaches

3.1 An interior point algorithm

The *ILP* feasibility problem obtained from *SAT* as described in the previous section is solved with an *interior point algorithm* in [56, 57]. In the *interior point algorithm* one applies a function minimization method based on *continuous* mathematics to the inherently discrete *SAT* problem.

In [57] the application is to a problem of *inductive inference*, in which one aims at identifying a hidden Boolean function using outputs obtained by applying a limited number of random inputs to the hidden function. The task is formulated as a *SAT* problem, which is in turn formulated as an integer linear program:

$$A^T y \leq c \, , \, y \in \{-1, 1\}^n \qquad (1)$$

where A^T is an $m \times n$ real matrix and c a real m vector.

The interior point algorithm is based on finding a local minimum in the box $-1 \leq y_j \leq 1$ of the *potential function*:

$$\phi(y) = \log \left\{ \frac{n - y^T y}{\prod_{k=1}^{m}(c_k - a_k^T y)^{1/m}} \right\} \qquad (2)$$

by an iterative method. The denominator of the argument of the log is the geometric mean of the *slacks* (a_k is the k-th column of matrix A). It is shown that, if the integer linear program has a solution, y^* is a global minimum of this potential function if and only if y^* solves the integer program. The next iterate y^{k+1} (interior point solution, i.e., such that $A^T y < c$) is obtained by moving in a descent direction Δy from the current iterate y^k such that $\phi(y^{k+1}) = \phi(y^k + \alpha \Delta y) < \phi(y^k)$. Each iteration in [57] is based on the *trust region approach* of continuous optimization where the Riemannian metric used for defining the search region is dynamically modified. The feasibility of the approach for inductive inference is demonstrated in [57].

3.2 Continuous unconstrained optimization

In some techniques the *MAX-SAT* (or *SAT*) problem is transformed into an unconstrained optimization problem on the real space R^n and solved by using existing global optimization techniques.

Some examples of this approach include the UNISAT models [43] and the *neural network* approaches [54, 19]. In general, these techniques do not have performance guarantees because they assure only the local convergence to a locally optimal point, not necessarily the global optimum.

The local convergence properties of some optimization algorithms are considered in [44]. The main results are that, for any CNF formula, if y^* is a solution point of the objective function f defined on R^n associated to the problem, i.e., $f(y^*) = 0$, the *Hessian* matrix $H(y^*)$ is positive definite and therefore the convergence ratios of the steepest descent, Newton's method and coordinate descent methods can be derived, see [44] for the details. Let us note that, to obtain these results, one assumes that the initial solution is "sufficiently close" to the optimal solution.

4 Approximate algorithms

The present section presents the first important approximate algorithms for *MAX-SAT*. However, in order to evaluate the goodness of the algorithms, one needs to define the meaning of *approximation algorithm* with a "guaranteed" quality of approximation.

In the following it is assumed that the reader is familiar with the concepts of the complexity classes classes $\mathcal{P}$ and $\mathcal{NP}$ and with elementary concepts from probability theory.

4.1 Definitions and basic results

First of all, let us present a general definition of optimization problem.

Definition 4.1 *An optimization problem $P = (I, sol, m, opt)$ belongs to the class $\mathcal{NPO}$ if the following holds:*

1. *the set of instances I is recognizable in polynomial time,*

2. *given an instance $x \in I$, $sol(x)$ is the set of the feasible solutions of x; moreover there exists a polynomial q such that, given an instance $x \in I$, for any $y \in sol(x)$, $|y| < q(|x|)$ and, besides, for any y such that $|y| < q(|x|)$, it is decidable in polynomial time whether $y \in sol(x)$,*

3. *given an instance $x \in I$, a feasible solution y of x, $m(x, y)$ is the objective function and is computable in (deterministic) polynomial time.*

4. *$opt \in \{max, min\}$ specifies whether one has a maximization or minimization problem.*

Finally $m^*(x)$ will denote the optimal value of instance x. When it is clear from the context, one will use simply m^*.

A problem belonging to $\mathcal{NPO}$ will be called an $\mathcal{NPO}$ problem.

Note that the difficulty of solving an $\mathcal{NPO}$ problem is based on the fact that, in many cases, the set of feasible solutions is exponentially large.

Even if not explicitly stated, there is a nondeterministic polynomial time computation model underlying this definition. The nondeterministic machine of polynomial complexity may run in the following way: in non-deterministic polynomial time, all strings y such that $|y| < q(|x|)$ are generated. Afterwards any string is tested for membership in $sol(x)$ in polynomial time. If the test is positive, $m(x, y)$ is computed (again in polynomial time) and both y and $m(x, y)$ are returned.

The definition of the class $\mathcal{NPO}$ formalizes the notion of optimization problem with an associated decision version which is in $\mathcal{NP}$. In addition, let as define as $\mathcal{PO}$ the subclass of $\mathcal{NPO}$ formed by problems that can be solved in polynomial time. Many classical combinatorial problems belong to the class $\mathcal{NPO}$; for instance, the traveling salesperson problem, the knapsack problem, the minimal covering of a graph and so on.

MAX W-SAT is another important example of $\mathcal{NPO}$ problem. In this case one has

1. I = sets U of Boolean variables and a collection $\mathbf{C} = C_1, \ldots, C_m$ of clauses over U, a set $W = w_1, \ldots, w_m$ of integers (weights) associated to the clauses.

2. Given an instance x of I, $sol(x)$ = set of truth assignments U to the variables in the problem. Moreover $|U| < (|x|)$ and it is possible to decide in polynomial time whether a string is a truth assignment for the formula;

3. Given an instance x of I and a feasible solution y of x, $m(x,y)$ = sum of the weights associated to the satisfied clauses; m is trivially computable in polynomial time

4. opt= max.

A subset is given by the *MAX-SAT* problem, obtained when all weights are equal to one. Note that, in this definition, the set of Boolean variables and a truth assignment are denoted with the same symbol U; even if formally questionable, this identification will allow to simplify the presentation of many results. However, when this abuse of notation could raise problems to the reader, different symbols will be used.

Because an $\mathcal{NPO}$ problem that is not in $\mathcal{PO}$ cannot be solved in polynomial time unless $\mathcal{P} = \mathcal{NP}$, a natural approach consists of looking for "good" approximate solutions.

Definition 4.2 *Given an $\mathcal{NPO}$ problem $P = (I, sol, m, opt)$, an algorithm A is an* approximation algorithm *if, for any given instance $x \in I$, it returns an* approximate solution, *that is a feasible solution $A(x) \in sol(x)$.*

Because the present work is dedicated to *MAX-SAT* the following definitions will be restricted to the case of maximization problems.

An approximation algorithm can be usefully applied only if it achieves approximate solutions whose values are "near" to the optimum value. Therefore one is interested in determining how far from the optimal value is the value of the achieved solution.

Definition 4.3 *Given an $\mathcal{NPO}$ problem P, an instance x and a feasible solution y, the* performance ratio *of y is*

$$R(x, y) = \frac{m(x, y)}{m^*(x)}.$$

When the performance ratio is close to 1, the value of the approximate solution is close to the optimum one.

Definition 4.4 *Given an $\mathcal{NPO}$ problem P and an approximation algorithm A, A is said to be an ε-approximate algorithm if, given any input instance x, the performance ratio of the approximate solution $A(x)$ verifies the following relation:*

$$R(x, A(x)) \geq \varepsilon.$$

In other words, the solution provided by the algorithm must guarantee at least a value $\varepsilon m^*(x)$.

Definition 4.5 *An $\mathcal{NPO}$ problem P is ε-approximable if there exists a polynomial-time ε-approximate algorithm for P, with $0 < \varepsilon < 1$*

Definition 4.6 *$\mathcal{APX}$ is the class of all $\mathcal{NPO}$ problems that are ε-approximable.*

For a problem to join $\mathcal{APX}$ it is sufficient that the performance ratio is greater than or equal to ε for a particular value of ε. Of course, the goodness of the approximation algorithm strictly depends on how near ε is to 1.

In fact, in the class $\mathcal{APX}$ there are problems with different values of ε and therefore with different approximation properties. After taking this fact into account, the definition of ε-approximate algorithm can be strengthened in the following way:

Definition 4.7 *Let P be an $\mathcal{NPO}$ problem. An algorithm A is said to be a polynomial-time approximation scheme ($\mathcal{PTAS}$) for P if, for any instance x of P and any rational value $0 < \varepsilon < 1$, $A(x, \varepsilon)$ returns an ε-approximate solution of x in time polynomial in the size of the instance x.*

Definition 4.8 *$\mathcal{PTAS}$ is the class of $\mathcal{NPO}$ problems that allow a polynomial-time approximation scheme.*

The following theorem holds:

Theorem 4.1
- *MAX W–SAT belongs to the class $\mathcal{APX}$.*

- *MAX W–SAT does not belong to the class $\mathcal{PTAS}$ unless $\mathcal{P} = \mathcal{NP}$.*

The first part of the theorem will be demonstrated in the sequel, while citations will be given for the second part.

Finally, let us conclude this Subsection by introducing the important notion of *completeness in an approximation class*. As it is done in the $\mathcal{NP}$-completeness theory, one is interested in finding the "hardest" problem in the classes $\mathcal{APX}$ and $\mathcal{PTAS}$ that is, the most difficult ones from a computational point of view. One looks for problems which cannot have stronger approximation properties unless $\mathcal{P} = \mathcal{NP}$.

In complexity theory, the notion of hardest problem in a class is equivalent to saying that a problem is complete with respect to a suitable *reduction*. The same approach can be followed for approximation classes. Therefore, a definition of approximation–preserving reduction is presented and one will be able to define a complete problem. Actually, many different reductions have been proposed. In this paper we consider the reduction presented in [8] which has the relevant advantage that it can be used for defining the notion of completeness both in $\mathcal{APX}$ and in $\mathcal{PTAS}$.

Intuitively, in order to map an optimization problem P_1 into another optimization problem P_2, we need not only a function f mapping instances of P_1 into instances of P_2 but also a second function g mapping back feasible solutions of P_2 into feasible solutions of P_1, see also Fig. 5.

Definition 4.9 *Let P_1 and P_2 be two $\mathcal{NPO}$ problems. P_1 is said to be $\mathcal{PTAS}$-reducible to P_2 ($P_1 \leq_{\mathcal{PTAS}} P_2$) if three functions f, g, and c exist such that*

1. *For any $x \in I_{P_1}$, $f(x) \in I_{P_2}$ is computable in polynomial time.*

2. *For any $x \in I_{P_1}$, for any $y \in sol_{P_2}(f(x))$, and for any $\epsilon \in (0,1)^Q$ (set of positive rational numbers smaller than 1), $g(x,y,\epsilon) \in sol_{P_1}(x)$ is computable in time polynomial with respect to both $\mid x \mid$ and $\mid y \mid$.*

3. *$c : (0,1)^Q \to (0,1)^Q$ is computable and surjective.*

4. *For any $x \in I_{P_1}$, for any $y \in sol_{P_2}(f(x))$, and for any $\epsilon \in (0,1)^Q$,*
   $$1 - c(\epsilon) \leq R_{P_2}((f(x),y)) \text{ implies } 1 - \epsilon \leq R_{P_1}(x, g(x,y,\epsilon)).$$

The triple (f,g,c) is said to be a $\mathcal{PTAS}$-reduction from P_1 to P_2.

It is easy to demonstrate the following Lemma:

Lemma 4.2 *If $P_1 \leq_{\mathcal{PTAS}} P_2$ and $P_2 \in \mathcal{APX}$ (respectively, $P_2 \in \mathcal{PTAS}$), then $P_1 \in \mathcal{APX}$ (respectively, $P_1 \in \mathcal{PTAS}$).*

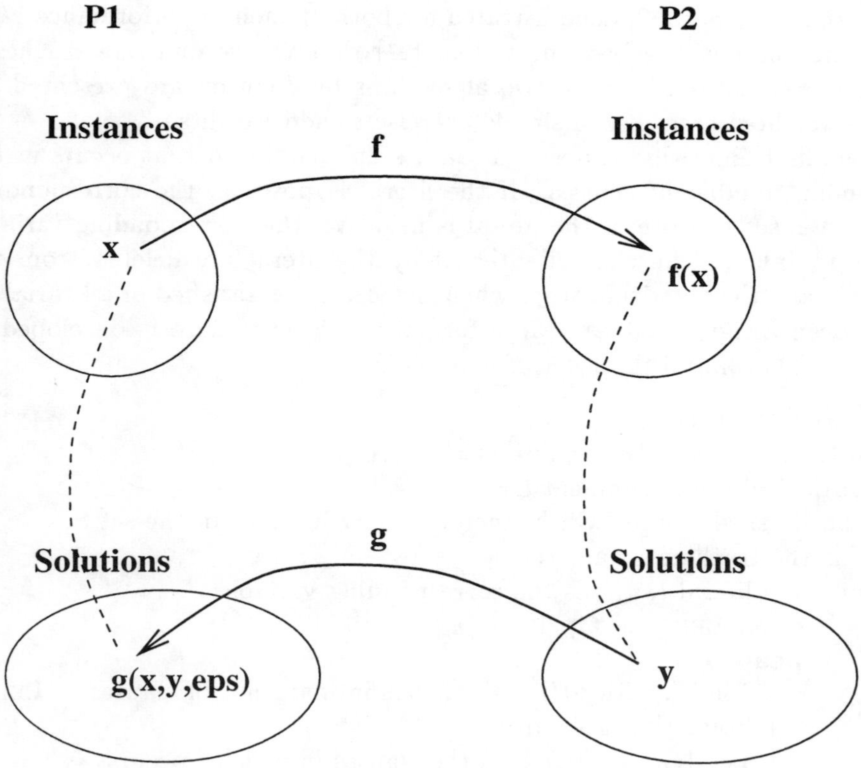

[Figure 5: Approximation preserving reduction.]

For a proof see [8].

Definition 4.10 *A problem $P_1 \in \mathcal{NPO}$ (respectively, $P_1 \in \mathcal{APX}$) is $\mathcal{NPO}$-complete (respectively $\mathcal{APX}$-complete) if, for any $P_2 \in \mathcal{NPO}$ (respectively, $P_2 \in \mathcal{APX}$), $P_2 \leq_{\mathcal{PTAS}} P_1$.*

The above Lemma shows that the reduction we have introduced is really capable of preserving the level of approximability. Moreover, a consequence of the definition of $\mathcal{NPO}$-completeness (respectively $\mathcal{APX}$-completeness) is that an $\mathcal{NPO}$-complete (respectively $\mathcal{APX}$-complete) problem does not belong to $\mathcal{APX}$ (respectively to $\mathcal{PTAS}$).

4.2 Johnson's approximate algorithms

Let us now present the two first approximate algorithms for *MAX W-SAT*. They were proposed by Johnson [52] and use *greedy* construction strategies.

The original paper [52] demonstrated for both of them a performance ratio 1/2. Recently it has been proved in [21] that the second one reaches a performance ratio 2/3. The two algorithms by Johnson are presented for the unweighted case: it is a simple exercise to add weights.

The first algorithm chooses, at each step, the literal that occurs in the maximum number of clauses. If the literal is positive, the corresponding variable is set to **true**; if the literal is negative, the corresponding variable is set to **false**. The clauses satisfied by the literal are deleted from the formula and the algorithm stops when the formula is satisfied or all variables have been assigned values. More formally, this procedure is developed in algorithm GREEDYJOHNSON1 of Fig. 6.

GREEDYJOHNSON1
  Input: Boolean CNF formula $\mathbf{C} = \{C_1, C_2, \ldots, C_m\}$;
  Output: Truth assignment U;
  △ The satisfied clauses will be incrementally inserted in the set $\mathbf{S}$;
  △ U is the truth assignment;
  △ for every literal l, $u(l)$ is the corresponding variable;
  1    $\mathbf{S} \leftarrow \emptyset$; LEFT $\leftarrow \mathbf{C}$; $V \leftarrow \{u \mid u \text{ variable in } \mathbf{C}\}$;
  2   **repeat**
  3   Find l, with $u(l) \in V$, that is in max. no. of clauses in LEFT
  4   Solve ties arbitrarily
  5   Let $\{C_{l_1}, \ldots, C_{l_k}\}$ be the clauses in which l occurs
  6        $\mathbf{S} \leftarrow \mathbf{S} \cup \{C_{l_1}, \ldots, C_{l_k}\}$
  7   LEFT $\leftarrow$ LEFT $\setminus \{C_{l_1}, \ldots, C_{l_k}\}$
  8   **if** l is positive **then** $u(l) \leftarrow$ **true else** $u(l) \leftarrow$ **false**
  9        $V \leftarrow V \setminus \{u(l)\}$
  10   **until** no literal l with $u(l) \in V$ is contained in any clause of LEFT
  11   **if** $V \neq \emptyset$ **then forall** $u \in V$ **do** $u \leftarrow$ **true**
  12   **return** U

Figure 6: The GREEDYJOHNSON1 algorithm, a $k/(k+1)$-approximate algorithm.

Theorem 4.3 *Algorithm* GREEDYJOHNSON1 *is a polynomial time 1/2-approximate algorithm for MAX–SAT.*

Proof. One can prove that, given a formula with m clauses, algorithm GREEDYJOHNSON1 always satisfies at least $m/2$ clauses, by induction on the number of variables. Because no optimal solution can be larger than

Approximate Algorithms for MAX-SAT

m, the theorem follows. The result is trivially true in the case of one variable. Let us assume that it is true in the case of $i - 1$ variables ($i > 1$) and let us consider the case in which one has i variables. Let u be the last variable to which a truth value has been assigned. We can suppose that u appears positive in k_1 clauses, negative in k_2 clauses and does not appear in $m - k_1 - k_2$ clauses. Without loss of generality suppose that $k_1 \geq k_2$. Then, by inductive hypothesis, algorithm GREEDYJOHNSON1 allows us to choose suitable values for the remaining $i - 1$ variables in such a way to satisfy at least $(m - k_1 - k_2)/2$ clauses; if according to the algorithm we now choose $u = $ **true** we satisfy
$$\frac{m - k_1 - k_2}{2} + k_1 \geq \frac{m}{2}$$
clauses.
∎

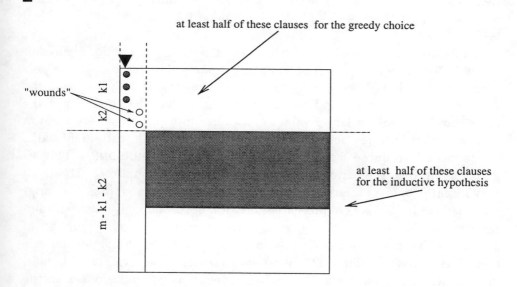

Figure 7: Illustration of the GREEDYJOHNSON1 algorithm.

Let us note that one does not use the fact that the chosen literal occurs in the *maximum* number of clauses for the above proof. What is required is that, given an unset variable that appears in at least an unsatisfied clause, the variable is set to **true** or **false** in a way that maximizes the number of newly satisfied clauses.

This result can be made more specific by considering the number of variables in a clause.

Theorem 4.4 *Let k be the minimum number of variables occurring in any clause of the formula. For any integer $k \geq 1$, algorithm* GREEDYJOHNSON1 *achieves a feasible solution y of an instance x such that*

$$\frac{m(x,y)}{m^*(x)} \geq 1 - \frac{1}{k+1}.$$

Proof. Because of the greediness, when literal l is picked in line 3 of Fig. 6, the number of newly satisfied clauses is at least as large as the number of new *wounds*, defined as the number of occurrences of literal $\bar{l}$ in clauses of LEFT that will never be matched in the future steps, given the choice of l, see Fig. 7. When the algorithm halts, the only clauses remaining in LEFT are those that have a number of wounds equal to the number of their literals, and hence are *dead*. This means that, when the algorithm halts, there are at least $k|\text{LEFT}|$ wounds, and therefore $|S| \geq k|\text{LEFT}|$. Thus $m^* \leq m = |S| + |\text{LEFT}| \leq \frac{(k+1)}{k}|S|$. The bound follows. ∎

Note that, according to the definition of performance ratio, algorithm GREEDYJOHNSON1 is $\frac{k}{k+1}$-approximate. In particular, for $k = 1$, the performance ratio is $1/2$, for $k = 2$ the performance ratio is $2/3$, for $k = 3$ the performance ratio is $3/4$ and so on. This means that the goodness of the algorithm improves for larger values of k. Therefore the worst case is given by $k = 1$, that is, when one has unit clauses (clauses with just one literal).

Johnson introduced a second algorithm (GREEDYJOHNSON2). This algorithm improves the performance ratio and obtains a bound $2/3$ [21]. Until very recently, only a performance ratio $1/2$ was demonstrated [52]. The original theorem in [52] is here presented, because of its simplicity and paradigmatic nature and because it gives a better performance as a function of k, the minimum number of literals in some clause. In the algorithm one associates a *mass* $w(C_i) = 2^{-|C_i|}$ to each clause. The term *mass* is used instead of the original term "weight" in order to avoid confusions with the clause *weight* in the *MAX W–SAT* problem. The mass will be proportional to the weight in the version of the algorithm for the *MAX W–SAT* problem ($w(C_i) = w_i \, 2^{-|C_i|}$). In [52] the analysis of the performance of algorithm GREEDYJOHNSON2 leads to the following:

Theorem 4.5 *Let k be the minimum number of clauses occurring in any clause of the formula. For any integer $k \geq 1$, algorithm* GREEDYJOHNSON2 *achieves a feasible solution y of an instance x such that*

$$\frac{m(x,y)}{m^*(x)} \geq 1 - \frac{1}{2^k}.$$

GREEDYJOHNSON2
Input: Boolean CNF formula $\mathbf{C} = \{C_1, C_2, \ldots, C_m\}$;
Output: Truth assignment U;
△ The satisfied clauses will be incrementally inserted in the set $\mathbf{S}$;
△ U is the truth assignment;
△ for every literal l, let $u(l)$ be the corresponding variable;
1      $\mathbf{S} \leftarrow \emptyset$; LEFT $\leftarrow \mathbf{C}$; $V \leftarrow \{u \mid u$ variable in $\mathbf{C}\}$;
2   Assign to each clause C_i a mass $w(C_i) = 2^{-|C_i|}$
3   **repeat**
4   Determine $u \in V$, appearing in at least a clause $\in$ LEFT
5   Let **CT** be the clauses $\in$ LEFT cont. u, **CF** those cont. $\overline{u}$
6   **if** $\sum_{C_i \in \mathbf{CT}} w(C_i) \geq \sum_{C_i \in \mathbf{CF}} w(C_i)$ **then**
7              $u(l) \leftarrow$ **true**
8              $\mathbf{S} \leftarrow \mathbf{S} \cup \mathbf{CT}$
9   LEFT $\leftarrow$ LEFT $\setminus \mathbf{CT}$
10   **forall** $C_i \in \mathbf{CF}$ **do** $w(C_i) \leftarrow 2 \cdot w(C_i)$
11   **else**
12             $u(l) \leftarrow$ **false**
13             $\mathbf{S} \leftarrow \mathbf{S} \cup \mathbf{CF}$
14   LEFT $\leftarrow$ LEFT $\setminus \mathbf{CF}$
15   **forall** $C_i \in \mathbf{CT}$ **do** $w(C_i) \leftarrow 2 \cdot w(C_i)$
16   **until** no literal l in any clause of LEFT is such that $u(l)$ is in V
17   **if** $V \neq \emptyset$ **then forall** $u \in V$ **do** $u \leftarrow$ **true**
18   **return** U

Figure 8: The GREEDYJOHNSON2 algorithm, a $(1 - 1/2^k)$-approximate algorithm.

Proof. Initially, because each clause has at least k literals, the total mass of all the clauses in LEFT cannot exceed $m/2^k$. During each iteration, the total mass of the clauses in LEFT cannot increase. In fact, the mass removed from LEFT is at least as large as the mass added to those remaining clauses which receive new *wounds*, see lines 6–15 of Fig. 8. Therefore, when the algorithm halts, the total mass still cannot exceed $m/2^k$. But each of the *dead* clauses in LEFT when the algorithm halts must have been wounded as many times as it had literals, hence must have had its mass doubled that many times, and so must have final mass equal to one. Therefore $|\text{LEFT}| \leq m/2^k$, and so $|S| \geq m(1 - 1/2^k)$ and the bound follows.
∎

Again, for larger values of k, algorithm GREEDYJOHNSON2 obtains better performance ratios and, generally speaking, because $1 - \frac{1}{2^k} > 1 - \frac{1}{k+1}$ for any integer $k \geq 2$, algorithm GREEDYJOHNSON2 has a better performance than that of algorithm GREEDYJOHNSON1.

The performance ratio 2/3 has been proved in a paper by Chen, Friesen, and Zheng [21]. Because they consider the *MAX W–SAT* problem, line 2 in Fig. 8 must be modified to take the weights w_i into account: the *mass* becomes $w(C_i) = w_i \, 2^{-|C_i|}$. The preceding bound 1/2 depends on the fact that the only upper bound used in the above proofs was given by the total weight of the clauses; of course this upper bound can be far from the optimal value. The novelty of the approach of [21] is that the performance ratio can be derived by using the correct value of the optimal solution. In order to prove that algorithm GREEDYJOHNSON2 has this better performance ratio let us introduce a generalization of the algorithm. It is important to stress that this generalization is introduced to perform a more accurate analysis of the performance ratio and it is used in the following as a theoretical tool.

The difference between GREEDYJOHNSON2 and its generalization is rather subtle. The generalized algorithm, that we denote as GENJOHNSON2, considers an arbitrary Boolean array $b[1..n]$ of size n as additional input, and examines b to decide what to do if an equality is present in line 6 of Fig. 8. Let us assume that the variable one is considering is u_j. In line 6 of GREEDYJOHNSON2 in Fig. 8, when $\sum_{C_i \in \mathbf{CT}} w(C_i) = \sum_{C_i \in \mathbf{CF}} w(C_i)$, the **if** condition is true and u_j is set to **true**. Now, instead, when one obtains an equality one considers two different cases: if the variable $b[j]$ is **true** u_j is set to **true**; if the variable $b[j]$ is **false** u_j is set to **false**.

This generalized algorithm is then used in the proof with this Boolean array equal to the optimal assignment. Of course the optimal assignment cannot be derived in polynomial time but here we are not interested in running an algorithm but in performing a theoretical analysis.

We will prove that GENJOHNSON2 has a performance ratio 2/3 and this fact will imply that also GREEDYJOHNSON2 has performance ratio 2/3.

Let us give some definitions needed in the proof.

Definition 4.11 • *A literal is positive if it is a Boolean variable u_i for some i.*

• *A literal is negative if it the negation $\bar{u}_i$ of a Boolean variable for some i.*

Definition 4.12 *Assume that algorithm* GENJOHNSON2 *is applied to a formula* **C** *and consider a fixed moment in the execution.*

Approximate Algorithms for MAX-SAT

- A literal l is active if it has not been assigned a truth value yet.
- A clause C_j is killed if all literals in C_j are assigned value **false**.
- A clause C_j is negative if it is neither satisfied nor killed, and all active literals in C_j are negative literals.

Definition 4.13 Let $0 \leq t \leq n$. Assume that in GENJOHNSON2 the t-th iteration has been completed (a truth assignment has been given to t variables). Then $\mathbf{S}^t$ denotes the set of satisfied clauses, $\mathbf{K}^t$ denotes the set of killed clauses, $\mathbf{N}_i^t$ denotes the set of negative clauses with exactly i active liberals.

Without less of generality, one assumes that each clause in the formula has at most r literals. The proof of the performance ratio $2/3$ depends on the following Lemma.

Given a set of clauses $\mathbf{C}$, let us define as $w(\mathbf{C})$ the sum of the weights of all clauses of $\mathbf{C}$.

Lemma 4.6 For any formula $\mathbf{C}$ of MAX W-SAT and for any Boolean array $b[1..n]$, when the algorithm GENJOHNSON2 is applied on $\mathbf{C}$ the following inequality holds at all iterations $0 \leq t \leq n$:

$$w(\mathbf{S}^t) \geq 2w(\mathbf{K}^t) + \sum_{i=1}^{r} \frac{1}{2^{i-1}} w(\mathbf{N}_i^t) - A_0 \tag{3}$$

where $A_0 = \sum_{i=1}^{r} \frac{1}{2^{i-1}} w(\mathbf{N}_i^0)$

The proof of the Lemma proceeds by induction on t and can be found in [21].

Theorem 4.7 The performance ratio of algorithm GREEDYJOHNSON2 is $2/3$.

Proof. Let $\mathbf{C}$ be an instance of MAX W-SAT and let U_0 an optimal truth assignment for $\mathbf{C}$. Now one considers another formula $\mathbf{C}'$ that is derived from $\mathbf{C}$ as follows. If $U_0(u_t) = \mathbf{false}$ for a variable u_t then one negates u_t (u_t and $\bar{u}_t$ are interchanged) in $\mathbf{C}'$. No change on the weights is done. Therefore there exists a one-to-one correspondence between the set of clauses in $\mathbf{C}$ and the set of clauses in $\mathbf{C}'$; moreover the corresponding clauses have the same weight. In addition, the Boolean array $b[1..n]$ is constructed such that $b[j] = \mathbf{false}$ if and only if $U_0(u_j) = \mathbf{false}$.

It is easy to see (for the details, see again [21]) that

- the weight of an optimal assignment to $\mathbf{C'}$ is equal to the weight of an optimal assignment to $\mathbf{C}$.

- the truth assignment for $\mathbf{C}$ found by GREEDYJOHNSON2 and the truth assignment for $\mathbf{C'}$ found by GENJOHNSON2 have the same weight.

This means that, if we prove that GENJOHNSON2 has a performance ratio $2/3$ on the formula $\mathbf{C'}$, the theorem is shown.

Note that the truth assignment U_0' for $\mathbf{C'}$ that gives value **true** to all variables corresponds to the optimal truth assignment U_0 for $\mathbf{C}$. Therefore U_0' is optimal for $\mathbf{C'}$.

When GENJOHNSON2 stops, that is, for $t = n$, $\mathbf{S}^n$ is the set satisfied by the algorithm and $\mathbf{K}^n$ is the set of clauses not satisfied. $\mathbf{N}_i^n$ is the empty set for any i.

Applying the inequality 3 of Lemma 4.6 to this case, one obtains:

$$w(\mathbf{S}^n) \geq 2w(\mathbf{K}^n) - A_0. \tag{4}$$

On the other hand, A_0 can be upperbounded in the following way:

$$A_0 = \sum_{i=1}^{r} \frac{1}{2^{i-1}} w(\mathbf{N}_i^0) \leq \sum_{i=1}^{r} w(\mathbf{N}_i^0) \leq 2 \sum_{i=1}^{r} w(\mathbf{N}_i^0). \tag{5}$$

From inequalities 4 and 5 one has:

$$\frac{3}{2} w(\mathbf{S}^n) \geq w(\mathbf{S}^n) + w(\mathbf{K}^n) - \sum_{i=1}^{r} w(\mathbf{N}_i^0) \tag{6}$$

Note that, on one hand, $w(\mathbf{S}^n)$ is the weight of the truth assignment found by GENJOHNSON2. On the other hand, $\mathbf{S}^n \cup \mathbf{K}^n$ is the whole set of clauses in $\mathbf{C'}$ and the optimal truth assignment U_0' for $\mathbf{C'}$ that gives value **true** to all variables satisfies all clauses in $\mathbf{C'}$ except those belonging to $\mathbf{N}_i^0$ for $i = 1, 2, \ldots, r$.

Therefore an optimal truth assignment for $\mathbf{C'}$ has weight exactly

$$w(\mathbf{S}^n) + w(\mathbf{K}^n) - \sum_{i=1}^{r} w(\mathbf{N}_i^0).$$

Then the inequality 6 says that the weight of the truth assignment found by GENJOHNSON2 is at least $2/3$ of the weight of an optimal assignment

to **C'**. In consequence, the weight of the assignment constructed by the original GREEDYJOHNSON2 algorithm for the instance **C** is at least 2/3 of the weight of an optimal assignment to **C**, thus proving the theorem.
∎

Finally, it is worthwhile to note this performance ratio 2/3 is tight. There are formulae for which GREEDYJOHNSON2 finds a truth assignment such that the ratio is 2/3. Therefore this bound cannot be improved. In [21] a set of formulae with this characteristic has been presented.

Let us consider the following formula $\mathbf{C}_h$ formed by $3h$ clauses with h integer greater than 0 where all the clauses have the same weight:

$$\mathbf{C}_h = \{(u_{3k+1} \vee u_{3k+2}), (u_{3k+1} \vee u_{3k+3}), (\overline{u}_{3k+1}) | 0 \leq k \leq h-1\}$$

GREEDYJOHNSON2 gives value **true** to all variables so satisfying $2h$ clauses, that is $(u_{3k+1} \vee u_{3k+2})$, $(u_{3k+1} \vee u_{3k+3})$ for $0 \leq k \leq h-1$. On the other hand, the truth assignment u_{3k+1} =**false**, $u_{3k+2} = u_{3k+3}$ =**true** for $0 \leq k \leq h-1$ satisfies all the $3h$ clauses of the formula.

4.3 Randomized algorithms for *MAX W–SAT*

4.3.1 A randomized 1/2–approximate algorithm for *MAX W–SAT*

One of the most interesting approaches in the design of new algorithms is the use of *randomization*. During the computation, random bits are generated and used to influence the algorithm process.

In many cases randomization allows to obtain better (expected) performance or to to simplify the construction of the algorithm. Particularly in the field of approximation, randomized algorithms are widely used and, for many problems, the algorithm can be "derandomized" in polynomial time while preserving the approximation ratio. However, it is important to note that, often, the derandomization leads to algorithms which are very complicated in practice.

Let us now use this approach to present more efficient approximate algorithms for *MAX W–SAT*. More precisely, this section introduces two different randomized algorithms that achieve a performance ratio of 3/4. Moreover, it is possible to derandomize these algorithms, that is, to obtain deterministic algorithms that preserve the same bound 3/4 for every instance.

The derandomization is based on the *method of conditional probabilities* that has revealed its usefulness in numerous cases and is a general technique that often permits to obtain a deterministic algorithm from a randomized one while preserving the quality of approximation.

Let us first present the algorithm RANDOM, a simple randomized algorithm, that, while just achieving a performance ratio 1/2, will be used in the following subsections as an ingredient to reach the performance ratio 3/4.

RANDOM
 Input: Set C of weighted clauses in conjunctive normal form
 Output: Truth assignment U, $\mathbf{C}'$, $\sum_{C_j \in \mathbf{C}'} w_j$
 1   Independently set each variable u_i to **true** with probability 1/2
 2   Compute $\mathbf{C}' = \{C_j \in \mathbf{C} : C_j \text{ is satisfied }\}$
 3   Compute $\sum_{C_j \in \mathbf{C}'} w_j$

Figure 9: The RANDOM algorithm, a randomized $(1 - 1/2^k)$-approximate algorithm.

Because the algorithm is randomized, one is interested in the *expected* performance when the algorithm is run with different sequences of random bits (i.e., with different random assignments).

Lemma 4.8 *Given an instance of MAX W–SAT in which all clauses have at least k literals, the expected weight W of the solution found by algorithm* RANDOM *is such that*

$$W \geq (1 - \frac{1}{2^k}) \sum_{C_j \in \mathbf{C}} w_j.$$

Proof. The probability that any clause with k literals is not satisfied by the assignment found by the algorithm is 2^{-k} (all possible k matches must fail). Therefore the probability that a clause is satisfied is $1 - 2^{-k}$. Then

$$W = (1 - \frac{1}{2^k}) \sum_{C_j \in \mathbf{C}} w_j.$$

∎

As an immediate consequence of Lemma 4.8, one obtains the following Corollary.

Corollary 4.9 *Algorithm* RANDOM *finds a solution for MAX W–SAT whose expected value is at least one half of the optimum value.*

The performance of algorithm RANDOM is the same, in a probabilistic setting, as that of algorithm GREEDYJOHNSON2.

Actually it is possible to show that, by applying the method of conditional probabilities to algorithm RANDOM, one essentially obtains algorithm GREEDYJOHNSON2. In this way the algorithm RANDOM is derandomized.

Let us consider the following greedy algorithm that it is described informally and divided in two phases:

First phase (initialization). Assuming (as in algorithm RANDOM) that every variable u_i is **true** with probability $1/2$, for every clause C_i compute the probability d_i that C_i is not satisfied. According to Lemma 4.8 this probability is $\frac{1}{2^k}$ where k is the number of literals occurring in C_i.

Second phase Given a variable u_j that has not been assigned a value, let **CT** be the set of remaining clauses that contain u_j and let **CF** be the set of remaining clauses that contain $\overline{u_j}$.

If $\sum_{C_i \in \mathbf{CT}} w_i d_i \geq \sum_{C_i \in \mathbf{CF}} w_i d_i$ then assign to u_j **true**, otherwise assign **false**. In the first case remove all clauses of **CT** from the formula and double d_i for all clauses C_i of **CF**; in the second case remove all clauses of **CF** from the formula and double d_i for all clauses C_i of **CT**.

It is immediate to note that the truth assignment computed by such an algorithm has weight at least equal to the expected value W of the solution found by algorithm RANDOM. Moreover the algorithm is deterministic.

On the other hand, the two phases correspond to what is made in algorithm GREEDYJOHNSON2.

The approach shown for derandomizing RANDOM can be applied to many randomized algorithms. For the sake of simplicity, let us assume that the input is given by n Boolean variables. Now let E be the expected value of the solutions achieved by a randomized algorithm A. One is now interested in finding a deterministic algorithm B that achieves a solution of value E in polynomial time. In such a way A has been efficiently derandomized.

The algorithm B consists of n iterations and, at each iteration, the value of a variable is determined. The Boolean value of the i-th variable is found in the following way: given the values of variables $u_1, \ldots, u_{i-1}$, we set $u_i = 1$ and we compute the expected weight of clauses satisfied by that truth assignment; then we compute the expected weight of clauses satisfied by the truth assignment in which one has $u_i = 0$, given the current assignment to $u_1, \ldots, u_{i-1}$. We assign u_i the value that maximizes the conditional expectation.

After n iterations, a truth assignment is found deterministically. If we are able to compute each conditional expectation in polynomial time, the algorithm runs in polynomial time and has found an approximate solution whose value is at least E.

Coming back to algorithm RANDOM, one has that, for $k = 1$, the algorithm achieves an expected performance ratio $1/2$. The performance of the algorithm improves if we increase the number of literals. In particular, for $k = 2$, that is for formulae which do not contain unit clauses, one obtains an expected value which is at least $3/4$ of the optimal value. Therefore if one could discard unit clauses, one would already have a $3/4$-approximate algorithm for *MAX W–SAT*, after applying the derandomization. This observation will reveal its usefulness in the following.

4.3.2 A randomized 3/4-approximate algorithm for *MAX W–SAT*

This subsection presents an algorithm that considerably improves the performance of algorithm RANDOM, and obtains a performance ratio $3/4$.

First of all we consider a generalization of algorithm RANDOM. In the previous case the value of every variable was chosen randomly and uniformly, that is with probability $1/2$; now the value of variable u_i is chosen with probability p_i, obtaining algorithm GENRANDOM.

GENRANDOM
  Input: Set **C** of weighted clauses in conjunctive normal form
  Output: Truth assignment U, $\mathbf{C}'$, $\sum_{C_j \in \mathbf{C}'} w_j$
1   Independently set each variable u_i to **true** with probability p_i
2   Compute $\mathbf{C}' = \{C_j \in \mathbf{C} : C_j \text{ is satisfied }\}$
3   Compute $\sum_{C_j \in \mathbf{C}'} w_j$

Figure 10: The GENRANDOM algorithm.

The expected number of clauses satisfied by algorithm GENRANDOM can be immediately computed as a function of p_i.

Lemma 4.10 *The expected weight W of the set of clauses **C** is:*

$$W = \sum_{C_j \in C} w_j (1 - \prod_{i \in U_j^+} (1 - p_i) \prod_{i \in U_j^-} p_i)$$

where U_j^+ (U_j^-) denotes the set of indices of the variables appearing unnegated (negated) in the clause C_j.

Proof. It is an obvious generalization of the proof given in the particular case $p_i = 1/2$.
■

Approximate Algorithms for MAX-SAT

Now, if one manages to find suitable values p_i such that $W \geq 3/4 \, m^*(\mathbf{C})$ for every formula $\mathbf{C}$, one would obtain a 3/4-approximate randomized algorithm.

To aim at this result, let us consider the representation of the instances of *MAX W-SAT* as instances of an integer linear programming problem (*ILP*) already presented in Section 2:

$$max \sum_{C_j \in \mathbf{C}} w_j z_j$$

subject to :

$$\sum_{i \in U_j^+} y_i + \sum_{i \in U_j^-} (1 - y_i) \geq z_j, \forall C_j \in \mathbf{C}$$

$$y_i \in \{0, 1\}, 1 \leq i \leq n$$

$$z_j \in \{0, 1\}, \forall C_j \in \mathbf{C}$$

Let $u_1, \ldots, u_n$ be the Boolean variables appearing in the formula. An instance of *MAX W-SAT* is equivalent to an instance of *ILP* if we choose the following conditions:

- $y_i = 1$ iff variable u_i is **true**;
- $y_i = 0$ iff variable u_i is **false**;
- $z_j = 1$ iff clause C_j is satisfied;
- $z_j = 0$ iff clause C_j is not satisfied.

The linear inequality states the fact that a clause can be satisfied ($z_j = 1$) only if at least one of its literals is matched.

One cannot compute the optimal value in polynomial time because *ILP* is $\mathcal{NP}$-complete. However let us consider the *LP* relaxation (by *relaxation* one means that the set of admissible solution increases with respect to that of the original problem) in which one relaxes the conditions $y_i, z_j \in \{0, 1\}$ with the new constraints $0 \leq y_i, z_j \leq 1$. It is known that *LP* can be solved in polynomial time finding a solution

$$(y^* = (y_1^*, \ldots, y_n^*), z^* = (z_1^*, \ldots, z_m^*))$$

with value $m^*_{LP}(x) \geq m^*_{ILP}(x)$, for every instance x, where $m^*_{LP}(x)$ and $m^*_{ILP}(x)$ denote the optimal value of the *LP* and *ILP* instances, respectively. The upper bound is obvious given that the set of admissible solutions is enlarged by the relaxation.

Let us consider algorithm GENAPPROX, see Fig. 11, that works as follows: first it solves the linear programming relaxation and so computes the optimal values (y^*, z^*); then, given a function g to be specified later, it computes, for each i, $i = 1, \ldots, n$, the probabilities $p_i = g(y^*_i)$. By Lemma 4.10 we know that a solution of weight:

$$W = \sum_{C_j \in \mathbf{C}} w_j(1 - \prod_{i \in U_j^+}(1 - p_i) \prod_{i \in U_j^-} p_i)$$

must exist; by applying the method of conditional probabilities, such solution can be deterministically found.

GENAPPROX
Input: Set $\mathbf{C}$ of clauses in disjunctive normal form
Output: Set $\mathbf{C}'$ of clauses, $W = \sum_{C_j \in \mathbf{C}'} w_j$
1   Express the input $\mathbf{C}$ as an equivalent instance x of ILP
2   Find the optimum value y^*, z^* of x in the linear relaxation
3   Choose $p_i \leftarrow g(y^*_i)$, $i = 1, 2, \ldots, n$, for a suitable function g
4    $W \leftarrow \sum_{C_j \in \mathbf{C}} w_j(1 - \prod_{i \in X_j^+}(1 - p_i) \prod_{i \in X_j^-} p_i)$
5   Apply the method of conditional probabilities to find
6   a feasible solution $\mathbf{C}' = \{C_j \in \mathbf{C} : C_j \text{ is satisfied}\}$ of value W

Figure 11: The GENAPPROX algorithm, deterministic version.

If the function g can be computed in polynomial time then algorithm GENAPPROX runs in polynomial time. In fact the linear relaxation can be solved efficiently and the computation of the feasible solution can be computed in polynomial time with the method of conditional probabilities explained before.

The quality of approximation naturally depends on the choice of the function g. Let us suppose that this function finds suitable values such that:

$$(1 - \prod_{i \in U_j^+}(1 - p_i) \prod_{i \in U_j^-} p_i) \geq \frac{3}{4} z_j^*.$$

If this inequality is satisfied, then the algorithm is a 3/4-approximate algorithm for *MAX W–SAT*. In fact one has :

$$W = \sum_{C_j \in C} w_j (1 - \prod_{i \in U_j^+}(1-p_i) \prod_{i \in U_j^-} p_i) \geq \frac{3}{4} \sum_{C_j \in C} w_j z_j^* =$$

$$= \frac{3}{4} m_{LP}^*(x) \geq \frac{3}{4} m_{ILP}^*(x)$$

More generally if one has :

$$(1 - \prod_{i \in U_j^+}(1-p_i) \prod_{i \in U_j^-} p_i) \geq \alpha z_j^*$$

one obtains a α-approximate algorithm.

A first interesting way of choosing the function g consists of applying the following technique, called *Randomized Rounding*, to get an integral solution from a linear programming relaxation. In order to get integer values one rounds the fractional values, that is each variable y_i is independently set to 1 (corresponding to the Boolean variable u_i being set to **true**) with probability y_i^*, for each $i = 1, 2, \ldots, n$. Hence the use of the randomized rounding technique is equivalent to choosing $p_i = g(y_i^*) = y_i^*, i = 1, 2, \ldots, n$.

Lemma 4.11 *Given the optimal values (y^*, z^*) to LP and given any clause C_j with k literals, one has*

$$(1 - \prod_{i \in U_j^+}(1-y_i^*) \prod_{i \in U_j^-} y_i^*) \geq \alpha_k z_j^*$$

where

$$\alpha_k = 1 - \left(1 - \frac{1}{k}\right)^k.$$

Proof. Let us consider a clause C_j and, for the sake of simplicity, let us assume that every variable is unnegated. If a variable u_i would appear negated in C_j, one could substitute u_i by its negation $\bar{u}_i$ in every clause and also replace y_i by $1 - y_i$. So we can assume $C_j = u_1 \vee \cdots \vee u_k$ with the associated condition $y_1^* + \cdots + y_k^* \geq z_j^*$. The Lemma is proved by showing that:

$$1 - \prod_{i=1}^{k}(1 - y_i^*) \geq \alpha_k z_j^*.$$

In the proof we exploit the geometric inequality based on the properties of the arithmetic mean: given a finite set of nonnegative numbers $\{a_1, \ldots, a_k\}$,

$$\frac{a_1 + \cdots + a_k}{k} \geq \sqrt[k]{a_1 a_2 \cdots a_k}.$$

Now we apply the geometric inequality to the set $\{1 - y_1^*, \ldots, 1 - y_k^*\}$. Because $\sum_{i=1}^{k} \frac{1-y_i^*}{k} = 1 - \frac{\sum_{i=1}^{k} y_i^*}{k}$, one has

$$1 - \prod_{i=1}^{k}(1 - y_i^*) \geq 1 - (1 - \frac{\sum_{i=1}^{k} y_i^*}{k})^k \geq 1 - (1 - \frac{z_j^*}{k})^k.$$

We note that the function $g(z_j^*) = 1 - (1 - \frac{z_j^*}{k})^k$ is concave in the interval $[0, 1]$; hence it is sufficient to prove that $g(z_j^*) \geq \alpha_k z_j^*$ at the extremal points of the interval. Because one has

$$g(0) = 0 \text{ and } g(1) = \alpha_k$$

the Lemma is shown.

∎

One can conclude that algorithm GENAPPROX with the choice $p_i = y_i^*$ reaches an approximation ratio equal to α_k. In particular for $k = 2$, the ratio is 3/4. Note that, because α_k is decreasing with k, algorithm GENAPPROX is an α_k-approximation algorithm for formulae with *at most* k literals per clause.

Moreover, it is well known that $\lim_{k \to \infty}(1 - \frac{1}{k})^k = \frac{1}{e}$; hence for arbitrary formulae one finds approximate solutions whose value is at least $1 - \frac{1}{e}$ times the optimal value. Because $1 - \frac{1}{e} = 0.632\ldots$, the randomized rounding obtains a better performance than RANDOM, but it looks as if one is far from achieving a 3/4-approximation ratio.

Luckily, with a suitable merging of the above algorithm with RANDOM one obtains the desired performance ratio. Firstly let us recall that RANDOM is a 3/4-approximation algorithm if all clauses have *at least* two literals. On the other hand, GENAPPROX is a 3/4-approximation algorithm if we work with clauses with *at most* two literals. One algorithm is good for large clauses, the other for short ones. A simple combination consists of running both algorithm and choosing the best truth assignment obtained. Let us now consider the expected value obtained from the combination.

Theorem 4.12 *Let W_1 be the expected weight corresponding to $p_i = 1/2$ and let W_2 be the expected weight corresponding to $p_i = y_i^*$, $i = 1, 2, \ldots, n$. Then one has :*

$$\max(W_1, W_2) \geq \frac{3}{4} m_{LP}^*(x), \text{ for any instance } x.$$

Proof. Because $\max(W_1, W_2) \geq \frac{W_1+W_2}{2}$, it is sufficient to show that $\frac{W_1+W_2}{2} \geq \frac{3}{4} m^*_{LP}(x)$ for any x. Let us denote by $\mathbf{C}^k$ the set of clauses with exactly k literals. By Lemma 4.8, because $0 \leq z_j^* \leq 1$ one has

$$W_1 = \sum_{k \geq 1} \sum_{C_j \in \mathbf{C}^k} \gamma_k w_j \geq \sum_{k \geq 1} \sum_{C_j \in \mathbf{C}^k} \gamma_k w_j z_j^* \qquad (7)$$

where $\gamma_k = (1 - \frac{1}{2^k})$.

Moreover, by applying Lemma 4.11, one obtains:

$$W_2 \geq \sum_{k \geq 1} \sum_{C_j \in \mathbf{C}^k} \alpha_k w_j z_j^*. \qquad (8)$$

Summing 7 and 8 one has :

$$\frac{W_1 + W_2}{2} \geq \sum_{k \geq 1} \sum_{C_j \in \mathbf{C}^k} \frac{\gamma_k + \alpha_k}{2} w_j z_j^*.$$

We note that $\gamma_1 + \alpha_1 = \gamma_2 + \alpha_2 = 3/2$ and for $k \geq 3$ one has that $\gamma_k + \alpha_k \geq 7/8 + 1 - \frac{1}{e} \geq 3/2$; Therefore:

$$\frac{W_1 + W_2}{2} \geq \sum_{k \geq 1} \sum_{C_j \in \mathbf{C}^k} \frac{3}{4} w_j z_j^* = \frac{3}{4} m^*_{LP}(x).$$

∎

Note that it is not necessary to separately apply the two algorithms but it is sufficient to randomly choose one of the two algorithms with probability 1/2, as it is done in algorithm 3/4–APPROXIMATE SAT.

Corollary 4.13 *Algorithm* 3/4–APPROXIMATE SAT *is a 3/4-approximation algorithm for MAX W–SAT.*

Proof. The proof derives from the above theorem and from the use of the method of conditional probabilities.

∎

4.3.3 A variant of the randomized rounding technique

This subsection shows that it possible to directly design a 3/4-approximate algorithm for *MAX W–SAT* based on randomized rounding. However, in order to reach this aim, one needs to apply some modifications to the standard technique.

3/4-APPROXIMATE SAT
Input: Set **C** of clauses in conjunctive normal form
Output: Set **C**' of clauses, $W = \sum_{C_j \in \mathbf{C}'} w_j$
1   Express the input **C** as an equivalent instance x of *ILP*
2   Find the optimum value (y^*, z^*) of x in the linear relaxation
3   With probability 1/2 choose $p_i = 1/2$ or $p_i = y_i^*, i = 1, 2, \ldots, n$
4     $W \leftarrow \sum_{C_j \in \mathbf{C}} w_j (1 - \prod_{i \in U_j^+}(1 - p_i) \prod_{i \in U_j^-} p_i)$
5   Apply the method of conditional probabilities to find a feasible
6   solution $\mathbf{C}' = \{C_j \in \mathbf{C} : C_j \text{ is satisfied}\}$ of value W

Figure 12: The 3/4-APPROXIMATE SAT algorithm: deterministic with performance ratio 3/4.

Let us start again from algorithm GENAPPROX. One has already seen that, by choosing $g(y_i^*) = y_i^*$, one cannot obtain a performance ratio 3/4. Therefore a different choice of g is necessary.

Let us consider the following definition:

Definition 4.14 *A function* $g : [0,1] \longrightarrow [0,1]$ *has property 3/4 if*

$$1 - \prod_{i=1}^{l}(1 - g(y_i)) \prod_{i=l+1}^{k} g(y_i) \geq \frac{3}{4} \min(1, \sum_{i=1}^{l} y_i + \sum_{i=l+1}^{k}(1 - y_i)).$$

for any integers k,l with $k \geq l$ and any $y_1, \ldots, y_k \in [0,1]$.

By Lemma 4.11 if a function g with property 3/4 is found, then algorithm GENAPPROX becomes a 3/4-approximate algorithm. In order to prove the existence of functions with property 3/4 one needs the following lemma:

Lemma 4.14 *A function* $g : [0,1] \longrightarrow [0,1]$ *has property 3/4 if satisfies the following conditions:*

i) $1 - \prod_{i=1}^{k}(1 - g(y_i)) \geq \frac{3}{4} \min(1, \sum_{i=1}^{k} y_i)$, $\forall k$ *and* $\forall y_i \in [0,1]$
  for any integer k and $y_i \in [0,1], i = 1, 2, \ldots, n$.

ii) $g(y) \leq 1 - g(1 - y)$.

Proof. Given integers k, l with $k \geq l$, let $y_i' = y_i$ for $i = 1 \ldots, l$ and $y_i' = 1 - y_i$

for $i = l+1, \ldots, k$. one has

$$1 - \prod_{i=1}^{l}(1-g(y_i)) \prod_{i=l+1}^{k} g(y_i) \geq 1 - \prod_{i=1}^{l}(1-g(y_i)) \prod_{i=l+1}^{k}(1-g(1-y_i))$$

$$= 1 - \prod_{i=1}^{k}(1-g(y'_i)) \geq \frac{3}{4}\min(1, \sum_{i=1}^{k} y'_i)$$

$$= \frac{3}{4}\min(1, \sum_{i=1}^{l} y_i + \sum_{i=l+1}^{k}(1-y_i)).$$

∎

Lemma 4.15 *The following function g_α verifies property 3/4:*

$$g_\alpha(y) = \alpha + (1-2\alpha)y$$

where $2 - \frac{3}{\sqrt[3]{4}} \leq \alpha \leq \frac{1}{4}$

Proof. It is immediate to verify that g_α satisfies (ii). In order to prove that also the condition (i) of the lemma 4.14 is verified, one has

$$1 - \prod_{i=1}^{k}(1-g_\alpha(y_i)) = 1 - \prod_{i=1}^{k}(1 - \alpha - (1-2\alpha)y_i)$$

$$\geq 1 - \left(1 - \alpha - (1-2\alpha)\frac{\sum_{i=1}^{k} y_i}{k}\right)^k$$

where one exploits the fact that the arithmetic mean is greater or equal than the geometric mean of k numbers. Let us define $Y = \frac{\sum_{i=1}^{k} y_i}{k}$. It is sufficient to prove that:

$$h_k(Y) = 1 - (1 - \alpha - (1-2\alpha)Y)^k \geq \frac{3}{4}\min(1, kY) \; \forall Y \in [0,1].$$

Let us first prove the result in the interval $[0, 1/k]$. Because the function h_k is concave, the minimum value is reached at one of the extremal points of the interval. This means that it is sufficient to check that the inequality holds for $Y = 0$ and $Y = 1/k$. This fact is immediately true for $Y = 0$. In the other case it is sufficient to prove that:

$$1 - (1 - \alpha - (1-2\alpha)\frac{1}{k})^k \geq 3/4, \; \forall k \geq 1.$$

For $k = 1$ the inequality is satisfied if $\alpha \leq 1/4$. On the other side for $k = 2$ the inequality becomes an identity and therefore it is always satisfied. For $k \geq 3$, by easy algebraic steps, one needs to show that

$$\alpha \geq \frac{k - 1 - k4^{-1/k}}{k - 2}. \tag{9}$$

Note that the right-hand-side of the inequality is a function decreasing in k. Moreover, for $k = 3$, inequality 9 holds for $\alpha \geq 2 - \frac{3}{\sqrt[3]{4}}$. Therefore the proof is completed for the interval $[0, \frac{1}{k}]$.

To finish the proof of the lemma it is sufficient to observe that, for any given k, the function $h_k(Y)$ is increasing in the interval $(\frac{1}{k}, 1]$.
∎

Lemmata 4.11 and 4.15 imply the following theorem:

Theorem 4.16 *Given α such that $2 - \frac{3}{\sqrt[3]{4}} \leq \alpha \leq \frac{1}{4}$, algorithm* GENAPPROX *with the choice $p_i = \alpha + (1 - 2\alpha)y_i^*$ is a 3/4-approximate algorithm for MAX W–SAT.*

4.4 Another $\frac{3}{4}$–approximate algorithm by Yannakakis

It is possible to achieve a performance ratio $\frac{3}{4}$ also by using a very different approach. In fact Yannakakis [84] introduced an algorithm that exploits network flow techniques and again obtains the performance bound $\frac{3}{4}$.

Because of the complexity of the proofs, we will limit ourselves to consider the case of MAX W–2–SAT, by showing that MAX W–2–SAT can be approximated with a performance ratio $\frac{3}{4}$. After generalizing the techniques used for MAX W–2–SAT to an arbitrary formula, it is still possible to obtain the same bound. About the notation: in this section we consider the MAX W–2–SAT problem with clauses containing *one or two* literals. As already shown, if every clause has at least two literals, a simple greedy algorithm finds the ratio $\frac{3}{4}$. Therefore, if one can *eliminate the unit clauses* from instances of MAX W–2–SAT, one obtains the desired bound.

More precisely, given a set **C** of clauses with at most two literals per clause, one fixes the truth value for a subset of the variables and builds a set **C'** of clauses with exactly two literals per clause in which the remaining variables occur. **C'** is constructed in such a way that a truth assignment for **C'** with approximation ratio R gives a truth assignment for **C** (when combined with the truth values of fixed variables) with an approximation ratio at least R.

Approximate Algorithms for MAX–SAT

Formally, one has the following theorem in which we use this notation: $w(\mathbf{C}, \rho)$ is the weight of the formula $\mathbf{C}$ with respect the truth assignment ρ. In this Subsection, for the sake of clarity, we will use different symbols for the set of Boolean variables and the truth assignments.

Theorem 4.17 *Let $\mathbf{C}$ be an instance of MAX W–2–SAT defined over a set U of Boolean variables. It is possible to find in polynomial time a subset V of variables, a truth assignment σ for V, a nonnegative constant h, and a set $\mathbf{C}'$ of clauses with exactly two literals per clause in which only variables belonging to the set $U - V$ occur such that:*

1. *For every truth assignment θ to the set $U - V$ of variables one has $w(\mathbf{C}, \sigma \cup \theta) = w(\mathbf{C}', \theta) + h$ where $\sigma \cup \theta$ is the global truth assignment obtained applying σ to V and θ to $U - V$.*

2. *For every truth assignment ρ to U with restriction θ to $U - V$ one has*
   $w(\mathbf{C}, \rho) \leq w(\mathbf{C}', \theta) + h$

This theorem says that one does not lose in the level of approximation by choosing the truth assignment σ for the variables in V; an optimal (or near-optimal) truth assignment for $U - V$ together with σ for V gives an optimal (near-optimal) assignment for the entire set U of variables.

Because one knows how to approximate formulae with exactly two literals per clause with a performance ratio 3/4, the above theorem implies the following:

Corollary 4.18 *If MAX W–2–SAT with exactly two literals per clause can be approximated with performance ratio R, then the general MAX W–2–SAT problem can be approximated with performance ratio R.*
In particular, MAX W–2–SAT can be approximated with performance ratio 3/4.

Proof. Assume that it is possible to achieve an approximation ratio R for the formula $\mathbf{C}'$, that is $w(\mathbf{C}', \theta) \geq R m^*(\mathbf{C}')$. Then the same ratio holds for $\mathbf{C}$ according to the following calculation. By Part 2 of the above Theorem, one has $m^*(\mathbf{C}) \leq m^*(\mathbf{C}') + h$. Moreover, applying Part 1 and because $h \geq 0$ and $R \leq 1$ one obtains $w(\mathbf{C}, \sigma \cup \theta) = w(\mathbf{C}', \theta) + h \geq R m^*(\mathbf{C}') + h \geq R(m^*(\mathbf{C}') + h) \geq R m^*(\mathbf{C})$.

Intuitively, the proof of theorem 4.17 is based on the idea of finding a correspondence between formulae and networks so that the weight of a formula is evaluated by computing the maximum flow of a network.

In the sequel we will assume that some basic notions of network flow theory are known. For a clear introduction to this area see, for instance, [73].

Given a formula **C**, a network $N(\mathbf{C})$ is built in the following way: Every literal in **C** becomes a node in $N(\mathbf{C})$; moreover two other nodes are introduced, that is, s (which is the source of the network) and t (which is the sink of the network).

The arcs are defined as follows. First of all, two arcs (u,v) and (w,z) are said to *correspond* to each other if $u = \overline{z}$ and $v = \overline{w}$. In this approach $\overline{\overline{a}} = a$ for an arbitrary literal a and $\overline{s} = t$. Now, given a clause C_i of **C** with weight w_i, C_i is associated to two corresponding arcs of $N(\mathbf{C})$, each having capacity $w_i/2$. If $C_i = a$ is a unit clause then its associated arcs are (s,a) and $(\overline{a},t)$; if $C_i = a \vee b$ is a clause of length two, its associated arcs are $(\overline{a},b)$ and $(\overline{b},a)$. Finally note that the source node stands for the constant **true** and the sink node for **false**.

Some other definition is needed.

Definition 4.15 • *A network is symmetric if corresponding arcs have the same capacity.*

• *A flow is symmetric if corresponding arcs have the same flow.*

According to such definition $N(\mathbf{C})$ is symmetric.

Let f^* be the maximum flow in $N(\mathbf{C})$. Now let us consider a new flow f in $N(\mathbf{C})$. If e_{i_1} and e_{i_2} are the two arcs associated to a clause C_i, then f is defined in the following way: $f(e_{i_1}) = f(e_{i_2}) = \frac{f^*(e_{i_1}) + f^*(e_{i_2})}{2}$.

Two introductory lemmata are now presented. The proofs can be found in [84].

Lemma 4.19 *The flow f satisfies the capacity and flow conservation constraints and has maximum value.*

Definition 4.16 *Let G and G' be two formulae defined over the same set of variables. G and G' are said to be equivalent if every truth assignment gives the same weight to the two formulae.*

In the following lemma one will assume that all the considered clauses have the same weight.

Lemma 4.20 1. *Let us consider the following two formulae:*
$G = \{\overline{u}_i \vee u_{i+1} | i = 0, \ldots, k\}$ *and* $G' = \{\overline{u}_{i+1} \vee u_i | i = 0, \ldots, k\}$. *Then G and G' are equivalent.*

2. Let us consider the following two formulae:
   $H = \{u_1\} \cup \{\overline{u}_i \vee u_{i+1} | i = 1, \ldots, k-1\}$ and $H' = \{u_k\} \cup \{\overline{u}_{i+1} \vee u_i | i = 1, \ldots, k-1\}$. Then also H and H' are equivalent.

Let us define the *residual network* M with respect to the flow f. Given any arc $e = (u, v)$ of $N(\mathbf{C})$, M contains e with capacity $c(e) - f(e)$ if e is not saturated, where c denotes the capacity in $N(\mathbf{C})$; moreover M contains the reverse arc (v, u) with capacity $f(e)$, again if the capacity is not saturated. No arc going into source s or going out of t is included.

The reversal of two corresponding arcs gives two arcs that are still corresponding to each other. Furthermore M is symmetric because the network $N(\mathbf{C})$ and the flow f are symmetric. Finally note that M is the network $N(\tilde{\mathbf{C}})$ of a formula $\tilde{\mathbf{C}}$ on the same set of variables. If M has an arc (s, l) and hence an arc $(\bar{l}, t)$ of weight w, then $\tilde{\mathbf{C}}$ has a unit clause l with weight $2w$. If M contains the corresponding arcs (a, b) and $(\bar{b}, \bar{a})$ of weight w, then $\tilde{\mathbf{C}}$ contains the clause $\bar{a} \vee b$ with weight $2w$.

Now one is ready to state the following important Lemma:

Lemma 4.21 *Let f_{opt} be the value of the maximum flow f. For any truth assignment θ, one has*

$$w(\mathbf{C}, \theta) = w(\tilde{\mathbf{C}}, \theta) + f_{opt}.$$

Proof. A flow can be decomposed into a set of simple paths $P_1, \ldots, P_l$ from the source to the sink and into a set of cycles $K_1, \ldots, K_m$. Given an arbitrary arc e, the flow $f(e)$ through e is equal to the sum of the weights associated with the paths and cycles containing e. Moreover f_{opt} is equal to the sum of the weights of the paths. For this decomposition, in the case of the residual network M, one has to reverse every path and cycle from $N(\mathbf{C})$. This operation is performed by subtracting the associated weight from the capacities of all the arcs of the path or cycle and summing the weight to the capacities of the arcs in the reverse path or cycle (except for the arcs going into the source or out of the sink).

Let us consider what happens to the clauses in $\mathbf{C}$ after applying this reversal operation. By reversing a cycle K_j with weight w_j, w_j is subtracted from all the clauses that correspond to arcs of the cycle and w_j is added to the clauses corresponding to arcs of the reverse cycle. Part 1 of Lemma 4.20 guarantees that a set of equivalent clauses is obtained. Considering the paths, assume that an arbitrary path P_j consists of arcs $(s, u_1), \ldots, (u_{k-1}, u_k), (u_k, t)$. By Part 2 of Lemma 4.20, the corresponding set of clauses

$\{u_1\} \cup \{\overline{u_i} \vee u_{i+1} | i = 1, \ldots, k-1\} \vee \{\overline{u_k}\}$ is equivalent to the set $\{u_k, \overline{u_k}\} \cup \{\overline{u_{i+1}} \vee u_i | i = 1, \ldots, k-1\}$. $\{u_k, \overline{u_k}\}$ is equivalent to the constant clause **true** while $\{\overline{u_{i+1}} \vee u_i | i = 1, \ldots, k-1\}$ corresponds to the reverse path of P_j. Hence, by reversing a path with weight w_j, the weight is subtracted from the corresponding clauses and added to the clauses of the reverse path in M; moreover, the weight w_j is given to constant clause **true**, so preserving the equivalence.

Globally speaking, one obtains an equivalent set of clauses that consists of $\tilde{C}$ and the clause **true** with weight equal to the sum of the weights of the paths, that is, f_{opt}.
∎

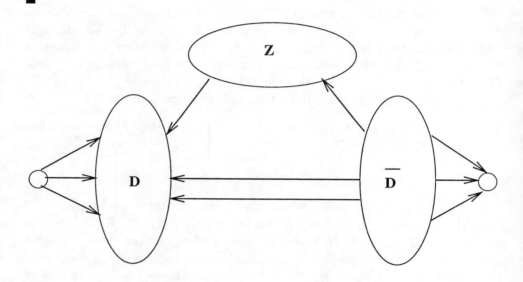

Figure 13: The residual network M.

Note that the residual network M does not contain any path from s to t, because f is a maximum flow (see [73]). Let D be the set of nodes reachable from s in M. The symmetry of M implies that there exists a path from s to a node a if and only if there exists a path from $\overline{a}$ to t. One can conclude that D does not contain any complementary literals because otherwise M would contain a path from s to t. Again, by symmetry, the set of nodes that can reach t is given by the set $\overline{D} = \{\overline{a} | a \in D\}$. The set Z is given by the the remaining nodes of M, that is, by the nodes that do not belong to D or $\overline{D}$. By construction, there are no arcs coming out of D and, by symmetry of M, no arcs going into $\overline{D}$, see Fig. 13.

Approximate Algorithms for MAX–SAT

So one has obtained the following Lemma.

Lemma 4.22 *The set D does not contain any complementary literals. Every clause of $\tilde{C}$ that contains the negation $\bar{a}$ of a literal a in D, also contains (positively) a literal b in D.*

Exploiting this lemma, if one sets every literal in D to **true**, all the clauses in $\tilde{C}$ involving such literals and their negations are satisfied. Let d^* be the total weight of these clauses. Lemma 4.21 implies Theorem 4.17 if one does these choices: V is given by those variables with a literal in D, σ is the above truth assignment for V, C' is the set of remaining clauses of $\tilde{C}$ that involve only literals from Z and $h = f_{opt} + d^*$. As desired, C' does not have unit clauses because M does not contain any arc from s to nodes outside D.

Finally it is worthwhile to note that, because the weights of the clauses of C are integers, it follows that, in general, the weights of $\tilde{C}$ and C', are half-integers. Then these weights can become integers if they are doubled. This multiplication by two does not change the problem.

The proof of Theorem 4.17 is therefore completed.

A generalization of this approach allows to eliminate the unit clauses in the case of formulae in which there are clauses of length 3 or more.

4.5 Approximate solution of *MAX W–SAT*: improvements

The approximation ratio 3/4 can be slightly improved by applying different relation techniques. More precisely, Goemans and Williamson [40] introduce a new way to approximate another classical optimization problem: *MAX–CUT*.

Definition 4.17 *MAX–CUT is the following $\mathcal{NPO}$ problem: Given a graph $G = (V, E)$ and a weight $w(e)$ for each $e \in E$, one looks for a subset V' of V that maximizes the sum of the weights of the edges from E that have one endpoint in V' and one endpoint in $V - V'$.*

MAX–CUT can be easily approximated with a performance ratio 1/2 (see, for instance, [84]). For many years it was impossible to improve this bound. Goemans and Williamson devised a new approach. The problem can be represented as a quadratic programming problem. As in the case of *MAX W–SAT*, one can look for some relaxation in order to find a good approximate algorithm. Actually, in the case of *MAX–CUT*, a relaxation based on the so called *semidefined programming* allows to design an approximation ratio .87856, therefore strongly improving the bound 1/2. In fact, by applying similar techniques, it is possible to reach better results also for *MAX W-SAT*, although the improvement is very small in this case:

Theorem 4.23 *There exists a polynomial time approximate algorithm for MAX W-SAT with a performance ratio .7584.*

For the restricted case of *MAX 2-SAT*, one can obtain a more substantial improvement with the technique of Feige and Goemans [32]. Actually they have obtained a performance ratio 0.931. Coming back to the general problem some other small improvements have been given. Asano [5] (following [6]) has improved the bound to .77. If one considers only *satisfiable MAX W-SAT* instances, Trevisan [83] obtains a 0.8 approximation factor, while Karloff and Zwick [58] claim a 0.875 performance ratio for satisfiable instances of *MAX W-3-SAT*.

4.6 Negative results about approximability

Until now one has shown how to approximate *MAX W-SAT*, obtaining better and better approximation ratios. It is natural, in this framework, to wonder whether it is possible to further improve the approximation properties of *MAX W-SAT*, by showing that the problem belongs to $\mathcal{PTAS}$. We recall that if *MAX W-SAT* belongs to $\mathcal{PTAS}$, one would be able to introduce a polynomial time ε-approximate algorithm for every ε between 0 and 1. Unfortunately it is possible to prove that, unless $\mathcal{P} = \mathcal{NP}$, this is not true.

The negative results for *MAX W-SAT* and, more generally, for $\mathcal{NPO}$ optimization problems are based on the theory of probabilistic checkable proof (for short, $\mathcal{PCP}$). This theory was developed in a different context but, surprisingly, can be applied to the field of approximation algorithms allowing to find very interesting negative results. Because of its complexity it is impossible to present this theory. The reader interested in understanding this nice relationship can read the pioneering papers that introduced $\mathcal{PCP}$, for instance [3] and [4]. Exploiting this theory many negative results for *MAX W-SAT* have been given. We limit ourselves to present the strongest one which is due to Håstad [50].

Theorem 4.24 *Unless* $\mathcal{P} = \mathcal{NP}$ *MAX W-SAT cannot be approximated in polynomial time within a performance ratio greater than 7/8.*

More precisely, this result has been obtained for the *MAX W-SAT* problem in which each clause is of length exactly three. Since this version is a particular case of *MAX W-SAT*, of course the result holds in general.

5 A different *MAX-SAT* problem and completeness results

In this section we present another optimization problem again having *SAT* as associated recognition problem. While in *MAX W-SAT* we associate a weight to each clause, now we associate a weight to each variable. More formally we introduce *MAX-VAR SAT*.

MAX-VAR SAT is an $\mathcal{NPO}$ problem in which (I, sol, m, opt) are defined in the following way:

1. I = sets $U = u_1, \ldots, u_n$ of Boolean variables and a collection **C** of clauses over U, a set $W = w_1, \ldots, w_n$ of integers (weights) associated to the variables.

2. Given an instance x of I, $sol(x)$ = set of truth assignments to the variables in U that satisfy all clauses in **C**.

3. Given an instance x of I and a feasible solution τ of x,
   $m(x, \tau)$ =max(1, sum of weights associated to the variables that are **true** in τ).

4. opt= max.

We note that, in the case of this new problem, the feasible solutions are restricted to those truth assignments that satisfy the formula completely. Formally, in the definition, the measure is found by determining a maximum between 1 and the sum of the weights associated to the variables that are **true** in the truth assignment, because the formula could be not satisfiable. In this case we directly assume that the optimum value is 1, to define the optimization problem for every instance.

A first important result due to [9] and independently to [71] is the following:

Theorem 5.1 *MAX-VAR SAT is $\mathcal{NPO}$-complete.*

Proof. Let P be an $\mathcal{NPO}$ problem and let us consider the corresponding non deterministic Turing machine M associated to P and that was presented in Section 4.1.

According to Cook's Theorem, for any instance x, one can find a Boolean formula whose satisfying truth assignments are in one-to-one correspondence with the halting computation paths of $M(x)$. Let $y_1, y_2, \ldots, y_r$ be the Boolean variables describing the feasible solution y of x and let $m_1, \ldots, m_s$ be the Boolean variables that correspond to the tape cells on which M prints the value $m_P(x, y)$. Then a zero weight is assigned to every variable except the m_i's which are given weight 2^{s-i}.

Given a satisfying truth assignment, one is able to find a solution for P just by looking at the values of the y_i's. From the construction $m_P(x, y)$ is equal to the sum of the weights of the **true** variables. Therefore it has been proved that $P \leq_{\mathcal{PTAS}}$ MAX-VAR SAT with, in this case $c(\epsilon) = \epsilon$.
∎

Considering a particular version of MAX-VAR SAT one can exhibit an example of a problem which is $\mathcal{APX}$-complete. Let us consider the problem MAX-VAR BOUNDED SAT in which the total sum of the weights is between Z and $2Z$, where Z is an integer given in input. Consequently the measure is changed in the following way: $\max(Z, \sum_{i=1}^{n} w_i \tau(u_i))$ if the formula is satisfied, $m(x, \tau) = Z$ otherwise

Theorem 5.2 *MAX-VAR BOUNDED SAT is $\mathcal{APX}$-complete.*

The proof of the theorem can be found in [27].

Historically MAX-VAR BOUNDED SAT is the first example of a problem that is $\mathcal{APX}$-complete. However, by combining together different techniques (including PCP) it would be possible to prove the following:

Theorem 5.3 *MAX W-SAT is $\mathcal{APX}$-complete.*

A presentation of the proof can be found in [8]. Let us note that this theorem is another way of stating that MAX W-SAT does not belong to $\mathcal{PTAS}$.

6 Local search

According to [73] "local search is based on what is perhaps the oldest optimization method – trial and error." The idea is simple and natural and it is surprising to see how successful local search has been on a variety of difficult problems. MAX-SAT is among the problems for which local search has been very effective: different variations of local search with randomness techniques have been proposed for SAT and MAX-SAT starting from the late eighties, see for example [42, 81], motivated by previous applications of "min-conflicts" heuristics in the area of Artificial Intelligence [66].

The general scheme is based on generating a starting point in the set of admissible solution and trying to improve it through the application of simple *basic* moves. If a move ("trial") is successful one accepts it, otherwise ("error") one keeps the current point. Of course, the successfulness of a local search technique depends on the neighborhood chosen and there are often trade-offs between the size of the neighborhood (and the related computational requirements to calculate it) and the quality of the obtained local optima.

In addition, as it will be demonstrated in Sec. 6.2, the use of a guiding function different from the original one can in some cases guarantee local optima of better quality.

Because this presentation is dedicated to the *MAX-SAT* problem, the search space that we consider is given by all possible truth assignments. Of course, a truth assignment can be represented by a binary string. For this presentation, let us consider the elementary changes to the current assignment obtained by changing a single truth value. The definitions are as follows.

Let $\mathcal{U}$ be the discrete search space: $\mathcal{U} = \{0,1\}^n$, and let $f : \mathcal{U} \longrightarrow R$ (R are the real numbers) be the function to be maximized, i.e., in our case, the number of satisfied clauses. In addition, let $U^{(t)} \in \mathcal{U}$ be the current configuration along the *search trajectory* at iteration t, and $N(U^{(t)})$ the neighborhood of point $U^{(t)}$, obtained by applying a set of basic moves μ_i ($1 \leq i \leq n$), where μ_i complements the i-th bit u_i of the string: $\mu_i (u_1, u_2, ..., u_i, ..., u_n) = (u_1, u_2, ..., 1 - u_i, ..., u_n)$. Clearly, these moves are idempotent ($\mu_i^{-1} = \mu_i$).

$$N(U^{(t)}) = \{U \in \mathcal{U} \text{ such that } U = \mu_i U^{(t)}, i = 1, ..., n\}$$

The version of *local search* (LS) that we consider starts from a random initial configuration $U^{(0)} \in \mathcal{U}$ and generates a search trajectory as follows:

$$V = \text{Best-Neighbor}(\, N(U^{(t)}) \,) \qquad (10)$$

$$U^{(t+1)} = \begin{cases} V & \text{if} \quad f(V) > f(U^{(t)}) \\ U^{(t)} & \text{if} \quad f(V) \leq f(U^{(t)}) \end{cases} \qquad (11)$$

where Best-Neighbor selects $V \in N(U^{(t)})$ with the best f value and ties are broken randomly. V in turn becomes the new current configuration if f improves. Other versions are satisfied with an improving (or non-worsening) neighbor, not necessarily the best one. Clearly, local search stops as soon as the first local optimum point is encountered, when no improving moves are available, see eqn. 11. Let us define as LS^+ a modification of LS where a specified number of iterations are executed and the candidate move obtained by BEST-NEIGHBOR is *always* accepted even if the f value remains equal or worsens.

6.1 Quality of local optima

Let m^* be the optimum value and k the minimum number of literals contained in the problem clauses.

For the following discussion it is useful to consider the different degree of *coverage* of the various clause for a given assignment. Precisely, let us define as Cov_s the subset of clauses that have exactly s literals matched by the current assignment, and by $\text{Cov}_s(l)$ the number of clauses in Cov_s that contain literal l.

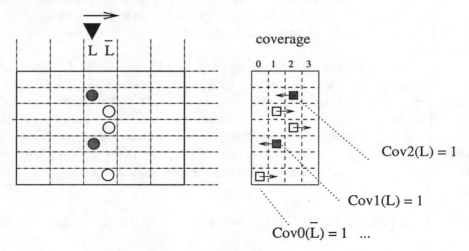

Figure 14: Literal L is changed from **true** to **false**.

One has the following theorem [48]:

Theorem 6.1 *Let m_{loc} be the number of satisfied clauses at a local optimum of any instance of MAX-SAT with at least k literals per clause. m_{loc} satisfies the following bound*

$$m_{loc} \geq \frac{k}{k+1} m$$

Approximate Algorithms for MAX-SAT

and the bound is sharp.

Proof. By definition, if the assignment U is a local optimum, one cannot flip the truth value of a variable (from **true** to **false** or *vice versa*) and obtain a net increase in the number of satisfied clauses f. Now, let $(\Delta f)_i$ by the increase in f if variable u_i is flipped. By using the above introduced quantities one verifies that:

$$(\Delta f)_i = -\text{Cov}_1(u_i) + \text{Cov}_0(\overline{u_i}) \leq 0 \qquad (12)$$

In fact, when u_i is flipped one looses the clauses that contain u_i as the single matched literal, i.e., $\text{Cov}_1(u_i)$ and gains the clauses that have no matched literal and that contain $\overline{u_i}$, i.e., $\text{Cov}_0(\overline{u_i})$.

After summing over all variables:

$$\sum_{i=1}^{n} \text{Cov}_0(\overline{u_i}) \leq \sum_{i=1}^{n} \text{Cov}_1(u_i) \qquad (13)$$

$$k|\text{Cov}_0| \leq |\text{Cov}_1| \leq m_{loc} \qquad (14)$$

where the equality $\sum_{i=1}^{n} \text{Cov}_0(\overline{u_i}) = k|\text{Cov}_0|$ and $\sum_{i=1}^{n} \text{Cov}_1(u_i) = |\text{Cov}_1|$ have been used. The equality are demonstrated by counting how many times a clause in Cov_0 (or Cov_1) is uncountered during the sum. For example, because all literals are unmatched for the clauses in Cov_0, each of them will be encountered k times during the sum.

The conclusion is immediate:

$$m = m_{loc} + |\text{Cov}_0| \leq (1 + \frac{1}{k})m_{loc} = \frac{k+1}{k}m_{loc} \qquad (15)$$

∎

The intuitive explanation is as follows: if there are too many clauses in Cov_0, because each of them has k unmatched literals, there will be at least one variable whose flipping will satisfy so many of these clauses to lead to a net increase in the number of satisfied clauses.

There is therefore a very simple local search algorithm that reaches the same bound as the GREEDYJOHNSON1 algorithm. One starts from a truth assignments and keeps flipping variables that cause a net increase of satisfied clauses, until a local optimum is encountered. Of course, because one gains at least one clause at each step, there is an upper bound of m on the total number of steps executed before reaching the local optimum.

The following corollary is immediate:

Corollary 6.2 *If m_{loc} is the number of satisfied clauses at a local optimum, then:*

$$m_{loc} \geq \frac{k}{k+1} m^* \tag{16}$$

Besides *MAX-SAT*, many important optimization problems share the property that the ratio between the value of the local optimum and the optimal value is bounded by a constant. It is possible to define a class $\mathcal{GLO}$ composed of these problems. It is of interest to note that the closure of $\mathcal{GLO}$ coincides with $\mathcal{APX}$ [10].

6.2 Non-oblivious local optima

In the design of efficient approximation algorithms for *MAX-SAT* a recent approach of interest is based on the use of *non-oblivious functions* independently introduced in [2] and in [59].

Let us consider the classical local search algorithm LS for *MAX-SAT*, here redefined as *oblivious* local search (LS-OB). Clearly, the feasible solution found by LS-OB typically is only a *local* and not a *global* optimum.

Now, a different type of local search can be obtained by using a *different* objective function to direct the search, i.e., to select the best neighbor at each iteration. Local optima of the standard objective function f are not necessarily local optima of the different objective function. In this event, the second function causes an *escape* from a given local optimum. Interestingly enough, suitable *non-oblivious* functions f_{NOB} improve the performance of LS if one considers both the worst-case performance ratio and, as it has been shown in [13], the actual average results obtained on benchmark instances.

Let us mention a theoretical result for *MAX 2-SAT*. The d-neighborhood of a given truth assignment is defined as the set of all assignment where the values of at most d variables are changed. The theoretically-derived non-oblivious function for *MAX 2-SAT* is:

$$f_{NOB}(U) = \frac{3}{2}|\text{Cov}_1| + 2|\text{Cov}_2|$$

Theorems 7-8 of [59] state that:

Theorem 6.3 *The performance ratio for any oblivious local search algorithm with a d-neighborhood for MAX 2-SAT is 2/3 for any $d = o(n)$. Non-oblivious local search with an 1-neighborhood achieves a performance ratio 3/4 for MAX 2-SAT.*

Proof. While one is referred to the cited papers for the complete details, let us only demonstrate the second part of the theorem. The proof is a generalization of that for Theorem 6.1. Let the non-oblivious function be a weighted linear combination of the number of clauses with one and two matched literals:

$$f_{NOB} = a|\text{Cov}_1| + b|\text{Cov}_2|$$

Let $(\Delta f)_i$ by the increase in f if variable u_i is flipped. By using the definition of local optimum and the quantities introduced in Sec. 6.1 one has that $(\Delta f)_i \leq 0$ for each possible flip of a variable u_i. After expressing $(\Delta f)_i$ by using the above introduced quantities, one obtains:

$$-a|\text{Cov}_1(u_i)| - (b-a)|\text{Cov}_2(u_i)| + a|\text{Cov}_0(\overline{u_i})| + (b-a)|\text{Cov}_1(\overline{u_i})| \leq 0 \tag{17}$$

In fact, when u_i is flipped, all clauses that contain it decrease their coverage by one, while the clauses that contain $\overline{u_i}$ increase it by one, see also Fig. 14. As usual, let us assume that no clause contains both a literal and its negation.

After summing over all variables and collecting the sizes of the sets Cov_i one obtains:

$$\sum_{i=1}^{n}(\Delta f)_i \leq 0 \tag{18}$$

$$\frac{b-a}{a}|\text{Cov}_2| + \frac{2a-b}{2a}|\text{Cov}_1| \geq |\text{Cov}_0| \tag{19}$$

Now one can fix the relative size of the values a and b in order to get the best possible bound. This occurs when the coefficients of the terms $|\text{Cov}_2|$ and $|\text{Cov}_1|$ in equation 19 are equal, that is, for $b = \frac{4}{3}a$.

For these values one obtains the following bound:

$$|\text{Cov}_2| + |\text{Cov}_1| \geq 3|\text{Cov}_0| \tag{20}$$

The number of satisfied clauses must be larger than three times the number of unsatisfied ones, which implies that $|\text{Cov}_0| \leq \frac{1}{4}m$, or $m_{loc} \geq \frac{3}{4}m$. ∎

Therefore LS-NOB, by using a function that weights in different ways the satisfied clauses according to the number of matched literals, improves considerably the performance ratio, even if the search is restricted to a much smaller neighborhood. In particular the "standard" neighborhood where all possible flips are tried is sufficient.

With a suitable generalization the above result can be extended: LS-NOB achieves a performance ratio $1 - \frac{1}{2^k}$ for $MAX\text{-}k\text{-}SAT$. The oblivious function for $MAX\text{-}k\text{-}SAT$ is of the form:

$$f_{NOB}(U) = \sum_{i=1}^{k} c_i |\text{Cov}_i|$$

and the above given performance ratio is obtained if the quantities $\Delta_i = c_{i+1} - c_i$ satisfy:

$$\Delta_i = \frac{1}{(k-i+1)\binom{k}{i-1}} \left[\sum_{j=0}^{k-i} \binom{k}{j} \right]$$

Because the positive factors c_i that multiply $|\text{Cov}_i|$ in the function f_{NOB} are strictly increasing with i, the approximations obtained through f_{NOB} tend to be characterized by a "redundant" satisfaction of many clauses. Better approximations, at the price of a limited number of additional iterations, can be obtained by a two-phase local search algorithm (NOB&OB): after a random start f_{NOB} guides the search until a local optimum is encountered [13]. As soon as this happens a second phase of LS is started where the move evaluation is based on f. A further reduction in the number of unsatisfied clauses can be obtained by a "plateau search" phase following NOB&OB: the search is continued for a certain number of iterations after the local optimum of OB is encountered, by using LS$^+$, with f as guiding function [13].

6.2.1 An example of non-oblivious search

Let us consider the following task with number of variables $n = 5$, and clauses $m = m^* = 4$, see also Fig. 15:

$$(\overline{u}_1 \vee \overline{u}_2 \vee u_3) \wedge (\overline{u}_1 \vee \overline{u}_2 \vee u_4) \wedge (\overline{u}_1 \vee \overline{u}_2 \vee u_5) \wedge (\overline{u}_3 \vee \overline{u}_4 \vee \overline{u}_5)$$

Let us assume that the assignment $U = (11111)$ is reached by OB local search. It is immediate to check that $U = (11111)$ is an oblivious local optimum with one unsatisfied clause (clause-4). While OB stops here, a possible sequence to reach the global optimum starting from U is the following: i)

Approximate Algorithms for MAX-SAT

u_1 is set to **false**, ii) u_3 is set to **false**. Now, the first move does not change the number of satisfied clauses, but it changes the "amount of redundancy" (in clause-1 two literals are now satisfied, i.e., clause-1 enters COV_2) and the move *is* a possible choice for a selection based on the *non-oblivious* function. The *oblivious* plateau has been eliminated and the search can continue toward the globally optimal point $U = (01011)$.

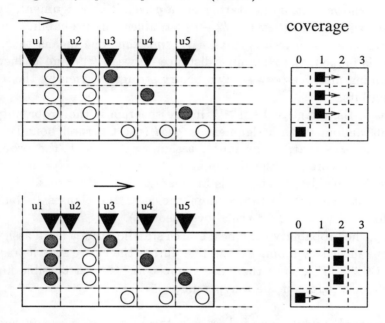

Figure 15: Non-oblivious search takes the different coverage into account.

6.3 Local search satisfies most *3-SAT* formulae

An intriguing result by Koutsoupias and Papadimitriou [63] shows that, for the vast majority of satisfiable *3-SAT* formulae, the local search heuristic that starts at a random truth assignments and repeatedly flips a variable that improves the number of satisfied clauses, almost always succeeds in discovering a satisfying truth assignment.

Let us consider all clauses that are satisfied by a given truth assignment $\hat{U}$ and let us pick each of them with probability $p = 1/2$ to build a *3-SAT* formula. The following theorem [63] is demonstrated:

Theorem 6.4 *Let $0 < \varepsilon < 1/2$. Then there exists c,*
$c \approx \left(1 - \sqrt{1 - (1/2 - \varepsilon)^2}\right)^2 / 6$, *such that for all but a fraction of at most*

$n2^n e^{-cn^2/2}$ satisfiable 3-SAT formulae with n variables, the probability that local search succeeds in discovering a truth assignment in each independent trial from a random start is at least $1 - e^{-\varepsilon^2 n}$.

Proof. Let us focus on the structure of the proof, without giving the technical details. One assumes that there is an assignment $\hat{U}$ that satisfies all clauses and shows that, if one starts from a *good* initial assignment, i.e., one that agrees with $\hat{U}$ in at least $(1/2 - \varepsilon)$ variables, the probability that the local search is ever *mislead* is small. By "*mislead*" one means that, when a variable is flipped, the Hamming distance between $U^{(t)}$ and $\hat{U}$ increases. The Hamming distance between two binary strings is given by the number of differing bits.

In detail, the quantity $1 - e^{-\varepsilon^2 n}$ in the theorem is the probability that the initial random truth assignment is *good* (use Chernoff bound). Then one demonstrates that, if the initial assignment is good, the probability that one does not reduce the Hamming distance between $U^{(t)}$ and $\hat{U}$ when an improving neighbor is chosen is at most $2e^{-cpn^2}$, the probability being measured with respect to the random choice of the clauses to build the original formula ($p = 1/2$ for the above theorem).

Finally, the probability that local search starting from a good assignment will ever be misled by flipping a variable during the entire search trajectory is at most $n2^n e^{-cpn^2}$, since there are at most $n2^{n-1}$ such possible flippings – the number of edges of the n-hypercube.
∎

The original formulation of the above theorem is for a *greedy* version of local search, using the function BEST-NEIGHBOR described in eqn. 10, but the authors note that greediness is not required for the theorem to hold, although it may be important in practice.

Let us finally note that the result, while of theoretical interest, is valid for formulae with many clauses (p must be such that the expected number of clauses is $\Omega(n^2)$), while the most difficult formulae have a number of clauses that is linear in n, see also Sec 9.2.

6.4 Randomized search for *2-SAT* (Markov processes)

A "natural" polynomial-time *randomized* search algorithm for *2-SAT* is presented in [72]. While it has long been known that *2-SAT* is a polynomially solvable problem, the algorithm is of interest because of its simplicity and is summarized here also because it motivated the GSAT-WITH-WALK algorithm of [78], see also Sec. 7.2.

In its "standard" form, local search is guided by the number of satisfied clauses and the basic criterion is that of accepting a neighbor only if more clauses are satisfied. The paper by Papadimitriou [72] changes the perspective by concentrating the attention to the *unsatisfied* clauses.

The algorithm for *2-SAT*, is extremely simple:

MARKOVSEARCH
1   Start with any truth assignment
2   **while**   there are unsatisfied clauses **do**
3   pick one of them and flip a random literal in it

Figure 16: The MARKOVSEARCH randomized algorithm for *2-SAT*.

Let us note that worsening moves, leading to a lower number of satisfied clause, can be accepted during the search.
One can prove that:

Theorem 6.5 *The* MARKOVSEARCH *randomized algorithm for 2-SAT, if the instance is satisfiable, finds a satisfying assignment in $O(n^2)$ expected number of steps.*

Proof. The proof involves an aggregation of the states of the Markov chain so that the chain is mapped to the *gambler's ruin* chain. A sketch of the proof is derived from [68]. Given an instance with a satisfying assignment $\hat{U}$, and the current assignment $U^{(t)}$, the progress of the algorithm can be represented by a particle moving between the integers $\{0, 1, ..., n\}$ on the real line. The position of the particle indicates how many variables in $U^{(t)}$ agree with those of $\hat{U}$. At each iteration the particle's position can change only by one, from the current position i to $i+1$ or $i-1$ for $0 < i < n$. A particle at 0 can move only to 1, and the algorithm terminates when the particle reaches position n, although it may terminate at some other position with a satisfying assignment different from $\hat{U}$. The crucial fact is that, in an unsatisfied clause, at least one of the *two* literals has an incorrect value and therefore, with probability at least 1/2, the number of correct variables increases by one when a randomized step is executed.

The random walk on the line is one of the most extensively studied stochastic processes. In particular, the above process is a version of the "gambler's ruin" chain with reflecting barrier (that is, the house cannot lose its last dollar). Average number of steps for the gambler to be ruined is $O(n^2)$.
∎

7 Memory-less Local Search Heuristics

State-of-the-art heuristics for *MAX–SAT* are obtained by complementing local search with schemes that are capable of producing better approximations beyond the locally optimal points. In some cases, these schemes generate a sequence of points in the set of admissible solutions in a way that is fixed before the search starts. An example is given by *multiple runs* of local search starting from different random points. The algorithm does not take into account the *history* of the previous phase of the search when the next points are generated. The term *memory-less* denotes this lack of feedback from the search history.

In addition to the cited *multiple-run* local search, these techniques are based on Markov processes (Simulated Annealing), see Sec. 7.1, "plateau" search and "random noise" strategies, see Sec. 7.2, or combinations of randomized constructions and local search, see Sec. 7.3.

7.1 Simulated Annealing

The use of a Markov process (Simulated Annealing or SA for short) to generate a stochastic search trajectory is adopted for example in [82].

```
SA
1    for tries ← 1 to MAX-TRIES
2        U ← random truth assignment ; iter ← 0
3        forever
4            if U satisfies all clause then return U
5            temperature ← MAX-TEMP × e^(−iter×decay_rate)
6            if temperature < MIN-TEMP then exit loop
7            for i ← 1 to n
8                δ ← increase of satisfied clauses if u_i is flipped
9                FLIP(u_i) with probability 1/(1 + e^(−δ/temperature))
10           iter ← iter + 1
```

Figure 17: The Simulated Annealing algorithm for *SAT*.

The main structure of the algorithm is illustrated in Fig. 17, adapted from [82]. For a certain number of tries, a random truth assignment is generated (line 2) and the *temperature* parameter is set to MAX-TEMP. In the inner loop, new assignments are generated by probabilistically flipping each variable based on the improvement δ in the number of satisfied clauses that would occur after the flip. Of course, the improvement can be negative.

Approximate Algorithms for MAX-SAT

The probability to flip is given by a logistic function that penalizes smaller or negative improvements (line 9). The inner loop controls the *annealing schedule*: when *iter* increases the *temperature* slowly decreases (line 5) until a minimum of MIN-TEMP is reached and the control exits the loop (line 6) Let us note that, when the *temperature* is large, the moves are similar to those produced by a random walk, while, when the *temperature* is low the acceptance criterion of the moves is that of local search and the algorithm resembles GSAT, that will be introduced in Sec. 7.2. Implementation details, the addition of a "random walk" modification inspired by [78], and experimental results are described in the cited paper.

7.2 GSAT with "random noise" strategies

SAT is of special concern to Artificial Intelligence because of its connection to *reasoning*. In particular, deductive reasoning is the complement of satisfiability: from a collection of base facts A one should deduce a sentence F if and only if $A \cup \overline{F}$ is *not* satisfiable, see also Sec. 2.2. The popular and effective algorithm GSAT was proposed in [81] as a *model-finding* procedure, i.e., to find an interpretation of the variables under which the formula comes out **true**. GSAT consists of multiple runs of LS$^+$, each run consisting of a number of iterations that is typically proportional to the problem dimension n. The experiments in [81] show that GSAT can be used to solve hard (see sec. 9.2) randomly generated problems that are an order of magnitude larger than those that can be solved by more traditional approaches like Davis-Putnam or resolution. Of course, GSAT is an incomplete procedure: it could fail to find an optimal assignment. An extensive empirical analysis of GSAT is presented in [37, 36].

Different "noise" strategies to escape from attraction basins are added to GSAT in [78, 80]. In particular, the GSAT-WITH-WALK algorithm has been tested in [80] on the Hansen-Jaumard benchmark of [48], where a better performance with respect to SAMD is demonstrated, although requiring much longer CPU times. See Sec. 8.1 for the definition of SAMD.

The algorithm is briefly summarized in Fig. 18. A certain number of tries (MAX-TRIES) is executed, where each try consists of a number of iterations (MAX-FLIPS). At each iteration a variable is chosen by two possible criteria and then flipped by the function FLIP, i.e., U_i becomes equal to $(1 - U_i)$. One criterion, active with "noise" probability p, selects a variable occurring in some unsatisfied clause with uniform probability over these variables, the other one is the standard method based on the function f given by the number of satisfied clauses. The first criterion was motivated

GSAT-WITH-WALK
```
1    for i ← 1 to MAX-TRIES
2        U ← random truth assignment
3        for j ← 1 to MAX-FLIPS
4            if RANDOMNUMBER < p then
5                u ← any variable occurring in some unsat. clause
6            else
7                u ← any variable with largest Δf
8            FLIP(u)
```

Figure 18: The GSAT-WITH-WALK algorithm. RANDOMNUMBER generates random numbers in the range $[0, 1]$.

by [72], see also Sec. 6.4. For a generic move μ, the quantity $\Delta_\mu f$ (or Δf for short) is defined as $f(\mu\, U^{(t)}) - f(U^{(t)})$. The straightforward book-keeping part of the algorithm is not shown. In particular, the best assignment found during all trials is saved and reported at the end of the run. In addition, the run is terminated immediately if an assignment is found that satisfies all clauses. The original GSAT algorithm can be obtained by setting $p = 0$ in the GSAT-WITH-WALK algorithm of Fig. 18.

7.3 Randomized Greedy and Local Search (GRASP)

A hybrid algorithm that combines a randomized greedy construction phase to generate initial candidate solutions, followed be a local improvement phase is the GRASP scheme proposed in [75] for the *SAT* and generalized for the *MAX W–SAT* problem in [76], a work that is briefly summarized in this section.

GRASP is an iterative process, with each iteration consisting of two phases, a construction phase and a local search phase.

During each construction, all possible choices are ordered in a candidate list with respect to a greedy function measuring the (myopic) benefit of selecting it. The algorithm is randomized because one picks in a random way one of the best candidates in the list, not necessarily the top candidate. In this way different solutions are obtained at the end of the construction phase.

Because these solutions are not guaranteed to be locally optimal with respect to simple neighborhoods, it is usually beneficial to apply a local search to attempt to improve each constructed solution.

A high-level description of the GRASP algorithm is presented in Fig. 19,

Approximate Algorithms for MAX-SAT

GRASP(*RCLSize, MaxIter, RandomSeed*)
1   △ Input instance and initialize data structures
2   for $i \leftarrow 1$ to *MaxIter*
3         $U \leftarrow$ CONSTRUCTGREEDYRAND(*RCLSize, RandomSeed*)
4         $U \leftarrow$ LOCALSEARCH(U)

CONSTRUCTGREEDYRAND(*RCLSize, RandomSeed*)
1   for $k \leftarrow 1$ to n
2   MAKERCL(*RCLSize*)
3         $s \leftarrow$ SELECTINDEX(*RandomSeed*)
4   ASSIGNVARIABLE(s)
5   ADAPTGREEDYFUNCTION(s)

Figure 19: The GRASP algorithm (above) and the randomized greedy construction (below).

a summarized version of the more detailed description in [76]. After reading the instance and initializing the data structures one repeats for *MaxIter* iterations the construction of an assignment U and the application of local search starting from U to produce a possibly better assignment (lines 2–4). Of course, the best assignment found during all iterations is saved and reported at the end. In addition to *MaxIter*, the parameters are *RCLSize*, the size of the restricted candidate list of moves out of which a random selection is executed, and a random seed used by the random number generator. In detail, see function CONSTRUCTGREEDYRAND in Fig. 19, the restricted candidate list of assignments is created by MAKERCL, the index of the next variable to be assigned a truth value is chosen by SELECTINDEX, the truth value is assigned by ASSIGNVARIABLE and the greedy function that guides the construction is changed by ADAPTGREEDYFUNCTION to reflect the assignment just made.

The remaining details about the greedy function (designed to maximize the total weight of yet-unsatisfied clauses that become satisfied after a given assignment), the creation of the restricted candidate list, and local search (based on the *1-flip* neighborhood) are presented in [76], together with experimental results.

8 History-sensitive Heuristics

Different history-sensitive heuristics have been proposed to continue local search schemes beyond local optimality. These schemes aim at intensifying the search in promising regions and at diversifying the search into uncharted territories by using the information collected from the previous phase (the *history*) of the search. The *history* at iteration t is formally defined as the set of ordered couples (U, s) such that $0 \leq s \leq t$ and $U = U^{(s)}$.

Because of the internal feedback mechanism, some algorithm parameters can be modified and tuned in an *on-line* manner, to reflect the characteristics of the *task* to be solved and the *local* properties of the configuration space in the neighborhood of the current point. This tuning has to be contrasted with the *off-line* tuning of an algorithm, where some parameters or choices are determined for a given problem in a preliminary phase and they remain fixed when the algorithm runs on a specific instance.

8.1 Prohibition-based Search: TS and SAMD

Tabu Search (TS) is a *history-sensitive* heuristic proposed by F. Glover [38] and, independently, by Hansen and Jaumard, that used the term SAMD ("steepest ascent mildest descent") and applied it to the *MAX–SAT* problem in [48]. The main mechanism by which the history influences the search in TS is that, at a given iteration, some neighbors are *prohibited*, only a non-empty subset $N_A(U^{(t)}) \subset N(U^{(t)})$ of them is *allowed*. The general way of generating the search trajectory that we consider is given by:

$$N_A(U^{(t)}) = \text{ALLOW}(N(U^{(t)}), U^{(0)}, ..., U^{(t)}) \qquad (21)$$
$$U^{(t+1)} = \text{BEST-NEIGHBOR}(\ N_A(U^{(t)})\ ) \qquad (22)$$

The set-valued function ALLOW selects a non-empty subset of $N(U^{(t)})$ in a manner that depends on the entire previous history of the search $U^{(0)}, ..., U^{(t)}$. Let us note that worsening moves *can* be produced by eqn. 22, as it must be in order to exit local optima.

The introduction of algorithm SAMD is motivated in [48] by contrasting the technique with Simulated Annealing (SA) [60] for maximization. The directions of local changes are little explored by SA: for example, if the objective function increases, the change is always accepted however small it may be. On the contrary, it is desirable to exploit the information on the direction of *steepest* ascent and yet to retain the property of not being blocked at the first local optimum found. SAMD performs local changes in the direction of steepest ascent until a local optimum is encountered, then a local change along the direction of mildest descent takes place and the reverse move is *forbidden* for a given number of iterations to avoid cycling with a high probability. The details of the SAMD technique as well as additional specialized devices for detecting and breaking cycles are outlined in [48]. A computational comparison with SA and with Johnson's two algorithms is also presented. A specialized Tabu Search heuristic is used in [51] to speed up the search for a solution (if the problem is satisfiable) as part of a branch-and-bound algorithm for SAT, that adopts both a relaxation and a decomposition scheme by using polynomial instances, i.e., 2-SAT and Horn SAT.

8.2 HSAT and "clause weighting"

In addition to the already cited SAMD [48] heuristic that uses the temporary prohibitions of recently executed moves, let us mention two variations of GSAT that make use of the previous history.

HSAT [37] introduces a tie-breaking rule into GSAT: if more moves produce the same (best) Δf, the preferred move is the one that has not been applied for the longest span. HSAT can be seen as a "soft" version of Tabu Search: while TS prohibits recently-applied moves, HSAT discourages recent moves if the same Δf can be obtained with moves that have been "inactive" for a longer time. It is remarkable to see how this innocent variation of GSAT can increase its performance on some SAT benchmark tasks [37].

Clause–weighting has been proposed in [79] in order to increase the effectiveness of GSAT for problems characterized by strong asymmetries. In this algorithm a positive weight is associated to each clause to determine how often the clause should be counted when determining which variable to flip. The weights are dynamically modified during problem solving and the qualitative effect is that of "filling in" local optima while the search proceeds. Clause–weighting can be considered as a "reactive" technique where a repulsion from a given local optimum is generated in order to induce an escape from a given attraction basin.

8.3 Reactive Search

Different methods to generate prohibitions produce qualitatively different *search trajectories*, i.e., sequences of visited configurations $U^{(t)}$. In particular, prohibitions based on a list of *moves* lead to a faster escape from a locally optimal point than prohibitions based on a list of visited *configurations* [11]. In this method prohibitions are determined by the last moves applied. In detail, the ALLOW function can be specified by introducing a *prohibition parameter* T (also called *list size*) that determines how long a move will remain prohibited after its execution. The FIXED-TS algorithm is obtained by fixing T throughout the search [38]. A neighbor is allowed if and only if it is obtained from the current point by applying a move that has not been used during the last T iterations. In detail, if $\text{LASTUSED}(\mu)$ is the last usage time of move μ ($\text{LASTUSED}(\mu) = -\infty$ at the beginning):

$$N_A(U^{(t)}) = \{U = \mu\, U^{(t)} \text{ such that } \text{LASTUSED}(\mu) < (t - T)\} \quad (23)$$

The Reactive Tabu Search algorithm [14], REACTIVE-TS for short, defines simple rules to determine the prohibition parameter by reacting to the repetition of previously-visited configurations. One has a repetition if $U^{(t+R)} = U^{(t)}$, for $R \geq 1$. The prohibition period T depends on the iteration t (therefore the notation is $T^{(t)}$), and the discrete dynamical system that generates the search trajectory comprises an additional evolution equation for $T^{(t)}$, that is specified through the function REACT, see eqn. 24 below. The dynamical system becomes:

$$T^{(t)} = \text{REACT}(T^{(t-1)}, U^{(0)}, ..., U^{(t)}) \quad (24)$$
$$N_A(U^{(t)}) = \{U = \mu\, U^{(t)} \text{ such that } \text{LASTUSED}(\mu) < (t - T^{(t)})\} \quad (25)$$
$$U^{(t+1)} = \text{BEST-NEIGHBOR}(N_A(U^{(t)})) \quad (26)$$

While the reader is referred to [14] for the details, the design principles of REACTIVE-TS are that $T^{(t)}$ (in the range $1 \leq T^{(t)} \leq n - 2$) increases when repetitions happen, and decreases when repetitions disappear for a sufficiently long search period. For convenience, let us introduce a "fractional prohibition" T_f, such that the prohibition is obtained by setting $T = \lfloor T_f\, n \rfloor$. T_f ranges between zero and one, with bounds inherited from those on T. Larger T values imply larger diversification, in particular the relationship between T and the diversification is as follows:

- The Hamming distance H between a starting point and successive point along the trajectory is strictly increasing for $T + 1$ steps.

$$H(U^{(t+\tau)}, U^{(t)}) = \tau \quad \text{for} \quad \tau \leq T + 1$$

- The minimum repetition interval R along the trajectory is $2(T + 1)$.

$$U^{(t+R)} = U^{(t)} \Rightarrow R \geq 2(T + 1)$$

REACTIVE-TS has been applied to various problems with competitive performance with respect to alternative heuristics like FIXED-TS, SA, Neural Networks, and Genetic Algorithms, see the review in [11].

8.3.1 The Hamming-Reactive Tabu Search (H-RTS) algorithm

An algorithm that combines the previously described techniques of local search (oblivious and non-oblivious), the use of prohibitions (see TS and SAMD), and a reactive scheme to determine the prohibition parameter is presented in [12]. The algorithm is called HAMMING-REACTIVE-TS algorithm, and its core is illustrated in Fig. 20.

HAMMING-REACTIVE-TS
1 repeat
2      $t_r \leftarrow t$
3      $U \leftarrow$ random truth assignment
4      $T \leftarrow \lfloor T_f\, n \rfloor$

5   repeat   { *NOB local search* }
6   [$U \leftarrow$ BEST-MOVE(LS, f_{NOB})
7   until largest $\Delta f_{NOB} = 0$

8   repeat
9   repeat   { *local search* }
10   [$U \leftarrow$ BEST-MOVE(LS, f_{OB})
11   until largest $\Delta f_{OB} = 0$
12         $U_I \leftarrow U$

13   for $2(T+1)$ *iterations*   { *reactive tabu search* }
14   [$U \leftarrow$ BEST-MOVE(TS, f_{OB})
15         $U_F \leftarrow U$

16         $T \leftarrow$ REACT(T_f, U_F, U_I)
17   until $(t - t_r) > 10\,n$
18 until solution is acceptable or max. number of iterations reached

Figure 20: The *H-RTS* algorithm.

The initial truth assignment is generated in a random way, and non-oblivious local search (LS-NOB) is applied until the first local optimum of f_{NOB} is encountered. LS-NOB obtains local minima of better average quality than LS-OB, but then the guiding function becomes the standard oblivious one. This choice was motivated by the success of the NOB & OB combination [13] and by the poor diversification properties of NOB alone, see [12].

The search proceeds by repeating phases of local search followed by phases of TS (lines 8–17 in Fig. 20), until a suitable number of iterations are accumulated after starting from the random initial truth assignment (see line 17 in Fig. 20). A single elementary move is applied at each iteration. The variable t, initialized to zero, identifies the current iteration and increases after a local move is applied, while t_r identifies the iteration when the last random assignment was generated. Some trivial bookkeeping details (like the increase of t) are not shown in the figure.

During each combined phase, first the local optimum of f is reached, then $2(T + 1)$ moves of Tabu Search are executed. The design principle underlying this choice is that prohibitions are necessary for diversifying the search only after LS reaches a local optimum. The fractional prohibition T_f is changed during the run by the function REACT to obtain a proper balance of diversification and bias [12].

The random restart executed after $10\,n$ moves guarantees that the search trajectory is not confined in a localized portion of the search space.

Being an heuristic algorithm, there is not a natural termination criterion. In its practical application, the algorithm is therefore run until either the solution is acceptable, or a maximum number of moves (and therefore CPU time) has elapsed. What is demonstrated in the computational experiments in [12] is that, given a fixed number of iterations, HAMMING-REACTIVE-TS achieves much better average results with respect to competitive algorithms (GSAT and GSAT-WITH-WALK). Because, to a good approximation, the actual running time is proportional to the number of iterations, HAMMING-REACTIVE-TS should therefore be used to obtain better approximations in a given allotted number of iterations, or equivalent approximations in a much smaller number of iterations.

9 Experimental analysis and threshold effects

Given the hardness of the problem and the relevancy for applications in different fields, the emphasis on the experimental analysis of algorithms for the MAX-SAT problem has been growing in recent years.

In some cases the experimental comparisons have been executed in the framework of "challenges," with support of electronic collection and distribution of software, problem generators and test instances. An example is the the Second DIMACS Algorithm Implementation Challenge on Cliques, Coloring and Satisfiability, whose results have been published in [53]. The archive is currently available from:

http://dimacs.rutgers.edu/Challenges/.
Practical and industrial *MAX–SAT* problems and benchmarks, with significant case studies are also presented in [30], see also the contained review [45].

9.1 Models

Let us describe some basic problem models that are considered both in theoretical and in experimental studies of *MAX–SAT* algorithms [45].

- **k-SAT model**, also called **fixed length clause model**. A randomly generated CNF formula consists of independently generated random clauses, where each clause contains exactly k literals. Each literal is chosen uniformly from $U = \{u_1, ..., u_n\}$ without replacement, and negated with probability p. The default value for p is $1/2$.

- **average k-SAT model**, also called **random clause model**. A randomly generated CNF formula consists of independently generated random clauses. Each literal has ha probability p of being part of a clause. In detail, each of the n variables occurs positively with probability $p(1-p)$, negatively with probability $p(1-p)$, both positively and negatively with probability p^2, and is absent with probability $(1-p)^2$.

Both models have many variations depending on whether the clauses are required to be different, whether a variable and its negation can be present in the same clause, etc.

Although superficially similar, the two models differ in the *difficulty* to solve the obtained formulae and in the mathematical analysis. In particular, when the initial formula comes from the average k-SAT model, a step that fixes the value of a variable produces a set of clauses from the same model, while if the same step is executed in the k-SAT model, the resulting clauses do not necessarily have the same length and therefore do not come from the k-SAT model.

Other *structured* problem models are derived from the mapping of instances of different problems, like *coloring*, n-queens, etc. [43]. The performance of algorithms on these more structured models tend to have little correlation with the performance tested on the above introduced random problems. Unfortunately, the theoretical analysis of these more structured problems is very hard. The situation worsens if one considers "real-world" practical applications, where one is typically confronted with a few instances and little can be derived about the "average" performance, both because the probability distribution is not known and because the number of instances tends to be very small.

A compromise can be reached by having parametrized generators that capture some of the relevant structure of the "real-world" problems of interest.

9.2 Hardness and threshold effects

Different algorithms demonstrate a different degree of effort, measured by number of elementary steps or CPU time, when solving different kinds of instances. For example, Mitchell et al. [67] found that some distributions used in past experiments are of little interest because the generated formulae are almost always very easy to satisfy. They also reported that one can generate very hard instances of k-SAT, for $k \geq 3$. In addition, they report the following observed behavior for random fixed length *3-SAT* formulae: if r is the ratio r of clauses to variables ($r = m/n$), almost all formulae are satisfiable if $r < 4$, almost all formulae are unsatisfiable if $r > 4.5$. A rapid transition seems to appear for $r \approx 4.2$, the same point where the computational complexity for solving the generated instances is maximized, see [61, 26] for reviews of experimental results.

A series of theoretical analyses aim at approximating the unsatisfiability threshold of random formulae. Let us define the notation and summarize some results obtained.

Let **C** be a random k-SAT formula. The research problem that has been considered, see for example [62], is to compute the least real number κ such that, if r is larger than κ, then the probability of **C** being satisfiable converges to 0 as n tends to infinity. In this case one says that **C** is asymptotically almost certainly satisfiable. Experimentally, κ is a threshold value marking a "sudden" change from probabilistically certain satisfiability to probabilistically certain unsatisfiability. More precisely [1], given a sequence of events $\mathcal{E}_i$, one says that $\mathcal{E}_n$ occurs almost surely (a.s.) if $\lim_{n\to\infty} \mathbf{Pr}\left[\mathcal{E}_n\right] = 1$, where $\mathbf{Pr}\left[event\right]$ denotes the probability of an event. The behavior observed in experiments with random k-SAT leads to the following conjecture:

For every $k \geq 2$, there exist r_k such that for any $\varepsilon > 0$, random instances of k-SAT with $(r_k - \varepsilon)n$ clauses are a.s. satisfiable and random instances with $(r_k + \varepsilon)n$ clauses are a.s. unsatisfiable

For $k = 2$ (i.e., for the polynomially solvable *2-SAT*) the conjecture was proved [23, 41], in fact showing that $r_2 = 1$. For $k = 3$ much less progress has been made: neither the existence of r_3 nor its value has been determined.

In the fixed-length *3-SAT* model, the total number of all possible clauses is $8\binom{n}{3}$ and the probability that a random clause is satisfied by a truth assignment U is 7/8.

Let $\mathcal{U}_n$ be the set of all truth assignments on n variables, and let $\mathcal{S}_n$ be the set of assignments that satisfy the random formula **C**. Therefore the cardinality $|\mathcal{S}_n|$ is a random variable. Given **C**, let $|\mathcal{S}_n(\mathbf{C})|$ be the number of assignments satisfying **C**.

The expected value of the number of satisfying truth assignments of a random formula, $\mathbf{E}\left[|\mathcal{S}_n|\right]$, is defined as:

$$\mathbf{E}\left[|\mathcal{S}_n|\right] = \sum_{\mathbf{C}} (\mathbf{Pr}\left[\mathbf{C}\right]|\mathcal{S}_n(\mathbf{C})|) \qquad (27)$$

The probability that a random formula is satisfiable is:

$$\mathbf{Pr}\left[\text{the random formula is satisfiable}\right] = \sum_{\mathbf{C}} (\mathbf{Pr}\left[\mathbf{C}\right] I_{\mathbf{C}}) \qquad (28)$$

where $I_{\mathbf{C}}$ is 1 if **C** is satisfiable, 0 otherwise.

From equations (27) and (28) the following Markov's inequality follows:

$$\mathbf{Pr}\left[\text{the random formula is satisfiable}\right] \leq \mathbf{E}\left[|\mathcal{S}_n|\right] \qquad (29)$$

Let us now consider the "first moment" argument to obtain an upper bound for κ in the *3-SAT* model. First one observes that the expected number of truth assignments that satisfy **C** is $2^n (7/8)^{rn}$, then one lets this expected value converge to zero and uses the above Markov's inequality. From this one obtains

$$\kappa \leq \log_{8/7} 2 = 5.191$$

This result has been found independently by many people, including [33] and [24].

The weakness of the above technique is that, in the right-hand side of equation (27) one can have small probabilities multiplied by large cardinalities, therefore the condition may be unnecessarily strong to ensure only that **C** is almost certainly satisfiable. In [62], instead of considering the random class $\mathcal{S}_n$ that may have a large cardinality for a formula with small probability, one considers a *subset* of it obtained by considering truth assignments that satisfy a local maximality condition. In particular, one considers the subset $\mathcal{S}_n^\#$ defined as the random class of assignments U satisfying **C** such that any assignment obtained from U by changing exactly one **false** value of U to **true** does not satisfy **C**.

It is demonstrated in the cited paper that the expected value $\mathbf{E}\left[\mathcal{S}_n^{\#}\right]$ is at most $(7/8)^{rn}(2 - e^{-3r/7} + o(1))^n$. It follows that the unique positive solution of the equation

$$(7/8)^{rn}(2 - e^{-3r/7})^n = 1$$

is an upper bound for κ. This solution is less than 4.667. Better bounds can be obtained by increasing the range of locality when selecting the local maxima that represent $\mathcal{S}_n$. A previous best bound of 4.758 had been obtained in [55] by non-elementary means. Independently, Dubois and Boufkhad [31] obtained an upper bound of 4.64.

Unlike upper bounds, which are based on probabilistic counting arguments, all known lower bounds for r_3 are algorithmic. The UNIT CLAUSE algorithm for *3-SAT* is considered in [20], where it is shown that, for $r < 2.9$ or $r < 8/3$, depending on the presence or absence of a "majority rule," it finds a satisfying assignment with positive probability instead of a.s. (therefore this does not imply that $r_3 \geq 2.9$). The PURE LITERAL algorithm succeeds a.s. for $r < 1.63$, see [17]. A generalization of the analysis [34] shows that the GUC algorithm succeeds a.s. for $r < 3.003$, giving the best known lower bound for r_3. Additional recent developments are presented in [1].

References

[1] D. Achlioptas, L. M. Kirousis, E. Kranakis, and D. Krinzac, *Rigorous results for random (2+p)-SAT*, Proc. Work. on Randomized Algorithms in Sequential, Parallel and Distributed Computing, Santorini, Greece, 1997.

[2] P. Alimonti, *New local search approximation techniques for maximum generalized satisfiability problems*, Proc. Second Italian Conf. on Algorithms and Complexity, Rome, 1994, pp. 40–53.

[3] S. Arora, C. Lund, R. Motwani, M. Sudan, and M. Szegedy, *Proof verification and hardness of approximation problems*, Proc. 33rd Annual IEEE Symp. on Foundations of Computer Science, IEEE Computer Society, 1992, pp. 14–23.

[4] S. Arora and S. Safra, *Probabilistic checking of proofs: a new characterization of NP*, Proc. 33rd Annual IEEE Symp. on Foundations of Computer Science, IEEE Computer Society, 1992, pp. 2–13.

[5] T. Asano, *Approximation algorithms for MAX-SAT: Yannakakis vs. Goemans-Williamson*, Proc. 3rd Israel Symp. on the Theory of Computing and Systems, Ramat Gan, Israel, 1997, pp. 24–37.

[6] T. Asano, T. Ono, and T. Hirata, *Approximation algorithms for the maximum satisfiability problem*, Proc. 5th Scandinavian Work. on Algorithms Theory, 1996, pp. 110–111.

[7] P. Asirelli, M. de Santis, and A. Martelli, *Integrity constraints in logic databases*, Journal of Logic Programming **3** (1985), 221–232.

[8] G. Ausiello, P. Crescenzi, and M. Protasi, *Approximate solution of NP optimization problems*, Theoretical Computer Science **150** (1995), 1–55.

[9] G. Ausiello, A. D'Atri, and M. Protasi, *Lattice theoretic properties of NP-complete problems*, Fundamenta Informaticae **4** (1981), 83–94.

[10] G. Ausiello and M. Protasi, *Local search, reducibility and approximability of NP-optimization problems*, Information Processing Letters **54** (1995), 73–79.

[11] R. Battiti, *Reactive search: Toward self-tuning heuristics*, Modern Heuristic Search Methods (V. J. Rayward-Smith, I. H. Osman, C. R. Reeves, and G. D. Smith, eds.), John Wiley and Sons, 1996, pp. 61–83.

[12] R. Battiti and M. Protasi, *Reactive search, a history-sensitive heuristic for MAX-SAT*, ACM Journal of Experimental Algorithmics **2** (1997), no. 2, http://www.jea.acm.org/.

[13] _____, *Solving MAX-SAT with non-oblivious functions and history-based heuristics*, Satisfiability Problem: Theory and Applications, DIMACS: Series in Discrete Mathematics and Theoretical Computer Science, no. 35, AMS and ACM Press, 1997.

[14] R. Battiti and G. Tecchiolli, *The reactive tabu search*, ORSA Journal on Computing **6** (1994), no. 2, 126–140.

[15] C.E. Blair, R.G. Jeroslow, and J.K. Lowe, *Some results and experiments in programming for propositional logic*, Computers and Operations Research **13** (1986), no. 5, 633–645.

[16] M. Boehm and E. Speckenmeyer, *A fast parallel sat solver - efficient workload balancing*, Annals of Mathematics and Artificial Intelligence **17** (1996), 381–400.

[17] A. Broder, A. Frieze, and E. Upfal, *On the satisfiability and maximum satisfiability of random 3-CNF formulas*, Proc. of the 4th Annual ACM-SIAM Symp. on Discrete Algorithms, 1993.

[18] M. Buro and H. Kleine Buening, *Report on a SAT competition*, EATCS Bulletin **49** (1993), 143–151.

[19] S. Chakradar, V. Agrawal, and M. Bushnell, *Neural net and boolean satisfiability model of logic circuits*, IEEE Design and Test of Computers (1990), 54–57.

[20] M.-T. Chao and J. Franco, *Probabilistic analysis of two heuristics for the 3-satisfiability problem*, SIAM Journal on Computing **15** (1986), 1106–1118.

[21] J. Chen, D. Friesen, and H. Zheng, *Tight bound on Johnson's algorithm for MAX-SAT*, Proc. 12th Annual IEEE Conf. on Computational Complexity, Ulm, Germany, 1997, pp. 274–281.

[22] J. Cheriyan, W. H. Cunningham, T. Tuncel, and Y. Wang, *A linear programming and rounding approach to MAX 2-SAT*, Proc. of the Second DIMACS Algorithm Implementation Challenge on Cliques, Coloring and Satisfiability (M. Trick and D. S. Johson, eds.), DIMACS Series on Discrete Mathematics and Theoretical Computer Science, no. 26, 1996, pp. 395–414.

[23] V. Chvatal and B. Reed, *Mick gets some (the odds are on his side)*, Proc. 33th Ann. IEEE Symp. on Foundations of Computer Science, IEEE Computer Society, 1992, pp. 620–627.

[24] V. Chvátal and E. Szemerédi, *Many hard examples for resolution*, Journal of the ACM **35** (1988), 759–768.

[25] S.A. Cook, *The complexity of theorem-proving procedures*, Proc. of the Third Annual ACM Symp. on the Theory of Computing, 1971, pp. 151–158.

[26] S.A. Cook and D.G. Mitchell, *Finding hard instances of the satisfiability problem: a survey*, Satisfiability Problem: Theory and Applications

(D.-Z. Du, J. Gu, and P.M. Pardalos, eds.), DIMACS Series in Discrete Mathematics and Theoretical Computer Science, vol. 35, AMS and ACM Press, 1997.

[27] P. Crescenzi and A. Panconesi, *Completeness in approximation classes*, Information and Computation **93** (1991), 241–262.

[28] M. Davis, G. Logemann, and D. Loveland, *A machine program for theorem proving*, Communications of the ACM **5** (1962), 394–397.

[29] M. Davis and H. Putnam, *A computing procedure for quantification theory*, Journal of the ACM **7** (1960), 201–215.

[30] D. Du, J. Gu, and P.M. Pardalos (Eds.), *Satisfiability problem: Theory and applications*, DIMACS Series in Discrete Mathematics and Theoretical Computer Science, vol. 35, AMS and ACM Press, 1997.

[31] O. Dubois and Y. Boufkhad, *A general upper bound for the satisfiability threshold of random r-SAT formulas*, Tech. report, LAFORIA, CNRS-Univ. Paris 6, 1996.

[32] U. Feige and M.X. Goemans, *Approximating the value of two proper proof systems, with applications to MAX-2SAT and MAX-DICUT*, Proc. of the Third Israel Symp. on Theory of Computing and Systems, 1995, pp. 182–189.

[33] J. Franco and M. Paull, *Probabilistic analysis of the davis-putnam procedure for solving the satisfiability problem*, Discrete Applied Mathematics **5** (1983), 77–87.

[34] A. Frieze and S. Suen, *Analysis of two simple heuristics on a random instance of k-SAT*, Journal of Algorithms **20** (1996), 312–355.

[35] H. Gallaire, J. Minker, and J. M. Nicolas, *Logic and databases: a deductive approach*, Computing Surveys **16** (1984), no. 2, 153–185.

[36] I.P. Gent and T. Walsh, *An empirical analysis of search in gsat*, Journal of Artificial Intelligence Research **1** (1993), 47–59.

[37] _____, *Towards an understanding of hill-climbing procedures for SAT*, Proc. of the Eleventh National Conf. on Artificial Intelligence, AAAI Press / The MIT Press, 1993, pp. 28–33.

[38] F. Glover, *Tabu search - part I*, ORSA Journal on Computing **1** (1989), no. 3, 190–260.

[39] M.X. Goemans and D.P. Williamson, *New 3/4-approximation algorithms for the maximum satisfiability problem*, SIAM Journal on Discrete Mathematics **7** (1994), no. 4, 656–666.

[40] _____, *Improved approximation algorithms for maximum cut and satisfiability problems using semidefinite programming*, Journal of the ACM **42** (1995), no. 6, 1115–1145.

[41] A. Goerdt, *A threshold for unsatisfiability*, Journal of Computer and System Sciences **53** (1996), 469–486.

[42] J. Gu, *Efficient local search for very large-scale satisfiability problem*, ACM SIGART Bulletin **3** (1992), no. 1, 8–12.

[43] _____, *Global optimization for satisfiability (SAT) problem*, IEEE Transactions on Data and Knowledge Engineering **6** (1994), no. 3, 361–381.

[44] J. Gu, Q.-P. Gu, and D.-Z.Du, *Convergence properties of optimization algorithms for the SAT problem*, IEEE Transactions on Computers **45** (1996), no. 2, 209–219.

[45] J. Gu, P.W. Purdom, J. Franco, and B.W. Wah, *Algorithms for the satisfiability (SAT) problem: A survey*, Satisfiability Problem: Theory and Applications (D.-Z. Du, J. Gu, and P.M. Pardalos, eds.), DIMACS Series in Discrete Mathematics and Theoretical Computer Science, vol. 35, AMS and ACM Press, 1997.

[46] J. Gu and R. Puri, *Asynchronous circuit synthesis with boolean satisfiability*, IEEE Transactions on Computer-Aided Design of Integrated Circuits **14** (1995), no. 8, 961–973.

[47] P.L. Hammer, P. Hansen, and B. Simeone, *Roof duality, complementation and persistency in quadratic 0-1 optimization*, Mathematical Programming **28** (1984), 121–155.

[48] P. Hansen and B. Jaumard, *Algorithms for the maximum satisfiability problem*, Computing **44** (1990), 279–303.

[49] J.N. Hooker, *Resolution vs. cutting plane solution of inference problems: some computational experience*, Operations Research Letters **7** (1988), no. 1, 1–7.

[50] J. Håstad, *Some optimal inapproximability results*, Proc. 28th Annual ACM Symp. on Theory of Computing, El Paso, Texas, 1997, pp. 1–10.

[51] B. Jaumard, M. Stan, and J. Desrosiers, *Tabu search and a quadratic relaxation for the satisfiability problem*, Proc. of the Second DIMACS Algorithm Implementation Challenge on Cliques, Coloring and Satisfiability (M. Trick and D. S. Johson, eds.), DIMACS Series on Discrete Mathematics and Theoretical Computer Science, no. 26, 1996, pp. 457–477.

[52] D.S. Johnson, *Approximation algorithms for combinatorial problems*, Journal of Computer and System Sciences **9** (1974), 256–278.

[53] D.S. Johnson and M. Trick (Eds.), *Cliques, coloring, and satisfiability: Second DIMACS implementation challenge*, vol. 26, DIMACS Series in Discrete Mathematics and Theoretical Computer Science, no. 26, AMS, 1996.

[54] J.L. Johnson, *A neural network approach to the 3-satisfiability problem*, Journal of Parallel and Distributed Computing **6** (1989), 435–449.

[55] A. Kamath, R. Motwani, K. Palem, and P. Spirakis, *Tail bounds for occupancy and the satisfiability threshold conjecture*, Random Structures and Algorithms **7** (1995), 59–80.

[56] A.P. Kamath, N.K. Karmarkar, K.G. Ramakrishnan, and M.G. Resende, *Computational exprience with an interior point algorithm on the satisfiability problem*, Annals of Operations Research **25** (1990), 43–58.

[57] _____, *A continuous approach to inductive inference*, Mathematical programming **57** (1992), 215–238.

[58] H. Karloff and U. Zwick, *A 7/8-approximation algorithm for MAX 3SAT?*, Proc. of the 38th Annual IEEE Symp. on Foundations of Computer Science, IEEE Computer Society, 1997, in press.

[59] S. Khanna, R.Motwani, M.Sudan, and U.Vazirani, *On syntactic versus computational views of approximability*, Proc. 35th Ann. IEEE Symp. on Foundations of Computer Science, IEEE Computer Society, 1994, pp. 819–836.

[60] S. Kirkpatrick, C.D. Gelatt Jr., and M.P. Vecchi, *Optimization by simulated annealing*, Science **220** (1983), 671–680.

[61] S. Kirkpatrick and B. Selman, *Critical behavior in the satisfiability of random boolean expressions*, Science **264** (1994), 1297–1301.

[62] L. M. Kirousis, E. Kranakis, and D. Krizanc, *Approximating the unsatisfiability threshold of random formulas*, Proc. of the Fourth Annual European Symp. on Algorithms (Barcelona), Springer-Verlag, September 1996, pp. 27–38.

[63] E. Koutsoupias and C.H. Papadimitriou, *On the greedy algorithm for satisfiability*, Information Processing Letters **43** (1992), 53–55.

[64] O. Kullmann and H. Luckhardt, *Deciding propositional tautologies: Algorithms and their complexity*, Tech. Report 1596, JohannWolfgang Goethe-Univ., Fachbereich Mathematik, Frankfurt, Germany, January 1997.

[65] D.W. Loveland, *Automated theorem proving: A logical basis*, North-Holland, 1978.

[66] S. Minton, M. D. Johnston, A. B. Philips, and P. Laird, *Solving large-scale constraint satisfaction and scheduling problems using a heuristic repair method*, Proc. of the 8th National Conf. on Artificial Intelligence (AAAI-90), 1990, pp. 17–24.

[67] D. Mitchell, B. Selman, and H. Levesque, *Hard and easy distributions of SAT problems*, Proc. of the 10th National Conf. on Artificial Intelligence (AAAI-92) (San Jose, Ca), July 1992, pp. 459–465.

[68] R. Motwani and P. Raghavan, *Randomized algorithms*, Cambridge University Press, New York, 1995.

[69] T.A. Nguyen, W.A. Perkins, T.J. Laffrey, and D. Pecora, *Checking an expert system knowledge base for consistency and completeness*, Proc. of the International Joint Conf. on Artificial Intelligence (Los Altos, CA), 1985, pp. 375–378.

[70] P. Nobili and A. Sassano, *Strengthening lagrangian bounds for the MAX-SAT problem*, Tech. Report 96-230, Institut fuer Informatik, Koln Univ., Germany, 1996, Proc. of the Work. on the Satisfiability Problem, Siena, Italy (J. Franco and G. Gallo and H. Kleine Buening, Eds.).

[71] P. Orponen and H. Mannila, *On approximation preserving reductions: complete problems and robust measures*, Tech. Report C-1987-28, Dept. of Computer Science, Univ. of Helsinki, 1987.

[72] C. H. Papadimitriou, *On selecting a satisfying truth assignment (extended abstract)*, Proc. of the 32th Annual Symp. on Foundations of Computer Science, 1991, pp. 163–169.

[73] C.H. Papadimitriou and K. Steiglitz, *Combinatorial optimization, algorithms and complexity*, Prentice-Hall, NJ, 1982.

[74] R. Puri and J. Gu, *A BDD SAT solver for satisfiability testing: an industrial case study*, Annals of Mathematics and Artificial Intelligence **17** (1996), no. 3-4, 315–337.

[75] M.G.C. Resende and T. A. Feo, *A grasp for satisfiability*, Proc. of the Second DIMACS Algorithm Implementation Challenge on Cliques, Coloring and Satisfiability (M. Trick and D. S. Johson, eds.), DIMACS Series on Discrete Mathematics and Theoretical Computer Science, no. 26, 1996, pp. 499–520.

[76] M.G.C. Resende, L.S. Pitsoulis, and P.M. Pardalos, *Approximate solution of weighted MAX-SAT problems using GRASP*, Satisfiability Problem: Theory and Applications, DIMACS: Series in Discrete Mathematics and Theoretical Computer Sc ience, no. 35, 1997.

[77] J. A. Robinson, *A machine-oriented logic based on the resolution principle*, Journal of the ACM **12** (1965), 23–41.

[78] B. Selman and H. Kautz, *Domain-independent extensions to GSAT: Solving large structured satisfiability problems*, Proc. of the International Joint Conf. on Artificial Intelligence, 1993, pp. 290–295.

[79] B. Selman and H.A. Kautz, *An empirical study of greedy local search for satisfiability testing*, Proc. of the 11th National Conf. on Artificial Intelligence (AAAI-93) (Washington, D. C.), 1993.

[80] B. Selman, H.A. Kautz, and B. Cohen, *Local search strategies for satisfiability testing*, Proc. of the Second DIMACS Algorithm Implementation Challenge on Cliques, Coloring and Satisfiability (M. Trick and D. S. Johson, eds.), DIMACS Series on Discrete Mathematics and Theoretical Computer Science, no. 26, 1996, pp. 521–531.

[81] B. Selman, H. Levesque, and D. Mitchell, *A new method for solving hard satisfiability problems*, Proc. of the 10th National Conf. on Artificial Intelligence (AAAI-92) (San Jose, Ca), July 1992, pp. 440–446.

[82] W.M. Spears, *Simulated annealing for hard satisfiability problems*, Proc. of the Second DIMACS Algorithm Implementation Challenge on Cliques, Coloring and Satisfiability (M. Trick and D. S. Johnson, eds.), DIMACS Series on Discrete Mathematics and Theoretical Computer Science, no. 26, 1996, pp. 533–555.

[83] L. Trevisan, *Approximating satisfiable satisfiability problems*, Proc. of the 5th Annual European Symp. on Algorithms, Graz, Springer Verlag, 1997, pp. 472–485.

[84] M. Yannakakis, *On the approximation of maximum satisfiability*, Journal of Algorithms **17** (1994), 475–502.

HANDBOOK OF COMBINATORIAL OPTIMIZATION (VOL. 1)
D.-Z. Du and P.M. Pardalos (Eds.) pp. 149-188
©1998 Kluwer Academic Publishers

Connections between Nonlinear Programming and Discrete Optimization[1]

Franco Giannessi
Department of Mathematics,
Università di Pisa
Via Buonarroti 2, 56127 Pisa, Italy
E-mail: gianness@dm.unipi.it

Fabio Tardella
Department of Mathematics
Faculty of Economics
University of Rome "La Sapienza"
Via del Castro Laurenziano 9, 00161 Roma, Italy
E-mail: tardella@imc.pi.cnr.it

Contents

1 Introduction 150

2 Equivalence, via Relaxation–Penalization, of two Constrained Extremum Problems 152

3 Equivalence between Integer and Real Optimization 157

4 Functions attaining the Minimum on the Vertices of a Polyhedron 161
   4.1 The Box Case 166
   4.2 The Polymatroid Case 167

5 Piecewise Concave Functions and Minimax 168

6 Location Problems 172

---
[1] Sect.s 2,3 are due to F. Giannessi; all the other Sections are due to F. Tardella.

7 Convex-Concave Problems	176
8 Discretization, Extension and Relaxation	178
9 Necessary and Sufficient Optimality Conditions	180
10 Conclusions and Remarks on Further Developments	181
References	

1 Introduction

Given a set X, a function $f : X \to \mathbb{R}$ and a subset S of X we consider the problem:

$$min f(x) \quad s.t. \quad x \in S \tag{1}$$

Problem (1) is usually called a *combinatorial optimization problem* when S is finite and a *discrete optimization problem* when the points of S are isolated in some topology, i.e., every point of S has a neighbourhood which does not contain other points of S. Obviously, all combinatorial optimization problems are also discrete optimization problems but the converse is not true. A simple example is the problem of minimizing a function on the set of integer points contained in an unbounded polyhedron.

Even though some discrete and combinatorial optimization problems have been studied since ancient times, the increase of their importance and their development have been very fast in the last few decades thanks to the possibility of practically solving them with modern computers and because of the several applications that they have found in many fields.

In most cases, discrete and combinatorial optimization can be formulated as linear or nonlinear integer programs and are solved by means of methods which exploit in a crucial way the finiteness or discreteness of the feasible region.

In the case where the feasible set S is defined by a family of equalities and inequalities in $\mathbb{R}^n$ and at least one of the constraining or objective functions is nonlinear, problem (1) is called a *nonlinear program*. Problems of this kind have been studied for at least four centuries with tools that exploit the differential, geometric and topological properties of the functions and sets involved.

Although the methods employed in integer and nonlinear programming are quite different in general, they can be complementary in several cases. In fact, it is often possible to restrict the search for a global solution of a nonlinear program to a finite set of points. On the other hand, most discrete optimization problems can be reformulated equivalently as nonlinear programs. Clearly, the restriction to a finite set of points or the nonlinear reformulation do not always lead to practical solution methods. Nevertheless, these approaches provide useful tools for many important classes of problems, as will be shown in the sequel.

It should be clear that we do not attempt here to make an exhaustive survey of all connections between discrete optimization and nonlinear programming. We do not cover, e.g., the many types of continuous relaxation introduced for integer programming - including the recent and very promising semidefinite relaxation (see, e.g., [1, 55, 56, 70]) - nor the several continuous approaches to graph problems already described in [53].

In Section 2 we present a general result on the equivalence between the problem of minimizing a function on a set and the problem of minimizing a penalized function on a larger set. This result is then specialized in Section 3 and used to reformulate integer programs as concave programs or as linear complementarity problems. In Section 4 we address the problem of establishing conditions under which a function achieves its minimum on the vertices of a polyhedron. The class of piecewise concave functions is introduced in Section 5 where it is proved that such functions achieve their minimum on a set of points that is often finite. This result is then applied to some minimax problems thereby strengthening known recent results. Location problems are discussed in Section 6. It is shown that the optimal solutions for many continuous problems of this type are attained in a finite subset of the feasible set. In Section 7 we consider a generalization of bilinear programming problems. Extension and relaxation of a function from a discrete set to a larger continuous set are briefly discussed in Section 8 together with the opposite approach of discretizing a function defined on a polyhedron. Finally, in Section 9, some connections between global optimality conditions for continuous and discrete problems are presented.

2 Equivalence, via Relaxation–Penalization, of two Constrained Extremum Problems

In this section we will consider a constrained extremum problem $\mathcal{P}$ in the following format:

$$\min f(x) \ , \quad s.t. \quad x \in R \cap Z, \tag{2}$$

where $f : \mathbb{R}^n \to \mathbb{R}$, $R \subseteq \mathbb{R}^n$, $Z \subset \mathbb{R}^n$. Assume we are given a set $X \subset \mathbb{R}^n$ such that $Z \subseteq X$. Replacement of $R \cap Z$ with $R \cap X$ is known as relaxation of (2); it leads to a lower bound for the minimum of problem (2), which generally does not equal the minimum. Equality between them may be forced by means of a suitable penalization of the objective function of (2). To this end let us introduce a function $\varphi : \mathbb{R}^n \to \mathbb{R}$, and consider the family $\{\mathcal{P}(\mu)\}_{\mu \in \mathbb{R}}$ of problems, where $\mathcal{P}(\mu)$ is defined as follows

$$\min[f(x) + \mu \varphi(x)], \quad s.t. \quad x \in R \cap X, \tag{3}$$

and shows, with respect to (2), a relaxation of the feasible region and a penalization of the objective function.

We want to state conditions under which (2) and (3) are equivalent in the sense that they have the same minimum (or infimum; $+\infty$, if $R \cap Z = \emptyset$) and the same set of minimum points. If no assumption is made on φ and if f is bounded, then the answer is trivial: it is sufficient to choose:

$$\varphi(x) \doteq \begin{cases} 0, & \text{if } x \in Z, \\ 1, & \text{if } x \in X \backslash Z, \end{cases}$$

and

$$\mu > \sup_{x \in X} f(x) - \inf_{x \in X} f(x)$$

to guarantee the equivalence between (2) and (3). Indeed, with the above choice of φ, the equivalence holds iff

$$\mu > \inf_{x \in R \cap Z} f(x) - \inf_{x \in R \cap (X \backslash Z)} f(x)$$

Of course, the function φ above is discontinuous; thus the equivalence is of no much interest. The following theorem [23] gives a condition under which the above equivalence is achieved within the class of continuous functions φ. Fig.s 1 and 2 illustrate the sets which appear in Theorem 2.1 by means of two examples.

Nonlinear Programming and Discrete Optimization

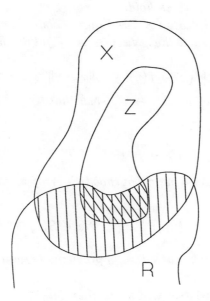

Fig.1

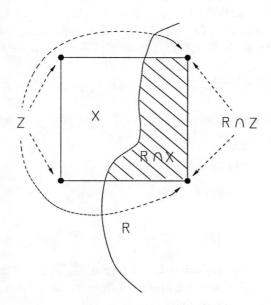

Fig.2

Theorem 2.1 *Let $R \subseteq \mathbb{R}^n$ be closed, $Z \subseteq X \subset \mathbb{R}^n$, Z and X be compact,* and let the following hypotheses hold:

(H_1) $f : \mathbb{R}^n \to \mathbb{R}$ is bounded on X, and there exists an open set $\mathcal{A} \supset Z$ and real numbers $\alpha, L > 0$, such that, $\forall x, y \in \mathcal{A}$, f fulfils the following Hölder condition:
$$|f(x) - f(y)| \leq L\|x - y\|^\alpha.$$

(H_2) It is possible to find $\varphi : \mathbb{R}^n \to \mathbb{R}$, such that:

(i) φ is continuous on X;

(ii) $\varphi(x) = 0 \; \forall x \in Z$, $\varphi(x) > 0 \; \forall x \in X \setminus Z$;

(iii) $\forall z \in Z$, there exists a neighbourhood $S(z)$ of z, and a real $\epsilon(z) > 0$, such that:
$$\varphi(x) \geq \epsilon(z)\|x - z\|^\alpha, \quad \forall x \in S(z) \cap (X \setminus Z).$$

Then, a real μ_0 exists, such that $\forall \mu > \mu_0$ problems (2) and (3) are equivalent.

Proof. To prove the thesis we will show that $\exists \mu_0 \in \mathbb{R}$, such that $\forall \mu > \mu_0$ the minimum of $f(x) + \mu\varphi(x)$ on $R \cap X$ is achieved necessarily at a point $\bar{z} \in R \cap Z$. Since $\varphi(\bar{z}) = 0$ for every $\bar{z} \in R \cap Z$, we can then conclude that the solution sets of problems (2) and (3) are the same whenever $\mu > \mu_0$. Let us introduce the sets $\bar{X} \doteq R \cap X$ and $\bar{Z} \doteq R \cap Z$. It will be shown that the function
$$F_{\bar{z}}(x) \doteq \frac{f(\bar{z}) - f(x)}{\varphi(x)}, \quad x \in \mathcal{A} \cap (\bar{X} \setminus \bar{Z}), \; \bar{z} \in \bar{Z},$$
is bounded in some neighbourhood of $\bar{z}$. To see this, consider $\bar{S}(\bar{z}) \doteq \mathcal{A} \cap S(\bar{z})$. For every $x \in \bar{S}(\bar{z}) \cap (\bar{X} \setminus \bar{Z})$ we have:
$$\varphi(x) \geq \epsilon(\bar{z})\|x - \bar{z}\|^\alpha, \quad \epsilon(\bar{z}) > 0,$$
$$|f(x) - f(\bar{z})| \leq L\|x - \bar{z}\|^\alpha,$$
so that
$$|F_{\bar{z}}(x)| < \frac{L}{\epsilon(\bar{z})} < +\infty.$$

The family $\{\bar{S}(z), z \in \bar{Z}\}$ is obviously a cover of $\bar{Z}$. Since Z is compact and $\bar{Z}$ is a closed subset of Z, there is a finite subfamily, say $\{\bar{S}(z_i), \; i = 1, \ldots, k\}$, which is a cover of $\bar{Z}$. Consider the set:
$$S \doteq \left[\bigcup_{i=1}^{k} \bar{S}(z_i)\right] \cap \bar{X} \supseteq \bar{Z}.$$

Nonlinear Programming and Discrete Optimization

It is clear that $\mu > \beta \doteq \max\{\frac{L}{\epsilon(z_i)},\ i=1,\ldots,k\}$ implies:

$$f(x) + \mu\varphi(x) > f(z_i)\ ,\quad \forall x \in S\backslash \bar{Z},\ i=1,\ldots,k. \tag{4}$$

On the other side, the set

$$X_0 \doteq \bar{X}\backslash S = \bar{X} \cap \left[\mathbb{R}^n \backslash \bigcup_{i=1}^{k} \bar{S}(z_i)\right]$$

is compact, and we have:

$$\bar{X} = X_0 \cup (S\backslash \bar{Z}) \cup \bar{Z},\ X_0 \cap \bar{Z} = \emptyset,\ X_0 \cap (S\backslash \bar{Z}) = \emptyset,\ (S\backslash \bar{Z}) \cap \bar{Z} = \emptyset.$$

Besides, f is bounded on $\bar{X}$ and thus

$$M_f \doteq \inf_{x \in \bar{X}} f(x) > -\infty;$$

φ is continue and positive on X_0, and thus

$$M_\varphi \doteq \inf_{x \in X_0} \varphi(x) = \min_{x \in X_0} \varphi(x) > 0.$$

Since f is bounded on $\bar{X}$, we have:

$$\lambda_0 \doteq \sup_{x \in \bar{X}} \frac{f(x) - M_f}{M_\varphi} < +\infty$$

and, of course, $\lambda_0 \geq 0$. If $\lambda > \lambda_0$, then, by the definition of λ_0, for any $z \in \bar{Z}$ and $x \in X_0$

$$f(x) + \lambda\varphi(x) > M_f + \lambda_0 M_\varphi \geq f(z). \tag{5}$$

Inequalities (4) and (5) hold if

$$\mu > \mu_0 \doteq \max\{\beta, \lambda_0\}.$$

Therefore, $f(x) + \mu\varphi(x)$ cannot have its minimum neither at a point in X_0, for this would not agree with (5), nor at one in $S\backslash\bar{Z}$, for this would not agree with (4). This completes the proof. □

In some applications of Theorem 2.1, where $f(x) + \mu\varphi(x)$ cannot be convex, it is useful to be able to choose φ strictly concave; this happens, e.g., in the case of 0–1 programs, as it will be shown later. An extensive treatment of both the theory and methods of concave minimization problems can be found in [4, 40, 54]. If f is not concave, then (3) may require the minimization of an indefinite form, which may be undesirable. The following theorem states a condition under which the objective function in (3) is strictly concave. Consider the case where

$$X = X_Q \doteq \{x \in \mathbb{R}^n : 0 \leq x \leq e\}, \quad \varphi(x) = x^T(e-x),$$

with $e \doteq (1,\ldots,1)^T$. Such a special case is interesting because it represents the classic relaxation for 0–1 extremum problems.

Theorem 2.2 *If $f \in C^2(X_Q)$ and $Z \subset X_Q$, then there exists a real μ_1 such that for all $\mu > \mu_1$ problems (2) and (3) with $X = X_Q$ are equivalent and (3) has a strictly concave objective function.*

Proof. Let $H(x)$ and $\hat{H}(x)$ be the Hessian matrices of f and of $f + \mu x^T(e-x)$, respectively. We have:

$$\hat{H}(x) = H(x) - 2\mu I_n.$$

Moreover, H is continuous and, because of a well known property [2], the same is true for its eigenvalues, say $\lambda_1(x),\ldots,\lambda_n(x)$; these are bounded since X is compact. Thus,

$$\tilde{\lambda} \doteq \max_{i=1,\ldots,n} \sup_{x \in X} |\lambda_i(x)| < +\infty.$$

For every $x \in X$, ν is a (real) eigenvalue of the (symmetric) matrix $\hat{H}(x)$ iff $det\,[\hat{H}(x) - \nu I_n] = det\,[H(x) - (2\mu + \nu)I_n] = 0$. Thus, ν is an eigenvalue of $\hat{H}(x)$ iff $\lambda \doteq \nu + 2\mu$ is an eigenvalue of $H(x)$. If λ_0 and μ_0 are defined as in the proof of Theorem 2.1, and

$$\mu \geq \mu_1 \doteq \max\{\frac{1}{2}\tilde{\lambda}, \mu_0\},$$

then (2) and (3) are equivalent and, furthermore,

$$\nu = \lambda - 2\mu \leq \lambda - \lambda_0 < 0.$$

Hence, $\hat{H}(x)$ is negative definite. This completes the proof. □

The closeness of X_Q and the assumption $f \in C^2(X_Q)$ in Theorem 2.2 cannot be weakened, as the following example shows.

---

[2] Let M_n be the space of $n \times n$ real matrices, and let $T : \mathbb{R}^n \to M_n$ be a continuous function. The function $\lambda : \mathbb{R}^n \to \mathbb{C}$, where $\lambda(y)$ is any eigenvalue of $M_n(y)$, is continuous. This is a straightforward consequence of a known theorem on linear continuous operators in a normed space [15].

Example 2.3 Set $n = 1$, $X_Q = [0,1]$, $\varphi(x) = x(1-x)$, and

$$f(x) = \begin{cases} -x^3 \sin \frac{1}{x}, & \text{if } x \neq 0, \\ 0, & \text{if } x = 0, \end{cases}$$

so that

$$f'(x) = \begin{cases} -3x^2 \sin \frac{1}{x} + x \cos \frac{1}{x}, & \text{if } x \neq 0, \\ 0, & \text{if } x = 0, \end{cases}$$

$$f''(x) = \begin{cases} (\frac{1}{x} - 6x) \sin \frac{1}{x} + 4 \cos \frac{1}{x}, & \text{if } x \neq 0, \\ \text{undefined}, & \text{if } x = 0, \end{cases}$$

$$f''(x) + \mu \varphi''(x) = \begin{cases} (\frac{1}{x} - 6x) \sin \frac{1}{x} + 4 \cos \frac{1}{x} - 2\mu, & \text{if } x \neq 0, \\ \text{undefined}, & \text{if } x = 0. \end{cases}$$

For every $\mu \in \mathbb{R}_+ \setminus \{0\}$ it is possible to find $x \in]0,1]$ (close enough to zero) such that $f''(x) + \mu \varphi''(x) > 0$. Indeed, it is sufficient[3] to choose $x = \frac{2}{(8\lceil \mu \rceil + 13)\pi}$, where $\lceil \mu \rceil$ denotes the upper integer part of μ. Hence, $f(x) + \mu \varphi(x)$ is not concave in $[0,1]$ even if $f \in C^1([0,1])$ and f has continuous second derivative in $]0,1]$.

3 Equivalence between Integer and Real Optimization

We now consider a special case of (2) which, however, embraces most combinatorial extremum problems. Let

$$R = \{x \in \mathbb{R}^n : g(x) \geq 0\} \text{ and } Z = \mathbb{B}^n, \tag{6}$$

where $g : \mathbb{R}^n \to \mathbb{R}^m$, $\mathbb{B} \doteq \{0,1\}$ and $\mathbb{B}^n \doteq \mathbb{B} \times \ldots \times \mathbb{B}$ (n times). Thus, (2) becomes:

$$\min f(x), \text{ s.t. } g(x) \geq 0, \ x \in \mathbb{B}^n. \tag{7}$$

---

[3] In this case the above inequality becomes ($\lceil \mu \rceil$ denotes the upper integer part of μ):

$$\pi^2 (8\lceil \mu \rceil + 13)^2 - 4\pi (8\lceil \mu \rceil + 13)\mu - 24 > 0,$$

and is easily verified by $\mu \geq 1$.

The case where, in (7), $x \in \mathbb{B}^n$ is replaced by $x \in \mathbb{Z}^n$ can be reduced to (7) by means of well known devices like binary expansion. The natural relaxation of $Z = \mathbb{B}^n$ and the penalization when R is defined by (6) are the hypercube X_Q and $\varphi(x) = x^T(e - x)$ of the preceding section, respectively; with this choice (3) becomes:

$$\min[f(x) + \mu x^T(e - x)], \text{ s.t. } g(x) \geq 0, \; 0 \leq x \leq e. \tag{8}$$

Theorem 2.1 becomes here:

Theorem 3.1 *Let f verify assumption (H_1) of Theorem 2.1 with $\alpha = 1$, i.e., is bounded on $X_Q = \{x \in \mathbb{R}^n : 0 \leq x \leq e\}$ and Lipschitz continuous on an open set $\mathcal{A} \supset Z = \mathbb{B}^n$. Then, there exists $\mu_0 \in \mathbb{R}$ such that for every $\mu \geq \mu_0$ problems (7) and (8) are equivalent.*

Proof. We only need to prove that $\varphi(x) = x^T(e - x)$ satisfies assumption (H_2) of Theorem 2.1. Note that (i) and (ii) are trivially true. We will now prove that (iii) holds with $S(z) = \{x \in \mathbb{R}^n : \|x - z\| \leq \bar{\rho} < 1\}$ and $\epsilon(z) = 1 - \bar{\rho}$. To see this, consider $\rho \in \mathbb{R}$ and $u = (u_1, \ldots, u_n)$ satisfying

$$\rho \doteq \|x - z\| \leq \bar{\rho},$$

$$u \doteq \frac{1}{\rho}(x - z).$$

Then

$$\varphi(x) = \sum_{j=1}^{n}(\rho u_j + z_j)(1 - z_j - \rho u_j). \tag{9}$$

Since $z + \rho u = x \in X_Q$, $u_j > 0$ implies $z_j = 0$ and $u_j < 0$ implies $z_j = 1$; therefore, from (9) we deduce:

$$\varphi(x) = \sum_{u_j>0} \rho u_j(1 - \rho u_j) + \sum_{u_j<0}(1 + \rho u_j)(-\rho u_j) =$$

$$\sum_{j=1}^{n} \rho|u_j|(1 - \rho|u_j|) = \rho \sum_{j=1}^{n}|u_j| - \rho^2 \sum_{j=1}^{n}|u_j|^2. \tag{10}$$

Since

$$\rho \leq \bar{\rho} < 1, \; \sum_{j=1}^{n}|u_j|^2 = \|u\|^2 = 1, \; \sum_{j=1}^{n}|u_j| \geq \|u\| = 1,$$

from (10) we obtain:
$$\varphi(x) \geq \rho(1 - \bar{\rho}) = \epsilon(z)\|x - z\|.$$

This completes the proof. □

When f is linear (or quadratic) and g affine, Theorem 3.1 states an equivalence between (7) – called 0–1 linear (or quadratic) programming problem – and (8) which, because of Theorem 2.2, is a strictly concave quadratic real program, if μ is large enough. An analogous remark can be made in the more general case $f \in C^2(X_Q)$. This condition is not redundant, as it may be shown by Example 2.3.

It is well known that the satisfiability problem in logic and many problems in Graph Theory can be formulated as $0-1$ programming problems. Therefore, such problems can also be formulated as strictly concave programs in a continuous setting as shown in Theorem 3.1. Other formulations as nonlinear programs for this type of problems can be found in [53].

When the equivalence between (7) and (8) holds, properties and methods valid for one of the two problems can be transferred to the other one. As an instance of this, consider the case

$$f(x) = c^T x + \frac{1}{2}x^T C x \,, \quad g(x) = Ax - b,$$

where $b \in \mathbb{R}^m$, $c \in \mathbb{R}^n$, $A \in \mathbb{R}^{m \times n}$ and $C \in \mathbb{R}^{n \times n}$ is a symmetric matrix.

Theorem 3.2 *If $\mu \in \mathbb{R}$ is large enough, then the 0–1 quadratic programming problem:*

$$\min(c^T x + \frac{1}{2}x^T C x), \text{ s.t. } Ax \geq b \,, \ x \in \mathbb{B}^n \tag{11}$$

is equivalent to the linear complementarity problem:

$$\min \bar{c}^T \xi \,, \text{ s.t. } \bar{A}\xi + \eta = \bar{b}, \ \xi \geq 0, \ \eta \geq 0, \ \xi^T \eta = 0, \tag{12}$$

where $\xi, \eta \in \mathbb{R}^{2n+m}$ and

$$\bar{c}^T \doteq \frac{1}{2}(c^T + \mu e^T + e^T, b^T, e^T) \,, \quad \bar{b}^T \doteq (c^T, b^T, e^T),$$

$$\bar{A} \doteq \begin{pmatrix} -C + 2\mu I_n & A^T & -I_n \\ A & 0 & 0 \\ I_n & 0 & 0 \end{pmatrix}.$$

Proof. Because of Theorem 3.1, whose hypotheses are trivially satisfied, (11) is equivalent to the quadratic problem:

$$\min[(c^T + \mu e^T)x + \frac{1}{2}x^T(C - 2\mu I_n)x], \ s.t. \ Ax \geq b, \ 0 \leq x \leq e, \quad (13)$$

if μ is large enough. The well known Karush–Kuhn–Tucker necessary condition for (13) is:

$$c^T + \mu e^T + (C - 2\mu I_n)x - A^T y + t - u = 0; \quad (14)$$

$$Ax - v = b \ ; \ x + w = e \ ; \ x, y, t, u, v, w \geq 0; \quad (15)$$

$$x^T u = y^T v = t^T w = 0; \quad (16)$$

where y, t, n are vectors of multipliers associated to the inequalities $Ax \geq b$, $x \leq e$, $x \geq 0$, respectively; and v, w are slack variables. Solving (13) is equivalent to finding, among the solutions of the complementarity system (14-16) (stationary points), those which minimize the function in square brackets of (13). Such a function, evaluated at the stationary points, *"becomes linear"*:

$$\frac{1}{2}[(c^T + \mu e^T)x + (b^T y + e^T t)].$$

In fact, (14) implies

$$Cx - 2\mu x = -(c + \mu e) + A^T y - t + u;$$

from (16) we have:

$$\begin{aligned} x^T u &= 0, \\ 0 &= y^T v = y^T b - y^T Ax, \\ 0 &= t^T w = t^T e - t^T x, \end{aligned}$$

and therefore:

$$y^T b = y^T Ax,$$
$$t^T e = t^T x.$$

Now, to achieve the thesis it is sufficient to set

$$\xi^T = (x^T, y^T, t^T) \ , \ \eta^T = (u^T, v^T, w^T).$$

This completes the proof. □

Note that no assumption has been made on C, so that the convex case, as well as the nonconvex one, have been considered. See also [53] for a reduction of the mixed integer feasibility problem to a linear complementarity system.

4 Functions attaining the Minimum on the Vertices of a Polyhedron

In many cases it is possible to restrict the search for the global solution of a nonlinear program to a finite set of points. One way of doing this consists in considering only the set of points that satisfy some kind of first or second order necessary condition for local optimality. This set is often finite, but in general it is not practical to compute all its elements or to minimize the objective function over it. Another important case where the search for a solution of a nonlinear program can be restrained to a finite set is when at least one solution is guaranteed to be in the set of extreme points of the feasible region and such set is finite. We recall here the standard definition of extreme point of a set and we introduce a slightly more restrictive property which will be used in Section 5.

Definition 4.1 *Let X be a subset of $\mathbb{R}^n$. A point $x \in X$ is called an* extreme point *of X iff $y, z \in X$, $\alpha \in]0, 1[$ and $x = \alpha y + (1 - \alpha)z$ imply $y = z$. A point $x \in X$ is called a* convex hull extreme point *of X iff it is an extreme point of the* convex hull *(denoted by $co(X)$) of X.*

Note that the set of extreme points of a polyhedron coincides with the set of its vertices and is finite. Let $\mathcal{E}(X)$ denote the set of extreme points of a set X. Then the set of convex hull extreme points is given by $\mathcal{E}(co(X))$. Since $X \subseteq co(X)$, it follows from the above definition that $\mathcal{E}(co(X)) \subseteq \mathcal{E}(X) \subseteq X$.

It is well known (see, e.g., [58]) that the convex hull of a compact set is compact. Furthermore, the Krein-Millman Theorem states that for a nonempty convex compact set Y the equality $Y = co(\mathcal{E}(Y))$ holds. Hence, for any nonempty compact set X one has

$$co(X) = co(\mathcal{E}(co(X))) \subseteq co(\mathcal{E}(X)) \subseteq co(X),$$

which trivially implies that

$$\mathcal{E}(co(X)) \neq \emptyset \text{ and } co(\mathcal{E}(co(X))) = co(\mathcal{E}(X)).$$

One of the basic properties of Linear Programming, which underlies the validity of the Simplex Algorithm, is the following:

(A) *If in (1) f attains its global minimum on S and S is a polyhedron, then at least one global minimum point must be in a vertex of S.*

This property, which establishes an equivalence between a continuous problem and a combinatorial problem, actually holds also for classes of feasible sets S and of objective functions f that are considerably larger than the classes of polyhedra and of linear functions. Indeed, it is well-known (see, e.g., [37, 69]) that the same property holds for concave functions on closed convex sets and for quasi-concave functions on compact convex sets. Note however that Property (A) does not hold in general for quasi-concave functions on *unbounded* closed convex sets. A simple counterexample is provided by the quasi-concave (and convex) function $f : X = [0,+\infty[\to \mathbb{R}$ defined by $f(x) = -x$, for $x \in [0,1]$, and $f(x) = -1$, for $x \in]1,+\infty[$.

We now state a definition of concavity and quasi-concavity for functions defined on a subset X of $\mathbb{R}^n$ without the usual assumption of convexity of X. This greater generality is required for the results of Section 5.

Definition 4.2 *Let X be a subset of $\mathbb{R}^n$. A function $f : X \to \mathbb{R}$ is concave (resp. quasi-concave) on X iff for every $x \in X$ and for every set of points $\{x^1, \ldots, x^m\} \subseteq X$ and of coefficients $\alpha_1, \ldots, \alpha_m \in]0,1[$ such that $\sum_{i=1}^m \alpha_i = 1$ and $x = \sum_{i=1}^m \alpha_i x^i$ one has*

$$f(x) \geq \sum_{i=1}^m \alpha_i f(x^i)$$
$$(resp.\ f(x) \geq \min_i f(x^i)) \tag{17}$$

A function is called strictly concave *or* strictly quasi-concave *iff the above relations are satisfied with a strict inequality whenever $x^i \neq x$ for at least one index i.*

Note that if the set X coincides with $\mathcal{E}(co(X))$ (like e.g. in the case of a circle), then every function is strictly concave on X with the above definition. However, when X is convex Definition 4.2 is equivalent to the standard definition of concavity and strict concavity of a function.

In this section we restrict our attention to the minimization of a function on a polyhedron and, in order to avoid trivial cases, we also assume that all (unbounded) polyhedra have at least one vertex.

It is easily seen that the class of quasi-concave functions is the most general class of functions for which Property (A) holds for *every* bounded polyhedron S. Indeed, by applying Property (A) to the special polyhedron that coincides with the line segment joining two points x^1 and x^2, one trivially obtains the inequality defining quasi-concave functions.

For a given polyhedron, or class of polyhedra, it is however possible to find classes of functions which properly include quasi-concave functions and satisfy Property (A). We recall here some results obtained in [64] and extend them to the case of unbounded polyhedra and of polymatroids.

In the sequel we denote by riX and $rbdX$ the *relative interior* and the *relative boundary* of a set $X \subset \mathbb{R}^n$ respectively, i.e., the interior and the boundary of X with respect to the topology induced on the smallest affine manifold containing X. Furthermore, we denote by $\dim(X)$ the dimension of the smallest affine manifold containing X.

Definition 4.3 *Let X be a subset of $\mathbb{R}^n$ and let f be a function from X into $\mathbb{R}$. We say that f satisfies the* Weak Minimum Principle *(WMP for short) on X iff, whenever $x^* \in X$ and $f(x^*) \leq f(x)$ for every $x \in X$, either $f(x) = f(x^*)$ for every $x \in X$ or $x^* \in rbdX$.*

Lemma 4.4 *Let f be a real function on a polyhedron P and assume that f satisfies WMP on all faces of P belonging to a subset Ω of the set of all faces of P. Then, the set P_{min} of global minimum points of f over P satisfies the following relation*

$$P_{min} \cap (V_P \cup \bigcup_{F \in \Omega^c}) \neq \emptyset \iff P_{min} \neq \emptyset \tag{18}$$

Proof. Assume that $P_{min} \neq \emptyset$ and let F^* denote a face of P of minimal dimension among the faces F satisfying the relation $P_{min} \cap F \neq \emptyset$. If $F^* \in \Omega^c$ or $dim(F^*) = 0$, then (18) holds. Hence assume that $F^* \in \Omega$ and $dim(F^*) > 0$ and let $x \in P_{min} \cap F^*$. By the WMP one can then find $y \in rbdF^*$ such that $y \in P_{min}$, contradicting the minimality assumption on F^*. □

By choosing Ω equal to the set of all faces of dimension greater than m we obtain the following:

Theorem 4.5 *If f satisfies WMP on all faces of P of dimension greater than m, then, if $P_{min} \neq \emptyset$, at least one global minimum point lies in an m-dimensional face of P. In particular, if $m = 0$, then Property (A) holds.*

Theorem 4.5 provides a fairly general condition for the validity of Property (A); however, in general it is not easy to check whether a function f satisfies WMP on some or all faces of P. The following corollaries provide useful sufficient conditions for this hypothesis to hold.

Corollary 4.6 ([30]) *If $f \in C^2(P)$ and the Hessian matrix $H_f(x)$ of f has least $n - m$ negative eigenvalues at any $x \in P$ then, if $P_{min} \neq \emptyset$, at least one global minimum point lies in an m-dimensional face of P.*

Proof. The assumption implies that f cannot satisfy the second order necessary condition for optimality at any point in the relative interior of a face of dimension greater than m. Hence, WMP holds on all faces of P of dimension greater than m. □

Given a point $s \in \mathbb{R}^n$ define $P_s(x) = \{z \in P : z = x + \lambda s, \lambda \in \mathbb{R}\}$. Given a face F of P define also $I_F = \{h : s^h$ is parallel to some edge of $F\}$. Furthermore, let $H^1 = \{1, 2, \ldots, q^1\}, H^2 = \{1, 2, \ldots, q^2\}$ and $H = H^1 \cup H^2$, and let $\{s^h : h \in H\}$ be a set of vectors in $\mathbb{R}^n$ such that each bounded edge of P is parallel to some s^h with $h \in H^1$ and each unbounded edge of P is parallel to some s^h with $h \in H^2$.

Lemma 4.7 *If $P_{min} \neq \emptyset$, f is quasi-concave on $P_{s^h}(x)$ for all $h \in K^1 \subset H^1$ and $x \in P$, and concave or strictly quasi-concave on $P_{s^h}(x)$ for all $h \in K^2 \subset H^2$ and $x \in P$, then a global minimum point of f over P belongs to a face F satisfying $I_F \cap (K^1 \cup K^2) = \emptyset$*

Proof. Let $\Omega = \{F : F$ is a face of P and $I_F \cap (K^1 \cup K^2) \neq \emptyset\}$. Then by Lemma 4.4 we only need to prove that f satisfies WMP on every face $F \in \Omega$. To this end, assume that $F \in \Omega$ and $x^* \in F$ is a global minimum point for f over F. If $x^* \in rbdF$, then WMP trivially holds on F. So suppose that $x^* \in riF$ and let $h \in I_F \cap (K^1 \cup K^2)$. Then $x^* \in riP_{s^h}(x^*)$. If $h \in H^1$, then $P_{s^h}(x^*)$ is bounded and its extreme points x^1 and x^2 belong to $rbdF$. Furthermore, quasi-concavity of f on $P_{s^h}(x^*)$ implies that either x^1 or x^2 is also a global minimum point. Hence, WMP holds. If $h \in H^2$, then $P_{s^h}(x^*)$ is unbounded. In this case let x^1 be the only extreme point of $P_{s^h}(x^*)$ (which exists because P has vertices) and take any point x^2 on $P_{s^h}(x^*)$ such that x^* is in the interior of the line segment joining x^1 and x^2. If f is concave on $P_{s^h}(x)$, then $f(x^1) = f(x^*) = f(x^2)$ which establishes WMP. On the other hand, if f is strictly quasi-concave on $P_{s^h}(x)$, then one should have $f(x^*) > min\{f(x^1), f(x^2)\} \geq f(x^*)$, which is a contradiction. □

Since the only faces of P that satisfy $I_F \cap H = \emptyset$ are the vertices of P, we immediately obtain the following:

Theorem 4.8 *If f is quasi-concave on $P_{sh}(x)$ for all $h \in H^1$ and $x \in P$, and concave or strictly quasi-concave on $P_{sh}(x)$ for all $h \in H^2$ and $x \in P$, then Property (A) holds.*

In order to apply the above proposition one needs to know the directions of all the edges of P. When such directions are not explicitly known, one can still guarantee the validity of Property (A) by making some more restrictive assumptions on f.

Definition 4.9 *Let f be a real function defined on a convex set $X \subset \mathbb{R}^n$ and let $m \leq n$. We say that f is m-concave (m-quasi-concave) on X iff*

$$f(\alpha x^1 + (1-\alpha)x^2)) \geq \alpha f(x^1) + (1-\alpha)f(x^2)$$

$$(resp. \ f(\alpha x^1 + (1-\alpha)x^2)) \geq min\{f(x^1), f(x^2)\}),$$

for every $\alpha \in [0,1]$ and for every $x^1, x^2 \in X$ such that $x_i^1 = x_i^2$ for at least $n - m$ indices i in $\{1, \ldots, n\}$.

Assume now that the polyhedron P is expressed in one of the following ways:

$$P = \{x \in \mathbb{R}^n : Ax = b, \ell \leq x \leq u\}$$

or (19)

$$P = \{x \in \mathbb{R}^n : Ax \leq b, \ell \leq x \leq u\},$$

where $A \in \mathbb{R}^{m \times n}, b \in \mathbb{R}^m, \ell, u \in \mathbb{R}^n, \ell = (\ell_1, \ldots, \ell_n), u = (u_1, \ldots, u_n)$, $-\infty \leq \ell_i < u_i \leq +\infty$ and $min\{|\ell_i|, |u_i|\} < +\infty$. Note that the assumptions on ℓ and u imply that, if $P \neq \emptyset$, then P has at least one vertex.

Theorem 4.10 *Let P be a polyhedron defined by (19) and assume that A has rank $m - 1$. If f is m-concave on P or f is m-quasi-concave on P and P is bounded, then Property (A) holds.*

Proof. Let $s = (s_1, \ldots, s_n)$ be any vector parallel to an edge of P. Observe that every edge of P lies in the intersection of $n - 1$ linearly independent hyperplanes taken from among those defining P in (19). Since $rank(A) = m - 1$, we have $s_i = 0$ for at least $n - m$ indices i. From the m(-quasi)-concavity of f we then derive that f is m(-quasi)-concave on the sets $P_s(x)$ for every $x \in P$. Hence the conclusion follows from Theorem 4.8. □

4.1 The Box Case

We now consider the case where P is a box, i.e., it is defined by $P = \{x \in \mathbb{R}^n : \ell \leq x \leq u\}$ where $\ell, u \in \mathbb{R}^n, \ell = (\ell_1, \ldots, \ell_n), u = (u_1, \ldots, u_n), -\infty \leq \ell_i < u_i \leq +\infty$ and $min\{|\ell_i|, |u_i|\} < +\infty$. In this case, it is easy to see that a set of vectors parallel to the edges of P is the canonical basis of $\mathbb{R}^n$ denoted by $e^1, \ldots, e^n$.

Consider the sets of indices $J^1 = \{i : \ell_i = -\infty\}, J^2 = \{i : u_i = +\infty\}$ and $J = \{1, \ldots, n\} \setminus (J^1 \cup J^2)$ and apply the transformation $x_i = (u_i - \ell_i)y_i + \ell_i$ for $i \in J$, $x_i = u_i$ for $i \in J^1$ and $x_i = \ell_i$ for $i \in J^2$. The problem of minimizing f on the vertices V_P of P is then equivalent to the problem of minimizing the transformed function $\tilde{f}(y)$ on the $0-1$ hypercube $\mathbb{B}^{n'} = \{0, 1\}^{n'}$, where $n' = card(J)$. Hence, when Property (A) holds, the problem of minimizing f over a box can be reduced to a problem of (nonlinear) $0-1$ programming for which several solution methods are available (see, e.g., [31, 33]).

In [19] the equivalence between minimization on P and on V_P in the box case is exploited to efficiently perform a stability test for a system of linear differential equations with uncertain real parameters.

Taking into account the directions of the edges of P, Theorem 4.8 can be reformulated as follows:

Theorem 4.11 *If f is quasi-concave with respect to each coordinate x_i such that $i \in J$, and concave or strictly quasi-concave with respect to each coordinate x_i such that $i \in J^1 \cup J^2$, then Property (A) holds.*

By exploiting the above reduction to the 0–1 case, a function f can be efficiently minimized over a box P if it is *submodular* on V_P in addition to satisfying Property (A).

We recall that a subset $X \subset \mathbb{R}^n$ is a sublattice of $\mathbb{R}^n$ iff

$$x^1 \vee x^2, x^1 \wedge x^2 \in X, \quad \forall x^1, x^2 \in X,$$

where

$$(x^1 \vee x^2)_i = max\{x_i^1, x_i^2\} \quad \text{and} \quad (x^1 \wedge x^2)_i = min\{x_i^1, x_i^2\}.$$

Note that a box is trivially a sublattice of $\mathbb{R}^n$. Furthermore, a real-valued function g is submodular on a sublattice X if

$$g(x^1 \vee x^2) + g(x^1 \wedge x^2) \leq g(x^1) + g(x^2) \text{ for all } x^1, x^2 \in X.$$

Nonlinear Programming and Discrete Optimization

A well-known result in $0-1$ optimization is the fact that the problem of minimizing a submodular function over the $0-1$ hypercube can be solved in polynomial time [27].

In the case where $f \in C^2$ it has been proved [67] that f is submodular on a box Q iff the second order mixed partial derivatives $f_{x_i x_j}(x)$ are nonpositive for all $x \in Q$. Note that f is concave with respect to each coordinate iff $f_{x_i x_i}(x) \leq 0$ for all i and x. Hence, by choosing Q equal to the convex hull of V_P we obtain the following:

Theorem 4.12 *If $f_{x_i x_i}(x) \leq 0$ for every index i and for all $x \in P$, and $f_{x_i x_j}(x) \leq 0$ for all $i, j \in J$ with $i \neq j$, then the problem of minimizing f over P can be solved in polynomial time.*

4.2 The Polymatroid Case

A more complex type of polyhedra for which a simple characterization of the edges is available are polymatroids and base polyhedra. These polyhedra, originally introduced by Edmonds [16], have attracted considerable interest due to their combinatorial structure, their connections with submodular function minimization and the possibility of efficiently minimizing linear and separable functions over them. A *polymatroid* is a polyhedron P in $\mathbb{R}^n$ described by the following inequalities:

$$\sum_{i \in I} x_i \leq g(I) \text{ for all } I \subset N = \{1, \ldots, n\},$$

$$x_i \geq 0 \text{ for all } i \in N,$$

where the function $g : 2^N \to \mathbb{R}$ is
isotone: $\quad g(S) \leq g(T)$ for all $S, T \subset N$,
submodular: $\quad g(S \cap T) + g(S \cup T) \leq g(S) + g(T)$ for all $S, T \subset N$,
normalized: $\quad g(\emptyset) = 0$.

A *base polyhedron* is the facet of a polymatroid determined by the additional constraint $\sum_{i \in N} x_i = g(N)$. From known results on the characterization of adjacent vertices of polymatroids and base polyhedra (see [18, 25, 68]) one can easily deduce that every edge of a polymatroid is parallel to a vector of the set $\{e^i : i \in N\} \cup \{e^i - e^j : i, j \in N, i \neq j\}$, while every edge of a base polyhedron is parallel to a vector of the set $\{e^i - e^j : i, j \in N, i \neq j\}$.

In the case of a base polyhedron and of a twice differentiable function f a sufficient condition for the validity of Property (A) is then the following:

$$f_{x_i x_i}(x) + f_{x_j x_j}(x) - 2f_{x_i x_j}(x) \leq 0, \forall x \text{ and } \forall i, j \in N, \text{ with } i \neq j.$$

For a polymatroid one must add the condition

$$f_{x_i x_i}(x) \leq 0, \forall x \text{ and } \forall i \in N.$$

Note that the polyhedra described by non-negativity constraints together with a single "knapsack" constraint of the type $\sum_i a_i x_i \leq b$ or $\sum_i a_i x_i = b$, with $a_i > 0$, can be transformed into special cases of polymatroids and base polyhedra by setting $y_i = a_i x_i$ and choosing $g(S) = b$ for all $S \neq \emptyset$. This type of polyhedra arise naturally in problems of resource allocation (see [38, 43]).

5 Piecewise Concave Functions and Minimax

In this section we introduce the classes of piecewise concave and piecewise quasi-concave functions on a set X, i.e., those functions which are concave or quasi-concave on each element of a family of subsets of X whose union covers X. These classes of functions arise naturally from several applications including location theory and various types of minimax problems. We show here that for these functions the search for a global minimum point can be restricted to a special subset of X that, under suitable assumptions, is guaranteed to be finite.

Let $X \subset \mathbb{R}^n$ and assume that $X = \bigcup_{i \in I} X_i$, where each X_i is a closed subset of $\mathbb{R}^n$. Consider a family $\{f_i\}_{i \in I}$ of functions with $f_i : X_i \to \mathbb{R}$ and $f_i(x) = f_j(x)$ for every $x \in X_i \cap X_j$ and for all $i, j \in I$. Define a function $g : X \to \mathbb{R}$ by setting

$$g(x) \doteq f_i(x), \qquad \text{if } x \in X_i. \tag{20}$$

The function g defined above is called *(strictly) piecewise concave* or *piecewise quasi-concave* iff all the functions f_i are (strictly) concave or quasi-concave respectively. We point out that the sets X_i are not required to be convex and that the notions of concavity and quasi-concavity on X_i are those introduced in Definition 4.2. In fact, the sets X_i are not necessarily convex, e.g., in the important problem of minimizing a function defined as the maximum of a family of concave or quasi-concave functions on a convex set. Clearly, *piecewise convexity* or *piecewise quasi-convexity* of a function can be defined in a similar manner.

Consider the subset $\mathcal{E}$ of X containing all the convex hull extreme points of all sets X_i and its subset $D \subseteq \mathcal{E}$ formed by those points that are convex hull extreme points of every set X_i to which they belong. Formally we set:

$$\mathcal{E} = \bigcup_{i \in I} \mathcal{E}(co(X_i)) \quad \text{and} \quad D = \{x \in \mathcal{E} : x \in X_i \Rightarrow x \in \mathcal{E}(co(X_i))\} \qquad (21)$$

Denoting by X_{min} the set of global minimum points of g over X, we can now state the main result of this section which generalizes and strengthens some well-known results for piecewise concave functions and for minimax problems [66].

Theorem 5.1 (i) *If, for all $i \in I$, f_i is quasi-concave on X_i and X_i is compact, then*
$$X_{min} \neq \emptyset \Leftrightarrow X_{min} \cap \mathcal{E} \neq \emptyset.$$

(ii) *If, for all $i \in I$, f_i is strictly quasi-concave on X_i and X_i is compact, then*
$$X_{min} \neq \emptyset \Leftrightarrow X_{min} \cap D \neq \emptyset.$$

(iii) *If X is compact, g is lower semicontinuous on X and f_i is concave on X_i for all $i \in I$, then*
$$X_{min} \cap D \neq \emptyset \quad \text{and} \quad \mathcal{E}(co(X_{min})) \subseteq D.$$

Proof.

(i) Let x^* be a global minimum point for g over X and assume that $x^* \in X_i$. Since X_i is compact, $co(X_i)$ is also compact and hence $co(X_i) = co(\mathcal{E}(co(X_i)))$, by the Krein-Millman Theorem. Hence $X_i \subseteq co(\mathcal{E}(co(X_i)))$, so that there exist points $x^1, \ldots, x^m \in \mathcal{E}(co(X_i))$ and coefficients $\alpha_1, \ldots, \alpha_m \in]0, 1[$ such that $x^* = \sum_{i=1}^m \alpha_i x^i$ and $\sum_{i=1}^m \alpha_i = 1$. Then, by the quasi-concavity of f_i, at least one of the points $x^1, \ldots, x^m$ must be a global minimum point for g.

(ii) Let x^* be a global minimum point for g over X and assume that $x^* \in X_i$. With an argument similar to the one employed in the proof of (i) if $x^* \notin \mathcal{E}(co(X_i))$ we obtain $f_i(x^*) > \min\{f_i(x^1), \ldots, f_i(x^m)\}$ which contradicts the minimality of x^*.

(iii) First note that, by the lower semicontinuity of g, the set X_{min} is nonempty, closed and hence compact since it is contained in the compact set X. Hence, $\mathcal{E}(co(X_{min})) \neq \emptyset$ and $co(X_{min}) = co(\mathcal{E}(co(X_{min})))$ by the Krein-Millman Theorem. To complete the proof, we show that if $x^* \in \mathcal{E}(co(X_{min}))$, then $x^* \in \mathcal{E}(co(X_i))$ for every i such that $x^* \in X_i$.

Ab absurdo, if $x^* \in X_i \setminus \mathcal{E}(co(X_i))$, then there exist $y, z \in co(X_i)$ and $\lambda \in]0,1[$ such that $x^* = \lambda y + (1-\lambda)z$. Then, by definition of $co(X_i)$, there exist $y^1, \ldots, y^{m_1}$ and $z^1, \ldots, z^{m_2}$ in X_i and coefficients $\alpha_1, \ldots, \alpha_{m_1}$, $\beta_1, \ldots, \beta_{m_2} \in]0,1[$ such that $y = \sum_{i=1}^{m_1} \alpha_i y^i$, $z = \sum_{i=1}^{m_2} \beta_i z^i$, $\sum_{i=1}^{m_1} \alpha_i = 1$ and $\sum_{i=1}^{m_2} \beta_i = 1$. Setting $\gamma_i = \lambda \alpha_i$ for $i = 1, \ldots, m_1$ and $\delta_i = (1-\lambda)\beta_i$ for $i = 1, \ldots, m_2$ one has $x^* = \sum_{i=1}^{m_1} \gamma_i y^i + \sum_{i=1}^{m_2} \delta_i z^i$ and $\sum_{i=1}^{m_1} \gamma_i + \sum_{i=1}^{m_2} \delta_i = 1$. From the concavity of f_i, it then follows that $f_i(x^*) = f_i(y^i) = f_i(z^i)$ for all indices i. Hence, $g(x^*) = g(y) = g(z)$ and therefore $y, z \in X_{min}$, contradicting $x^* \in \mathcal{E}(co(X_{min}))$. □

Remark 5.2 *Note that if $co(X_i)$ is a polytope, then, taking into account Theorem 4.8, the assumption of (i) in Theorem 5.1 can be weakened by requiring only quasi-concavity of f_i on all line segments in $co(X_i)$ which are parallel to some edge of $co(X_i)$.*

Remark 5.3 *The result of Theorem 5.1(i) holds also when some set X_i is an unbounded polyhedron. In this case, taking into account Theorem 4.8, the assumption of (i) in Theorem 5.1 can be replaced by the requirement of quasi-concavity of f_i on all line segments in X_i which are parallel to some bounded edge of X_i and concavity or strict quasi-concavity of f_i on all line segments in X_i which are parallel to some unbounded edge of X_i.*

In several applications (see, e.g., [11, 13]) one has to solve minimax problems of the form

$$\min_{x \in X} \max_{i \in I} f_i(x), \qquad (22)$$

where X is a compact subset of $\mathbb{R}^n$ (often a polytope) and each f_i is (quasi-)concave on X. In this case Theorem 5.1 can be applied to the piecewise (quasi-)concave function $g(x) \doteq \max_{i \in I} f_i(x)$ which is (quasi-)concave on the sets $X_i \doteq \{x \in X : g(x) = f_i(x)\}$. Thus we obtain the following result first stated by Zangwill [71] for the concave case with a somewhat incomplete proof:

Theorem 5.4 *If X is a compact subset of $\mathbb{R}^n$ and the functions $f_i : X \to \mathbb{R}$ are continuous and (quasi-)concave, then at least one solution of (22) belongs to the set D (resp. $\mathcal{E}$).*

Suppose now that X is a polytope described by a set of linear inequalities $a_j^T x \leq b_j$, $j \in J$ and, for every $x \in X$ define $I(x) = \{i \in I : g(x) = f_i(x)\}$

and $J(x) = \{j \in J : a_j^T x = b_j\}$. It can be easily verified that a point $x \in X$ is a vertex (extreme point) of X iff $I(x)$ is maximal, i.e., there does not exist any $y \in X$ such that $I(x)$ is a proper subset of $I(y)$. Analogously, a point $x \in X$ is called a *g-vertex* of X iff $I(x) \cup J(x)$ is maximal. This notion has been introduced by Du and Hwang [12] (see also [11]). They also proved the following result, which has been extended to the case of an infinite index set I by Du and Pardalos [14]:

Theorem 5.5 *If X is a polytope in $\mathbb{R}^n$ and the functions $f_i : X \to \mathbb{R}$ are continuous and concave, then at least one solution of (22) belongs to the set G of g-vertices of X.*

Theorems 5.4 and 5.5 establish an interesting equivalence between the continuous problem (22) and the problem of minimizing $g(x)$ over the sets $D, \mathcal{E}$ or G which are often finite. Note however that Theorem 5.4 provides a stronger restriction for global minimum points than Theorem 5.5. Indeed, it is clear that $D \subseteq \mathcal{E}$ and it has been proved in [66] that $D \subseteq G$. Furthermore, simple examples, like the following one, show that there are cases where the inclusions $D \subset \mathcal{E} \subset G$ can be strict.

Example 5.6 *Consider the interval $X = [a, e] \subset \mathbb{R}$ and the functions $f_1, f_2 : X \to \mathbb{R}$ illustrated in fig.3. Then we have $X_1 = [a, b] \cup [c, d]$, $X_2 = [b, c] \cup [d, e]$, $D = \{a, e\}$, $\mathcal{E} = \{a, b, d, e\}$ and $G = \{a, b, c, d, e\}$.*

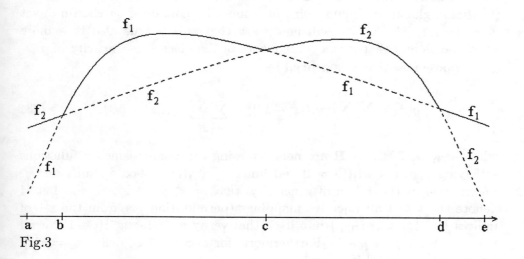

Fig.3

6 Location Problems

Piecewise concavity has some important applications in location theory. Consider, e.g., the rectilinear distance facility location problem which consists in finding the coordinates of m new facilities $x^1,\ldots,x^m$ in an n-dimensional space so as to minimize a weighted sum of the rectilinear (or L^1) distances among them and between them and a set of p existing facilities located in the points $a^1,\ldots,a^p$. Formally, we seek to minimize over $\mathbb{R}^{n\times m}$ the function

$$g(x^1,\ldots,x^n) \doteq \sum_{k=1}^{n}\sum_{i=1}^{m}\sum_{j=1}^{m} w_{ijk}|x_k^i - x_k^j| + \sum_{k=1}^{n}\sum_{i=1}^{m}\sum_{j=1}^{p} v_{ijk}|x_k^i - a_k^j|,$$

where w_{ijk} and v_{ijk} are non-negative weights. Note that the function g is separable, i.e., it can be written in the form $g(x) = \sum_{k=1}^{n} g_k(x_k^1,\ldots,x_k^n)$, where

$$g_k(x_k^1,\ldots,x_k^n) = \sum_{i=1}^{m}\sum_{j=1}^{m} w_{ijk}|x_k^i - x_k^j| + \sum_{i=1}^{m}\sum_{j=1}^{p} v_{ijk}|x_k^i - a_k^j|.$$

Hence, the problem of minimizing g over $\mathbb{R}^{n\times m}$ can be solved by minimizing each function g_k over $\mathbb{R}^m$ separately. For this reason we will restrict our attention to the one-dimensional case where $x^i, a^i \in \mathbb{R}$. It is well-known [5] that a global minimum point of g must be contained in the finite set $A \doteq \{a^1,\ldots,a^p\}^m$. We will now show that this is true also in a more general case and that it is a consequence of the piecewise concavity of g.

Suppose now that g is defined by

$$g(x) = \sum_{i=1}^{m}\sum_{j=1}^{m} \varphi_{ij}(|x^i - x^j|) + \sum_{i=1}^{m}\sum_{j=1}^{p} \psi_{ij}(|x^i - a^j|), \tag{23}$$

where $\varphi_{ij}, \psi_{ij} : \mathbb{R}_+ \to \mathbb{R}$ are nondecreasing continuous concave functions satisfying $\varphi_{ij}(0) = \psi_{ij}(0) = 0$ and $\lim_{x\to+\infty}\psi_{ij}(t) = +\infty$ for all i and j. We assume, without loss of generality, that $a^1 < a^2 < \ldots < a^p$. Let H denote the set of all bijective mappings (permutations) π from the set of indices $L = \{1,\ldots,m+p\}$ into itself that verify $\pi^{-1}(i) < \pi^{-1}(i+1)$ for all $i = m+1,\ldots,m+p-1$. Furthermore, for every $i \in L$, define $y_i = x^i$ if $i \leq m$ and $y_i = a^{i-m}$ if $i > m$.

Lemma 6.1 *The function g defined in (23) is concave on all the sets $X_\pi \doteq \{x \in \mathbb{R}^n : y_{\pi(i)} \leq y_{\pi(i+1)}, i = 1, \ldots, m+p-1\}$.*

Proof. Indeed, on X_π one has $|y_i - y_j| = y_i - y_j$ if $\pi^{-1}(i) < \pi^{-1}(j)$ and $|y_i - y_j| = y_j - y_i$ if $\pi^{-1}(i) > \pi^{-1}(j)$. Hence, on X_π the functions $\varphi_{ij}(|x^i - x^j|)$ and $\psi_{ij}(|x^i - a^j|)$ are concave for all i, j, since they are obtained as the composition of a concave and a linear function. Therefore, the function g is concave on X_π because it is a sum of concave functions. □

Theorem 6.2 *At least one global minimum point of g on $\mathbb{R}^n$ is achieved at a point in A.*

Proof. Note that the sets X_π are polyhedra with vertices in A and that $\mathbb{R}^n = \cup_{\pi \in H} X_\pi$. Furthermore, since $\lim_{t \to +\infty} \psi_{ij}(t) = +\infty$ for all i and j and g is continuous, there must be a global minimum point of g in a bounded neighbourhood of 0. Hence, by Theorem 5.1, a global minimum point of g must be contained in A. □

Another continuous location problem which can be solved by restricting the search to a finite set of candidate solutions [65] is the problem of locating some undesirable facilities in locations $x^1, \ldots, x^m \in \mathbb{R}^n$ in a way that minimizes the maximum of some decreasing nuisance cost functions of the distances among them and between them and a given set of points $b^1, \ldots, b^p \in \mathbb{R}^n$ (see [17] for a survey on this type of problems). Formally, we wish to minimize the function:

$$g(x) = g(x^1, \ldots, x^m) \doteq \max\{\max_{i \in I, j \in J} \alpha_{ij}(\|x^i - b^j\|), \max_{i \neq j \in I} \beta_{ij}(\|x^i - x^j\|)\},$$

where α_{ij} and β_{ij} are continuous nonincreasing functions, $I = \{1, \ldots, m\}$, $J = \{1, \ldots, p\}$ and $\|\cdot\|$ is any norm in $\mathbb{R}^n$. In this case the function g is piecewise quasi-concave, since it is the maximum of a finite family of continuous quasi-concave functions. Clearly, the infimum of g on $\mathbb{R}^{n \times m}$ is obtained when $\|x^i\| \to \infty$ for all i. Hence, this location problem becomes meaningful only when x is required to be in a compact set X. Under this assumption, Theorem 5.4 can be applied to establish the following result.

Theorem 6.3 *At least one global minimum point of g on a compact set X is achieved on the set $\mathcal{E} = \bigcup_{i,j \in I} \mathcal{E}(co(X_{ij})) \cup \bigcup_{i \in I, j \in J} \mathcal{E}(co(Y_{ij}))$, where*

$$X_{ij} \doteq \{x \in X : \alpha_{ij}(\|x^i - b^j\|) \geq$$

$$\geq \max\{\max_{h \in I, k \in J} \alpha_{hk}(\|x^h - b^k\|)\}, \max_{h \neq k \in I} \beta_{hk}(\|x^h - x^k\|)\}\}.$$

$$Y_{ij} \doteq \{x \in X : \beta_{ij}(\|x^i - x^j\|) \geq$$

$$\geq \max\{\max_{h \in I, k \in J} \alpha_{hk}(\|x^h - b^k\|)\}, \max_{h \neq k \in I} \beta_{hk}(\|x^h - x^k\|)\}\}.$$

When only one new undesirable facility has to be established, the objective function $g : \mathbb{R}^n \to \mathbb{R}$ becomes

$$g(x) = \max_{j \in J} \alpha_j(\|x - b^j\|).$$

In this case, Theorem 6.3 implies that a global minimum point of g on a compact set X is attained at a point in $\mathcal{E} = \bigcup_{j \in J} \mathcal{E}(co(X \cap X_j))$, where $X_j = \{x \in \mathbb{R}^n : \alpha_j(\|x - b^j\|) \leq \alpha_k(\|x - b^k\|) \text{ for all } k \neq i\}$. The family of sets X_j forms the (generalized) Voronoi diagram with respect to the functions $\alpha_j(\|x\|)$. This geometric structure has received much attention in recent years especially in the case where $\alpha_j(\|x\|) = \|x\|$ or $\alpha_j(\|x\|) = \lambda_j \|x\|$ and $\|\cdot\|$ is the Euclidean norm (see [2] and references therein). In particular, when $\alpha_j(\|x\|) = \|x\|$ and $\|\cdot\|$ is the Euclidean norm, the sets X_j are polyhedra described by:

$$X_j = \{x \in \mathbb{R}^n : (b^k - b^j)^T (x - \frac{b^j + b^k}{2}) \leq 0 \text{ for all } k \neq j\}.$$

In this case, the maximum number N of points in $\mathcal{E} = \bigcup_{j \in J} \mathcal{E}(X_j)$ satisfies the following bounds [45]:

$$n/2!\, p^{n/2} \leq N \leq 2(n/2!\, p^{n/2}) \quad \text{for } n \text{ even}$$

and

$$\frac{(\lceil n/2 \rceil - 1)!}{e} p^{\lceil n/2 \rceil} \leq N \leq 2(\lceil n/2 \rceil!\, p^{\lceil n/2 \rceil}) \quad \text{for } n \text{ odd}.$$

Furthermore, the points of $\mathcal{E}$ can be computed in $O(p^{\lceil n/2 \rceil + 1})$ time in the general case [3], and in time $O(p \log p)$ for $n = 2$. If we wish to minimize g on a polyhedron X, then a global solution can be found by simple enumeration of all vertices of the sets $X_i \cap X$. In this way we always obtain a finite algorithm, while in [62] an algorithm is presented which requires some additional assumptions to solve the same problem in finitely many iterations.

Facility location on networks is an important field in location theory where much research has concentrated in identifying a *finite* set of points that necessarily contain an optimal solution. Good surveys on this topic are [29, 47, 50].

In order to formulate location problems on networks, we introduce some definitions and notations taken from [47]. An *edge* of length $\ell > 0$ is the image of an interval $[0, \ell]$ by a continuous mapping f form $[0, \ell]$ to $\mathbb{R}^d$ such that $f(\theta) \neq f(\theta')$ for any $\theta \neq \theta'$ in $[0, \ell]$. A *network* is defined as a subset N of $\mathbb{R}^d$ that satisfies the following conditions: (i) N is the union of a finite number of edges; (ii) any two edges intersect at most at their extremities; (iii) N is connected. The *vertices* of the network are the extremities of the edges defining N and are denoted by $V = \{v^1, \ldots, v^n\}$. The set of edges defining the network is denoted by E. For every edge $[v^i, v^j] \in E$ let f_{ij} from $[0, \ell_{ij}]$ to $\mathbb{R}^d$ be the defining mapping and denote by θ_{ij} the inverse of f_{ij} which maps $[v^i, v^j]$ into $[0, \ell_{ij}]$. Given two points $x^1, x^2 \in [v^i, v^j]$, the subset of points of $[v^i, v^j]$ between and including x^1, x^2 is a subedge $[x^1, x^2]$. The length of $[x^1, x^2]$ is given by $|\theta_{ij}(x^1) - \theta_{ij}(x^2)|$. A *path* joining two points $x^1 \in N$ and $x^2 \in N$ is a minimal connected subset of N containing x^1 and x^2. The length of a path is equal to the sum of all its constituent edges and subedges. A metric d on N is defined by setting $d(x^1, x^2)$ equal to the length of a shortest path joining x^1 and x^2.

For any point $z \in N$ consider the function on $[0, \ell_{ij}]$ defined by $d(z, f_{ij}(\theta))$. Note that, by the definition of the distance, one has

$$d(z, f_{ij}(\theta)) = \min\{d(z, v^i) + \theta, d(z, v^j) + \ell_{ij} - \theta\}.$$

Hence, $d(z, f_{ij}(\theta))$ is the minimum of two linear functions in θ. Therefore, the following properties hold [47]:

(i) $d(z, f_{ij}(\theta))$ is continuous and concave on $[0, \ell_{ij}]$;

(ii) $d(z, f_{ij}(\theta))$ is linearly increasing with slope $+1$ on $[0, \theta_{ij}(z)[$ and linearly decreasing with slope -1 on $]\theta_{ij}(z), \ell_{ij}]$, where $\theta_{ij}(z) = \frac{1}{2}[\ell_{ij} + d(z, v^j) - d(z, v^i)]$.

The (single) median problem on the network N consists of finding a point in N that minimizes the function

$$F(x) = \sum_{v^i \in V} w_i((d(v^i, x))),$$

where $w_i(t)$ are concave nondecreasing functions from $\mathbb{R}_+$ to $\mathbb{R}$. In 1964 Hakimi [28] showed that at least one solution of this problem belongs to V when $w_i(t) = \lambda_i t$ with $\lambda_i \geq 0$. This result was extended in [50] to the case where the $w_i(t)$ are concave nondecreasing functions. The proof is based on the remark that, if $w_i(t)$ is concave nondecreasing, then the function $w_i((d(v^i, f_{ij}(\theta))))$ is concave on $[0, \ell_{ij}]$ and hence $F(f_{ij}(\theta))$ is also concave on $[0, \ell_{ij}]$. Therefore, the global minimum of F must be attained at one of the points $f_{ij}(0)$ or $f_{ij}(\ell_{ij})$, i.e., at a vertex of N.

In the (single) center problem on the network N one seeks to minimize the function

$$G(x) = \max_{v^i \in V}(d(v^i, x)).$$

In this case the function $G(f_{ij}(\theta))$ is not concave on the whole segment $[0, \ell_{ij}]$, but only on subsegments thereof. However, the number of such subsegments is bounded by $|V|^2|E|$ and their extremities can be explicitly determined (see [47] for details). Hence, also the center problem on a network can be solved by enumerating a finite (and polynomially bounded) number of points.

7 Convex-Concave Problems

Consider a problem of the form:

$$\min f(x, y) \quad s.t. \ (x, y) \in S \times T, \tag{24}$$

where $S, T \subset \mathbb{R}^n$ and $f : S \times T \to \mathbb{R}$.

When $f(x, y) = c^T x + x^T Q y + d^T y$, with $Q \in \mathbb{R}^{p \times q}$, $c, x \in \mathbb{R}^p$, $d, y \in \mathbb{R}^q$ and S and T are polyhedra, problem (24) is called a *bilinear programming problem*. This kind of problems, which can be reduced to concave minimization problems, have been studied by several authors (see, e.g., [4, 40, 54] and references therein). It is well known that at least one solution of a bilinear programming problem is achieved at a vertex of its feasible region. We now show that this property holds for a more general class of functions.

It is straightforward to prove that, if $\min_{x \in S} f(x, y)$ exists for every $y \in T$, then problem (24) is equivalent to the following:

$$\min_{y \in T} \varphi(y), \qquad (25)$$

where $\varphi(y) \doteq \min_{x \in S} f(x, y)$.

Hence, the solution of problem (24) may be reduced to the solution of the two subproblems $\min_{x \in S} f(x, y)$ and $\min_{y \in T} \varphi(y)$. In the case where T is a polyhedron Theorem 4.8 can be applied to ensure that $\varphi(y)$ attains its minimum at one of the vertices of T.

Theorem 7.1 *If T is a polyhedron and the function $y \mapsto f(x, y)$ is quasi-concave on all directions parallel to bounded edges of T and concave or strictly quasi-concave on all directions parallel to unbounded edges of T, then, if the function $\varphi(y)$ attains its minimum on T, at least one global minimum point lies in a vertex of T.*

Proof. It is sufficient to notice that, since $\varphi(y)$ is defined as a minimum of functions that are concave or quasi-concave on the directions parallel to the edges of T, φ shares the same properties of such functions and hence Theorem 4.8 may be employed to complete the proof. □

Clearly, if $\min_{x \in S} f(x, y)$ and $\min_{y \in T} \varphi(y)$ can be evaluated efficiently, then problem (24) can also be solved in an efficient manner. We now show that this is the case for a special class of indefinite quadratic programs. Consider the problem

$$\begin{cases} \min f(x, y) = x^T B x + x^T C y + y^T D y + c^T x + d^T y \\ x \in S = \{x : A_1 x \leq b_1\} \\ y \in T = \{y : A_2 y \leq b_2\}, \end{cases} \qquad (26)$$

where $B \in \mathbb{R}^{p \times p}, C \in \mathbb{R}^{p \times q}, D \in \mathbb{R}^{q \times q}, A_1 \in \mathbb{R}^{m_1 \times p}, A_2 \in \mathbb{R}^{m_2 \times q}, c, x \in \mathbb{R}^p, d, y \in \mathbb{R}^q, b_1 \in \mathbb{R}^{m_1}$ and $b_2 \in \mathbb{R}^{m_2}$. Note that, if B is positive semidefinite, then $\varphi(y) = \min_{x \in S} f(x, y)$ can be evaluated in polynomial time [46] and, if $s^T D s \leq 0$ for all vectors s parallel to an edge of T, then $\varphi(y)$ achieves its minimum on a vertex of T. Hence, problem (26) can also be solved in polynomial time if the set V_T of vertices of T is small or if $\varphi(y)$ can be minimized in polynomial time over V_T. This is the case, e.g., if $T = \{y \in \mathbb{R}^n : \ell \leq y \leq u\}$ and $b_{ij} \leq 0$, $c_{ij} \leq 0$, $d_{ij} \leq 0$ for all $i \neq j$. In fact, these assumptions imply submodularity of $f(x, y)$ which in turn implies submodularity of $\varphi(y)$ by a result of Topkis [67]. We have thus proved the following result:

Theorem 7.2 *Problem (26) can be solved in polynomial time if* $T = \{y \in \mathbb{R}^n : \ell \leq y \leq u\}$, B *is positive semidefinite,* $b_{ij}, d_{ij} \leq 0$ *for all* $i \neq j$ *and* $c_{ij} \leq 0$ *for all* i *and* j.

8 Discretization, Extension and Relaxation

A function f defined on a subset X of $\mathbb{Z}^n$ can be extended in several ways to a piecewise linear function $\bar{f}$ on the convex hull $co(X)$ of X. In [20, 49, 63] some extensions of this type are presented which also satisfy the condition that the global minimum of f over X and of $\bar{f}$ over $co(X)$ coincide. When $\bar{f}$ is convex, the problem of minimizing f on X can be efficiently solved by minimizing $\bar{f}$ on $co(X)$. This approach has been exploited in [49] to prove that the problem of minimizing a submodular function over the 0–1 hypercube $\mathbb{B}^n$ can be solved in polynomial time and in [20] to show that submodular functions for which a particular extension is convex can be minimized in polynomial time over any box in $\mathbb{Z}^n$.

A typical method for finding lower bounds for problems of the form

$$\min f(x) \quad s.t. \quad x \in S \cap \mathbb{Z}^n \qquad (27)$$

consists in solving the following continuous relaxation obtained by dropping the integrality constraint:

$$\min f(x) \quad s.t. \quad x \in S. \qquad (28)$$

This approach is frequently adopted when f is linear and S is a polyhedron since, in this case, problem (28) is considerably easier to solve than (27). However, even in the linear case, the distance $\|x^* - x'\|$ between an optimal solution x^* of (28) and any optimal solution x' of (27) can be quite large and the same thing can happen to the difference $f(x') - f(x^*)$ between the optimal objective values. Some upper bounds on the distance between the optimal solutions of (27) and (28) have been established in [9] for the linear case and in [26] for the quadratic case.

Recently, much attention has been devoted to polyhedral methods for integer programming. The key step of these methods is the addition of valid constraints to the set S in order to find a smaller feasible region $S' \subseteq S$ which satisfies $S' \cap \mathbb{Z}^n = S \cap \mathbb{Z}^n$. The final aim of polyhedral methods is that of finding a polyhedral representation S' of the convex hull of the integer points in S, i.e., $S' = co(S \cap \mathbb{Z}^n)$, so that problem (27) can be solved by minimizing f on S'. See, e.g., [51] for an introduction to these methods and [60, 61] for a procedure which constructs a hierarchy of sets $S_0 = S \subseteq S_1 \subseteq \ldots \subseteq S_n = co(S \cap \mathbb{Z}^n)$.

A different type of relaxation for 0-1 programming, which gives tighter bounds than standard continuous relaxation, has been introduced and analyzed in the last few years. The main idea consists in reformulating the 0-1 program as a linear program in a space with n^2 variables with the addition of the condition that a symmetric matrix formed with the n^2 variables is positive definite. Such problems are called *semidefinite programs* and can be solved efficiently both in theory and in practice. See [1, 55, 56, 70] for an introduction to these problems and for the connections with 0-1 and combinatorial optimization.

Another active field of research consists in the investigation of conditions that guarantee integrality of the vertices of S. Several classes of constraint matrices for which this property holds have been identified and the problem of recognizing them has also been addressed. The book of Schrijver [59] contains many results of this type, while [8] is a very recent survey on this topic.

When f is linear (or concave) and equality between the optimal solutions of (27) and (28) does not hold, problem (27) is often solved by means of a branch and bound procedure which exploits the bounds obtained from the solution of several continuous relaxations of type (28). When f is not linear, however, the roles might be exchanged. If f is separable and convex and S is a polyhedron [39] or f is separable and S is a polymatroid [38], then problem (27) can be solved efficiently and problem (28) may then be solved to within any prescribed accuracy by solving a sequence of discretized problems of the form

$$\min f(x) \quad s.t. \quad x \in S \cap \lambda \mathbb{Z}^n, \tag{29}$$

where λ is a positive scaling factor which enlarges the grid of integer points when $\lambda > 1$ and shrinks it when $\lambda < 1$. The methods developed in [38, 39] are based on proximity results between the optimal solutions of problems (28) and (29) which allow to efficiently improve the scaling factor λ in order to obtain a point that is at distance at most ϵ from an optimal solution of problem (28).

9 Necessary and Sufficient Optimality Conditions

Some of the main tools in nonlinear programming are necessary and/or sufficient optimality conditions. However, unless some kind of convexity assumption is made, such conditions only concern local solutions and one is then faced with the (combinatorial) problem of finding the global optimal solution among them.

Recently, some new *global* optimality conditions have been proposed for nonlinear minimization problems, which do not require any convexity assumptions (see [10, 34, 35, 52]). Since all 0–1 programs and many other discrete optimization problems can be equivalently reformulated as nonlinear programs, it seems natural to try to exploit global optimality conditions in the nonlinear setting to obtain analogous conditions in the discrete case.

A first step in this direction has been made by Hiriart-Urruty and Lemarechal [36] who observed that, since 0–1 quadratic minimization problems can be equivalently formulated as *concave* quadratic minimization problems on the hypercube $[0,1]^n$ (see Sect. 3), one could deduce from the general global optimality condition established in [34] for concave minimization problems a global optimality condition for the 0–1 quadratic case. This condition has been further analyzed in [6, 7], where some computational results are provided.

Another global optimality condition for concave minimization problems which might be fruitfully employed in discrete optimization has been introduced by A. Strekalovski and refined by Y. Ledyaev (see [35]). When f is a concave differentiable function on a convex set S this condition can be formulated as follows:

Theorem 9.1 *A point $\bar{x} \in S$ is a global minimum point for f over S if and only if*

$$(s-x)^T \nabla f(x) \geq 0, \qquad (30)$$

for all $s \in S$ and for all $x \in S$ satisfying $f(x) = f(\bar{x})$. Furthermore, if $\tilde{x} \in S$ satisfies $f(\tilde{x}) = f(\bar{x})$ and $(\bar{s} - \tilde{x})^T \nabla f(\tilde{x}) < 0$, for some $\bar{s} \in S$, then $f(\bar{s}) < f(\bar{x})$.

Note that the first part of the Theorem 9.1 provides an optimality test for a candidate solution $\bar{x}$. The second part suggests a method for improving the candidate solution, if the optimality test (30) fails. This is illustrated in the following example.

Example 9.2 *Consider the problem of minimizing the function $f(x_1, x_2) = -(x - \frac{1}{2})^2 - (x_2 - \frac{1}{4})^2$ on the 0–1 square $\mathbb{B}^2 = \{0,1\}^2$. Since f is concave, this is equivalent to the problem of minimizing f over $S = co(\mathbb{B}^2) = [0,1]^2$. The point $\bar{x} = (0,1)$ is a local minimum point for f over S since it satisfies (30) for every $s \in S$. However, the point $\tilde{x} = (1, \frac{1}{2})$ satisfies $f(\tilde{x}) = f(\bar{x}) = \frac{-5}{16}$ and $\min_{s \in S} s^T \nabla f(\tilde{x}) = (1,1)^T \nabla f(\tilde{x}) < \tilde{x}^T \nabla f(\tilde{x})$. Hence, $\bar{x}$ is not a global minimum point, by the first part of Theorem 9.1. Furthermore, a better solution is provided by the point $\bar{s} = (1,1)$. Since $\bar{s}$ satisfies (30) and the only other point in S satisfying $f(s) = f(\bar{s})$ is $(1,0)$, that also satisfies condition (30), we can conclude that $(1,1)$ and $(1,0)$ are global minimum points for f over S and hence, a fortiori, over $\mathbb{B}^2$.*

The previous example gives the flavor of how global optimality conditions for nonlinear programs can be used for developing global solution methods for continuous and discrete optimization problems. However, we feel that the development of global optimality conditions in nonlinear programming and their application to discrete optimization is still at a fairly early stage and deserves further investigation.

10 Conclusions and Remarks on Further Developments

It is a common belief that in order to find a good solution to a problem it is advisable to look at it from various angles. In this paper we presented several ways of approaching discrete optimization problems from a continuous viewpoint and viceversa. These connections have been already exploited in several efficient solution methods. We now discuss some ideas for further developments in this direction.

Some of the assumptions of Theorem 2.1 might be weakened in the general case or, at least, in special cases such as that of Section 3. In the general case $\mathbb{R}^n$ might be replaced with a metric space. For instance, isoperimetric problems or optimal control problems, where the feasible region has only a discrete or even finite number of curves, might be equivalently transformed into a classic differentiable problem which admits Euler equation as necessary condition. To this end compactness might be usefully replaced with bicompactness, continuity with lower semicontinuity.

In Section 3 we have noted that integer programs can be reduced to (7). However, since such a reduction implies, in general, an enormous increase of the number of variables. Hence, it is interesting to extend the analysis to the case where $x \in \mathbb{B}^n$ is replaced by $x \in \mathbb{Z}^n$. In this case, functions like $\sum_{j=1}^n \sin^2(\pi x_j)$ or $\sum_{j=1}^n \prod_{s=0}^{L_j}(x_j - s)^2$, with L_j upper bound of x_j, might be used as a penalization function $\varphi(x)$. Furthermore, for problem (7) – or, more generally, when Z is finite – it is interesting to investigate other types of functions φ, e.g., in the field of gauge functions.

The idea of resolvent for a Boolean problem [41] might be transferred to a concave minimization problem through the equivalence established in Sections 3 and 4.

Theorem 3.2 states a connection between 0–1 programming problems and complementarity problems which are a special case (when the domain is a cone) of Variational Inequalities. Hence, we have the possibility of connecting two fields very far from each other. For instance, fixed–point methods or gap function methods, which have been designed for Variational Inequalities, may be adapted to 0–1 problems. On the other hand, cutting plane methods and the related group theory, as well as other methods conceived for integer programs, may be transferred to Variational Inequalities.

It is well known that the concept of saddle point of the Lagrangian function associated to a problem like (8) can be used to achieve sufficient optimality conditions (besides necessary ones). Through the equivalence such a concept can be introduced into 0–1 programs.

The equivalence between nonlinear and 0–1 programs can be used also in the context of duality. For instance, it can be used to close the duality gap when Lagrangian duality is applied to a facial constraint [48].

Since the optimality of a constrained extremum problem can be put in terms of the impossibility of a suitable system of inequalities, it might be useful to extend the connections between continuous and discrete problems to a system of inequalities, recovering constrained extremum problems as special cases. This might let us embrace vector extremum problems.

In Section 4 we described general classes of functions which attain their minimum on the vertices of a polyhedron. It is interesting to investigate the possibility of extending the methods employed for concave minimization problems (see, e.g., [4, 40, 54]) to this more general case.

References

[1] F. Alizadeh, Interior point methods in semidefinite programming with applications to combinatorial optimization, *SIAM Journal Optim.* Vol.5 (1995) pp. 13–51.

[2] F. Aurenhammer, Voronoi diagrams – A survey of a fundamental geometric data structure, *ACM Computing Surveys* Vol.23 (1991) pp. 345–405.

[3] D. Avis and B.K. Bhattacharya, Algorithms for computing d–dimensional Voronoi diagrams and their duals, *Adv. Comput. Res.* Vol.1 (1983) pp. 159–180.

[4] H.P. Benson, Concave minimization: theory, applications and algorithms, in R. Horst and P.M. Pardalos (eds.), *Handbook of Global Optimization*, (Boston, Kluwer Academic Publisher, 1995) pp. 43–148.

[5] V. Cabot, R.L. Francis and M.A. Stuart, A network flow solution to a rectilinear distance facility location problem, *AIIE Trans.* Vol.2 (1970) pp. 132–141.

[6] P. Carraresi, F. Farinaccio and F. Malucelli, Testing optimality for quadratic 0–1 problems, Technical Report 11/95 (1995), Dept. of Comp. Sci., Univ. of Pisa.

[7] P. Carraresi, F. Malucelli and M. Pappalardo, Testing optimality for quadratic 0–1 problems, *ZOR Mathematical Methods of Op. Res.* Vol.42 (1995) pp. 295–312.

[8] M. Conforti, G. Cornuéjols, A. Kapoor, and K. Vušković, Perfect, ideal and balanced matrices, In M. Dell'Amico, F. Maffioli and S. Martello (eds.), *Annotated bibliographies in combinatorial optimization*, (New York, Wiley, 1997).

[9] W. Cook, A.M. Gerards, A. Schrijver and É. Tardos, Sensitivity theorems in linear integer programming, *Mathematical Programming* Vol.34 (1986) pp. 251–264.

[10] G. Danninger, Role of copositivity in optimality criteria for nonconvex optimization problems, *Journal of Optimization Problems and Applications* Vol.75 (1992) pp. 535–558.

[11] D.-Z. Du, Minimax and its applications, in R. Horst and P.M. Pardalos (eds.), *Handbook of Global Optimization*, (Boston, Kluwer Academic Publisher, 1995), pp. 339–367.

[12] D.-Z. Du and F.K. Hwang, A proof of Gilbert-Pollak conjecture on the Steiner ratio, *Algoritmica* Vol.7 (1992) pp. 121–135.

[13] D.-Z. Du and P.M. Pardalos (eds.), *Minimax and applications*, (Boston, Kluwer Academic Publisher, 1995).

[14] D.-Z. Du and P.M. Pardalos, A continuous version of a result of Du and Hwang, *Journal of Global Optimization* Vol.5 (1994) pp. 127–130.

[15] N. Dunford and J.T. Schwartz, *Linear operators. Part I*, (New York, Interscience, 1958).

[16] J. Edmonds, Submodular functions, matroids and certain polyhedra, in R. Guy et al. (eds.), *Combinatorial Structures and their Applications (Proceedings Calgary International Conference, 1969)*, (New York, Gordon and Breach, 1970).

[17] E. Erkut and S. Neumann, Analytical models for locating undesirable facilities, *European Journal of Operational Research* Vol.40 (1989) pp. 275–291.

[18] S. Fujishige, *Submodular Functions and Optimization*, (Amsterdam, North Holland, 1991).

[19] P. Gahinet, P. Apkarian and M. Chilali, Parameter-dependent Lyapunov functions for real parametric uncertainty, *IEEE Trans. Automat. Contr.* Vol.41 (1996) pp. 436–442.

[20] P. Favati and F. Tardella, Convexity in nonlinear integer programming, *Ricerca Operativa* Vol.53 (1990) pp. 3–44.

[21] A. Frank and É Tardos, Generalized polymatroids and submodular flows, *Mathematical Programming* Vol.42 (1988) pp. 489–563.

[22] A. Frank, Matroids and submodular functions, In M. Dell'Amico, F. Maffioli and S. Martello (eds.), *Annotated bibliographies in combinatorial optimization*, (New York, Wiley, 1997).

[23] F. Giannessi and F. Niccolucci, Connections between nonlinear and integer programming problems, in *Symposia Mathematica*, Vol. XIX, (London, Academic Press, 1976) pp. 161–176.

[24] F. Giannessi, On some connections among variational inequalities, combinatorial and continuous optimization, *Annals of Operations Research* Vol.58 (1995) pp. 181–200.

[25] E. Girlich and M. Kovalev, Classification of polyhedral matroids, *ZOR Mathematical Methods of Op. Res.* Vol.43 (1996) pp. 143–160.

[26] F. Granot and J. Skorin-Kapov, Some proximity and sensitivity results in quadratic integer programming. *Mathematical Programming* Vol.47 (1990) pp. 259–268.

[27] M. Grötschel, l. Lovász and A. Schrijver, The ellipsoid algorithm and its consequences in combinatorial optimization, *Combinatorica* Vol.1 (1981) pp. 169–197.

[28] S.L. Hakimi, Optimum location of switching centers and the absolute centers and medians of a graph, *Operations Research* Vol.12 (1964) pp. 450–459.

[29] G.Y. Handler and P.B. Mirchandani, *Location on Networks: Theory and Algorithms*, (Cambridge, MIT Press, 1979).

[30] W.W. Hager, P.M. Pardalos, I.M. Roussos, and H.D. Sahinoglou, Active constraints, indefinite quadratic test problems, and complexity, *Journal of Optimization Theory and Applications* Vol. 68 (1991) pp. 499–511.

[31] P. Hansen, Methods of nonlinear 0–1 programming, *Annals of Discrete Math.* Vol.5 (1979) pp. 53–70.

[32] P. Hansen, D. Peeters and J.-F. Thisse, On the location of an obnoxious facility, *Sistemi Urbani* Vol.3 (1981) pp. 299–317.

[33] P. Hansen, B. Jaumard and V. Mathon, Constrained nonlinear 0–1 programming, Rutcor Research Report 47-89 (1989), Rutgers University.

[34] J.-B. Hiriart-Urruty, From convex optimiztion to nonconvex optimization. Part I: necessary and sufficient conditions for global optimality, in R. Horst and P.M. Pardalos (eds.), *Nonsmooth Optimization and Related Topics*, (New York, Plenum Press, 1995) pp. 219–239.

[35] J.-B. Hiriart-Urruty, Conditions for global optimality, in R. Horst and P.M. Pardalos (eds.), *Handbook of Global Optimization*, (Boston, Kluwer Academic Publisher, 1995) pp. 1–26.

[36] J.-B. Hiriart-Urruty and C. Lemarechal, Testing necessary and sufficient conditions for global optimality in the problem of maximizing a convex quadratic function over a convex polyhedron, Tech. Report (1990), Univ. Paul Sabatier of Toulouse, Seminar of Numerical Analysis.

[37] W.M. Hirsch and A.J. Hoffman, Extreme varieties, concave functions and the fixed charge problem, *Communications on Pure and Applied Mathematics* Vol.14 (1961) pp. 355–369.

[38] D.S. Hochbaum, Lower and upper bounds for the allocation problem and other nonlinear optimization problems, *Mathematics of Operations Research* Vol.19 (1994) pp. 309–409.

[39] D.S Hochbaum and J.G. Shanthikumar, Convex separable optimization is not much harder than linear optimization, *J. Assoc. Comput. Mach.* Vol.37 (1990) pp. 843–862.

[40] R. Horst and H. Tuy, *Global Optimization. Deterministic approaches*, (Berlin, Springer-Verlag, 1990).

[41] P.L. Hammer and S. Rudeanu, *Méthods Booléennes en Recherche Opérationelle*, (Paris, Dunod, 1970).

[42] F.K. Hwang and U.G. Rothblum, Directional quasi-convexity, asymmetric Schur-convexity and optimality of consecutive partitions, *Mathematics of Operations Research* Vol.21 (1996) pp. 540–554.

[43] T. Ibaraki and N. Katoh, *Resource Allocation Problems: Algorithmic Approaches*, (Boston, MIT Press, 1988).

[44] B. Kalantari and J.B. Rosen, Penalty for zero–one integer equivalent problem, *Mathematical Programming* Vol.24 (1982) pp. 229–232.

[45] V. Klee, On the complexity of d–dimensional Voronoi diagrams, *Arkiv der Mathematik* Vol.34 (1980) pp. 75–80.

[46] M.K. Kozolov, S.P. Tarasov and L.G. Hačijan, Polynomial solvability of convex quadratic programs, *Soviet Math. Dokl.* Vol.20 (1979) pp. 1108–1111.

[47] M. Labbé, D. Peeters and J.F. Thisse, Location on networks, in M.O. Ball et al. (eds.) *Handbooks in OR and MS* Vol.8, (Amsterdam, Elsevier, 1995) pp. 551–624.

[48] C. Larsen and J. Tind, Lagrangean duality for facial programs with applications to integer and complementarity problems, *Operations Research Letters* Vol.11 (1992) pp. 293–302.

[49] L. Lovász, Submodular functions and convexity, in A. Bachem et al. (eds.), *Mathematical Programming – The State of the Art*, (Berlin, Springer, 1983) pp. 235–257.

[50] P.B. Mirchandani and R.L. Francis (eds.), *Discrete Location Theory*, (New York, Wiley, 1990).

[51] G.L. Nemhauser and L.A. Wolsey, *Integer and Combinatorial Optimization*, (New York, Wiley, 1988).

[52] A. Neumaier, Second-order sufficient optimality conditions for local and global nonlinear programming, *Journal of Global Optimization* Vol.9 (1996) pp. 141–151.

[53] P.M. Pardalos, Continuous approaches to discrete optimization problems, in G. Di Pillo and F. Giannessi (eds.), *Nonlinear Optimization and Applications*, (New York, Plenum Press, 1996) pp. 313–328.

[54] P.M. Pardalos and J.B. Rosen, *Constrained Global Optimization: Algorithms and Applications*, (Berlin, Springer, 1987).

[55] P.M. Pardalos and H. Wolkowicz (eds.), *Topics in Semidefinite and Interior-Point Methods*, Fields Institute Communications Series, American Mathematical Society (in press 1997).

[56] S. Poljak, F. Rendl, and H. Wolkowicz, A Recipe for Best Semidefinite Relaxation for (0,1)-Quadratic Programming, *Journal of Global Optimization* Vol.7 (1995) pp. 51–73.

[57] M. Ragavachari, On the connections between zero-one integer programming and concave programming under linear constraints, *Operations Research* Vol.17 (1969) pp. 680–683.

[58] R.T. Rockafellar, *Convex Analysis*, (Princeton, Princeton University Press, 1970).

[59] A. Schrijver, *Theory of Linear and Integer Programming*, (New York, Wiley, 1986).

[60] H.D. Sherali and W.P. Adams, A hierarchy of relaxations between the continuous and convex hull representations for zero-one programming problems, *SIAM J. Discrete Math.* Vol.3 (1990) pp. 411–430.

[61] H.D. Sherali and W.P. Adams, A hierarchy of relaxations and convex hull characterizations for mixed-integer zero-one programming problems., *Discrete Applied Mathematics* Vol.52 (1994) pp. 83–106.

[62] J. Shi and Y. Yoshitsugu, A D.C. approach to the largest empty sphere problem in higher dimension, in C.A. Floudas and P.M. Pardalos (eds.), *State of the Art in Global Optimization*, (Boston, Kluwer Academic Publisher, 1996) pp. 395–411.

[63] I. Singer, Extension of functions of 0–1 variables and applications to combinatorial optimization, *Numer. Funct. Anal. and Optimiz.* Vol.7 (1984-85) pp. 23–62.

[64] F. Tardella, On the equivalence between some discrete and continuous optimization problems, Rutcor Research Report 30–90 (1990), Rutgers University. Published in *Annals of Operations Research* Vol.25 (1990) pp. 291–300.

[65] F. Tardella, Discretization of continuous location problems, manuscript (1997).

[66] F. Tardella, Piecewise concavity and minimax problems, manuscript (1997).

[67] D.M. Topkis, Minimizing a submodular function on a lattice, *Operations Research* Vol.26 (1978) pp. 305–321.

[68] D.M. Topkis, Adjacency on polymatroids, *Mathematical Programming* Vol.30 (1984) pp. 229–237.

[69] H. Tuy, Concave programming under linear constraints, *Soviet Math. Dokl.* Vol.5 (1964) pp. 1437–1440.

[70] L. Vandenberghe and S. Boyd, Semidefinite Programming, *SIAM Review* Vol.38 (1996) pp. 49-95.

[71] W.I. Zangwill, The piecewise concave function, *Management Science* Vol.13 (1967) pp. 900–912.

# Interior Point Methods for Combinatorial Optimization

John E. Mitchell
Mathematical Sciences
Renssaeler Polytechnic Institute, Troy, NY 12180 USA
E-mail: mitchj@rpi.edu

Panos M. Pardalos
Center for Applied Optimization, ISE Department
University of Florida, Gainesville, FL 32611 USA
E-mail: pardalos@ufl.edu

Mauricio G. C. Resende
Information Sciences Research
AT&T Labs Research, Florham Park, NJ 07932 USA
E-mail: mgcr@research.att.com

Contents

1 Introduction   **191**

2 **Combinatorial optimization**   **191**
   2.1 Examples of combinatorial optimization problems 191
   2.2 Scope and computational efficiency 196

3 **Solution techniques**   **202**
   3.1 Combinatorial approach 202
   3.2 Continuous approach 203
     3.2.1 Examples of embedding 204
     3.2.2 Global approximation 205
     3.2.3 Continuous trajectories 205
     3.2.4 Topological properties 207

4 Interior point methods for linear and network programming   **208**
    4.1   Linear programming 208
    4.2   Network programming 212
    4.3   Components of interior point network flow methods 217
         4.3.1   The dual affine scaling algorithm 217
         4.3.2   Computing the direction 218
         4.3.3   Network preconditioners for conjugate gradient method . . . 221
         4.3.4   Identifying the optimal partition 225
         4.3.5   Recovering the optimal flow 229

5 Branch and bound methods   **233**
    5.1   General concepts 233
    5.2   An example: The QAP 236

6 Branch and cut methods   **242**
    6.1   Interior point cutting plane methods 246
    6.2   Solving the relaxations approximately 249
    6.3   Restarting 251
    6.4   Primal heuristics and termination 252
    6.5   Separation routines 253
    6.6   Fixing variables 253
    6.7   Dropping constraints 254
    6.8   The complete algorithm 254
    6.9   Some computational results 254
    6.10   Combining interior point and simplex cutting plane algorithms . . . 256
    6.11   Interior point column generation methods for other problems 257
    6.12   Theoretical issues and future directions 258

7 Nonconvex potential function minimization   **259**
    7.1   Computing the descent direction 265
    7.2   Some computational considerations 269
    7.3   Application to combinatorial optimization 270

8 A lower bounding technique   **272**

9 Semidefinite Programming Relaxations   **276**

10 Concluding Remarks   **280**

11 Acknowledgement   **282**

    References

1 Introduction

Interior-point methods, originally invented in the context of linear programming, have found a much broader range of applications, including *discrete* problems that arise in computer science and operations research as well as continuous computational problems arising in the natural sciences and engineering. This chapter describes the conceptual basis and applications of interior-point methods for discrete problems in computing.

The chapter is organized as follows. Section 2 explains the nature and scope of combinatorial optimization problems and illustrates the use of interior point approaches for these problems. Section 3 contrasts the combinatorial and continuous approaches for solving discrete problems and elaborates on the main ideas underlying the latter approach. The continuous approach constitutes the conceptual foundation of interior-point methods. Section 4 is dedicated to interior point algorithms for linear and network optimization. Sections 5 and 6 discuss branch-and-bound and branch-and-cut methods based on interior point approaches. Sections 7 and 8 discuss the application of interior point techniques to minimize nonconvex potential functions to find good feasible solutions to combinatorial optimization problems as well as good lower bounds. In Section 9, a brief introduction to semidefinite programming techniques and their application to combinatorial optimization is presented. We conclude the paper in Section 10 by observing the central role played by optimization in both natural and man-made sciences. We provide selected pointers to web sites constaining up-to-date information on interior point methods and their applications to combinatorial optimaization.

2 Combinatorial optimization

In this section, we discuss several examples of combinatorial optimization problems and illustrate the application of interior point techniques to the development of algorithms for these problems.

2.1 Examples of combinatorial optimization problems

As a typical real-life example of combinatorial optimization, consider the problem of operating a flight schedule of an airline at minimum cost. A flight schedule consists of many flights connecting many cities, with specified arrival and departure times. There are several operating constraints. Each plane must fly a round trip route. Each pilot must also fly a round trip route, but not necessarily the same route taken by the plane, since at each airport the pilot can change planes. There are obvious timing constraints interlocking the schedules of pilots, planes and flights. There must be adequate rest built into the pilot schedule and periodic maintenance built into the plane schedule. Only certain crews are qualified to operate certain types of planes. The operating cost consists of many components, some of them more subtle than others. For example, in an imperfect schedule, a pilot may have to fly as a passenger on some flight. This results in lost revenue not only because a passenger seat is taken up but also because the pilot has to be paid even when riding as a passenger. How does one make an operating plan for an airline that minimizes the total cost while meeting all the constraints? A problem of this type is called a *combinatorial optimization* problem, since there are only a finite number of combinations possible, and in principle, one can enumerate all of them, eliminate the ones that do not meet the conditions and among those that do, select the one that has the least operating cost. Needless to say, one needs to be more clever than simple enumeration, due to the vast number of combinations involved.

As another example, consider a communication network consisting of switches interconnected by trunks (e.g. terrestrial, oceanic, satellite) in a particular topology. A telephone call originating in one switch can take many different paths (of switches) to terminate in another switch, using up trunks along the path. The problem is to design a minimum cost network that can carry the expected traffic. After a network is designed and implemented, operating the network involves various other combinatorial optimization problems, e.g. dynamic routing of calls.

As a third example, consider inductive inference, a central problem in artificial intelligence and machine learning. Inductive inference is the process of hypothesizing a general rule from examples. Inductive inference involves the following steps: (i) Inferring rules from examples, finding compact abstract models of data or hidden patterns in the data; (ii) Making predictions based on abstractions; (iii) Learning, i.e. modifying the abstraction based on comparing predictions with actual results; (iv) Designing questions to generate new examples. Consider the first step of the above process, i.e. discovering patterns in data. For example, given the sequence $2, 4, 6, 8, \ldots$, we may ask, "What comes next?" One could pick any number and justify it by fitting a fourth degree polynomial through the 5 points. However, the answer "10" is considered the most "intelligent." That is so because it is based on the first-order polynomial $2n$, which is linear and hence *simpler* than a fourth degree polynomial. The answer to an inductive inference problem is not unique. In inductive inference, one wants a *simple* explanation that fits a given set of observations. Simpler answers are considered better answers. One therefore needs a way to measure simplicity. For example, in finite automaton inference, the number of states could be a measure of simplicity. In logic circuit inference, the measure could be the number of gates and wires. Inductive inference, in fact leads to a discrete optimization problem, where one wants to maximize simplicity, or find a model, or set of rules, no more complex than some specified measure, consistent with already known data.

As a further example, consider the linear ordering problem, an important problem in economics, the social sciences, and also archaeology. In this problem, we are given several objects that we wish to place in order. There is a cost associated with placing object i before object j and a cost for placing object j before object i. The objective is to order the objects to minimize the total cost. There are methods for ranking sports teams that can be formulated as linear ordering problems: if team A beats team B then team A should go ahead of team B in the ranking, but it may be that team B beat team C, who in turn beat team A, so the determination of the "best" ordering is a non-trivial task, usually depending on the margin of victory.

Even though the four examples given above come from four different facets of life and look superficially to be quite different, they all have a common mathematical structure and can be described in a common mathematical notation called *integer programming*. In integer programming, the unknowns are represented by variables that take on a finite or discrete set of values. The various constraints or conditions on the problem are captured by algebraic expressions of these variables. For example, in the airline crew assignment problem discussed above, let us denote by variable x_{ij} the decision quantity that assigns crew i to flight j. Let there be m crew and n flights. If the variable x_{ij} takes on a value 1, then we say that crew i is assigned to flight j, and the cost of that assignment is c_{ij}. If the value is 0, then crew i is *not* assigned to flight j. Thus, the total crew-scheduling cost for the airline is given by the expression

$$\sum_{i=1}^{m} \sum_{j=1}^{n} c_{ij} x_{ij} \qquad (1)$$

that must be minimized. The condition that every flight should have exactly one crew is expressed by the equations

$$\sum_{i=1}^{m} x_{ij} = 1, \text{ for every flight } j = 1, \ldots, n. \qquad (2)$$

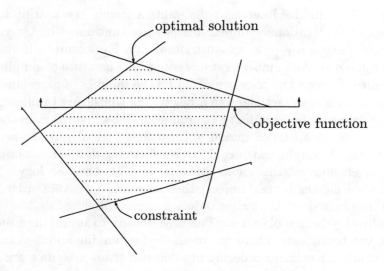

Figure 1: Geometric view of linear programming

We should also stipulate that the variables should take on only values 0 or 1. This condition is denoted by the notation

$$x_{ij} \in \{0,1\}, \ 1 \leq i \leq m; \ 1 \leq j \leq n. \tag{3}$$

Other conditions on the crew can be expressed in a similar fashion. Thus, an integer programming formulation of the airline crew assignment problem is to minimize the operating cost given by (1) subject to various conditions given by other algebraic equations and inequalities. The formulations of the network design problem, the inductive inference problem, as well as the linear operdering problem, look mathematically similar to the above problem.

Linear programming is a special and simpler type of combinatorial optimization problem in which the integrality constraints of the type (3) are absent and we are given a linear objective function to be minimized subject to linear inequalities and equalities. A standard form of linear program is stated as follows:

$$\min_{x \in \mathrm{R}^n} \{c^\top x | Ax \leq b; \ l \leq x \leq u\}, \tag{4}$$

where $c, u, l, x \in \mathrm{R}^n, b \in \mathrm{R}^m$ and $A \in \mathrm{R}^{m \times n}$. In (4), x is the vector of decision variables, $Ax \leq b$ and $l \leq x \leq u$ represent constraints on the decision variables, and $c^\mathsf{T} x$ is the linear objective function to be minimized. Figure 1 shows a geometric interpretation of a linear program on the Euclidean plane. Each linear inequality is represented by a line that partitions the plane into two *half-spaces*. Each inequality requires that for a solution to be feasible, it must lie in one of the half-spaces. The feasible region is the intersection of the half-spaces and is represented in the figure by the hashed area. The objective function, that must be minimized over the feasible region, is represented by a sliding line. This line intersects the feasible region in a set of points, all having the same objective value. As the line is swept across the feasible region in the direction of improvement, its objective value decreases. The set of points determined by the intersection of the sliding line with the feasible region that attains the best objective function value is called the *optimal* solution. In the example of the figure there is a unique optimal solution. In fact, a fundamental theorem in linear programming states that the optimal solution of a linear program occurs at a vertex of the polytope defined by the constraints of the linear program. This result gives linear programming its combinatorial nature. Even though the linear programming decision variables are continuous in nature, this result states that only a discrete and finite number of points in the solution space need to be examined.

Linear programming has a wide range of applications, including personnel assignment, production planning and distribution, refinery planning, target assignment, medical imaging, control systems, circuit simulation, weather forecasting, signal processing and financial engineering. Many polynomial-time solvable combinatorial problems are special cases of linear programming (e.g. matching and maximum flow). Linear programming has also been the source of many theoretical developments, in fields as diverse such as economics and queueing theory.

Combinatorial problems occur in diverse areas. These include graph theory (e.g. graph partitioning, network flows, graph coloring), linear inequalities (e.g. linear and integer programming), number theory (e.g. factoring, primality testing, discrete logarithm), group theory (e.g. graph isomorphism, group intersection), lattice theory (e.g. basis reduction), and logic and artificial intelligence (e.g. satisfiability, inductive and deductive inference boolean function minimization). All these problems, when abstracted mathematically have a commonality of discreteness. The solution approaches for solving these problems also have a great deal in common. In fact, attempts to come up with solution techniques revealed more commonality of the problems than was revealed from just the problem formulation. The solution of combinatorial problems has been the subject of much research. There is a continuously evolving body of knowledge, both theoretical and practical, for solving these problems.

2.2 Scope and computational efficiency

We illustrate with some examples the broad scope of applications of the interior-point techniques and their computational effectiveness. Since the most widely applied combinatorial optimization problem is linear programming, we begin with this problem. Each step of the interior-point method as applied to linear programming involves the solution of a linear system of equations. While a straightforward implementation of solving these linear systems can still outperform the Simplex Method, more sophisticated implementations have achieved orders of magnitude improvement over the Simplex Method. These advanced implementations make use of techniques from many disciplines such as linear algebra, numerical analysis, computer architecture, advanced data structures, and differential geometry. Tables 1–2 show the performance comparison between implementations of the Simplex (CPLEX) and interior-point (ADP [79]) methods on a class of linear programming relaxations of the quadratic assignment problems [127, 122]. Similar relative performances have been observed in problems drawn from disciplines such as operations research, electrical engineering, computer science, and statistics [79]. As the table shows, the relative superiority of interior-point method over the Simplex Method grows as the problem size grows and the speed-up factor can exceed 1000. Larger problems in the table could only be solved by the interior-point method because of impracticality of running the Simplex Method. In fact, the main practical contribution of the interior-point method has been to enable the solution of many large-scale real-life problems in fields such as telecommunication, transportation and defense, that could not be solved earlier by the Simplex Method.

From the point of view of efficient implementation, interior-point methods have another important property: they can exploit parallelism rather well [60, 131]. A parallel architecture based on multi-dimensional finite projective geometries, particularly well suited for interior-point methods, has been proposed [75].

We now illustrate computational experience with an interior point based heuristic for integer programming [74, 77]. Here again, the main computational task at each iteration, is the solution of one or more systems of linear equations. These systems have a structure similar to the system solved in each iteration of interior point algorithms for linear programming

Table 1: LP relaxations of QAP integer programming formulation

name	LP relaxation rows	LP relaxation vars	simplex itr	simplex time	int. pt. itr	int. pt. time	time ratio
nug05	210	225	103	0.2s	14	1.6s	0.1
nug06	372	486	551	2.3s	17	2.6s	0.9
nug07	602	931	2813	22.0s	19	6.2s	3.5
nug08	912	1632	5960	91.3s	18	9.5s	9.6
nug12	3192	8856	57524	9959.1s	29	754.1s	13.2
nug15	6330	22275	239918	192895.2s	36	5203.8s	37.1
nug20	15240	72600	est. time: > 2 months		31	6745.5s	-
nug30	52260	379350	did not run		36	35058.0s	-

Table 2: CPLEX 3.0 and ADP runs on selected QAPLIB instances

prob	LP relaxation rows	LP relaxation vars	primal simplex itr	primal simplex time	dual simplex itr	dual simplex time	ADP itr	ADP time
nug05	1410	825	265	1.7s	370	1.1s	48	3.2s
nug06	3972	2886	7222	604.3s	1872	22.2s	55	12.2s
nug07	9422	8281	39830	47970.3s	6057	720.3s	59	43.3s
nug08	19728	20448	did not run		16034	37577.1s	63	139.1s
nug12	177432	299256	did not run		did not run		91	6504.2s

and therefore software developed for linear programming can be reused in integer programming implementations.

Consider, as a first example, the Satisfiability (SAT) Problem in propositional calculus, a central problem in mathematical logic. During the last decade, a variety of heuristics have been proposed for this problem [61, 30]. A Boolean variable x can assume only values 0 or 1. Boolean variables can be combined by the logical connectives **or** ($\vee$), **and** ($\wedge$) and **not** ($\bar{x}$) to form Boolean formulae (e.g. $x_1 \wedge \bar{x}_2 \vee x_3$). A variable or a single negation of the variable is called a *literal*. A Boolean formula consisting of only literals combined by the $\vee$ operator is called a *clause*. SAT can be stated as follows: Given m clauses $C_1, \ldots, C_m$ involving n variables $x_1, \ldots, x_n$, does the formula $C_1 \wedge \cdots \wedge C_m$ evaluate to 1 for some Boolean input vector $[x_1, \ldots, x_n]$? If so, the formula is said to be satisfiable. Otherwise it is unsatisfiable.

SAT can be formulated as the integer programming feasibility problem

$$\sum_{j \in \mathcal{I}_C} x_j - \sum_{j \in \mathcal{J}_C} x_j \geq 1 - |\mathcal{J}_c|, \ C = C_1, \ldots, C_m, \tag{5}$$

where

$$\mathcal{I}_C = \{j \mid \text{literal } x_j \text{ appears in clause } C\}$$
$$\mathcal{J}_C = \{j \mid \text{literal } \bar{x}_j \text{ appears in clause } C\}.$$

If an integer vector $x \in \{0, 1\}^n$ is produced satisfying (5), the corresponding SAT problem is said to be satisfiable.

An interior point implementation was compared with an approach based on the Simplex Method to prove satisfiability of randomly generated instances of SAT [65]. Instances with up to 1000 variables and 32,000 clauses were solved. Compared with the Simplex Method approach on small problems (Table 3), speedups of over two orders of magnitude were observed. Furthermore, the interior point approach was successful in proving satisfiability in over 250 instances that the Simplex Method approach failed.

As a second example, consider inductive inference. The interior point approach was applied to a basic model of inductive inference [68]. In this model there is a black box (Figure 2) with n Boolean input variables $x_1, \ldots, x_n$ and a single Boolean output variable y. The black box contains a hidden Boolean function $\mathcal{F} : \{0, 1\}^n \to \{0, 1\}$ that maps inputs to outputs. Given a limited number of inputs and corresponding outputs, we ask: Does there exist an algebraic sum-of-products expression with no more than K product terms that matches this behavior? If so, what is it? It turns out that this problem can be formulated as a SAT problem.

Table 3: SAT: Comparison of Simplex and interior point methods

SAT Problem Size			Speed
Variables	Clauses ($|\mathcal{C}|$)	Avg Lits/Clause	Up
50	100	5	5
100	200	5	22
200	400	7	66
400	800	10	319

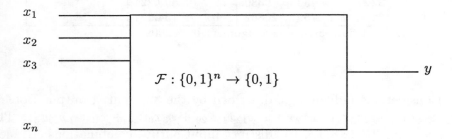

Figure 2: Black box with hidden logic

Table 4: Inductive inference SAT problems: 32-variable hidden logic

I/O Samples	SAT Size Vars	$|\mathcal{C}|$	itr	CPU time	Inferred Logic	Prediction Accuracy
50	332	2703	49	66s	$y = \bar{x}_{22}x_{28}\bar{x}_{29} + x_{12}\bar{x}_{17}\bar{x}_{25}x_{27} +$ $\bar{x}_3 x_9 x_{20} + x_{11}x_{12}\bar{x}_{16}\bar{x}_{32}$	.74
100	404	5153	78	178s	$y = x_9 x_{11}\bar{x}_{22}\bar{x}_{29} + x_4 x_{11}\bar{x}_{22} +$ $\bar{x}_3 x_9 x_{20} + x_{12}\bar{x}_{15}\bar{x}_{16}\bar{x}_{29}$	.91
400	824	19478	147	1227s	$y = x_4 x_{11}\bar{x}_{22} + \bar{x}_{10}x_{11}\bar{x}_{29}x_{32} +$ $\bar{x}_3 x_9 x_{20} + x_2 x_{12}\bar{x}_{15}\bar{x}_{29}$	exact

Table 5: Efficiency on inductive inference problems: interior point and combinatorial approaches

Variables Hidden Logic	SAT Problem vars	$|\mathcal{C}|$	Interior Method itr	time	Combinatorial Method (time)
8	396	2798	1	9.33s	43.05s
8	930	6547	13	45.72s	11.78s
8	1068	8214	33	122.62s	9.48s
16	532	7825	89	375.83s	20449.20s
16	924	13803	98	520.60s	*
16	1602	23281	78	607.80s	*
32	228	1374	1	5.02s	159.68s
32	249	2182	1	9.38s	176.32s
32	267	2746	1	9.76s	144.40s
32	450	9380	71	390.22s	*
32	759	20862	1	154.62s	*

* Did not find satisfiable assignment in 43200s.

Consider the hidden logic described by the 32-input, 1-output Boolean expression $y = x_4 x_{11} x_{15} \bar{x}_{22} + x_2 x_{12} \bar{x}_{15} \bar{x}_{29} + \bar{x}_3 x_9 x_{20} + \bar{x}_{10} x_{11} \bar{x}_{29} x_{32}$. This function has $2^{32} \simeq 4.3 \times 10^9$ distinct input-output combinations. Table 4 summarizes the computational results for this instance, where subsets of input-output examples of size 50, 100 and 400 were considered and the number of terms in the expression to be synthesized was fixed at $K = 4$. In all instances, the interior point algorithm synthesized a function that described completely the behavior of the sample. With a sample of only 400 input-output patterns the approach succeeded in exactly describing the hidden logic. The prediction accuracy given in the table was computed with Monte Carlo simulation, where 10,000 random vectors were input to the black box and to the inferred logic and their outputs compared. Table 5 illustrates the efficiency of the interior-point method compared to the combinatorial Davis-Putnam Method [24].

As another example of an application of the continuous approach to combinatorial problems, consider the wire routing problem for gate arrays, an important subproblem arising in VLSI design. As shown in Figure 3, a gate array can be abstracted mathematically as a grid graph. Input to the wire routing problem consists of a list of wires specified by end points on a rectangular grid. Each edge of the graph, also known as a channel, has a pre-

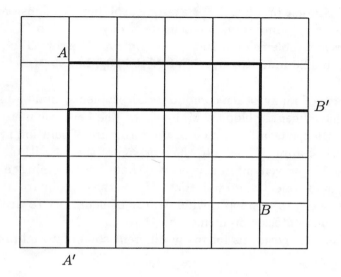

Figure 3: Wire routing

specified capacity representing the maximum number of wires it can carry. The combinatorial problem is to find a wiring pattern without exceeding capacity of horizontal and vertical channels. This problem can be formulated as an integer programming problem. The interior-point approach has successfully obtained provably optimal global solutions to large-scale problems of this type having more than 20,000 wires [109]. On the other hand, combinatorial heuristics, such as simulated annealing are not comparable either in the quality of the solution they can find or in terms of the computational cost.

A further example of the successful application of interior point methods to solve combinatorial optimization problems comes from statistical physics. The problem of finding the ground state of an Ising spin glass is related to the magnetism of materials. Finding the ground state can be modelled as the problem of finding the maximum cut in a graph whose vertices and edges are those of a grid on a torus. It can be formulated as an integer programming problem and solved using a cutting plane approach. If the weights on the edges are ± 1 then the linear programs suffer from degeneracy, which limits the size of problems that can be solved efficiently using the Simplex Method. The use of an interior point algorithm to solve the relaxations allows the solution of far larger problems. For example, solving problems on a 70 × 70 toroidal grid using simplex required up to a day on a Sun SPARCstation 10 [26], whereas problems on a 100 × 100 grid could be solved in an average of about 3 hours 20 minutes on a Sun SPARC 20/71 when using an interior point code [98].

In the case of many other combinatorial problems, numerous heuristic approaches have been developed. Many times, the heuristic merely encodes the prior knowledge or anticipation of the structure of solution to a specific class of practical applications into the working of the algorithm. This may make a limited improvement in efficiency without really coming to grips with the problem of exponential growth that plagues the combinatorial approach. Besides, one needs to develop a wide variety of heuristics to deal with different situations. Interior-point methods have provided a unified approach to create efficient algorithms for many different combinatorial problems.

3 Solution techniques

Solution techniques for combinatorial problems can be classified into two groups: combinatorial and continuous approaches. In this section, we contrast these approaches.

3.1 Combinatorial approach

The combinatorial approach creates a sequence of states drawn from a discrete and finite set. Each state represents a suboptimal solution or a partial solution to the original problem. It may be a graph, a vertex of a polytope, a collection of subsets of a finite set or some other combinatorial object. At each major step of the algorithm, the next state is chosen in an attempt to improve the current state. The improvement may be in the quality of the solution measured in terms of the objective function, or it may be in making the partial solution more feasible. In any case, the improvement is guided by *local* search. By local search we mean that the solution procedure only examines a neighboring set of configurations and greedily selects one that improves the current solution. Thus, local search is quite myopic, with no consideration given to evaluate whether this move may make any sense globally. Indeed, a combinatorial approach often lacks the information needed for making such an evaluation. In many cases, the greedy local improvement may trap the solution in a local minimum that is qualitatively much worse than a true global minimum. To escape from a local minimum, the combinatorial approach needs to resort to techniques such as backtracking or abandoning the sequences of states created so far altogether and restarting with a different initial state. Most combinatorial problems suffer from the property of having a large number of local minima when the search space is confined to a discrete set. For a majority of combinatorial optimization problems, the phenomenon of multiple local minima may create a problem for the combinatorial approach.

On the other hand, for a limited class of problems, one can rule out the possibility of local minima and show that local improvement also leads to global improvement. For many problems in this class, polynomial-time algorithms (i.e. algorithms whose running time can be proven to be bounded from above by polynomial functions of the lengths of the problems) have been known for a long time. Examples of problems in this class are bipartite matching and network flows. It turns out that many of these problems are special cases of linear programming, which is also a polynomial-time problem. However, the Simplex Method, which employs a combinatorial approach to solving linear programs, has been shown to be an exponential-time algorithm. In contrast, all polynomial-time algorithms for solving the general linear programming problem employ a continuous approach. These algorithms use either the Ellipsoid Method [81] or one of the variants of the Karmarkar Method.

3.2 Continuous approach

In the continuous approach to solving discrete problems, the set of candidate solutions to a given combinatorial problem is embedded in a larger continuous space. The topological and geometric properties of the continuous space play an essential role in the construction of the algorithm as well as in the analysis of its efficiency. The algorithm involves the creation of a sequence of points in the enlarged space that converges to the solution of the original combinatorial problem. At each major step of the algorithm, the next point in the sequence is obtained from the current point by making a good *global* approximation to the entire set of relevant solutions and solving it. Usually it is also possible to associate a continuous trajectory or a set of trajectories with the limiting case of the discrete algorithm obtained by taking infinitesimal steps. Topological properties of the underlying continuous space such as connectivity of the level sets of the function being optimized are used for bounding the number of local minima and choosing an effective formulation

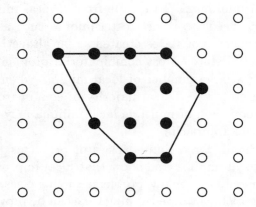

Figure 4: Discrete set embedded into continuous set

of the continuous optimization problem. The geometric properties such as distance, volume and curvature of trajectories are used for analyzing the rate of convergence of an algorithm, whereas the topological properties help determine if a proposed algorithm will converge at all. We now elaborate further on each of the main concepts involved in the continuous approach.

3.2.1 Examples of embedding

Suppose the candidate solutions to a discrete problem are represented as points in the n-dimensional real space R^n. This solution set can be embedded into a larger continuous space by forming the convex hull of these points (Figure 4). This is the most common form of continuous embedding and is used for solving linear and integer programming problems. As another example, consider a discrete problem whose candidate solution set is a finite cyclic group. This can be embedded in a continuous Lie group $\{e^{i\theta}|0 \leq \theta < 2\pi\}$ [20]. A Lie group embedding is useful for the problem of graph isomorphism or automorphism. In this problem, let A denote the adjacency matrix of the graph. Then the discrete solution set of the automorphism problem is the permutation group given by $\{P|AP = PA;\ P$ is a permutation matrix $\}$. This can be embedded in a larger continuous group given by $\{U|AU = UA;\ U$ is a complex unitary matrix$\}$.

3.2.2 Global approximation

At each major step of the algorithm, a subproblem is solved to obtain the next point in the sequence. The subproblem should satisfy two properties: (i) it should be a *global* approximation to the original problem; and (ii) should be efficiently solvable. In the context of linear programming, the Karmarkar Method contains a way of making a global approximation having both of the above desirable properties and is based on the following theorem.

Theorem 3.1 *Given any polytope P and an interior point $\mathbf{x} \in P$, there exists a projective transformation T, that transforms P to P' and $\mathbf{x}$ to $\mathbf{x}' \in P'$ so that it is possible to find in the transformed space a circumscribing ball $B(\mathbf{x}', R) \supseteq P'$, of radius R, center $\mathbf{x}'$, containing P' and a inscribing ball $B(\mathbf{x}', r) \subseteq P'$ of radius r, center $\mathbf{x}'$ contained in P' such that the ratio R/r is at most n.*

The inverse image (under T) of the inscribed ball is used as the optimization space for the subproblem and satisfies the two properties stated above, leading to a polynomial-time algorithm for linear programming. The effectiveness of this global approximation is also borne out in practice since Karmarkar's Method and its variants take very few approximation steps to find the global optimum of the original problem. Extension of this global approximation step to integer programming have led to new algorithms for solving NP-complete problems, as we shall see later.

3.2.3 Continuous trajectories

Suppose an interior-point method produces an iteration of the following type,

$$\mathbf{x}^{(k+1)} \leftarrow \mathbf{x}^{(k)} + \alpha \mathbf{f}^{(k)} + \mathcal{O}(\alpha^2), \tag{6}$$

where $\mathbf{x}^{(k)}$ is the k-th iterate, $\mathbf{f}^{(k)}$ is the k-th direction of improvement, and α is the step-length parameter. Then by taking the limit as $\alpha \to 0$, we get the *infinitesimal version* of the algorithm whose continuous trajectories are given by the differential equation

$$\frac{d\mathbf{x}}{d\alpha} = \mathbf{f}(\mathbf{x}), \tag{7}$$

where $\mathbf{f}(\mathbf{x})$ defines a vector field. Thus, the infinitesimal version of the algorithm can be thought of as a nonlinear dynamical system. For the projective method for linear programming, the differential equation is given by

$$\frac{d\mathbf{x}}{dt} = -[D - \mathbf{x}\mathbf{x}^T]P_{AD} \cdot D\mathbf{c},$$

where

$$P_{AD} = I - DA^T(AD^2A^T)^{-1}AD,$$

$$D = \text{diag}\{x_1, x_2, \cdots, x_n\}.$$

Similarly, continuous trajectories and the corresponding differential equations can be derived for other interior-point methods. These trajectories have a rich mathematical structure in them. Many times they also have algebraic descriptions and alternative interpretations. The continuous trajectory given above for the linear programming problem converges to an optimal solution of the problem corresponding to the objective function vector **c**. Note that the vector field depends on **c** in a smooth way and as the vector **c** is varied one can get to each vertex of the polytope as limit of some continuous trajectory. If one were to attempt a direct combinatorial description of the discrete solution set of a linear programming problem, it would become enormously complex since the number of solutions can be exponential with respect to the size of the problem. In contrast, the simple differential equation given above implicitly encodes the complex structure of the solution set. Another important fact to be noticed is that the differential equation is written in terms of the *original* input matrix A defining the problem. Viewing combinatorial objects as limiting cases of continuous objects often makes them more accessible to mathematical reasoning and also permits construction of more efficient algorithms.

The power of the language of differential equations in describing complex phenomena is rather well-known in the natural sciences. For example, if one were to attempt a direct description of the trajectories involved in planetary motion, it would be enormously complex. However, a small set of differential equations written in terms of the *original* parameters of the problem is able to describe the same motion. One of the most important accomplishments of Newtonian mechanics was finding a *simple* description of the apparently complex phenomena of planetary motion, in the form of differential equations.

In the context of combinatorial optimization, the structure of the solution set of a discrete problem is often rather complex. As a result, a straightforward combinatorial approach to solving these problems has not succeeded in many cases and has led to a belief that these problems are intractable.

Even for linear programming, which is one of the simplest combinatorial optimization problems, the best known method, in both theory and practice, is based on the continuous approach rather than the combinatorial approach. Underlying this continuous approach is a small set of differential equations, capable of encoding the complicated combinatorial structure of the solution set. As this approach is extended and generalized, one hopes to find new and efficient algorithms for many other combinatorial problems.

3.2.4 Topological properties

There are many ways to formulate a given discrete problem as a continuous optimization problem, and it is rather easy to make a formulation that would be difficult to solve even by means of continuous trajectories. How does one make a formulation that is solvable? The most well-known class of continuous solvable problems is the class of convex minimization problems. This leads to a natural question: Is convexity the characteristic property that separates the class of efficiently solvable minimization problems from the rest? To explore this question we need to look at topological properties. Topology is the study of properties invariant under any continuous, one-to-one transformation of space having a continuous inverse.

Suppose we have a continuous optimization problem that is solvable by means of continuous trajectories. It may be a convex problem, for example. Suppose we apply a nonlinear transformation to the space that is a diffeomorphism. The transformed problem need not be convex, but it will continue to be solvable by means of continuous trajectories. In fact, the image of the continuous trajectories in the original space, obtained by applying the *same* diffeomorphism gives us a way of solving the transformed problem. Conversely, if the original problem was unsolvable, it could not be converted into a solvable problem by any such transformation. Hence any diffeomorphism maps solvable problems onto solvable problems and unsolvable problems onto unsolvable problems. This argument suggests that the property characterizing the solvable class may be a topological property and not simply a geometric property such as convexity.

The simplest topological property relevant to the performance of interior-point methods is connectivity of the level sets of the function being optimized. Intuitively, a subset of continuous space is connected if any two points of the subset can be joined by a continuous path lying entirely in the subset. In the context of function minimization, the significance of connectivity lies in the fact that functions having connected level sets do not have spurious local minima. In other words every local minimum is necessarily a global minimum. A continuous formulation of NP-complete problems having such desirable topological properties is given in [74]. The approach described there provides a theoretical foundation for constructing efficient algorithms for discrete problems on the basis of a common principle. Algorithms for many practical problems can now be developed which differ mainly in the way the combinatorial structure of he problem is exploited to gain additional computational efficiency.

4 Interior point methods for linear and network programming

4.1 Linear programming

Interior point methods were first described by Dikin [29] in the mid 1960s, and current interest in them started with Karmarkar's algorithm in the mid 1980s [72]. As the name suggests, these algorithms generate a sequence of iterates which moves through the relative interior of the feasible region, in marked contrast to the simplex method [22], where each iterate is an extreme point. Like the ellipsoid method [81], many interior point methods have polynomial complexity, whereas every known variant of the simplex method can take an exponential number of iterations in the worst case. Computationally, interior point methods usually require far less time than their worst-case bounds, and they appear to be superior to the simplex method, at least for problems with a large number of constraints and variables (say, more than one thousand). Recent books discussing interior point methods for linear programming include [132, 140, 146, 148].

The dual affine scaling method is similar to Dikin's original method and is discussed in section 4.3.1. In this section, we consider a slightly more complicated interior point method, namely the primal-dual predictor-corrector method PDPCM [92, 90]. This is perhaps the most popular and widely implemented interior point method. The basic idea with an interior point method is to enable the method to take long steps, by choosing directions that do not immediately run into the boundary. With the PDPCM this is achieved by considering a modification of the original problem, with a penalty term for approaching the boundary. Thus, for the standard form linear programming problem

$$\begin{array}{ll} \min & c^T x \\ \text{subject to} & Ax = b \\ & x \geq 0 \end{array}$$

where c and x are n-vectors, b is an m-vector, and A is dimensioned appropriately, the barrier function subproblem is constructed:

$$\begin{array}{ll} \min & c^T x - \mu \sum_{i=1}^{n} \log(x_i) \\ \text{subject to} & Ax = b \\ & x \geq 0 \end{array}$$

Here, log denotes the natural logarithm, and μ is a positive constant. Note that if x_i approaches zero for any component, then the objective function value approaches ∞.

If the original linear program has a compact set of optimal solutions then each barrier subproblem will have a unique optimal solution. The set of all these optimal solutions is known as the *central trajectory*. The limit point of the central trajectory as μ tends to zero is an optimal point for the original linear program. If the original linear program has more than one optimal solution, then the limit point is in the relative interior of the optimal face.

Fiacco and McCormick [34] suggested following the central trajectory to the optimal solution. This requires solving a barrier subproblem for a particular choice of μ, decreasing μ, and repeating. The hope is that knowing the solution to one subproblem will make it easy to solve the next one. It also suffices to only solve the subproblems approximately, both theoretically and practically. Monteiro and Adler [102] showed that if μ is decreased by a sufficiently small amount then an approximate solution to one subproblem can be used to obtain an approximate solution to the next one in just one iteration, leading to an algorithm that requires $O(n^{1/2})$ iterations. With a more aggressive reduction in μ (for example, μ is halved at each iteration), more iterations are required to obtain an approximate solution to the new subproblem, and the best complexity result that has been proved for such algorithms is that they require $O(n)$ iterations.

There are several issues that need to be resolved in order to specify the algorithm, including the choice of values of μ, methods for solving the subproblems, and the desired accuracy of the solutions of the subproblems. Many different choices have been investigated, at least theoretically. Much of the successful computational work has focussed on the PDPCM, and so we now describe that algorithm.

The optimality conditions for the subproblems can be written:

$$Ax = b \tag{8}$$
$$A^T y + z = c \tag{9}$$
$$XZe = \mu e \tag{10}$$

where e denotes the vector of all ones of appropriate dimension, y is an m-vector, z is a nonnegative n-vector, and X and Z are $n \times n$ diagonal matrices with $X_{ii} = x_i$ and $Z_{ii} = z_i$ for $i = 1, \ldots, n$. Equation (9) together with the nonnegativity restriction on z corresponds to dual feasibility. Notice that the last condition is equivalent to saying that $x_i z_i = \mu$ for each component i. Complementary slackness for the original linear program would require $x_i z_i = 0$ for each component i. The duality gap is $x^T z$, so if a point satisfies the optimality conditions (8–10) then the duality gap will be $n\mu$.

We assume we have a strictly positive feasible solution x and a dual feasible solution (y, z) satisfying $A^T y + z = c$, $z > 0$. If these assumptions are not satisfied, the algorithm can be modified appropriately; see, for example, [90] or Zhang [150].

An iteration of PDPCM consists of three parts:

- A Newton step to move towards the solution of the linear program is calculated (but not taken). This is known as the *predictor step*.

- This predictor step is used to update μ.

- A *corrector step* is taken, which combines the decrease of the predictor step with a step that tries to stay close to the central trajectory.

The predictor step gives an immediate return on the value of the objective function, and the corrector step brings the iterate back towards the central trajectory, making it easier to obtain a good decrease on future steps.

The calculation of the predictor step requires solving the following system of equations to obtain the search directions $\Delta^p x$, $\Delta^p y$, and $\Delta^p z$:

$$A\Delta^p x = 0 \tag{11}$$
$$A^T \Delta^p y + \Delta^p z = 0 \tag{12}$$
$$Z\Delta^p x + X\Delta^p z = -XZe. \tag{13}$$

One method to solve this system is to notice that it requires

$$AZ^{-1}XA^T \Delta^p y = -AXe. \tag{14}$$

The Cholesky factorization of the matrix $AZ^{-1}XA^T$ can be calculated, and this can be used to obtain $\Delta^p y$. The vectors $\Delta^p z$ and $\Delta^p x$ can then be calculated from the equations (12) and (13).

It can be shown that if we choose a step of length α then the duality is reduced by a factor of α, so if we took a step of length one then the duality gap would be reduced to zero. However, it is not usually possible to take such a large step and still remain feasible. Thus, we calculate α_P^p and α_D^p, the maximum possible step lengths to maintain primal and dual feasibility.

We use these steplengths to aid in the adaptive update of μ. If the steplengths are close to one, then the duality gap can be decreased dramatically, and the new μ should be considerably smaller than the old value. Conversely, if the steplengths are short then the iterates are close to the boundary, so μ should only be decreased slightly and the iterates should be pushed back towards the central trajectory. This leads to one possible update of μ as

$$\mu^+ = (g_p/x^T z)^2 g_p/n \qquad (15)$$

where g_p is the duality gap that would result if primal and dual steps of lengths α_P^p and α_D^p, respectively, were taken.

A corrector step $(\Delta x, \Delta y, \Delta z)$ is used to bring the iterates back towards the central trajectory. This involves solving the system of equations

$$A\Delta x = 0 \qquad (16)$$
$$A^T \Delta y + \Delta z = 0 \qquad (17)$$
$$Z\Delta x + X\Delta z = \mu^+ e - XZe - v^p \qquad (18)$$

where v^p is an n-vector, with components $v_i^p = \Delta^p x_i \Delta^p z_i$. This system can be solved using the Cholesky factors of the matrix $AZ^{-1}XA^T$, which were already formed when calculating the predictor step.

Once the direction has been calculated, the primal and dual step lengths α_P and α_D are chosen to ensure that the next iterate has $x^+ > 0$ and $z^+ > 0$. The iterates are updated as:

$$x^+ = x + \alpha_P \Delta x \qquad (19)$$
$$y^+ = y + \alpha_D \Delta y \qquad (20)$$
$$z^+ = z + \alpha_D \Delta z. \qquad (21)$$

Typically, α_P and α_D are chosen to move the iterates as much as 99.95% of the way to the boundary.

The predictor-corrector method [103, 92] can be thought of as finding a direction by using a second order approximation to the central trajectory. A second order solution to (8–10) would require that the direction satisfy

$$Z\Delta x + X\Delta z + v = \mu^+ e - XZe$$

where v is a vector to be determined with $v_i = \Delta x_i \Delta z_i$. It is not possible to easily solve this equation together with (16–17), so it is approximated by the system of equations (16–18).

The method makes just one iteration for each reduction in μ, but typically μ is decreased very quickly, by perhaps at least a constant factor at each iteration. The duality gap usually close to $n\mu$, so the algorithm is terminated when the duality gap drops below some tolerance. The algorithm typically takes only 40 or so iterations even for problems with thousands of constraints and/or variables.

The computational work in an iteration is dominated by the factorization of the matrix $AZ^{-1}XA^T$. The first step in the factorization is usually to permute the rows of A to reduce the number of nonzeroes in the Cholesky factors — this step need only be performed once in the algorithm. Once the ordering is set, a numerical factorization of the matrix is performed at each iteration. These factors are then used to calculate the directions by means of backward and forward substitutions. The use of the correction direction was found to decrease the number of iterations required to solve a linear program by enough to justify the extra work at each iteration of calculating an extra direction. Higher order approximations [13, 14, 3, 76] have proven useful for some problems where the cost of the factorization is far greater than the cost of backward and forward substitutions — see [148]. For a discussion of the computational issues involed in implementing an interior point algorithm, see, for example, Adler et al. [3] and Andersen *et al.* [7].

4.2 Network programming

A large number of problems in transportation, communications, and manufacturing can be modeled as network flow problems. In these problems one seeks to find the most efficient, or optimal, way to move flow (e.g. materials, information, buses, electrical currents) on a network (e.g. postal network, computer network, transportation grid, power grid). Among these optimization problems, many are special classes of linear programming problems, with combinatorial properties that enable development of efficient solution techniques. In this section, we limit our discussion to these linear network

Interior Point Methods for Combinatorial Optimization

flow problems. For a treatment of classes of nonlinear network flow problems, the reader is referred to [31, 50, 51, 114] and references therein.

Given a directed graph $G = (\mathcal{N}, \mathcal{A})$, where $\mathcal{N}$ is a set of m nodes and $\mathcal{A}$ a set of n arcs, let (i,j) denote a directed arc from node i to node j. Every node is classified in one of the following three categories. *Source* nodes produce more flow than they consume. *Sink* nodes consume more flow than they produce. *Transshipment* nodes produce as much flow as they consume. Without loss of generality, one can assume that the total flow produced in the network equals the total flow consumed. Each arc has associated with it an origination node and a destination node, implying a direction for flow to follow. Arcs have limitations (often called capacities or bounds) on how much flow can move through them. The flow on arc (i,j) must be no less than l_{ij} and can be no greater than u_{ij}. To set up the problem in the framework of an optimization problem, a unit flow cost c_{ij}, incurred by each unit of flow moving through arc (i,j), must be defined. Besides being restricted by lower and upper bounds at each arc, flows must satisfy another important condition, known as Kirchhoff's Law (conservation of flow), which states that for every node in the network, the sum of all incoming flow together with the flow produced at the node must equal the sum of all outgoing flow and the flow consumed at the node. The objective of the *minimum cost network flow problem* is to determine the flow on each arc of the network, such that all of the flow produced in the network is moved from the source nodes to the sink nodes in the most cost-effective way, while not violating Kirchhoff's Law and flow limitations on the arcs. The minimum cost network flow problem can be formulated as the following linear program:

$$\min \sum_{(i,j) \in \mathcal{A}} c_{ij} x_{ij} \qquad (22)$$

subject to:

$$\sum_{(j,k) \in \mathcal{A}} x_{jk} - \sum_{(k,j) \in \mathcal{A}} x_{kj} = b_j, \quad j \in \mathcal{N} \qquad (23)$$

$$l_{ij} \leq x_{ij} \leq u_{ij}, \quad (i,j) \in \mathcal{A}. \qquad (24)$$

In this formulation, x_{ij} denotes the flow on arc (i,j) and c_{ij} is the cost of transporting one unit of flow on arc (i,j). For each node $j \in \mathcal{N}$, let b_j denote a quantity associated with node j that indicates how much flow is produced or consumed at the node. If $b_j > 0$, node j is a source. If $b_j < 0$, node j is a sink. Otherwise ($b_j = 0$), node j is a transshipment node. For each arc $(i,j) \in \mathcal{A}$, as before, let l_{ij} and u_{ij} denote, respectively, the lower and upper bounds on flow on arc (i, j). The case where $u_{ij} = \infty$, for all $(i, j) \in \mathcal{A}$, gives rise to the *uncapacitated* network flow problem. Without loss of generality, l_{ij} can be set to zero. Most often, the problem data (i.e. c_{ij}, u_{ij}, l_{ij}, for $(i, j) \in \mathcal{A}$ and b_j, for $j \in \mathcal{N}$) are assumed to be integer, and many codes adopt this assumption. However, there can exist applications where the data are real numbers, and algorithms should be capable of handling problems with real data.

Constraints of type (23) are referred to as the flow conservation equations, while constraints of type (24) are called the flow capacity constraints. In matrix notation, the above network flow problem can be formulated as a linear program of the special form

$$\min \{c^T x \mid Ax = b, \ l \leq x \leq u\},$$

where A is the $m \times n$ *node-arc incidence matrix* of the graph $G = (\mathcal{N}, \mathcal{A})$, i.e. for each arc (i, j) in $\mathcal{A}$ there is an associated column in matrix A with exactly two nonzero entries: an entry 1 in row i and an entry -1 in row j. Note that from the mn entries of A, only $2n$ are nonzero and because of this, the node-arc incidence matrix is not a space-efficient representation of the network. There are many other ways to represent a network. A popular representation is the *node-node adjacency* matrix B. This is an $m \times m$ matrix with an entry 1 in position (i, j) if arc $(i, j) \in \mathcal{A}$ and 0 otherwise. Such a representation is efficient for dense networks, but is inefficient for sparse networks. A more efficient representation of sparse networks is the *adjacency list*, where for each node $i \in \mathcal{N}$ there exists a list of arcs emanating from node i, i.e. a list of nodes j such that $(i, j) \in \mathcal{A}$. The *forward star* representation is a multi-array implementation of the adjacency list data structure. The adjacency list enables easy access to the arcs emanating from a given node, but not the incoming arcs. The *reverse star* representation enables easy access to the list of arcs incoming into i. Another representation that is much used in interior point network flow implementations is a simple *arc list*, where the arcs are stored in a linear array. The complexity of an algorithm for solving network flow problems depends greatly on the network representation and the data structures used for maintaining and updating the intermediate computations.

We denote the i-th column of A by A_i, the i-th row of A by $A_{\cdot i}$ and a submatrix of A formed by columns with indices in set S by A_S. If graph G is disconnected and has p connected components, there are exactly p redundant flow conservation constraints, which are sometimes removed from the problem formulation. We rule out a trivially infeasible problem by assuming

$$\sum_{j \in \mathcal{N}^k} b_j = 0, \quad k = 1, \ldots, p, \qquad (25)$$

where $\mathcal{N}^k$ is the set of nodes for the k-th component of G.

Often it is further required that the flow x_{ij} be integer, i.e. we replace (24) with

$$l_{ij} \leq x_{ij} \leq u_{ij}, \; x_{ij} \text{ integer}, \; (i,j) \in \mathcal{A}. \qquad (26)$$

Since the node-arc incidence matrix A is totally unimodular, when the data is integer all vertex solutions of the linear program are integer. An algorithm that finds a vertex solution, such as the simplex method, will necessarily produce an integer optimal flow. In certain types of network flow problems, such as the assignment problem, one may be only interested in solutions having integer flows, since fractional flows do not have a logical interpretation.

In the remainder of this section we assume, without loss of generality, that $l_{ij} = 0$ for all $(i,j) \in \mathcal{A}$ and that $c \neq 0$. A simple change of variables can transform the original problem into an equivalent one with $l_{ij} = 0$ for all $(i,j) \in \mathcal{A}$. The case where $c = 0$ is a simple feasibility problem, and can be handled by solving a maximum flow problem [4].

Many important combinatorial optimization problems are special cases of the minimum cost network flow problem. Such problems include the linear assignment and transportation problems, and the maximum flow and shortest path problems. In the transportation problem, the underlying graph is bipartite, i.e. there exist two sets $\mathcal{S}$ and $\mathcal{T}$ such that $\mathcal{S} \cup \mathcal{T} = \mathcal{N}$ and $\mathcal{S} \cap \mathcal{T} = \emptyset$ and arcs occur only from nodes of $\mathcal{S}$ to nodes of $\mathcal{T}$. Set $\mathcal{S}$ is usually called the set of source nodes and set $\mathcal{T}$ is the set of sink nodes. For the transportation problem, the right hand side vector in (23) is given by

$$b_j = \begin{cases} s_j & \text{if } j \in \mathcal{S} \\ -t_j & \text{if } j \in \mathcal{T}, \end{cases}$$

where s_j is the supply at node $j \in \mathcal{S}$ and t_j is the demand at node $j \in \mathcal{T}$. The assignment problem is a special case of the transportation problem, in which $s_j = 1$ for all $j \in \mathcal{S}$ and $t_j = 1$ for all $j \in \mathcal{T}$.

The computation of the maximum flow from node s to node t in $G = (\mathcal{N}, \mathcal{A})$ can be done by computing a minimum cost flow in $G' = (\mathcal{N}', \mathcal{A}')$, where $\mathcal{N}' = \mathcal{N}$ and $\mathcal{A}' = \mathcal{A} \cup (t,s)$, where

$$c_{ij} = \begin{cases} 0 & \text{if } (i,j) \in \mathcal{A} \\ -1 & \text{if } (i,j) = (t,s), \end{cases}$$

and
$$u_{ij} = \begin{cases} \text{cap}(i,j) & \text{if } (i,j) \in \mathcal{A} \\ \infty & \text{if } (i,j) = (t,s), \end{cases}$$

where $\text{cap}(i,j)$ is the capacity of arc (i,j) in the maximum flow problem.

The shortest paths from node s to all nodes in $\mathcal{N} \setminus \{s\}$ can be computed by solving an uncapacitated minimum cost network flow problem in which c_{ij} is the length of arc (i,j) and the right hand side vector in (23) is given by

$$b_j = \begin{cases} m-1 & \text{if } j = s \\ -1 & \text{if } j \in \mathcal{N} \setminus \{s\}. \end{cases}$$

Although all of the above combinatorial optimization problems are formulated as minimum cost network flow problems, several specialized algorithms have been devised for solving them efficiently.

In many practical applications, flows in networks with more than one commodity need to be optimized. In the multicommodity network flow problem, k commodities are to be moved in the network. The set of commodities is denoted by $\mathcal{K}$. Let x_{ij}^k denote the flow of commodity k in arc (i,j). The multicommodity network flow problem can be formulated as the following linear program:

$$\min \sum_{k \in \mathcal{K}} \sum_{(i,j) \in \mathcal{A}} c_{ij}^k x_{ij}^k \tag{27}$$

subject to:
$$\sum_{(j,l) \in \mathcal{A}} x_{jl}^k - \sum_{(l,j) \in \mathcal{A}} x_{lj}^k = b_j^k, \quad j \in \mathcal{N}, \; k \in \mathcal{K} \tag{28}$$

$$\sum_{k \in \mathcal{K}} x_{ij}^k \leq u_{ij}, \quad (i,j) \in \mathcal{A}, \tag{29}$$

$$x_{ij}^k \geq 0, \quad (i,j) \in \mathcal{A}, \; k \in \mathcal{K}. \tag{30}$$

The minimum cost network flow problem is a special case of the multicommodity network flow problem, in which there is only one commodity.

In the 1940s, Hitchcock [58] proposed an algorithm for solving the transportation problem and later Dantzig [23] developed the Simplex Method for linear programming problems. In the 1950s, Kruskal [83] developed a minimum spanning tree algorithm and Prim [118] devised an algorithm for the shortest path problem. During that decade, commercial digital computers were introduced widely. The first book on network flows was published by

Ford and Fulkerson [37] in 1962. Since then, active research produced a variety of algorithms, data structures, and software for solving network flow problems. For an introduction to network flow problems and applications, see the books [4, 15, 31, 37, 80, 84, 133, 137].

4.3 Components of interior point network flow methods

Since Karmarkar's breakthrough in 1984, many variants of his algorithm, including the dual affine scaling, with and without centering, reduced dual affine scaling, primal path following (method of centers), primal-dual path following, predictor-corrector primal-dual path following, and the infeasible primal-dual path following, have been used to solve network flow problems. Though these algorithms are, in some sense, different, they share many of the same computational requirements. The key ingredients for efficient implementation of these algorithms are:

1. The solution of the linear system $ADA^\mathsf{T} u = t$, where D is a diagonal $n \times n$ scaling matrix, and u and t are m-vectors. This requires an iterative algorithm for computing approximate directions, preconditioners, stopping criteria for the iterative algorithm, etc.

2. The recovery of the desired optimal solution. This may depend on how the problem is presented (integer data or real data), and what type of solution is required (fractional or integer solution, ϵ-optimal or exact solution, primal optimal or primal-dual optimal solution, etc.).

In this subsection, we present in detail these components, illustrating their implementation in the dual affine scaling network flow algorithm DLNET of Resende and Veiga [130].

4.3.1 The dual affine scaling algorithm

The dual affine scaling (DAS) algorithm [12, 29, 135, 141] was one of the first interior point methods to be shown to be competive computationally with the simplex method [2, 3]. As before, let A be an $m \times n$ matrix, c, u, and x be n-dimensional vectors and b an m-dimensional vector. The DAS algorithm solves the linear program

$$\min \{c^\mathsf{T} x \mid Ax = b,\ 0 \leq x \leq u\}$$

indirectly, by solving its dual

$$\max \{b^\mathsf{T} y - u^\mathsf{T} z \mid A^\mathsf{T} y - z + s = c,\ z \geq 0, s \geq 0\}, \tag{31}$$

where z and s are an n-dimensional vectors and y is an m-dimensional vector. The algorithm starts with an initial interior solution $\{y^0, z^0, s^0\}$ such that

$$A^\top y^0 - z^0 + s^0 = c, \ z^0 > 0, \ s^0 > 0,$$

and iterates according to

$$\{y^{k+1}, z^{k+1}, s^{k+1}\} = \{y^k, z^k, s^k\} + \alpha \{\Delta y, \Delta z, \Delta s\},$$

where the search directions $\Delta y, \Delta z,$ and Δs satisfy

$$\begin{aligned} A(Z_k^2 + S_k^2)^{-1} A^\top \Delta y &= b - A Z_k^2 (Z_k^2 + S_k^2)^{-1} u, \\ \Delta z &= Z_k^2 (Z_k^2 + S_k^2)^{-1} (A^\top \Delta y - S_k^2 u), \\ \Delta s &= \Delta z - A^\top \Delta y, \end{aligned}$$

where

$$Z_k = \mathrm{diag}(z_1^k, \ldots, z_n^k) \text{ and } S_k = \mathrm{diag}(s_1^k, \ldots, s_n^k)$$

and α is such that $z^{k+1} > 0$ and $s^{k+1} > 0$, i.e. $\alpha = \gamma \times \min\{\alpha_z, \alpha_s\}$, where $0 < \gamma < 1$ and

$$\alpha_z = \min\{-z_i^k / (\Delta z)_i \mid (\Delta z)_i < 0, \ i = 1, \ldots, n\}$$
$$\alpha_s = \min\{-s_i^k / (\Delta s)_i \mid (\Delta s)_i < 0, \ i = 1, \ldots, n\}.$$

The dual problem (31) has a readily available initial interior point solution:

$$\begin{aligned} y_i^0 &= 0, \ i = 1, \ldots, n \\ s_i^0 &= c_i + \lambda, \ i = 1, \ldots, n \\ z_i^0 &= \lambda, \ i = 1, \ldots, n, \end{aligned}$$

where λ is a scalar such that $\lambda > 0$ and $\lambda > -c_i, \ i = 1, \ldots, n$. The algorithm described above has two important parameters, γ and λ. For example, in DLNET, $\gamma = 0.95$ and $\lambda = 2 \|c\|_2$.

4.3.2 Computing the direction

The computational efficiency of interior point network flow methods relies heavily on a preconditioned conjugate gradient algorithm to solve the direction finding system at each iteration. The preconditioned conjugate gradient algorithm is used to solve

$$M^{-1}(AD_k A^\top)\Delta y = M^{-1} \bar{b} \tag{32}$$

```
procedure pcg(A, D_k, b̄, ε_cg, Δy)
1    Δy_0 := 0;
2    r_0 := b̄;
3    z_0 := M^{-1} r_0;
4    p_0 := z_0;
5    i := 0;
6    do stopping criterion not satisfied →
7        q_i := A D_k A^T p_i;
8        α_i := z_i^T r_i / p_i^T q_i;
9        Δy_{i+1} := Δy_i + α_i p_i;
10       r_{i+1} := r_i - α_i q_i;
11       z_{i+1} := M^{-1} r_{i+1};
12       β_i := z_{i+1}^T r_{i+1} / z_i^T r_i;
13       p_{i+1} := z_{i+1} + β_i p_i;
14       i := i + 1
15   od;
16   Δy := Δy_i
end pcg;
```

Figure 5: The preconditioned conjugate gradient algorithm

where M is a positive definite matrix and, in the case of the DAS algorithm, $\bar{b} = b - AZ_k^2 D_k u$, and $D_k = (Z_k^2 + S_k^2)^{-1}$ is a diagonal matrix of positive elements. The objective is to make the preconditioned matrix

$$M^{-1}(AD_k A^T) \tag{33}$$

less ill-conditioned than $AD_k A^T$, and improve the convergence of the conjugate gradient algorithm.

The pseudo-code for the preconditioned conjugate gradient algorithm is presented in Figure 5. The computationally intensive steps in the preconditioned conjugate gradient algorithm are lines 3, 7 and 11 of the pseudo-code. These lines correspond to a matrix-vector multiplication (7) and solving linear systems of equations (3 and 11). Line 3 is computed once and lines 7 and 11 are computed once every conjugate gradient iteration. The matrix-vector multiplications are of the form $AD_k A^T p_i$, carried out without forming $AD_k A^T$ explicitly. One way to compute the above matrix-vector multiplication is to decompose it into three sparse matrix-vector multiplications.

Let
$$\zeta' = A^\top p_i \quad \text{and} \quad \zeta'' = D_k \zeta'.$$
Then
$$(A (D_k (A^\top p_i))) = A\zeta''.$$

The complexity of this matrix-vector multiplication is $O(n)$, involving n additions, $2n$ subtractions and n floating point multiplications.

The preconditioned residual is computed in lines 3 and 11 when the system of linear equations
$$M z_{i+1} = r_{i+1}, \qquad (34)$$
is solved, where M is a positive definite matrix. An efficient implementation requires a preconditioner that can make (34) easy to solve. On the other hand, one needs a preconditioner that makes (33) well conditioned. In the next subsection, we show several preconditioners that satisfy, to some extent, these two criteria.

To determine when the approximate direction Δy_i produced by the conjugate gradient algorithm is satisfactory, one can compute the angle θ between $(AD_k A^\top)\Delta y_i$ and $\bar{b}$ and stop when $|1-\cos\theta| < \epsilon_{cos}$, where ϵ_{cos} is some small tolerance. In practice, one can initially use $\epsilon_{cos} = 10^{-3}$ and tighten the tolerance as the interior point iterations proceed, as $\epsilon_{cos} = \epsilon_{cos} \times 0.95$. The exact computation of
$$\cos\theta = \frac{|\bar{b}^\top (AD_k A^\top)\Delta y_i|}{\|\bar{b}\|_2 \cdot \|(AD_k A^\top)\Delta y_i\|_2}$$
has the complexity of one conjugate gradient iteration and is therefore expensive if computed at each conjugate gradient iteration. One way to proceed is to compute the cosine every l_{cos} conjugate gradient iterations. A more efficient procedure [116] follows from the observation that $(AD_k A^\top)\Delta y_i$ is approximately equal to $\bar{b} - r_i$, where r_i is the estimate of the residual at the i-th conjugate gradient iteration. Using this approximation, the cosine can be estimated by
$$\cos\theta = \frac{|\bar{b}^\top (\bar{b} - r_i)|}{\|\bar{b}\|_2 \cdot \|(\bar{b} - r_i)\|_2}.$$
Since, in practice, the conjugate gradient method finds good directions in few iterations, this estimate has been shown to be effective and can be computed at each conjugate gradient iteration.

4.3.3 Network preconditioners for conjugate gradient method

A useful preconditioner for the conjugate gradient algorithm must be one that allows the efficient solution of (34), while at the same time causing the number of conjugate gradient iterations to be small. Five preconditioners have been found useful in conjugate gradient based interior point network flow methods: diagonal, maximum weighted spanning tree, incomplete QR decomposition, the Karmarkar-Ramakrishnan preconditioner for general linear programming, and the approximate Cholesky decomposition preconditioner [93].

A diagonal matrix constitutes the most straightforward preconditioner used in conjunction with the conjugate gradient algorithm [45]. They are simple to compute, taking $O(n)$ double precision operations, and can be effective [129, 131, 149]. In the diagonal preconditioner, $M = \text{diag}(AD_kA^T)$, and the preconditioned residue systems of lines 3 and 11 of the conjugate gradient pseudo-code in Figure 5 can each be solved in $O(m)$ double precision divisions.

A preconditioner that is observed to improve with the number of interior point iterations is the maximum weighted spanning tree preconditioner. Since the underlying graph G is not necessarily connected, one can identify a maximal forest using as weights the diagonal elements of the current scaling matrix,

$$w = D_k e, \qquad (35)$$

where e is a unit n-vector. In practice, Kruskal's and Prim's algorithm have been used to compute the maximal forest. Kruskal's algorithm, implemented with the data structures in [137] has been applied to arcs, ordered approximately with a bucket sort [62, 130], or exactly using a hybrid QuickSort [64]. Prim's algorithm is implemented in [116] using the data structures presented in [4].

At the k-th interior point iteration, let $\mathcal{S}_k$ be the submatrix of A with columns corresponding to arcs in the maximal forest, $t_1, \ldots, t_q$. The preconditioner can be written as

$$M = \mathcal{S}_k \mathcal{D}_k \mathcal{S}_k^T,$$

where, for example, in the DAS algorithm

$$\mathcal{D}_k = \text{diag}(1/z_{t_1}^2 + 1/s_{t_1}^2, \ldots, 1/z_{t_q}^2 + 1/s_{t_q}^2).$$

For simplicity of notation, we include in $\mathcal{S}_k$ the linear dependent rows corresponding to the redundant flow conservation constraints. At each conjugate gradient iteration, the preconditioned residue system

$$(\mathcal{S}_k \mathcal{D}_k \mathcal{S}_k^\top) z_{i+1} = r_{i+1} \qquad (36)$$

is solved with the variables corresponding to redundant constraints set to zero. As with the diagonal preconditioner, (36) can be solved in $O(m)$ time, as the system coefficient matrix can be ordered into a block triangular form.

Portugal et al. [116] introduced a preconditioner based on an incomplete QR decomposition (IQRD) for use in interior point methods to solve transportation problems. They showed empirically, for that class of problems, that this preconditioner mimics the diagonal preconditioner during the initial iterations of the interior point method, and the spanning tree preconditioner in the final interior point method iterations, while causing the conjugate gradient method to take fewer iterations than either method during the intermediate iterations. In [117], the use of this preconditioner is extended to general minimum cost network flow problems. In the following discussion, we omit the iteration index k from notation for the sake of simplicity. Let $\bar{\mathcal{T}} = \{1, 2, \ldots, n\} \setminus \mathcal{T}$ be the index set of the arcs not in the computed maximal spanning tree, and let

$$D = \begin{bmatrix} D_\mathcal{T} & \\ & D_{\bar{\mathcal{T}}} \end{bmatrix},$$

where $D_\mathcal{T} \in \mathsf{R}^{q \times q}$ is the diagonal matrix with the arc weights of the maximal spanning tree and $D_{\bar{\mathcal{T}}} \in \mathsf{R}^{(n-q) \times (n-q)}$ is the diagonal matrix with weights of the arcs not in the maximal spanning tree. Then

$$\begin{aligned} ADA^\top &= \begin{bmatrix} A_\mathcal{T} & A_{\bar{\mathcal{T}}} \end{bmatrix} \begin{bmatrix} D_\mathcal{T} & \\ & D_{\bar{\mathcal{T}}} \end{bmatrix} \begin{bmatrix} A_\mathcal{T}^\top \\ A_{\bar{\mathcal{T}}}^\top \end{bmatrix} \\ &= \begin{bmatrix} A_\mathcal{T} D_\mathcal{T}^{\frac{1}{2}} & A_{\bar{\mathcal{T}}} D_{\bar{\mathcal{T}}}^{\frac{1}{2}} \end{bmatrix} \begin{bmatrix} D_\mathcal{T}^{\frac{1}{2}} A_\mathcal{T}^\top \\ D_{\bar{\mathcal{T}}}^{\frac{1}{2}} A_{\bar{\mathcal{T}}}^\top \end{bmatrix}. \end{aligned}$$

The Cholesky factorization of $ADA^\top$ can be found by simply computing the QR factorization of

$$\bar{A} = \begin{bmatrix} D_\mathcal{T}^{\frac{1}{2}} A_\mathcal{T}^\top \\ D_{\bar{\mathcal{T}}}^{\frac{1}{2}} A_{\bar{\mathcal{T}}}^\top \end{bmatrix}. \qquad (37)$$

In fact, if $Q\bar{A} = R$, then

$$ADA^\top = \bar{A}^\top \bar{A} = R^\top Q^\top Q R = R^\top R.$$

The computation of the QR factorization is not recommended here, since besides being more expensive than a Cholesky factorization, it also destroys the sparsity of the matrix $\bar{A}$. Instead, Portugal et al. [116] propose an *incomplete QR decomposition* of $\bar{A}$. Applying Givens rotations [39] to $\bar{A}$, using the diagonal elements of $D_\mathcal{T}^{\frac{1}{2}} A_\mathcal{T}^\top$, the elements of $D_{\bar{\mathcal{T}}}^{\frac{1}{2}} A_{\bar{\mathcal{T}}}^\top$ become null. No fill-in is incurred in this factorization. See [116] for an example illustrating this procedure. After the factorization, we have the preconditioner

$$M = F\mathcal{D}F^\top,$$

where F is a matrix with a diagonal of ones that can be reordered to triangular form, and $\mathcal{D}$ is a diagonal matrix with positive elements.

To avoid square root operations, $\mathcal{D}$ and F are obtained without explicitly computing $\mathcal{D}^{\frac{1}{2}}F^\top$. Suppose that the maximum spanning tree is rooted at node r, corresponding to the flow conservation equation that has been removed from the formulation. Furthermore, let $\mathcal{A}_\mathcal{T}$ denote the subset of arcs belonging to the tree and let ρ_i represent the predecessor of node i in the tree. The procedure used to compute the nonzero elements of $\mathcal{D}$ and the off-diagonal nonzero elements of F is presented in the pseudo-code in Figure 6.

The computation of the preconditioned residual with $F\mathcal{D}F^\top$ requires $O(m)$ divisions, multiplications, and subtractions, since $\mathcal{D}$ is a diagonal matrix and F can be permuted into a triangular matrix with diagonal elements equal to one. The construction of F and $\mathcal{D}$, that constitute the preconditioner, requires $O(n)$ additions and $O(m)$ divisions.

In practice, the diagonal preconditioner is effective during the initial iterations of the DAS algorithm. As the DAS iterations progress, the spanning tree preconditioner is more effective as it becomes a better approximation of matrix $AD_k A^\top$. Arguments as to why this preconditioner is effective are given in [62, 116]. The DLNET implementation begins with the diagonal preconditioner and monitors the number of iterations required by the conjugate gradient algorithm. When the conjugate gradient takes more than $\beta\sqrt{m}$ iterations, where $\beta > 0$, DLNET switches to the spanning tree preconditioner. Upper and lower limits to the number of DAS iterations using a diagonal preconditioned conjugate gradient are specified.

```
procedure iqrd(𝒯, 𝒯̄, 𝒩, D, 𝒟, F)
1      do i ∈ 𝒩 \ {r} →
2          j = ρ_i;
3          if (i,j) ∈ 𝒜_𝒯 → 𝒟_{ii} = D_{ij} fi;
4          if (j,i) ∈ 𝒜_𝒯 → 𝒟_{ii} = D_{ji} fi;
5      od;
6      do (i,j) ∈ 𝒜_𝒯̄ →
7          if i ∈ 𝒩 \ {r} → 𝒟_{ii} = 𝒟_{ii} + D_{ij} fi;
8          if j ∈ 𝒩 \ {r} → 𝒟_{jj} = 𝒟_{jj} + D_{ij} fi;
9      od;
10     do i ∈ 𝒩 \ {r} →
11         j = ρ_i;
12         if j ∈ 𝒩 \ {r} →
13             if (i,j) ∈ 𝒜_𝒯 → F_{ij} = D_{ij}/𝒟_{ii}
fi;
14             if (j,i) ∈ 𝒜_𝒯 → F_{ji} = D_{ji}/𝒟_{ii}
fi;
15         fi;
16     od;
end iqrd;
```

Figure 6: Computing the F and $\mathcal{D}$ matrices in IQRD

In [93], is proposed a Cholesky decomposition of an approximation of the matrix $A\Theta A^\top$ (CDAM) as preconditioner. This preconditioner has the form
$$M = LL^\top, \tag{38}$$
whith L the lower triangular Cholesky factor of the matrix
$$B\Theta_B B^\top + \rho \times \text{diag}(N\Theta_N N^\top), \tag{39}$$
where B and N are such that $A = [B \; N]$ with B a basis matrix, Θ_B and Θ_N are the diagonal submatrices of Θ corresponding to B and N, respectively, and ρ is a parameter.

Another preconditioner used in an interior point implementation is the one for general linear programming, developed by Karmarkar and Ramakrishnan and used in [79, 120]. This preconditioner is based on a dynamic scheme to drop elements of the original scaled constraint matrix DA, as well as the from the factors of the matrix $ADA^\top$ of the linear system, and use the incomplete Cholesky factors as the preconditioner. Because of the way elements are dropped, this preconditioner mimics the diagonal preconditioner in the initial iterations and the tree preconditioner in the final iterations of the interior point algorithm.

4.3.4 Identifying the optimal partition

One way to stop an interior point algorithm before the (fractional) interior point iterates converge is to estimate (or guess) the optimal partition of arcs at each iteration, attempt to recover the flow from the partition and, if a feasible flow is produced, test if that flow is optimal. In the discussion that follows we describe a strategy to partition the set of arcs in the dual affine scaling algorithm. The discussion follows [128] closely, using a dual affine scaling method for uncapacitated networks to illustrate the procedure.

Let $A \in \mathsf{R}^{m \times n}$, $c, x, s \in \mathsf{R}^n$ and $b, y \in \mathsf{R}^m$. Consider the linear programming problem
$$\begin{aligned} \text{minimize} \quad & c^\top x \\ \text{subject to} \quad & Ax = b, \quad x \geq 0 \end{aligned} \tag{40}$$
and its dual
$$\begin{aligned} \text{maximize} \quad & b^\top y \\ \text{subject to} \quad & A^\top y + s = c, \quad s \geq 0. \end{aligned} \tag{41}$$

The dual affine scaling algorithm starts with an initial dual solution $y^0 \in \{y : s = c - A^\top y > 0\}$ and obtains iterate y^{k+1} from y^k according to $y^{k+1} = y^k + \alpha^k d_y^k$, where the search direction d_y is $d_y^k = (AD_k^{-2}A^\top)^{-1}b$ and $D_k = \mathrm{diag}(s_1^k, ..., s_n^k)$. A step moving a fraction γ of the way to the boundary of the feasible region is taken at each iteration, namely,

$$\alpha^k = \gamma \times \min\{-s_i^k/(d_s^k)_i : (d_s^k)_i < 0, \ i = 1, ..., n\}, \tag{42}$$

where $d_s^k = -A^\top d_y^k$ is a unit displacement vector in the space of slack variables. At each iteration, a tentative primal solution is computed by $x^k = D_k^{-2}A^\top(AD_k^{-2}A^\top)^{-1}b$. The set of optimal solutions is referred to as the *optimal face*. We use the index set N_* for the always-active index set on the optimal face of the primal, and B_* for its complement. It is well-known that B_* is the always-active index set on the optimal face of the dual, and N_* is its complement. An *indicator* is a quantity to detect whether an index belongs to N_* or B_*. We next describe three indicators that can be implemented in the DAS algorithm. For pointers to other indicators, see [33].

Under a very weak condition, the iterative sequence of the DAS algorithm converges to a relative interior point of a face on which the objective function is constant, i.e. the sequence $\{y^k\}$ converges to an interior point of a face on which the objective function is constant. Let B be the always-active index set on the face and N be its complement, and let b^∞ be the limiting objective function value. There exists a constant $C_0 > 0$ such that

$$\limsup_{k \to \infty} \frac{s_i^k}{b^\infty - b^\top y^k} \leq C_0 \tag{43}$$

for all $i \in B$, while

$$\frac{s_i^k}{b^\infty - b^\top y^k} \tag{44}$$

diverges to infinity for all $i \in N$. Denote by s^∞ the limiting slack vector. Then $s_N^\infty > 0$ and $s_B^\infty = 0$. The vector

$$u^k \equiv \frac{(D^k)^{-1}d_s^k}{b^\infty - b^\top y^k} = \frac{D^k x^k}{b^\infty - b^\top y^k} \tag{45}$$

plays an important role, since

$$\lim_{k \to \infty}(u^k)^\top e = \lim_{k \to \infty} \frac{(s^k)^\top x^k}{b^\infty - b^\top y^k} = 1. \tag{46}$$

Consequently, in the limit $b^\infty - b^\top y^k$ can be estimated by $(s^k)^\top x^k$ asymptotically, and (43) can be stated as

$$\lim_{k\to\infty} \sup \frac{s_i^k}{(s^k)^\top x^k} \leq C_0.$$

Then, if $i \in B$, for any β such that $0 < \beta < 1$,

$$\lim_{k\to\infty} \sup \frac{s_i^k}{((s^k)^\top x^k)^\beta} = 0,$$

since $((s^k)^\top x^k)^\beta$ converges to zero at a slower rate than $((s^k)^\top x^k)$ for any β such that $0 < \beta < 1$. Therefore, if $\beta = 1/2$, the following indicator has the property that $\lim_{k\to\infty} N^k = N_*$.

Indicator 1: Let $C_1 > 0$ be any constant, and define

$$N^k = \{i \in E \,:\, s_i^k \leq C_1 \sqrt{(s^k)^\top x^k}\}. \tag{47}$$

This indicator is available under very weak assumptions, so it can be used to detect B_* and N_* without any substantial restriction on step-size. On the other hand, it gives the correct partition only if the limit point y^∞ happens to be a relative interior point of the optimal face of the dual and thus lacks a firm theoretical justification. However, since we know by experience that y^∞ usually lies in the relative interior of the optimal face, we may expect that it should work well in practice. Another potential problem with this indicator is that it is not scaling invariant, so that it will behave differently if the scaling of the problem is changed.

Now we assume that the step-size γ is asymptotically less than or equal to $2/3$. Then the limiting point exists in the interior of the optimal face and b^∞ is the optimal value. Specifically, $\{y^k\}$ converges to an interior point of the optimal face of the dual problem, $\{x^k\}$ converges to the analytic center of the optimal face of the primal problem, and $\{b^\top y^k\}$ converges linearly to the optimal value b^∞ asymptotically, where the (asymptotic) reduction rate is exactly $1 - \gamma$. Furthermore, one can show that

$$\lim_{k\to\infty} u_i^k = 1/|B_*| \quad \text{for } i \in B_* \tag{48}$$

$$\lim_{k\to\infty} u_i^k = 0 \quad \text{otherwise.} \tag{49}$$

The vector u^k is not available because the exact optimal value is unknown a priori, but $b^\infty - b^\top y^k$ can be estimated by $(s^k)^\top x^k$ to obtain

$$\lim_{k \to \infty} \frac{s_i^k x_i^k}{(s^k)^\top x^k} = 1/|B_*| \quad \text{for } i \in B_* \tag{50}$$

$$\lim_{k \to \infty} \frac{s_i^k x_i^k}{(s^k)^\top x^k} = 0 \quad \text{otherwise.} \tag{51}$$

On the basis of this fact, the following procedure to construct N^k, which asymptotically coincides with N_*:

Indicator 2: Let δ be a constant between 0 and 1. We obtain N^k according to the following procedure:

- Step 1: Sort $g_i^k = s_i^k x_i^k/(s^k)^\top x^k$ according to its order of magnitude. Denote i_l the index for the l-th largest component.

- Step 2: For $p := 1, 2, \ldots$ compare g_{i_p} and δ/p, and let p^* be the first number such that $g_{i_{p^*}} \leq \delta/p^*$. Then set

$$N^k = \{i_1, i_2, \ldots, i_{p^*-1}\}. \tag{52}$$

To state the third, and most practical indicator, let us turn our attention to the asymptotic behavior of s_i^{k+1}/s_i^k. If $i \in N_*$, then s_i^k converges to a positive value, and hence

$$\lim_{k \to \infty} \frac{s_i^{k+1}}{s_i^k} = 1. \tag{53}$$

If $i \in B_*$, s_i^k converges to zero. Since

$$\lim_{k \to \infty} \frac{s_i^k x_i^k}{b^\infty - b^\top y^k} = \frac{1}{|B_*|}, \tag{54}$$

x_i^k converges to a positive number, and the objective function reduces with a rate of $1 - \gamma$, then

$$\lim_{k \to \infty} \frac{s_i^{k+1}}{s_i^k} = 1 - \gamma, \tag{55}$$

which leads to the following indicator:

Indicator 3: Take a constant η such that $1 - \gamma < \eta < 1$. Then let

$$N^k = \{i : \frac{s_i^{k+1}}{s_i^k} \geq \eta\} \tag{56}$$

be defined as the index set. Then $N^k = N_*$ holds asymptotically.

Of the three indicators described here, Indicators 2 and 3 stand on the firmest theoretical basis. Furthermore, unlike Indicator 1, both are scaling invariant. The above discussion can be easily extended for the case of capacitated network flow problems. DLNET uses Indicator 3 to identify the set of active arcs defining the optimal face by examining the ratio between subsequent iterates of each dual slack. At the optimum, the flow on each arc can be classified as being at its upper bound, lower bound, or as active. From the discussion above, if the flow on arc i converges to its upper bound,

$$\lim_{k\to\infty} s_i^k/s_i^{k-1} = 1-\gamma \quad \text{and} \quad \lim_{k\to\infty} z_i^k/z_i^{k-1} = 1.$$

If the flow on arc i converges to its lower bound,

$$\lim_{k\to\infty} s_i^k/s_i^{k-1} = 1 \quad \text{and} \quad \lim_{k\to\infty} z_i^k/z_i^{k-1} = 1-\gamma.$$

If the flow on arc i is active,

$$\lim_{k\to\infty} s_i^k/s_i^{k-1} = 1-\gamma \quad \text{and} \quad \lim_{k\to\infty} z_i^k/z_i^{k-1} = 1-\gamma.$$

From a practical point of view, scale invariance is the most interesting feature of this indicator. An implementable version can use constants which depend only on the step size factor γ. Let $\kappa_0 = .7$ and $\kappa_1 = .9$. At each iteration of DLNET, the arcs are classified as follows:

- If $s_i^k/s_i^{k-1} < \kappa_0$ and $z_i^k/z_i^{k-1} > \kappa_1$, arc i is set to its upper bound.
- If $s_i^k/s_i^{k-1} > \kappa_1$ and $z_i^k/z_i^{k-1} < \kappa_0$, arc i is set to its lower bound.
- Otherwise, arc i is set active, defining the tentative optimal face.

4.3.5 Recovering the optimal flow

The simplex method restricts the sequence of solutions it generates to nodes of the linear programming polytope. Since the matrix A of the network linear program is totally unimodular, when a simplex variant is applied to a network flow problem with integer data, the optimal solution is also integer. On the other hand, an interior point algorithm generates a sequence of interior point (fractional) solutions. Unless the primal optimal solution is unique, the primal solution that an interior point algorithm converges to is not guaranteed to be integer. In an implementation of an interior point network flow method, one would like to be capable of recovering an integer flow even when the problem has multiple optima. We discuss below the stopping strategies implemented in DLNET and used to recover an integer optimal solution.

Besides the indicator described in subsection 4.3.4, DLNET uses the arcs of the spanning forest of the tree preconditioner as an indicator. If there exists a unique optimal flow, this indicator correctly identifies an optimal primal basic sequence, and an integer flow can be easily recovered by solving a triangular system of linear equations. In general, however, the arc indices do not converge to a basic sequence. Let $\mathcal{T} = \{t_1, \ldots, t_q\}$ denote the set of arc indices in the spanning forest. To obtain a tentative primal basic solution, first set flow on arcs not in the forest to either their upper or lower bound, i.e. for all $i \in \mathcal{A} \setminus \mathcal{T}$:

$$x_i^* = \begin{cases} 0 & \text{if } s_i^k > z_i^k \\ u_i & \text{otherwise,} \end{cases}$$

where s^k and z^k are the current iterates of the dual slack vectors as defined in (31). The remaining basic arcs have flows that satisfy the linear system

$$A_{\mathcal{T}} x_{\mathcal{T}}^* = b - \sum_{i \in \Omega^-} u_i A_i, \qquad (57)$$

where $\Omega^- = \{i \in \mathcal{A} \setminus \mathcal{T} : s_i^k \leq z_i^k\}$. Because $A_{\mathcal{T}}$ can be reordered in a triangular form, (57) can be solved in $\mathcal{O}(m)$ operations. If $u_{\mathcal{T}} \geq x_{\mathcal{T}}^* \geq 0$ then the primal solution is feasible and optimality can be tested.

Optimality can be verified by producing a dual feasible solution (y^*, s^*, z^*) that is either complementary or that implies a duality gap less than 1. The first step to build a tentative optimal dual solution is identify the set of dual constraints defining the supporting affine space of the dual face complementary to x^*,

$$\mathcal{F} = \{i \in \mathcal{T} : 0 < x_i^* < u_i\},$$

i.e. the set of arcs with zero dual slacks. Since, in general, x^* is not feasible, $\mathcal{F}$ is usually determined by the indicators of subsection 4.3.4, as the indexset of active arcs. To ensure a complementary primal-dual pair, the current dual interior vector y^k is projected orthogonally onto this affine space. The solution y^* of the least squares problem

$$\min_{y^* \in \mathbb{R}^m} \{ \|y^* - y^k\|_2 \ : \ A_{\mathcal{F}}^\top y^* = c_{\mathcal{F}} \} \qquad (58)$$

is the projected dual iterate.

Let $G_\mathcal{F} = (\mathcal{N}, \mathcal{F})$ be the subgraph of G with $\mathcal{F}$ as its set of arcs. Since this subgraph is a forest, its incidence matrix, $A_\mathcal{F}$, can be reordered into a block triangular form, with each block corresponding to a tree in the forest. Assume $G_\mathcal{F}$ has p components, with $T_1, \ldots, T_p$ as the sets of arcs in each component tree. After reordering, the incidence matrix can be represented as

$$A_\mathcal{F} = \begin{bmatrix} A_{T_1} & & \\ & \ddots & \\ & & A_{T_p} \end{bmatrix}.$$

The supporting affine space of the dual face can be expressed as the sum of orthogonal one-dimensional subspaces. The operation in (58) can be performed by computing the orthogonal projections onto each individual subspace independently, and therefore can be completed in $\mathcal{O}(m)$ time. For $i = 1, \ldots, p$, denote the number of arcs in T_i by m_i, and the set of nodes spanned by those arcs by $\mathcal{N}_i$. A_{T_i} is an $(m_i + 1) \times m_i$ matrix and each subspace

$$\Psi_i = \{y_{\mathcal{N}_i} \in \mathsf{R}^{m_i+1} : A_{T_i}^\top y_{\mathcal{N}_i} = c_{T_i}\}$$

has dimension one. For all $y_{\mathcal{N}_i} \in \Psi_i$,

$$y_{\mathcal{N}_i} = y_{\mathcal{N}_i}^0 + \alpha_i y_{\mathcal{N}_i}^h, \tag{59}$$

where $y_{\mathcal{N}_i}^0$ is a given solution in Ψ_i and $y_{\mathcal{N}_i}^h$ is a solution of the homogeneous system $A_{T_i}^\top y_{\mathcal{N}_i} = 0$. Since A_{T_i} is the incidence matrix of a tree, the unit vector is a homogeneous solution. The given solution $y_{\mathcal{N}_i}^0$ can be computed by selecting $v \in \mathcal{N}_i$, setting $y_v^0 = 0$, removing the row corresponding to node v from matrix A_{T_i} and solving the resulting triangular system

$$\tilde{A}_{T_i}^\top y_{\mathcal{N}_i \setminus \{v\}} = c_{T_i}.$$

With the representation in (59), the orthogonal projection of $y_{\mathcal{N}_i}$ onto subspace Ψ_i is

$$y_{\mathcal{N}_i}^* = y_{\mathcal{N}_i}^0 + \frac{e_{\mathcal{N}_i}^\top (y_{\mathcal{N}_i} - y_{\mathcal{N}_i}^0)}{(m_i + 1)} e_{\mathcal{N}_i}$$

where e is the unit vector. The orthogonal projection, as indicated in (58), is obtained by combining the projections onto each subspace,

$$y^* = (y_{\mathcal{N}_i}^*, \ldots, y_{\mathcal{N}_q}^*).$$

A feasible dual solution is built by computing the slacks as

$$z_i^* = \begin{cases} -\delta_i & \text{if } \delta_i < 0 \\ 0 & \text{otherwise,} \end{cases} \qquad s_i^* = \begin{cases} 0 & \text{if } \delta_i < 0 \\ \delta_i & \text{otherwise,} \end{cases}$$

where $\delta_i = c_i - A^\top y^*$.

If the solution of (57) is feasible, optimality can be checked at this point, using the projected dual solution as a lower bound on the optimal flow. The primal and dual solutions, x^* and (y^*, s^*, z^*), are optimal if complementary slackness is satisfied, i.e. if for all $i \in \mathcal{A} \setminus \mathcal{T}$ either $s_i^* > 0$ and $x_i^* = 0$ or $z_i^* > 0$ and $x_i^* = u_i$. Otherwise, the primal solution, x^*, is still optimal if the duality gap is less than 1, i.e. if $c^\top x^* - b^\top y^* + u^\top z^* < 1$.

However, in general, the method proceeds attempting to find a feasible flow x^* that is complementary to the projected dual solution y^*. Based on the projected dual solution y^*, a refined tentative optimal face is selected by redefining the set of active arcs as

$$\tilde{\mathcal{F}} = \{i \in \mathcal{A} : |c_i - A_{\cdot i}^\top y^*| < \epsilon\}.$$

Next, the method attempts to build a primal feasible solution, x^*, complementary to the tentative dual optimal solution by setting the inactive arcs to lower or upper bounds, i.e., for $i \in \mathcal{A} \setminus \tilde{\mathcal{F}}$,

$$x_i^* = \begin{cases} 0 & \text{if } i \in \Omega^+ = \{i \in \mathcal{A} \setminus \tilde{\mathcal{F}} : c_i - A_{\cdot i}^\top y^* > 0\} \\ u_i & \text{if } i \in \Omega^- = \{i \in \mathcal{A} \setminus \tilde{\mathcal{F}} : c_i - A_{\cdot i}^\top y^* < 0\}. \end{cases}$$

By considering only the active arcs, a *restricted network* is built, represented by the constraint set

$$A_{\tilde{\mathcal{F}}} x_{\tilde{\mathcal{F}}} = \tilde{b} = b - \sum_{i \in \Omega^-} u_i A_i, \qquad (60)$$

$$0 \leq x_i \leq u_i, \quad i \in \tilde{\mathcal{F}}. \qquad (61)$$

Clearly, from the flow balance constraints (60), if a feasible flow $x_{\tilde{\mathcal{F}}}^*$ for the restricted network exists, it defines, along with $x_{\Omega^+}^*$ and $x_{\Omega^-}^*$, a primal feasible solution complementary to y^*. A feasible flow for the restricted network can be determined by solving a maximum flow problem on the *augmented network* defined by underlying graph $\tilde{G} = (\tilde{\mathcal{N}}, \tilde{\mathcal{A}})$, where

$$\tilde{\mathcal{N}} = \{\sigma\} \cup \{\theta\} \cup \mathcal{N}$$

Interior Point Methods for Combinatorial Optimization 233

and
$$\tilde{\mathcal{A}} = \Sigma \cup \Theta \cup \tilde{\mathcal{F}}.$$

In addition, for each arc $(i,j) \in \tilde{\mathcal{F}}$ there is an associated capacity u_{ij}. The additional arcs are such that

$$\Sigma = \{(\sigma, i) \ : \ i \in \mathcal{N}^+\},$$

with associated capacity $\tilde{b}_i$ for each arc (σ, i), and

$$\Theta = \{(i, \theta) \ : \ i \in \mathcal{N}^-\},$$

with associated capacity $-\tilde{b}_i$ for each arc (i, θ), where $\mathcal{N}^+ = \{i \in \mathcal{N} \ : \ \tilde{b}_i > 0\}$ and $\mathcal{N}^- = \{i \in \mathcal{N} \ : \ \tilde{b}_i < 0\}$. It can be shown that if $\mathcal{M}_{\sigma,\theta}$ is the maximum flow value from σ to θ, and $\tilde{x}$ is a maximal flow on the augmented network, then $\mathcal{M}_{\sigma,\theta} = \sum_{i \in \mathcal{N}^+} \tilde{b}_i$ if and only if $\tilde{x}_{\tilde{\mathcal{F}}}$ is a feasible flow for the restricted network. Therefore, finding a feasible flow for the restricted network involves the solution a maximum flow problem. Furthermore, this feasible flow is integer, as we can select a maximum flow algorithm [4] that provides an integer solution.

5 Branch and bound methods

Branch and bound methods are exact algorithms for integer programming problems — given enough time, they are guaranteed to find an optimal solution. If there is not enough time available to solve a given problem exactly, a branch and bound algorithm can still be used to provide a bound on the optimal value. These methods can be used in conjunction with a heuristic algorithm such as local search, tabu search, simulated annealing, GRASP, genetic algorithms, or more specialized algorithms, to give a good solution to a problem, with a guarantee on the maximum possible improvement available over this good solution. Branch and bound algorithms work by solving relaxations of the integer programming problem, and selectively partitioning the feasible region to eventually find the optimal solution.

5.1 General concepts

Consider an integer programming problem of the form

$$\begin{aligned} \min \quad & c^T x \\ \text{subject to} \quad & Ax \leq b \\ & x \geq 0, \text{ integer}, \end{aligned}$$

where A is an $m \times n$ matrix, c and x are n-vectors, and b is an m-vector. The linear programming relaxation (*LP relaxation*) of this problem is

$$\begin{array}{ll} \min & c^T x \\ \text{subject to} & Ax \leq b \\ & x \geq 0. \end{array}$$

If the optimal solution x^* to the LP relaxation is integral then it solves the integer programming problem also. Generally, the optimal solution to the LP relaxation will not be an integral point. In this case, the value of the LP relaxation provides a lower bound on the optimal value of the integer program, and we attempt to improve the relaxation.

In a branch and bound method, the relaxation is improved by dividing the relaxation into two subproblems, where one of the variables is restricted to take certain values. For example, if $x_i^* = 0.4$, we may set up one subproblem where x_i must be zero and another subproblem where x_i is restricted to take a value of at least one. We think of the subproblems as forming a tree, rooted at the initial relaxation.

If the solution to the relaxation of one of the subproblems in the tree is integral then it provides an upper bound on the optimal value of the complete integer program. If the solution to the relaxation of another subproblem has value larger than this upper bound, then that subproblem can be pruned, as no feasible solution for it can be optimal for the complete problem. If the relaxation of the subproblem is infeasible then the subproblem itself is infeasible and can be pruned. The only other possibility at a node of the tree is that the solution to the relaxation is fractional, with value less than that of the best known integral solution. In this case, we further subdivide the subproblem. There are many techniques available for choosing the branching variable and for choosing the next subproblem to examine; for more details, see, for example, Parker and Rardin [115].

Interior point methods are good for linear programming problems with a large number of variables, so they should also be useful for large integer programming problems. Unfortunately, large integer programming problems are often intractable for a general purpose method like branch and bound, because the tree becomes prohibitively large. Branch and bound interior point methods have proven successful for problems such as capacitated facility location problems [19, 25], where the integer variables correspond to the decision as to whether to build a facility at a particular location, and there are a large number of continuous variables corresponding to transporting goods from the facilities to customers. For these problems, the LP relaxations can be large even for instances with only a few integer variables.

As with interior point cutting plane methods (see section 6), the most important technique for making an interior point branch and bound method competitive is early termination. There are four possible outcomes at each node of the branch and bound tree; for three of these, it suffices to solve the relaxation approximately. The first outcome is that the relaxation has value greater than the known upper bound on the optimal value, so the node can be pruned by bounds. Usually, an interior point method will get close to the optimal value quickly, so the possibility of pruning by bounds can be detected early. The second possible outcome is that the relaxation is infeasible. Even if this is not detected quickly, we can usually iterate with a dual feasible algorithm (as with interior point cutting plane algorithms), so if the dual value becomes larger than the known bound we can prune. The third possible outcome is that the optimal solution to the relaxation is fractional. In this case, there are methods (including the Tapia indicator [33] and other indicators discussed in section 4.3.4) for detecting whether a variable is converging to a fractional value, and these can be used before optimality is reached. The final possible outcome is that the optimal solution to the relaxation is integral. In this situation, we can prune the node, perhaps resulting in an improvement in the value of the best known integral solution. Thus, we are able to prune in the only situation where it is necessary to solve the relaxation to optimality.

If the optimal solution to the relaxation is fractional, then the subproblem must be subdivided. The iterate for the parent problem will be dual feasible but primal infeasible for the child problems. The solution process can be restarted at these child problems either by using an infeasible interior point method or by using methods similar to those described for interior point cuting plane methods in section 6. For very large or degenerate problems, the interior point method has proven superior to simplex even when the interior point code is started from scratch at each node.

The first interior point branch and bound code was due to Borchers and Mitchell [19]. This method was adapted by De Silva and Abramson [25] specifically for facility location problems. Ramakrishnan *et al.* [119] have developed a branch and bound algorithm for the quadratic assignment problem. The linear programming relaxations at the nodes of the tree for this problem are so large that it was necessary to use an interior point method to solve them. Lee and Mitchell have been developing a parallel interior point branch and cut algorithm for mixed integer nonlinear programming problems [85].

5.2 An example: The QAP

The quadratic assignment problem (QAP) can be stated as

$$\min_{p \in \Pi} \sum_{i=1}^{n} \sum_{j=1}^{n} a_{ij} b_{p(i)p(j)},$$

where Π is the set of all permutations of $\{1, 2, \ldots, n\}$, $A = (a_{ij}) \in \mathsf{R}^{n \times n}$, $B = (b_{ij}) \in \mathsf{R}^{n \times n}$.

Resende, Ramakrishnan, and Drezner [127] consider the following linear program as a lower bound (see also [1]) to the optimal solution of a QAP.

$$\min \sum_{i \in I} \overset{(i<j)}{\sum_{r \in I}} \overset{(r \neq s)}{\sum_{j \in I} \sum_{s \in I}} (a_{ij} b_{rs} + a_{ji} b_{sr}) y_{irjs}$$

subject to:

$$\overset{(j>i)}{\sum_{j \in I}} y_{irjs} + \overset{(j<i)}{\sum_{j \in I}} y_{jsir} = x_{ir}, \quad i \in I, r \in I, s \in I \ (s \neq r),$$

$$\overset{(r \neq s)}{\sum_{s \in I}} y_{irjs} = x_{ir}, \quad i \in I, r \in I, j \in I \ (j > i),$$

$$\overset{(r \neq s)}{\sum_{s \in I}} y_{jsir} = x_{ir}, \quad i \in I, r \in I, j \in I \ (j < i),$$

$$\sum_{i \in I} x_{ir} = 1, \quad r \in I,$$

$$\sum_{r \in I} x_{ir} = 1, \quad i \in I,$$

$$0 \leq x_{ir} \leq 1, \quad i \in I, r \in I,$$

$$0 \leq y_{irjs} \leq 1, \quad i \in I, r \in I, j \in I, s \in I,$$

where the set $I = \{1, 2, \ldots, n\}$. This linear program has $n^2(n-1)^2/2 + n^2$ variables and $2n^2(n-1) + 2n$ constraints. Table 6 shows the dimension of theses linear programs for several values of n.

Table 6: Dimension of lower bound linear programs

n	constraints	variables
2	12	6
3	42	27
4	104	88
5	210	225
6	372	486
7	602	931
8	912	1632
9	1314	2673
10	1820	4150
11	2442	6171
12	3192	8856
13	4082	12337
14	5124	16758

The linear programs were solved with ADP [79], a dual interior point algorithm (see Subsection 2.2). The solver produces a sequence of lower bounds (dual interior solutions), each of which can be compared with the best upper bound to decide if pruning of the search tree can be done at the node on which the lower bound is computed. Figure 7 illustrates the sequence of lower bounds produced by ADP, compared to the sequence of feasible primal solutions produced by the primal simplex code of CPLEX on QAPLIB test problem nug15. The figure suggests that the algorithm can be stopped many iterations prior to convergence to the optimal value and still be close in value to the optimal solution. This is important in branch and bound codes, where often a lower bound needed to prune the search tree is less than the value of the best lower bound.

Pardalos, Ramakrishnan, Resende, and Li [110] describe a branch and bound algorithm used to study the effectiveness of a variance reduction based lower bound proposed by Li, Pardalos, Ramakrishnan, and Resende [87]. This branch and bound algorithm is used by Ramakrishnan, Pardalos, and Resende [121] in conjunction with the LP-based lower bound described earlier.

In the first step, an initial upper bound is computed and an initial branch-and-bound search tree is set up. The branch and bound tree is a

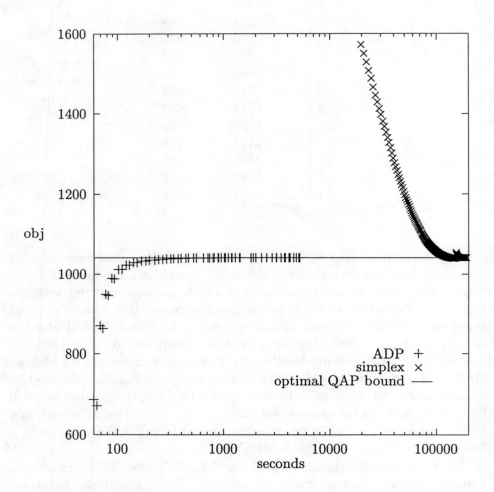

Figure 7: CPLEX simplex and ADP iterates on nug15

binary tree, each node having a left and right child. For the purpose of describing the branching process, denote, at any node of the branch and bound tree, S_A to be the set of already assigned facilities in the partial assignment, S_E the facilities that will never be in the partial assignment in any node of the subtree rooted at the current node. Let S_A^l, S_E^l and S_A^r, S_E^r be the corresponding sets for the left and right children of the current node. Let q denote the partial assignment at the current node. Each node of the branch and bound tree is organized as a heap with a key that is equal to the lower bound on the solution to the original QAP obtainable by any node in the subtree rooted at this node. The binary tree is organized in maximum order, i.e. the node with the largest lower bound is first.

The initial best known upper bound is computed by the GRASP heuristic described in [88, 126]. The initial search tree consists of n nodes with $S_A = \{i\}$ and $S_E = \emptyset$ for $i = 1, \ldots, n$, and $q(i) = p(i)$, where p is the permutation obtained by the GRASP and for $k \neq i$, $q(k) = 0$ and a key of 0.

In the second step, the four procedures of the branch-and-bound as described earlier are:

- *Selection*: The selection procedure simply chooses the partial permutation stored in the root of the heap, i.e. we pick the node with the maximum key.

- *Branching*: The branching procedure creates two children, the left and right children, as follows:

$$\begin{aligned} \text{pick}\ i &\notin S_A \\ S_A^l &= S_A \\ S_E^l &= S_E \cup \{i\} \\ S_A^r &= S_A \cup \{i\} \\ S_E^r &= \emptyset \\ q^l &= q \\ q^r &= q\ \text{and}\ q(i) = p(i), \text{where}\ p\ \text{is the incumbent}, \end{aligned}$$

and the key of left child is the same as the key of the current node and the key of the right child is the newly computed lower bound.

- *Elimination*: The elimination procedure compares the newly computed lower bound of the right child to the incumbent and deletes the right child if its key is greater than the incumbent, thus pruning the entire subtree rooted at the right child.

Table 7: Branch and bound algorithm on nug05

node	UB	LB	permutation				
1	52	58	1	-	-	-	-
2	52	55	2	-	-	-	-
3	52	52	5	-	-	-	-
4	52	57	3	-	-	-	-
5	52	50	4	-	-	-	-
6	52	57	4	3	-	-	-
7	52	50	4	5	-	-	-
8	52	56	4	5	3	-	-
9	52	56	4	5	2	-	-
10	52	50	4	5	1	-	-
11	52	60	4	5	1	3	-
12	52	50	4	5	1	2	-
	50	-	4	5	1	2	3
13	50	56	4	2	-	-	-
14	50	50	4	1	-	-	-

- *Termination Test*: The algorithm stops if, and only if, the heap is empty.

In the final step, a best permutation found is taken as the global optimal permutation.

As an example of the branch and bound algorithm, consider the QAPLIB instance nug05. The iterations of the branch and bound algorithm are summarized in Table 7. The GRASP approximation algorithm produced a solution (UB) having cost 52. The branch and bound algorithm examined 14 nodes of the search tree. In the first five nodes, each facility was fixed to location 1 and the lower bounds of each branch computed. The lower bounds corresponding to branches rooted at nodes 1 through 4 were all greater than or equal to the upper bound, and thus those branches of the tree could be pruned. At node 6 a level-2 branching begins with a lower bound less than the upper bound produced at node 7. Deeper branchings are done at nodes 8, 11, and 12, at which point a new upper bound is computed having value 50. Nodes 13 and 14 complete the search. The same branch and bound algorithm using the GLB scans 44 nodes of the tree to prove optimality.

We tested the codes on several instances from the QAP library QAPLIB.

Table 8: QAP test instances: LP-based vs. GLB-based B&B algorithms

problem	dim	LP-based B&B nodes	LP-based B&B time	GLB-based B&B nodes	GLB-based B&B time	time ratio	nodes ratio
nug05	5	12	11.7	44	0.1	117.0	3.7
nug06	6	6	9.5	82	0.1	95.0	13.7
nug07	7	7	16.6	115	0.1	166.0	16.4
nug08	8	8	35.1	895	0.2	175.5	111.9
nug12	12	220	5238.2	49063	14.6	358.8	223.0
nug15	15	1195	87085.7	1794507	912.4	95.4	1501.7
scr10	10	19	202.1	1494	0.6	336.8	78.6
scr12	12	252	5118.7	12918	4.8	1066.4	51.3
scr15	15	228	3043.3	506360	274.7	11.1	2220.9
rou10	10	52	275.7	2683	0.8	344.6	51.6
rou12	12	152	2715.9	37982	12.3	220.8	249.9
rou15	15	991	30811.7	4846805	2240.3	13.8	4890.8
esc08a	8	8	37.4	57464	7.0	5.3	7183.0
esc08b	8	208	491.1	7352	0.7	701.6	35.3
esc08c	8	8	42.7	2552	0.3	142.3	319.0
esc08d	8	8	38.1	2216	0.3	127.0	277.0
esc08e	8	64	251.0	10376	1.0	251.0	162.1
esc08f	8	8	37.6	1520	0.3	125.3	190.0
chr12a	12	12	312.0	672	0.7	445.7	56.0
chr12b	12	12	289.4	318	0.6	482.3	26.5
chr12c	12	12	386.1	3214	1.5	257.4	267.8
chr15a	15	15	1495.9	413825	235.5	6.4	27588.3
chr15b	15	15	1831.9	396255	217.8	8.4	26417.0
chr15c	15	15	1908.5	428722	240.0	8.0	28581.5
chr18a	18	35	1600.0	$> 1.6 \times 10^9$	$> 10^6$	$< 648.0^{-1}$	$> 45 \times 10^6$

Table 8 summarizes the runs on both algorithms. For each instance it displays the name and dimension of the problem, as well as the solution times and number of branch and bound search tree nodes examined by each of the algorithms. The ratio of CPU times is also displayed.

The number of GRASP iterations was set to 100,000 for all runs.

Table 9 shows statistics for the LP-based algorithm. For each run, the table lists the number of nodes examined, the number of nodes on which the lower bound obtained was greater than the best upper bound at that moment, the number of nodes on which the lower bound obtained was less than or equal to the best upper bound at that moment, and the percentage of nodes examined that were of levels 1, 2, 3, 4, and 5 or greater.

6 Branch and cut methods

For some problems, branch and bound algorithms can be improved by refining the relaxations solved at each node of the tree, so that the relaxation becomes a better and better approximation to the set of integral feasible solutions. In a general branch and cut method, many linear programming relaxations are solved at each node of the tree. Like branch and bound, a branch and cut method is an exact algorithm for an integer programming problem.

In a cutting plane method, extra constraints are added to the relaxation. These extra constraints are satisfied by all feasible solutions to the integer programming problem, but they are violated by the optimal solution to the LP relaxation, so we call them *cutting planes*. As the name suggests, a branch and cut method combines a cutting plane approach with a branch and bound method, attacking the subproblems at the nodes of the tree using a cutting plane method until it appears that no further progress can be made in a reasonable amount of time.

Consider, for example, the integer programming problem

$$\begin{aligned} \min \quad & -2x_1 - x_2 \\ \text{s.t.} \quad & x_1 + 2x_2 \leq 7 \\ & 2x_1 - x_2 \leq 3 \\ & x_1, x_2 \geq 0, \text{ integer.} \end{aligned}$$

This problem is illustrated in figure 8. The feasible integer points are indicated. The LP relaxation is obtained by ignoring the integrality restrictions; this is given by the polyhedron contained in the solid lines. The boundary

Table 9: QAP test instances: B&B tree search

problem	nodes of B&B tree			percentage of nodes of level				
	scan	good	bad	1	2	3	4	≥ 5
nug05	14	10	4	35.7	28.6	21.4	14.3	0.0
nug06	6	6	0	100.0	0.0	0.0	0.0	0.0
nug07	7	7	0	100.0	0.0	0.0	0.0	0.0
nug08	8	8	0	100.0	0.0	0.0	0.0	0.0
nug12	220	200	20	5.5	45.0	45.5	4.1	0.0
nug15	1195	1103	92	1.3	17.6	56.6	21.1	3.5
scr10	19	18	1	52.6	47.4	0.0	0.0	0.0
scr12	252	228	24	4.8	43.7	23.8	21.4	6.3
scr15	228	211	17	6.6	49.1	11.4	10.5	22.4
rou10	54	46	8	18.5	16.7	14.8	13.0	37.0
rou12	154	137	17	7.8	57.1	6.5	5.8	22.7
rou15	991	912	79	1.5	21.2	69.5	1.2	6.6
esc08a	8	8	0	100.0	0.0	0.0	0.0	0.0
esc08b	208	176	32	3.8	26.9	69.2	0.0	0.0
esc08c	8	8	0	100.0	0.0	0.0	0.0	0.0
esc08d	8	8	0	100.0	0.0	0.0	0.0	0.0
esc08e	64	56	8	12.5	87.5	0.0	0.0	0.0
esc08f	8	8	0	100.0	0.0	0.0	0.0	0.0
chr12a	12	12	0	100.0	0.0	0.0	0.0	0.0
chr12b	12	12	0	100.0	0.0	0.0	0.0	0.0
chr12c	12	12	0	100.0	0.0	0.0	0.0	0.0
chr15a	15	15	0	100.0	0.0	0.0	0.0	0.0
chr15b	15	15	0	100.0	0.0	0.0	0.0	0.0
chr15c	15	15	0	100.0	0.0	0.0	0.0	0.0
chr18a	35	17	18	51.4	48.6	0.0	0.0	0.0

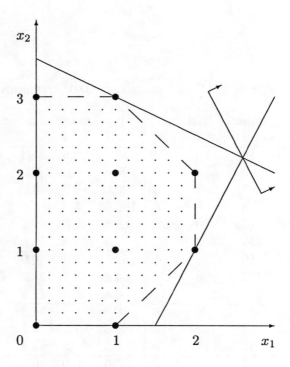

Figure 8: A cutting plane example

of the *convex hull* of the feasible integer points is indicated by dashed lines and can be described by the inequalities

$$\begin{aligned} x_1 - x_2 &\leq 1 \\ x_1 &\leq 2 \\ x_1 + x_2 &\leq 4 \\ x_2 &\leq 3 \\ x_1, x_2 &\geq 0. \end{aligned}$$

When solving this problem using a cutting plane algorithm, the linear programming relaxation is first solved, giving the point $x_1 = 2.6$, $x_2 = 2.2$, which has value -7.4. The inequalities $x_1 + x_2 \leq 4$ and $x_1 \leq 2$ are satisfied by all the feasible integer points but they are violated by the point $(2.6, 2.2)$. Thus, these two inequalities are valid *cutting planes*. Adding these two inequalities to the relaxation and solving again gives the point $x_1 = 2$, $x_2 = 2$, with value -6. Notice that this point is feasible in the original integer program, so it must actually be optimal for that problem, since it is optimal for a relaxation of the integer program.

If instead of adding both inequalities, we had just added the inequality $x_1 \leq 2$, the optimal solution to the new relaxation would have been $x_1 = 2$, $x_2 = 2.5$, with value -6.5. We could then have looked for a cutting plane that separates this point from the convex hull, for example $x_1 + x_2 \leq 4$, added this to the relaxation and solved the new relaxation. This illustrates the basic structure of a cutting plane algorithm:

- Solve the linear programming relaxation.

- If the solution to the relaxation is feasible in the integer programming problem, STOP with optimality.

- Else, find one or more cutting planes that separate the optimal solution to the relaxation from the convex hull of feasible integral points, and add a subset of these constraints to the relaxation.

- Return to the first step.

Notice that the values of the relaxations provide lower bounds on the optimal value of the integer program. These lower bounds can be used to measure progress towards optimality, and to give performance guarantees on integral solutions. None of these constraints can be omitted from the description of the convex hull, and they are called *facets* of the convex hull. Cutting planes that define facets are the strongest possible cutting planes, and they should be added to the relaxation in preference to non-facet defining inequalities, if possible. Families of facet defining inequalities are known for many classes of integer programming problems (for example, the traveling salesman problem [47, 108], the matching problem [32], the linear ordering problem [48], and the maximum cut problem [27, 28]). Jünger et al. [63] contains a survey of cutting plane methods for various integer programming problems. Nemhauser and Wolsey [105] gives more background on cutting plane methods for integer programming problems.

Traditionally, Gomory cutting planes [46] were used to improve the relaxation. These cuts are formed from the optimal tableau for the LP relaxation of the integer program. Cutting plane methods fell out of favour for many years because algorithms using Gomory cuts showed slow convergence. The resurgence of interest in these methods is due to the use of specialized methods that search for facets, enabling the algorithm to converge far more rapidly. The cutting planes are determined using a *separation routine*, which is usually very problem specific. General integer programming problems have been solved by using cutting planes based on facets of the *knapsack problem* $\min\{c^T x : a^T x \leq b, x \geq 0, x \text{ integer}\}$: each constraint of the general problem can be treated as a knapsack constraint [59]. Other general cutting plane techniques include lift-and-project methods [10]. Gomory cutting planes have also been the subject of a recent investigation [11]. It appears that they are not as bad as originally thought, and they in fact work quite well if certain modifications are made, such as adding many constraints at once.

The *separation problem* for the problem $\min\{c^T x : Ax \leq b, x \text{ integer}\}$ can be defined as:

Given a point $\bar{x}$, either determine that $\bar{x}$ is in the convex hull Q of the feasible integer points, or find a cutting plane that separates $\bar{x}$ from the convex hull.

Grötschel et al. [49] used the ellipsoid algorithm to show that if the separation problem can be solved in polynomial time then the problem (IP) itself can also be solved in polynomial time. It follows that the separation problem for an NP-hard problem cannot be solved in polynomial time, unless $P = NP$. Many of the separation routines in the literature are heuristics designed to find cutting planes belonging to certain families of facets; there are many undiscovered families of facets, and, for NP-hard problems, it is unlikely that a complete description of the facets of the convex hull will be discovered. Such a description would certainly contain an exponential number of facets (provided $P \neq NP$). Even small problems can have many facets. For example, the convex hull of the travelling salesman problem with only nine cities has over 42 million facets [21].

6.1 Interior point cutting plane methods

We now assume that our integer programming problem takes the form

$$
\begin{array}{rl}
\min & c^T x \\
\text{subject to} & Ax = b \\
& 0 \leq x \leq u \\
& x_i \text{ binary for } i \text{ in } I \\
& x \text{ satisfies some additional conditions}
\end{array}
\qquad (IP)
$$

where A is an $m \times n$ matrix of rank m, c, u, and x are n-vectors, b is an m-vector, and I is a subset of $\{1,\ldots,n\}$. We assume that these additional conditions can be represented by linear constraints, perhaps by an exponential number of such constraints. For example, the traveling salesman problem can be represented in this form, with the additional conditions being the subtour elimination constraints [47, 108], and the conditions $Ax = b$ representing the degree constraints that the tour must enter and leave each vertex exactly once. It is also possible that the problem does not need any such additional conditions. Of course, problems with inequality constraints can be written in this form by including slack variables. Note that we allow a mixture of integer and continuous variables. In this section, we describe cutting plane methods to solve (IP) where the LP relaxations are solved using *interior point methods*. Computational experience with interior point cutting plane methods is described in [99, 96, 101]. Previous surveys on interior point cutting plane methods include [94, 95].

It has been observed that interior point algorithms do not work very well when started from close to a nonoptimal extreme point. Of course, this is exactly what we will have to do if we solve the LP relaxation to optimality, since the fractional optimal solution to the relaxation will be a nonoptimal infeasible extreme point after adding a cutting plane. The principal method used to overcome this drawback is to only solve the relaxation approximately. We use this approximate solution to generate an integral feasible point that is, with luck, close to the optimal integral solution. The best integral solution found so far gives an upper bound on the optimal value of (IP) and the value of the dual solution gives a lower bound. A conceptual interior point cutting plane algorithm is given in Figure 9.

To make this algorithm practical, we have to decide how accurately to solve the relaxations. Notice that if the entries in c are integral then it is sufficient to reduce the gap between the integral solution and the lower bound to be less than one. Other refinements include methods for choosing which cuts to add, generating good integral solutions, dropping unimportant constraints, and fixing variables at their bounds. We discuss all of these issues, and conclude by presenting the complete algorithm, and some typical computational results.

In what follows, we refer several times to the linear ordering problem and to finding the ground state of an Ising spin glass with no external force, which we call the Ising spin glass problem. We now define those problems.

The linear ordering problem:

> 1. Solve the current relaxation of (IP) approximately using an interior point method.
>
> 2. Generate an integral feasible solution from the approximate primal solution.
>
> 3. If the gap between the best integral solution found so far and the best lower bound provided by a dual solution is sufficiently small, STOP with an optimal solution to the original problem.
>
> 4. Otherwise, use a separation routine to generate cutting planes, add these constraints to the LP relaxation, and return to Step 1.

Figure 9: A conceptual interior point cutting plane algorithm

Given p sectors with costs g_{ij} for placing sector i before sector j for each pair i and j of sectors, find a permutation of the sectors with minimum total cost.

This problem can be represented algebraically as follows:

$$\begin{array}{ll} \min & \sum_{1\leq i\leq p, 1\leq j\leq p, i\neq j} g_{ij} x_{ij} \\ \text{subject to} & x_{ij} + x_{ji} = 1 \text{ for all pairs } i \text{ and } j \\ & x \text{ binary} \\ & x \text{ satisfies the triangle inequalities,} \end{array}$$

where the triangle inequalities require that

$$x_{ij} + x_{jk} + x_{ki} \leq 2$$

for each triple (i, j, k). When this problem is solved using a cutting plane approach, the triangle inequalities are used as cuts. They define facets of the convex hull of feasible solutions. Other facets are known (see Grötschel et al. [48]), but these prove to be unnecessary for many problems.

The Ising spin glass problem:
Given a grid of points on a torus, and given interaction forces c_{ij} between each point and each of its neighbours, partition the vertices into

two sets to minimize the total cost, where the total cost is the sum of all interaction forces between vertices that are in different sets.

The physical interpretation of this problem is that each point possesses either a positive or a negative charge, and the interaction force will be either $+1$ or -1 depending on the charges on the neighbours. The interactions between the points can be measured, but the charges at the points cannot and need to be determined. The Ising spin glass problem is a special case of the *maximum cut problem*:

Given a graph $G = (V, E)$ and edge weights w_{ij}, partition the vertices into two sets to maximize the value of the cut, that is, the sum of the weights of edges where the two ends of the edge are in opposite sets of the partition.

This problem can be represented as an integer program, where x_e indicates whether e is in the cut:

$$\begin{aligned} \min \quad & c^T x \\ \text{subject to} \quad & x \text{ is binary} \\ & x \text{ satisfies the cycle/cut inequalities.} \end{aligned}$$

The cycle/cut inequalities exploit the fact that every cycle and every cut intersect in an even number of edges. They can be stated as

$$x(F) - x(C \setminus F) \leq |F| - 1$$

for sets F of odd cardinality, where C is a cycle in the graph, and $x(S) := \sum_{e \in S} x_e$ for any subset S of the edges. An inequality of this form defines a facet if the cycle C is chordless.

6.2 Solving the relaxations approximately

The principle technique used to make an interior point cutting plane algorithm practical is early termination: the current relaxation is only solved approximately. Typically, the relaxations are solved more exactly as the algorithm proceeds.

There are two main, related, advantages to early termination. In the first place, iterations are saved on the current relaxation and the early solution is usually good enough to enable the efficient detection of cutting planes, so solving the relaxation to optimality would not provide any additional information but would require additional computational effort. Secondly, the approximate solution provides a better starting point for the method on the next relaxation, because it is more centered than the optimal solution.

The disadvantages result from the fact that the solution to the current relaxation may be the optimal solution to the integer program. It is possible that the approximate solution is not in the convex hull of feasible integral solutions, even though the optimal solution is in this set, and so cutting planes may be generated and the relaxation may be modified unnecessarily. On the other hand, if the approximate solution is in the convex hull, the separation routines will not find cutting planes, but time will be wasted in trying to find cuts. The effect of the first drawback can be mitigated by initializing the relaxation with a point that is not too far from the center of the convex hull, and by solving the relaxation to optimality occasionally, for example on every tenth relaxation. This last technique proved to be very useful in the experiments on Ising spin glass problems described in [96].

One way to reduce the cost of the first drawback is to control how accurately the relaxations are solved by using a dynamically adjusted tolerance for the duality gap: one searches for cutting planes once the duality gap falls below this tolerance. If many cutting planes are found, then perhaps one did not need to solve the current relaxation so accurately, so one can increase this tolerance. On the other hand, if only a few cutting planes are found then the tolerance should be decreased. In most of those experiments, the tolerance was initiliazed with a value of 0.3 on the relative duality gap and then was modified by multiplying by a power of 1.1, with the power depending on the number of cuts found and on how badly these cuts were violated by the current iterate.

Other ways to control the accuracy include requiring that the current primal solution should have better value than the best known integral solution, and that the dual solution should be better than the best known lower bound. Perhaps surprisingly, it was observed that the condition based on the dual value is in general too restrictive, forcing the algorithm to perform more iterations than necessary on the current relaxation without resulting in a reduction in the total number of relaxations solved. Mitchell has even had mixed results with the condition on the primal solution: for the linear ordering problem, this condition resulted in an increase in computational times, but it improved runtimes for Ising spin glass problems. (The Ising spin glass problems are harder than the linear ordering problems.)

A more sophisticated condition is to require that the relaxations be solved to an accuracy such that it appears that the optimal value to the relaxation would not be sufficiently good to enable the algorithm to terminate, unless it provided an optimal integral solution. For example, one can require that the average of the primal and dual values should be at least one less than the best known integral value, if all the data is integral. If one solves such a relaxation to optimality, the lower bound would not be sufficient to prove optimality with the current best known integral solution. A similar condition was helpful for the Ising spin glass problems.

6.3 Restarting

When cutting planes are added, the current primal iterate is no longer feasible, and the algorithm must be restarted. It is possible to restart from the current iterate using a primal-dual infeasible interior point method, perhaps with an initial centering step, but it has been observed that other techniques have proved superior in practice.

After adding cutting planes, the primal relaxation becomes

$$\begin{array}{rl}
\min & c^T x \\
\text{subject to} & Ax = b \\
& A^o x + x^o = b^o \quad (LPnew) \\
& 0 \leq x \leq u \\
& 0 \leq x^o \leq u^o,
\end{array}$$

where x_o is the vector of slack variables for the cutting planes given by $A^o x \leq b^o$. The dual ($LDnew$) to this problem can be written as

$$\begin{array}{rl}
\max & b^T y + b^{oT} y^o - u^T w - u^{oT} w^o \\
\text{subject to} & \\
& A^T y + A^{oT} y^o + z - w = c \\
& y^o + z^o - w^o = 0 \\
& z, z^o \quad w, w^o \geq 0.
\end{array}$$

Since one uses an interior point method, and did not solve the last relaxation to optimality, the last iterate is a primal dual pair $\bar{x}$, $(\bar{y}, \bar{z}, \bar{w})$ satisfying $A\bar{x} = b$, $0 < \bar{x} < u$, $A^T \bar{y} + \bar{z} - \bar{w} = c$, $\bar{z} > 0$, $\bar{w} > 0$.

A feasible interior solution to ($LDnew$) is obtained by setting $y^o = 0$ and $z^o = w^o = \epsilon$ for any positive ϵ (a typical value is 10^{-3}). It is often beneficial to update the dual solution to an older iterate than $(\bar{y}, \bar{z}, \bar{w})$, which will be more centered. It is also useful to increase any small components of w or z up to ϵ if necessary; if w_i is increased, then z_i is also increased to maintain dual feasibility, and vice versa.

Primal feasibility is harder to maintain, since $A^o\bar{x} > b$. Possible updating schemes are based upon knowing a point x^Q in the interior of the convex hull of feasible integer points. Of course, such a point will be an interior point in (*LPnew*). One can either update to this point, or to an appropriate convex combination of this point and $\bar{x}$. It is often straightforward to initialize x^Q: for the linear ordering problem and for the maximum cut problem, one can take x^Q to be the vector of halves; for the undirected traveling salesman problem on a complete graph with n cities, one can take each component of x^Q to be $2/(n-1)$ (each component corresponds to an edge; an edge e is in the tour if and only if $x_e = 1$). The point x^Q can be updated by moving towards the current primal iterate or by moving towards the best integral solution found so far. For the Ising spin glass problem, Mitchell found it best not to update x^Q, but to restart by taking a convex combination of x^Q and $\bar{x}$ which was 95% of the way from x^Q to the boundary of new the relaxation. On the other hand, updating x^Q by moving towards the current primal iterate worked well for the linear ordering problem. Another possible restarting scheme is to store earlier iterates, and take the most recent iterate that is feasible in the current relaxation. This works well on some instances, but it is generally outperformed by methods based on x^Q.

6.4 Primal heuristics and termination

A good primal heuristic can save many iterations and stages, especially if the objective function data is integral. The algorithm terminates when the gap between the best known integral solution and the best lower bound drops below some tolerance. If the data is integer, then a tolerance of one is sufficient, so it is not necessary to refine the relaxation to such a degree that it completely describes the convex hull in the region of the optimal integral solution.

The importance of the primal heuristic varies from problem to problem. Mitchell found that his runtimes improved dramatically for the Ising spin glass problem when he implemented a good local search heuristic, even though the heuristic itself required as much as 60% of the total runtime on some large instances. The primal heuristic was not nearly so important for the linear ordering problem, where it was relatively easy to generate the optimal ordering from a very good fractional solution. Another indication of the difference in the importance of the primal heuristic for these two problems could be observed when Mitchell tried to solve them so that the gap between the best integral solution and the lower bound was less than, say, 10^{-6}. The linear ordering problems could be solved almost as easily as before, but the larger spin glass problems became computationally intractable.

6.5 Separation routines

Separation routines are problem specific. Good routines for simplex based cutting plane algorithms can usually be adapted to interior point cutting plane methods. Because the iterates generated by the interior point approach are more centered, it may be possible to find deeper cuts and cutting planes that are more important. This is a topic that warrants further investigation.

One issue that is specific to separation routines for interior point cutting plane algorithms is the effect of the cutting planes on the sparsity of the matrix AA^T. (Here, A represents the whole constraint matrix.) If the structure of this matrix is unfavourable, then a simplex method will outperform an interior point method based on Cholesky factorization, even for the linear programming relaxation (see, for example, [90]). For this reason, it is often useful to add cuts that are variable-disjoint, that is, a particular x_i appears in just one of the constraints added at a particular stage.

6.6 Fixing variables

When using a simplex cutting plane algorithm, it is well known that a variable can be fixed at zero or one if the corresponding reduced cost is sufficiently large (see, for example [108]). The dual variables can be used for the same purpose if an interior point method is used.

When using a cutting plane algorithm, an upper bound v_U on the optimal value is provided by a feasible integral solution. Let $\bar{v}$ be the value of the current dual iterate $(\bar{y}, \bar{z}, \bar{w})$. It was shown in [97] that if $z_i > v_U - \bar{v}$ then x_i must be zero in any optimal integral solution. Similarly, if $\bar{w}_i > v_U - \bar{v}$ then x_i must be one in any optimal solution.

These techniques can be very useful for reducing the size of the relaxations. They are most useful when the objective function data is fractional, since the gap between the upper and lower bounds has to become small in order to prove optimality, so many of the dual variables will eventually be large enough that the integral variables can be fixed.

Notice that if a variable is fixed, and thus eliminated from the relaxation, the point x^Q is no longer feasible. Therefore care has to be taken when restarting the algorithm. In particular, it is useful to examine the logical implications of fixing a variable; it may be possible to fix further variables, or to impose constraints on the remaining variables. For example, when solving a maximum cut problem, if one fixes two of the edges of a cycle of length 3 then the third edge can also be fixed. If one fixes one edge of a cycle of length 3, then the variables for the other two edges can be constrained to either take the same values as each other, or to take opposite values, depending on the value of the fixed edge. Fixing variables and adding these logical constraints can worsen the conditioning of the constraint matrix, perhaps introducing rank deficiencies. Thus, care must be exercised.

6.7 Dropping constraints

Dropping unimportant constraints reduces the size of the relaxation and so enables the relaxation to be solved more quickly. It is possible to develop tests based on ellipsoids to determine when a constraint can be dropped, but the cost of these tests outweighs the computational savings. Therefore, an implementation will generally drop a constraint based on the simple test that its slack variable is large. Of course, it is undesirable to have a constraint be repeatedly added and dropped; a possible remedy is to insist that a constraint cannot be dropped for several stages.

The development of efficient, rigorous tests for dropping constraints would be useful.

6.8 The complete algorithm

The complete algorithm is contained in figure 10.

If a primal feasible solution is known, v^U can be initialized in Step 1 to take the value of that solution; otherwise v^U should be some large number. If all the objective function coefficients c_i correspond to binary variables, then the lower bound v^L can be initialized to be $\sum_i min\{c_i, 0\}$; otherwise, the lower bound can be taken to be a negative number with large absolute value.

6.9 Some computational results

We present some computational results for Ising spin glass problems on grid of sizes up to 100×100 in Table 10. For comparison, De Simone et al. [26] have solved problems of size up to 70×70 with a simplex cutting plane algorithm using CPLEX3.0 on a Sun Sparc 10 workstation, requiring up to a day for each problem. The results in Table 10 were obtained on a Sun

1. **Initialize:** Read in the problem. Set up the initial relaxation. Find initial interior feasible primal and dual points. Find a point x^Q in the interior of the convex hull of feasible integral solutions. Choose a tolerance τ on optimality for the integer program. Choose a tolerance ρ on the duality gap for the relaxation. Initialize the upper and lower bounds v^U and v^L on the optimal value appropriately.

2. **Iterate:** Take a primal-dual predictor-corrector step from the current iterate.

3. **Add cuts?** If the relative duality gap δ is smaller than ρ (and perhaps if other conditions on the primal and dual values are met), then go to Step 4; otherwise, return to Step 2.

4. **Primal heuristic:** Search for a good integral solution, starting from the current primal iterate. Update v^U if a solution is found which is better than this bound.

5. **Check for optimality:** If $v^U - v^L < \tau$, STOP: the best integer solution found so far is optimal.

6. **Search for cutting planes:** Use the separation routines to find cutting planes. If cutting planes are found, go to Step 7. If none are found and $\delta \geq 10^{-8}$, reduce ρ and return to Step 2. If none are found and $\delta < 10^{-8}$ then STOP with a nonoptimal solution; use branch and bound to find the optimal solution.

7. **Modify the relaxation:** Add an appropriate subset of the violated constraints to the relaxation. Increase ρ if it appears that the relaxations do not need to be solved so accurately. Decrease ρ if it appears that the relaxations need to be solved more accurately. Fix any variables if possible, and add any resulting constraints. Drop unimportant constraints.

8. **Restart:** Update the primal and dual solutions to give feasible interior points in the new relaxation. Return to Step 2.

Figure 10: An interior point cutting plane algorithm

L	Sample Size	Mean	Std Dev	Minimum	Maximum
10	100	0.42	0.20	0.17	1.17
20	100	4.87	2.01	1.30	12.48
30	100	24.32	11.84	7.42	87.00
40	100	88.46	43.68	32.50	259.02
50	100	272.86	151.59	96.35	795.50
60	100	860.57	969.79	227.38	7450.18
70	100	1946.14	1286.13	593.57	8370.37
80	100	5504.11	4981.00	1403.27	32470.40
90	100	10984.82	6683.37	2474.20	28785.30
100	100	12030.69	3879.55	3855.02	21922.60

Table 10: Time (seconds) to solve Ising spin glass problems

Sparc 20/71, and are taken from [98]. As can be seen, even the largest problems were solved in an average of less than $3\frac{1}{2}$ hours. They needed approximately nine iterations per relaxation — the later relaxations required more iterations and the earlier relaxations fewer. The primal heuristic took approximately 40% of the total runtime.

6.10 Combining interior point and simplex cutting plane algorithms

Practical experience with interior point cutting plane algorithms has shown that often initially they add a large number of constraints at a time (hundreds or even thousands), and the number of added constraints decreases to just a handful at a time towards the end. The number of iterations to reoptimize increases slightly as optimality is approached, because the relaxations are solved to a higher degree of accuracy.

When a simplex method is used to solve the relaxations, the number of iterations to reoptimize depends greatly on the number of added constraints. Initially, when many constraints are added, the dual simplex method can take a long time to reoptimize, but towards the end it can reoptimize in very few iterations, perhaps as few as ten.

Because of the time required for an iteration of an interior point method, it is very hard to compete with the speed of simplex for solving these last few relaxations. Conversely, the interior point method is considerably faster for the first few stages. The interior point method may also make a better selection of cutting planes in these initial stages, because it is cutting off an interior point that is well-centered, a property that is intensified because it is looking for cutting planes before termination.

Mitchell and Borchers [100] investigated solving linear ordering problems with a cutting plane code that uses an interior point method for the first few stages and a dual simplex method for the last few stages. Computational results are contained in table 11. These problems have up to 250 sectors,

n	% zeros	Interior	Simplex	Combined
150	0	206	75	68
200	0	755	385	209
250	0	4492	3797	592
100	20%	1405	1296	230
150	10%	2247	1294	208
200	10%	N/A	9984	879

Table 11: Preliminary Results on Linear Ordering Problems.

with a percentage of the cost entries zeroed out. The nonzero costs above the diagonal were uniformly distributed between 0 and 99, and those below the diagonal were uniformly distributed between 0 and 39. The table contains runtimes in seconds on a Sun SPARC 20/71 for an interior point cutting plane code, a simplex cutting plane code using CPLEX 4.0, and a combined cutting plane code. The interior point code was unable to solve the problems with 200 sectors and 20% of the entries zeroed out because of space limitations. As can be seen the combined code is more than 10 times faster than the simplex code on the largest problems, and the interior point and simplex codes require similar amounts of time, at least on the harder problems.

6.11 Interior point column generation methods for other problems

A cutting plane method can be regarded as a column generation method applied to the dual problem. Interior point methods have been successfully applied in several other situations amenable to solution by a column generation approach. Goffin et al. [42] have solved nondifferentiable optimization problems. Bahn et al. [9] have used an interior point method within the L-shaped decomposition method of Van Slyke and Wets [136] for stochastic programming problems. Goffin et al. [41] have also solved multicommodity network flow problems using an interior point column generation approach. In this method, the columns correspond to different paths from an origin to a destination, and they are generated by solving a shortest path problem with an appropriate cost vector.

6.12 Theoretical issues and future directions

As mentioned earlier, the ellipsoid algorithm can be used to solve an integer programming problem in polynomial time if the separation problem can be solved in polynomial time. It is not currently known how to use an interior point method in an exactly analogous way. Atkinson and Vaidya [8] developed an interior point algorithm for this process, but their algorithm requires that unimportant constraints be dropped, unlike the ellipsoid algorithm. Vaidya later obtained a similar result for an algorithm that used the volumetric center [138]. Goffin et al. [44] have proposed a fully polynomial algorithm that does not require that unimportant constraints be removed. It is an interesting open theoretical question to find an interior point algorithm that does not require that unimportant constraints be removed, and also solves the optimization problem in polynomial time provided the separation problem can be solved in polynomial time.

The algorithms proposed in [8, 138, 44] required that only a single constraint be added at a time, and that the constraint be added far from the current iterate. These algorithms have been extended to situations where many cuts are added at once, and the constraints are added right through the current iterate, with no great increase in the complexity bound [124, 125, 43]. It has been shown that if p constraints are added through the analytic center then the analytic center of the new feasible region can be found in $O(\sqrt{p})$ iterations [124].

There are several open computational questions with interior point cutting plane methods. Combining interior point methods with the simplex algorithm needs to be investigated further. When a direct method is used to calculate the Newton direction, it is necessary to choose an ordering of the columns of AA^T to reduce fill in the Cholesky factor; it would be interesting to see if the ordering from one stage can be efficiently modified to give an ordering for the next stage, rather than calculating an ordering from scratch. When the constraint matrix contains many dense columns, it becomes expensive to use a direct method to calculate the Newton direction; it would be interesting to examine whether it is efficient to switch to a preconditioned conjugate gradient method in the later stages.

7 Nonconvex potential function minimization

Consider the problem of maximizing a convex quadratic function defined as

$$\max w^T w = \sum_{i=1}^{m} w_i^2 \qquad (62)$$

subject to

$$A^T w \le b. \qquad (63)$$

The significance of this optimization problem is that many combinatorial optimization problems can be formulated as above with the additional requirement that the variables are binary.

In [73, 77] a new affine scaling algorithm was proposed for solving the above problem using a logarithmic potential function. Consider the nonconvex optimization problem

$$\min \{\varphi(w) \mid A^T w \le b\}, \qquad (64)$$

where

$$\varphi(w) = \log(m - w^T w)^{1/2} - \frac{1}{n}\sum_{i=1}^{n} \log d_i(w) \qquad (65)$$

$$= \log\left\{\frac{m - w^T w}{2\prod_{i=1}^{n} d_i(w)^{1/n}}\right\} \qquad (66)$$

and where

$$d_i(w) = b_i - a_i^T w, \quad i = 1, \ldots, n, \qquad (67)$$

are the slacks. The denominator of the log term of $\varphi(w)$ is the geometric mean of the slacks and is maximized at the analytic center of the polytope defined by

$$\mathcal{L} = \left\{w \in \mathbb{R}^m \mid A^T w \le b\right\}.$$

To find a local (perhaps global) solution of (64), an approach similar to the classical Levenberg-Marquardt methods [86, 91] is used. Let

$$w^0 \in \mathcal{L}^0 = \left\{w \in \mathbb{R}^m \mid A^T w < b\right\}$$

be a given initial interior point. The algorithm generates a sequence of interior points of $\mathcal{L}$.

Let $w^k \in \mathcal{L}^0$ be the k-th iterate. Around w^k a quadratic approximation of the potential function is set up. Let $D = \text{diag}(d_1(w), \ldots, d_n(w))$, $e = (1, \ldots, 1)$, $f_0 = m - w^T w$ and C be a constant. The quadratic approximation of $\varphi(w)$ around w^k is given by

$$Q(w) = \frac{1}{2}(w - w^k)^T H(w - w^k) + h^T(w - w^k) + C \tag{68}$$

where the Hessian is

$$H = -\frac{1}{f_0} I - \frac{2}{f_0^2} w^k w^{kT} + \frac{1}{n} A D^{-2} A^T \tag{69}$$

and the gradient is

$$h = -\frac{1}{f_0} w^k + \frac{1}{n} A D^{-1} e. \tag{70}$$

Recall that minimizing (68) over a polytope is NP-complete. However, if the polytope is substituted by an inscribed ellipsoid, the resulting approximate problem can be solved in polynomial time [147]. Since preliminary implementations of this algorithm indicate that trust region methods are more efficient for solving these problems, in the discussion that follows we consider a trust region approach.

Consider the ellipsoid

$$\mathcal{E}(r) = \left\{ w \in \mathbb{R}^m \mid (w - w^k)^T A D^{-2} A^T (w - w^k) \leq r^2 \right\}.$$

To see that the ellipsoid $\mathcal{E}(r)$ is inscribed in the polytope $\mathcal{L}$, assume that $r = 1$ and let $y \in \mathcal{E}(1)$. Then

$$(y - w^k)^T A D^{-2} A^T (y - w^k) \leq 1$$

and consequently

$$D^{-1} A^T (y - w^k) \leq e,$$

where $w^k \in \mathcal{L}^0$. Denoting the i-th row of A^T by $a_{i\cdot}^T$, we have

$$\frac{1}{b_i - a_{i\cdot}^T w^k} a_{i\cdot}^T (y - w^k) \leq 1, \; \forall i = 1, \ldots, n.$$

Hence,

$$a_{i\cdot}^T (y - w^k) \leq b_i - a_{i\cdot}^T w^k, \; \forall i = 1, \ldots, n,$$

Interior Point Methods for Combinatorial Optimization

and consequently
$$a_{i.}^T y \leq b_i, \quad \forall i = 1, \ldots, n,$$

i.e. $A^T y \leq b$, showing that $y \in \mathcal{L}$. This shows that $\mathcal{E}(1) \subset \mathcal{L}$ and since $\mathcal{E}(r) \subset \mathcal{E}(1)$, for $0 \leq r < 1$, then $\mathcal{E}(r) \subset \mathcal{L}$, i.e. $\mathcal{E}(r)$ is an inscribed ellipsoid in $\mathcal{L}$.

Substituting the polytope by the appropriate inscribed ellipsoid and letting $\Delta w \equiv w - w^k$ results in the minimization of a quadratic function over an ellipsoid, i.e.

$$\min \frac{1}{2}(\Delta w)^T H \Delta w + h^T \Delta w \tag{71}$$

subject to

$$(\Delta w)^T A D^{-2} A^T (\Delta w) \leq r^2. \tag{72}$$

The optimal solution Δw^* to (71–72) is a descent direction of $Q(w)$ from w^k. For a given radius $r > 0$, the value of the original potential function $\varphi(w)$ may increase by moving in the direction Δw^*, because of the higher order terms ignored in the approximation. It can be easily verified, however, that if the radius is decreased sufficiently, the value of the potential function will decrease by moving in the new Δw^* direction. We shall say a *local minimum* to (64) has been found if the radius must be reduced below a tolerance ϵ to achieve a reduction in the value of the potential function.

The following result, proved in [73], characterizes the optimal solution of (71–72). Using a linear transformation, the problem is transformed into the minimization of a quadratic function over a sphere.

Consider the optimization problem

$$\min \frac{1}{2} x^T Q x + c^T x \tag{73}$$

subject to

$$x^T x \leq r^2, \tag{74}$$

where $Q \in \mathbb{R}^{m \times m}$ is symmetric and indefinite, $x, c \in \mathbb{R}^m$ and $0 < r \in \mathbb{R}$. Let $u_1, \ldots, u_m$ denote a full set of orthonormal eigenvectors spanning $\mathbb{R}^m$ and let $\lambda_1, \ldots, \lambda_m$ be the corresponding eigenvalues ordered so that $\lambda_1 \leq \lambda_2 \leq \cdots \leq \lambda_{m-1} \leq \lambda_m$. Denote $0 > \lambda_{min} = \min\{\lambda_1, \ldots, \lambda_m\}$ and u_{min} the corresponding eigenvector. Furthermore, let q be such that $\lambda_{min} = \lambda_1 = \cdots = \lambda_q < \lambda_{q+1}$. To describe the solution to (73–74) consider two cases:

Case 1: Assume $\sum_{i=1}^{q}(c^T u_i)^2 > 0$. Let the scalar $\lambda \in (-\infty, \lambda_{min})$ and consider the parametric family of vectors

$$x(\lambda) = -\sum_{i=1}^{m} \frac{(c^T u_i) u_i}{\lambda_i - \lambda}.$$

For any $r > 0$, denote by $\lambda(r)$ the unique solution of the equation $x(\lambda)^T x(\lambda) = r^2$ in λ. Then $x(\lambda(r))$ is the unique optimal solution of (73-74).

Case 2: Assume $c^T u_i = 0, \forall i = 1, \ldots, q$. Let the scalar $\lambda \in (-\infty, \lambda_{min})$ and consider the parametric family of vectors

$$x(\lambda) = -\sum_{i=q+1}^{m} \frac{(c^T u_i) u_i}{\lambda_i - \lambda}. \tag{75}$$

Let

$$r_{max} = \|x(\lambda_{min})\|_2.$$

If $r < r_{max}$ then for any $0 < r < r_{max}$, denote by $\lambda(r)$ the unique solution of the equation $x(\lambda)^T x(\lambda) = r^2$ in λ. Then $x(\lambda(r))$ is the unique optimal solution of (73-74).

If $r \geq r_{max}$, then let $\alpha_1, \alpha_2, \ldots, \alpha_q$ be any real scalars such that

$$\sum_{i=1}^{q} \alpha_i^2 = r^2 - r_{max}^2.$$

Then

$$x = \sum_{i}^{q} \alpha_i u_i - \sum_{i=q+1}^{m} \frac{(c^T u_i) u_i}{(\lambda_i - \lambda_{min})}$$

is an optimal solution of (73-74). Since the choice of α_i's is arbitrary, this solution is not unique.

This shows the existence of a unique optimal solution to (73-74) if $r < r_{max}$. The proof of this result is based on another fact, used to develop the algorithm described in [73, 77], that we state next.

Let the length of $x(\lambda)$ be

$$l(x(\lambda)) \equiv \|x(\lambda)\|_2^2 = x(\lambda)^T x(\lambda),$$

then $l(x(\lambda))$ is monotonically increasing in λ in the interval $\lambda \in (-\infty, \lambda_{min})$. To see this is so, consider two cases. First, assume $\sum_{i=1}^{q}(c^T u_i)^2 > 0$. Consider the parametric family of vectors

$$x(\lambda) = -\sum_{i=1}^{m} \frac{(c^T u_i) u_i}{\lambda_i - \lambda},$$

for $\lambda \in (-\infty, \lambda_{min})$. Now, assume that $c^T u_i = 0, \forall i = 1, \ldots, q$ and consider the parametric family of vectors

$$x(\lambda) = -\sum_{i=q+1}^{m} \frac{(c^T u_i) u_i}{\lambda_i - \lambda}, \qquad (76)$$

for $\lambda \in (-\infty, \lambda_{min})$. Furthermore, assume

$$r < \|x(\lambda_{min})\|_2.$$

Then $l(x(\lambda))$ is monotonically increasing in λ in the interval $\lambda \in (-\infty, \lambda_{min})$.

The above result suggests an approach to solve the nonconvex optimization problem (64). At each iteration, a quadratic approximation of the potential function $\varphi(w)$ around the iterate w^k is minimized over an ellipsoid inscribed in the polytope $\{w \in R^m | A^T w \le b\}$ and centered at w^k. Either a descent direction Δw^* of $\varphi(w)$ is produced or w^k is said to be a local minimum. A new iterate w^{k+1} is computed by moving from w^k in the direction Δw^* such that $\varphi(w^{k+1}) < \varphi(w^k)$. This can be done by moving a fixed step α in the direction Δw^* or by doing a line search to find α that minimizes the potential function $\varphi(w^k + \alpha \Delta w^*)$ [134].

Figure 11 shows a pseudo-code procedure cmq, for finding a local minimum of the convex quadratic maximization problem. Procedure cmq takes as input the problem dimension n, the A matrix, the b right hand side vector, an initial estimate μ_0 of parameter μ and initial lower and upper bounds on the acceptable length, $\underline{l}_0$ and $\bar{l}_0$, respectively. In line 2, get_start_point returns a strict interior point of the polytope under consideration, i.e. $w^k \in \mathcal{L}^0$.

The algorithm iterates in the loop between lines 3 and 13, terminating when a local optimum is found. At each iteration, a descent direction of the potential function $\varphi(w)$ is produced in lines 4 through 8. In line 4, the minimization of a quadratic function over an ellipsoid (71–72) is solved. Because of higher order terms the direction returned by descent_direction may not be a descent direction for $\varphi(w)$. In this case, loop 5 to 8 is repeated until an improving direction for the potential function is produced or the largest acceptable length falls below a given tolerance ϵ.

If an improving direction for $\varphi(w)$ is found, a new point w^{k+1} is defined (in line 10) by moving from the current iterate w^k in the direction Δw^* by a step length $\alpha < 1$.

```
procedure cmq(n, A, b, μ₀, l₀, l̄₀)
1     k = 0;  γ = 1/(μ₀ + 1/n);  l = l₀;  l̄ = l̄₀;  K = 0;
2     wᵏ = get_start_point(A, b);
3     do l̄ > ε →
4         Δw* = descent_direction(γ, wᵏ, l, l̄);
5         do φ(wᵏ + αΔw*) ≥ φ(wᵏ) and l̄ > ε →
6             l̄ = l̄/l̄ᵣ;
7             Δw* = descent_direction(γ, wᵏ, l, l̄);
8         od;
9         if φ(wᵏ + αΔw*) < φ(wᵏ) →
10            wᵏ⁺¹ = wᵏ + αΔw*;
11            k = k + 1;
12        fi;
13    od;
end cmq;
```

Figure 11: Procedure cmq: Algorithm for nonconvex potential function minimization

7.1 Computing the descent direction

Now consider in more detail the computation of the descent direction for the potential function. The algorithm described in this section is similar to the trust region method described in Moré and Sorensen [104].

As discussed previously, the algorithm solves the optimization problem

$$\min \frac{1}{2}(\Delta w)^T H \Delta w + h^T \Delta w \qquad (77)$$

subject to

$$(\Delta w)^T A D^{-2} A^T \Delta w \leq r^2 \leq 1 \qquad (78)$$

to produce a descent direction Δw^* for the potential function $\varphi(w)$. A solution $\Delta w^* \in R^m$ to (77–78) is optimal if and only if there exists $\mu \geq 0$ such that

$$\left(H + \mu A D^{-2} A^T\right) \Delta w^* = -h \qquad (79)$$

$$\mu \left((\Delta w^*)^T A D^{-2} A^T \Delta w^* - r^2\right) = 0 \qquad (80)$$

$$H + \mu A D^{-2} A^T \text{ is positive semidefinite.} \qquad (81)$$

With the change of variables $\gamma = 1/(\mu + 1/n)$ and substituting the Hessian (69) and the gradient (70) into (79) we obtain

$$\Delta w^* = -\left(AD^{-2}A^T - \frac{2\gamma}{f_0^2} w^k w^{kT} - \frac{\gamma}{f_0} I\right)^{-1} \times \\ \gamma \left(-\frac{1}{f_0} w^k + \frac{1}{n} AD^{-1} e\right) \qquad (82)$$

that satisfies (79). Note that r does not appear in (82) and that (82) is not defined for all values of r. However, if the radius r of the ellipsoid (78) is kept within a certain range, then there exists an interval $0 \leq \gamma \leq \gamma_{max}$ such that

$$AD^{-2}A^T - \frac{2\gamma}{f_0^2} w^k w^{kT} - \frac{\gamma}{f_0} I \qquad (83)$$

is nonsingular. Next, we show that for γ small enough Δw^* is a descent direction of $\varphi(w)$. Note that

$$\Delta w^* = -\left(AD^{-2}A^T - \frac{2\gamma}{f_0^2} w^k w^{kT} - \frac{1\gamma}{f_0} I\right)^{-1} \gamma \left(-\frac{1}{f_0} w^k + \frac{1}{n} AD^{-1} e\right)$$

$$= -\left[AD^{-2}A^T\left\{I - \gamma(AD^{-2}A^T)^{-1}\left(-\frac{2}{f_0^2}w^k w^{kT} - \frac{1}{f_0}I\right)\right\}\right]^{-1} \times$$
$$\gamma\left(-\frac{1}{f_0}w^k + \frac{1}{n}AD^{-1}e\right)$$
$$= -\gamma\left[I + \gamma(AD^{-2}A^T)^{-1}\left(\frac{2}{f_0^2}w^k w^{kT} + \frac{1}{f_0}I\right)\right]^{-1}(AD^{-2}A^T)^{-1} \times$$
$$\left(-\frac{1}{f_0}w^k + \frac{1}{n}AD^{-1}e\right)$$
$$= \gamma\left[I + \gamma(AD^{-2}A^T)^{-1}\left(\frac{2}{f_0^2}w^k w^{kT} + \frac{1}{f_0}I\right)\right]^{-1} \times$$
$$(AD^{-2}A^T)^{-1}(-h). \tag{84}$$

Let $\gamma = \epsilon > 0$ and consider $\lim_{\epsilon \to 0^+} h^T \Delta w^*$. Since

$$\lim_{\epsilon \to 0^+} \Delta w^* = \epsilon\,(AD^{-2}A^T)^{-1}(-h)$$

then

$$\lim_{\epsilon \to 0^+} h^T \Delta w^* = -\epsilon\, h^T(AD^{-2}A^T)^{-1}h.$$

Since, by assumption, $\epsilon > 0$ and $h^T(AD^{-2}A^T)^{-1}h > 0$ then

$$\lim_{\epsilon \to 0^+} h^T \Delta w^* < 0,$$

showing that there exists $\gamma > 0$ such that the direction Δw^*, given in (82), is a descent direction of $\varphi(w)$.

The idea of the algorithm is to solve (77–78), more than once if necessary, with the radius r as a variable. Parameter γ is varied until r takes a value in some given interval. Each iteration of this algorithm is comprised of two tasks. To simplify notation, let

$$H_c = AD^{-2}A^T \tag{85}$$

$$H_o = -\frac{2}{f_0^2}w^k w^{kT} - \frac{1}{f_0}I \tag{86}$$

and define

$$M = H_c + \gamma H_o.$$

Given the current iterate w^k, we first seek a value of γ such that $M\Delta w = \gamma h$ has a solution Δw^*. This can be done by binary search, as we will see shortly. Once such a parameter γ is found, the linear system

$$M\Delta w^* = \gamma h \tag{87}$$

is solved for $\Delta w^* \equiv \Delta w^*(\gamma(r))$. As was shown previously, the length $l(\Delta w^*(\gamma))$ is a monotonically increasing function of γ in the interval $0 \leq \gamma \leq \gamma_{max}$. Optimality condition (80) implies that $r = \sqrt{l(\Delta w^*(\gamma))}$ if $\mu > 0$. Small lengths result in small changes in the potential function, since r is small and the optimal solution lies on the surface of the ellipsoid. A length that is too large may not correspond to an optimal solution of (77–78), since this may require $r > 1$. An interval $(\underline{l}, \overline{l})$ called the *acceptable length region*, is defined such that a length $l(\Delta w^*(\gamma))$ is accepted if $\underline{l} \leq l(\Delta w^*(\gamma)) \leq \overline{l}$. If $l(\Delta w^*(\gamma)) < \underline{l}$, γ is increased and (87) is resolved with the new M matrix and h vector. On the other hand, if $l(\Delta w^*(\gamma)) > \overline{l}$, γ is reduced and (87) is resolved. Once an acceptable length is produced we use $\Delta w^*(\gamma)$ as the descent direction.

Figure 12 presents pseudo-code for procedure descent_direction, where (77–78) is optimized. As input, procedure descent_direction is given an estimate for parameter γ, the current iterate w^k around which the inscribing ellipsoid is to be constructed and the current acceptable length region defined by $\underline{l}$ and $\overline{l}$. The value of γ passed to descent_direction at minor iteration k of cmq is the value returned by descent_direction at minor iteration $k - 1$. It returns a descent direction Δw^* of the quadratic approximation of the potential function $Q(w)$ from w^k, the next estimate for parameter γ and the current lower bound of the acceptable length region $\underline{l}$.

In line 1, the length l is set to a large number and several logical keys are initialized: LD_{key} is **true** if a linear dependency in the rows of M is ever found during the solution of the linear system (87) and is **false** otherwise; $\overline{\gamma}_{key}$ ($\underline{\gamma}_{key}$) is **true** if an upper (lower) bound for an acceptable γ has been found and **false** otherwise.

The problem of minimizing a nonconvex quadratic function over an ellipsoid is carried out in the loop going from line 2 to 19. The loop is repeated until either a length l is found such that $\underline{l} \leq l \leq \overline{l}$ or $l \leq \underline{l}$ due to a linear dependency found during the solution of (87), i.e. if $LD_{key} =$ **true**. Lines 3 to 8 produce a descent direction that may not necessarily have an acceptable length. In line 3 the matrix M and the right hand side vector b are formed. The linear system (87) is tentatively solved in line 4. The solution procedure may not be successful, i.e. M may be singular. This implies that parameter γ is too large and parameter γ is reduced in line 5 of loop 4–7, which is repeated until a nonsingular matrix M is produced.

Once a nonsingular M matrix is available, a descent direction Δw^* is computed in line 8 along with its corresponding length l. Three cases can

```
procedure descent_direction(γ, w^k, l̲, l̄)
1    l = ∞;  LD_key = false;  γ̄_key = false;  γ̲_key = false;
2    do l > l̄ or (l < l̲ and LD_key = false) →
3        M = H_c + γH_o;  b = γh;
4        do MΔw = b has no solution →
5            γ = γ/γ_r;  LD_key = true;
6            M = H_c + γH_o;  b = γh;
7        od;
8        Δw* = M^{-1}b;  l = (Δw*)^T AD^{-2} A^T Δw*;
9        if l < l̲ and LD_key = false →
10           γ̲ = γ;  γ̲_key = true;
11           if γ̄_key = true → γ = √(γ γ̄) fi;
12           if γ̄_key = false → γ = γ · γ_r fi;
13       fi;
14       if l > l̄ →
15           γ̄ = γ;  γ̄_key = true;
16           if γ̲_key = true → γ = √(γ γ̲) fi;
17           if γ̲_key = false → γ = γ/γ_r fi;
18       fi;
19   od;
20   do l < l̲ and LD_key = true → l̲ = l̲/l_r od;
21   return(Δw*);
end descent_direction;
```

Figure 12: Procedure descent_direction: Algorithm to compute descent direction in nonconvex potential function minimization

occur: (*i*) - the length is too small even though no linear dependency was detected in the factorization; (*ii*) - the length is too large; or (*iii*) - the length is acceptable. Case (*iii*) is the termination condition for the main loop 2-19. In lines 9-13 the first case is considered. The value of γ is a lower bound on an acceptable value of γ and is recorded in line 10 and the corresponding logical key is set. If an upper bound $\overline{\gamma}$ for an acceptable value of γ has been found the new estimate for γ is set to the geometric mean of $\underline{\gamma}$ and $\overline{\gamma}$ in line 11. Otherwise γ is increased by a fixed factor in line 12.

Similar to the treatment of case (*i*), case (*ii*) is handled in lines 14-18. The current value of γ is an upper bound on an acceptable value of γ and is recorded in line 15 and the corresponding logical key is set. If a lower bound $\underline{\gamma}$ for an acceptable value of γ has been found the new estimate for γ is set to the geometric mean of $\underline{\gamma}$ and $\overline{\gamma}$ in line 16. Otherwise γ is decreased by a fixed factor in line 17.

Finally, in line 20, the lower bound $\underline{l}$ may have to be adjusted if $l < \underline{l}$ and LD_{key} = **true**. Note that the key LD_{key} is used only to allow the adjustment in the range of the acceptable length, so that the range returned contains the current length l.

7.2 Some computational considerations

The density of the linear system solved at each iteration of `descent_direction` is determined by the density of the Hessian matrix. Using the potential function described in the previous section, this Hessian,

$$M = AD^{-2}A^T - \frac{2}{f_0^2}w^k w^{kT} - \frac{1}{f_0}I,$$

is totally dense, because of the rank one component $\frac{2}{f_0^2}w^k w^{kT}$. Consequently, direct factorization solution techniques must be ruled out for large instances. However, in the case where the matrix A is sparse, iterative methods can be applied to approximately solve the linear system. In [71], a preconditioned conjugate gradient algorithm, using diagonal preconditioning, was used to solve the system efficiently taking advantage of the special structure of the coefficient matrix. In this approach, the main computational effort is the multiplication of a dense vector ξ and the coefficient matrix M, i.e. $M\xi$. This multiplication can be done efficiently, by considering fact that M is the sum of three matrices, each of which has special structure. The first

multiplication,
$$\frac{1}{f_0}I\xi$$
is simply a scaling of ξ. The second product,
$$\frac{2}{f_0^2}w^k w^{k^T}\xi$$
is done in two steps. First, an inner product $w^{k^T}\xi$ is computed. Then, the vector $\frac{4}{f_0^2}w^k$ is scaled by the inner product. The third product,
$$AD^{-2}A^T\xi$$
is done in three steps. First the product $A^T\xi$ is carried out. The resulting vector is scaled by D^{-2} and multiplies A. Therefore, if A is sparse, the entire matrix vector multiplication can be done efficiently.

In a recent study, Warners et al. [142] describe a new potential function
$$\phi_\rho(w) = m - w^T w - \sum_{i=1}^n \rho_i \log d_i(w),$$
whose gradient and Hessian are given by
$$h = -2w + AD^{-1}\rho,$$
and
$$H = -2I + AD^{-1}PD^{-1}A^T,$$
where $\rho = (\rho_1, \ldots, \rho_n)$ and $P = \text{diag}(\rho)$. Note that the density of the Hessian depends only on the density of AA^T. Consequently, direct factorization methods can be used efficiently when the density of AA^T is small.

7.3 Application to combinatorial optimization

The algorithms discussed in this section have been applied to the following integer programming problem: Given $A' \in \mathsf{R}^{m \times n'}$ and $b' \in \mathsf{R}^{n'}$, find $w \in \mathsf{R}^m$ such that:
$$A'^T w \leq b' \tag{88}$$
$$w_i = \{-1, 1\}, \; i = 1, \ldots, m. \tag{89}$$

The more common form of integer programming, where variables x_i take on (0,1) values, can be converted to the above form with the change of variables

$$x_i = \frac{1 + w_i}{2}, \; i = 1, \ldots, m.$$

More specifically, let I denote an $m \times m$ identity matrix,

$$A = \begin{bmatrix} A' \vdots I \vdots -I \end{bmatrix} \in \mathsf{R}^{m \times n}$$

and

$$b = \begin{bmatrix} b' \\ 1 \\ \vdots \\ 1 \end{bmatrix} \in \mathsf{R}^n$$

and let

$$\mathcal{I} = \left\{ w \in \mathsf{R}^m \mid A^T w \leq b \text{ and } w_i = \{-1, 1\} \right\}.$$

With this notation, we can state the integer programming problem as: Find $w \in \mathcal{I}$.

As before, let

$$\mathcal{L} = \left\{ w \in \mathsf{R}^m \mid A^T w \leq b \right\}$$

and consider the linear programming relaxation of (88–89), i.e. find $w \in \mathcal{L}$. One way of selecting ±1 integer solutions over fractional solutions in linear programming is to introduce the quadratic objective function,

$$\max \; w^T w = \sum_{i=1}^{m} w_i^2$$

and solve the nonconvex quadratic programming problem (62–63). Note that $w^T w \leq m$, with the equality only occurring when $w_j = \pm 1$, $j = 1, \ldots, m$. Furthermore, if $w \in \mathcal{I}$ then $w \in \mathcal{L}$ and $w_i = \pm 1$, $i = 1, \ldots, m$ and therefore $w^T w = m$. Hence, if w is the optimal solution to (62–63) then $w \in \mathcal{L}$. If $w^T w = m$ then $w_i = \pm 1$, $i = 1, \ldots, m$ and therefore $w \in \mathcal{I}$. Consequently, this shows that if $w \in \mathcal{L}$ then $w \in \mathcal{I}$ if and only if $w^T w = m$.

In place of (62–63), one solves the nonconvex potential function minimization

$$\min \; \{\varphi(w) \mid A^T w \leq b\}, \tag{90}$$

where $\varphi(w)$ is given by (65–67). The generally applied scheme rounds each iterate to an integer solution, terminating if a feasible integer solution is produced. If the algorithm converges to a nonglobal local minimum of (90), then the problem is modified by adding a cut and the algorithm is applied to the augmented problem. Let v be the integer solution rounded off from the local minimum. A valid cut is

$$v^T w \leq m - 2. \qquad (91)$$

Observe that if $w = v$ then $v^T w = m$. Otherwise, $v^T w \leq m - 2$. Therefore, the cut (91) excludes v but does not exclude any other feasible integral solution of (88–89).

We note that adding a cut of the type above will not, theoretically, prevent the algorithm from converging to the same local minimum twice. In practice [77], the addition of the cut changes the objective function, consequently altering the trajectory followed by the algorithm.

Most combinatorial optimization problems have very natural equivalent integer and quadratic programming formulations [113]. The algorithms described in this section have been applied to a variety of problems, including maximum independent set [78], set covering [77], satisfiability [71, 134], inductive inference [69, 70], and frequency assignment in cellular telephone systems [143].

8 A lower bounding technique

A lower bound for the globally optimal solution of the quadratic program

$$\min\ q(x) = \frac{1}{2} x^T Q x + c^T x \qquad (92)$$

subject to

$$x \in \mathcal{P} = \{x \in R^n \mid Ax = b,\ x \geq 0\}, \qquad (93)$$

where $Q \in \mathsf{R}^{n \times n}$, $A \in \mathsf{R}^{m \times n}$, $c \in \mathsf{R}^n$, and $b \in \mathsf{R}^m$, can be obtained by minimizing the objective function over the largest ellipsoid inscribed in $\mathcal{P}$. This technique can be applied to quadratic integer programming, a problem that is NP-hard in the general case. Kamath and Karmarkar [66] proposed a polynomial time interior point algorithm for computing these bounds. This is one of the first computational approaches to solve semidefinite programming relaxations. The problem is solved as a minimization of the trace of a matrix subject to positive definiteness conditions. The algorithm takes no more than $O(nL)$ iterations (where L is the the number of bits required to represent the input). The algorithm does two matrix inversions per iteration.

Consider the quadratic integer program

$$\min f(x) = x^T Q x \tag{94}$$

subject to

$$x \in S = \{-1, 1\}^n, \tag{95}$$

where $Q \in \mathsf{R}^{n \times n}$ is symmetric. Let f_{min} be the value of the optimal solution of (94–95).

Consider the problem of finding good lower bounds on f_{min}. To apply an interior point method to this problem, one needs to embed the discrete set S in a continuous set $T \supseteq S$. Clearly, the minimum of $f(x)$ over T is a lower bound on f_{min}.

A commonly used approach is to choose the continuous set to be the box

$$B = \{x \in \mathsf{R}^n \mid -1 \le x_i \le 1, i = 1, \ldots, n\}.$$

However, if $f(x)$ is not convex, the problem of minimizing $f(x)$ over B is NP-hard. Consider this difficult case, and therefore assume that Q has at least one negative eigenvalue. Since optimizing over a box can be hard, instead enclose the box in an ellipsoid E. Let

$$U = \{w = (w_1, \ldots, w_n) \in \mathsf{R}^n \mid \sum_{i=1}^n w_i = 1 \text{ and } w_i > 0,\, i = 1, \ldots, n, \},$$

and consider the parameterized ellipsoid

$$E(w) = \{x \in \mathsf{R}^n \mid x^T W x \le 1\},$$

where $w \in U$ and $W = \text{diag}(w)$.

Clearly, the set S is contained in $E(w)$. If $\lambda_{min}(w)$ is the minimum eigenvalue of $W^{-1/2} Q W^{-1/2}$, then

$$\min \frac{x^T Q x}{x^T W x} = \min \frac{x^T W^{-1/2} Q W^{-1/2} x}{x^T x} = \lambda_{min}(w),$$

and therefore

$$x^T Q x \ge \lambda_{min}(w),\ \forall x \in E(w).$$

Hence, the minimum value of $f(x)$ over $E(w)$ can be obtained by simply computing the minimum eigenvalue of $W^{-1/2}QW^{-1/2}$. To further improve the bound on f_{min} requires that $\lambda_{min}(w)$ be maximized over the set U. Therefore, the problem of finding a better lower bound is transformed into the optimization problem

$$\max \mu$$

subject to

$$\frac{x^T Q x}{x^T W x} \geq \mu, \forall x \in \mathsf{R}^n \setminus \{0\} \text{ and } w \in U.$$

One can further simplify the problem by defining $d = (d_1, \ldots, d_n) \in \mathsf{R}^n$ such that $\sum_{i=1}^n d_i = 0$. Let $D = \text{diag}(d)$. If

$$\frac{x^T (Q - D) x}{x^T W x} \geq \mu,$$

then, since $\sum_{i=1}^n w_i = 1$ and $\sum_{i=1}^n d_i = 0$,

$$x^T Q x \geq \mu x^T W x + x^T D x = \mu,$$

for $x \in S$. Now, define $z = \mu w + d$ and let $Z = \text{diag}(z)$. For all $x \in S$,

$$x^T Z x = e^T z = \mu,$$

and therefore the problem becomes

$$\max e^T z$$

subject to

$$x^T (Q - Z) x \geq 0.$$

Let $M(z) = Q - Z$. Observe that solving the above problem amounts to minimizing the trace of $M(z)$ while keeping $M(z)$ positive semidefinite. Since $M(z)$ is real and symmetric, it has n real eigenvalues $\lambda_i(M(z)), i = 1, \ldots, n$. To ensure positive definiteness, the eigenvalues of $M(z)$ must be nonnegative. Hence, the above problem is reformulated as

$$\min \text{tr}(M(z))$$

subject to

$$\lambda_i(M(z)) \geq 0, i = 1, \ldots, n.$$

```
procedure qplb(Q, ε, z, opt)
1     z^(0) = (λ_min(Q) - 1)e;
2     v^(0) = 0;
3     M(z^(0)) = Q - Z^(0);  k = 0;
4     do tr(M(z^(k))) - v^(k) ≥ ε →
5         Construct H^(k) where H_{ij}^(k) = (e_i^T M(z^(k))^{-1} e_j)^2;
6         f^(k)(z) = 2n ln(tr(M(z^(k))) - v^(k)) - ln det M(z^(k));
7         g^(k) = ∇f^(k)(z^(k));
8         β = 0.5/√(g^(k)T H^(k)-1 g^(k));
9         Solve H^(k) Δz = -β g^(k);
10        if g^(k)T Δz < 0.5 →
11            Increase v^(k) until g^(k)T Δz = 0.5;
12        fi;
13        z^(k+1) = z^(k) + Δz;  k = k + 1;
14    od;
15    z = z^(k); opt = tr(Q) - v^(k);
end qplb;
```

Figure 13: Procedure qplb: Interior point algorithm for computing lower bounds

Kamath and Karmarkar [66, 67] proposed an interior point approach to solve the above trace minimization problem, that takes no more that $O(nL)$ iterations, having two matrix inversions per iteration. Figure 13 shows a pseudo code for this algorithm.

To analyze the algorithm, consider the parametric family of potential functions given by

$$g(z,v) = 2n\ln(\text{tr}(M(z)) - v) - \ln\det(M(z)),$$

where $v \in \mathbb{R}$ is a parameter. This algorithm will generate a monotonically increasing sequence of parameters $v^{(k)}$ that converges to the optimal value v^*. The sequence $v^{(k)}$ is constructed together with the sequence $z^{(k)}$ of interior points, as shown in the pseudo code in Figure 13. Since $Q - Z^*$ is a positive definite matrix, $v^{(0)} = 0 \leq v^*$ is used as the initial point in the sequence.

Let $g_1^{(k)}(z,v)$ be the linear approximation of $g(z,v)$ at $z^{(k)}$. Then

$$g_1^{(k)}(z,v) = -\frac{2n}{\text{tr}(M(z^{(k)})) - v}e^T z + \nabla \ln\det(M(z^{(k)})^T z + C,$$

where C is a constant. Kamath and Karmarkar show how $g_1^{(k)}(z,v)$ can be reduced by a constant amount at each iteration. They prove that it is possible to compute $v^{(k+1)} \in \mathbb{R}$ and a point $z^{(k+1)}$ in a closed ball of radius α centered at $z^{(k)}$ such that $v^{(k)} \leq v^{(k+1)} \leq v^*$ and

$$g_1^{(k)}(z^{(k+1)}, v^{(k+1)}) - g_1^{(k)}(z^{(k)}, v^{(k+1)}) \leq -\alpha.$$

Using this fact, they show that, if $z^{(k)}$ is the current interior point and $v^{(k)} \leq v^*$ is the current estimate of the optimal value, then

$$g(z^{(k+1)}, v^{(k+1)}) - g(z^{(k)}, v^{(k)}) \leq -\alpha + \frac{\alpha^2}{2(1-\alpha)},$$

where $z^{(k+1)}$ and $v^{(k+1)}$ are the new interior point and new estimate, respectively. This proves polynomial-time complexity for the algorithm.

9 Semidefinite Programming Relaxations

There has been a great deal of interest recently in solving semidefinite programming relaxations of combinatorial optimization problems [5, 6, 40, 144,

145, 152, 54, 53, 57, 112, 123, 111]. The semidefinite relaxations are solved by an interior point approach. These papers have shown the strength of the relaxations, and some of these papers have discussed cutting plane and branch and cut approaches using these relaxations. The bounds obtained from semidefinite relaxations are often better than those obtained using linear programming relaxations, but they are also usually more expensive to compute.

Semidefinite programming relaxations of some integer programming problems have proven to be very powerful, and they can often provide better bounds than those given by linear programming relaxations. There has been interest in semidefinite programming relaxations since at least the seventies (see Lovász [89]). These were regarded as being purely of theoretical interest until the recent development of interior point methods for semidefinite programming problems [6, 106, 56, 107, 82, 139]. Interest was increased further by Goemans and Williamson [40], who showed that the bounds generated by semidefinite programming relaxations for the maximum cut and satisfiability problems were considerably better than those that could be obtained from a linear programming relaxation, in the worst case, and that the solutions to these relaxations could be exploited to generate good integer solutions.

For an example of a semidefinite programming relaxation, consider the quadratic integer programming problem

$$\min \ f(x) = x^T Q x$$

subject to

$$x \in S = \{-1, 1\}^n,$$

where $Q \in \mathbb{R}^{n \times n}$ is symmetric, first discussed in equations (94) and (95) in section 8.

We let trace(M) denote the trace of a square matrix. By exploiting the fact that trace(AB)=trace(BA), we can rewrite the product $x^T Q x$:

$$x^T Q x = \text{trace}(x^T Q x) = \text{trace}(Q x x^T)$$

This lets us reformulate the quadratic program as

$$\begin{aligned} \min \quad & \text{trace}(QX) \\ \text{subject to} \quad & X = xx^T \\ & X_{ii} = 1 \quad i = 1, \ldots, n. \end{aligned}$$

The constraint that $X = xx^T$ is equivalent to saying that X must have rank equal to one. This is a hard constraint to enforce, so it is relaxed to the constraint that X is positive semi-definite, written $X \succeq 0$. This gives the following semidefinite programming relaxation:

$$\begin{array}{ll} \min & \text{trace}(QX) \\ \text{subject to} & X_{ii} = 1 \quad i = 1, \ldots, n \\ & X \succeq 0. \end{array}$$

Once we have a good relaxation, we can then (in principle) use a branch and bound (or branch and cut) method to solve the problem to optimality. Helmberg et al. [53] showed that the semidefinite programming relaxations of general constrained $0-1$ quadratic programming problems could be strengthened by using valid inequalities of the cut-polytope. There are a large number of such inequalities, and in [54], a branch and cut approach using semidefinite relaxations is used to solve quadratic $\{-1, 1\}$ problems with dense cost matrices. They branch using the criterion that two variables either take the same value or they take opposite values. This splits the current SDP relaxation into two SDP subproblems, each corresponding to quadratic $\{-1, 1\}$ problems of dimension one less. They are able to solve problems with up to about 100 variables in a reasonable amount of time.

Helmberg et al. [57] contains a nice discussion of different families of constraints for semidefinite relaxations of the quadratic knapsack problem. They derive semidefinite constraints from both the objective function and from the knapsack constraint. Many of the semidefinite constraints derived from the objective function are manipulations of the linear constraints for the Boolean quadric polytope. Similarly, they derive semidefinite constraints from known facets of the knapsack polytope. They attempt to determine the relative importance of different families of constraints.

The bottleneck with the branch and cut approach is the time required to solve each relaxation, and in particular to calculate the interior point directions. One way to reduce this time is to fix variables at -1 or 1, in much the same way that variables with large reduced costs can be fixed when we use a branch and cut algorithm that solves linear programming relaxations at each node. Helmberg [52] has proposed a method to determine whether a variable can be fixed when solving an SDP relaxation. This method examines the dual to the SDP relaxation. If it appears that a variable should be fixed at 1, say, then the effect of adding an explicit constraint that the variable should take the value -1 is examined. The change in the dual value that would result is then bounded; if this change is large enough then the variable can be fixed at 1.

The papers [54, 53, 57, 18] all contain semidefinite relaxations of quadratic programming problems with at most one constraint. By contrast, Wolkowicz and Zhao [144, 145] and Zhao et al. [152] have looked at semidefinite relaxations of more complicated integer programming problems. This required the development of some techniques that appear to be widely applicable. For these problems, the primal semidefinite programming relaxation does not have an interior feasible point, that is, there is no positive definite matrix that satisfies all the constraints. This implies that the dual problem will have an unbounded optimal face, so the problem is computationally intractable for an interior point method. To overcome this difficulty, the authors recast the problem in a lower dimensional space, where the barycenter of the known integer solutions corresponds to an interior point. In particular, if X is the matrix of variables for the original semidefinite formulation, a constant matrix V is determined so that the problem can be recast in terms of a matrix Z of variables, with $X = VZV^T$ and Z is of smaller dimension than X. To ensure that the new problem corresponds to the original problem, a *gangster operator* is used, which forces some components of VZV^T to be zero. With this reformulation, an interior point method can be used successfully to solve the semidefinite relaxations. An extension of the gangster operator may make it possible to use these relaxations in a branch and cut approach.

Another interesting aspect of [145] is the development of an alternative, slightly weaker, semidefinite relaxation that allows the exploitation of some sparsity in the original matrix for the set covering problem. The resulting relaxation contains both semidefinite constraints and linear constraints. This may make the semidefinite approach viable for problems of this type which are far larger than those previously tackled with semidefinite programming approaches. Whether this approach can be extended to other problems is an interesting question.

Some work on attempting to exploit sparsity in the general setting has been performed by Fujisawa et al. [38], and by Helmberg et al. [56] in their MATLAB implementation. Zhao et al. [152, 151] propose using a preconditioned conjugate gradient method to calculate the directions for the quadratic assignment problem (QAP) within a primal-infeasible dual-feasible variant of the method proposed in [56]. In the setting of solving a QAP, the semidefinite relaxation is used to obtain a lower bound on the optimal value; this bound is provided by the dual solution. Thus, only dual feasibility is needed to get a lower bound, and so primal feasibility is not as important, and it is possible to solve the Newton system of equations only approximately while still maintaining dual feasibility. It should be possible to extend this approach to other problems.

Zhao et al. [152] also developed another relaxation for the QAP which contains a large number of constraints. This second relaxation is stronger than the relaxation that uses the gangster operator, but because of the number of constraints, they could only use it in a cutting plane algorithm. Due to memory limitations, the gangster approach provided a better lower bound than the other relaxation for larger problems.

One way to solve sparse problems using semidefinite programming techniques is to look at the dual problem. Benson et al. [16] and Helmberg and Rendl [55] have both recently proposed methods that obtain very good bounds and sometimes optimal solutions for sparse combinatorial optimization problems by looking at the dual problem or relaxations of the dual.

There are several freely available implementations of SDP methods. Many of these codes are written in MATLAB. One of the major costs in an iteration of an SDP algorithm is constructing the Newton system of equations, with a series of `for` loops. MATLAB does not appear to handle this well, because of the slowness of its interpreted loops: in compiled C code, each iteration of these loops takes half a dozen machine language instructions, while in the interpreted code, each pass through one of these loops takes 100 or more instructions. For details of a freely available C implementation, see [17].

10 Concluding Remarks

Optimization is of central importance in both the natural sciences, such as physics, chemistry and biology, as well as artificial or man-made sciences, such as computer science and operations research. Nature inherently seeks optimal solutions. For instance, crystalline structure is the minimum energy state for a set of atoms, and light travels through the shortest path. The behavior of nature can often be explained on the basis of variational principles. Laws of nature then simply become optimality conditions. Concepts from continuous mathematics have always played a central role in the description of these optimality conditions and in analysis of the structure of their solutions. On the other hand, in artificial sciences, the problems are stated using the language of discrete mathematics or logic, and a simple-minded search for their solution confines one to a discrete solution set.

With the advent of interior point methods, the picture is changing, because these methods do not confine their working to a discrete solution set, but instead view combinatorial objects as limiting cases of continuous objects and exploit the topological and geometric properties of the continuous space. As a result, the number of efficiently solvable combinatorial problems is expanding. Also, the interior-point methods have revealed that the mathematical structure relevant to optimization in the natural and artificial sciences have a great deal in common. Recent conferences on global optimization (e.g. [35, 36]) are attracting researchers from diverse fields, ranging from computer science to molecular biology, thus merging the development paths of natural and artificial sciences. The phenomena of multiple solutions to combinatorial problems is intimately related to multiple configurations a complex molecule can assume. Thus, understanding the structure of solution sets of nonlinear problems is a common challenge faced by both natural and artificial sciences, to explain natural phenomena in the former case and to create more efficient interior-point algorithms in the latter case.

In the last decade we have witnessed computational breakthroughs in the approximate solution of large scale combinatorial optimization problems. Many of these breakthroughs are due to the development of interior point algorithms and implementations. Starting with linear programming in 1984 [72], these developments have spanned a wide range of problems, including network flows, graph problems, and integer programming. There is a continuing activity with new papers and codes being announced almost on daily basis. The interested reader can consult the following web sites:

1. http://www.mcs.anl.gov:80/home/otc/InteriorPoint
   is an archive of technical reports and papers on interior-point methods, maintained by S.J. Wright at Argonne National Laboratory.

2. http://www.zib.de/helmberg/semidef.html
   contains a special home page for semidefinite programming organized by C. Helmberg, at Berlin Center for Scientific Computing, Konrad Zuse Zentrum fur Informationstechnik, Berlin.

3. ftp://orion.uwaterloo.ca/pub/henry/reports/psd.bib.gz
   contains a bib file with papers related to SDP.

Acknowledgement

The first author acknowledges support in part by ONR grant N00014-94-1-0391, and by a grant from the Dutch NWO and by Delft University of Technology for 1997-98, while visiting TWI/SSOR at Delft University of Technology. The second author acknowledges support in part by NSF grant BIR-9505913 and U.S. Department of Air Force grant F08635-92-C-0032.

References

[1] W.P. Adams and T.A. Johnson. Improved linear programming-based lower bounds for the quadratic assignment problem. In P.M. Pardalos and H. Wolkowicz, editors, *Quadratic assignment and related problems*, volume 16 of *DIMACS Series on Discrete Mathematics and Theoretical Computer Science*, pages 43-75. American Mathematical Society, 1994.

[2] I. Adler, N. Karmarkar, M.G.C. Resende, and G. Veiga. Data structures and programming techniques for the implementation of Karmarkar's algorithm. *ORSA Journal on Computing*, 1:84-106, 1989.

[3] I. Adler, N. Karmarkar, M.G.C. Resende, and G. Veiga. An implementation of Karmarkar's algorithm for linear programming. *Mathematical Programming*, 44:297-335, 1989.

[4] Navindra K. Ahuja, Thomas L. Magnanti, and James B. Orlin. *Network Flows*. Prentice Hall, Englewood Cliffs, NJ, 1993.

[5] F. Alizadeh. Optimization over positive semi-definite cone: Interior-point methods and combinatorial applications. In P.M. Pardalos, editor, *Advances in Optimization and Parallel Computing*, pages 1-25. North-Holland, Amsterdam, 1992.

[6] F. Alizadeh. Interior point methods in semidefinite programming with applications to combinatorial optimization. *SIAM Journal on Optimization*, 5(1):13-51, 1995.

[7] E. D. Andersen, J. Gondzio, C. Mészáros, and X. Xu. Implementation of interior point methods for large scale linear programming. In T. Terlaky, editor, *Interior Point Methods in Mathematical Programming*, chapter 6. Kluwer Academic Publishers, 1996.

[8] D. S. Atkinson and P. M. Vaidya. A cutting plane algorithm for convex programming that uses analytic centers. *Mathematical Programming*, 69:1–43, 1995.

[9] O. Bahn, O. Du Merle, J. L. Goffin, and J. P. Vial. A cutting plane method from analytic centers for stochastic programming. *Mathematical Programming*, 69:45–73, 1995.

[10] E. Balas, S. Ceria, and G. Cornuéjols. A lift-and-project cutting plane algorithm for mixed 0-1 programs. *Mathematical Programming*, 58:295–324, 1993.

[11] E. Balas, S. Ceria, G. Cornuéjols, and N. Natraj. Gomory cuts revisited. *Operations Research Letters*, 19:1–9, 1996.

[12] E.R. Barnes. A variation on Karmarkar's algorithm for solving linear programming problems. *Mathematical Programming*, 36:174–182, 1986.

[13] D. A. Bayer and J. C. Lagarias. The nonlinear geometry of linear programming, I. Affine and projective scaling trajectories. *Transactions of the American Mathematical Society*, 314:499–526, 1989.

[14] D. A. Bayer and J. C. Lagarias. The nonlinear geometry of linear programming, II. Legendre transform coordinates and central trajectories. *Transactions of the American Mathematical Society*, 314:527–581, 1989.

[15] M.S. Bazaraa, J.J. Jarvis, and H.D. Sherali. *Linear Programming and Network Flows*. Wiley, New York, NY, 1990.

[16] S. J. Benson, Y. Ye, and X. Zhang. Solving large-scale sparse semidefinite programs for combinatorial optimization. Technical report, Department of Management Sciences, University of Iowa, Iowa City, Iowa 52242, September 1997.

[17] B. Borchers. CSDP, a C library for semidefinite programming. Technical report, Mathematics Department, New Mexico Tech, Socorro, NM 87801, March 1997.

[18] B. Borchers, S. Joy, and J. E. Mitchell. Three methods to the exact solution of max-sat problems. Talk given at *IN-*

FORMS Conference, Atlanta, 1996. Slides available from the URL http://www.nmt.edu/~borchers/atlslides.ps, 1996.

[19] B. Borchers and J. E. Mitchell. Using an interior point method in a branch and bound algorithm for integer programming. Technical Report 195, Mathematical Sciences, Rensselaer Polytechnic Institute, Troy, NY 12180, March 1991. Revised July 7, 1992.

[20] C. Chevalley. *Theory of Lie Groups.* Princeton University Press, Princeton, New Jersey, 1946.

[21] T. Christof and G. Reinelt. Parallel cutting plane generation for the tsp (extended abstract). Technical report, IWR Heidelberg, Germany, 1995.

[22] G. B. Dantzig. Maximization of a linear function of variables subject to linear inequalities. In Tj. C. Koopmans, editor, *Activity Analysis of Production and Allocation*, pages 339–347. Wiley, New York, 1951.

[23] G.B. Dantzig. Application of the simplex method to a transportation problem. In T.C. Koopsmans, editor, *Activity Analysis of Production and Allocation.* John Wiley and Sons, 1951.

[24] M. Davis and H. Putnam. A computing procedure for quantification theory. *Journal of the ACM*, 7:201–215, 1960.

[25] A. de Silva and D. Abramson. A parallel interior point method and its application to facility location problems. Technical report, School of Computing and Information Technology, Griffith University, Nathan, QLD 4111, Australia, 1995.

[26] C. De Simone, M. Diehl, M. Jünger, P. Mutzel, G. Reinelt, and G. Rinaldi. Exact ground states of two-dimensional $\pm J$ Ising spin glasses. *Journal of Statistical Physics*, 84:1363–1371, 1996.

[27] M. Deza and M. Laurent. Facets for the cut cone I. *Mathematical Programming*, 56:121–160, 1992.

[28] M. Deza and M. Laurent. Facets for the cut cone II: Clique-web inequalities. *Mathematical Programming*, 56:161–188, 1992.

[29] I. I. Dikin. Iterative solution of problems of linear and quadratic programming. *Doklady Akademiia Nauk SSSR*, 174:747–748, 1967. English Translation: *Soviet Mathematics Doklady*, 1967, Volume 8, pp. 674–675.

[30] D. Z. Du, J. Gu, and P. M. Pardalos, editors. *Satisfiability Problem: Theory and Applications*, volume 35 of *DIMACS Series on Discrete Mathematics and Theoretical Computer Science*. American Mathematical Society, 1997.

[31] Ding-Zhu Du and Panos M. Pardalos, editors. *Network Optimization Problems: Algorithms, Applications and Complexity*. World Scientific, 1993.

[32] J. Edmonds. Maximum matching and a polyhedron with 0, 1 vertices. *Journal of Research National Bureau of Standards*, 69B:125–130, 1965.

[33] A. S. El-Bakry, R. A. Tapia, and Y. Zhang. A study of indicators for identifying zero variables in interior–point methods. *SIAM Review*, 36:45–72, 1994.

[34] A. V. Fiacco and G. P. McCormick. *Nonlinear Programming: Sequential Unconstrained Minimization Techniques*. John Wiley and Sons, New York, 1968. Reprinted as Volume 4 of the SIAM Classics in Applied Mathematics Series, 1990.

[35] C. Floudas and P. Pardalos. *Recent Advances in Global Optimization*. Princeton Series in Computer Science. Princeton University Press, 1992.

[36] C. Floudas and P. Pardalos. *State of the Art in Global Optimization: Computational Methods and Applications*. Kluwer Academic Publishers, 1996.

[37] L.R. Ford and D.R. Fulkerson. *Flows in Networks*. Princeton University Press, Princeton, NJ, 1990.

[38] K. Fujisawa, M. Kojima, and K. Nakata. Exploiting sparsity in primal-dual interior-point methods for semidefinite programming. Technical report, Department of Mathematical and Computing Sciences, Tokyo Institute of Technology, 2-12-1 Oh-Okayama, Meguro-ku, Tokyo 152, Japan, January 1997.

[39] A. George and M. Heath. Solution of sparse linear least squares problems using Givens rotations. *Linear Algebra and Its Applications*, 34:69–83, 1980.

[40] Michel X. Goemans and David P. Williamson. Improved Approximation Algorithms for Maximum Cut and Satisfiability Problems Using Semidefinite Programming. *J. Assoc. Comput. Mach.*, 42:1115–1145, 1995.

[41] J.-L. Goffin, J. Gondzio, R. Sarkissian, and J.-P. Vial. Solving nonlinear multicommodity network flow problems by the analytic center cutting plane method. *Mathematical Programming*, 76:131–154, 1997.

[42] J.-L. Goffin, A. Haurie, and J.-P. Vial. Decomposition and nondifferentiable optimization with the projective algorithm. *Management Science*, 38:284–302, 1992.

[43] J.-L. Goffin, Z.-Q. Luo, and Y. Ye. Further complexity analysis of a primal-dual column generation algorithm for convex or quasiconvex feasibility problems. Technical report, Faculty of Management, McGill University, Montréal, Québec, Canada, November 1993.

[44] J.-L. Goffin, Z.-Q. Luo, and Y. Ye. On the complexity of a column generation algorithm for convex or quasiconvex problems. In *Large Scale Optimization: The State of the Art*. Kluwer Academic Publishers, 1993.

[45] G.H. Golub and C.F. van Loan. *Matrix Computations*. The Johns Hopkins University Press, Baltimore, MD, 1983.

[46] R. E. Gomory. Outline of an algorithm for integer solutions to linear programs. *Bulletin of the American Mathematical Society*, 64:275–278, 1958.

[47] M. Grötschel and O. Holland. Solution of large-scale travelling salesman problems. *Mathematical Programming*, 51(2):141–202, 1991.

[48] M. Grötschel, M. Jünger, and G. Reinelt. A cutting plane algorithm for the linear ordering problem. *Operations Research*, 32:1195–1220, 1984.

[49] M. Grötschel, L. Lovasz, and A. Schrijver. *Geometric Algorithms and Combinatorial Optimization*. Springer-Verlag, Berlin, Germany, 1988.

[50] G.M. Guisewite. Network problems. In Reiner Horst and Panos M. Pardalos, editors, *Handbook of global optimization*. Kluwer Academic Publishers, 1994.

[51] G.M. Guisewite and P.M. Pardalos. Minimum concave cost network flow problems: Applications, complexity, and algorithms. *Annals of Operations Research*, 25:75–100, 1990.

[52] C. Helmberg. Fixing variables in semidefinite relaxations. Technical Report SC-96-43, Konrad-Zuse-Zentrum fuer Informationstechnik, Berlin, December 1996.

[53] C. Helmberg, S. Poljak, F. Rendl, and H. Wolkowicz. Combining semidefinite and polyhedral relaxations for integer programs. In E. Balas and J. Clausen, editors, *Integer Programming and Combinatorial Optimization, Lecture Notes in Computer Science*, volume 920, pages 124–134. Springer, 1995.

[54] C. Helmberg and F. Rendl. Solving quadratic (0,1)-problems by semidefinite programs and cutting planes. Technical Report SC-95-35, Konrad-Zuse-Zentrum fuer Informationstechnik, Berlin, 1995.

[55] C. Helmberg and F. Rendl. A spectral bundle method for semidefinite programming. Technical Report SC-97-37, Konrad-Zuse-Zentrum fuer Informationstechnik, Berlin, August 1997. Revised: October 1997.

[56] C. Helmberg, F. Rendl, R. J. Vanderbei, and H. Wolkowicz. An interior point method for semidefinite programming. *SIAM Journal on Optimization*, 6:342–361, 1996.

[57] C. Helmberg, F. Rendl, and R. Weismantel. Quadratic knapsack relaxations using cutting planes and semidefinite programming. In W. H. Cunningham, S. T. McCormick, and M. Queyranne, editors, *Integer Programming and Combinatorial Optimization, Lecture Notes in Computer Science*, volume 1084, pages 175–189. Springer, 1996.

[58] F.L. Hitchcock. The distribution of product from several sources to numerous facilities. *Journal of Mathematical Physics*, 20:224–230, 1941.

[59] K. L. Hoffman and M. Padberg. Improving LP-representation of zero-one linear programs for branch-and-cut. *ORSA Journal on Computing*, 3(2):121–134, 1991.

[60] E. Housos, C. Huang, and L. Liu. Parallel algorithms for the AT&T KORBX System. *AT&T Technical Journal*, 68:37–47, 1989.

[61] D. S. Johnson and M. A. Trick, editors. *Cliques, Coloring, and Satisfiability: Second DIMACS Implementation Challenge*, volume 26 of *DIMACS Series on Discrete Mathematics and Theoretical Computer Science*. American Mathematical Society, 1996.

[62] A. Joshi, A.S. Goldstein, and P.M. Vaidya. A fast implementation of a path-following algorithm for maximizing a linear function over a network polytope. In David S. Johnson and Catherine C. McGeoch, editors, *Network Flows and Matching: First DIMACS Implementation Challenge*, volume 12 of *DIMACS Series in Discrete Mathematics and Theoretical Computer Science*, pages 267–298. American Mathematical Society, 1993.

[63] M. Jünger, G. Reinelt, and S. Thienel. Practical problem solving with cutting plane algorithms in combinatorial optimization. In *Combinatorial Optimization: DIMACS Series in Discrete Mathematics and Theoretical Computer Science*, pages 111–152. AMS, 1995.

[64] J.A. Kaliski and Y. Ye. A decomposition variant of the potential reduction algorithm for linear programming. *Management Science*, 39:757–776, 1993.

[65] A. P. Kamath, N. K. Karmarkar, K. G. Ramakrishnan, and M. G. C. Resende. Computational experience with an interior point algorithm on the Satisfiability problem. *Annals of Operations Research*, 25:43–58, 1990.

[66] A.P. Kamath and N. Karmarkar. A continuous method for computing bounds in integer quadratic optimization problems. *Journal of Global Optimization*, 2:229–241, 1992.

[67] A.P. Kamath and N. Karmarkar. An $O(nL)$ iteration algorithm for computing bounds in quadratic optimization problems. In P.M. Pardalos, editor, *Complexity in Numerical Optimization*, pages 254–268. World Scientific, Singapore, 1993.

[68] A.P. Kamath, N. Karmarkar, K.G. Ramakrishnan, and M.G.C. Resende. A continuous approach to inductive inference. *Mathematical Programming*, 57:215–238, 1992.

[69] A.P. Kamath, N. Karmarkar, K.G. Ramakrishnan, and M.G.C. Resende. A continuous approach to inductive inference. *Mathematical Programming*, 57:215–238, 1992.

[70] A.P. Kamath, N. Karmarkar, K.G. Ramakrishnan, and M.G.C. Resende. An interior point approach to Boolean vector function synthesis. In *Proceedings of the 36th MSCAS*, pages 185–189, 1993.

[71] A.P. Kamath, N. Karmarkar, N. Ramakrishnan, and M.G.C. Resende. Computational experience with an interior point algorithm on the Satisfiability problem. *Annals of Operations Research*, 25:43–58, 1990.

[72] N. Karmarkar. A new polynomial-time algorithm for linear programming. *Combinatorica*, 4:373–395, 1984.

[73] N. Karmarkar. An interior-point approach for NP-complete problems. *Contemporary Mathematics*, 114:297–308, 1990.

[74] N. Karmarkar. An interior-point approach to NP-complete problems. In *Proceedings of the First Integer Programming and Combinatorial Optimization Conference*, pages 351–366, University of Waterloo, 1990.

[75] N. Karmarkar. A new parallel architecture for sparse matrix computation based on finite projective geometries. In *Proceedings of Supercomputing '91*, pages 358–369. IEEE Computer Society, 1991.

[76] N. Karmarkar, J. Lagarias, L. Slutsman, and P. Wang. Power series variants of Karmarkar-type algorithms. *AT&T Technical Journal*, 68:20–36, 1989.

[77] N. Karmarkar, M.G.C. Resende, and K. Ramakrishnan. An interior point algorithm to solve computationally difficult set covering problems. *Mathematical Programming*, 52:597–618, 1991.

[78] N. Karmarkar, M.G.C. Resende, and K.G. Ramakrishnan. An interior point approach to the maximum independent set problem in dense random graphs. In *Proceedings of the XIII Latin American Conference on Informatics*, volume 1, pages 241–260, Santiago, Chile, July 1989.

[79] N. K. Karmarkar and K. G. Ramakrishnan. Computational results of an interior point algorithm for large scale linear programming. *Mathematical Programming*, 52:555–586, 1991.

[80] J.L. Kennington and R.V. Helgason. *Algorithms for network programming*. John Wiley and Sons, New York, NY, 1980.

[81] L. G. Khachiyan. A polynomial algorithm in linear programming. *Doklady Akademiia Nauk SSSR*, 224:1093–1096, 1979. English Translation: *Soviet Mathematics Doklady*, Volume 20, pp. 1093–1096.

[82] M. Kojima, S. Shindoh, and S. Hara. Interior point methods for the monotone semidefinite linear complementarity problem in symmetric matrices. *SIAM Journal on Optimization*, 7:86–125, 1997.

[83] J.B. Kruskal. On the shortest spanning tree of graph and the traveling salesman problem. *Proceedings of the American Mathematical Society*, 7:48–50, 1956.

[84] Eugene Lawler. *Combinatorial Optimization: Networks and Matroids*. Holt, Rinehart and Winston, 1976.

[85] E. K. Lee and J. E. Mitchell. Computational experience in nonlinear mixed integer programming. In *Proceedings of Symposium on Operations Research, August 1996, Braunschweig, Germany*, pages 95–100. Springer-Verlag, 1996.

[86] K. Levenberg. A method for the solution of certain problems in least squares. *Quart. Appl. Math.*, 2:164–168, 1944.

[87] Y. Li, P.M. Pardalos, K.G. Ramakrishnan, and M.G.C. Resende. Lower bounds for the quadratic assignment problem. *Annals of Operations Research*, 50:387–410, 1994.

[88] Y. Li, P.M. Pardalos, and M.G.C. Resende. A greedy randomized adaptive search procedure for the quadratic assignment problem. In P.M. Pardalos and H. Wolkowicz, editors, *Quadratic assignment and related problems*, volume 16 of *DIMACS Series on Discrete Mathematics and Theoretical Computer Science*, pages 237–262. American Mathematical Society, 1994.

[89] L. Lovász. On the Shannon capacity of a graph. *IEEE Transactions on Information Theory*, 25:1–7, 1979.

[90] I. J. Lustig, R. E. Marsten, and D. F. Shanno. Interior point methods for linear programming: Computational state of the art. *ORSA*

Journal on Computing, 6(1):1–14, 1994. See also the following commentaries and rejoinder.

[91] D. Marquardt. An algorithm for least-squares estimation of nonlinear parameters. *SIAM J. Appl. Math.*, 11:431–441, 1963.

[92] S. Mehrotra. On the implementation of a (primal–dual) interior point method. *SIAM Journal on Optimization*, 2(4):575–601, 1992.

[93] S. Mehrotra and J. Wang. Conjugate gradient based implementation of interior point methods for network flow problems. In L. Adams and J. Nazareth, editors, *Linear and Nonlinear Conjugate Gradient Related Methods*. SIAM, 1995.

[94] J. E. Mitchell. Interior point algorithms for integer programming. In J. E. Beasley, editor, *Advances in Linear and Integer Programming*, chapter 6, pages 223–248. Oxford University Press, 1996.

[95] J. E. Mitchell. Interior point methods for combinatorial optimization. In Tamás Terlaky, editor, *Interior Point Methods in Mathematical Programming*, chapter 11, pages 417–466. Kluwer Academic Publishers, 1996.

[96] J. E. Mitchell. Computational experience with an interior point cutting plane algorithm. Technical report, Mathematical Sciences, Rensselaer Polytechnic Institute, Troy, NY 12180–3590, February 1997. Revised: April 1997.

[97] J. E. Mitchell. Fixing variables and generating classical cutting planes when using an interior point branch and cut method to solve integer programming problems. *European Journal of Operational Research*, 97:139–148, 1997.

[98] J. E. Mitchell. An interior point cutting plane algorithm for Ising spin glass problems. Technical report, Mathematical Sciences, Rensselaer Polytechnic Institute, Troy, NY 12180–3590, July 1997.

[99] J. E. Mitchell and B. Borchers. Solving real-world linear ordering problems using a primal-dual interior point cutting plane method. *Annals of Operations Research*, 62:253–276, 1996.

[100] J. E. Mitchell and B. Borchers. Solving linear ordering problems with a combined interior point/simplex cutting plane algorithm. Technical report, Mathematical Sciences, Rensselaer Polytechnic Institute, Troy, NY 12180-3590, September 1997.

[101] J. E. Mitchell and M. J. Todd. Solving combinatorial optimization problems using Karmarkar's algorithm. *Mathematical Programming*, 56:245-284, 1992.

[102] R. D. C. Monteiro and I. Adler. Interior path following primal-dual algorithms. Part I: Linear programming. *Mathematical Programming*, 44(1):27-41, 1989.

[103] R. D. C. Monteiro, I. Adler, and M. G. C. Resende. A polynomial-time primal-dual affine scaling algorithm for linear and convex quadratic programming and its power series extension. *Mathematics of Operations Research*, 15(2):191-214, 1990.

[104] J.J. Moré and D.C. Sorenson. Computing a trust region step. *SIAM J. of Stat. Sci. Comput.*, 4:553-572, 1983.

[105] G. L. Nemhauser and L. A. Wolsey. *Integer and Combinatorial Optimization*. John Wiley, New York, 1988.

[106] Y. E. Nesterov and A. S. Nemirovsky. *Interior Point Polynomial Methods in Convex Programming : Theory and Algorithms*. SIAM Publications. SIAM, Philadelphia, USA, 1993.

[107] Y. E. Nesterov and M. J. Todd. Self-scaled barriers and interior-point methods for convex programming. *Mathematics of Operations Research*, 22:1-42, 1997.

[108] M. W. Padberg and G. Rinaldi. A branch-and-cut algorithm for the resolution of large-scale symmetric traveling salesman problems. *SIAM Review*, 33(1):60-100, 1991.

[109] R. Pai, N. K. Karmarkar, and S. S. S. P. Rao. A global router for gate-arrays based on Karmarkar's interior point methods. In *Proceedings of the Third International Workshop on VLSI System Design*, pages 73-82, 1990.

[110] P. M. Pardalos, K. G. Ramakrishnan, M. G. C. Resende, and Y. Li. Implementation of a variance reduction-based lower bound in a branch-and-bound algorithm for the quadratic assignment problem. *SIAM Journal on Optimization*, 7:280–294, 1997.

[111] P. M. Pardalos and M.G.C. Resende. Interior point methods for global optimization problems. In T. Terlaky, editor, *Interior Point Methods of Mathematical Programming*, pages 467–500. Kluwer Academic Publishers, 1996.

[112] P. M. Pardalos and H. Wolkowicz, editors. *Topics in Semidefinite and Interior-Point Methods*. Fields Institute Communications Series. American Mathematical Society, New Providence, Rhode Island, 1997.

[113] P.M. Pardalos. Continuous approaches to discrete optimization problems. In G. Di Pillo and F. Giannessi, editors, *Nonlinear optimization and applications*. Plenum Publishing, 1996.

[114] P.M. Pardalos and H. Wolkowicz, editors. *Quadratic assignment and related problems*, volume 16 of *DIMACS Series in Discrete Mathematics and Theoretical Computer Science*. American Mathematical Society, 1994.

[115] R. G. Parker and R. L. Rardin. *Discrete Optimization*. Academic Press, San Diego, CA 92101, 1988.

[116] L. Portugal, F. Bastos, J. Júdice, J. Paixão, and T. Terlaky. An investigation of interior point algorithms for the linear transportation problem. *SIAM J. Sci. Computing*, 17:1202–1223, 1996.

[117] L. Portugal, M.G.C. Resende, G. Veiga, and J. Júdice. An efficient implementation of an infeasible primal-dual network flow method. Technical report, AT&T Bell Laboratories, Murray Hill, New Jersey, 1994.

[118] R.C. Prim. Shortest connection networks and some generalizations. *Bell System Technical Journal*, 36:1389–1401, 1957.

[119] K. G. Ramakrishnan, M. G. C. Resende, and P. M. Pardalos. A branch and bound algorithm for the quadratic assignment problem using a lower bound based on linear programming. In C. Floudas and P.M. Pardalos, editors, *State of the Art in Global Optimization: Computational Methods and Applications*. Kluwer Academic Publishers, 1995.

[120] K.G. Ramakrishnan, N.K. Karmarkar, and A.P. Kamath. An approximate dual projective algorithm for solving assignment problems. In David S. Johnson and Catherine C. McGeoch, editors, *Network Flows and Matching: First DIMACS Implementation Challenge*, volume 12 of *DIMACS Series in Discrete Mathematics and Theoretical Computer Science*, pages 431–451. American Mathematical Society, 1993.

[121] K.G. Ramakrishnan, M.G.C. Resende, and P.M. Pardalos. A branch and bound algorithm for the quadratic assignment problem using a lower bound based on linear programming. In *State of the Art in Global Optimization: Computational Methods and Applications*, pages 57–73. Kluwer Academic Publishers, 1996.

[122] K.G. Ramakrishnan, M.G.C. Resende, B. Ramachandran, and J.F. Pekny. Tight QAP bounds vias linear programming. In *From Local to Global Optimization*. Kluwer Academic Publishers, 1998. To appear.

[123] M. Ramana and P. M. Pardalos. Semidefinite programming. In T. Terlaky, editor, *Interior Point Methods of Mathematical Programming*, pages 369–398. Kluwer Academic Publishers, 1996.

[124] S. Ramaswamy and J. E. Mitchell. On updating the analytic center after the addition of multiple cuts. Technical Report 37-94-423, DSES, Rensselaer Polytechnic Institute, Troy, NY 12180, October 1994.

[125] S. Ramaswamy and J. E. Mitchell. A long step cutting plane algorithm that uses the volumetric barrier. Technical report, DSES, Rensselaer Polytechnic Institute, Troy, NY 12180, June 1995.

[126] M.G.C. Resende, P.M. Pardalos, and Y. Li. FORTRAN subroutines for approximate solution of dense quadratic assignment problems using GRASP. *ACM Transactions on Mathematical Software*, To appear.

[127] M.G.C. Resende, K.G. Ramakrishnan, and Z. Drezner. Computing lower bounds for the quadratic assignment problem with an interior point algorithm for linear programming. *Operations Research*, 43(5):781–791, 1995.

[128] M.G.C. Resende, T. Tsuchiya, and G. Veiga. Identifying the optimal face of a network linear program with a globally convergent interior

point method. In W.W. Hager, D.W. Hearn, and P.M. Pardalos, editors, *Large scale optimization: State of the art*, pages 362–387. Kluwer Academic Publishers, 1994.

[129] M.G.C. Resende and G. Veiga. Computing the projection in an interior point algorithm: An experimental comparison. *Investigación Operativa*, 3:81–92, 1993.

[130] M.G.C. Resende and G. Veiga. An efficient implementation of a network interior point method. In David S. Johnson and Catherine C. McGeoch, editors, *Network Flows and Matching: First DIMACS Implementation Challenge*, volume 12 of *DIMACS Series in Discrete Mathematics and Theoretical Computer Science*, pages 299–348. American Mathematical Society, 1993.

[131] M.G.C. Resende and G. Veiga. An implementation of the dual affine scaling algorithm for minimum cost flow on bipartite uncapaciated networks. *SIAM Journal on Optimization*, 3:516–537, 1993.

[132] C. Roos, T. Terlaky, and J.-Ph. Vial. *Theory and Algorithms for Linear Optimization: An Interior Point Approach*. John Wiley, Chichester, 1997.

[133] Günther Ruhe. *Algorithmic Aspects of Flows in Networks*. Kluwer Academic Publishers, Boston, MA, 1991.

[134] C.-J. Shi, A. Vannelli, and J. Vlach. An improvement on Karmarkar's algorithm for integer programming. In P.M. Pardalos and M.G.C. Resende, editors, *COAL Bulletin – Special issue on Computational Aspects of Combinatorial Optimization*, volume 21, pages 23–28. Mathematical Programming Society, 1992.

[135] L.P. Sinha, B.A. Freedman, N.K. Karmarkar, A. Putcha, and K.G. Ramakrishnan. Overseas network planning. In *Proceedings of the Third International Network Planning Symposium – NETWORKS'86*, pages 8.2.1–8.2.4, June 1986.

[136] R. Van Slyke and R. Wets. L-shaped linear programs with applications to optimal control and stochastic linear programs. *SIAM Journal on Applied Mathematics*, 17:638–663, 1969.

[137] R.E. Tarjan. *Data Structures and Network Algorithms*. Society for Industrial and Applied Mathematics, Philadelphia, PA, 1983.

[138] P. M. Vaidya. A new algorithm for minimizing convex functions over convex sets. *Mathematical Programming*, 73:291–341, 1996.

[139] L. Vandenberghe and S. Boyd. Semidefinite programming. *SIAM Review*, 38:49–95, 1996.

[140] R. J. Vanderbei. *Linear Programming: Foundations and Extensions*. Kluwer Academic Publishers, Boston, 1996.

[141] R.J. Vanderbei, M.S. Meketon, and B.A. Freedman. A modification of Karmarkar's linear programming algorithm. *Algorithmica*, 1:395–407, 1986.

[142] J.P. Warners, T. Terlaky, C. Roos, and B. Jansen. Potential reduction algorithms for structured combinatorial optimization problems. *Operations Research Letters*, 21:55–64, 1997.

[143] J.P. Warners, T. Terlaky, C. Roos, and B. Jansen. A potential reduction approach to the frequency assignment problem. *Discrete Applied Mathematics*, 78:251–282, 1997.

[144] H. Wolkowicz and Q. Zhao. Semidefinite programming relaxations for the graph partitioning problem. Technical report, Combinatorics and Optimization, University of Waterloo, Waterloo, Ontario, N2L 3G1 Canada, October 1996.

[145] H. Wolkowicz and Q. Zhao. Semidefinite programming relaxations for the set partitioning problem. Technical report, Combinatorics and Optimization, University of Waterloo, Waterloo, Ontario, N2L 3G1 Canada, October 1996.

[146] S. Wright. *Primal-dual interior point methods*. SIAM, Philadelphia, 1996.

[147] Y. Ye. On affine scaling algorithms for nonconvex quadratic programming. *Mathematical Programming*, 56:285–300, 1992.

[148] Y. Ye. *Interior Point Algorithms: Theory and Analysis*. John Wiley, New York, 1997.

[149] Quey-Jen Yeh. *A reduced dual affine scaling algorithm for solving assignment and transportation problems*. PhD thesis, Columbia University, New York, NY, 1989.

[150] Y. Zhang. On the convergence of a class of infeasible interior-point methods for the horizontal linear complementarity problem. *SIAM Journal on Optimization*, 4(1):208–227, 1994.

[151] Q. Zhao. *Semidefinite programming for assignment and partitioning problems*. PhD thesis, Combinatorics and Optimization, University of Waterloo, Waterloo, Ontario, N2L 3G1, Canada, 1996.

[152] Q. Zhao, S. E. Karisch, F. Rendl, and H. Wolkowicz. Semidefinite programming relaxations for the quadratic assignment problem. Technical Report 95-27, Combinatorics and Optimization, University of Waterloo, Waterloo, Ontario, N2L 3G1, Canada, September 1996.

Knapsack Problems

David Pisinger
DIKU, University of Copenhagen
Universitetsparken 1
DK-2100 Copenhagen
E-mail: pisinger@diku.dk

Paolo Toth
DEIS, University of Bologna
Viale Risorgimento 2
I-40136 Bologna
E-mail: paolo@deis.unibo.it

Contents

1 **Introduction**   **302**
   1.1 Historical Overview 302
   1.2 Applications 304
   1.3 The Problems 306
   1.4 $\mathcal{NP}$-hardness and Solvability 309
   1.5 Fundamental Properties of the Knapsack Problems 310
   1.6 Experimental Comparisons 314
   1.7 Notation 316
   1.8 Overview of the Chapter 317

2 **0-1 Knapsack Problem**   **318**
   2.1 Upper Bounds 320
       2.1.1 Lagrangian Relaxation 322
       2.1.2 Tighter Bounds 323
       2.1.3 Bounds from Minimum and Maximum Cardinality 324
   2.2 Heuristics 329
   2.3 Reduction 331

	2.4	Branch-and-bound Algorithms	332
	2.5	Dynamic Programming Algorithms	336
		2.5.1 Primal-dual Dynamic Programming	336
		2.5.2 Horowitz and Sahni Subdivision	337
		2.5.3 Other Dynamic Programming Algorithms	337
	2.6	Solution of Large-sized Problems	338
		2.6.1 Deriving a Core	339
		2.6.2 Fixed-core Algorithms	342
		2.6.3 Expanding-core Algorithms	343
		2.6.4 Inconveniences in Core Problems	344
	2.7	Solution of Hard Problems	345
	2.8	Approximation Schemes	346
	2.9	Computational Experiments	348
3	**Subset-sum Problem**		**351**
	3.1	Upper Bounds	353
	3.2	Dynamic Programming Algorithms	354
		3.2.1 Horowitz and Sahni Decomposition	355
		3.2.2 Balancing	356
	3.3	Hybrid Algorithms	359
	3.4	Solution of Large-sized Instances	361
	3.5	Computational Experiments	362
4	**Multiple-choice Knapsack Problem**		**364**
	4.1	Upper Bounds	367
		4.1.1 Linear Time Algorithms for the Continuous Problem	368
		4.1.2 Bounds from Lagrangian Relaxation	371
		4.1.3 Other Bounds	372
	4.2	Heuristics	372
	4.3	Class Reduction	373
	4.4	Branch-and-bound Algorithms	373
	4.5	Dynamic Programming Algorithms	374
	4.6	Reduction of States	376
	4.7	Solution of Large-sized Instances	377
	4.8	Computational Experiments	379
5	**Bounded Knapsack Problem**		**382**
	5.1	Upper Bounds	383
	5.2	Heuristics	385
	5.3	Reduction	386
	5.4	Branch-and-bound Algorithms	387
	5.5	Dynamic Programming	388
	5.6	Reduction of States	390
	5.7	Solution of Large-sized Problems	391

5.8	Computational Experiments	392

6 Unbounded Knapsack Problem — 394
6.1	Upper Bounds	395
6.2	Heuristics	396
6.3	Dynamic Programming	397
6.4	Branch-and-bound	397
6.5	Reduction Algorithms	398
	6.5.1 Reductions from Bounds	399
6.6	Solution of Large-sized Instances	400
6.7	Computational Experiments	400

7 Multiple Knapsack Problem — 402
7.1	Upper Bounds	403
7.2	Tightening Constraints	406
7.3	Reduction Algorithms	407
7.4	Heuristics and Approximate Algorithms	408
7.5	Dynamic Programming	409
7.6	Branch-and-bound Algorithms	410
	7.6.1 The mtm Algorithm	410
	7.6.2 The mulknap Algorithm	412
7.7	Computational Experiments	413

8 Conclusion and Future Trends — 417

References

Abstract

Knapsack Problems are the simplest $\mathcal{NP}$-hard problems in Combinatorial Optimization, as they maximize an objective function subject to a single resource constraint. Several variants of the classical 0-1 Knapsack Problem will be considered with respect to relaxations, bounds, reductions and other algorithmic techniques for the exact solution. Computational results are presented to compare the actual performance of the most effective algorithms published.

1 Introduction

Knapsack Problems have been intensively studied since the emergence of Combinatorial Optimization, both because of their immediate applications in industry and financial management, and even more for theoretical reasons, as Knapsack Problems often occur by relaxation of different integer programming problems. In such applications, we need to solve a Knapsack Problem each time a bounding function is derived, demanding extremely fast solution times of the Knapsack algorithm.

The family of Knapsack Problems all consider a set of items, each item j having an associated profit p_j and weight w_j. The problem is then to choose a subset of the given items such that the corresponding profit sum is maximized without exceeding the capacity c of the knapsack(s). Different types of Knapsack Problems occur depending on the distribution of items and knapsacks: In the *0-1 Knapsack Problem* each item may be chosen at most once, while in the *Bounded Knapsack Problem* we have a bounded amount of each item type. The *Multiple-choice Knapsack Problem* occurs when the items should be chosen from disjoint classes and, if several knapsacks are to be filled simultaneously we get the *Multiple Knapsack Problem*. The most general form is the *Multi-constrained Knapsack Problem*, which is basically a general *Integer Programming* (IP) Problem with non-negative coefficients.

Although Knapsack Problems, from a theoretical point of view, are almost intractable as they belong to the family of $\mathcal{NP}-hard$ problems, several of the problems may be solved to optimality in fractions of a second. This surprising result is the outcome of several decades of research which has exposed the special structural properties of Knapsack Problems that make the problems so easy to solve. The intention of this chapter is to state several of these properties and show how they influence the solution methods.

1.1 Historical Overview

Knapsack Problems are some of the most intensively studied discrete optimization problems. The reason for such interest basically derives from three facts: (a) they can be viewed as the simplest *Integer Linear Programming* problem; (b) they appear as a subproblem in many more complex problems; (c) they represent many practical situations.

Recently, Knapsack Problems have been used for generating minimal cover induced constraints (see, e.g., Crowder, Johnson and Padberg [16]) and in several coefficient reduction procedures for strengthening LP bounds in general integer programming (see, e.g., Dietrich and Escudero [19, 20]). During the last few decades, Knapsack Problems have been studied through different approaches, according to the theoretical development of *Combinatorial Optimization*.

In the fifties, Bellman's *dynamic programming* theory produced the first algorithms to exactly solve the 0-1 Knapsack Problem. In 1957 Dantzig gave an elegant and efficient method to determine the solution of the continuous relaxation of the problem, and hence a *bound* on the integer solution value. The bound was used in the following twenty years in almost all studies on KP.

In the sixties, the dynamic programming approach to the KP and other knapsack-type problems was deeply investigated by Gilmore and Gomory. In 1967 Kolesar experimented with the first *branch-and-bound* algorithm for the problem.

In the seventies, the branch-and-bound approach was further developed, proving to be capable of solving problems with a large number of variables. The most well-known algorithm of this period was developed by Horowitz and Sahni. In 1973 Ingargiola and Korsh presented the first *reduction procedure*, a preprocessing algorithm which significantly reduces the number of variables. In 1974 Johnson gave the first *polynomial-time approximation scheme* for the subset-sum problem; the result was extended by Sahni to the 0-1 Knapsack Problem. The first *fully polynomial-time approximation scheme* was obtained by Ibarra and Kim in 1975. In 1977 Martello and Toth proposed the first upper bound dominating the value of the continuous relaxation.

The main results of the eighties concern the solution of large-sized problems, for which sorting of the variables (required to solve the continuous solution) takes a very high percentage of the running time. In 1980 Balas and Zemel presented a new approach to solve the problem by sorting, in many cases, only a small subset of the variables (the *core problem*). Several efficient algorithms were designed based on this idea.

The current decade has been focused on solving difficult instances of reasonable size instead of extremely large easy instances. Martello and Toth showed in 1993 how upper bounds derived from Lagrangian relaxation of cardinality bounds may help solving several difficult problems.

Dynamic Programming has also been accepted as an efficient solution technique for two reasons: by incorporating bounding rules in the enumeration, the number of states can often be held at a reasonable level, and new improved recursions can focus the enumeration on those items which are most interesting.

The outcome of the latest research has been some algorithms with very stable overall performance. On the theoretical frontier the nineties have brought algorithms with improved worst-case time bounds. In 1995 Pisinger was the first to present a dynamic programming recursion for the subset-sum problem which has better worst-case complexity than Bellman's classical recursion, and many of these results can be generalized to other Knapsack Problems.

Each year numerous papers on Knapsack Problems are presented, and several new variants of the classical problem are considered. Since every problem with a single weight constraint can be seen as some kind of Knapsack Problem, recent papers have considered quadratic versions of the KP, as well as collapsing, nested, bottleneck, graph and tree Knapsack Problems to mention only the names. The techniques applied for these more sophisticated problems may vary a lot, but it is interesting to see that many ideas from the 0-1 Knapsack Problem are applicable even in the generalized versions.

Where the nineties have brought efficient algorithms to solve difficult problems in reasonable time, the theoretical work is far behind. For Knapsack Problems there is a very large gap between the theoretical worst-case performance of the best algorithms and the practical ability to solve large and difficult problems in reasonable time. Thus, there are still several theoretical and practical problems to be solved.

1.2 Applications

Knapsack Problems have numerous applications in theory as well as in practice. From a theoretical point of view, the simple structure calls for exploitation of numerous interesting properties, that can make the problems easier to solve. Knapsack Problems also arise as subproblems in several more complex algorithms in combinatorial optimization, and these algorithms will benefit from any improvement in this field.

Despite the name, practical applications of Knapsack Problems are not limited to packing problems: Assume that you may invest in n projects, each giving the profit p_j. It costs w_j to invest in project j, and you have only c dollars available. The best projects for investment may be found by solving a 0-1 Knapsack Problem.

Another application appears in a restaurant, where a person has to choose k courses, without surpassing the amount of c calories, that his diet prescribes. Assume that there are N_i dishes to choose from for each course $i = 1, \ldots, k$, w_{ij} is the nutritive value and p_{ij} is a rating saying how good each dish tastes. Then an optimal meal may be found by solving the corresponding Multiple-choice Knapsack Problem [109].

A two-processor scheduling problem, where a number of jobs have to be divided among two processors such that the completion time is minimized, may be solved as a Subset-sum Problem [77]. Cassette recorders have been introduced which are able to select a number of songs from a CD such that the longest possible play time will be recorded. This algorithm also solves some kinds of Subset-sum Problem.

Apart from these simple illustrations, Knapsack Problems are frequently used in the following industrial fields: Problems in cargo loading, cutting stock, budget control, and financial management may be formulated as Knapsack Problems, where the specific model depends on the side constraints present. Sinha and Zoltners [109] proposed using Multiple-choice Knapsack Problems to select which components should be linked in series in order to maximize fault tolerance. In several two- and three-dimensional cutting and packing problems, the Knapsack Problem is used heavily as a subproblem to find an optimal partition in layers or strips (Gilmore and Gomory [43]); Diffe and Hellman [21] designed a public cryptographic scheme whose security relies on the difficulty of solving the Subset-sum Problem.

The more theoretical applications appear either where a general problem is transformed into a Knapsack Problem, or where the Knapsack Problem appears as a subproblem, e.g. for deriving bounds in a branch-and-bound algorithm intended to solve a more complex problem. In the first category Mathews [84] a century ago showed how several constraints may be aggregated to one single knapsack constraint, making it possible to solve any IP Problem as a 0-1 Knapsack Problem. Moreover Nauss [89] proposed transforming nonlinear Knapsack Problems into Multiple-choice Knapsack Problems. In the second category we should mention that the 0-1 Knapsack Problem appears as a subproblem when solving the Generalized Assignment Problem, which is also heavily used when solving Vehicle Routing Problems [70]. Knapsack Problems also occur as a subproblem in airline scheduling problems [53], production planning problems [108], clustering and graph partitioning problems [32] and in the design of some electronic circuits [31].

1.3 The Problems

All Knapsack Problems consider a set of *items* with associated *profit* p_j and *weight* w_j. A subset of the items is to be chosen such that the weight sum does not exceed the *capacity* c of the knapsack, and such that the largest possible profit sum is obtained. We will assume that all coefficients p_j, w_j, c are positive integers, although weaker assumptions may sometimes be handled in the individual problems.

The *0-1 Knapsack Problem* is the problem of choosing some of the n items such that the corresponding profit sum is maximized without the weight sum exceeding the capacity c. Thus, it may be formulated as the following maximization problem:

$$\begin{aligned} \text{maximize} \quad & \sum_{j=1}^{n} p_j x_j \\ \text{subject to} \quad & \sum_{j=1}^{n} w_j x_j \leq c, \\ & x_j \in \{0,1\}, \quad j = 1, \ldots, n, \end{aligned} \quad (1)$$

where x_j is a binary variable having value 1 if item j should be included in the knapsack, and 0 otherwise. If we have a bounded amount m_j of each item type j, then the *Bounded Knapsack Problem* appears:

$$\begin{aligned} \text{maximize} \quad & \sum_{j=1}^{n} p_j x_j \\ \text{subject to} \quad & \sum_{j=1}^{n} w_j x_j \leq c, \\ & x_j \in \{0, 1, \ldots, m_j\}, \quad j = 1, \ldots, n. \end{aligned} \quad (2)$$

Here x_j gives the amount of each item type that should be included in the knapsack in order to obtain the largest objective value. The *Unbounded Knapsack Problem* is a special case of the Bounded Knapsack Problem, since an unlimited amount of each item type is available:

$$\begin{aligned} \text{maximize} \quad & \sum_{j=1}^{n} p_j x_j \\ \text{subject to} \quad & \sum_{j=1}^{n} w_j x_j \leq c, \\ & x_j \geq 0 \text{ integer}, \quad j = 1, \ldots, n. \end{aligned} \quad (3)$$

Actually, any variable x_j of an Unbounded Knapsack Problem will be bounded by the capacity c, as the weight of each item is at least one. But generally there is no benefit by transforming an Unbounded Knapsack Problem into its bounded version.

Another generalization of the 0-1 Knapsack problem is to choose exactly one item j from each class N_i, $i = 1, \ldots, k$ such that the profit sum is maximized. This gives the *Multiple-choice Knapsack Problem* which is defined as

$$\begin{aligned}
\text{maximize} \quad & \sum_{i=1}^{k} \sum_{j \in N_i} p_{ij} x_{ij} \\
\text{subject to} \quad & \sum_{i=1}^{k} \sum_{j \in N_i} w_{ij} x_{ij} \leq c, \\
& \sum_{j \in N_i} x_{ij} = 1, \qquad i = 1, \ldots, k, \\
& x_{ij} \in \{0,1\}, \qquad i = 1, \ldots, k, \ j \in N_i.
\end{aligned} \qquad (4)$$

Here p_{ij} and w_{ij} are the profit and weight of item j in class i, while the binary variable x_{ij} is one if item j was chosen in class i, and zero otherwise. The constraint $\sum_{j \in N_i} x_{ij} = 1$, $i = 1, \ldots, k$ ensures that exactly one item is chosen from each class.

If the profit p_j equals the weight w_j for each item j in a 0-1 Knapsack Problem, then we obtain the *Subset-sum Problem*, which may be formulated as:

$$\begin{aligned}
\text{maximize} \quad & \sum_{j=1}^{n} w_j x_j \\
\text{subject to} \quad & \sum_{j=1}^{n} w_j x_j \leq c, \\
& x_j \in \{0,1\}, \qquad j = 1, \ldots, n.
\end{aligned} \qquad (5)$$

The name indicates that it may also be seen as the problem of choosing a subset of the values $w_1, \ldots, w_n$ such that the sum is the largest possible without exceeding c.

Now imagine a cashier who has to give back an amount of money c by using the least possible number of coins of values $w_1, \ldots, w_n$. The *Change-making Problem* is then defined as

$$\begin{aligned}
\text{minimize} \quad & \sum_{j=1}^{n} x_j \\
\text{subject to} \quad & \sum_{j=1}^{n} w_j x_j = c, \\
& x_j \geq 0 \text{ integer}, \qquad j = 1, \ldots, n,
\end{aligned} \qquad (6)$$

where w_j is the face value of coin j, and we assume that an unlimited amount of each coin is available. The optimal number of each coin j that should be used is then given by x_j.

If we choose some of n items to pack in m knapsacks of (maybe) different capacity c_i, such that the largest possible profit sum is obtained, then we get the *Multiple Knapsack Problem*:

$$\begin{aligned}
\text{maximize} \quad & \sum_{i=1}^{m} \sum_{j=1}^{n} p_j x_{ij} \\
\text{subject to} \quad & \sum_{j=1}^{n} w_j x_{ij} \leq c_i, \qquad i = 1, \ldots, m, \\
& \sum_{i=1}^{m} x_{ij} \leq 1, \qquad j = 1, \ldots, n, \\
& x_{ij} \in \{0,1\}, \qquad i = 1, \ldots, m, \ j = 1, \ldots, n.
\end{aligned} \qquad (7)$$

Here $x_{ij} = 1$ says that item j should be packed into knapsack i ($x_{ij} = 0$ otherwise), while the constraint $\sum_{j=1}^{n} w_j x_{ij} \leq c_i$ ensures that the capacity constraint of each knapsack i is respected. The constraint $\sum_{i=1}^{m} x_{ij} \leq 1$ ensures that each item j is chosen at most once.

A very useful model is the *Bin-packing Problem* where all the n items should be packed in a number of equally large *bins*, such that the number of bins used is the smallest possible. Thus we have

$$\begin{aligned}
\text{minimize} \quad & \sum_{i=1}^{n} y_i \\
\text{subject to} \quad & \sum_{j=1}^{n} w_j x_{ij} \leq c y_i, \quad i = 1, \ldots, n, \\
& \sum_{i=1}^{n} x_{ij} = 1, \quad j = 1, \ldots, n, \\
& y_i \in \{0, 1\}, \quad i = 1, \ldots, n, \\
& x_{ij} \in \{0, 1\}, \quad i = 1, \ldots, n, \ j = 1, \ldots, n,
\end{aligned} \quad (8)$$

where $y_i = 1$ indicates that bin i is used ($y_i = 0$ otherwise), and x_{ij} says that item j should be packed in bin i. The constraint $\sum_{i=1}^{n} x_{ij} = 1$ ensures that every item j is packed exactly once, while inequalities $\sum_{j=1}^{n} w_j x_{ij} \leq c y_i$ ensure that the capacity constraint is respected for all bins that are actually used.

The most general form of a Knapsack Problem is the *Multi-constrained Knapsack Problem*, which is basically a general Integer Programming Problem where all coefficients p_j, w_{ij} and c_i are nonnegative integers. Thus, it may be formulated as

$$\begin{aligned}
\text{maximize} \quad & \sum_{j=1}^{n} p_j x_j \\
\text{subject to} \quad & \sum_{j=1}^{n} w_{ij} x_j \leq c_i, \quad i = 1, \ldots, m, \\
& x_j \geq 0 \text{ integer}, \quad j = 1, \ldots, n.
\end{aligned} \quad (9)$$

Gavish and Pirkul [41] consider different relaxations of this problem, and propose an exact algorithm. Approximation algorithms are considered in Frieze and Clarke [38] as well as Plotkin, Shmoys and Tardos [107]. If the number of constraints is $m = 2$ then the *Bidimensional Knapsack Problem* appears. Exact solution techniques for this problem are presented in e.g. Fréville and Plateau [37].

The *Quadratic Knapsack Problem* presented by Gallo, Hammer and Simeone [39] is an example of a Knapsack Problem with a quadratic objective function. It may be stated as

$$\begin{aligned}
\text{maximize} \quad & \sum_{j=1}^{n} \sum_{i=1}^{n} p_{ij} x_i x_j \\
\text{subject to} \quad & \sum_{j=1}^{n} w_j x_j \leq c, \\
& x_j \in \{0, 1\}, \quad j = 1, \ldots, n.
\end{aligned} \quad (10)$$

Here p_{ij} is the profit obtained if *both* items i and j are chosen, while w_j is the weight of item j. The Quadratic Knapsack Problem is a knapsack counterpart to the Quadratic Assignment Problem, and the problem has several applications in telecommunication and hydrological studies. Caprara, Pisinger and Toth [11] have recently presented an exact algorithm for large instances with up to $n = 400$ variables.

Other related problems within the family of Knapsack Problems are: The Collapsing Knapsack Problem which is considered in Fayard and Plateau [30] and Pferschy, Pisinger, Woeginger [95], and the Nested Knapsack Problem, treated in Dudzinski and Walukiewicz [24] together with several generalizations. Morin and Marsten [85] as well as Hochbaum [52] report some results on Nonlinear Knapsack Problems. Burkard and Pferschy consider the Inverse-parametric Knapsack Problem in [10]. Bottleneck versions of the 0-1 Knapsack Problem are considered in Martello and Toth [82] where polynomial algorithms are proposed for their solution. Although not usually grouped as a Knapsack Problem, Martello and Toth treat the Generalized Assignment Problem in [80] using the terminology of Knapsack Problems.

Numerous other Knapsack Problems appear by combining the above constraints in some way, e.g. the Bounded Multiple-choice Knapsack Problem [104], the Multiple-choice Subset-sum Problem [99], the Multiple-choice Nested Knapsack Problem [24] etc.

We have presented the above Knapsack Problems in the maximization form (with the change-making problem as an important exception), although equivalent minimization versions can be defined. For several of the problems the maximization problem may however be transformed into an equivalent minimization problem and vice versa. We will describe these transformations when dealing with each problem.

For all the problems (apart from change-making), a feasible solution may be obtained in polynomial time. If the capacity constraint is however changed to demanded equality, even finding a feasible solution becomes an $\mathcal{NP}$-hard problem.

1.4 $\mathcal{NP}$-hardness and Solvability

For most of the Knapsack Problems it is easy to derive a recursive formulation, expressing the objective value as the maximum of the optimal solution values for a number of subproblems. Since the solution space for several of these subproblems overlaps, dynamic programming yields an effective solution technique. Basically, the recursion is evaluated in an iterative way, saving the intermediate solutions (so-called *states*) in a table. By dynamic programming, many of the Knapsack Problems can be solved in pseudo-polynomial time, i.e. in a time bounded by a polynomium in the size of the instance and the magnitude of the coefficients. For the 0-1 Knapsack Problem a straightforward dynamic programming algorithm yields a time bound of $O(nc)$.

Since most Knapsack Problems are pseudo-polynomially solvable, we are balancing at the border between $\mathcal{NP}$-hard problems and Polynomially solvable problems $\mathcal{P}$. One could claim that since c in all practical applications is bounded by the word-length of a computer, and thus by a constant, all Knapsack Problems considered in practice are polynomially solvable. This does not however give a clue to efficient solution techniques, since the word-length may be large.

Most of the test-instances considered in this chapter have coefficients bounded by a constant, and thus we are basically dealing with polynomially solvable problems. But even for coefficients of modest magnitude, instances can be constructed which tend to be difficult. Avis [4] found a class of problems where the magnitude of the weights is bounded by $O(n^2)$, but any branch-and-bound algorithm not using dominance and cuts will perform badly. Chvátal [14] proved that if all coefficients of a Knapsack Problem are exponentially growing, and if the profit equals the weight for each item, then no bounding and no dominance relations will stop the enumerative process before at least $(2^{n/10})$ nodes have been enumerated, thus implying strictly exponentially growing computational times. Lagarias and Odlyzko [69] however showed that more general algorithms can handle such instances in a better way.

Recently Krass, Sethi and Sorger [68] investigated properties that make a Knapsack Problem hard to solve. Although considering only Knapsack Problems where the weights are elements of a given subset S of the positive integers, they show that the family of Knapsack Problems obtained by varying the parameter S in the power set of the positive integers Z^+ contains polynomially solvable problems and $\mathcal{NP}$-complete problems, even when S is restricted to the class of polynomially recognizable sets.

1.5 Fundamental Properties of the Knapsack Problems

Knapsack Problems are highly structured, which fortunately implies that several instances may be solved in fractions of a second despite the worst-case complexity. These structural properties give rise to different techniques which either lead to algorithms with known worst-case performance, or which are able to limit the solution space considerably such that real life problems are generally easily solved despite bad worst-case behavior. It is important to distinguish between these two directions of research, since when dealing with $\mathcal{NP}$-hard problems it often happens that solution techniques with good worst-case performance are outperformed by less mathematically founded techniques when practical applications are considered.

The perhaps most important property of the Knapsack Problems is that the *continuous relaxation* of the problems, where the constraints on the variables $x_j \in \{0, \ldots, m_j\}$ are changed to $0 \leq x_j \leq m_j$, are so fast to solve. Back in 1957, Dantzig [17] showed an elegant solution for the continuous 0-1 Knapsack Problem, by ordering the items according to their profit-to-weight ratio,

$$\frac{p_1}{w_1} \geq \frac{p_2}{w_2} \geq \ldots \geq \frac{p_n}{w_n}, \qquad (11)$$

and using a greedy algorithm for filling the knapsack. The first item which does not fit into the knapsack is denoted the *break item* b (also called *critical item*) and an optimal solution is found by choosing all items $j < b$ plus a fraction of item b corresponding to the residual capacity. The ordering (11) takes $O(n \log n)$ time, but Balas and Zemel [6] showed that the continuous 0-1 Knapsack Problem may be solved in linear time without sorting, since b may be found as a weighted median. This result has later been generalized to several of the Knapsack Problems, such that we now know linear time algorithms for the continuous version of problems (1) to (8).

The existence of quickly obtainable *upper bounds* makes it possible to solve large-sized problems through branch-and-bound. Since branch-and-bound techniques are basically based on a complete enumeration, their performance strictly depends on the quality of the bounds applied. For many applications, the bounds obtained by continuous relaxation are sufficient, but instance classes exist where these bounds overestimate the integer solution value, and thus do not efficiently cut off unpromising branches in the search tree. Recently, much effort has been made to improve the bounds in order to obtain branch-and-bound algorithms with stable performance for a large class of instances.

Another essential property is that, having solved the continuous relaxed problem generally only a few decision variables need to be changed in order to obtain the optimal solution. Fig. 1 shows a typical solution to a 0-1 Knapsack Problem compared to the continuous solution. It can be seen that most of the solution values are the same, while the differing variables are generally close to the break item. To be more specific, Fig. 2 shows how often the integer solution differs from the continuous solution as average over 1000 "easy" instances constructed such that b is the same for all instances. The figure shows that the only differences are close to b, while other variables maintain their solution value from the continuous problem.

solution	$x_i (1 \leq i \leq 6)$	x_7	x_8	x_9	x_{10}	x_{11}	x_{12}	x_{13}	x_{14}	x_{15}	x_{16}
continuous	1		1	1	1	0.35	0	0	0	0	0	0
IP-optimal	1		0	0	1	1	0	1	0	0	0	0

Figure 1: A typical solution to the 0-1 Knapsack Problem compared to the continuous solution. The break item is $b = 10$, and it is seen that those variables, where the two solution values differ, are generally close to b

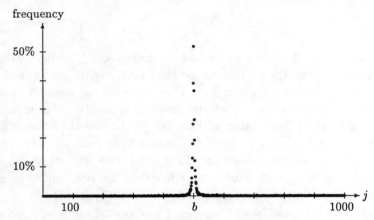

Figure 2: Frequency of items j where the optimal solution x_j^* differs from the continuous solution $\overline{x}_j$

This behavior motivated Balas and Zemel [6] to propose that only a few variables around b are considered in order to solve the Knapsack Problem to optimality. This problem was denoted the *core problem* and has been an essential part of all efficient algorithms for Knapsack Problems. Recent results have, however, shown that it is important how a core is chosen, as degeneration may occur such that all item weights in the core will be almost the same. Obviously this makes it very difficult to obtain a solution which fills the knapsack well, and thus a tight lower bound.

For all the Knapsack Problems, efficient *reduction algorithms* have been developed, which enable one to fix some decision variables at their optimal values before the problem is solved, thus considerably decreasing the size of an instance. Basically, these tests may be viewed as a special case of the branch-and-bound technique; for each 0-1 variable, we test both branches, fathoming one of them if a bounding test shows that a better solution cannot be found.

Already in the seventies Balas [5], Padberg [93] and Wolsey [115] investigated *polyhedral properties* of the knapsack polytope. Among their results we find that if the weights are ordered in increasing order and we let $\alpha = \min\{h : \sum_{j=1}^{h} w_j > c\}$ then the inequality

$$\sum_{j=1}^{n} x_j \leq \alpha - 1 \qquad (12)$$

defines a facet of the knapsack polytope. Balas and Zemel [6] made the first experiment with adding such constraints to the formulation in order to get solutions for the continuous problem which are closer to the integer optimum. But only recently have these cardinality constraints been used efficiently in an enumerative algorithm by relaxation with the original weight constraint. This technique often closes the gap between the integer and continuous solutions of *ill-conditioned* problems (i.e. problems where the continuous solution is far from the integer solution).

Several Knapsack Problems are solvable in pseudo-polynomial time by dynamic programming. We do in fact know that problems (1) to (6) are pseudo-polynomially solvable, while the remaining problems are $\mathcal{NP}$-hard in the strong sense. The dominance relations are generally very efficient, making it possible to fathom several unpromising states. By incorporating bounding tests in the dynamic programming, very efficient algorithms may be developed.

Another important property of Knapsack Problems is that they are *separable*, as observed by Horowitz and Sahni [54], which means that a 0-1 Knapsack Problem may be solved in $O(\sqrt{2^n})$ worst-case time. The idea is to divide the problem into two equally sized parts, and enumerate each subproblem through dynamic programming. The two tables of optimal profit sums that may be obtained for each capacity are easily combined in time proportional to the size of the sets by considering pairs of states which have the largest possible weight sum not exceeding c. The technique gives an improvement over a complete enumeration by a factor of a square-root. Although this bound is still exponential, the consequence of this observation is that we may solve a 0-1 Knapsack Problem through parallel computation by recursively dividing the problem into two parts. The resulting algorithm runs in $O(\log n \log c)$, which is probably the best one can hope for, as mentioned in Kindervater and Lenstra [65], but the number of processors required is huge: $O(nc^2/(\log n(\log c)^2))$.

A final technique to limit the search is *balancing*. A balanced solution is loosely speaking a solution which is sufficiently filled, i.e. the weight sum does not differ from the capacity by more than the weight of a single item. Obviously, an optimal solution is balanced, and thus the enumeration may be limited to only consider balanced states. Using balancing, several problems from the knapsack family are solvable in linear time provided that the weights w_j are bounded by a constant r [99]. For the subset-sum problem one gets the attractive solution time of $O(nr)$, which is an improvement over the Bellman recursion running in $O(nc) = O(n^2 r)$. For 0-1 Knapsack problems balancing yields solution times of $O(nr^2)$, which can only compete with the Bellman recursion for large-sized problems having small weights.

Finally, fully polynomial *approximation schemes* have been derived for some of the Knapsack Problems. The algorithms are generally based on dynamic programming where we make some kind of scaling of the profit sums in order to limit the number of states in the dynamic programming. The scaling introduces some kind of relative error for each state; thus two techniques are used for limiting the error: 1) The items are divided into "large" and "small" items, where the dynamic programming is done for the large items only. 2) At each stage of the dynamic programming, states are deleted which are "sufficiently" close to each other. In practice fully polynomial approximation schemes do, however, have quite disappointing performance compared to well-designed heuristics with no formal performance guarantee [80].

1.6 Experimental Comparisons

Due to the large gap between the worst-case performance of algorithms and their practical solution times, one cannot study algorithms for Knapsack Problems without performing some experiments.

In the following sections we will end each presentation of a problem with comprehensive computational comparisons showing the performance of the best algorithms available. The experimental work is also intended to expose other properties which may be relevant for the design of new algorithms, such as measuring the quality of the upper bounds, showing the size of the minimal core, etc.

For the experimental comparisons we will apply different groups of ran-

Knapsack Problems

domly generated instances, which have been constructed to reflect special properties that may influence the solution process. Thus, we will discuss the nature of each group of instances. In all instances the weights are randomly distributed in a given interval while the profits are expressed as a function of the weights, yielding the specific properties of each group. The instance groups are graphically illustrated in Fig. 3.

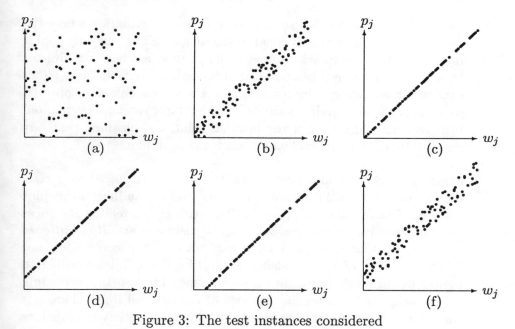

Figure 3: The test instances considered

a *Uncorrelated data instances:* In these instances there is no correlation between the profit and weight of an item. Such instances illustrate those situations where it is reasonable to assume that the profit does not depend on the weight. Uncorrelated instances are generally easy to solve, as there is a large variation between the weights, making it easy to obtain a filled knapsack. Moreover, it is easy to fathom numerous variables by upper bound tests or by dominance relations.

b *Weakly correlated instances:* Despite their name, weakly correlated instances have a very high correlation between the profit and weight of an item. Typically the profit only differs from the weight by a couple of percent. Such instances are perhaps the most realistic in management, as it is well-known that the return of an investment is generally proportional to the sum invested within some small variations. The high correlation means that it is generally difficult to fathom variables by upper bound tests. In spite of this, weakly correlated instances are generally easy to solve, since there is a large variation in the weights, making it easy to obtain a filled knapsack, and filled solutions are generally very close to an optimal solution due to the correlation.

c *Subset-sum instances:* These instances reflect the situation where the profit of each item is proportional to the weight. Thus, our only goal is to obtain a filled knapsack. Subset-sum instances are however difficult to solve, as any upper bound returns the same trivial value c, so we cannot use bounding rules for cutting off branches before an optimal solution has been found. On the other hand, large randomly generated instances generally have many optimal solutions, meaning that any permutation algorithm will easily reach the optimum.

d *Strongly correlated instances:* Such instances correspond to a real-life situation where the return is proportional to the investment plus some fixed charge for each project. The strongly correlated instances are hard to solve for two reasons: 1) The instances are *ill-conditioned* in the sense that there is a large gap between the continuous and integer solution of the problem. 2) Sorting the items according to nonincreasing profit-to-weight ratios correspond to a sorting according to the weights. Thus for any interval of the ordered items there is a small variation in the weights, making it difficult to satisfy the capacity constraint with equality.

e *Inverse strongly correlated instances:* These instances are like strongly correlated instances, but the fixed charge is negative.

f *Almost strongly correlated instances:* Are a kind of fixed-charge problem with variance. Thus they reflect the properties of both strongly and weakly correlated instances.

1.7 Notation

For the following presentation we define $x(\text{P})$ and $z(\text{P})$ to represent the optimal solution of a problem P and its objective value, respectively. The continuous relaxation of a problem P will be denoted by $C(\text{P})$ while the

$L(P, \lambda)$ denotes the Lagrangian relaxation of the same problem using multiplier λ. Finally, $S(P, \lambda)$ denotes the surrogate relaxed problem of P using multiplier λ. Other symbols used during the presentation are:

x^*, z^*	optimal solution and the corresponding objective value
$\hat{x}, \hat{p}, \hat{w}$	break solution and the corresponding profit and weight sum
λ	surrogate or Lagrangian multiplier
δ	core size
γ	median, middle value

In the following sections, we will consider several upper and lower bounds for the Knapsack Problem. The worst-case performance ratio $\rho(U)$ of an upper bound U is defined as the smallest real number such that

$$\frac{U(P)}{z^*} \leq \rho(U) \qquad (13)$$

for a problem P with optimal solution z^*. Similarly assume that a lower bound z^h on P has been obtained with an approximation algorithm H. Then the worst-case performance ratio of algorithm H is defined as the largest real number $r(H)$ such that

$$\frac{z^h(P)}{z^*} \geq r(H) \qquad (14)$$

1.8 Overview of the Chapter

In this chapter we will consider theoretical aspects of several problems from the knapsack family, giving outlines of the most efficient algorithms. Several new and unpublished results will be presented, and old results are presented in a new and systematical way, using simplified proofs and algorithms where possible. All the codes presented have been tested on the same computer (HP9000 model 735/99, performance index according to SPEC [110]: 3.27 specint95, 3.98 specfp95). This gives an up-to-date performance index of the different algorithms, as well as making it possible to compare algorithms across sections.

Numerous results have been presented for Knapsack Problems, thus due to the limited space, we have chosen to focus on those techniques which lead to effective solution methods. We will however, as far as possible, also give references to important theoretical results, which have not yet been applied in the best algorithms.

Since the family of Knapsack Problems is so wide, only some of the problems can be covered in this chapter. We have chosen to focus on the most fundamental Knapsack Problems, since most research goes on in this field, and since the solution techniques presented may also be used in some of the more general cases. Thus in Section 2 we consider the 0-1 Knapsack Problem and show several fundamental properties. Upper and lower bounds are presented, and these are combined to show the foremost solution techniques from the literature. The following sections generalize these results to other kinds of knapsack problems. First, the Subset-sum Problem is considered in Section 3. Although this problem is a special case of the 0-1 Knapsack Problem, specialized solution techniques are developed to fully exploit the structure of the problem. In Section 4 we consider the knapsack problem with multiple-choice constraints. Since it is a generalization of the 0-1 Knapsack Problem, it is interesting to see which principles are adaptable to the more general case, and which techniques are difficult to utilize. Sections 5 and 6 consider the Bounded Knapsack Problem and Unbounded Knapsack Problem, respectively. Despite being quite similar in their formulation, these problems have very different properties and thus solution techniques. Finally, Section 7 considers the Multiple Knapsack Problem, which is $\mathcal{NP}$-hard in the strong sense, and thus differs from the previous pseudopolynomially solvable problems.

The text-book by Martello and Toth [80] is the most comprehensive work on Knapsack Problems, and the reader is referred to it for additional details. Other surveys published recently are: Martello and Toth [74] consider exact algorithms for the 0-1 Knapsack Problem and their average computational performance; the study is extended to the other linear Knapsack Problems and to approximate algorithms in Martello and Toth [78]. Dudzinski and Walukiewicz [24] analyze dual methods for solving Lagrangian and linear programming relaxations. In addition, almost all books on Integer Programming contain a section on Knapsack Problems: Papadimitriou and Steiglitz [94], Syslo, Deo and Kowalik [111], Nemhauser and Wolsey [92] and Ibaraki [56, 57] are some of the more recent ones.

2 0-1 Knapsack Problem

The *0-1 Knapsack Problem* (KP) is the problem of choosing some of the n items such that the corresponding profit sum is maximized without the weight sum exceeding the capacity c. Thus it may be formulated as the following maximization problem:

$$\text{maximize} \quad \sum_{j=1}^{n} p_j x_j$$
$$\text{subject to} \quad \sum_{j=1}^{n} w_j x_j \leq c, \qquad (15)$$
$$x_j \in \{0,1\}, \quad j = 1,\ldots,n,$$

where x_j is a binary variable having value 1 if item j should be included in the knapsack, and 0 otherwise.

Many industrial problems can be formulated as 0-1 Knapsack Problems: Cargo loading, cutting stock, project selection, and budget control to mention a few examples. Several combinatorial problems can be reduced to KP, and KP has important applications as a subproblem of several algorithms of integer linear programming.

To avoid trivial cases we assume that $w_j \leq c$ for all $j = 1,\ldots,n$ and $\sum_{j=1}^{n} w_j > c$. Items violating the first constraint will not be present in any solution, and thus we can set $x_j = 0$ for such items. Problems violating the second constraint have a trivial solution with all items chosen.

Without loss of generality, we may also assume that p_j, w_j and c are positive integers, since fractions can be handled by multiplying the coefficients by a proper factor, while instances with nonpositive coefficients can be treated as described in Glover [45]: Let

$$\begin{array}{ll} N^0 = \{j \in N : p_j \leq 0,\ w_j \geq 0\} & N^1 = \{j \in N : p_j \geq 0,\ w_j \leq 0\} \\ N^- = \{j \in N : p_j \leq 0,\ w_j \leq 0\} & N^+ = \{j \in N : p_j \geq 0,\ w_j \geq 0\} \end{array} \qquad (16)$$

All items $j \in N^0$ will never be chosen, thus $x_j = 0$ for these items. Similarly, all items $j \in N^1$ must be present in an optimal solution, thus for all these items $x_j = 1$. The remaining items are transformed as follows

$$\begin{array}{lll} \overline{p}_j = -p_j, & \overline{w}_j = -w_j, & \overline{x}_j = 1 - x_j \quad \text{for} \quad j \in N^- \\ \overline{p}_j = p_j, & \overline{w}_j = w_j, & \overline{x}_j = x_j \quad \text{for} \quad j \in N^+ \end{array} \qquad (17)$$

The transformed problem satisfies the nonnegativity of the coefficients and thus may be solved as

$$\text{maximize} \quad \sum_{j \in N^- \cup N^+} \overline{p}_j \overline{x}_j + \sum_{j \in N^1 \cup N^-} p_j$$
$$\text{subject to} \quad \sum_{j \in N^- \cup N^+} \overline{w}_j \overline{x}_j \leq c - \sum_{j \in N^1 \cup N^-} w_j \qquad (18)$$
$$\overline{x}_j \in \{0,1\}, \quad j \in N^- \cup N^+$$

Throughout this section we will consider KP in the maximization form as it is equivalent to the minimization form

$$\begin{aligned}\text{minimize} \quad & \sum_{j=1}^{n} p_j y_j \\ \text{subject to} \quad & \sum_{j=1}^{n} w_j y_j \geq d, \\ & y_j \in \{0,1\}, \quad j=1,\ldots,n,\end{aligned} \quad (19)$$

by setting $y_j = 1 - x_j$ and $d = \sum_{j=1}^{n} w_j - c$. Intuitively, we here minimize the profit of the items *not inserted* in the knapsack.

In Section 2.1 we will present different upper bounds for the knapsack problem and show how they may be tightened by addition of cardinality bounds. Heuristic solutions for KP are discussed in Section 2.2, and Section 2.3 shows how the size of a KP may be reduced by using some upper bound tests. Enumerative algorithms are presented in Sections 2.4 and 2.5, considering respectively branch-and-bound algorithms and dynamic programming algorithms. In Section 2.6 we consider the solution of large-sized instances, where a *core* is used to focus the enumeration on the most interesting items. In Section 2.7 we consider the solution of hard problems, where basically all the previous techniques are applied to obtain a stable and well-performing algorithm. Although this text is mainly focused on exact solutions of Knapsack Problems, we show a few results on approximation schemes in Section 2.8. Finally, comprehensive computational experiments are presented in Section 2.9.

2.1 Upper Bounds

The continuous relaxation $C(\text{KP})$ of (15) leads to the so-called Dantzig bound [17], which may be derived as follows: Order the items according to nonincreasing profit-to-weight ratios

$$\frac{p_1}{w_1} \geq \frac{p_2}{w_2} \geq \ldots \geq \frac{p_n}{w_n} \quad (20)$$

and then using the *greedy principle* items $j = 1, 2, \ldots$ are added to the knapsack as long as

$$\sum_{h=1}^{j} w_h \leq c. \quad (21)$$

The first item b which cannot be included in the knapsack is denoted the *break item* and an optimal solution $\bar{x}$ to the continuous problem is given by $\bar{x}_j = 1$ for $j = 1, \ldots, b-1$ and $\bar{x}_j = 0$ for $j = b+1, \ldots, n$ while $\bar{x}_b$ is given by

$$\bar{x}_b = \left(c - \sum_{i=1}^{b-1} w_i\right) / w_b. \tag{22}$$

The continuous solution has objective value

$$U_1 = z(C(\text{KP})) = \sum_{j=1}^{b-1} p_j + \left(c - \sum_{j=1}^{b-1} w_j\right) \frac{p_b}{w_b} \tag{23}$$

Since any integer solution will have an integral objective value, we may actually tighten this bound to $\lfloor U_1 \rfloor$. An integer solution may be obtained by truncating $\bar{x}_b$ to zero. The solution will be denoted the *break solution* $\hat{x}$ and the profit sum of $\hat{x}$ is $\hat{p} = \sum_{j=1}^{b-1} p_j$ while the weight sum is $\hat{w} = \sum_{j=1}^{b-1} w_j$. Notice that if $\bar{x}_b = 0$ in the continuous solution, then the break solution $\hat{x}$ is an optimal solution x^* to (15).

Balas and Zemel [6], independently from Fayard and Plateau [28], showed that the Dantzig bound may be found in $O(n)$ time through a partitioning algorithm. Basically the problem is to find a break item b which satisfies the following criteria:

$$\begin{array}{ll} p_j/w_j \geq p_b/w_b & \text{for } j = 1, \ldots, b-1 \\ p_j/w_j \leq p_b/w_b & \text{for } j = b+1, \ldots, n \\ \sum_{j=1}^{b-1} w_j \leq c < \sum_{j=1}^{b} w_j & \end{array} \tag{24}$$

Thus finding b may be seen as a kind of weighted median problem, which can be solved as follows: Let $[s, t]$ (initially equal to $[1, n]$) be the current interval of items. Choose γ as the median of $p_s/w_s, \ldots, p_t/w_t$. Partition the items into two intervals $[s, i-1]$ and $[i, t]$ such that $p_j/w_j \geq \gamma$ for $j \in [s, i-1]$ and $p_j/w_j \leq \gamma$ for $j \in [i, t]$. Let $\tilde{w} = \sum_{h=1}^{i-1} w_j$. If $\tilde{w} > c$ then the break item b cannot be in $[i, t]$, thus this interval is discarded, while otherwise b cannot be in $[s, i-1]$ and we discard the latter. The process is repeated until $[s, t]$ contains only one item, which is obviously the break item.

algorithm bal_zem;
$s \leftarrow 1; t \leftarrow n; \overline{w} \leftarrow 0;$
while $(s < t)$ **do**
    $\gamma \leftarrow \text{median}(p_s/w_s, \ldots, p_t/w_t);$

$i \leftarrow s;\ j \leftarrow t;\ \tilde{w} \leftarrow \overline{w};$
repeat
  while $(p_i/w_i \geq \gamma)$ **and** $(i < j)$ **do** $\tilde{w} \leftarrow \tilde{w} + w_i;\ i \leftarrow i + 1;$
  while $(p_j/w_j \leq \gamma)$ **and** $(i < j)$ **do** $j \leftarrow j - 1;$
  if $(i < j)$ **then** $\text{swap}(i, j);$
until $(i \geq j);$
if $(\tilde{w} > c)$ **then** $t \leftarrow i - 1$ **else** $s \leftarrow i;\ \overline{w} \leftarrow \tilde{w};$
$b \leftarrow t;$

The partitioning of the interval $[s, t]$ is done in the same way as in most sorting algorithms: Search forward from s until an item i is met which does not satisfy the ordering $p_i/w_i \geq \gamma$. Similarly, starting from t and searching backward find the first item j violating $p_j/w_j \leq \gamma$. By swapping the items i and j the ordering is satisfied, and the search can continue.

When γ is chosen as the median of the remaining items, each iteration will discard half of the items, thus leading to a computational effort of $O(n)$. Balas and Zemel, however, experimentally showed that better average-case performance can be obtained by choosing γ as a median-of-three.

2.1.1 Lagrangian Relaxation

The Lagrangian relaxation of KP with a nonnegative multiplier λ leads to the problem $L(\text{KP}, \lambda)$:

$$\text{maximize} \quad \sum_{j=1}^{n} p_j x_j + \lambda \left(c - \sum_{j=1}^{n} w_j x_j \right) \tag{25}$$
$$\text{subject to} \quad x_j \in \{0, 1\}, \qquad j = 1, \ldots, n,$$

The objective function can be restated as $z(L(\text{KP}, \lambda)) = \max \sum_{j=1}^{n} \tilde{p}_j x_j + \lambda c$, where $\tilde{p}_j = p_j - \lambda w_j$ for $j = 1, \ldots, n$, and the optimal solution to (25) is easily determined in $O(n)$ time as

$$x_j = \begin{cases} 1 & \text{if } \tilde{p}_j > 0, \\ 0 & \text{if } \tilde{p}_j < 0, \end{cases} \tag{26}$$

where x_j may be chosen to any of the two values when $\tilde{p}_j = 0$. The corresponding objective value is

$$z(L(\text{KP}, \lambda)) = \sum_{\{j : p_j/w_j > \lambda\}} \tilde{p}_j + \lambda c\ . \tag{27}$$

Knapsack Problems

Notice that the continuous relaxation of (25) will produce the same (integer) solution as above, thus we have

$$z(L(\text{KP}, \lambda)) = z(C(L(\text{KP}, \lambda))) \geq z(C(\text{KP})) \geq z(\text{KP}) . \qquad (28)$$

This means that the bound obtained by Lagrangian relaxation will never be tighter than the continuous bound $z(C(\text{KP}))$. The value of λ producing the minimal value of $z(L(\text{KP}, \lambda))$ is $\lambda^* = p_b/w_b$, in which case we exactly obtain the continuous bound, thus $z(L(\text{KP}, \lambda^*)) = z(C(\text{KP})) = U_1$ [80].

2.1.2 Tighter Bounds

Martello and Toth [72] derived a tighter bound on KP than the continuous bound by applying the integrality of x_b. Obviously, in an optimal solution either $x_b = 0$ or $x_b = 1$. Thus first assuming $x_b = 0$ a valid upper bound on KP is:

$$U' = \left\lfloor \hat{p} + (c - \hat{w}) \frac{p_{b+1}}{w_{b+1}} \right\rfloor, \qquad (29)$$

and with $x_b = 1$ we have the bound

$$U'' = \left\lfloor \hat{p} + p_b + (c - \hat{w} - w_b) \frac{p_{b-1}}{w_{b-1}} \right\rfloor, \qquad (30)$$

where $\hat{p}$ and $\hat{w}$ are the profit and weight sums of the break solution. Both bounds have been derived by relaxing the constraints on x_{b-1} and x_{b+1} to $x_{b-1} > 0$ resp. $x_{b+1} > 0$. Now the *Martello and Toth upper bound* is given by

$$U_2 = \max\{U', U''\} \qquad (31)$$

Obviously U_2 can be derived in $O(n)$ through the bal_zem algorithm, and it is easy to see that $U_2 \leq U_1$ [72].

Fayard and Plateau [29] used the same dichotomy to derive the bound

$$U_3 = \max\left\{z(C(\text{KP}, x_b = 1)), \ z(C(\text{KP}, x_b = 0))\right\} \qquad (32)$$

where the continuous KP with additional constraint $x_b = \alpha$ can be solved in $O(n)$ time by using the bal_zem algorithm. Obviously $U_3 \leq U_2 \leq U_1$.

Generalizing the above principle, Martello and Toth [79] proposed the derivation of arbitrarily tight upper bounds by using *partial enumeration*. Let $M \subseteq N$ be a subset of the items, and $X_M = \{x_j \in \{0, 1\}, \ j \in M\}$.

Then every optimal solution to KP must follow one of the paths through X_M, and an upper bound on KP is thus given by

$$U_M = \max_{\tilde{x} \in X_M} U(\tilde{x}), \qquad (33)$$

where $U(\tilde{x})$ is any upper bound on (15) with additional constraint $x_j = \tilde{x}_j$ for $j \in M$. For $M = N$ we obtain the upper bound $U_M = z^*$ i.e. the optimal solution value. The computational effort is however large: $O(2^n)$. A good trade-off between tightness and computational effort may however be obtained by choosing M as a small subset of items with profit-to-weight ratios close to that of the break item.

Müller-Merbach [86] proposed a bound based on the fact that, if the continuous solution given by (22) becomes integral, then either $\bar{x}_b$ is truncated to zero, or a variable $\bar{x}_j$, $j \neq b$ changes value to $1 - \bar{x}_j$. Thus we have

$$U_4 = \max\left\{\hat{p},\ \max_{j<b}\lfloor u'_j \rfloor,\ \max_{j>b}\lfloor u''_j \rfloor\right\} \qquad (34)$$

The first term comes from assuming that $\bar{x}_b = 0$, in which case we get the objective value $\hat{p}$. If we set $\bar{x}_j = 0$ for $j < b$ then an upper bound is given by $u'_j = \hat{p} - p_j + (c - \hat{w} + w_j)p_b/w_b$. Similarly, if we set $\bar{x}_j = 1$ for $j > b$ then an upper bound is given by $u''_j = \hat{p} + p_j + (c - \hat{w} - w_j)p_b/w_b$. Both bounds can be derived by relaxing the integrality constraint on the break item to $x_b \in \Re$.

The Müller-Merbach bound is derived in $O(n)$ time, and we have $U_4 \leq U_1$ but no dominance exist between U_3 and U_4. Dudzinski and Walukiewicz [24] further exploited the above technique, obtaining a bound which dominates all of the above, and which can be derived in $O(n)$ time.

2.1.3 Bounds from Minimum and Maximum Cardinality

Polyhedral theory has recently found application in the field of Knapsack Problem where it has led to some efficient solution techniques for classes of hard problems. The study of facets of the knapsack polytope however dates back to 1975 when Balas [5], Hammer, Johnson and Peled [50] and Wolsey [115] gave necessary and sufficient conditions for a canonical inequality to be a facet of the knapsack polytope. Already in 1980 Balas and Zemel [6] made the first experiments adding facets to the formulation of strongly correlated knapsack problems, but the solution times were too large to show the benefits of this approach. For a recent treatment of Knapsack Polyhedra, see e.g. Weismantel [113].

Valid inequalities do not exclude any integer solution, and are thus redundant for the integer problem. The continuous relaxation is however strengthened by addition of valid inequalities, leading to tighter continuous bounds. We will here consider different inequalities which may be obtained from minimum and maximum cardinality of an optimal solution. Inequalities from minimum and maximum cardinality are very tight, and professional mixed-integer programming solvers like CPLEX automatically impose cardinality constraints for each Boolean inequality.

Cardinality bounds as presented by e.g. Balas [5] and Martello and Toth [83] are derived as follows: Assume that the weights are ordered according to nondecreasing weights, and define k as

$$k = \min\left\{h : \sum_{j=1}^{h} w_j > c\right\} - 1 \ . \tag{35}$$

Since items $1,\ldots,k$ have the smallest weights, every feasible solution will comprise no more than k items, and thus we can impose a *maximum cardinality constraint* of the form

$$\sum_{j=1}^{n} x_j \leq k \tag{36}$$

without excluding any feasible solution. However, adding a constraint which is not violated will not tighten the formulation, thus for problem (15) we will only add a maximum cardinality constraint if $k = b - 1$. Notice that since the break solution is a feasible solution, we must always have $k \geq b-1$.

In a similar way we may define *minimum cardinality constraints*. Assume that the current lower bound is given by z and that the items are ordered according to nonincreasing profits. We set

$$k = \max\left\{h : \sum_{j=1}^{h} p_j \leq z\right\} + 1 \tag{37}$$

and thus we have the constraint

$$\sum_{j=1}^{n} x_j \geq k \tag{38}$$

for any solution with objective value larger than z. As before, there is no reason to add this constraint to (15) if it is not violated, thus we will only use the constraint when $k = b$. Notice that in every case $k \leq b$, since an ordering according to (20) gives $k = b$ in (37) and ordering according to nonincreasing profits can only decrease this value.

Adding the constraints (36) or (38) to our model leads to a two-constraint knapsack problem which may be difficult to solve, as it is $\mathcal{NP}$-hard in the strong sense. The continuous relaxation may be solved by any LP-solver which is however time-consuming for large instances. Thus Martello and Toth [83] presented a specialized algorithm for the continuous solution as follows:

We consider the KP with maximum cardinality constraint added, thus having the problem (KP′)

$$\begin{aligned}
\text{maximize} \quad & \sum_{j=1}^{n} p_j x_j \\
\text{subject to} \quad & \sum_{j=1}^{n} w_j x_j \leq c, \\
& \sum_{j=1}^{n} x_j \leq k \\
& x_j \in \{0,1\}, \quad j = 1, \ldots, n,
\end{aligned} \tag{39}$$

By Lagrangian relaxing the cardinality constraint using a nonnegative multiplier $\lambda \geq 0$, and by relaxing the integrality constraints, one gets the problem $L(\text{KP}', \lambda)$ given by

$$\begin{aligned}
\text{maximize} \quad & \sum_{j=1}^{n} p_j x_j - \lambda \left(\sum_{j=1}^{n} x_j - k \right) = \sum_{j=1}^{n} \tilde{p} x_j + \lambda k \\
\text{subject to} \quad & \sum_{j=1}^{n} w_j x_j \leq c, \\
& 0 \leq x_j \leq 1, \quad j = 1, \ldots, n,
\end{aligned} \tag{40}$$

which is a continuous KP where the items have profits $\tilde{p} = p_j - \lambda$ that may be nonpositive. The above problem is easily solved for each value of λ by using transformation (16) for the nonpositive values of $\tilde{p}_j$, and then sorting the remaining items according to

$$\frac{p_1 - \lambda}{w_1} \geq \frac{p_2 - \lambda}{w_2} \geq \ldots \geq \frac{p_n - \lambda}{w_n} \tag{41}$$

in order to find the Dantzig bound. This can be done in $O(n)$ using the bal_zem algorithm. Let $K(x, \lambda)$ denote the cardinality of the solution x to $L(\text{KP}', \lambda)$.

We are interested in deriving the multiplier λ^* which leads to the smallest objective value of $L(\text{KP}', \lambda)$, i.e.

$$z(L(\text{KP}', \lambda^*)) = \min_{\lambda \geq 0} \left\{ z(L(\text{KP}', \lambda)) \right\} \tag{42}$$

or, equivalently

$$|K(x, \lambda^*) - k| = \min_{\lambda \geq 0} |K(x, \lambda) - k| \tag{43}$$

since in this way the cardinality constraint (36) has the greates effect on the continuous solution. It can be proved that a multiplier λ^* exists such that $K(x, \lambda^*) = k$. Equation (43) may be solved efficiently due to the following

Theorem 2.1 (Martello and Toth [83]) $K(x, \lambda)$ *is monotonically non-increasing as λ is increasing.*

Proof. First note that if we have given two items i, j and two multipliers λ', λ'' then if

$$\frac{p_i - \lambda'}{w_i} \geq \frac{p_j - \lambda'}{w_j} \quad \text{and} \quad \frac{p_i - \lambda''}{w_i} \leq \frac{p_j - \lambda''}{w_j} \tag{44}$$

then $w_j \geq w_i$, which can be seen by subtracting the second inequality from the first. But this means that when λ increases from λ' to λ'' then items i, j may change places in the ordering (41), but according to (44) bigger items will move to the first positions, thus never increasing the cardinality of the solution. □

The theorem means that we can use binary search to minimize (43). Thus, initially set $\lambda_1 = 0$ and $\lambda_2 = k$'th largest p_j value. Now, repeatedly derive $\lambda = (\lambda_1 + \lambda_2)/2$, and solve (40). If $K(x, \lambda) = k$, we have found the optimal multiplier $\lambda^* = \lambda$. Otherwise, if $K(x, \lambda) > k$ set $\lambda_1 = \lambda$ and if $K(x, \lambda) < k$ set $\lambda_2 = \lambda$.

For each value of λ we need to solve (40), which can however be done in $O(n)$ time. Martello and Toth [83] investigated further properties of $L(\text{KP}', \lambda)$, and derived an algorithm which finds the optimal multiplier λ^* in $O(n^2)$ time. Obviously the derived bound

$$U_5 = z(L(\text{KP}', \lambda^*)) = z(C(\text{KP}')) \tag{45}$$

is tighter than the continuous bound U_1 since it contains U_1 as a special case for $\lambda = 0$.

It is interesting to note that if instead we Lagrangian relax the capacity constraint with $\lambda \geq 0$ in (39), without relaxing the integrality constraint, we get

$$\begin{aligned}
\text{maximize} \quad & \sum_{j=1}^{n}(p_j - \lambda w_j)x_j + \lambda c \\
\text{subject to} \quad & \sum_{j=1}^{n} x_j \leq k, \\
& x_j \in \{0,1\}, \qquad j = 1,\ldots,n \ .
\end{aligned} \qquad (46)$$

which can be solved in $O(n)$ time by selecting the items corresponding to the k largest positive Lagrangian profits $(p_j - \lambda w_j)$. It can be easily proved, however, that the best upper bound which can be obtained through this relaxation is equal to U_5.

Martello and Toth [83] derive symmetrical results for bounds from minimum cardinality, thus the reader is referred to this paper for more details.

Surrogate relaxation of the capacity constraint with maximum cardinality constraint (36) using multiplier $\lambda \geq 0$ gives the problem

$$\begin{aligned}
\text{maximize} \quad & \sum_{j=1}^{n} p_j x_j \\
\text{subject to} \quad & \sum_{j=1}^{n}(w_j + \lambda)x_j \leq c + k\lambda, \\
& x_j \in \{0,1\}, \qquad j = 1,\ldots,n \ .
\end{aligned} \qquad (47)$$

which is also a Knapsack Problem. The latter problem may, however, be easier to solve, since with a well-chosen multiplier λ, the continuous bounds are tighter than for the original problem. In general, it is however sufficient to look at the continuous relaxation $C(S(\text{KP}))$

$$\begin{aligned}
\text{maximize} \quad & \sum_{j=1}^{n} p_j x_j \\
\text{subject to} \quad & \sum_{j=1}^{n}(w_j + \lambda)x_j \leq c + k\lambda, \\
& 0 \leq x_j \leq 1, \qquad j = 1,\ldots,n \ .
\end{aligned} \qquad (48)$$

in order to derive a tight bound on KP. For $C(S(\text{KP}))$ Pisinger [103] analytically proved that the best choice of multiplier is $\lambda = \alpha$ for strongly correlated Knapsack Problems with $p_j = w_j + \alpha$. In this way the problem becomes a Subset-sum Problem, and specialized algorithms can be used for its solution. For less structured instances, the surrogate multiplier λ must be determined through iterative techniques. Monotonicity of the continuous solution to (47) can however be proved as in [83], thus making it possible to use a binary search to efficiently derive the optimal multiplier. With the best multiplier λ for $C(S(\text{KP}))$ one however gets the same bound U_5 as found in (45).

2.2 Heuristics

In section 2.1 we derived a feasible solution from the continuous solution by truncating x_b to zero. The corresponding break solution $\hat{x}$ has objective value

$$z' = \hat{p} = \sum_{j=1}^{b-1} p_j. \qquad (49)$$

On average z' is quite close to the optimal solution value z. In fact $z' \leq z \leq U_1 \leq z' + p_b$, i.e. the absolute error is bounded by p_b. The worst-case performance ratio, however, is arbitrarily bad. This is shown by the series of problems with $n = 2$, $p_1 = w_1 = 1$, $p_2 = w_2 = k$ and $c = k$, for which $z' = 1$ and $z = k$, so the ratio z'/z is arbitrarily close to 0 for sufficiently large k.

Noting that the above pathology occurs when p_b is relatively large, we can obtain a heuristic with improved performance ratio by also considering the feasible solution given by the break item alone and taking the best of the two solution values, i.e.

$$z^h = \max\{z', p_b\}. \qquad (50)$$

The worst-case performance ratio of the new heuristic is $\frac{1}{2}$. We have already noted, in fact, that $z \leq z' + p_b$, so, from (50) $z \leq 2z^h$. To see that $\frac{1}{2}$ is tight, consider the series of problems with $n = 3$, $p_1 = w_1 = 1$, $p_2 = w_2 = p_3 = w_3 = k$ and $c = 2k$: we have $z^h = k+1$ and $z = 2k$, so z^h/z is arbitrarily close to $\frac{1}{2}$ for sufficiently large k.

The computation of z^h requires $O(n)$ time, once the break item is known. If the items are sorted as in (20), a more effective algorithm is to start from the break solution $\hat{x}$ and repeatedly insert the next item $j \geq b$ if it fits. The algorithm is known as the *Greedy Algorithm* and the obtained solution $\tilde{x}$ will have an objective value that is not smaller than z' since we start from the break solution. The worst-case performance can however again be as bad as 0 (take e.g. the series of problems introduced for z').

The performance ratio of the greedy algorithm can be improved to $\frac{1}{2}$ if we also consider the solution given by the item of maximum profit alone. Assume that the items are ordered according to (20) and let $\ell = \arg\max\{p_j, j = 1, \ldots, n\}$ be the index of the most profitable item. Then the greedy lower bound is given by

$$z^g = \max\left\{\sum_{j=1}^n p_j \tilde{x}_j, p_\ell\right\} \tag{51}$$

The worst-case performance ratio is $\frac{1}{2}$ since: (a) $p_\ell \geq p_b$, so $z^g \geq z^h$; (b) the series of problems introduced for z^h proves the tightness. The time complexity is $O(n)$, plus $O(n \log n)$ for the initial sorting.

When a 0-1 knapsack problem in minimization form (see Section 2) is heuristically solved by applying the *Greedy Algorithm* to its equivalent maximization instance, we of course obtain a feasible solution, but the worst-case performance is not preserved. Consider, in fact, the series of minimization problems with $n = 3$, $p_1 = w_1 = k$, $p_2 = w_2 = 1$, $p_3 = w_3 = k$ and $q = 1$, for which the optimal solution value is 1. Applying the Greedy Algorithm to the maximization version (with $c = 2k$), we get $z^g = k + 1$ and hence an arbitrarily bad heuristic solution of value k for the minimization problem.

Fixed-cardinality heuristics have been proposed in [96]. These are especially well-suited for strongly correlated problems, since it can be shown that optimal solutions of strongly correlated problems are characterized by choosing the largest possible number of items and filling the knapsack as much as possible.

The *forward greedy solution* is the best value of the objective function when adding one item to the break solution:

$$z^f = \max_{j=b,\ldots,n} \{\hat{p} + p_j \,:\, \hat{w} + w_j \leq c\}, \tag{52}$$

and the *backward greedy solution* is the best value of the objective function when adding the break item to the break solution and removing another item:

$$z^b = \max_{j=1,\ldots,b-1} \{\hat{p} + p_b - p_j \,:\, \hat{w} + w_b - w_j \leq c\}. \tag{53}$$

The time complexity of bounds z^f, z^b is $O(n)$ but the performance ratio is arbitrarily close to zero.

The 0-1 Knapsack Problem accepts a fully polynomial approximation scheme, as will be described in Section 2.8.

2.3 Reduction

The size of a Knapsack Problem may be reduced by applying a reduction algorithm to fix variables at their optimal value. For the following discussion assume that a lower bound z has been found through any kind of heuristic and that the corresponding solution vector x has been saved. The reason for this assumption is that it allows us to use tighter reductions.

Ingargiola and Korsch [59] presented the first reduction algorithm for the 0-1 Knapsack Problem. This reduction algorithm and all similar algorithms are based on the dichotomy of each variable x_j. If one of the two branches cannot lead to an improved solution, we can fix the variable at the opposite value.

Thus let u_j^0 be an upper bound for (15) with additional constraint $x_j = 0$. Similarly let u_j^1 be an upper bound for (15) with constraint $x_j = 1$. Then the set of items which can be fixed at optimal value $x_j = 1$ is

$$N^1 = \{j \in N : u_j^0 < z + 1\} \tag{54}$$

In a similar way the set of items which can be fixed at $x_j = 0$ is

$$N^0 = \{j \in N : u_j^1 < z + 1\} \tag{55}$$

the remaining items $F = N \setminus (N^1 \cup N^0)$ are called free, and a solution with objective value better than z may be found by solving the reduced problem

$$\begin{aligned}
\text{maximize} \quad & z' = \sum_{j \in F} x_j + P \\
\text{subject to} \quad & \sum_{j \in F} w_j x_j \leq c - W, \\
& x_j \in \{0, 1\}, \qquad j \in F,
\end{aligned} \tag{56}$$

where P is the profit of items fixed at one: $P = \sum_{j \in N^1} p_j$ and W is the corresponding weight sum $W = \sum_{j \in N^1} w_j$. The optimal solution value is now found as $z^* = \max\{z, z'\}$, where z is the lower bound used in the reduction.

The time complexity of the reduction depends on the complexity of the applied bound. Dembo and Hammer [18] used the bounds

$$u_j^0 = \hat{p} - p_j + \left(c - \hat{w} + w_j\right)\frac{p_b}{w_b} \qquad u_j^1 = \hat{p} + p_j + \left(c - \hat{w} - w_j\right)\frac{p_b}{w_b} \tag{57}$$

which, as we saw in (34) are both obtained by relaxing the integrality constraint on the break item to $x_b \in \Re$. Since each bound is derived in constant time, the whole reduction may be performed in $O(n)$.

Tighter reductions have been proposed by Ingargiola and Korsh [59]. Here u_j^0 and u_j^1 are derived by using the Dantzig bound for (15) with additional constraint $x_j = 0$, resp. $x_j = 1$. Since the break item of the constrained problem may differ from b, a linear search is used to find the new break item b_j^1 (resp. b_j^0) for each of the problems. This means that each bound is found in $O(n)$ time, demanding totally $O(n^2)$ for the reduction. If the weights are randomly distributed, the new break item will be close to b meaning that the whole reduction can be done in $O(n)$ on average.

Notice that there is no reason for deriving u_j^1 for $j < b$ since for these items $u_j^1 = U_1$ which cannot be used for fixing x_j at 0. Similarly u_j^0 should not be derived for $j > b$ as $u_j^0 = U_1$.

A further improvement proposed by Ingargiola and Korsh is trying to improve the lower bound z during the reduction. Since each break item corresponds to a greedy solution, we may set $z \leftarrow \max\{z, z^0, z^1\}$ with

$$z^0 = \max_{j<b} \left(\sum_{i=1}^{b_j^0} p_i - p_j \right) \qquad z^1 = \max_{j>b} \left(\sum_{i=1}^{b_j^1} p_i + p_j \right). \qquad (58)$$

Martello and Toth [79] further improved the approach by using the bound U_2 instead of the Dantzig upper bound U_1, and by using a binary search to find the break item. The latter needs a preprocessing of the items where we set $\overline{p}_j = \sum_{i=1}^{j} p_i$ resp. $\overline{w}_j = \sum_{i=1}^{j} w_i$. Then for any capacity c' one may find the break item b' in $O(\log n)$ time as the index which satisfies

$$\overline{w}_{b'-1} \leq c' < \overline{w}_{b'}. \qquad (59)$$

where $c' = c - w_j$ if u_j^0 is computed and $c' = c + w_j$ for the opposite bound. The Martello and Toth reduction algorithm also improves the sequel of the execution by first finding the break item for all additionally constrained problems, and during this process improving the lower bound z. Having found and saved all break items, reductions (54) and (55) are performed. This gives a complexity of $O(n \log n)$, and generally the reductions are more effective than those of the Ingargiola and Korsh scheme, since a better lower bound is generally applied.

2.4 Branch-and-bound Algorithms

The first branch-and-bound algorithm for the 0-1 Knapsack Problem was presented by Kolesar [67] back in 1967. Since then several variants have emerged. The most efficient of these are based on a depth-first search to limit the number of life nodes to $O(n)$ at any stage of the enumeration. Moreover, a greedy strategy for choosing the next item for branching seems to lead to best solution times. Among the greedy depth-first algorithms we may distinguish between two variants: A primal method which builds feasible solutions from scratch until the weight constraint is exceeded, and a primal-dual approach which accepts infeasible solutions in a transition stage.

The primal method dates back to Horowitz and Sahni [54] and several improved versions have emerged since then. In a recursive formulation, each iteration corresponds to a branch on x_i where $x_i = 1$ is investigated before $x_i = 0$ according to the greedy principle. The profit and weight sum of the currently fixed variables $x_j, j < i$ is $\overline{p}$ resp. $\overline{w}$, thus if $\overline{w} > c$ for a given node, we may backtrack. Also, if the Dantzig upper bound U_1 for the remaining problem is less or equal to the current lower bound z, we may backtrack. The lower bound z is improved during the search while the optimal solution vector $x = x^*$ is defined during the backtracking.

algorithm hs_branch$(i, \overline{p}, \overline{w})$: boolean;
var improved;
if $(\overline{w} > c)$ **then return false**;
if $(\overline{p} > z)$ **then** $z \leftarrow \overline{p}$; improved $\leftarrow$ **true**; **else** improved $\leftarrow$ **false**;
if $(i > n)$ **or** $(c - \overline{w} < \underline{w}_i)$ **or** $(\overline{p} + (c - \overline{w})p_i/w_i < z + 1)$
then return improved;
if hs_branch$(i + 1, \overline{p} + p_i, \overline{w} + w_i)$ **then** $x_i \leftarrow 1$; improved $\leftarrow$ **true**;
if hs_branch$(i + 1, \overline{p}, \overline{w})$ **then** $x_i \leftarrow 0$; improved $\leftarrow$ **true**;
return improved;

The original formulation of the Horowitz and Sahni algorithm may insert more than one item for each branching node, but from a computational point of view the above formulation is equivalent. To execute the algorithm properly, all decision variables x_i must be set to 0 and the lower bound set to $z = 0$ before calling hs_branch$(0, 0, 0)$.

Notice how the optimal solution vector x is defined during the backtracking of the algorithm. This idea of lazy updating of the solution vector is due to Martello and Toth [72]. Their mt1 algorithm is based on the same frame as above, but the Martello-Toth upper bound is used instead of the Dantzig bound, and a table is used for storing the last derived profit and weight sums at each branching node. This is implicitly done by the above recursive formulation. The use of a minimal weight table $\underline{w}_i = \min_{j \geq i} w_j$ is also due to Martello and Toth. Obviously, we may backtrack whenever no item $j \geq i$ fits into the knapsack. Finally, the original mt1 algorithm contains a special dominance step; whenever an item is removed, only specific items can replace it in order to obtain a better solution.

A comprehensive comparison of algorithms built over the above frame is presented in Martello and Toth [81] showing that there are only minor differences among the performance of the algorithms, for easy instances.

A different approach is what we could call the *primal-dual* algorithm by Pisinger [96]. The algorithm starts from the break item b and repeatedly inserts an item if the current weight sum is less than c and otherwise removes an item. In this way one only needs to consider solutions which are appropriately filled, but on the other hand infeasible solutions must be considered in a transition stage. The variables fixed at some value are $x_{s+1}, \ldots, x_{t-1}$ where in the profit and weight sum $\overline{p}, \overline{w}$, we implicitly assume that all items $j \leq s$ also have been chosen. Thus we insert an item when $\overline{w} \leq c$ and remove an item when $\overline{w} > c$. As opposed to the hs_branch algorithm we cannot backtrack whenever $\overline{w} > c$, as infeasible solutions may become feasible by removal of some items, but must rely on the Dantzig bound U_1 to state when backtracking should occur.

```
algorithm pi_branch(s, t, p̄, w̄): boolean;
var i, improved;
improved ← false;
if (w̄ ≤ c) then { insert some item j ≥ t }
    if (p̄ > z) then z ← p̄; improved ← true; A ← {};
    for i ← t to n
        if (p̄ + (c − w̄)pᵢ/wᵢ < z + 1) then return improved;
        if pi_branch(s, i + 1, p̄ + pᵢ, w̄ + wᵢ)
            then A ← A ∪ {xᵢ}; improved ← true;
else
    for i ← s downto 1
        if (p̄ + (c − w̄)pᵢ/wᵢ < z + 1) then return improved;
        if pi_branch(i − 1, t, p̄ − pᵢ, w̄ − wᵢ)
            then A ← A ∪ {xᵢ}; improved ← true;
return improved;
```

The algorithm is called pi_branch($b, b, \hat{p}, \hat{w}$), and initially the lower bound must be set to $z = 0$.

As opposed to the hs_branch algorithm we cannot directly define the solution vector during the backtracking. Instead, all changes performed are pushed to a stack A. Then the solution vector x is easily defined by first setting $x_j = 1$ for $j = 1,\ldots,b-1$ resp. $x_j = 0$ for $j = b,\ldots,n$, and then changing

$$x_j \leftarrow 1 - x_j \text{ for } j \in A \qquad (60)$$

Notice that only solutions with weight sum close to the capacity c are considered:

Definition 2.1 *A balanced filling is a solution x obtained from the break solution through balanced operations as follows:*

- *The break solution $\hat{x}$ is a balanced filling.*

- *Balanced insert: If we have a balanced filling x with $\sum_{j=1}^n w_j x_j \leq c$ and change a variable x_t, $(t \geq b)$ from $x_t = 0$ to $x_t = 1$ then the new filling is also balanced.*

- *Balanced remove: If we have a balanced filling x with $\sum_{j=1}^n w_j x_j > c$ and change a variable x_s, $(s < b)$ from $x_s = 1$ to $x_s = 0$ then the new filling is balanced.*

Theorem 2.2 (Pisinger [99]) *An optimal solution to (15) is a balanced filling, i.e. it may be obtained through balanced operations.*

Proof. Assume that the optimal solution is given by x^*. Let $s_1,\ldots,s_\alpha$ be the indices $s_i < b$ where $x^*_{s_i} = 0$, and $t_1,\ldots,t_\beta$ be the indices $t_i \geq b$ where $x^*_{t_i} = 1$. Order the indices such that $s_\alpha < \cdots < s_1 < b \leq t_1 < \cdots < t_\beta$. Starting from the break solution $x = \hat{x}$ we perform balanced operations in order to reach x^*. As the break solution satisfies $\sum_{j=1}^n w_j x_j \leq c$ we must insert an item t_1, thus set $x_{t_1} = 1$. If the hereby obtained weight sum $\sum_{j=1}^n w_j x_j$ is greater than c we remove item s_1 by setting $x_{s_1} = 0$, otherwise we insert the next item t_2. Continue this way till one of the following three situations occur: (1) All the changes corresponding to $\{s_1,\ldots,s_\alpha\}$ and $\{t_1,\ldots,t_\beta\}$ have been made, meaning that we have reached the optimal solution x^* through balanced operations. (2) We reach a situation where $\sum_{j=1}^n w_j x_j > c$ and all indices $\{s_i\}$ have been used but some $\{t_i\}$ have not been used. This however implies that x^* could not be a feasible solution from the start as the knapsack is filled and we still have to insert items. (3) A similar situation occurs when $\sum_{j=1}^n w_j x_j \leq c$ is reached and all indices $\{t_i\}$ have been used, but some $\{s_i\}$ are missing. This implies that x^* cannot be an optimal solution, as a better feasible solution can be obtained by *not* removing the remaining items s_i. □

2.5 Dynamic Programming Algorithms

Bellman [8] presented the first exact solution technique for the Knapsack Problem by using Dynamic Programming. The solution technique is based on a recursive formulation of the problem, which is then evaluated in an iterative way. Since several subsolutions overlap, these can be saved in a table of size $O(c)$, thus reaching the pseudo-polynomial complexity $O(nc)$ for solving KP.

At any stage we let $f_i(\tilde{c})$ for $0 \leq i \leq n$ and $0 \leq \tilde{c} \leq c$ be an optimal solution value to the subproblem of KP, which is defined on items $1, \ldots, i$ with capacity $\tilde{c}$. Then the recursion by Bellman is given by

$$f_i(\tilde{c}) = \begin{cases} f_{i-1}(\tilde{c}) & \text{for } \tilde{c} = 0, \ldots, w_i - 1 \\ \max\{f_{i-1}(\tilde{c}), \ f_{i-1}(\tilde{c} - w_i) + p_i\} & \text{for } \tilde{c} = w_i, \ldots, c \end{cases} \quad (61)$$

while we set $f_0(\tilde{c}) = 0$ for $\tilde{c} = 0, \ldots, c$.

Toth [112] improved the Bellman recursion in several respects: First of all dynamic programming by reaching is used (see Ibaraki [57] for a definition) and dominance is used to delete unpromising states. Moreover, if at the ith stage a state has weight sum μ satisfying

$$\mu < c - \sum_{j=i+1}^{n} w_j \quad (62)$$

then the state will never reach a filled solution with $\mu = c$ and thus can be fathomed. Moreover, if

$$c - \min_{i < j \leq n} w_j < \mu < c \quad (63)$$

then the state cannot be improved further and can thus be fathomed.

2.5.1 Primal-dual Dynamic Programming

The Bellman recursion builds the optimal solution from scratch by repeatedly adding a new item to the problem. A more efficient recursion [99] should however take into account that generally only a few items around b need be changed from their LP-optimal values in order to obtain the IP-optimal values. Thus, assume that the items are ordered according to (20), and let $f_{s,t}(\tilde{c})$, $(s \leq b,\ t \geq b - 1,\ 0 \leq \tilde{c} \leq 2c)$ be an optimal solution to the

problem:
$$f_{s,t}(\tilde{c}) = \max \begin{cases} \sum_{j=1}^{s-1} p_j + \sum_{j=s}^{t} p_j x_j : \\ \sum_{j=1}^{s-1} w_j + \sum_{j=s}^{t} w_j x_j \leq \tilde{c}, \\ x_j \in \{0,1\} \text{ for } j = s, \ldots, t \end{cases}. \quad (64)$$

We may use the following recursion for the enumeration

$$f_{s,t}(\tilde{c}) = \max \begin{cases} f_{s,t-1}(\tilde{c}) & \text{if } t \geq b \\ f_{s,t-1}(\tilde{c} - w_t) + p_t & \text{if } t \geq b,\ \tilde{c} - w_t \geq 0 \\ f_{s+1,t}(\tilde{c}) & \text{if } s < b \\ f_{s+1,t}(\tilde{c} + w_s) - p_s & \text{if } s < b,\ \tilde{c} + w_s \leq 2c \end{cases} \quad (65)$$

setting $f_{b,b-1}(\tilde{c}) = -\infty$ for $\tilde{c} = 0, \ldots, \overline{w} - 1$ and $f_{b,b-1}(\tilde{c}) = \overline{p}$ for $\tilde{c} = \overline{w}, \ldots, 2c$. Thus the enumeration starts at $(s,t) = (b, b-1)$ and continues by either removing an item s from the knapsack, or inserting an item t into the knapsack. An optimal solution to KP is found as $f_{1,n}(c)$.

Since generally far less states than $2c$ need be considered at each stage, the algorithm may be modified to dynamic programming *by reaching*, obtaining a time bound $O(2^{t-s+1})$ for enumerating an interval $[s, t]$ of items around the break item. The pseudo-polynomial time bound is also valid giving a different bound $O(c(t - s + 1))$.

2.5.2 Horowitz and Sahni Subdivision

Horowitz and Sahni [54] presented a different recursion based on the subdivision of the original problem of n variables into two subproblems of $n/2$ size. For each of the subproblems a normal dynamic programming recursion (61) is used to enumerate all states and the two sets are easily combined in time proportional to the set sizes by considering pairs of states which have the largest possible weight sum not exceeding c.

The worst-case complexity of the Horowitz and Sahni algorithm is $O(2^{n/2})$ as well as the pseudo-polynomial bound $O(nc)$. Thus, for difficult problems the recursion improves by a square-root over a complete enumeration of magnitude $O(2^n)$.

2.5.3 Other Dynamic Programming Algorithms

A final recursion can be obtained based on the principle of balanced solutions described in Definition 2.1. Pisinger [99] presented a dynamic programming algorithm running in $O(nr_1 r_2)$ where $r_1 = \max w_j$ and $r_2 \leq \max p_j$ is the gap between any upper and lower bounds on the problem. The algorithm is however nontrivial, so a simplified version will be presented in the section on Subset-sum Problems.

A topic which has not yet been dealt with is how the solution vector x corresponding to the optimal solution value z should be found. Bellman gave the simple answer that one should follow the states backwards and monitor those choices that lead to the optimal solution. This however means that all states must be saved during the solution process, which may be quite a hindrance for large-sized problems.

A different approach is to save only the last two levels of the recursion at any stage, thus reducing the space complexity to $O(c)$. In this way one can get information about the optimal value of the last variable x_n, and previous solution values can be found by repeatedly solving the problem with the last variable removed. In this way, the computational effort however increases to $O(n^2 c)$.

Munro and Ramirez [87] improved this technique further. In their version, every state $f_i(\tilde{c})$ of recursion (61) with $i > n/2$ should save the corresponding profit and weight sum at level $i = n/2$. When the optimal solution value is found as $f_n(c)$, then the problem of finding the solution vector may be decomposed into two equally large subproblems of $n/2$ items each. Recursively applying this approach one gets the space bound $O(c)$ and a time bound $O(cn \log n)$.

The Munro and Ramirez technique cannot be directly used for primal-dual dynamic programming algorithms like recursion (65), since these are designed to terminate the enumeration before all items are considered, and thus the subdivision at $n/2$ is not optimal. Instead, for each state $f_{s,t}(\tilde{c})$, one should save the corresponding profit and weight sum when the core size $t - s + 1$ most recently became a power of two. This basically means that each time $t - s + 1 = 2^k$ then the current profit and weight sums are saved as part of all states, and these values follow all succeeding states. The space bound is again $O(c)$ and the time bound for deriving an optimal solution becomes $O(c|C| \log |C|)$ for a final core C.

2.6 Solution of Large-sized Problems

When large-sized knapsack problems are solved it may be beneficial to consider a *core problem* — basically a problem defined on a subset of the items where there is a high probability of finding an optimal solution. The first algorithm of this kind was presented by Balas and Zemel [6] where the core problem was used to avoid a complete sorting of the items according to (20).

The concept of solving a core problem unfortunately has some drawbacks, as pointed out by Pisinger [97] since it tends to "count the chickens before they are hatched". However, the technique is important as it allows one to focus the search on those items that are most interesting according to general knowledge. In Fig. 2 we saw that an optimal solution generally corresponds to the break solution apart from some items close to the break item which have been changed. Thus obviously it is more interesting to start the enumeration around the break item than, say, at item 1.

Balas and Zemel [6] presented it as follows: Assume that the items are ordered according to nonincreasing profit-to-weight ratios as in (20), and let an optimal solution be given by x^*. Now let

$$j_1 = \min\{j : x_j^* = 0\} \qquad j_2 = \max\{j : x_j^* = 1\} \qquad (66)$$

then the *core* is given by the items in the interval $C = [j_1, j_2]$ and the *core problem* is defined as

$$\begin{aligned} \text{maximize} \quad & \sum_{j \in C} p_j x_j + P \\ \text{subject to} \quad & \sum_{j \in C} w_j x_j \leq c - W, \\ & x_j \in \{0,1\}, \qquad j \in C \end{aligned} \qquad (67)$$

where $P = \sum_{j<j_1} p_j$ and $W = \sum_{j<j_1} w_j$. For several classes of instances, with strongly correlated problems as an important exception, the size of the core is a small fraction of n. Hence if we knew a priori the values of j_1 and j_2 we could easily solve the complete problem by setting $x_j^* = 1$ for $j < j_1$ and $x_j^* = 0$ for $j > j_2$ and simply solving the core problem through any enumerative algorithm.

2.6.1 Deriving a Core

The definition of the core C is based on the knowledge of the optimal solution x^*. Since this knowledge is obviously not present, most algorithms rely on an approximated core, basically choosing 2δ items around the break item.

The following algorithm finds the break item b and sorts 2δ items around b such that the core is given by $C = [s, t] = [b - \delta, b + \delta]$. The algorithm is a `quicksort` algorithm with slight modifications such that the interval containing b is always partitioned first. This means that b is well-defined when the second interval is considered upon backtracking, and thus it is trivial to check whether the interval is outside the wanted core C.

>**algorithm** findcore$(s, t, \overline{w})$;
>**comment** $[s, t]$ *is an interval to be partitioned, and* $\overline{w} = \sum_{j=1}^{s-1} w_j$
>As in bal_zem partition $[s, t]$ in two intervals satisfying $p_j/w_j \geq \gamma$
> for $j \in [s, i-1]$ and $p_j/w_j \leq \gamma$ for $j \in [i, t]$.
> Moreover set $\tilde{w} = \sum_{j=1}^{i-1} w_j$.
>**if** $(\tilde{w} \leq c)$ **and** $(c < \tilde{w} + w_i)$ **then** $b \leftarrow i$;
>**if** $(\tilde{w} > c)$ **then**
>   findcore$(s, i-1, \overline{w})$;
>   **if** $(i \leq b + \delta)$ **then** findcore$(i, t, \tilde{w})$ **else** $L \leftarrow L \cup (i, t)$
>**else**
>   findcore$(i, t, \tilde{w})$;
>   **if** $(i - 1 \geq b - \delta)$ **then** findcore$(s, i-1, \overline{w})$ **else** $H \leftarrow H \cup (s, i-1)$

The procedure is called findcore$(1, n, 0)$ and it returns with items in $C = [b - \delta, b + \delta]$ sorted while all preceding items have the profit-to-weight ratios larger than that of the items in the core, and succeeding items have smaller profit-to-weight ratios.

As already seen, the findcore algorithm is a normal sorting algorithm apart from the fact that only intervals containing the break item are partitioned further. Thus all discarded intervals represent a partial ordering of the items according to p_j/w_j which may be used later in an enumerative algorithm. Hence, Pisinger [96] proposed to push the delimiters of discarded intervals into two stacks H and L as sketched in the algorithm.

Several proposals have been made as to how large a core should be chosen. Balas and Zemel [6] claimed that the size of the core is constant $|C| = 25$. Martello and Toth [79] however chose $|C| = \sqrt{n}$ since more items are needed in order to prove optimality. Pisinger [105] found the minimal core size according to definition (66) for several problems. These are given in Table 1 as average values of 100 instances. For comparison Table 2 gives the minimal core size which is necessary in order to prove optimality of the solution by a branch-and-bound algorithm using continuous bounds. The instances are generated with weights distributed in $[1, R]$, having different problem sizes n. Notice that difficult problems demand a very large core to find the optimal solution, and even more items need to be enumerated before optimality can be proved. The entries considered are average values for different capacities c and there may be a large variation in the core size

Table 1: Minimal core size (number of items). Average of 100 instances.

$n \setminus R$	uncorr. 10^3	uncorr. 10^4	weak.corr 10^3	weak.corr 10^4	str.corr 10^3	str.corr 10^4	inv.str.corr 10^3	inv.str.corr 10^4	al.str.corr 10^3	al.str.corr 10^4	subs.sum 10^3	subs.sum 10^4	sim.w 10^5
50	4	3	8	8	10	12	7	9	12	12	11	15	0
100	5	5	10	12	12	13	11	12	17	20	10	14	0
200	6	7	12	14	18	18	14	17	24	26	10	13	0
500	8	8	15	17	25	25	21	23	32	41	10	13	0
1000	9	11	14	17	34	36	31	32	41	59	10	13	0
2000	11	13	13	20	46	44	43	40	53	72	10	13	0
5000	12	14	11	21	76	79	65	62	68	97	10	13	0
10000	11	17	16	25	105	104	93	88	71	106	10	13	0

Table 2: Core size needed to prove optimality (number of items). Average of 100 instances.

$n \setminus R$	uncorr. 10^3	uncorr. 10^4	weak.corr 10^3	weak.corr 10^4	str.corr 10^3	str.corr 10^4	inv.str.corr 10^3	inv.str.corr 10^4	al.str.corr 10^3	al.str.corr 10^4	subs.sum 10^3	subs.sum 10^4	sim.w 10^5
50	5	5	11	11	20	21	19	20	20	19	12	16	15
100	7	7	13	14	38	39	35	37	38	40	12	15	32
200	9	9	16	16	76	81	68	80	73	70	11	15	64
500	12	12	17	22	192	194	180	197	159	170	11	15	153
1000	13	16	17	24	392	407	369	393	261	332	11	15	302
2000	15	18	16	27	739	693	722	649	439	526	11	14	595
5000	15	21	12	28	1956	2034	1774	1876	731	998	11	15	1460
10000	14	24	12	29	3889	3768	3633	3926	804	1301	11	15	2585

depending on c.

Goldberg and Marchetti-Spaccamela [46] claimed that the core size increases with the size of n and thus the proposed core size by Balas and Zemel of $|C| = 25$ is not correct. Table 1 does not give a unique answer to this question, as the core size apparently depends on n as well as on the instance type and on the range R of the weights. Apart from the strongly correlated instances, choosing $|C| = 50$ should be sufficient.

2.6.2 Fixed-core Algorithms

Algorithms which solve a fixed core have been presented by Balas and Zemel [6], Fayard and Plateau [29] and Martello and Toth [79]. The latter, which is the most general of the algorithms, may be sketched as follows:

algorithm mt2

1. Determine an approximate core $C = [b - \delta, b + \delta]$ using the findcore algorithm. Sort the items in the core and solve the core problem to optimality, deriving a heuristic solution z_C.

2. Derive the enumerative bound U_C given by (33). If $U_C \leq \sum_{j=1}^{b-\delta-1} p_j + z_C$ then the core solution is optimal, and we may terminate.

3. Reduce the problem by using the Martello-Toth version of reduction (58).

4. If all variables $j \notin C$ have been fixed at their optimal value, the core solution is optimal and we may terminate.

5. Sort the items not fixed at their optimal value, and solve the remaining problem to optimality.

The core size in Step 1 is chosen as $\delta = \sqrt{n}$ for $n \geq 100$ while $|C| = n$ for smaller instances, meaning that all items are considered for small problems. The enumerative algorithm used in step 1 and 5 is the mt1 algorithm described in Section 2.4.

The algorithm by Balas and Zemel [6] differs from the above frame, by only finding an approximate solution to the core problem in Step 1, and by choosing a core size of $|C| = 50$. The Dantzig upper bound is used in Step 2 instead of the tighter enumerative bound. Finally the enumeration in Step 5 is based on a branch-and-bound algorithm by Zoltners [118].

Knapsack Problems

The Fayard and Plateau algorithm [29] basically solves a core of size one, since only the break item is derived using algorithm bal_zem. However, knowing b, a heuristic solution may be derived by using the greedy algorithm without sorting. The items are then reduced using the Dembo and Hammer bounds (57). Finally the remaining problem is enumerated by using a branch-and-bound algorithm specialized for this problem.

A detailed description of all the above algorithms can be found in Martello and Toth [80] including some computational comparisons of the algorithms.

2.6.3 Expanding-core Algorithms

Realizing that the core is only a guess on which items may be the most interesting to consider, a different principle is to use expanding-core algorithms, which simply start with an empty core and add more items when needed. Using branch-and-bound for the enumeration, one gets the expknap algorithm presented by Pisinger [96]:

procedure expknap

1. Find core $C = [b, b]$ through algorithm findcore pushing all discarded intervals into two stacks H and L containing respectively items with higher or lower profit-to-weight ratios than that of the break item b.

2. Find a heuristic solution z using the forward and backward heuristics (52) and (53).

3. Run algorithm pi_branch; for each step testing if the set $[s, t]$ is within the current core C. If this is not the case, the next interval from H or L is popped, and the items are reduced using (57) for fixing variables at their optimal values. The remaining variables are sorted according to nonincreasing profit-to-weight ratios, and added to the core C.

Since the sorting and reduction is done as they are needed, no more effort will be made than absolutely necessary. Also the reductions are tighter when postponed in this way, since one may expect that a better lower bound is found the later a reduction test is performed in an enumeration.

The branch-and-bound algorithm however may follow an unpromising branch to completion, thus basically extending the core with all items, even if the instance could be solved with a smaller core by following a different branch. This inadequate behavior is avoided by using dynamic programming since the breadth-first search basically ensures that all variations of the solution vector have been considered before a new item is added to the core. This idea is due to [105] and can be sketched as:

algorithm minknap

1. Find an initial core $C = [b, b]$ as in the expknap algorithm.

2. Run the dynamic programming recursion (65) alternately inserting an item t or removing an item s. Check if s, t is within the current core C, otherwise pick the next interval from H or L, reduce and sort the items, finally adding the remaining items to the core.

3. The process stops when all states in the dynamic programming have been fathomed due to an upper bound test, or all items $j \notin C$ have been fixed at their optimal value.

Both algorithms have the property that the core size adapts to the problem, but due to the breadth first-search of the dynamic programming, one can show that

Theorem 2.3 (Pisinger [105]) *The enumerated core by algorithm* minknap *is minimal.*

The meaning of "minimal" is in this context, that one cannot find a smaller core, which has symmetrical size around the break item b, such that a fixed-core algorithm like mt2 will terminate before reaching the last enumeration in Step 5. Tables 1 and 2 are immediate results of the above theorem.

2.6.4 Inconveniences in Core Problems

For large-sized problems it is generally beneficial to focus the search on those items which are most interesting, i.e. solve a core problem. But in some situation this focusing makes the problem more difficult than the original complete KP. This behavior has been studied by Pisinger [97].

A core is by definition a collection of items where the profit-to-weight ratio is close to that of the break item b. Pisinger's experiments however demonstrated that in some cases this definition of a core results in many items with similar or proportional weights. This degeneration of the core results in a difficult problem, since with many proportional weights it is difficult to obtain a solution which fills the knapsack. With this lack of a good lower bound, many algorithms get stuck in the enumeration of the core. The paradox is that in most situations one can find an item which fills the knapsack to completion by considering items outsides the core, but the algorithm which solves the core problem cannot "see" this item.

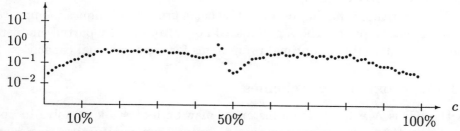

Figure 4: Average log computational times for mt2 in seconds, as a function of the capacity c. Weakly correlated instances, $n = 3000, R = 100$. The core has size $|C| = 100$

A computational comparison in [97] showed that branch-and-bound algorithms are sensitive to the choice of the core, while dynamic programming algorithms are generally not affected in their performance, since similar weights are handled efficiently by dominance criteria. The degeneration of a core however only occurs for large-sized instances and for *some* choices of the capacity, as illustrated in Fig. 4. When c is about 45% of the total weight sum $\sum_{j=1}^{n} w_j$ the core problem is 10 times more difficult to solve than when $c = 50\%$ of the weight sum. Methods to avoid such inadequate behavior are either choosing a larger core, or putting a time limit on the enumeration of the core (see e.g. Martello and Toth [83]).

2.7 Solution of Hard Problems

Martello and Toth [83] recently published an algorithm mthard specially designed for hard problems, which uses cardinality bounds to effectively limit the enumeration.

Before starting the branch-and-bound algorithm with cardinality bounds, an attempt is made to solve the problem by less demanding techniques. Thus the mt2 algorithm is initially called with a limit on the number of branching steps it may perform. If the mt2 terminates with an optimal solution z, the algorithm stops.

If the mt2 algorithm halts without proving optimality of the solution found, the branching process starts. The upper bounds are derived by using the cardinality bounds described in Section 2.1.3. If the upper bound u

derived satisfies $u > z$ then a branching step is performed by exploring the dichotomy on the next binary variable x_j. Let $\bar{x}_j$ be the value of x_j in a continuous solution to KP. Then the branch $x_j = \lceil \bar{x}_j \rceil$ is made first, thus following the greedy principle.

The algorithm is further improved by using greedy techniques to improve z, dominance steps to fathom dominated nodes, and finally partial enumeration to build a table of optimal profit sums for small residual capacities.

2.8 Approximation Schemes

As the KP is $\mathcal{NP}$-hard, some instances may be impossible to solve to optimality within a reasonable amount of time. In such situations one may be interested in an approximate solution with objective value z, where the relative error is bounded from above by a certain constant ϵ, i.e.

$$\frac{z - z^*}{z^*} \leq \epsilon, \tag{68}$$

where z^* is the optimal objective value. *Fully polynomial approximation schemes* must satisfy the condition that for any $\epsilon > 0$ they find a feasible solution satisfying (68) in time polynomially bounded by the size of the problem and by $1/\epsilon$.

Ibarra and Kim [58] presented the first fully polynomial approximation scheme for the 0-1 Knapsack problem. The algorithm is based on dynamic programming, where state space relaxation (see Ibaraki [57] for a thorough treatment of this field) is used in order to limit the number of possible states. Since the relative error by scaling profits is larger for the small profits, Ibarra and Kim divide the items into those with large profits, and those with small profits. The first group of items is enumerated through dynamic programming, and a greedy algorithm is used to improve the enumerated states by adding some additional items from the second group.

algorithm iba_kim
Assume that the items are ordered according to nonincreasing profit-to-weight ratios (20) and let the break item be defined by $b = \min\{h : \sum_{j=1}^{h} w_i > c\}$. If $\sum_{j=1}^{b-1} w_j = c$ then the break solution $z = \hat{p}$ is optimal, and the algorithm stops. Otherwise let $\tilde{z} = \sum_{j=1}^{b} p_j$ be an upper bound on the objective value. Obviously an optimal solution z^* must satisfy

$$\tilde{z}/2 \leq z^* < \tilde{z}, \tag{69}$$

since $z^* \geq \max\{\tilde{z} - p_b, p_b\}$, thus $2z^* \geq \tilde{z}$. The upper bound $\tilde{z}$ is used to partition the items into two groups such that

$$p_j > \tilde{z}\epsilon/3, \quad \text{for} \quad j = 1,\ldots,s,$$
$$p_j \leq \tilde{z}\epsilon/3, \quad \text{for} \quad j = s+1,\ldots,n \qquad (70)$$

still preserving the ordering (20) on each of the intervals.

State space relaxation is used for the dynamic programming algorithm, thus scale the profits with a factor $\delta = \tilde{z}(\epsilon/3)^2$, obtaining $\overline{p}_j = \lfloor p_j/\delta \rfloor$, for $j = 1,\ldots,s$. As $\tilde{z}$ is an upper bound on the objective value, it is not possible to obtain profits larger than $q = \lfloor \tilde{z}/\delta \rfloor = \lfloor (3/\epsilon)^2 \rfloor$ in the dynamic programming. Let $f_i(\pi)$, $(0 \leq \pi \leq q, 0 \leq i \leq s)$ be the smallest weight sum, such that a solution with scaled profit sum equal to π can be obtained on the variables $j = 1,\ldots,i$. Thus

$$f_i(\pi) = \min\left\{\sum_{j=1}^{i} w_j : \sum_{j=1}^{i} \overline{p}_j x_j = \pi, \; x_j \in \{0,1\}, j = 1,\ldots,i\right\}. \qquad (71)$$

The normal Bellman recursion can be used to evaluate f_i as

$$f_i(\pi) = \begin{cases} f_{i-1}(\pi) & \text{for} \quad \pi = 0,\ldots,\overline{p}_i - 1 \\ \min\{f_{i-1}(\pi), f_{i-1}(\pi - \overline{p}_i) + w_i\} & \text{for} \quad \pi = \overline{p}_i,\ldots,q \end{cases}, \qquad (72)$$

setting $f_0(\pi) = \infty$ for $\pi = 1,\ldots,q$ and $f_0(0) = 0$. Now for all states $f_s(\pi)$, $\pi = 0,\ldots,q$ where $f_s(\pi) \neq \infty$ a greedy solution is found by inserting some items $j = s+1,\ldots,n$ into the knapsack to fill the residual capacity $c - f_s(\pi)$. Let z be the objective value of the best heuristic solution obtained in this way.

Theorem 2.4 *The space and time complexity of* iba_kim *is $O(n/\epsilon^2)$ and hence polynomial in n and $1/\epsilon$.*

Proof. The dynamic programming part considers $q = \lfloor (3/\epsilon)^2 \rfloor$ states at each stage of $i = 1,\ldots,s$, which gives the space complexity $O(nq)$, i.e. $O(n/\epsilon^2)$.

The time complexity is $O(nq)$ for the dynamic programming, while the heuristic filling demands considering $n - s$ items for each value of $\pi = 0,\ldots,q$, giving the complexity $O(nq)$. Thus the time bound becomes $O(n/\epsilon^2)$. Actually the time bound should also embrace the initial sorting, which is however obviously polynomial in n. □

Theorem 2.5 *For any instance of KP we have $(z^* - z)/z^* \leq \epsilon$ where z^* is the optimal solution value and z is the heuristic value returned by the above algorithm.*

See Ibarra and Kim [58] or Martello and Toth [80] for a proof.

2.9 Computational Experiments

We will compare the performance of minknap, mthard and mt2. With mt2 as a reference point, readers can go further back using [80].

We will consider how the algorithms behave for different problem sizes, instance types, and data ranges. Six types of randomly generated data instances are considered as sketched below. Each type will be tested with *data range* $R = 1000$ and $10\,000$ for different problem sizes n. We consider the following problems: *Uncorrelated instances* are generated by chosing p_j, w_j randomly in $[1, R]$. *Weakly correlated instances:* the weights w_j are distributed in $[1, R]$, and the profits p_j in $[w_j - R/10, w_j + R/10]$ such that $p_j \geq 1$. *Strongly correlated instances:* the weights w_j are distributed in $[1, R]$, and $p_j = w_j + R/10$. *Inverse strongly correlated instances:* the profits p_j are distributed in $[1, R]$, and $w_j = p_j + R/10$. *Almost strongly correlated instances:* the weights w_j are distributed in $[1, R]$, and the profits p_j in $[w_j + R/10 - R/500, w_j + R/10 + R/500]$. *Subset-sum problems:* the weights w_j are randomly distributed in $[1, R]$ and $p_j = w_j$. *Uncorrelated problems with similar weights:* the weights are distributed in $[100000, 100100]$ and the profits p_j in $[1, 1000]$.

We are considering series with $H = 100$ instances for each instance type. The capacity is in each instance chosen as $c = \frac{h}{H+1} \sum_{j=1}^{n} w_j$, for test h in a series of H tests. This is to smooth out variations due to the choice of capacity as described in Section 2.6.4. To respect the assumption that every item fits into the knapsack, the capacity is however chosen not smaller than the largest weight.

Tables 3 to 5 compare the solution times of the three algorithms. All tests were run on a HP9000/735, and a time limit of 5 hours was assigned to each instance type for all H instances. If some of the instances were not solved within the time limit, this is indicated by a dash in the table.

The oldest code, i.e. mt2, has the overall worst performance, although it is quite fast on easy instances like the uncorrelated and weakly correlated ones. Moreover it is the fastest code for the Subset-sum instances. For strongly correlated instances, mt2 performs badly, as it is able to solve only tiny instances.

The dynamic programming algorithm minknap has an overall stable performance, as it is able to solve all instances within reasonable time. It is the fastest code for uncorrelated and weakly correlated instances, and it has almost as good performance for the Subset-sum instances as mt2. The strongly correlated instances take considerably more time to be solved but

Table 3: Average solution times in seconds (mt2)

$n \backslash R$	uncorr.		weak.corr		str.corr		inv.str.corr		al.str.corr		subs.sum		sim.w
	10^3	10^4	10^3	10^4	10^3	10^4	10^3	10^4	10^3	10^4	10^3	10^4	10^5
50	0.00	0.00	0.00	0.00	0.06	0.04	0.01	0.02	0.03	0.03	0.00	0.01	0.02
100	0.00	0.00	0.00	0.00	26.26	24.78	4.44	—	5.90	16.02	0.00	0.01	3.28
200	0.00	0.00	0.00	0.00	—	—	—	—	—	—	0.00	0.02	—
500	0.00	0.00	0.01	0.01	—	—	—	—	—	—	0.00	0.02	—
1000	0.01	0.01	0.01	0.02	—	—	—	—	—	—	0.00	0.02	—
2000	0.01	0.01	0.01	0.04	—	—	—	—	—	—	0.01	0.02	—
5000	0.01	0.02	0.01	0.08	—	—	—	—	—	—	0.01	0.02	—
10000	0.02	0.05	0.02	0.13	—	—	—	—	—	—	0.01	0.03	—

Table 4: Average solution times in seconds (minknap)

$n \backslash R$	uncorr.		weak.corr		str.corr		inv.str.corr		al.str.corr		subs.sum		sim.w
	10^3	10^4	10^3	10^4	10^3	10^4	10^3	10^4	10^3	10^4	10^3	10^4	10^5
50	0.00	0.00	0.00	0.00	0.00	0.02	0.00	0.02	0.00	0.01	0.00	0.03	0.00
100	0.00	0.00	0.00	0.00	0.02	0.17	0.01	0.18	0.01	0.03	0.00	0.03	0.00
200	0.00	0.00	0.00	0.00	0.05	0.82	0.04	0.65	0.04	0.15	0.00	0.03	0.01
500	0.00	0.00	0.00	0.01	0.20	2.52	0.19	2.80	0.16	0.88	0.00	0.03	0.03
1000	0.00	0.00	0.01	0.01	0.48	8.30	0.45	7.59	0.37	3.18	0.00	0.03	0.10
2000	0.00	0.00	0.01	0.01	0.96	13.17	1.09	14.16	0.72	8.57	0.00	0.03	0.35
5000	0.00	0.01	0.01	0.02	3.73	54.11	3.20	54.66	1.63	26.57	0.01	0.04	1.32
10000	0.01	0.01	0.01	0.03	8.18	115.41	6.57	122.84	1.83	48.33	0.01	0.04	1.57

Table 5: Average solution times in seconds (mthard)

$n \setminus R$	uncorr. 10^3	uncorr. 10^4	weak.corr 10^3	weak.corr 10^4	str.corr 10^3	str.corr 10^4	inv.str.corr 10^3	inv.str.corr 10^4	al.str.corr 10^3	al.str.corr 10^4	subs.sum 10^3	subs.sum 10^4	sim.w 10^5
50	0.00	0.00	0.00	0.00	0.01	0.01	0.01	0.01	0.01	0.03	0.00	0.01	0.00
100	0.00	0.00	0.00	0.00	0.01	0.02	0.01	0.01	0.03	0.14	0.00	0.01	0.01
200	0.00	0.00	0.00	0.00	0.04	0.05	0.03	0.04	0.06	0.36	0.00	0.02	0.03
500	0.00	0.00	0.01	0.01	0.09	0.09	0.08	0.09	0.10	0.75	0.00	0.01	0.06
1000	0.01	0.01	0.01	0.02	0.15	0.23	0.14	0.16	0.19	1.01	0.00	0.02	0.11
2000	0.01	0.02	0.01	0.03	0.17	0.38	0.18	0.23	0.31	0.81	0.00	0.01	0.18
5000	0.02	0.04	0.02	0.06	0.17	1.75	0.33	0.66	2.55	1.46	0.00	0.02	0.24
10000	0.04	0.08	0.02	0.10	0.28	5.89	0.48	1.64	—	—	0.01	0.68	0.35

although minknap uses simple bounds from continuous relaxation, the pseudo-polynomial time complexity gets it safely through these instances.

The mthard algorithm has an excellent overall performance, as it is able to solve nearly all problems within seconds. The short solution times are mainly due to the cardinality bounds described in Section 2.1.3 which make it possible to terminate the branching after a few nodes have been explored. There are however a few anomalous entries for large-sized almost strongly correlated problems, where the cardinality bounds somehow fail. Also for some large-sized Subset-sum problems the mthard algorithm takes an unnecessarily long time.

3   Subset-sum Problem

Assume that a knapsack of *capacity* c is given, and that a subset of n items should be selected for the knapsack. Item j has *weight* w_j and we want to obtain the largest weight sum not exceeding c. Thus the *Subset-sum Problem* (SSP) can be stated:

$$
\begin{aligned}
\text{maximize} \quad & \sum_{j=1}^{n} w_j x_j \\
\text{subject to} \quad & \sum_{j=1}^{n} w_j x_j \leq c, \\
& x_j \in \{0,1\}, \quad j=1,\ldots,n.
\end{aligned}
\tag{73}
$$

where $x_j = 1$ if item j is selected and $x_j = 0$ otherwise. The Subset-sum Problem is related to the *diophantine equation*

$$
\begin{aligned}
\sum_{j=1}^{n} w_j x_j = c, \\
x_j \in \{0,1\}, \quad j=1,\ldots,n.
\end{aligned}
\tag{74}
$$

in the sense that the optimal solution of SSP is the largest c for which (74) has a solution. SSP is also called the *Value-independent Knapsack Problem* since it is a special case of the 0-1 Knapsack Problem arising when $p_j = w_j$ for all $j = 1,\ldots,n$. Hence without loss of generality we may assume that all weights w_j and the capacity c are nonnegative integers, that $\sum_{j=1}^{n} w_j > c$ and finally that $w_j < c$ for all $j = 1,\ldots,n$. Violations of these assumptions may be handled as described in Section 2.

The Subset-sum Problem appears in several Cutting problems where the most possible material should be cut out of the length c. SSP also appears as a subproblem in Multiple Knapsack Problems [101], and it can be used to strengthen constraints in the formulation of an integer-programming problem [19, 20, 27].

For SSP, all upper bounds based on some kind of continuous relaxation as described in Section 2.1–2.1.2 give the trivial bound $u = c$. Thus although SSP in principle can be solved by every KP algorithm, the lack of tight bounds may imply an unacceptably large computational effort. This is however seldom the case when the range of the weights is not large, as these instances generally have many solutions to (74), and thus the enumeration may be terminated when such a solution is found (See e.g. Tables 3-5 in Section 2.9).

Also, owing the lack of tight bounds, it may be necessary to apply heuristic techniques to obtain a reasonable solution within a limited time. Several approximate algorithms are considered in Martello and Toth [80]. Recently Gens and Levner [42] presented a fully polynomial approximation scheme with an improved time bound. The algorithm finds an approximate solution with relative error less than ϵ in time $O(\min\{n/\epsilon, n + 1/\epsilon^3\})$ and space $O(\min\{n/\epsilon, n + 1/\epsilon^2\})$.

Probabilistic results for SSP are considered in d'Atri and Puech [3]. Assuming that the weights are independently drawn from a uniform distribution over $\{1, \ldots, c(n)\}$ and the capacity from a uniform distribution over $\{1, \ldots, nc(n)\}$ where $c(n)$ is an upper bound on the weights value, they proved that a simple variant of the greedy algorithm solves SSP with probability tending to 1.

We will start the treatment of SSP by considering upper bounds tighter than $u = c$ in Section 3.1. Then Section 3.2 deals with different dynamic programming algorithms: If the coefficients are not too large, dynamic programming algorithms are generally able to solve Subset-sum Problems even where the diophantine equation has no solution. In Section 3.3 we will present some hybrid algorithms which combine branch-and-bound with dynamic programming. Section 3.4 shows how large-sized instances can be solved by defining a core problem, and finally Section 3.5 gives a computational comparison of the algorithms presented.

3.1 Upper Bounds

As previously mentioned, all upper bounds for SSP based on some kind of continuous relaxation, yield the trivial bound $u = c$. The set of feasible solutions for a KP and an SSP are identical, thus the same polyhedral properties apply. In particular, this means that minimum and maximum cardinality bounds are defined as in Section 2.1.3. Starting with the maximum cardinality bound, assume that the weights are ordered by $w_1 \leq w_2 \leq \ldots \leq w_n$ and let the break item b be defined by $b = \min\{h : \sum_{j=1}^{h} w_j > c\}$. We may now impose the maximum cardinality constraint to (73) as

$$\sum_{j=1}^{n} x_j \leq b - 1. \tag{75}$$

An obvious way of using this constraint is to choose the $b-1$ largest weights $\tilde{w} = \sum_{j=n-b+2}^{n} w_j$, getting the bound

$$u = \min\{c, \tilde{w}\}. \tag{76}$$

If we surrogate relax the cardinality constraint, using a nonnegative multiplier λ, we get the problem $S(\text{SSP}, \lambda)$ given by

$$\begin{aligned}
\text{maximize} \quad & \sum_{j=1}^{n} w_j x_j \\
\text{subject to} \quad & \sum_{j=1}^{n} (w_j + \lambda) x_j \leq c + \lambda(b-1), \\
& x_j \in \{0,1\}, \qquad\qquad j = 1, \ldots, n\,,
\end{aligned} \tag{77}$$

which is an inverse strongly correlated knapsack problem. Similarly Lagrangian relaxing the cardinality constraint leads to a strongly correlated knapsack problem. Since in Section 2 we saw that these instances may be difficult to solve, one may consider the continuous relaxation instead. For $C(S(\text{SSP}, \lambda))$ the items are already ordered according to nondecreasing ratios of $w_j/(w_j + \lambda)$, thus let the break item be defined by $b' = \max\{h : \sum_{j=h}^{n} w_j > c\}$. The continuous solution is then given by

$$z(C(S(\text{SSP}, \lambda))) = \sum_{j=b'+1}^{n} w_j + \left(c + \lambda(b-1) - \sum_{j=b'+1}^{n} (w_j + \lambda)\right) \frac{w_{b'}}{w_{b'} + \lambda} \tag{78}$$

Since $n - b' \leq b - 1$ we get

$$z(C(S(\text{SSP}, \lambda))) \geq \sum_{j=b'+1}^{n} w_j + \left(c - \sum_{j=b'+1}^{n} w_j\right) \frac{w_{b'}}{w_{b'} + \lambda} \quad (79)$$

$$= c \frac{w_{b'}}{w_{b'} + \lambda} + \sum_{j=b'+1}^{n} w_j \left(1 - \frac{w_{b'}}{w_{b'} + \lambda}\right)$$

which can be written $c\alpha + \tilde{w}(1 - \alpha)$ with $0 \leq \alpha = w_{b'}/(w_{b'} + \lambda) \leq 1$, thus $z(C(S(\text{SSP}, \lambda)))$ is a convex combination of c and $\tilde{w}$ and hence cannot be tighter than u given by (76).

If we instead Lagrangian relax the weight constraint of SSP using multiplier $\lambda > 0$, we get the problem $L(\text{SSP}, \lambda)$

$$\begin{aligned} \text{maximize} \quad & \sum_{j=1}^{n} w_j x_j - \lambda \left(\sum_{j=1}^{n} w_j x_j - c\right) \\ \text{subject to} \quad & \sum_{j=1}^{n} x_j \leq b - 1, \\ & x_j \in \{0, 1\}, \qquad\qquad j = 1, \ldots, n. \end{aligned} \quad (80)$$

The objective function may be written as

$$(1 - \lambda) \sum_{j=1}^{n} w_j x_j + \lambda c \quad (81)$$

For $\lambda \geq 1$ the optimal solution is found by not choosing any items, thus we get the value $z(L(\text{SSP}, \lambda)) = \lambda c \geq c$. For $0 \leq \lambda < 1$ the optimal value is found by choosing the $b - 1$ largest values w_j. The objective value is again a convex combination of $\tilde{w}$ and c, and it will never be tighter than the bound (76).

Thus the trivial bound (76) is the tightest one can get by simple relaxations of the cardinality constraint. Similar results are obtained for minimum cardinality bounds.

3.2 Dynamic Programming Algorithms

Due to the lack of tight bounds for SSP, dynamic programming may be the only way of obtaining an optimal solution in reasonable time when no solution to (74) exists. In addition, dynamic programming is used in branch-and-bound algorithms to avoid a repeated search for subsolutions with the same capacity. The dominance rule for SSP is extremely simple, since one state dominates another state if and only if they represent the same weight sum.

The Bellman recursion presented in Section 2.5 for the 0-1 Knapsack Problem is trivially generalized to the SSP: At any stage we let $f_i(\tilde{c})$, for $0 \leq i \leq n$ and $0 \leq \tilde{c} \leq c$, be an optimal solution value to the subproblem of SSP which is defined on items $1, \ldots, i$ with capacity $\tilde{c}$. Then the bellman recursion becomes

$$f_i(\tilde{c}) = \begin{cases} f_{i-1}(\tilde{c}) & \text{for } \tilde{c} = 0, \ldots, w_i - 1 \\ \max\{f_{i-1}(\tilde{c}), f_{i-1}(\tilde{c} - w_i) + w_i\} & \text{for } \tilde{c} = w_i, \ldots, c \end{cases} \quad (82)$$

where $f_0(\tilde{c}) = 0$ for $\tilde{c} = 0, \ldots, c$. This yields a time and space complexity of $O(nc)$. Using the improved techniques by Toth [112] as described in Section 2.5, the complexity may be brought down to $O(\min\{nc, 2^n\})$ by using dynamic programming by reaching.

3.2.1 Horowitz and Sahni Decomposition

Since SSP is a special case of KP it is straightforward to adapt the Horowitz and Sahni decomposition algorithm described in Section 2.5 to Subset-sum Problems.

Ahrens and Finke [1] proposed an algorithm where the decomposition technique is combined with a branch-and-bound algorithm to reduce the space requirements. Furthermore a *replacement technique* (Knuth [66]) is used in order to combine the dynamic programming lists, obtained by partitioning the variables into four subsets. The space bound of the Ahrens and Finke algorithm is $O(2^{n/4})$, making it best-suited for small but difficult instances.

Pisinger [101] presented a modified version of the Horowitz and Sahni decomposition, named decomp, which combines good worst-case properties with quick solution times for easy problems. Let b be the break item for (73) thus

$$b = \min\left\{h : \sum_{j=1}^{h} w_j > c\right\}. \quad (83)$$

The break solution $\hat{x}$ has weight sum $\hat{w} = \sum_{j=1}^{b-1} w_j$. Now let $f_t(\tilde{c})$, $(b - 1 \leq t \leq n, 0 \leq \tilde{c} \leq c)$ be the optimal solution value to (73) restricted to variables $b, \ldots, t$ as follows:

$$f_t(\tilde{c}) = \max\left\{\begin{array}{l} \sum_{j=b}^{t} w_j x_j : \sum_{j=b}^{t} w_j x_j \leq \tilde{c}; \\ x_j \in \{0, 1\}, \; j = b, \ldots, t. \end{array}\right\} \quad (84)$$

Let $g_s(\tilde{c})$, $(1 \leq s \leq b,\ 0 \leq \tilde{c} \leq c)$ be the optimal solution value to the problem defined on variables $s, \ldots, b-1$ with the additional constraint $x_j = 1$ for $j = 1, \ldots, s-1$, thus

$$g_s(\tilde{c}) = \max \left\{ \begin{array}{l} \sum_{j=1}^{s-1} w_j + \sum_{j=s}^{b-1} w_j x_j : \sum_{j=1}^{s-1} w_j + \sum_{j=s}^{b-1} w_j x_j \leq \tilde{c}; \\ x_j \in \{0,1\},\ j = s, \ldots, b-1. \end{array} \right\} \quad (85)$$

The recursion for f_t will repeatedly insert an item into the knapsack, while the recursion for g_s will remove one item, thus

$$f_t(\tilde{c}) = \begin{cases} f_{t-1}(\tilde{c}) & \text{for } \tilde{c} = 0, \ldots, w_t - 1 \\ \max\{f_{t-1}(\tilde{c}),\ f_{t-1}(\tilde{c} - w_t) + w_t\} & \text{for } \tilde{c} = w_t, \ldots, c \end{cases} \quad (86)$$

where $f_{b-1}(\tilde{c}) = 0$ for $\tilde{c} = 0, \ldots, c$. The corresponding recursion for g_s becomes

$$g_s(\tilde{c}) = \begin{cases} g_{s+1}(\tilde{c}) & \text{for } \tilde{c} = c - w_s + 1, \ldots, c \\ \max\{g_{s+1}(\tilde{c}),\ g_{s+1}(\tilde{c} + w_s) - w_s\} & \text{for } \tilde{c} = 0, \ldots, c - w_s \end{cases} \quad (87)$$

with initial values $g_b(\tilde{c}) = 0$ for $\tilde{c} \neq \hat{w}$ and $g_b(\tilde{c}) = \hat{w}$ for $\tilde{c} = \hat{w}$.

Now starting from $(s,t) = (b, b-1)$ the decomp algorithm repeatedly uses the above recursions, each time decreasing s and increasing t. At each iteration the two sets are merged in $O(c)$ time, in order to find the best current solution

$$z = \max_{\tilde{c} = 0, \ldots, c} f_t(\tilde{c}) + g_s(c - \tilde{c}), \quad (88)$$

and the process is terminated if $z = c$, or $(s,t) = (1,n)$ has been reached.

The algorithm may be improved in those cases where $v = \sum_{j=1}^n w_j - \hat{w} < c$. Since there is no need for removing more items $j < b$ than can be compensated for by inserting items $j \geq b$, the recursion g_s may be restricted to consider states $\tilde{c} = c - v, \ldots, c$ while recursion f_t only will consider states $\tilde{c} = 0, \ldots, v$.

The decomp algorithm has basically the same complexity $O(nc)$ as the Bellman recursion but if dynamic programming by reaching is used, only the live states need be saved thus giving the complexity $O(\min\{nc, nv, 2^b + 2^{n-b}\})$.

3.2.2 Balancing

If all weights w_j are bounded by a fixed constant r, the complexity of the Bellman recursion may be written $O(n^2 r)$, i.e. quadratic time for constant

Knapsack Problems

r. A linear time algorithm may however be derived by using balancing as defined in Section 2.4.

Due to the weight constraint in Theorem 2.2 in Section 2.4 let $f_{s,t}(\tilde{c})$, ($s \leq b$, $t \geq b-1$, $c-r < \tilde{c} \leq c+r$) be the optimal solution value to the subproblem of SSP which is defined on the variables $i = s, \ldots, t$ of the problem:

$$f_{s,t}(\tilde{c}) = \max \left\{ \begin{array}{l} \sum_{j=1}^{s-1} w_j + \sum_{j=s}^{t} w_j x_j : \sum_{j=1}^{s-1} w_j + \sum_{j=s}^{t} w_j x_j \leq \tilde{c}; \\ x_j \in \{0,1\} \text{ for } j = s, \ldots, t; \ x \text{ is a balanced filling} \end{array} \right\}. \tag{89}$$

We will only consider those states (s, t, μ) where $\mu = f_{s,t}(\mu)$, i.e. those weight sums μ which can be obtained by balanced operations on $x_s, \ldots, x_t$, applying the following (unusual) dominance relation:

Definition 3.1 *Given two states (s, t, μ) and (s', t', μ'). If $\mu = \mu'$, $s \geq s'$ and $t \leq t'$, then state (s, t, μ) dominates state (s', t', μ').*

If a state (s, t, μ) dominates another state (s', t', μ') then we may fathom the latter. Using the dominance rule, we will enumerate the states for t running from $b-1$ to n. Thus at each stage t and for each value of μ we will have only one index s, which is actually the largest s such that a balanced filling with weight sum μ can be obtained on the variables $x_s, \ldots, x_t$. Therefore let $s_t(\mu)$ for $t = b-1, \ldots, n$ and $c - r < \mu \leq c + r$ be defined as

$$s_t(\mu) = \max s \left\{ \begin{array}{l} \text{there exists a balanced filling } x \text{ which satisfies} \\ \sum_{j=1}^{s-1} w_j + \sum_{j=s}^{t} w_j x_j = \mu; \ x_j \in \{0,1\}, \ j = s, \ldots, t \end{array} \right. \tag{90}$$

where we set $s_t(\mu) = 0$ if no balanced filling exists. Notice that for $t = b - 1$ only one value of $s_t(\mu)$ is positive, namely $s_t(\hat{w}) = b$, as only the break solution is a balanced filling at this stage. An optimal solution to SSP is found as $z = \max\{\mu \leq c : s_n(\mu) > 0\}$.

After each iteration of t we will ensure that all states are feasible by removing a sufficient number of items $j < s_t(\mu)$ from those solutions where $\mu > c$. Thus only states $s_t(\mu)$ with $\mu \leq c$ need be saved, but in order to improve efficiency, we use $s_t(\mu)$ for $\mu > c$ to memorize that items $j < s_t(\mu)$ have been removed once before. We get:

1 **Algorithm** `balsub`
2   **for** $\mu \leftarrow c - r + 1$ **to** c **do** $s_{b-1}(\mu) \leftarrow 0$;

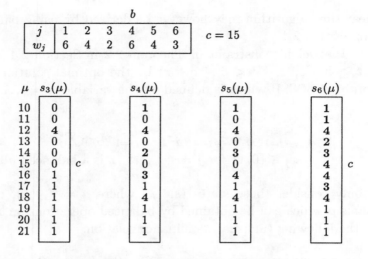

Figure 5: The items and table $s_t(\mu)$ for a given instance.

```
3      for μ ← c + 1 to c + r do s_{b-1}(μ) ← 1;
4      s_{b-1}(ŵ) ← b;
5      for t ← b to n do
6         for μ ← c − r + 1 to c + r do s_t(μ) ← s_{t-1}(μ);
7         for μ ← c − r + 1 to c do μ' ← μ + w_t;
                                     s_t(μ') ← max{s_t(μ'), s_{t-1}(μ)};
8         for μ ← c + w_t downto c + 1 do
9            for j ← s_t(μ) − 1 downto s_{t-1}(μ) do μ' ← μ − w_j;
                                     s_t(μ') ← max{s_t(μ'), j};
```

Algorithm balsub does the following (see Fig. 5 for an example): For $t = b - 1$ we only have one balanced solution, the break solution, thus $s_t(\mu)$ is initialized according to this in lines 2-4. Since the table $s_t(\mu)$ for $\mu > c$ is used for memorizing which items have been removed previously, we may set $s_t(\mu) = 1$ for $\mu > c$ as no items $j < s_t(\mu)$ have ever been removed.

Now we consider the items $t = b, \ldots, n$ in lines 5-9. In each iteration item t may be added to the knapsack or omitted. Line 6 corresponds to the latter case, thus the states $s_{t-1}(\mu)$ are copied to $s_t(\mu)$ without changes. Line 7 adds item t to each feasible state, obtaining the weight μ'. According to (90), $s_t(\mu')$ is the maximum of the previous value and the current balanced solution.

In lines 8-9 we complete the balanced operations by removing items

$j < s_t(\mu)$ from states with $\mu > c$. As it may be necessary to remove several items in order to maintain feasibility of the solution, we consider the states for decreasing μ, thus allowing for several removals.

Theorem 3.1 *Algorithm* balsub *finds the optimal solution* x^*.

Proof. We just need to show that the algorithm performs unrestricted balanced operations: 1) It starts from the break solution x'. 2) For each state with $\mu \leq c$ we perform a balanced insert, as each item t may be added or omitted. 3) For each state with $\mu > c$ we perform a balanced remove by removing an item $j < s_t(\mu)$. As the hereby obtained weight μ' satisfies $\mu' < \mu$, we must consider the weights μ in decreasing order in line 8 in order to allow multiple removals. This ensures that all states will be feasible after each iteration of t.

The only restriction in balanced operations is line 9, where we pass by items $j < s_{t-1}(\mu)$ when items are removed. But due to the memorizing we know that items $j < s_{t-1}(\mu)$ have been removed once before, meaning that $s_t(\mu - w_j) \geq j$ for $j = 1, \ldots, s_{t-1}(\mu)$. Thus repeating the same operations will not contribute to an increase in $s_t(\mu - w_j)$. □

Theorem 3.2 *The complexity of Algorithm* balsub *is* $O(nr)$ *in time and space.*

Proof. Space: The array $s_t(\mu)$ has size $(n - b + 1)(2r)$, thus $O(nr)$. Time: Lines 2-3 demand $2r$ operations. Line 6 is executed $2r(n - b + 1)$ times. Line 7 is executed $r(n - b + 1)$ times. Finally, for each $\mu > c$, line 9 is executed $s_n(\mu) \leq b$ times in all. Thus during the whole process, line 9 is executed at most rb times. □

3.3 Hybrid Algorithms

Despite the lack of tight bounds, branch-and-bound algorithms give excellent solution times in those cases where many solutions to the diophantine equation (74) exist. If no solutions to the diophantine equation exist, branch-and-bound algorithms yield very bad performance, and dynamic programming may be a better alternative. Several hybrid algorithms have been proposed which combine the best properties of dynamic programming and branch-and-bound. The most successful are those by Plateau and Elkihel [106] and Martello and Toth [77]. We will go into the latter in detail.

The mts algorithm assumes that the weights are ordered in nonincreasing order,
$$w_1 \geq w_2 \geq \ldots \geq w_n \qquad (91)$$
and apply dynamic programming to enumerate the last (small) weights, while branch-and-bound is used to search through combinations of the first (large) weights. The motivation is that during branch-and-bound a very great effort is used to search for solutions which fill a small residual capacity. Thus by building a table of such solutions through dynamic programming, repeated calculations can be avoided.

The mts algorithm actually builds two lists of partial solutions. First, the recursion (82) is used to enumerate items $\beta, \ldots, n$, for all weight sums not greater than c. Then the same recursion is used to enumerate items $\alpha, \ldots, n$, but only for weight sums up to $\bar{c}$. The constants $\alpha < \beta < n$ and $\bar{c} < c$ were experimentally determined to be

$$\alpha = n - \min\{2\log r,\ 0.7n\} \qquad \beta = n - \min\{2.5\log r,\ 0.8n\} \qquad \bar{c} = 1.3 w_\beta \qquad (92)$$

where $r = w_1$ is the largest weight. Thus we have two tables

$$f(\tilde{c}) = \max\left\{ \sum_{j=\alpha}^n w_j x_j : \sum_{j=\alpha}^n w_j x_j \leq \tilde{c}, j = \alpha, \ldots, n \right\}, \qquad \tilde{c} = 0, \ldots, \bar{c}$$

$$g(\tilde{c}) = \max\left\{ \sum_{j=\beta}^n w_j x_j : \sum_{j=\beta}^n w_j x_j \leq \tilde{c}, j = \beta, \ldots, n \right\}, \qquad \tilde{c} = 0, \ldots, c \qquad (93)$$

We will assume that if $\tilde{c} < 0$ the above tables return $-\infty$. The branch-and-bound part of mts repeatedly sets a variable x_i to 1 or 0, backtracking when either the current weight sum exceeds c or $i = \beta$. For each branching node, the dynamic programming lists are used to find the largest weight sum which fits into the residual capacity. Assuming that i is the next item to be inserted, and that $\overline{w}$ is the weight sum of the currently fixed items, a simplified sketch of the mts algorithm becomes:

algorithm mts$(i, \overline{w})$;
Use the dynamic programming lists to complete the solution:
$w' = \overline{w} + f(c - \overline{w});\ w'' = \overline{w} + g(c - \overline{w});$
if $(w' > z)$ **or** $(w'' > z)$ **then**
  $z \leftarrow \max\{w', w''\};$
  Construct solution vector by setting $x_j^* \leftarrow x_j$ for $j = 1, \ldots, i-1$ and
    remaining x_j^*'s as given by the dynamic programming tables.
if $(i < \beta)$ **and** $(c - \overline{w} \geq \underline{w}_i)$ **then**

if $(\overline{w} + w_i \leq c)$ then $x_i \leftarrow 1$; $\text{mts}(i+1, \overline{w} + w_i)$;
$x_i \leftarrow 0$; $\text{mts}(i+1, \overline{w})$;

The algorithm is called $\text{mts}(1,0)$ and it returns the optimal solution vector x^*. The minimum weight table must be initialized as $\underline{w}_i = \min_{j \geq i} w_j$.

Since dynamic programming by reaching is used for constructing the tables f and g, only weight sums which are obtainable will be saved. Thus a binary search method is used to look up values in the tables. This binary search algorithm also checks whether the solution vector x corresponding to $g(c - \overline{w})$ satisfies $x_j = 0$ for $j < i$, and returns $g(c - \overline{w}) = 0$ if this is not the case. This is a necessary restriction since the solution space of branch-and-bound and dynamic programming are not allowed to overlap.

The mts algorithm also allows several insertions in each iteration, such that the tables f and g are only accessed when a sufficiently filled solution has been obtained. A table $v_h = \sum_{j=h}^{n} w_j$ is used to avoid forward moves when $c - \overline{w} \geq v_i$, or when the insertion of all items $j \geq i$ will not lead to an improved solution.

3.4 Solution of Large-sized Instances

Many randomly generated instances have several optimal solutions which satisfy the weight constraint with equality. For such classes of instances one may expect that an optimal solution to large-sized instances can be obtained by considering only a relatively small subset of the items. Martello and Toth [77] presented such an algorithm which solves a *core problem* similar to the one for KP, although special properties for SSP make things much simpler. The mtsl algorithm may be outlined as follows:

For a given instance of SSP, let $b = \min\{h : \sum_{j=1}^{h} w_j > c\}$ be the break item, and define a *core* as the interval of items $[b - \delta, b + \delta]$, where δ is a appropriately chosen value. Martello and Toth experimentally found that $\delta = 45$ is a sufficiently large value. With this choice of the core, the core problem becomes:

$$\begin{aligned}
\text{maximize} \quad & \sum_{j=b-\delta}^{b+\delta} w_j x_j \\
\text{subject to} \quad & \sum_{j=b-\delta}^{b+\delta} w_j x_j \leq c - \sum_{j=1}^{b-\delta-1} w_j, \\
& x_j \in \{0,1\}, \qquad j = b-\delta, \ldots, b+\delta.
\end{aligned} \quad (94)$$

The above problem is solved using the mts algorithm. If a solution is obtained which satisfies the weight constraint with equality, we may obtain an optimal solution to the main problem by setting $x_j = 1$ for $j < b - \delta$ and $x_j = 0$ for $j > b + \delta$. Otherwise δ is increased to twice the size, and the process is repeated.

3.5 Computational Experiments

We will compare the presented algorithms mts1, balsub, bellman and decomp. A comparison between the Ahrens and Finke algorithm and mts1 can be found in [80]. Five types of data instances presented in Martello and Toth [80] are considered:

Problems *P3*: w_j randomly distributed in $[1, 10^3]$, and $c = \lfloor n10^3/4 \rfloor$. Problems *P6*: w_j randomly distributed in $[1, 10^6]$, and $c = \lfloor n10^6/4 \rfloor$. Problems *evenodd*: w_j even, randomly distributed in $[1, 10^3]$, and $c = 2\lfloor n10^3/8 \rfloor + 1$ (odd). Jeroslow [62] showed that every branch-and-bound algorithm enumerates an exponentially increasing number of nodes when solving *evenodd* problems. Problems *avis*: $w_j = n(n+1) + j$, and $c = n(n+1)\lfloor (n-1)/2 \rfloor + n(n-1)/2$. Avis [14] showed that any recursive algorithm which does not use dominance will perform poorly for the *avis* problems. Finally the *todd* problems: set $k = \lfloor \log_2 n \rfloor$ then $w_j = 2^{k+n+1} + 2^{k+j} + 1$, and $c = \lfloor \frac{1}{2} \sum_{j=1}^{n} w_j \rfloor$. Todd [14] constructed these problems such that any algorithm which uses upper bounding tests, dominance relations, and rudimentary divisibility arguments will have to enumerate an exponential number of states.

Each instance type was tested with 100 instances and a time limit of 10 hours was assigned for the solution of all problems in the series. For the *avis* and *todd* problems 100 permutations of the same problem were tested.

The running times of four different algorithms are compared in Table 6: The bellman recursion, the decomp dynamic programming algorithm, the balsub algorithm, and finally the mts1 algorithm. A dash indicates that the 100 instances could not be solved within the time limit.

For the randomly distributed problems *P3*, *P6* and *evenodd* we have r bounded by a (large) constant. Thus the bellman recursion runs in $O(n^2)$ time, while balsub has linear solution time. The problems *P3* and *P6* have the property that several solutions to $\sum_{j=1}^{n} w_j x_j = c$ exist when n is large, thus generally the algorithms may terminate before a complete enumeration has been performed. The bellman recursion has to enumerate all states up to at least $t = b$ before it can terminate. It is seen that the mts1 and decomp algorithms are superior for these problems.

Table 6: Solution times in seconds, as averages of 100 instances (hp9000/735).

algorithm	n	P3	P6	evenodd	avis	todd
bellman	10	0.00	0.00	0.00	0.00	0.00
	30	0.02	—	0.01	0.01	—*
	100	0.33	—	0.21	1.57	—*
	300	4.11	—	4.84	—	—*
	1000	52.86	—	69.34	—	—*
	3000	505.38	—	723.25	—*	—*
	10000	—	—*	—	—*	—*
	30000	—	—*	—	—*	—*
	100000	—	—*	—	—*	—*
mtsl	10	0.00	0.00	0.00	0.00	0.00
	30	0.00	0.01	3.84	12.39	—*
	100	0.00	0.00	—	—	—*
	300	0.00	0.00	—	—	—*
	1000	0.00	0.00	—	—	—*
	3000	0.00	0.01	—	—*	—*
	10000	0.00	—*	—	—*	—*
	30000	0.00	—*	—	—*	—*
	100000	0.02	—*	—	—*	—*
balsub	10	0.00	5.37	0.00	0.00	0.12
	30	0.00	8.68	0.01	0.01	—*
	100	0.00	4.21	0.02	0.26	—*
	300	0.00	2.62	0.07	15.21	—*
	1000	0.00	2.12	0.22	562.38	—*
	3000	0.00	2.11	0.66	—*	—*
	10000	0.00	—*	2.22	—*	—*
	30000	0.00	—*	6.66	—*	—*
	100000	0.02	—*	23.76	—*	—*
decomp	10	0.00	0.00	0.00	0.00	0.00
	30	0.00	0.00	0.01	0.00	—*
	100	0.00	0.00	0.24	0.25	—*
	300	0.00	0.00	2.90	23.94	—*
	1000	0.00	0.00	34.20	—	—*
	3000	0.00	0.00	311.28	—*	—*
	10000	0.00	—*	—	—*	—*
	30000	0.00	—*	—	—*	—*
	100000	0.02	—*	—	—*	—*

*Could not be generated on a 32-bit computer.

For the *evenodd* problems no solutions satisfying $\sum_{j=1}^{n} w_j x_j = c$ do exist, meaning that we get strict linear solution time for `balsub`. The `bellman` recursion has complexity $O(n^2)$, and thus cannot solve problems larger than $n = 3000$, and the same applies for the `decomp` algorithm. The `mtsl` algorithm cannot prove optimality of the solution, before an almost complete enumeration has been done in the branch-and-bound part, thus it is not able to solve problems larger than $n = 30$.

The *avis* problems have weights of magnitude $O(n^2)$ while the capacity is of magnitude $O(n^3)$, so the `bellman` and `decomp` algorithms demand $O(n^4)$ time, while `balsub` solves the problem in $O(n^3)$. Algorithm `mtsl` again needs a complete enumeration to prove optimality, and thus cannot solve problems larger than $n = 30$.

Finally the *todd* problems are considered. Due to the exponentially increasing weights, none of the algorithms are able to solve more than tiny instances. However, the `decomp` and `mtsl` algorithms are able to solve the largest instances, since they use some kind of decomposition to limit the enumeration.

The comparisons do not show a clear winner, since all algorithms have different properties, but the `bellman` recursion has worst solution times for all instances. Both `decomp` and `mtsl` are good algorithms for randomly generated instances *P3* and *P6*, but `decomp` is also able to solve some of the difficult problems. On the other hand `balsub` is excellent for the non-fill problems: *evenodd* and *avis*. Thus we may conclude that `decomp` dominates `mtsl` and `bellman`, while `balsub` dominates `bellman`. No dominance however exists between `decomp` and `balsub`.

4 Multiple-choice Knapsack Problem

Consider k classes $N_1, \ldots, N_k$ of *items* to be packed into a knapsack of *capacity* c. Each item $j \in N_i$ has a *profit* p_{ij} and a *weight* w_{ij}, and the problem is to choose one item from each class such that the profit sum is maximized without the weight sum exceeding c. The *Multiple-choice Knapsack Problem* (MCKP) may thus be formulated as:

$$\text{maximize} \quad z = \sum_{i=1}^{k} \sum_{j \in N_i} p_{ij} x_{ij}$$

$$\text{subject to} \quad \sum_{i=1}^{k} \sum_{j \in N_i} w_{ij} x_{ij} \leq c, \tag{95}$$

$$\sum_{j \in N_i} x_{ij} = 1, \quad i = 1, \ldots, k,$$

$$x_{ij} \in \{0, 1\}, \quad i = 1, \ldots, k, \quad j \in N_i,$$

where $x_{ij} = 1$ when item j is chosen in class N_i and $x_{ij} = 0$ otherwise. All coefficients p_{ij}, w_{ij}, and c are nonnegative integers, and the classes $N_1, \ldots, N_k$ are mutually disjoint, class N_i having *size* n_i. The *total* number of items is $n = \sum_{i=1}^{k} n_i$. The MCKP is also denoted as *Knapsack Problem with Generalized Upper Bound Constraints* or for short *Knapsack Problem with GUB*.

Negative coefficients p_{ij}, w_{ij} in (95) may be handled by adding a sufficiently large constant to all items in the corresponding class as well as to c. Fractions can be handled by multiplying through with an appropriate factor. If the multiple-choice constraints in (95) are replaced by $\sum_{j \in N_i} x_{ij} \leq 1$ as considered in [63] then this problem can be transformed into the above form by adding a dummy item $(p_{i,n_i+1}, w_{i,n_i+1}) = (0,0)$ to each class.

To avoid unsolvable or trivial situations we assume that

$$\sum_{i=1}^{k} \min_{j \in N_i} w_{ij} \leq c < \sum_{i=1}^{k} \max_{j \in N_i} w_{ij} \tag{96}$$

and moreover we assume that every item i', j' satisfies

$$w_{i',j'} + \sum_{i \neq i'} \min_{j \in N_i} w_{ij} \leq c \tag{97}$$

as otherwise it may be discarded.

If we relax the integrality constraint on x_{ij} in (95) we obtain the *Continuous Multiple-choice Knapsack Problem* C(MCKP). If each class has two items, where $(p_{i1}, w_{i1}) = (0,0)$, $i = 1, \ldots, k$, the problem (95) corresponds to the *0-1 Knapsack Problem* KP, and thus MCKP is $\mathcal{NP}$-hard.

The MCKP in a minimization form may be transformed into the above formulation (95) by finding for each class N_i the values $\overline{p}_i = \max_{j \in N_i} p_{ij}$, $\overline{w}_i = \max_{j \in N_i} w_{ij}$, and by setting $\tilde{p}_{ij} = \overline{p}_i - p_{ij}$ and $\tilde{w}_{ij} = \overline{w}_i - w_{ij}$ for $j \in N_i$ and $\tilde{c} = \sum_{i=1}^{k} \overline{w}_i - c$. Then the equivalent maximization problem is defined in $\tilde{p}, \tilde{w}$ and $\tilde{c}$.

The MCKP problem has a large range of applications: Nauss [89] mentions Capital Budgeting and transformation of nonlinear KP to MCKP as possible applications, while Sinha and Zoltners [109] propose MCKP used for Menu Planning or to determine which components should be linked in series in order to maximize fault tolerance. Witzgal [114] proposes MCKP used to accelerate ordinary LP/GUB problems by the dual simplex algorithm. Moreover MCKP appears by Lagrangian relaxation of several integer programming problems, as described by Fisher [36]. Recently MCKP has been used for solving Generalized Assignment Problems by Barcia and Jörnsten [7].

When dealing with Multiple-choice Knapsack Problems, two kinds of dominance appear:

Definition 4.1 *If two items r and s in the same class N_i satisfy*

$$w_{ir} \leq w_{is} \quad \text{and} \quad p_{ir} \geq p_{is}, \tag{98}$$

then we say that item r dominates item s. Similarly if some items $r, s, t \in N_i$ with $w_{ir} \leq w_{is} \leq w_{it}$ and $p_{ir} \leq p_{is} \leq p_{it}$ satisfy

$$\frac{p_{it} - p_{ir}}{w_{it} - w_{ir}} \geq \frac{p_{is} - p_{ir}}{w_{is} - w_{ir}} \tag{99}$$

then we say that item s is continuously dominated (for short LP-dominated) by items r and t.

Theorem 4.1 (Sinha and Zoltners [109]) *Given two items $r, s \in N_i$. If item r dominates item s then an optimal solution to MCKP with $x_{is} = 0$ exists. If two items $r, t \in N_i$ LP-dominate an item $s \in N_i$ then an optimal solution to $C(MCKP)$ with $x_{is} = 0$ exists.*

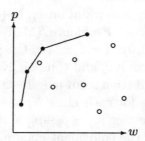

Figure 6: LP-undominated items R_i (black) form the upper convex hull of N_i.

In Section 4.1 we will use these dominance relations to derive different upper bounds, while Section 4.2 considers some heuristic solution methods.

Knapsack Problems

Reduction of classes is discussed in Section 4.3, and these results are incorporated into exact solution techniques presented in Section 4.4 and 4.5. Finally some reduction rules to fathom states in an enumerative algorithm are given in Section 4.6, and algorithms for large-sized instances are considered in Section 4.7. We end the section by bringing some computational results in Section 4.8.

4.1 Upper Bounds

The simplest bound is obtained from the continuous relaxation $C(\text{MCKP})$ of the formulation (95). In this case dominated and LP-dominated items may be fathomed, which is easily obtained by ordering the items in each class N_i according to increasing weights, and successively testing the items according to criteria (98) and (99). For each class N_i the reduction takes $O(n_i \log n_i)$ time due to the sorting. The remaining items R_i are called the *LP-undominated* items, and they form the upper convex hull of N_i, as illustrated in Fig. 6. We will assume that $w_{i1} < w_{i2} < \ldots < w_{i,|R_i|}$ in R_i.

The continuous bound $z(C(\text{MKP}))$ can now be derived in $O(n \log n)$ time by using the greedy algorithm:

algorithm greedy

1 Choose the lightest item from each class (i.e. set $x_{i1} \leftarrow 1$, $x_{ij} \leftarrow 0$ for $j = 2, \ldots, |R_i|$, $i = 1, \ldots, k$) and define the chosen weight and profit sum as $P \leftarrow \sum_{i=1}^{k} p_{i1}$, resp. $W \leftarrow \sum_{i=1}^{k} w_{i1}$. For all items $j \neq 1$ define the slope γ_{ij} as

$$\gamma_{ij} \leftarrow \frac{p_{ij} - p_{i,j-1}}{w_{ij} - w_{i,j-1}}, \quad i = 1, \ldots, k, \quad j = 2, \ldots, |R_i|. \qquad (100)$$

This slope is a measure of the profit-to-weight ratio obtained by choosing item j instead of item $j-1$ in class R_i. Using the greedy principle, order the slopes $\{\gamma_{ij}\}$ in nondecreasing order. With each value of γ_{ij} we associate the indices i, j during the sorting.

2 Assume that γ_{ij} is the next slope in $\{\gamma_{ij}\}$. If $W + w_{ij} > c$ go to Step 3. Otherwise set $x_{ij} \leftarrow 1$, $x_{i,j-1} \leftarrow 0$ and update the sums $P \leftarrow P + p_{ij} - p_{i,j-1}$, $W \leftarrow W + w_{ij} - w_{i,j-1}$. Repeat Step 2.

3 If $W = c$ we have an integer solution and the optimal objective value to $C(\text{MCKP})$ (and to MCKP) is $z^* = P$. Otherwise set $\gamma^* = \gamma_{ij}$.

We have two fractional variables $x_{ij} \leftarrow (c - W)/(w_{ij} - w_{i,j-1})$ and $x_{i,j-1} \leftarrow 1 - x_{ij}$, which both belong to the same class. The optimal objective value is

$$z^* = P + \gamma^*(c - W) . \tag{101}$$

The greedy algorithm is based on the transformation principles presented in Zemel [116]. The time complexity of the **greedy** algorithm is $O(n \log n)$ due to the sorting and preprocessing of each class.

The *LP-optimal choices* b_i obtained by the greedy algorithm are the variables for which $x_{ib_i} = 1$. The class containing two fractional variables in Step 3 will be denoted the *fractional class* N_a, and the *fractional variables* are x_{ab_a}, $x_{ab'_a}$ possibly with $x_{ab'_a} = 0$. An initial feasible solution to MCKP may be constructed by choosing the LP-optimal variables, i.e. by setting $x_{ib_i} = 1$ for $i = 1, \ldots, k$ and $x_{ij} = 0$ for $i = 1, \ldots, k$, $j \neq b_i$. The solution will be denoted the *break solution* and the corresponding weight and profit sums are $\hat{p} = P = \sum_{i=1}^k p_{ib_i}$ and $\hat{w} = W = \sum_{i=1}^k w_{ib_i}$, respectively. As a consequence of the greedy algorithm we have:

Theorem 4.2 *An optimal solution x^* to $C(MCKP)$ satisfies the following: 1) x^* has at most two fractional variables x_{ab_a} and $x_{ab'_a}$. 2) If x^* has two fractional variables they must be adjacent variables within the same class N_a. 3) If x^* has no fractional variables, then the break solution is an optimal solution to MCKP.*

4.1.1 Linear Time Algorithms for the Continuous Problem

Dyer [25] and Zemel [117] independently developed $O(n)$ algorithms for $C(\text{MCKP})$ which do not use the time-consuming preprocessing of classes N_i to R_i. Both algorithms are based on the convexity of the LP-dual problem to (95), which makes it possible to *pair* the dual line segments, so that at each iteration at least 1/6 of the line segments are deleted. The following primal algorithm is more intuitively appealing and can be seen as a generalization of Algorithm bal_zem in Section 2.1.

As a consequence of the greedy algorithm just described, an optimal solution to $C(\text{MCKP})$ is characterized by an optimal slope γ^* which satisfies:

$$\sum_{i=1}^k w_{i\phi_i} \leq c < \sum_{i=1}^k w_{i\psi_i} \tag{102}$$

Knapsack Problems

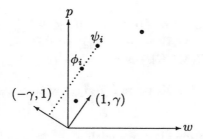

Figure 7: Projection of $(w_{ij}, p_{ij}) \in N_i$ on $(-\gamma, 1)$. Here we have $M_i = \{\phi_i, \psi_i\}$

where

$$\phi_i = \arg \min_{j \in M_i(\gamma^*)} w_{ij} \qquad \psi_i = \arg \max_{j \in M_i(\gamma^*)} w_{ij} \qquad (103)$$

and

$$M_i(\gamma) = \left\{ j \in N_i : (p_{ij} - \gamma w_{ij}) = \max_{\ell \in N_i}(p_{i\ell} - \gamma w_{i\ell}) \right\} \qquad (104)$$

Here (102) is the termination criteria from Step 2, while (103) and (104) ensure that γ^* is a slope corresponding to two items $j-1, j \in R_i$ as given by (100). This is graphically illustrated in Fig. 7. The optimal slope γ^* may be found by the following partitioning algorithm:

algorithm dye_zem

1. For all classes N_i: pair the items two by two as (ij_1, ij_2). Order each pair such that $w_{ij_1} \leq w_{ij_2}$ breaking ties such that $p_{ij_1} \geq p_{ij_2}$ when $w_{ij_1} = w_{ij_2}$. If item j_1 dominates item j_2 then delete item j_2 from N_i and pair item j_1 with another item from the class. Continue this process till all items in N_i have been paired (apart from one last item if $|N_i|$ is odd). Set $P \leftarrow 0, W \leftarrow 0$.

2. For all classes N_i: if the class has only one item j left, then set $P \leftarrow P + p_{ij}$, $W \leftarrow W + w_{ij}$, and fathom class N_i.

3. For all pairs (ij_1, ij_2) derive the slope $\gamma_{ij_1 j_2} = \frac{p_{ij_2} - p_{ij_1}}{w_{ij_2} - w_{ij_1}}$.

4. Let γ be the median of the slopes $\{\gamma_{ij_1 j_2}\}$.

5. Derive $M_i(\gamma)$ and ϕ_i, ψ_i for $i = 1, \ldots, k$ according to (104) and (103).

6 If γ is optimal according to (102), i.e. if $W + \sum_{i=1}^{k} w_{i\phi_i} \leq c < W + \sum_{i=1}^{k} w_{i\psi_i}$ then set $W \leftarrow W + \sum_{i=1}^{k} w_{i\phi_i}$ and $P \leftarrow P + \sum_{i=1}^{k} p_{i\phi_i}$. An optimal solution to $C(\text{MCKP})$ is $z^* = P + (c - W)\gamma$. Stop.

7 If $\sum_{i=1}^{k} w_{i\phi_i} \geq c$ then for all pairs (ij_1, ij_2) with $\gamma_{ij_1j_2} \leq \gamma$ delete item j_2.

8 If $\sum_{i=1}^{k} w_{i\psi_i} < c$ then for all pairs with $\gamma_{ij_1j_2} \geq \gamma$ delete item j_1.

9 Go to step 1.

In Step 7 we know that the chosen γ is too small according to the greedy algorithm, thus the validity of deleting j_2 can be stated as follows: If $j_1, j_2 \in R_i$, then since $\gamma_{ij_1j_2} \leq \gamma$ we may delete j_2. If $j_2 \notin R_i$ then j_2 cannot be in an optimal solution of $C(\text{MCKP})$, and thus can be deleted. Thus the final case is $j_1 \notin R_i$ while $j_2 \in R_i$. Let j_1' be the predecessor of j_2 in R_i. Since $\gamma_{ij_1'j_2} < \gamma_{ij_1j_2} \leq \gamma$ we may delete j_2. The validity of Step 8 is confirmed in a similar way.

Theorem 4.3 *The time complexity of the* dye_zem *algorithm is* $O(n)$.

Proof. Assume that all items and all classes are represented as lists, such that deletions can be made in $O(1)$. At any stage, n_i refers to the current number of items in class N_i and k is the current number of classes. Notice that each iteration 1-9 can be performed in $O(n)$ time, where n is the current number of items, since no step demands a higher complexity.

There are $\sum_{i=1}^{k} \lfloor n_i/2 \rfloor$ pairs of items (ij_1, ij_2). Since γ is the median of $\{\gamma_{ij_1j_2}\}$, half of the pairs will satisfy the criteria in step 7 or 8, and thus one item from these pairs will be deleted; i.e. at least $\frac{1}{2}\sum_{i=1}^{k}\lfloor n_i/2 \rfloor$ items are deleted out of $n = \sum_{i=1}^{k} n_i \leq \sum_{i=1}^{k}(2\lfloor n_i/2 \rfloor + 1)$. So each iteration deletes at least

$$\frac{\frac{1}{2}\sum_{i=1}^{k}\lfloor n_i/2 \rfloor}{\sum_{i=1}^{k}(2\lfloor n_i/2 \rfloor + 1)} \geq \frac{\sum_{i=1}^{k}\lfloor n_i/2 \rfloor}{2k + 4\sum_{i=1}^{k}\lfloor n_i/2 \rfloor} \geq \frac{1}{6} \qquad (105)$$

items since $\lfloor n_i/2 \rfloor \geq 1$, which gives the stated. □

Pisinger [102] proposed to improve the above algorithm by also deleting in Step 7 all items with $w_{ij} \geq w_{i\phi_i}$ for all classes N_i. In Step 8, all items with $p_{ij} \leq p_{i\psi_i}$ can be deleted. The validity of these reductions follows from dominance of items and the criterion (102).

4.1.2 Bounds from Lagrangian Relaxation

Two different bounds can be obtained from Lagrangian relaxation. If we relax the weight constraint with a multiplier $\lambda \geq 0$ we get $L_1(\text{MCKP}, \lambda)$ defined as

$$\text{maximize} \quad \sum_{i=1}^{k} \sum_{j \in N_i} p_{ij} x_{ij} + \lambda \left(c - \sum_{i=1}^{k} \sum_{j \in N_i} w_{ij} x_{ij} \right)$$
$$\text{subject to} \quad \sum_{j \in N_i} x_{ij} = 1, \quad i = 1, \ldots, k, \qquad (106)$$
$$x_{ij} \in \{0, 1\}, \quad i = 1, \ldots, k, \quad j \in N_i.$$

The objective function may be written

$$z(L_1(\text{MCKP}, \lambda)) = \lambda c + \sum_{i=1}^{k} \sum_{j \in N_i} \left(p_{ij} - \lambda w_{ij} \right) x_{ij} = \lambda c + \sum_{i=1}^{k} \max_{j \in N_i} \left(p_{ij} - \lambda w_{ij} \right) \qquad (107)$$

due to the multiple-choice constraint, thus the bound may be derived in $O(n)$ time. The corresponding solution vector is $x_{ij} = 1$ for $j = \arg\max_{j \in N_i} (p_{ij} - \lambda w_{ij})$ and 0 otherwise. Notice that an optimal solution to the continuously relaxed problem (106) will yield the same integer solution, thus we have

$$z(L_1(\text{MCKP}, \lambda)) = z(C(L_1(\text{MCKP}, \lambda))) \geq z(C(\text{MCKP})) \geq z(\text{MCKP}) \qquad (108)$$

A different relaxation can be obtained by relaxing the multiple-choice constraints. In this case, by using multipliers $\lambda_1, \ldots, \lambda_k$, we get $L_2(\text{MCKP}, \lambda_i)$:

$$\text{maximize} \quad \sum_{i=1}^{k} \sum_{j \in N_i} p_{ij} x_{ij} + \sum_{i=1}^{k} \lambda_i \left(1 - \sum_{j \in N_i} x_{ij} \right)$$
$$\text{subject to} \quad \sum_{i=1}^{k} \sum_{j \in N_i} w_{ij} x_{ij} \leq c \qquad (109)$$
$$x_{ij} \in \{0, 1\}, \quad i = 1, \ldots, k, \quad j \in N_i.$$

The objective function may be rewritten as

$$z(L_2(\text{MCKP}, \lambda_i)) = \sum_{i=1}^{k} \lambda_i + \sum_{i=1}^{k} \sum_{j \in N_i} (p_{ij} - \lambda_i) x_{ij} = \sum_{i=1}^{k} \lambda_i + \sum_{i=1}^{k} \sum_{j \in N_i} \tilde{p}_{ij} x_{ij} \qquad (110)$$

thus the relaxation leads to a 0-1 Knapsack Problem defined in profits $\tilde{p}_{ij} = p_{ij} - \lambda_i$. This means that any polynomial bounding procedure for KP as described in Section 2.1 can be used to obtain polynomial upper bounds for MCKP.

4.1.3 Other Bounds

Enumerative bounds (see Section 2.1.2) have been considered in [102]. Some work on the polyhedral properties of MCKP have been presented by Johnson and Padberg [63] as well as Ferreira, Martin and Weismantel [33] — although only for the case where the multiple-choice constraint is in the weaker form $\sum_{j \in N_i} x_{ij} \leq 1$. No bounds derived from polyhedral properties have however been presented.

4.2 Heuristics

The break solution $\hat{x}$, taking the LP-optimal choices x_{ib_i} in each class is generally a good heuristic solution of value $z' = \hat{p}$. The relative performance of the heuristic is however arbitrarily bad, as can be seen with the instance $k = 2$, $n_1 = n_2 = 2$, $c = d$ and items $(p_{11}, w_{11}) = (0,0)$, $(p_{12}, w_{12}) = (d, d)$, $(p_{21}, w_{21}) = (0, 0)$ and $(p_{22}, w_{22}) = (2, 1)$. The break solution is given by items 1,1 and 2,2 yielding $z' = \hat{p} = 2$ although the optimal objective value is $z^* = d$.

Dyer, Kayal and Walker [26] presented an improvement algorithm which runs in $O(\sum_{i=1}^k n_i) = O(n)$ time. Let $\beta_i = b_i$ for classes $i \neq a$ and $\beta_a = b'_a$. Thus, by setting $x_{i\beta_i} = 1$ in each class, we obtain an infeasible solution. Now, for each class $i = 1, \ldots, k$, find the item ℓ_i which when replacing item β_i give a feasible solution with the largest profit. Thus $\ell_i = \arg\max_{j \in N_i} \{p_{ij} : w_{ij} - w_{i\beta_i} + \hat{w} \leq c\}$ and let

$$z^r = \max_{i=1,\ldots,k} (\hat{p} + p_{i\ell_i} - p_{i\beta_i}). \tag{111}$$

This heuristic however also has an arbitrarily bad performance ratio which can be seen with the instance $N_1 = N_2 = N_3 = \{(0,0), (1,1), (d, d+1)\}$ and $k = 3$, $c = d + 1$. We have $z' = \hat{p} = 3$, and also $z^r = 3$ although $z^* = d$.

It is possible to obtain a heuristic solution with worst-case performance $\frac{1}{2}$ by setting

$$z^h = \max\{z', z^b\}, \tag{112}$$

where $z^b = p_{ab'_a} + \sum_{i \neq b} p_{i,\alpha_i}$ with $\alpha_i = \arg\min_{j \in N_i}\{w_{ij}\}$. Notice that z^b is a feasible solution according to (97). Obviously $z^* \leq z' + z^b$, thus $z^* \leq 2z^h$.

To see that the bound is tight, consider the KP instance given below (50) transformed to a MCKP.

A fully polynomial approximation scheme for MCKP has been presented by e.g. Chandra, Hirschberg and Wong [12].

4.3 Class Reduction

As for the 0-1 Knapsack Problem, several decision variables may be fixed a-priori at their optimal value through class reduction. Let u_{ij}^1 be an upper bound on MCKP with additional constraint $x_{ij} = 1$. If $u_{ij} < z+1$ then we may fix x_{ij} at zero. Similarly, if u_{ij}^0 is an upper bound on MCKP with additional constraint $x_{ij} = 0$ and $u_{ij}^0 < z+1$, then we may fix x_{ij} at one, and thus all other decision variables in the class at zero. Using continuous bounds for deriving u_{ij}^0 will however lead to the trivial bound $u_{ij}^0 = C(\text{MCKP})$, thus the latter test is seldom used.

A bound similar to the Dembo and Hammer bound for KP may be derived as follows [26]: Relax the constraint on the fractional variables $b_a, b_a' \in N_a$ in (95) to $x_{ab_a}, x_{ab_a'} \in \mathcal{R}$. Then the bound u_{ij}^1 may be derived in constant time as

$$u_{ij}^1 = \hat{p} - p_{ib_i} + p_{ij} + \gamma^*(c - \hat{w} + w_{ib_i} - w_{ij}), \qquad (113)$$

The complexity of reducing class N_i is $O(n_i)$, and if the reduced set has only one item left, say j, we fathom the class, fixing x_{ij} at 1.

Tighter, but more time-consuming bounds can be obtained from continuous relaxation. For these reductions it is appropriate to solve $C(\text{MCKP})$ through the zemel algorithm, as then it is easy to find the new break class by considering the γ_{ij} in decreasing order from γ. The worst-case complexity of the reduction becomes $O(n^2)$ which however in practice is $O(n)$ as only few γ_{ij} need be considered.

Dudzinski and Walukiewicz [24] propose bounds obtained from the Lagrangian relaxation $L_1(\text{MCKP})$ in (106) obtaining an $O(n)$ reduction.

4.4 Branch-and-bound Algorithms

Several enumerative algorithms for MCKP have been presented during the last two decades: Nauss [89], Sinha and Zoltners [109], Dyer, Kayal and Walker [26], Dudzinski and Walukiewicz [24]. Most of these algorithms start by solving $C(\text{MCKP})$ in order to obtain an upper bound for the problem. The $C(\text{MCKP})$ is solved in two steps: 1) The LP-dominated items are removed as described in Section 4.1. 2) The reduced $C(MCKP)$ is solved by a greedy algorithm. After these two initial steps, upper bound tests may be used to fix several variables in each class to their optimal value. The reduced MCKP problem is then solved to optimality through enumeration. We will go into the algorithm by Dyer, Kayal and Walker in detail, since this is a well-designed branch-and-bound algorithm. The algorithm may be sketched as follows:

procedure dyer_kay_wal

1. Remove LP-dominated items as described in Section 4.1. Solve the continuous relaxation $C(MCKP)$ in order to derive an upper bound. This is done by finding a good candidate $\bar{\gamma}$ for γ^* as an average value of the slopes γ_{ij}. Now, starting from $\gamma = \bar{\gamma}$ search forward or backward through the set $\{\gamma_{ij}\}$ as described in **greedy** until the weight constraint (102) is satisfied. Heaps are used to efficiently access the slopes $\{\gamma_{ij}\}$.

2. Reduce the classes as described in Section 4.3.

3. Solve the remaining problem through branch-and-bound: Bounds are derived by solving $C(MCKP)$ defined on the free variables. If $x_{ab_a}, x_{ab'_a}$ are the fractional variables of the continuous solution, branching is performed by first setting $x_{ab'_a} = 1$ and then $x_{ab'_a} = 0$. Backtracking is performed if either the upper bound $u = z(C(MCKP))$ is not larger than the current lower bound z, or if the MCKP problem with the current variables fixed is infeasible. The branch-and-bound algorithm follows a depth-first search to limit the space consumption.

Dyer, Kayal and Walker furthermore improve the lower bound z at every branching node by using the heuristic (111). For classes larger than $n_i \geq 25$ they also propose to use the class reduction from Section 4.3 at every branching node.

4.5 Dynamic Programming Algorithms

MCKP can be solved in pseudo-polynomial time through dynamic programming as follows [24]. Let $f_\ell(\tilde{c})$ be an optimal solution value to MCKP defined on the first ℓ classes and with restricted capacity $\tilde{c}$. Thus

$$f_\ell(\tilde{c}) = \max \left\{ \begin{array}{l} \sum_{i=1}^{\ell} \sum_{j \in N_i} p_{ij} x_{ij} : \sum_{i=1}^{\ell} \sum_{j \in N_i} w_{ij} x_{ij} \leq \tilde{c}; \\ \sum_{j \in N_i} x_{ij} = 1, \ i = 1, \ldots, \ell; \ x_{ij} \in \{0, 1\} \end{array} \right\}, \quad (114)$$

where we assume that $f_\ell(\tilde{c}) = -\infty$ if no solution exists. Initially we set $f_0(\tilde{c}) = 0$ for all $\tilde{c} = 0, \ldots, c$, Then for $\ell = 1, \ldots, k$ we can use the recursion

$$f_\ell(\tilde{c}) = \max \begin{cases} f_{\ell-1}(\tilde{c} - w_{\ell 1}) + p_{\ell 1} & \text{if } 0 \leq \tilde{c} - w_{\ell 1}, \\ f_{\ell-1}(\tilde{c} - w_{\ell 2}) + p_{\ell 2} & \text{if } 0 \leq \tilde{c} - w_{\ell 2}, \\ \vdots \\ f_{\ell-1}(\tilde{c} - w_{\ell n_\ell}) + p_{\ell n_\ell} & \text{if } 0 \leq \tilde{c} - w_{\ell n_\ell}, \end{cases} \quad (115)$$

where we assume that the maximum operator returns $-\infty$ if we are maximizing over an empty set. An optimal solution to MCKP is found as $z = f_k(c)$ and we obtain $z = -\infty$ if assumption (96) is violated. The space bound of the dynamic programming algorithm is $O(kc)$, while each iteration of (115) takes n_i time, demanding totally $O(c \sum_{i=1}^{k} n_i) = O(nc)$ operations.

The recursion presented has the drawback that an optimal solution is not reached before all classes have been enumerated, meaning that we have to pass through all $O(nc)$ steps. To avoid this problem, Pisinger [102] proposed a generalization of the primal-dual dynamic programming described in Section 2.5.1. Assume that the classes are reordered according to some global considerations, guessing that the last classes have a large probability for being fixed at their LP-optimal value.

Let $f_\ell(\tilde{c})$, $\tilde{c} = 0, \ldots, 2c$ be an optimal solution to the following MCKP problem defined on the first ℓ classes, and where variables in classes after ℓ are fixed at their LP-optimal values:

$$f_\ell(\tilde{c}) = \max \left\{ \begin{array}{l} \sum_{i=1}^{\ell} \sum_{j \in N_i} p_{ij} x_{ij} + \sum_{i=\ell+1}^{k} p_{ib_i} : \\ \sum_{i=1}^{\ell} \sum_{j \in N_i} w_{ij} x_{ij} + \sum_{i=\ell+1}^{k} w_{ib_i} \leq \tilde{c}; \\ \sum_{j \in N_i} x_{ij} = 1 \text{ for } i = 1, \ldots, \ell; \; x_{ij} \in \{0, 1\} \end{array} \right\}. \quad (116)$$

Initially we set $f_0(\tilde{c}) = \hat{p}$ for all $\tilde{c} \geq \hat{w}$, and $f_0(\tilde{c}) = -\infty$ for all $\tilde{c} < \hat{w}$. Then the following recursion is applied:

$$f_\ell(\tilde{c}) = \max \begin{cases} f_{\ell-1}(\tilde{c} - w_{\ell 1} + w_{\ell b_\ell}) + p_{\ell 1} - p_{\ell b_\ell} \\ \quad \text{if } 0 \leq \tilde{c} - w_{\ell 1} + w_{\ell b_\ell} \leq 2c, \\ f_{\ell-1}(\tilde{c} - w_{\ell 2} + w_{\ell b_\ell}) + p_{\ell 2} - p_{\ell b_\ell} \\ \quad \text{if } 0 \leq \tilde{c} - w_{\ell 2} + w_{\ell b_\ell} \leq 2c, \\ \vdots \\ f_{\ell-1}(\tilde{c} - w_{\ell n_\ell} + w_{\ell b_\ell}) + p_{\ell n_\ell} - p_{\ell b_\ell} \\ \quad \text{if } 0 \leq \tilde{c} - w_{\ell n_\ell} + w_{\ell b_\ell} \leq 2c. \end{cases} \quad (117)$$

An optimal solution to MCKP is found as $z = f_k(c)$, obtaining $z = -\infty$ if assumption (96) is violated. The recursion (117) demands $O(n_i)$ operations for each class in the core and for each capacity $\tilde{c}$, yielding the complexity $O(\sum_{i=1}^{k} 2cn_i) = O(nc)$ for a complete enumeration. The space complexity is $O(2kc)$. However if optimality of a state can be proved after enumerating classes up to N_ℓ, then we may terminate the process, having used the computational effort $O(c \sum_{i \leq \ell} n_i)$.

Other dynamic programming algorithms are: The separability property as presented by Horowitz and Sahni [54] is easily generalized to the MCKP. The classes should be separated in two classes such that $n_1 \cdot n_2 \cdots n_\ell$ and $n_{\ell+1} \cdot n_{\ell+2} \cdots n_k$ are of same magnitude and then use normal dynamic programming on each of the two problems. As for KP, the states are easily merged in linear time. This leads to a time bound of $O(\min\{nc, n_1 \cdots n_\ell, n_{\ell+1} \cdots n_k\})$.

No balanced algorithms for the MCKP have been published, but if $p_{ij} = w_{ij}$ for all items, we have a Multiple-choice Subset-sum Problem. For this problem Pisinger [99] presented a balanced algorithm which has complexity $O(nr)$ where $r = \max_{i,j} w_{ij}$.

As in dynamic programming algorithms for KP, it is convenient to represent states as a list of triples (π_i, μ_i, v_i), where $f_\ell(\mu_i) = \pi_i$ and v_i is a representation of the solution vector.

4.6 Reduction of States

During a branch-and-bound or dynamic programming algorithm, it is necessary to derive upper bounds for the considered branching nodes or states. Bounds could be derived in $O(n)$ using the Algorithm dye_zem, but specialized methods yield a better performance.

Dudzinski and Walukiewicz [23] derived an efficient bounding technique as follows: Initially the classes are reduced according to (98) and (99) obtaining classes R_i. Then, each time an upper bound should be derived, a median search algorithm is used in the sorted classes R_i, finding the continuous bound in $O(k \log^2(n'/k))$ where n' is the number of undominated items.

Weaker bounds, which are however derived in constant time, were proposed in Pisinger [102]. Assume that an expanding core is enumerated using recursion (117) up to class N_ℓ. Moreover define for each class i the extreme gradients

$$\Gamma_i^+ = \max_{\ell > i} \gamma_\ell^+ \qquad \Gamma_i^- = \min_{\ell > i} \gamma_\ell^- \qquad (118)$$

Then the bound on a state with profit π and weight μ may be found as

$$u(\pi,\mu) = \begin{cases} \pi + (c-\mu)\Gamma_\ell^+ & \text{if } \mu \leq c, \\ \pi + (c-\mu)\Gamma_\ell^- & \text{if } \mu > c. \end{cases} \quad (119)$$

Notice that the best bounds are obtained by ordering the classes such that the classes with γ_i^+, γ_i^- closest to γ^* are ordered first.

4.7 Solution of Large-sized Instances

As for other knapsack problems it may be beneficial to focus the enumeration on a *core* when dealing with large-sized problems. Basically one should distinguish between two kinds of core when dealing with MCKP: A core where only a subset $C \subset K = \{1,\ldots,k\}$ of the classes is enumerated, or a core where only some items $j \in C_i \subset N_i$ in each class are enumerated. The first approach is suitable for problems with many classes, while the second approach is more appropriate for problems with many items in each class. No work has however been published on the second approach, thus we will here consider a core consisting of a subset of the classes.

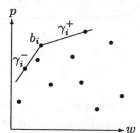

Figure 8: Gradients γ_i^+, γ_i^- in class N_i.

Pisinger [102] defined a core problem based on positive and negative gradients γ_i^+ and γ_i^- for each class N_i, $i \neq a$. The gradients are defined as (see Fig. 8):

$$\begin{aligned} \gamma_i^+ &= \max_{j \in N_i,\ w_{ij} > w_{ib_i}} \frac{p_{ij} - p_{ib_i}}{w_{ij} - w_{ib_i}}, \quad i = 1,\ldots,k,\ i \neq a, \\ \gamma_i^- &= \min_{j \in N_i,\ w_{ij} < w_{ib_i}} \frac{p_{ib_i} - p_{ij}}{w_{ib_i} - w_{ij}}, \quad i = 1,\ldots,k,\ i \neq a, \end{aligned} \quad (120)$$

and we set $\gamma_i^+ = 0$ (resp. $\gamma_i^- = \infty$) if the set we are maximizing (resp. minimizing) over is empty. Note that γ_i^+ and γ_i^- can be derived in $O(n_i)$

for each class N_i and they do not demand any preprocessing. The gradients are a measure of the expected gain (resp. loss) per weight unit by choosing a heavier (resp. lighter) item from N_i instead of the LP-optimal choice b_i.

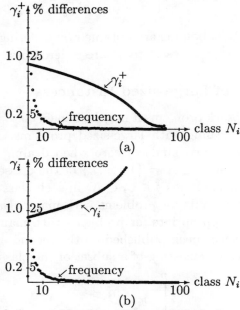

Figure 9: Frequency of classes N_i where IP-optimal choice differs from LP-optimal choice, compared to gradient γ_i^+.

In Fig. 9 (a) we have ordered the classes according to decreasing γ_i^+ and show how often the IP-optimal solution to MCKP differs from the LP-optimal choice in each class N_i. The figure is a result of 5000 randomly generated data instances ($k = 100$, $n_i = 10$), where we have measured how often the IP-optimal choice j (satisfying $w_{ij} > w_{ib_i}$ since we are considering forward gradients) differs from the LP-optimal choice b_i in each class N_i. It can be seen that when γ_i^+ decreases, so does the probability that b_i is not the IP-optimal choice. Similarly, in Fig. 9 (b) we have ordered the classes according to increasing γ_i^- to show how the probability for changes decreases with increased γ_i^-.

This observation motivates considering only a small number of the classes N_i, namely those classes where γ_i^+ or γ_i^- are sufficiently close to γ^* given by (102) to (104). Thus a fixed-core algorithm for MCKP can be derived by straightforward generalization of KP algorithms: Choose a subset $C \subset K$ of the classes where the gradients are close to γ. These classes may be obtained in $O(k)$ by choosing the δ classes with the largest value of γ_i^+. In a similar way additional δ classes with smallest value of γ_i^- are chosen. This gives a core C of *at most* 2δ classes, since some classes may have been chosen twice, and thus should be eliminated.

A fixed-core algorithm should enumerate the classes in the core, and then try to fix the remaining classes at their optimal values through the reduction rules described in Section 4.3. If all items in classes $N_i \not\in C$ can be fixed at their LP-optimal values, the solution to the core problem is optimal. Otherwise the reduced classes must be enumerated to completion either by branch-and-bound or dynamic programming. However no algorithm has presently been published which uses this approach.

An expanding core algorithm for MCKP has been published by Pisinger [102]: Initially the algorithm mcknap solves the continuous problem, using a simplified version of dye_zem. Then the classes are ordered such that classes N_i with gradient γ_i^+ or γ_i^- closest to γ^* are first. Initially only the break class N_a is enumerated, while other classes are introduced consecutively in recursion (117). Upper bound tests are used to fathom states which cannot lead to an optimal solution, hoping that optimality of the problem can be proved with classes $N_i \not\in C$ fixed at their LP-optimal value. Before a class is added to the core, it is reduced by using the tests (113), and if some items cannot be fixed at their LP-optimal values, these are sorted and then reduced according to (98) and (99).

4.8 Computational Experiments

We will compare the dyer_kay_wal and mcknap algorithms. Five types of randomly generated data instances are considered, each instance tested with data-range $R_1 = 1\,000$ or $R_2 = 10\,000$ for different number of classes k and sizes n_i as follows. *Uncorrelated data instances*: In each class we generate n_i items by choosing w_{ij} and p_{ij} uniformly in $[1, R]$. *Weakly correlated data instances*: In each class, w_{ij} is uniformly distributed in $[1, R]$ and p_{ij} is uniformly distributed in $[w_{ij} - 10, w_{ij} + 10]$, such that $p_{ij} \geq 1$. *Strongly correlated data instances*: For KP these instances are generated as w_j uniformly distributed in $[1, R]$ and $p_j = w_j + 10$, which are difficult instances for KP. These instances are however trivial for MCKP, since they degenerate to subset-sum data instances, but hard instances for MCKP may be constructed by cumulating strongly correlated KP-instances: For each class generate n_i items (w_j', p_j') as for KP, and order these by increasing weights. The data instance for MCKP is then $w_{ij} = \sum_{h=1}^{j} w_h'$, $p_{ij} = \sum_{h=1}^{j} p_h'$, $j =$

$1, \ldots, n_i$. Such instances have no dominated items, and form an upper convex hull. *Subset-sum data instances*: w_{ij} uniformly distributed in $[1, R]$ and $p_{ij} = w_{ij}$. *Zig-zag instances*: Sinha and Zoltners [109] consider some zig-zag instances with very few dominated items. For each class construct n_i items as (w'_j, p'_j) uniformly distributed in $[1, R]$. Order the profits and weights in increasing order, and set $w_{ij} = w'_j$, $p_{ij} = p'_j$, $j = 1, \ldots, n_i$.

The FORTRAN code dyer_kay_wal was obtained from Prof. Dyer. Both algorithms were run on a HP9000/735. The computational times obtained are given in Table 7 and 8. A time limit of one hour was given to each instance, and a dash in the tables indicates that some (or all) instances could not be solved within this limit. It follows that mcknap has an overall stable behavior, as it is able to solve all the instances. There are however several instances where the dyer_kay_wal algorithm is not able to solve the problem within the time limit, and in one situation, the algorithm does not find the optimal solution.

The dyer_kay_wal algorithm is sometimes better than the mcknap algorithm for small instances with very large classes N_i. This may be explained by the fact that mcknap is designed to enumerate a few classes instead of a few items. But when the classes are very large, this strategy demands a large enumeration. Other definitions of a core can however be applied in these situations as discussed at the end of Section 4.7.

Table 7: Total computing time (mcknap) in seconds. Averages of 100 instances.

k	n_i	Uncorrelated		Weakly corr.		Strongly corr.		Subset sum		Zig-zag	
		R_1	R_2	R_1	R_2	R_1	R_2	R_1	R_2	R_1	R_2
10	10	0.00	0.00	0.00	0.04	0.01	0.08	0.00	0.14	0.00	0.00
100	10	0.00	0.00	0.01	0.24	0.32	5.03	0.00	0.09	0.01	0.01
1000	10	0.02	0.02	0.02	0.19	6.35	90.37	0.01	0.07	0.02	0.04
10000	10	0.20	0.25	0.19	0.34	151.30	1461.16	0.13	0.13	0.27	0.33
10	100	0.00	0.00	0.02	0.50	0.02	0.18	0.05	0.92	0.01	0.01
100	100	0.01	0.01	0.02	0.46	0.27	6.68	0.01	0.59	0.03	0.05
1000	100	0.10	0.12	0.12	0.35	8.28	183.22	0.09	0.09	0.19	0.24
10	1000	0.02	0.02	0.09	2.38	1.96	0.11	0.02	11.26	0.15	0.61
100	1000	0.09	0.11	0.14	0.94	160.56	2.70	0.09	0.11	0.32	2.21

Table 8: Total computing time (dyer_kay_wal) in seconds. Averages of 100 instances.

k	n_i	Uncorrelated		Weakly corr.		Strongly corr.		Subset sum		Zig-zag	
		R_1	R_2	R_1	R_2	R_1	R_2	R_1	R_2	R_1	R_2
10	10	0.00	0.00	0.07	0.64	5.24	7.66	0.02	0.13	0.00	0.00
100	10	0.01	0.01	0.12	—	—	—	0.03	0.29	0.05	0.07
1000	10	0.11	0.21	0.09	—	—	—	0.04	0.24	0.15	1.22
10000	10	0.66	2.64	0.67	1.42	—	—	0.42	0.44	0.84	4.76
10	100	0.00	0.00	0.05	—	—	—	0.04	0.44	0.01	0.01
100	100	0.04	0.04	0.13	1.05	—	—	0.04	2.84	0.07	0.12
1000	100	0.41	0.95	0.53	9.57	—	—	0.37	*2.43	0.66	1.58
10	1000	0.03	0.03	0.08	—	34.52	—	0.04	2.31	0.09	0.33
100	1000	0.35	0.36	0.44	3.70	—	—	0.41	0.43	0.52	0.67

*Optimal solution not found in one instance.

5 Bounded Knapsack Problem

We consider the problem where a knapsack of *capacity* c should be filled by using n given item types, where type j has a *profit* p_j, a *weight* w_j, and a *bound* m_j on the availability. The problem is to select a number x_j ($0 \leq x_j \leq m_j$) of each item type j such that the profit sum of the included items is maximized without the weight sum exceeding c. The *Bounded Knapsack Problem (BKP)* may thus be defined as

$$\begin{aligned}\text{maximize} \quad & z = \sum_{j=1}^{n} p_j x_j \\ \text{subject to} \quad & \sum_{j=1}^{n} w_j x_j \leq c, \\ & x_j \in \{0, 1, \ldots, m_j\}, \quad j = 1, \ldots, n,\end{aligned} \qquad (121)$$

where all coefficients are positive integers. Without loss of generality we may assume that $w_j m_j \leq c$ for $j = 1, \ldots, n$ so all items available of a given type fit into the knapsack, and that $\sum_{j=1}^{n} w_j m_j > c$ to ensure a nontrivial problem. If the coefficients are negative or fractional, this can be handled by a straightforward adaptation of the Glover [45] method presented in Section 2. Also the transformation of a KP in minimization form into a KP in maximization form described in Section 2 can be immediately extended to BKP. If $m_j = 1$ for all item types, we get the ordinary KP and thus BKP is $\mathcal{NP}$-hard.

Several industrial problems which are usually solved as 0-1 Knapsack Problems may equally well be formulated as Bounded Knapsack Problems, thus taking advantage of the fact that most products come from series of identical item types. Many combinatorial problems can be reduced to BKP, and the problem arises also as a subproblem in several algorithms for Integer Linear Programming.

A close connection between BKP and KP is self-evident, so all the mathematical and algorithmic techniques analyzed in Section 2 could be extended to the present case. Only a few papers have however focused on generalizing the KP results to BKP, which may be due to the fact that BKP may be transformed into an equivalent KP, and thus solved effectively this way.

The transformation of BKP into an equivalent KP is based on a binary encoding of the bound m_j on each item type j (see Martello and Toth [80]). Each item type j is replaced by $\lfloor \log_2 m_j + 1 \rfloor$ items in the KP case, whose profits and weights are:

$$(p_j, w_j),\ (2p_j, 2w_j),\ (4p_j, 4w_j), \ldots, (2^{a-1}p_j, 2^{a-1}w_j),\ (dp_j, dw_j), \qquad (122)$$

where $a = \lfloor \log_2 m_j \rfloor$, and $d = m_j - \sum_{i=0}^{a-1} 2^i$. Thus the equivalent KP will contain $\sum_{j=1}^{n} \lfloor \log_2 m_j + 1 \rfloor$ variables. It can be seen that the transformation introduces 2^{a+1} binary combinations, i.e. $2^{a+1} - (m_j + 1)$ redundant representations of possible x_j values, since the values from d to $2^a - 1$ have a double representation. However, $a + 1$ is the minimum number of binary variables needed to represent the integers from 0 to m_j, thus any alternative transformation must introduce the same number of redundancies.

A different transformation into a Multiple-choice KP is possible, by having for each item type j a class of m_j items, each representing one of the m_j choices of x_j. This transformation however demands $\sum_{j=1}^{n} m_j$ variables, which is at first sight less attractive than the transformation to KP. But due to the simple structure of each class, it is not necessary to store all variables of the class, thus a simplified version of e.g. the mcknap algorithm may be used to solve the problem.

In the following we will present some upper bounds in Section 5.1, while lower bounds obtained through different heuristics and approximation algorithms are discussed in Section 5.2. Section 5.3 shows how the size of an instance may be reduced by applying the well-known reduction rules from KP, as well as specific techniques used for BKP. Branch-and-bound algorithms are considered in Section 5.4, while dynamic programming approaches are outlined in Section 5.5. Fathoming of states in enumerative algorithms is briefly considered in 5.6, while Section 5.7 deals with the solution of large-sized instances. Finally Section 5.8 gives a comparison of the best codes for BKP, considering specialized algorithms for BKP as well as approaches based on a transformation into KP.

5.1 Upper Bounds

The Continuous relaxation of BKP is easily solved by a greedy algorithm: Order the item types according to nonincreasing profit-to-weight ratios

$$\frac{p_1}{w_1} \geq \frac{p_2}{w_2} \geq \ldots \geq \frac{p_n}{w_n} \qquad (123)$$

and define the *break item type* as $b = \min\{j : \sum_{i=1}^{j} w_i m_i > c\}$. Then an optimal solution to $C(\text{BKP})$ is defined as $x_j = m_j$ for $j = 1, \ldots, b-1$, and

$x_j = 0$ for $j = b+1, \ldots, n$, while $x_b = (c - \sum_{j=1}^{b-1} w_j m_j)/w_b$. Thus the Dantzig bound for BKP becomes

$$U_1 = \left\lfloor \sum_{j=1}^{b-1} p_j m_j + \left(c - \sum_{j=1}^{b-1} w_j m_j\right) \frac{p_b}{w_b} \right\rfloor. \tag{124}$$

The break item type b may be found by adapting the bal_zem algorithm presented in Section 2.1, meaning that U_1 can be derived in $O(n)$ time. By truncating the continuous solution to $x_b = 0$ we obtain the *break solution* $\hat{x}$, which has profit sum $\hat{p} = \sum_{j=1}^{b-1} p_j m_j$ and weight sum $\hat{w} = \sum_{j=1}^{b-1} w_j m_j$.

A tighter bound than the Dantzig bound has been derived by Martello and Toth [73] by generalizing the results in Section 2.1.2. Let $\alpha = \lfloor (c - \hat{w})/w_b \rfloor$ be the number of items of type b which additionally fit into the break solution. Then the *Martello and Toth upper bound* is based on the fact that in an integer-optimal solution either $x_b \leq \alpha$ or $x_b \geq \alpha + 1$. A valid upper bound for BKP with the first constraint added is

$$U' = \left\lfloor \hat{p} + \alpha p_b + \left(c - \hat{w} - \alpha w_b\right) \frac{p_{b+1}}{w_{b+1}} \right\rfloor, \tag{125}$$

while an upper bound on BKP with constraint $x_b \geq \alpha + 1$ becomes

$$U'' = \left\lfloor \hat{p} + (\alpha+1)p_b + \left(c - \hat{w} - (\alpha+1)w_b\right) \frac{p_{b-1}}{w_{b-1}} \right\rfloor. \tag{126}$$

Hence
$$U_2 = \max\{U', U''\} \tag{127}$$

is an upper bound for BKP. Obviously U_2 satisfies $U_2 \leq U_1$ and thus the Martello and Toth upper bound is tighter than the Dantzig bound [73]. The bound U_2 may be found in $O(n)$ time, using the bal_zem algorithm for finding b.

The continuous solution of BKP in the transformed version to a 0-1 KP produces the same continuous bound U_1. This is however not the case for bound U_2, since U' and U'' are tighter than the corresponding values given by (29) and (30) in the KP version.

Enumerative bounds as presented in Section 2.1.2 have been considered in Pisinger [100]. Let $M \subset N$ be a subset of item types, and $X_M = \{x_j \in \{0, \ldots, m_j\}, j \in M\}$. Then an upper bound on BKP is given by

$$U_M = \max_{\tilde{x} \in X_M} U(\tilde{x}) \tag{128}$$

where $U(\tilde{x})$ is an upper bound on (121) with additional constraint $x_j = \tilde{x}_j$ for $j \in M$.

All the bounds introduced in Section 2.1 for the 0-1 Knapsack Problem can be generalized to obtain upper bounds for BKP. This could be done either in a straightforward way, by applying the formulae in Section 2.1 to BKP in 0-1 form (as was done for U_1) or, better, by exploiting the peculiar nature of the problem.

Maximum cardinality constraints are derived as follows: Assume that the weights are ordered according to nondecreasing weights and let $\beta = \min\{h : \sum_{j=1}^{h} w_j m_j > c\}$. Then setting

$$k = \sum_{j=1}^{\beta-1} m_j + \left\lfloor \frac{c - \sum_{j=1}^{\beta-1} w_j m_j}{w_\beta} \right\rfloor \qquad (129)$$

we may add the additional constraint

$$\sum_{j=1}^{n} x_j \leq k \qquad (130)$$

to the formulation (121) without excluding any integer solutions. An interesting observation is that if an instance of BKP is transformed into a KP defined in p', w', c according to (122), then the corresponding cardinality constraint $\sum_{j=1}^{n'} x'_j \leq k'$ will have $k' \leq k$ since the weights in the KP case are not smaller than those in the BKP case. Adding the cardinality constraint (130) to the BKP however leads to tighter bounds from continuous relaxation, than in the KP case. This can be seen with the instance $p_1 = w_1 = 10$, $p_2 = w_2 = 11$ with bounds $m_1 = m_2 = 3$ and capacity $c = 40$. We find $k = 3$ and thus we add the constraint $\sum_{j=1}^{2} x_j \leq 3$ to the problem, obtaining the continuous solution $z(C(\text{BKP})) = 33$. If we instead transform the instance into a KP, we get $p'_1 = w'_1 = 10$, $p'_2 = w'_2 = 20$, $p'_3 = w'_3 = 11$ and $p'_4 = w'_4 = 22$. Here $k' = 2$, and thus adding the constraint $\sum_{j=1}^{4} x'_j \leq 2$ to the problem gives the continuous solution $z(C(\text{KP})) = 40$.

5.2 Heuristics

The integer solution corresponding to bound U_1 is given by

$$z' = \sum_{j=1}^{b-1} p_j m_j + \left\lfloor \frac{c - \sum_{j=1}^{b-1} w_j m_j}{w_b} \right\rfloor p_b \qquad (131)$$

The absolute error $z^* - z'$ of this heuristic solution is bounded by p_b since $z' \leq z^* \leq U_1 \leq z' + p_b$, but the ratio z'/z can be arbitrarily close to 0 as can be proved by using the same instance as given below (49). The worst-case performance ratio, however, can be improved to $\frac{1}{2}$ by computing

$$z^h = \max\{z', p_b\} . \qquad (132)$$

The same instance as in (49) can be used to prove the performance ratio.

A greedy heuristic for BKP assumes that the items are ordered according to (123). Now repeatedly take the largest amount of each item type $j = 1, \ldots, n$ by setting $\tilde{x}_j = \min\{\lfloor \tilde{c}/w_j \rfloor, m_j\}$ where $\tilde{c} = c - \sum_{i=1}^{j-1} w_i \tilde{x}_i$. The performance of the greedy heuristic may be improved by deriving $\ell = \arg\max_{j=1,\ldots,n}\{p_j m_j\}$ and setting

$$z^g = \max\left\{\sum_{j=1}^n p_j \tilde{x}_j, \; p_\ell m_\ell\right\} . \qquad (133)$$

The relative performance ratio of z^g is $\frac{1}{2}$ which can be proved as in Section 2.2.

Transforming BKP into an equivalent 0-1 problem and then applying any of the *Polynomial-time* or *fully polynomial approximation schemes* of Section 2.8 leads to approximate solutions with worst-case bound as defined by such schemes.

5.3 Reduction

The reductions presented in Section 2.3 are easily generalized to BKP: An item type j may be fathomed if either all the available items of that type have to be included in the knapsack ($j < b$) or if none of the available items of that type can be included in the knapsack ($j \geq b$). Thus let u_j^0 be an upper bound on BKP with additional constraint $x_j \leq m_j - 1$, and u_j^1 be an upper bound with additional constraint $x_j \geq 1$. If $u_j^0 < z + 1$ then we may fix x_j at m_j and similarly if $u_j^1 < z + 1$ then we fix x_j at 0.

Pisinger [100] used a generalization of the Dembo and Hammer bound (57) getting the bounds

$$u_j^0 = \hat{p} - p_j + \left(c - \hat{w} + w_j\right)\frac{p_b}{w_b} \qquad u_j^1 = \hat{p} + p_j + \left(c - \hat{w} - w_j\right)\frac{p_b}{w_b} \qquad (134)$$

where u_j^0 is derived for $j < b$ and u_j^1 is derived for $j \geq b$. Both bounds may be derived in constant time for each item type.

Knapsack Problems

A different kind of reduction was presented in Pisinger [100], where some bounding tests are used to tighten the bound m_j on an item type j. In this way the bound $x_j \in \{0, \ldots, m_j\}$ may be restricted to $x_j \in \{0, \ldots, d_j\}$ when $j \geq b$ or to $x_j \in \{m_j - d_j, \ldots, m_j\}$ when $j < b$. If $d_j = 0$ then the decision variable x_j can only take one value, and thus we may fathom item type j.

It is only fruitful to insert d items of type $j \geq b$ in the knapsack if an upper bound $u_j(d)$ on the BKP with additional constraint $x_j = d$ exceeds the current lower bound z, i.e. if $u_j(d) \geq z + 1$. We use a generalization of the bound by Dembo and Hammer [18] for this purpose, obtaining the test

$$\hat{p} + dp_j + \left(c - \hat{w} - dw_j\right) \frac{p_b}{w_b} \geq z + 1. \tag{135}$$

From this inequality we may obtain the maximum number of an item type, which may be included into the knapsack, as:

$$d_j = \left\lfloor \frac{(z+1-\hat{p})w_b - (c-\hat{w})p_b}{p_j w_b - w_j p_b} \right\rfloor \quad \text{when } j > b, \tag{136}$$

where we set $d_j = m_j$ when $p_j/w_j = p_b/w_b$ or when the right side of equation (136) is larger than m_j. A similar result is obtained for items of type $j < b$, which have to be removed from the knapsack. Here an upper bound on the number of removed items is

$$d_j = \left\lfloor \frac{(z+1-\hat{p})w_b - (c-\hat{w})p_b}{-p_j w_b + w_j p_b} \right\rfloor \quad \text{when } j < b, \tag{137}$$

with the same conventions as for equation (136).

Pisinger [100] also derived a tighter reduction method than the above by using the enumerative bound (128). These bounds are derived in $O(|X_M|)$ and thus are computationally expensive to derive, but they are able to tighten the bound m_j on item type j by about 10% more than the bounds (136) and (137). In most applications the simple and cheaply evaluated bounds are however sufficient.

5.4 Branch-and-bound Algorithms

Martello and Toth [73] adapted the mt1 procedure of Section 2.4 to BKP. The obtained algorithm mtb could be sketched in simplified form as

algorithm mtb($i, \overline{p}, \overline{w}$): boolean;

var improved;
if ($\overline{w} > c$) **then return** false;
if ($\overline{p} > z$) **then** $z \leftarrow \overline{p}$; improved $\leftarrow$ **true**; **else** improved $\leftarrow$ **false**;
if ($i > n$) **or** ($c - \overline{w} < \underline{w}_i$) **or** ($U_2 < z+1$) **then return** improved;
for $a \leftarrow m_i$ **downto** 0 **do**
  if $\text{mtb}(i+1, \overline{p} + ap_i, \overline{w} + aw_i)$ **then** $x_i \leftarrow a$; improved $\leftarrow$ **true**;
return improved;

where the table $\underline{w}$ is given by $\underline{w}_i = \min_{j \geq i} w_j$. All decision variables x_j must be initialized to 0, and the lower bound set to $z = 0$ before calling $\text{mtb}(0,0,0)$. The Martello and Toth bound U_2 is derived by equation (127) restricted to variables $j \geq i$ and with reduced capacity $\tilde{c} = c - \overline{w}$.

Ingargiola and Korsh [61] presented a different branch-and-bound algorithm, using the reductions described in Section 5.3 to a-priori fix some variables at their optimal value. Bulfin, Parker and Shetty [9] presented a branch-and-bound algorithm where penalties were used to improve the bounding phase. According to Aittoniemi [2] these two branch-and-bound algorithms are outperformed by the mtb algorithm.

5.5 Dynamic Programming

Let $f_i(\tilde{c})$, ($0 \leq i \leq n$, $0 \leq \tilde{c} \leq c$) be the optimal solution value to the following subproblem of BKP, defined on the first i variables of the problem, and with capacity restricted to $\tilde{c}$

$$f_i(\tilde{c}) = \max\left\{ \sum_{j=1}^{i} p_j x_j : \sum_{j=1}^{i} w_j x_j \leq \tilde{c};\ x_j \in \{0, \ldots, m_j\},\ j = 1, \ldots, i \right\}. \tag{138}$$

Generalizing the results by Bellman [8] for the 0-1 Knapsack Problem, we obtain the following recursion

$$f_i(\tilde{c}) = \max \begin{cases} f_{i-1}(\tilde{c}) & \text{if } \tilde{c} \geq 0 \\ f_{i-1}(\tilde{c} - w_i) + p_i & \text{if } \tilde{c} - w_i \geq 0 \\ \vdots \\ f_{i-1}(\tilde{c} - w_i m_i) + p_i m_i & \text{if } \tilde{c} - w_i m_i \geq 0 \end{cases} \tag{139}$$

setting $f_0(\tilde{c}) = 0$ for $\tilde{c} = 0, \ldots, c$. Thus the Bellman recursion has time complexity $O(c \sum_{j=1}^{n} m_j)$ and space complexity $O(nc)$. Dynamic programming algorithms based on this recursion have been presented by Gilmore and Gomory [44] and Nemhauser and Ullmann [91].

In Section 2.5.1 we saw that a primal-dual dynamic programming algorithm for the 0-1 Knapsack Problem in general is more efficient than the Bellman recursion. A primal-dual algorithm for BKP assumes that the items are ordered according to nonincreasing efficiencies, and $f_{s,t}(\tilde{c})$, ($s \leq b$, $t \geq b-1$, $0 \leq \tilde{c} \leq 2c$) is the optimal solution value to the problem:

$$f_{s,t}(\tilde{c}) = \max \left\{ \begin{array}{l} \sum_{j=1}^{s-1} p_j m_j + \sum_{j=s}^{t} p_j x_j : \sum_{j=1}^{s-1} w_j m_j + \sum_{j=s}^{t} w_j x_j \leq \tilde{c}; \\ x_j \in \{0, \ldots, m_j\}, \; j = s, \ldots, t \end{array} \right\}. \tag{140}$$

We get the recursion

$$f_{s,t}(\tilde{c}) = \max \begin{cases} f_{s,t-1}(\tilde{c}) & \text{if } t \geq b \\ f_{s,t-1}(\tilde{c} - w_t) + p_t & \text{if } t \geq b, \; \tilde{c} - w_t \geq 0 \\ \;\;\vdots \\ f_{s,t-1}(\tilde{c} - w_t m_t) + p_t m_t & \text{if } t \geq b, \; \tilde{c} - w_t m_t \geq 0 \\ f_{s+1,t}(\tilde{c}) & \text{if } s < b \\ f_{s+1,t}(\tilde{c} + w_s) - p_s & \text{if } s < b, \; \tilde{c} + w_s \leq 2c \\ \;\;\vdots \\ f_{s+1,t}(\tilde{c} + w_s m_s) - p_s m_s & \text{if } s < b, \; \tilde{c} + w_s m_s \leq 2c \;. \end{cases} \tag{141}$$

If $\hat{p}$ and $\hat{w}$ are the profit and weight sums of the break solution, we may initially set $f_{b,b-1}(\tilde{c}) = \hat{p}$ for $\tilde{c} = \hat{w}, \ldots, 2c$ and $f_{b,b-1}(\tilde{c}) = -\infty$ for $\tilde{c} = 0, \ldots, \hat{w}-1$. Thus the enumeration starts at $(s,t) = (b, b-1)$ and continues by either removing some items of type s from the knapsack, or inserting some items of type t in the knapsack. An optimal solution to BKP is found as $f_{1,n}(c)$.

As for 0-1 Knapsack Problems, it is convenient to represent each state by the corresponding profit and weight sums (π, μ) where $\pi = f_{s,t}(\mu)$. Any state (π, μ) with $\mu > c + \sum_{j=1}^{s-1} w_j m_j$, may be fathomed, since even if we removed all items of types $j < s$ in the forthcoming iterations, the state would never become a feasible solution. This observation implies that no states with weights $\mu \geq 2c$ can occur, giving the algorithm space complexity $O(n2c) = O(nc)$.

An efficient iterative version of recursion (141) is obtained by applying the transformation (122) for the current item type, such that each insertion/removal of an item type results in $\lfloor \log_2 m_j + 1 \rfloor$ mergings of length $O(c)$ (see Pisinger [105] for details on the merging). This means that the time complexity for the dynamic programming is $O(c \sum_{j=1}^{n} \lfloor \log_2 m_j + 1 \rfloor)$, i.e. $O(nc \log c)$ in the worst case. For moderate core sizes $|C| = t - s + 1$ a tighter bound is obtained, as only $O(|C|c \log c)$ steps are necessary. Moreover by using dynamic programming by *reaching*, at most $O(m_s \cdot m_{s+1} \cdots m_t)$ states should be considered. Thus we obtain the time bound

$$O\left(\min\left\{\prod_{j=s}^{t} m_j, |C|c\log c\right\}\right), \tag{142}$$

on the enumeration of a core $C = [s, t]$.

A dynamic programming algorithm based on balancing has been presented in Pisinger [99]. The algorithm is based on transformation to KP, and since the solution times of balanced algorithms are bounded in the magnitude of the coefficients, a linear transformation is used instead of the binary transformation (122). Thus each item type j in BKP is replaced by m_j individual items in the KP case. The obtained solution times are $O(r_1 r_2 \sum_{j=1}^{n} m_j)$ where $r_1 = \max w_j$ and r_2 is a constant satisfying $r_2 \leq \max p_j$. Since the capacity c in the Bellman recursion is bounded by $c \leq \sum_{j=1}^{n} w_j m_j$, balancing becomes attractive when $r_1 r_2 \ll \sum_{j=1}^{n} w_j m_j$ and no bounding rules are able to terminate the process before complete enumeration.

Concerning the solution vector corresponding to an optimal solution value, the same principles as described in Section 2.5.3 may be applied to the above recursions.

5.6 Reduction of States

During dynamic programming or branch-and-bound it is necessary to derive upper bounds in an attempt to fathom the current node or state. Bounds can obviously be derived as described in Section 5.1, but specialized methods may often improve the performance.

Assume e.g. that items $[s,t]$, have been enumerated in (141), and that a given state i has the profit and weight sum (π, μ). We may then fathom the state if $u(\pi, \mu) < z + 1$ where the upper bound u is obtained by relaxing the constraints on x_{s-1} and x_{t+1} to $x_{s-1} \geq 0$ and $x_{t+1} \geq 0$, yielding:

$$u(\pi, \mu) = \begin{cases} \pi + (c-\mu)p_{t+1}/w_{t+1} & \text{if } \mu_i \leq c, \\ \pi + (c-\mu)p_{s-1}/w_{s-1} & \text{if } \mu_i > c. \end{cases} \tag{143}$$

This bound may also be used for deriving a global upper bound on (121) by applying u in equation (128).

5.7 Solution of Large-sized Problems

A core of the Bounded Knapsack Problem can be derived by adapting the principles described in Section 2.6. The `findcore` algorithm in Section 2.6.1 can be generalized in a straightforward way, returning a fixed-size core $C = [s, t] = [b - \delta, b + \delta]$. The core may then be solved using a specialized algorithm like `mtb`, or the core may be transformed into a 0-1 Knapsack Problem and then solved using one of the algorithms described in Section 2.4 or 2.5. If optimality of the solution can be proved, an optimal solution to the original problem is found by setting variables x_j with $j < s$ to m_j, and variables with $j > t$ to 0. If optimality cannot be proved, then the reduction algorithm in Section 5.5 is used, and the remaining problem is solved through enumeration. No algorithms have however been presented based on this approach.

Martello and Toth [80] presented an algorithm `mtb2` for BKP which transforms the instance into an equivalent 0-1 KP. The transformed instance is then solved through the `mt2` algorithm, which is designed for large-sized instances of KP.

Pisinger [100] presented an expanding-core algorithm for BKP, adapting the principles of Section 2.6.3. The `bouknap` algorithm is based on dynamic programming and can be sketched as follows:

algorithm bouknap

1. Find core $C = [b, b]$ through an algorithm similar to `findcore`, pushing all discarded intervals into two stacks H and L.

2. Run the dynamic programming recursion (141) alternately inserting an item type t or removing an item type s. Unpromising states with $u < z + 1$ are deleted, where u is obtained from (143).

3. Before an item type is added to the core, check if s, t is within the current core C, otherwise pick the next interval from H or L, and reduce the item types using (134). Those items which could not be fixed at their optimal values are sorted, and finally added to the core.

4. Before an item type is enumerated in the dynamic programming, tests (136) and (137) are used to tighten the bound on the item type.

5. The process stops when all states in the dynamic programming have been fathomed due to an upper bound test, or all item types $j \notin C$ have been fixed at their optimal value.

As in Section 2.6.3 it can be proved that the core, bouknap enumerates, is minimal.

5.8 Computational Experiments

In order to compare specialized appraoches for BKP with those based on a transformation to KP, we have chosen to focus on the algorithms bouknap and mtb2. The latter was tested in two versions: The published code mtb2 which solves the KP through algorithm mt2, and a new technique which solves KP through the mthard code. Using Martello and Toth's naming tradition we could call the last approach mtbhard.

The following data instances are considered: *Uncorrelated data instances*: p_j and w_j are randomly distributed in $[1, R]$. *Weakly correlated data instances*: w_j randomly distributed in $[1, R]$ and p_j randomly distributed in $[w_j - R/10, w_j + R/10]$ such that $p_j \geq 1$. *Strongly correlated data instances*: w_j randomly distributed in $[1, R]$ and $p_j = w_j + 10$. *Subset-sum data instances*: w_j randomly distributed in $[1, R]$ and $p_j = w_j$. The data range R will be tested with values $R = 100, 1000$ and $10\,000$, while the bounds m_j are randomly distributed in $[5, 10]$.

Each problem type is tested with a series of 200 instances, such that the capacity c of the knapsack varies from 1% to 99% of the total weight sum of the items. This approach takes into account that the computational times may depend on the chosen capacity. Since mtb2 transforms the BKP into an equivalent 0-1 KP, instances larger than $n = 50\,000$ cannot be solved due to memory limitations. Similarly the mtbhard code cannot solve instances larger than $n = 10\,000$. In the following tables a dash means that the 200 instances could not be solved in a total of 10 hours. The results have been achieved on a HP9000/735 computer.

The computational times are presented in Table 9. The mtb2 algorithm has substantial stability problems for low-ranged data instances even of relatively small size. Subset-sum data instances also show a few anomalous occurrences. The mtbhard algorithm generally has a more stable behavior, but it is interesting to note, that although mthard is excellent at solving strongly correlated KP instances, it is not able to solve strongly corrrelated instances when these are transformed from BKP into KP. But as mentioned in Section 5.1 the effect of cardinality bounds is weakened by the transformation. Finally bouknap has a very stable behavior, solving most instances within a fraction of a second. The strongly correlated instances demand more computational time, but the effort increases linearly with the problem

Table 9: Computing times in seconds, bouknap (top), mtb2 (mid) and mtbhard (bottom). Averages of 200 instances

$n \backslash R$	Uncorrelated			Weakly corr.			Strongly corr.			Subset-sum		
	100	1000	10000	100	1000	10000	100	1000	10000	100	1000	10000
100	0.00	0.00	0.00	0.00	0.00	0.00	0.02	0.16	1.55	0.00	0.02	0.27
300	0.00	0.00	0.00	0.00	0.01	0.01	0.06	0.62	7.36	0.00	0.02	0.27
1000	0.00	0.01	0.01	0.00	0.01	0.03	0.22	2.24	25.70	0.00	0.02	0.28
3000	0.00	0.01	0.02	0.01	0.01	0.09	0.81	9.37	109.67	0.01	0.03	0.30
10000	0.01	0.02	0.06	0.02	0.03	0.16	3.03	39.71	—	0.02	0.04	0.30
30000	0.04	0.05	0.12	0.08	0.08	0.20	13.59	116.96	—	0.05	0.06	0.31
100000	0.20	0.16	0.27	0.40	0.26	0.37	107.90	450.46	—	0.17	0.19	0.45
100	0.00	0.00	0.00	0.00	0.01	0.02	—	—	—	0.00	0.02	0.15
300	0.01	0.01	0.01	0.00	0.03	0.06	—	—	—	0.00	—	40.96
1000	0.01	0.03	0.04	0.01	0.05	0.26	—	—	—	0.00	0.01	0.09
3000	0.02	0.05	0.14	62.58	0.04	0.57	—	—	—	0.01	0.02	0.10
10000	14.72	0.11	0.43	—	0.17	0.98	—	—	—	0.05	0.06	0.14
30000	—	0.31	1.09	—	—	1.04	—	—	—	0.18	0.19	0.26
100	0.00	0.00	0.00	0.00	0.01	0.02	0.62	—	—	0.00	0.01	0.14
300	0.00	0.01	0.01	0.00	0.01	0.02	—	—	—	0.00	0.01	0.47
1000	0.01	0.03	0.03	0.01	0.03	0.17	—	—	—	0.00	0.02	3.32
3000	0.02	0.06	0.09	0.05	0.07	0.51	—	—	—	0.01	0.03	15.01
10000	0.11	0.22	0.43	0.22	0.19	1.05	—	—	—	0.05	0.13	39.15

size n and the data range R.

Thus the computational experiments indicate that specialized approaches for BKP are more efficient than those based on a transformation to KP, since tighter upper bounds can be derived, and specialized reductions can be applied as described in Section 5.3.

6 Unbounded Knapsack Problem

The *Unbounded Knapsack Problem* (*UKP*) is a special case of the Bounded Knapsack Problem, where there is an unlimited availability of each item type. Thus assume that n given item types are characterized by their *profit* p_j and *weight* w_j, and we want to choose a number x_j of each type j, such that the chosen profit sum is the largest possible, but such that the weight sum does not exceed a given *capacity* c. Thus the problem may be formulated as

$$\text{maximize} \quad \sum_{j=1}^{n} p_j x_j$$
$$\text{subject to} \quad \sum_{j=1}^{n} w_j x_j \leq c, \qquad (144)$$
$$x_j \geq 0, \text{ integer}, \quad j = 1, \ldots, n.$$

The problem has several applications in financial management, cargo loading and cutting stock, and it also appears as surrogate relaxation of IP problems with nonnegative coefficients. The Unbounded Knapsack Problem may be transformed into a Bounded Knapsack Problem by imposing the constraint $x_j \leq \lfloor c/w_j \rfloor$ for each variable considered, but according to Martello and Toth [80], algorithms for the BKP perform rather poorly for instances of this kind.

The problem is $\mathcal{NP}$-hard, as proved in Lueker [71] by transformation from SSP. However, it can be solved in polynomial time in the $n = 2$ case as proved by Hirschberg and Wong [51], and Kannan [64]. Notice that the result is not trivial, since a naive algorithm, testing $x_1 = i$, $x_2 = \lfloor (c - iw_1)/w_2 \rfloor$ for i taking on integer values from 0 to $\lfloor c/w_1 \rfloor$, would require a time $O(c)$, exponential in the input length.

Despite the fact that UKP is closely related to BKP, several unique techniques can be used to solve the unbounded case. Thus in Section 6.1 we present some upper bounds for UKP, while lower bounds are considered in Section 6.2. Exact solution methods are presented in the following two sections, first dealing with dynamic programming in Section 6.3, then branch-and-bound in Section 6.4. Reduction algorithms are considered in Section 6.5, concerning both reductions known from other Knapsack Problems, and also some reductions which are unique for UKP. Section 6.6 deals with the solution of large-sized problems, and we conclude the presentation of UKP in Section 6.7 by comparing the performance of some of the most efficient algorithms.

6.1 Upper Bounds

As usual assume that the item types are ordered according to

$$\frac{p_1}{w_1} \geq \frac{p_2}{w_2} \geq \ldots \geq \frac{p_n}{w_n} \qquad (145)$$

breaking ties such that $w_j \leq w_{j+1}$ when $p_j/w_j = p_{j+1}/w_{j+1}$. The continuous relaxation of UKP has the optimal solution $x_1 = c/w_1$ and $x_j = 0$ for $j = 2, \ldots, n$. This yields the trivial upper bound $U_0 = \lfloor c/w_1 \rfloor p_1$.

Since $x_1 \leq m = \lfloor c/w_1 \rfloor$ in any integer solution, imposing this constraint to UKP we get the continuous solution:

$$x_1 = m \qquad x_2 = r/w_2 \qquad x_j = 0, \ j = 3, \ldots, n \ . \qquad (146)$$

where the residual capacity is $r = c - mw_1$. We get the improved upper bound

$$U_1 = mp_1 + \left\lfloor r\frac{p_2}{w_2} \right\rfloor \qquad (147)$$

which is a counterpart to the Dantzig bound in Section 5.1. Note that the break item is always $b = 2$ and thus the profit and weight sum of the break solution is

$$\hat{p} = mp_1 + \lfloor r/w_2 \rfloor p_2 \qquad \hat{w} = mw_1 + \lfloor r/w_2 \rfloor w_2 \qquad (148)$$

Setting $\alpha = \lfloor r/w_2 \rfloor$ and noting that in every optimal solution either $x_2 \leq \alpha$ or $x_2 \geq \alpha + 1$, we get a counterpart to the Martello and Toth upper bound (31) as

$$U_2 = \max\{U', U''\} \qquad (149)$$

where

$$U' = \left\lfloor mp_1 + \alpha p_2 + (r - \alpha w_2)\frac{p_3}{w_3} \right\rfloor \qquad (150)$$

and
$$U'' = \left\lfloor mp_1 + (\alpha+1)p_2 + \left(r - (\alpha+1)w_2\right)\frac{p_1}{w_1}\right\rfloor. \quad (151)$$

In UKP we can exploit the fact that $b = 2$ to obtain a better bound. Since U'' is a bound on UKP with constraint $x_2 \geq \alpha+1$, this can only be obtained if at least $\beta = \lceil(r-(\alpha+1)w_2)/w_1\rceil$ items of type 1 are removed, and thus $r - (\alpha+1)w_2 + \beta w_1$ units of capacity are available for the items of type 2. Hence, a valid upper bound can be obtained by replacing U'' with

$$\overline{U}'' = \left\lfloor mp_1 + (\alpha+1)p_2 - \beta p_1 + \left(r - (\alpha+1)w_2 + \beta w_1\right)\frac{p_2}{w_2}\right\rfloor. \quad (152)$$

Furthermore, $\overline{U}'' \leq U''$ since we are "moving" some capacity units from items of type 1 to the less efficient items of type 2. Thus we have

Theorem 6.1 (Martello and Toth [81]) $U_3 = \max\{U', \overline{U}''\}$ *is an upper bound for UKP and,* $U_3 \leq U_2$.

The time complexity for the computation of U_0, U_1, U_2 and U_3 is $O(n)$, since only the three largest ratios p_j/w_j are needed.

The worst-case performance ratio of all four bounds is $\rho(U_0) = \rho(U_1) = \rho(U_2) = \rho(U_3) = 2$, since $U_3 \leq U_2 \leq U_1 \leq U_0 \leq \hat{p} + p_1 \leq 2z$. and the tightness is shown by considering a series of problems with $n = 3$, $p_j = w_j = k$ for all j, and $c = 2k - 1$. Here we get $U_0 = U_1 = U_2 = U_0 = 2k - 1$, and $z^* = k$, so the ratio U/z can be arbitrarily close to 2 for sufficiently large k.

6.2 Heuristics

The break solution of value $z' = \hat{p}$, is an obvious heuristic solution. But where the break solution for KP and BKP can provide an arbitrarily bad approximation of z^*, the situation is different for UKP, as we have $z'/z^* \geq \frac{1}{2}$. The proof is immediate by observing that $z^* - z' \leq p_1$ and from the assumption $w_j \leq c$ we get $z' \geq p_1$. The series of problems with $n = 2$, $p_1 = w_1 = k+1$, $p_w = w_2 = k$ and $c = 2k$ shows that $\frac{1}{2}$ is tight since $z'/z^* = (k+1)/(2k)$ and this ratio can be arbitrarily close to $\frac{1}{2}$ for sufficiently large k. The same property holds for the simpler heuristic $z'' = \lfloor c/w_1 \rfloor p_1$.

A counterpart to the greedy heuristic (133) thus gets the simpler form: Repeatedly take the largest amount of each item type $j = 1, \ldots, n$ by setting

$\tilde{x}_j = \lfloor \tilde{c}/w_j \rfloor$ where $\tilde{c} = c - \sum_{i=1}^{j-1} w_i \tilde{x}_i$. Thus

$$z^g = \sum_{j=1}^{n} p_j \tilde{x}_j \qquad (153)$$

UKP accepts a fully polynomial approximation scheme as shown by e.g. Ibarra and Kim [58].

6.3 Dynamic Programming

An immediate adaptation of the BKP recursion from Section 5.5 yields

$$f_i(\tilde{c}) = \max \left\{ f_{i-1}(\tilde{c} - \alpha w_i) + \alpha p_i : \alpha = 0, \ldots, \lfloor \tilde{c}/w_i \rfloor; \ \tilde{c} - \alpha w_i \geq 0 \right\} \qquad (154)$$

with $f_0(\tilde{c}) = 0$ for $\tilde{c} = 0, \ldots, c$. Since c/w_i may be of magnitude $O(c)$, the time complexity for determining $z = f_n(c)$ is $O(nc^2)$. Gilmore and Gomory [43] derived a better recursion as

$$f_i(\tilde{c}) = \max \begin{cases} f_{i-1}(\tilde{c}) & \text{if } \tilde{c} \geq 0 \\ f_i(\tilde{c} - w_i) + p_i & \text{if } \tilde{c} - w_i \geq 0 \end{cases} \qquad (155)$$

which reduces the time complexity to $O(nc)$. Specialized dynamic programming algorithms have been presented by e.g. Greenberg and Feldman [49], Greenberg [47, 48] to mention some of the more recent ones.

6.4 Branch-and-bound

Several branch-and-bound algorithms for UKP have been proposed, but Martello and Toth [73] showed that their algorithm mtu is the most efficient. Assuming that the items are sorted according to (145), let $\overline{p}$ and $\overline{w}$ be the profit and weight sum of the currently chosen items. Then the mtu algorithm can be sketched as:

algorithm mtu($i, \overline{p}, \overline{w}$): boolean;
var improved;
if ($\overline{w} > c$) **then return false**;
if ($\overline{p} > z$) **then** $z \leftarrow \overline{p}$; improved $\leftarrow$ **true**; **else** improved $\leftarrow$ **false**;
if ($i > n$) **or** ($c - \overline{w} < \underline{w}_i$) **or** ($\overline{p} + (c - \overline{w})p_i/w_i < z + 1$) **then return**
improved;
$m = \lfloor (c - \overline{w})/w_i \rfloor$;

```
for a ← m downto 0 do
    if mtb(i + 1, p̄ + ap_i, w̄ + aw_i) then x_i ← a; improved ← true;
return improved;
```

All decision variables x_j must be initialized to 0, and the lower bound set to $z = 0$ before calling mtu(0,0,0). The table $\underline{w}_j$ of minimal weights is initialized as $\underline{w}_j = \min_{i \geq j} w_i$.

6.5 Reduction Algorithms

An efficient way of solving UKP is to first apply some dominance relations to fathom unpromising item types, and then to solve the remaining problem through enumerative techniques. The general form of dominance (Pisinger [98]) is defined as follows

Definition 6.1 *Given an item type j, and a set of item types $i_1, \ldots, i_d$ where $i_k \neq j$ for $k = 1, \ldots, d$, but where the indices $\{i_k\}$ are not necessarily distinct. If*

$$p_{i_1} + \ldots + p_{i_d} \geq p_j \qquad w_{i_1} + \ldots + w_{i_d} \leq w_j \qquad (156)$$

then item type j is said to be dominated by item types $i_1, \ldots, i_d$.

Obviously a dominated item type j may be fathomed from the problem, as any optimal solution with $x_j > 0$ may be replaced by a solution where x_j items of types $i_1, \ldots, i_d$ are chosen instead.

A complete testing for dominated items may be performed in pseudo-polynomial time through dynamic programming, where recursion (155) is used to enumerate the left-hand sides in (156). Thus assume that the items are ordered according to (145). Now, running through the items $j = 1, \ldots, n$ recursion (155) is used to enumerate the states $f_j(\tilde{c})$. Before considering an item $j \geq 2$ let $\pi = f_{j-1}(w_j)$. If $\pi \geq p_j$ then item j is dominated by an integer combination of some items $i_1, \ldots, i_m < j$, and thus item j may be fathomed. Since we do not need to consider states with weight larger than $r = \max w_j$ in the reduction, the time complexity of this approach is $O(n + mr)$ where m is the number of undominated items.

Martello and Toth [81] consider a computationally cheaper dominance test, which assumes that the indices $i_i, \ldots, i_k$ correspond to the same item, say i, in equation (156). The tightest choice of d in this case is $d = \lfloor w_j/w_i \rfloor$ as this results in the largest profit sum not exceeding the weight w_j.

Theorem 6.2 (Martello and Toth [81]) *For two given item types i, j where $i \neq j$, we say that type j is dominated by type i if*

$$\left\lfloor \frac{w_j}{w_i} \right\rfloor p_i \geq p_j. \tag{157}$$

An optimal solution exists where $x_j = 0$ for a dominated item type j and thus item type j may be fathomed.

Martello and Toth gave a simple algorithm for reducing the dominated items by first sorting the items according to (145). Then an $O(n^2)$ algorithm is performed where each item type i attempts to dominate item types $j > i$. Dudziński [22] improved this reduction to $O(n \log n + mn)$ where m is the number of undominated items. Finally Pisinger [98] improved the reduction to $O(n \log n + \min\{mn, n\overline{w}/\underline{w}\})$ where $\overline{w} = \max w_j$ and $\underline{w} = \min w_j$.

6.5.1 Reductions from Bounds

The reductions presented in the previous section are unique to UKP, since they are based on the fact that any number of items can be chosen for each dominating subset. More common reduction techniques can however also be used for this problem. Thus, for $j \geq 3$ let u_j^1 be an upper bound for (144) with additional constraint $x_j \geq 1$. If $u_j^1 < z + 1$ then we may fix x_j at 0 and thus fathom the item type j.

Bounds derived in constant time can be obtained from a generalization of the Dembo and Hammer bound (57) getting

$$u_j^1 = \hat{p} + p_j + \left(c - \hat{w} - w_j\right) \frac{p_2}{w_2} \tag{158}$$

Martello and Toth [81] proposed using the tighter bound U_1 in the reduction. By adding the constraint $x_j \geq 1$ the capacity decreases to $\tilde{c} = c - w_j$ but the break item is still $b = 2$, thus U_1 can be derived in constant time. If it is not possible to fathom an item type by using bound U_1 then Martello and Toth proposed to use U_3 instead.

In Section 5.3 we saw that upper bound tests may be used to tighten the bound on an item type. Thus we may impose the bound on each item

$$m_j = \min\left\{ \left\lfloor \frac{c}{w_j} \right\rfloor, \left\lfloor \frac{(z+1-\hat{p})w_2 - (c-\hat{w})p_2}{p_j w_2 - w_j p_2} \right\rfloor \right\} \tag{159}$$

obtained directly from (136). By imposing the constraints $x_j \leq m_j$ on all item types, we get an ordinary BKP.

6.6 Solution of Large-sized Instances

The concept of core problem described in Section 2.6 can be directly extended to UKB by recalling that, in this case, the break item type b is always the second one. Hence a core may be defined as a collection of the δ first item types according to ordering (145).

The core $C = [1, \delta]$ can be derived in $O(n)$ time by a straightforward adaptation of the findcore algorithm described in Section 2.6.1: The break item type may initially be set to $b = 2$ and thus no weight sums need be maintained. Left intervals $[s, i - 1]$ are always partitioned further, while right intervals $[i, t]$ are only partitioned when $i \leq \delta$. Martello and Toth [81] experimentally found that $\delta' = \max\{100, \lfloor n/100 \rfloor\}$ is a good initial core size.

The only algorithm published for large sized UKP is the mtu2 algorithm by Martello and Toth [81]. It may be outlined as follows:

algorithm mtu2

1. Set $\delta = \delta'$.

2. Determine a core $C = [1, \delta]$ using a modified findcore algorithm.

3. Sort the item types in the core and use dominance rules (157) for all items in C to fathom unpromising item types.

4. Exactly solve the reduced core problem using algorithm mtu. If the obtained solution z equals the bound U_3 stop.

5. Reduce items $j \notin C$ by applying the reduction rules from previous section.

6. If all items $j \notin C$ were fathomed, then stop. Otherwise increase the core size to $\delta \leftarrow \delta + \delta'$ and go to Step 2.

In Step 2, the core C can also be defined in $O(n)$ time by looking for the δ'th largest element in a set (see e.g. Fischetti and Martello [34]).

6.7 Computational Experiments

We compare the solution times of the specialized algorithm mtu2 with those of the general algorithm bouknap designed for Bounded Knapsack Problems. The latter was run with trivial bounds on each item type $m_j = \lfloor c/w_j \rfloor$. Both

Table 10: Computing times in Seconds as average of 100 instances

R_1	n	Uncorrelated		Weakly corr.		Strongly corr.		Subset sum	
		mtu2	bouknap	mtu2	bouknap	mtu2	bouknap	mtu2	bouknap
10	100	0.00	0.00	0.00	0.00	0.00	0.00	0.01	0.09
10	300	0.00	0.00	0.00	0.00	0.00	0.01	0.01	0.31
10	1000	0.00	0.01	0.00	0.02	0.00	0.19	0.01	1.21
10	3000	0.01	0.02	0.01	0.02	0.01	2.63	0.01	−*
10	10000	0.02	0.08	0.02	0.05	0.02	37.87	0.02	−*
10	30000	0.07	0.16	0.07	−*	0.07	−*	0.07	−*
50	100	0.00	0.00	0.00	0.00	0.00	0.00	0.01	0.08
50	300	0.00	0.00	0.00	0.00	0.00	0.01	0.01	0.32
50	1000	0.00	0.00	0.00	0.00	0.01	0.05	0.01	1.47
50	3000	0.01	0.01	0.01	0.01	0.13	0.47	0.01	−*
50	10000	0.03	0.03	0.03	0.22	0.20	7.31	0.02	−*
50	30000	0.10	−*	0.10	−*	0.25	71.66	0.08	−*
100	100	0.00	0.00	0.00	0.00	0.00	0.01	0.01	0.12
100	300	0.00	0.00	0.00	0.01	0.01	0.06	0.01	0.42
100	1000	0.00	0.00	0.00	0.01	2.20	0.17	0.01	1.54
100	3000	0.01	0.01	0.01	0.04	—	0.51	0.01	6.77
100	10000	0.03	0.04	0.03	−*	—	4.86	0.02	−*
100	30000	0.12	0.13	0.16	−*	—	41.87	0.08	−*
500	100	0.00	0.00	0.00	0.00	—	0.16	0.01	0.13
500	300	0.00	0.01	0.00	0.02	—	0.72	0.01	0.63
500	1000	0.01	0.04	0.04	0.08	—	3.30	0.01	2.01
500	3000	0.04	0.31	15.35	0.77	—	13.11	0.01	7.04
500	10000	0.95	0.55	—	1.06	—	−*	0.02	−*
500	30000	4.22	−*	—	−*	—	−*	0.09	−*

*Insufficient space

algorithms were tested with a number of randomly generated instances as follows: The weights w_j are randomly distributed in $[R_1, R_2]$, while the distribution of the profits depend on the problem type. *Uncorrelated data instances*: p_j are randomly distributed in $[1, R_2]$. *Weakly correlated data instances*: p_j is randomly distributed in $[w_j - 100, w_j + 100]$ such that $p_j \geq 1$. *Strongly correlated data instances*: $p_j = w_j + 100$. *Subset-sum data instances*: $p_j = w_j$.

We consider instances with $R_2 = 1000$ and R_1 having values $10, 50, 100$ and 500. The capacity was chosen as $c = h/(H+1) \sum_{j=1}^{n} w_j$ for instance h in a series of H. Table 10 gives the computational times of mtu2 and bouknap. The results have been obtained on a HP9000/735, and a maximum time limit of 1 hour was assigned to each series of 100 instances.

It can be seen that mtu2 is able to quickly solve all instances for R_1 small, while when R_1 becomes larger it starts to have substantial problems for weakly and strongly correlated instances. This may be due to the fact that the reduction (157) for small values of R_1 is very effective, leaving only a few items. When R_1 becomes large, it is not possible to fathom so many items, and thus the enumerative part gets more difficult. Notice especially that for the strongly correlated instances, the core $[1, \delta]$ described in Section 6.6 will only contain items with weight close to R_1. Thus the enumeration gets stuck in replacing items of similar weight. The same happens in large-sized weakly correlated instances, as the ordering according to profit-to-weight ratios will also choose the smallest weights first.

The bouknap algorithm has generally worse computational times than mtu2, and in many cases it runs out of space in the dynamic programming part. This is due to the fact, that at least c/w_1 states will be enumerated already after considering the first item. But bouknap is able to solve some of the difficult weakly and strongly correlated problems, that mtu2 was not able to solve, which somehow confirms that dymamic programming algorithms are more tolerant to instances where many items have similar weights.

7 Multiple Knapsack Problem

We consider the problem where n given *items* should be packed in m knapsacks of distinct *capacities* c_i, $i = 1, \ldots, m$. Each item j has an associated *profit* p_j and a *weight* w_j, and the problem is to select m disjoint subsets of items, such that subset i fits into knapsack i and the total profit of the selected items is maximized. Thus the *0-1 Multiple Knapsack Problem (MKP)* may be defined as the optimization problem

$$\begin{aligned}
\text{maximize} \quad & \sum_{i=1}^{m} \sum_{j=1}^{n} p_j x_{ij} \\
\text{subject to} \quad & \sum_{j=1}^{n} w_j x_{ij} \leq c_i, \quad i = 1, \ldots, m, \\
& \sum_{i=1}^{m} x_{ij} \leq 1, \quad j = 1, \ldots, n, \\
& x_{ij} \in \{0, 1\}, \quad i = 1, \ldots, m, \; j = 1, \ldots, n,
\end{aligned} \quad (160)$$

where $x_{ij} = 1$ if item i is assigned to knapsack j and $x_{ij} = 0$ otherwise. It is usual to assume that the coefficients p_j, w_j and c_i are positive integers, as fractional values are handled by multiplying through by a proper factor, and items with $p_j \leq 0$ as well as knapsacks with $c_i \leq 0$ may be eliminated.

There is however no easy way of transforming an instance so as to handle negative weights, as described in Section 2.

To avoid trivial cases it is assumed that $w_j \leq \max_{i=1,\ldots,m} c_i$ for $j = 1, \ldots, n$, to ensure that item j fits into at least one knapsack as otherwise it may be removed from the problem. Also $c_i \geq \min_{j=1,\ldots,n} w_j$ for $i = 1, \ldots, m$ may be assumed, as any knapsack where no items fit into it, may be discounted. Finally it may be assumed that $\sum_{j=1}^{n} w_j > c_i$ for $i = 1, \ldots, m$ to avoid a trivial solution where all items fit into one of the knapsacks.

There are several applications for MKP, as the problem directly reflects a situation of loading m ships/containers or e.g. packing m envelopes. Martello and Toth [80] also proposed the problem used for deciding how to load m liquids into n tanks, when the liquids may not be mixed.

Several branch-and-bound algorithms for MKP have been presented during the last two decades, among which we should mention Hung and Fisk [55], Martello and Toth [75], Pisinger [101], Neebe and Dannenbring [90], and Christofides, Mingozzi and Toth [13]. The first three are best-suited for problems where several items are filled into each knapsack, while the last two are designed for problems with many knapsacks and few items. In the following we will mainly focus on the first kind of problems where the ratio n/m is assumed to be large.

In Section 7.1 we will consider different upper bounds for MKP, and see how these may be tightened by some polyhedral properties in Section 7.2. Reduction algorithms, attempting to fix some variables at their optimal values, are described in Section 7.3. Heuristics and approximate algorithms are considered in the following section, while Sections 7.5 and 7.6 deal with exact solution methods. A comparison of the most effective algorithms for MKP is finally presented in Section 7.7.

7.1 Upper Bounds

Different upper bounds for MKP may be obtained by either surrogate, Lagrangian or continuous relaxation. We will first consider *surrogate relaxation*. Let $\lambda_1, \ldots, \lambda_m$ be some nonnegative multipliers, then the surrogate relaxed problem $S(\text{MKP}, \lambda_i)$ becomes:

$$\begin{aligned}
\text{maximize} \quad & \sum_{i=1}^{m} \sum_{j=1}^{n} p_j x_{ij} \\
\text{subject to} \quad & \sum_{i=1}^{m} \lambda_i \sum_{j=1}^{n} w_j x_{ij} \leq \sum_{i=1}^{m} \lambda_i c_i, \\
& \sum_{i=1}^{m} x_{ij} \leq 1, & j = 1, \ldots, n, \\
& x_{ij} \in \{0,1\}, & i = 1, \ldots, m, \; j = 1, \ldots, n.
\end{aligned} \quad (161)$$

The best choice of multipliers λ_i are those producing the smallest objective value in (161). These can be analytically found as follows:

Theorem 7.1 (Martello and Toth [76]) *For any instance of MKP, the optimal choice of multipliers $\lambda_1, \ldots, \lambda_m$ for MKP is $\lambda_i = k$, $i = 1, \ldots, m$, where k is a positive constant.*

With this choice of multipliers $S(\text{MKP}, \lambda_i)$ becomes the following 0-1 Knapsack Problem

$$\begin{aligned}
\text{maximize} \quad & \sum_{j=1}^{n} p_j x'_j \\
\text{subject to} \quad & \sum_{j=1}^{n} w_j x'_j \leq c, \\
& x'_j \in \{0,1\}, \quad j = 1, \ldots, n,
\end{aligned} \quad (162)$$

where the introduced decision variables $x'_j = \sum_{i=1}^{m} x_{ij}$ indicate whether item j is chosen for any of the knapsacks $i = 1, \ldots, m$, and $c = \sum_{i=1}^{m} c_i$ may be seen as the capacity of the united knapsacks.

By *Continuous relaxation* of the variables x_{ij} the problem $C(\text{MKP})$ appears. Martello and Toth [80] proved that $z(C(\text{MKP})) = z(C(S(\text{MKP}, 1)))$ and thus the objective value of the continuously relaxed problem may be found in $O(n)$ as the Dantzig bound for (162), which however can never be less than $z(S(\text{MKP}, 1))$.

Two different *Lagrangian relaxations* of MKP are possible. By relaxing the constraints $\sum_{i=1}^{m} x_{ij} \leq 1$ using nonnegative multipliers $\lambda_1, \ldots, \lambda_n$ the problem $L_1(\text{MKP}, \lambda_j)$ becomes:

$$\begin{aligned}
\text{maximize} \quad & \sum_{i=1}^{m}\sum_{j=1}^{n} p_j x_{ij} - \sum_{j=1}^{n} \lambda_j \left(\sum_{i=1}^{m} x_{ij} - 1 \right) \\
\text{subject to} \quad & \sum_{j=1}^{n} w_j x_{ij} \leq c_i, \quad i = 1, \ldots, m, \\
& x_{ij} \in \{0, 1\}, \quad i = 1, \ldots, m, \ j = 1, \ldots, n.
\end{aligned} \quad (163)$$

By setting $\tilde{p}_j = p_j - \lambda_j$ for $j = 1, \ldots, n$ the relaxed problem can be decomposed into m independent 0-1 Knapsack Problems, where problem i has the form

$$\begin{aligned}
\text{maximize} \quad & z_i = \sum_{j=1}^{n} \tilde{p}_j x_{ij} \\
\text{subject to} \quad & \sum_{j=1}^{n} w_j x_{ij} \leq c_i, \\
& x_{ij} \in \{0, 1\}, \quad j = 1, \ldots, n,
\end{aligned} \quad (164)$$

for $i = 1, \ldots, m$. All the problems have similar profits and weights, thus only the capacity distinguishes the individual instances. An optimal solution to the $L_1(\text{MKP}, \lambda_i)$ is then $z = \sum_{i=1}^{m} z_i + \sum_{j=1}^{n} \lambda_j$.

As opposed to surrogate relaxation, there is however no optimal choice of multipliers λ_j for Lagrangian relaxation, thus approximate values may be found by subgradient optimization. Hung and Fisk [55] used predefined multipliers given by

$$\overline{\lambda}_j = \begin{cases} p_j - w_j p_b / w_b & \text{if } j < b \\ 0 & \text{if } j \geq b \end{cases} \quad (165)$$

where b is the break item of $S(\text{MKP}, 1)$. With this choice of λ_j, in (164) we have $\tilde{p}_j / w_j = p_b / w_b$ for $j \leq b$, and $\tilde{p}_j / w_j \leq p_b / w_b$ for $j > b$. Thus it follows that

$$z(C(L_1(\text{MKP}, \overline{\lambda}))) = \frac{p_b}{w_b} \sum_{i=1}^{m} c_i + \sum_{j=1}^{n} \overline{\lambda}_j = \sum_{j=1}^{b-1} p_j + \left(\sum_{i=1}^{m} c_i - \sum_{j=1}^{b-1} w_j \right) \frac{p_b}{w_b} \quad (166)$$

thus we get

$$z(C(L_1(\text{MKP}, \overline{\lambda}_j))) = z(C(S(\text{MKP}, 1))) = z(C(\text{MKP})) \quad (167)$$

i.e. the multipliers $\overline{\lambda}_j$ represent the best multipliers for $C(L_1(\text{MKP}, \lambda_j))$. In addition both $L_1(\text{MKP}, \overline{\lambda})$ and surrogate relaxation $S(\text{MKP}, 1)$ dominate continuous relaxation. There is however no dominance between Lagrangian and the surrogate relaxations.

Another Lagrangian relaxation $L_2(\text{MKP}, \lambda_i)$ appears by relaxing the weight constraints. Using positive multipliers $\lambda_1, \ldots, \lambda_m$, the relaxed problem becomes

$$\begin{aligned}\text{maximize} \quad & \sum_{i=1}^{m}\sum_{j=1}^{n} p_j x_{ij} - \sum_{i=1}^{m} \lambda_i \left(\sum_{j=1}^{n} w_j x_{ij} - c_i \right) \\ \text{subject to} \quad & \sum_{i=1}^{m} x_{ij} \leq 1 \\ & x_{ij} \in \{0,1\}, \quad i=1,\ldots,m, \ j=1,\ldots,n.\end{aligned} \quad (168)$$

It is important that $\lambda_i > 0$ for all $i = 1, \ldots, m$ as otherwise the above problem has a useless solution, where all items are put into the knapsack where $\lambda_i = 0$. The objective function can be rewritten as

$$\text{maximize} \quad \sum_{i=1}^{m}\sum_{j=1}^{n}(p_j - \lambda_i w_j)x_{ij} + \sum_{i=1}^{m}\lambda_i c_i \quad (169)$$

which shows that the optimal solution can be found by selecting the knapsack κ with the smallest associated value of λ_i and choosing all items with $p_j - \lambda_\kappa w_j > 0$ for this knapsack. Since this is also the optimal solution of the continuous relaxation of the same problem, i.e. $z(C(L_2(\text{MKP}, \lambda_i))) = z(L_2(\text{MKP}, \lambda_i)))$, we get $z(L_2(\text{MKP}, \lambda_i)) \geq z(C(\text{MKP}))$ and thus this Lagrangian relaxation cannot produce a bound tighter than the continuous one.

Note that any of the polynomially obtainable upper bounds presented in Section 2.1 can be used for the Knapsack Problems (162) and (164) to obtain upper bounds for MKP in polynomial time.

The most natural polynomially computable upper bound for MKP is

$$U_1 = \lfloor z(C(\text{MKP})) \rfloor = \lfloor z(C(S(\text{MKP}, 1))) \rfloor = \lfloor z(C(L_1(\text{MKP}, \overline{\lambda}))) \rfloor \quad (170)$$

Martello and Toth [80] have shown that the worst case performance of bound U_1 is $\rho(U_1) = m + 1$, where $\rho(U)$ is defined by (13).

7.2 Tightening Constraints

Polyhedral properties may be used to tighten the upper bound obtained by any of the above relaxations. Pisinger [101] used the following technique to tighten the constraints:

Knapsack Problems

Considering the capacity constraints in (160) for each knapsack separately, the largest possible filling $\hat{c}_i$ of knapsack i is found by solving the following Subset-sum Problem

$$\hat{c}_i = \max\left\{\sum_{j=1}^n w_j x_{ij} : \sum_{j=1}^n w_j x_{ij} \leq c_i;\ x_{ij} \in \{0,1\},\ j=1,\ldots,n\right\} \quad (171)$$

As no optimal solution x can have weight sum $\sum_{j=1}^n w_j x_{ij} > \hat{c}_i$, we may tighten the constraints in (160) to

$$\sum_{j=1}^n w_j x_{ij} \leq \hat{c}_i,\ i = 1, \ldots, m. \quad (172)$$

Since all problems (171) are defined for the same weights, dynamic programming may be used to derive all capacities $\hat{c}$, in $O(n\overline{c})$ time where $\overline{c} = \max_{i=1,\ldots,m} c_i$.

Within a branch-and-bound algorithm it is possible to use a tighter version of the above criterion, since only variables not fixed by the enumeration need be considered in (171). The tightening yields tighter upper bounds, and may also help in obtaining better lower bounds, as the capacities are more realistic. Pisinger [101] experimentally showed, that the tightening is able to decrease the capacity by several hundreds of weight units for a small problem $n = 50$, $m = 5$, $w_j \in [10, 10\,000]$ and so-called dissimilar capacities c_i given by (177).

Notice, for instance, that if knapsack i was tightened to capacity $\hat{c}_i$, then problem (164) with multipliers $\overline{\lambda}_j$ given by (165) is bounded by $z_i \leq \hat{c} p_b / w_b$ which is the continuous solution of (164).

Other polyhedral properties of MKP are considered in Ferreira, Martin and Weismantel [33].

7.3 Reduction Algorithms

The size of a Multiple Knapsack Problem may be reduced by preprocessing as described in Section 2.3: Assume that the solution vector corresponding to the current lower bound z has been saved. For a given item j let u_j^0 be any upper bound on (160) with the additional constraint $\sum_{i=1}^m x_{ij} = 0$. If $u_j^0 < z + 1$, then the constraint $\sum_{i=1}^m x_{ij} = 1$ may be added to the problem, i.e. in every improved solution to MKP, item j must be included in some knapsack. In a similar way let u_j^1 be any upper bound on (160) with the additional constraint $\sum_{i=1}^m x_{ij} = 1$. If $u_j^1 < z+1$, then x_{ij} may be fixed at 0 for $i = 1, \ldots, m$, thus in any improved solution, item j cannot be included in any of the knapsacks. Notice that the last equation is able to fix all variables x_{ij} to their optimal value for $i = 1, \ldots, m$, while the first equation only rules out one possibility out of $m + 1$. Pisinger [101] used the Dembo and Hammer [18] bound with respect to (162) for the reduction, getting an $O(n)$ reduction procedure. Only bounds u_j^1 were derived, attempting to exclude an item from all knapsacks. This reduction was performed at each branching node.

Ingargiola and Korsh [60] presented a reduction procedure based on dominance of items. An item j dominates another item k if $p_j \geq p_k$ and $w_j \leq w_k$. Whenever an item j is excluded from a given knapsack i during the branching process, all items dominated by j may also be excluded from knapsack i. Similarly, whenever an item j is included in a knapsack, all items dominating j must be included in one of the knapsacks.

7.4 Heuristics and Approximate Algorithms

Since MKP is $\mathcal{NP}$-hard in the strong sense, no fully polynomial approximation scheme can be found unless $\mathcal{P} = \mathcal{NP}$ [40]. The same conclusion holds for MKP in its minimization form. Martello and Toth [80] also proved that MKP in the minimization form cannot have a polynomial-time approximation scheme. It is however open whether the same holds for the maximization form, but currently no polynomial-time approximation scheme is known for MKP in the maximization form. For fixed m, Chandra, Hirschberg and Wong [12] however showed that a fully polynomial approximation scheme exists for MKP in the maximization form.

Greedy algorithms are considered in [80]. The most natural one is to use the continuous solution of $S(\text{MKP}, 1)$ to produce a feasible solution. Assume that the items are ordered according to

$$\frac{p_1}{w_1} \geq \frac{p_2}{w_2} \geq \ldots \geq \frac{p_n}{w_n}, \tag{173}$$

and let $b = \min\{h : \sum_{j=1}^{h} w_j > c\}$ be the break item. For each knapsack i define the ith break item b_i as $b_i = \min\{h : \sum_{j=b_{i-1}}^{h} w_j > c_i\}$ where $b_0 = 1$. Then each knapsack i can be filled with items $b_{i-1}, \ldots, b_i - 1$, and the objective value is $z' = \sum_{j=1}^{b_m - 1} p_j$. The absolute error of z' is less than $\sum_{j=b_m}^{b} p_j$ where $b - b_m < m$. The relative error is however arbitrarily bad.

A polynomial-time approximation algorithm was proposed by Martello and Toth [76]. The items are assumed to be sorted as above while the capacities are ordered $c_1 \leq c_2 \leq \ldots \leq c_m$. An initial feasible solution is determined by calling the greedy algorithm described in Section 2.2 for the first knapsack, removing the chosen items, and then calling it for the next knapsack until all m knapsacks have been considered. In the second phase, a number of exchanges are performed. If two items assigned to different knapsacks can be interchanged such that one more item fits into one of the knapsacks, then this exchange is made. The resulting algorithm runs in $O(n^2)$. Some computational results given in [80] show a good performance for randomly generated instances.

7.5 Dynamic Programming

The Multiple Knapsack Problem is $\mathcal{NP}$-hard in the strong sense, meaning that we cannot expect to find pseudo-polynomial algorithms for the problem except if $\mathcal{NP} = \mathcal{P}$. However if we only consider cases with a fixed number m of knapsacks, then the problem may in fact be solved in pseudo-polynomial time through dynamic programming. For instance, an MKP with two knapsacks ($m = 2$) may be solved in $O(nc_1c_2)$ time as follows: Let $f_k(\tilde{c}_1, \tilde{c}_2)$ be the optimal solution value to the problem defined on the first k items, with capacities $\tilde{c}_1, \tilde{c}_2$ i.e.

$$f_k(\tilde{c}_1, \tilde{c}_2) = \max \left\{ \begin{array}{l} \sum_{j=1}^{k} p_j(x_{1j} + x_{2j}) : \sum_{j=1}^{k} w_j x_{1j} \leq \tilde{c}_1; \\ \sum_{j=1}^{k} w_j x_{2j} \leq \tilde{c}_2; \\ x_{1j} + x_{2j} \leq 1, j = 1, \ldots, k; \\ x_{1j}, x_{2j} \in \{0,1\}, j = 1, \ldots, k \end{array} \right\} \quad (174)$$

The following recursion may be used to evaluate f_k:

$$f_k(\tilde{c}_1, \tilde{c}_2) = \max \left\{ \begin{array}{l} f_{k-1}(\tilde{c}_1, \tilde{c}_2) \\ f_{k-1}(\tilde{c}_1 - w_k, \tilde{c}_2) + p_k \\ f_{k-1}(\tilde{c}_1, \tilde{c}_2 - w_k) + p_k \end{array} \right. \quad (175)$$

with initial values $f_0(\tilde{c}_1, \tilde{c}_2) = 0$ for all $\tilde{c}_1, \tilde{c}_2$ and with $f_k(\tilde{c}_1, \tilde{c}_2) = -\infty$ if $\tilde{c}_1 < 0$ or $\tilde{c}_2 < 0$. An optimal solution is found as $f_n(c_1, c_2)$ by evaluating $f_k(\cdot, \cdot)$ for all values of $k = 1, \ldots, n$, yielding the complexity $O(nc_1c_2)$. The above may be generalized to a MKP problem with m knapsacks, where the time bound becomes $O(nm \prod_{i=1}^{m} c_i)$. Since this approach is not attractive for large values of m, most of the literature has been focused on branch-and-bound techniques, although Fischetti and Toth [35] used some kind of dominance tests to speed up the solution process.

7.6 Branch-and-bound Algorithms

Hung and Fisk [55] proposed a depth-first branch-and-bound algorithm where the upper bounds were derived by the Lagrangian relaxation, and branching was performed for the item which in the relaxed problem had been selected in most knapsacks. Each branching item was alternately assigned to the knapsacks in increasing index order, where the knapsacks were ordered in nonincreasing order $c_1 \geq c_2 \geq \ldots \geq c_m$. When all the knapsacks had been considered, a last branch was considered where the item was excluded from the problem.

A different branch-and-bound algorithm was proposed by Martello and Toth [75], where at each decision node, MKP was solved with constraint $\sum_{i=1}^{m} x_{ij} \leq 1$ omitted, and the branching item was chosen as an item which had been packed in $k > 1$ knapsacks of the relaxed problem. The branching operation generated k nodes by assigning the item to one of the corresponding $k-1$ knapsacks or excluding it from all of these. A parametric technique was used to speed up the computation of the upper bound at each node.

In a later work Martello and Toth [76] however focused on three aspects that make it difficult to solve MKP:

- Generally it is difficult to verify feasibility of an upper bound obtained either by surrogate relaxation or Lagrangian relaxation.

- The branch-and-bound algorithm needs good lower bounds for fathoming nodes in the enumeration.

- Some knowledge to guide the branching towards good feasible solutions is needed.

In order to avoid these problems, Martello and Toth proposed a *bound-and-bound* algorithm for MKP, where at each node of the branching tree not only an upper bound, but also a lower bound is derived. This technique is well-suited for problems where it is easy to find a fast heuristic solution which yields good lower bounds, and where it is difficult to verify feasibility of the upper bound.

7.6.1 The mtm Algorithm

The mtm bound-and-bound algorithm derives upper bounds by solving the surrogate relaxed problem (162) while lower bounds are found by solving m individual 0-1 Knapsack Problems as follows: The first knapsack $i = 1$

is filled optimally, the chosen variables are removed from the problem, and the next knapsack $i = 2$ may be filled. This process is continued until all the m knapsacks have been filled. The branching scheme follows this greedy solution, as a greedy solution should be better to guide the branching process than individual choices at each branching node. Thus at each node two branching nodes are generated, the first one assigning the next item j of a greedy solution to the chosen knapsack i, while the other branch excludes item j from knapsack i.

Martello and Toth assume that the capacities are ordered in nondecreasing order $c_1 \leq c_2 \leq \cdots \leq c_m$, while the items are ordered according to nonincreasing profit-to-weight ratios (173) so every branching takes place on the item with the largest profit-to-weight ratio. At any stage, knapsacks up to k have to be filled before knapsack $k+1$ is considered. The current solution vector is x, while the optimal solution vector is denoted x^*.

At any stage the list $T = \{(i_1, j_1), \ldots, (i_d, j_d)\}$ contains indices to the variables x_{ij} which have been fixed to either 0 or 1 during the branching process. The profit sum of the currently fixed variables is P while $c_1, \ldots, c_m$ refer to the current residual capacities.

algorithm mtm$(k, P, c_1, \ldots, c_m)$;
Find an upper bound u by solving $S(\text{MKP}, 1)$ defined on the free items and
    capacity $c = \sum_{i=1}^{m} c_i$.
Find a lower bound ℓ and the corresponding greedy solution y.
if $(P + \ell > z)$ **then** $x_{ij}^* \leftarrow x_{ij}$ for $(i,j) \in T$ and $x_{ij}^* \leftarrow y_{ij}$ for $(i,j) \notin T$.
    Set $z \leftarrow P + \ell$.
if $(P + u > z)$ **then**
    Choose the first item j where $y_{kj} = 1$ and item j has not been excluded
        from knapsack k, i.e. $(k, j) \notin T$.
    If no such item exists, increase k and repeat the search.
    Set $T \leftarrow T \cup (k, j)$;
    Set $x_{kj} \leftarrow 1$; { Assign item j to knapsack k }
    mtm$(k, P + p_j, c_1, \ldots, c_k - w_j, \ldots, c_m)$;
    Set $x_{kj} \leftarrow 0$; { Exclude item j from knapsack k }
    mtm$(k, P, c_1, \ldots, c_m)$;
    Set $T \leftarrow T \setminus (k, j)$;

Initially variables x_{ij}, x_{ij}^* are set to 0, and the lower bound is set to $z = 0$. Then the recursion mtm$(1, 0, 0, c_1, \ldots, c_m)$ is called. Martello and Toth used the mt1 algorithm to solve $S(\text{MKP}, 1)$ in the first part of the algorithm.

Since every forward move setting $x_{kj} = 1$ follows the greedy solution y, the lower bound ℓ and solution vector y will not change by this branching. Thus ℓ and y need only be determined after branches of the form $x_{kj} = 0$.

7.6.2 The mulknap Algorithm

Pisinger [101] noted that a solution to $S(\text{MKP}, 1)$ may be validated by solving a series of Subset-sum Problems, where the chosen items are attempted to be distributed among the m knapsacks. If this attempt succeeds, the lower bound equals the upper bound, and a backtracking occurs. Otherwise a feasible solution has been obtained which contains some (but not necessarily all) of the items selected by $S(MKP, 1)$. Thus the algorithm is merely an ordinary branch-and-bound approach since upper bounds yield a feasible solution. Pisinger also used Subset-sum Problems for tightening the capacity constraints corresponding to each knapsack as described in Section 7.2.

The branching scheme is based on a binary splitting where an item j is either assigned to knapsack i or excluded from the knapsack. The knapsacks are ordered in nondecreasing order $c_1 \leq c_2 \leq \cdots \leq c_m$ and the smallest knapsack is filled completely before starting to fill the next knapsack. At any stage, items $j \leq h$ have been fixed by the branching process, thus only items $j > h$ are considered when upper and lower bounds are determined. To keep track of which items are excluded from some knapsacks, a variable d_j for each item j indicates that the item may only be assigned to knapsacks $i \geq d_j$. The current solution vector is x while P is the profit sum of the currently fixed items.

algorithm mulknap$(h, P, c_1, \ldots, c_m)$;
Tighten the capacities c_i by solving m Subset-sum Problems defined on
  $h+1, \ldots, n$.
Solve the surrogate relaxed problem with capacity $c = \sum_{i=1}^{m} c_i$. Let $\hat{y}$ be
  the solution to this problem, with objective value u.
if $(P + u > z)$ **then**
  Split the solution $\hat{y}$ in the m knapsacks by solving a series of Subset-sum
    Problems defined on items with $\hat{y}_j = 1$. Let y_{ij} be the optimal filling
    of c_i with corresponding profit sum z_i.
  Improve the heuristic solution by greedy filling knapsacks with
    $\sum_{j=h+1}^{n} w_j y_{ij} < c_i$.
  if $(P + \sum_{i=1}^{m} z_i > z)$ **then** $x_{ij}^* = y_{ij}$ for $j \geq h$ and $x_{ij}^* = x_{ij}$ for $j < h$,

set $z \leftarrow P + \sum_{i=1}^{m} z_i$.
if $(P + u > z)$ then
  Reduce the items as described in Section 7.3, and swap the reduced
    items to the first positions, increasing h.
  Let i be the smallest knapsack with $c_i > 0$. Solve an ordinary 0-1
    Knapsack Problem with $c = c_i$ defined on the free variables.
    The solution vector is $\bar{y}$. Choose the branching item k
    as the item with largest profit-to-weight ratio among items $\bar{y}_j = 1$.
  Swap k to position $h + 1$ and set $j \leftarrow h + 1$.
  Set $x_{ij} \leftarrow 1$; { Assign item j to knapsack i }
  mulknap$(h + 1, P + p_j, w_j, c_1, \ldots, c_i - w_j, \ldots, c_m)$;
  Set $x_{ij} \leftarrow 0$; $d' \leftarrow d_j$; $d_j \leftarrow i + 1$; { Exclude item j from knapsack i }
  mulknap$(h, P, c_1, \ldots, c_m)$;
  Find j again, and set $d_j \leftarrow d'$.

The surrogate relaxed problem is solved using the minknap algorithm described in Section 2.6. The Subset-sum Problems are solved by the decomp algorithm described in Section 3.2.1. The algorithm does not demand any special ordering of the variables, as the minknap algorithm makes the necessary ordering itself. This however implies that items are permuted at each call to minknap thus the last line of mulknap cannot assume that item j is at the same position as before.

The main algorithm thus only has to order the capacities according to nondecreasing c_i, set $z = 0$ and initialize $d_j = 1$, and $x^*_{ij} = y_{ij} = 0$ for all i, j before calling mulknap$(0, 0, 0, c_1, \ldots, c_m)$.

7.7 Computational Experiments

We will compare the performance of the mtm algorithm with that of the mulknap algorithm. Four different types of randomly generated data instances are considered for different ranges $R = 100$, 1000 and $10\,000$. *Uncorrelated data instances*: p_j and w_j are randomly distributed in $[1, R]$. *Weakly correlated data instances*: w_j randomly distributed in $[1, R]$ and p_j randomly distributed in $[w_j - R/10, w_j + R/10]$ such that $p_j \geq 1$. *Strongly correlated data instances*: w_j randomly distributed in $[1, R]$ and $p_j = w_j + 10$. *Subset-sum data instances*: w_j randomly distributed in $[1, R]$ and $p_j = w_j$.

Table 11: Total computing time mulknap, small problems with $m = 5$.

$n \setminus R$	Uncorrelated			Weakly corr.			Strongly corr.			Subset-sum		
	100	1000	10000	100	1000	10000	100	1000	10000	100	1000	10000
25	0.57	0.70	0.44	5.02	15.29	9.49	0.10	2.45	3.85	0.00	1.82	13.16
50	0.00	0.00	0.02	0.00	0.00	15.67	0.00	0.01	0.12	0.00	0.01	0.10
75	0.00	0.00	0.01	0.00	0.00	0.01	0.01	0.02	0.22	0.00	0.01	0.04
100	0.00	0.00	0.00	0.00	0.00	0.01	0.01	0.03	0.24	0.00	0.01	0.04
200	0.00	0.00	0.00	0.00	0.00	0.01	0.01	0.11	1.17	0.00	0.01	0.03
25	0.01	0.12	0.03	0.20	0.21	0.87	0.00	0.13	0.47	0.00	0.12	1.29
50	0.00	0.00	0.01	0.00	0.34	2.80	0.00	0.02	0.69	0.00	0.01	0.38
75	0.00	0.00	0.01	0.00	0.01	0.10	0.01	0.08	1.70	0.00	0.02	0.28
100	0.00	0.00	0.01	0.01	0.03	0.25	0.01	0.03	0.44	0.00	0.01	0.16
200	0.00	0.00	0.01	0.00	0.01	0.02	0.02	0.10	1.37	0.00	0.01	0.12

Table 12: Total computing time mtm, small problems with $m = 5$.

$n \setminus R$	Uncorrelated			Weakly corr.			Strongly corr.			Subset-sum		
	100	1000	10000	100	1000	10000	100	1000	10000	100	1000	10000
25	0.34	0.80	0.75	3.96	7.87	4.42	0.37	9.74	29.12	0.00	19.64	111.36
50	0.06	2.13	6.21	0.63	62.06	181.67	3.02	6.68	522.53	0.00	0.01	0.06
75	0.05	1.19	13.38	0.41	21.04	123.00	—	126.91	322.34	0.00	0.00	0.05
100	0.05	1.27	10.35	0.11	21.36	242.46	—	623.90	—	0.00	0.00	0.05
200	0.03	1.26	5.81	0.02	16.10	278.77	—	—	—	0.00	0.01	0.05
25	0.02	0.04	0.06	0.17	0.14	0.80	0.03	0.36	1.49	0.00	0.67	5.82
50	0.02	0.13	1.06	0.18	2.82	22.60	0.16	2.58	151.76	0.00	0.28	14.47
75	0.02	0.29	1.17	0.02	3.54	21.49	8.68	111.48	169.50	0.00	0.08	4.97
100	0.01	0.23	2.63	0.03	4.17	21.71	—	197.25	212.37	0.00	0.02	1.22
200	0.01	0.09	0.80	0.01	11.73	153.46	—	—	—	0.00	0.01	0.41

Martello and Toth [80] proposed considering two different classes of capacities as follows: *Similar capacities* have the first $m-1$ capacities c_i randomly distributed in

$$\left[0.4\sum_{j=1}^{n}w_j/m,\ 0.6\sum_{j=1}^{n}w_j/m\right] \text{ for } i=1,\ldots,m-1, \qquad (176)$$

while *dissimilar capacities* have c_i distributed in

$$\left[0,\ 0.5\left(\sum_{j=1}^{n}w_j - \sum_{k=1}^{i-1}c_k\right)\right] \text{ for } i=1,\ldots,m-1. \qquad (177)$$

The last capacity c_m is in both classes chosen as $c_m = 0.5\sum_{j=1}^{n}w_j - \sum_{i=1}^{m-1}c_i$, to ensure that the sum of the capacities is half of the total weight sum. For each instance a check is made on whether the assumptions stated below (160) are respected, and a new instance is generated otherwise.

A maximum amount of 1 hour was given to each algorithm for solving the ten instances in each class, and a dash in the following tables indicates that the ten problems could not be solved within this time limit. The mtm algorithm is only designed for problems up to $n = 1000$ and cannot solve larger instances without a modification of the code. Thus no tests with $n > 1000$ have been run with mtm. Small data instances are tested with $m = 5$ knapsacks, while large instances have $m = 10$, as none of the algorithms are able to solve problems with small values of n/m. In each of the following tables, instances with similar capacities are considered in the upper part of the table, while the lower part of the table considers instances with dissimilar capacities.

Table 11 and 12 compare the solution times of the two algorithms for small instances up to $n = 200$. Both algorithms solve uncorrelated and subset-sum type problems in reasonable time, while mtm has considerable problems for large-sized and large-ranged problems of weak and strong correlation. If the individual entries are compared, it is seen that mulknap generally has faster solution times than mtm, and the larger the problems become the more efficient mulknap gets.

Table 13: Total computing time mulknap, large problems with $m = 10$.

$n \setminus R$	Uncorrelated			Weakly corr.			Strongly corr.			Subset-sum		
	100	1000	10000	100	1000	10000	100	1000	10000	100	1000	10000
100	0.00	0.01	0.03	0.00	0.01	0.04	0.01	0.04	0.32	0.00	0.01	0.11
300	0.00	0.01	0.02	0.01	0.01	0.03	0.02	0.23	1.75	0.01	0.01	0.08
1000	0.00	0.00	0.01	0.00	0.01	0.01	0.10	0.93	8.24	0.01	0.01	0.06
3000	0.01	0.01	0.01	0.01	0.01	0.03	0.23	3.04	29.60	0.01	0.01	0.06
10000	0.02	0.02	0.04	0.01	0.02	0.07	1.30	11.04	188.61	0.02	0.03	0.08
30000	0.05	0.05	0.10	0.05	0.06	0.09	4.63	48.79	684.22	0.06	0.07	0.14
100000	0.20	0.21	0.28	0.19	0.21	0.25	16.30	214.26	1623.33	0.24	0.25	0.36
100	0.04	0.01	0.87	0.02	0.02	27.88	0.01	2.51	16.10	0.00	0.39	10.35
300	0.01	0.01	0.07	0.01	0.02	0.29	0.03	4.00	4.99	0.01	0.25	0.47
1000	0.00	0.01	0.03	0.01	0.06	0.04	0.08	0.58	52.66	0.01	0.02	0.22
3000	0.01	0.01	0.03	0.01	0.01	0.11	0.21	2.48	31.08	0.01	0.02	0.10
10000	0.01	0.02	0.05	0.01	0.02	0.06	0.75	8.28	162.32	0.02	0.03	0.08
30000	0.04	0.04	0.10	0.04	0.04	0.08	3.23	38.55	477.18	0.05	0.07	0.12
100000	0.17	0.17	0.24	0.16	0.17	0.21	14.57	169.21	1287.07	0.20	0.21	0.30

Table 14: Total computing time mtm, large problems with $m = 10$.

$n \setminus R$	Uncorrelated			Weakly corr.			Strongly corr.			Subset-sum		
	100	1000	10000	100	1000	10000	100	1000	10000	100	1000	10000
100	1.78	337.84	—	54.25	—	—	—	—	—	0.00	0.01	0.08
300	0.12	31.71	502.31	0.11	391.93	—	—	—	—	0.00	0.01	0.10
1000	0.03	37.02	828.39	0.01	105.97	—	—	—	—	0.01	0.02	0.11
100	0.18	1.20	18.80	0.74	56.63	696.98	542.83	—	—	0.00	2.04	515.92
300	0.07	2.47	30.36	0.07	145.85	—	—	—	—	0.00	0.25	6.09
1000	0.08	2.49	75.87	0.02	24.76	—	—	—	—	0.01	0.04	4.79

The same situation appears for large instances with $n \geq 100$ as those presented in Table 13 and 14. Here mulknap is able to solve all of the large-sized instances while mtm only can solve low-ranged problems. For very large instances with $n \geq 10\,000$, mulknap is actually able to solve the problems in times comparable with the best solution times for the 0-1 Knapsack Problem. The computational results indicate that *large* Multiple Knapsack Problems, despite the $\mathcal{NP}$-hardness, are generally as easy to solve as ordinary 0-1 Knapsack Problems. For $n < 100$ and $m = 10$ both algorithms however have problems in solving the instances, as the n/m ratio becomes too small.

The performance of mtm can be improved by using a more recent algorithm for solving the 0-1 KP. Using mthard instead of mt1 will especially improve the performance for strongly correlated problems.

8 Conclusion and Future Trends

Knapsack Problems is a field of research where theoretical results and computational experience go hand in hand. The present chapter has shown that recent techniques make it possible to solve several large-sized difficult problems within reasonable time. But the computational results also show that several instances are impossible to solve, and thus require new solution techniques to be developed. One could mention real-valued knapsack problems, quadratic knapsack problems, non-fill problems, and knapsack problems with large weights as some of the fields where more research is required.

Where the 80s were mainly concerned with the solution of large-sized instances, the past decade has focused on developing algorithms which can solve a large spectrum of problems. In this development, one may notice that solving the *core problem* is not so attractive, since this technique is mainly applicable to easy problems. The core problem is however an elegant way of quickly obtaining a good lower bound, but for difficult instances it is necessary to involve other techniques: Facets lead to tight bounds for some difficult problems and dynamic programming gives useful worst-case bounds. A future research topic could be combining dynamic programming algorithms with the tight bounds developed for branch-and-bound algorithms, in order to derive effective algorithms with worst-case complexity given by the dynamic programming recursion.

It is well-known that a Bellman recursion for the 0-1 knapsack problem has some overheads: The problem is not only solved for a given capacity c, but for all capacities $\tilde{c} \leq c$. This indicates that the time bound $O(nc)$ of the Bellman recursion is not tight. One attempt to derive faster recursions is balancing, but we may see algorithms based on other properties, which obtain tighter worst-case time bounds.

Knapsack Problems have recently been applied [27] with great success to tighten the formulation of difficult real-life combinatorial problems. The problems that appear in this context are however not pure Knapsack Problems, but Knapsack Problems with some additional constraints like cardinality bounds, incompatibility-constraints, clique constraints, etc. Thus it is important that we find effective solution techniques for these combined problems.

When Knapsack Problems appear as subproblems in more general Combinatorial problems, it is not unusual that several Knapsack Problems are solved at every branching node of the master problem. It is however often the case that the knapsack instances solved are quite similar, e.g. only differ by a single item which has been removed or inserted. This motivates developing dynamic knapasack algorithms, which by maintaining some extra information about the previous problems, are able to solve the modified problem faster than if one started from scratch. Dynamic programming is an obvious technique to apply for such algorithms, but dynamic data structures may also be effective tools.

References

[1] J. H. Ahrens and G. Finke (1975), "Merging and Sorting Applied to the Zero-One Knapsack Problem", *Operations Research*, **23**, 1099–1109.

[2] L. Aittoniemi (1982), "Computational comparison of knapsack algorithms", presented at *XIth International Symposium on Mathematical Programming*, Bonn, August 23–27.

[3] G. d'Atri, C. Puech (1982), "Probabilistic analysis of the subset-sum problem", *Discrete Applied Mathematics*, **4**, 329–334.

[4] D. Avis (1980) Theorem 4. In V. Chvátal, "Hard knapsack problems", *Operations Research*, **28**, 1410-1411.

[5] E. Balas (1975), "Facets of the Knapsack Polytope", *Mathematical Programming*, **8**, 146–164.

[6] E. Balas and E. Zemel (1980), "An Algorithm for Large Zero-One Knapsack Problems", *Operations Research*, **28**, 1130–1154.

[7] P. Barcia and K. Jörnsten (1990), "Improved Lagrangean decomposition: An application to the generalized assignment problem", *European Journal of Operational Research*, **46**, 84–92.

[8] R. E. Bellman (1957), *Dynamic programming*, Princeton University Press, Princeton, NJ.

[9] R. L. Bulfin, R. G. Parker and C. M. Shetty (1979), "Computational results with a branch and bound algorithm for the general knapsack problem", *Naval Research Logistics Quarterly*, **26**, 41–46.

[10] R. E. Burkard and U. Pferschy (1995), "The Inverse-parametric Knapsack Problem", *European Journal of Operational Research*, **83** 376–393.

[11] A. Caprara, D. Pisinger, P. Toth (1997), "Exact Solution of Large Scale Quadratic Knapsack Problems", *Abstracts ISMP'97*, EPFL, Lausanne, 24–29 August 1997.

[12] A. K. Chandra, D. S. Hirschberg, C. K. Wong (1976), "Approximate algorithms for some generalized knapsack problems". *Theoretical Computer Science*, **3**, 293–304.

[13] N. Christofides, A. Mingozzi, P. Toth (1979). "Loading Problems". In N. Christofides, A. Mingozzi, P. Toth, C. Sandi (eds.), *Combinatorial Optimization*, Wiley, Chichester, 339–369.

[14] V. Chvátal (1980), "Hard Knapsack Problems", *Operations Research*, **28**, 1402–1411.

[15] T. H. Cormen, C. E. Leiserson and R. L. Rivest (1990), *Introduction to Algorithms*, MIT Press, Massachusetts.

[16] H. Crowder, E.L. Johnson, M.W. Padberg (1983), "Solving large-scale zero-one linear programming problems", *Operations Research*, **31**, 803–834.

[17] G. B. Dantzig (1957), "Discrete Variable Extremum Problems", *Operations Research*, **5**, 266–277.

[18] R. S. Dembo and P. L. Hammer (1980), "A Reduction Algorithm for Knapsack Problems", *Methods of Operations Research*, **36**, 49–60.

[19] B. L. Dietrich and L. F. Escudero (1989), "More coefficient reduction for knapsack-like constraints in 0-1 programs with variable upper bounds", *IBM T.J., Watson Research Center, RC-14389, Yorktown Heights N.Y.*

[20] B. L. Dietrich and L. F. Escudero (1989), "New procedures for preprocessing 0-1 models with knapsack-like constraints and conjunctive and/or disjunctive variable upper bounds", *IBM T.J., Watson Research Center, RC-14572, Yorktown Heights N.Y.*

[21] W. Diffe and M. E. Hellman (1976), "New directions in cryptography", *IEEE Trans. Inf. Theory*, **IT-36**, 644–654.

[22] K. Dudziński (1991), "A note on dominance relations in unbounded knapsack problems", *Operations Research Letters*, **10**, 417–419.

[23] K. Dudziński and S. Walukiewicz (1984), "A fast algorithm for the linear multiple-choice knapsack problem", *Operations Research Letters*, **3**, 205–209.

[24] K. Dudziński and S. Walukiewicz (1987), "Exact Methods for the Knapsack Problem and its Generalizations", *European Journal of Operational Research*, **28**, 3–21.

[25] M. E. Dyer (1984), "An $O(n)$ algorithm for the multiple-choice knapsack linear program", *Mathematical Programming*, **29**, 57-63.

[26] M. E. Dyer, N. Kayal and J. Walker (1984), "A branch and bound algorithm for solving the multiple choice knapsack problem", *Journal of Computational and Applied Mathematics*, **11**, 231–249.

[27] L. F. Escudero, S. Martello and P. Toth (1995), "A framework for tightening 0-1 programs based on an extension of pure 0-1 KP and SS problems", In: E. Balas, J. Clausen (eds.): *Integer Programming and Combinatorial Optimization, Fourth IPCO Conference. Lecture Notes in Computer Science*, **920**, 110-123.

[28] D. Fayard and G. Plateau (1977), "Reduction algorithm for single and multiple constraints 0-1 linear programming problems", Conference on Methods of Mathematical Programming, Zakopane (Poland).

[29] D. Fayard and G. Plateau (1982), "An Algorithm for the Solution of the 0-1 Knapsack Problem", *Computing*, **28**, 269–287.

[30] D. Fayard and G. Plateau (1994), "An exact algorithm for the 0-1 collapsing knapsack problem", *Discrete Applied Mathematics*, **49**, 175–187.

[31] C.E. Ferreira, M. Grötsche, S. Kiefl, C. Krispenz, A. Martin and R. Weismantel (1993), "Some integer programs arising in the design of mainframe computers", *ZOR*, **38** 77–100.

[32] C. E. Ferreira, A. Martin, C. de Souza, R. Weismantel and L. Wolsey (1994), "Formulations and valid inequalities for the node capacitated graph partitioning problem", *CORE discussion paper*, **9437**, Unversité Catholique de Louvain.

[33] C.E. Ferreira, A. Martin, R. Weismantel (1996), "Solving multiple knapsack problems by cutting planes", *SIAM Journal on Optimization*, **6**, 858–877.

[34] M. Fischetti and S. Martello (1988), "A hybrid algorithm for finding the kth smallest of n elements in $O(n)$ time. In B. Simeone, P. Toth, G. Gallo, F. Maffioli, S. Pallottino (eds), *Fortran codes for Network Optimization*, Annals of Operations Research, **13**, 401–419.

[35] M. Fischetti and P. Toth (1988), "A new dominance procedure for combinatorial optimization problems", *Operations Research Letters*, **7**, 181–187.

[36] M. L. Fisher (1981), "The Lagrangian Relaxation Method for Solving Integer Programming Problems", *Management Science*, **27**, 1–18.

[37] A. Fréville G. Plateau (1993), "An exact search for the solution of the surrogate dual for the 0-1 bidimensional knapsack problem", *European Journal of Operational Research*, **68**, 413–421.

[38] A.M. Frieze, M.R.B. Clarke (1984), "Approximation algorithms for the m-dimensional 0-1 knapsack problem: Worst-case and probabilistic analysis", *European Journal of Operational Research*, **15**, 100–109.

[39] G. Gallo, P.L. Hammer, B. Simeone (1980), "Quadratic knapsack problems", *Mathematical Programming* **12** 132–149.

[40] M. R. Garey and D. S. Johnson (1979), *Computers and Intractability: A Guide to the Theory of NP-Completeness*, Freeman, San Francisco.

[41] B. Gavish and H. Pirkul (1985), "Efficient algorithms for solving multiconstraint zero-one knapsack problems to optimality" *Mathematicial Programming*, **31**, 78–105.

[42] G. Gens and E. Levner (1994), "A fast approximation algorithm for the subset-sum problem", *INFOR*, **32**, 143–148.

[43] P. C. Gilmore and R. E. Gomory (1965), "Multistage cutting stock problems of two and more dimensions", *Operations Research*, **13**, 94–120.

[44] P. C. Gilmore and R. E. Gomory (1966), "The theory and computation of knapsack functions", *Operations Research*, **14**, 1045–1074.

[45] F. Glover (1965), "A multiphase dual algorithm for the zero-one integer programming problem", *Operations Research*, **13**, 879–919.

[46] A.V. Goldberg, A. Marchetti-Spaccamela (1984), "On finding the exact solution to a zero-one knapsack problem", *Proc. 16th Annual ACM Symposium Theory of Computing*, 359–368.

[47] H. Greenberg (1985), "An algorithm for the periodic solutions in the knapsack problem", *Journal of Mathematical Analysis and Applications*, **111**, 327–331.

[48] H. Greenberg (1986), "On equivalent knapsack problems", *Discrete Applied Mathematics*, **14**, 263–268.

[49] H. Greenberg, I. Feldman (1980), "A better-step-off algorithm for the knapsack problem", *Discrete Applied Mathematics*, **2**, 21–25.

[50] P.L. Hammer, E.L. Johnson, U.N. Peled (1975), "Facets of regular 0-1 polytopes", *Mathematical Programming*, **8** 179–206.

[51] D. S. Hirschberg and C. K. Wong (1976), "A polynomial time algorithm for the knapsack problem with two variables", *Journal of ACM*, **23** 147–154.

[52] D.S. Hochbaum (1995), "A Nonlinear Knapsack Problem", *Operations Research Letters*, **17**, 103–110.

[53] K. L. Hoffman and M. Padberg (1993), "Solving airline crew-scheduling problems by branch and cut", *Management Science*, **39**, 657–682.

[54] E. Horowitz and S. Sahni (1974), "Computing partitions with applications to the Knapsack Problem", *Journal of ACM*, **21**, 277–292.

[55] M. S. Hung and J. C. Fisk (1978), "An algorithm for 0-1 multiple knapsack problems", *Naval Research Logistics Quarterly*, **24**, 571–579.

[56] T. Ibaraki (1987), "Enumerative Approaches to Combinatorial Optimization – Part 1", *Annals of Operations Research*, **10**.

[57] T. Ibaraki (1987), "Enumerative Approaches to Combinatorial Optimization – Part 2", *Annals of Operations Research*, **11**.

[58] O. H. Ibarra and C. E. Kim (1975), "Fast approximation algorithms for the knapsack and sum of subset problem", *Journal of ACM*, **22**, 463–468.

[59] G. P. Ingargiola and J. F. Korsh (1973), "A Reduction Algorithm for Zero-One Single Knapsack Problems", *Management Science*, **20**, 460–463.

[60] G. P. Ingargiola and J. F. Korsh (1975), "An algorithm for the solution of 0-1 loading problems", *Operations Research*, **23**, 752–759.

[61] G. P. Ingargiola and J. F. Korsh (1977), "A general algorithm for the one-dimensional knapsack problem", *Operations Research*, **25**, 752–759.

[62] R. G. Jeroslow (1974), "Trivial Integer Programs Unsolvable by Branch-and-Bound", *Mathematical Programming*, **6**, 105–109.

[63] E. L. Johnson and M. W. Padberg (1981), "A note on the knapsack probem with special ordered sets", *Operations Research Letters*, **1**, 18–22.

[64] R. Kannan (1980), "A polynomial algorithm for the two-variables integer programming problem", *Journal of ACM*, **27**, 118–122.

[65] G. A. P. Kindervater and J. K. Lenstra (1986), "An introduction to parallelism in combinatorial optimization", *Discrete Applied Mathematics*, **14**, 135–156.

[66] D. E. Knuth (1973), *The art of Computer Programming, Vol. 3, Sorting and Searching*, Addison-Wesley, Reading, MA.

[67] P. J. Kolesar (1967), "A branch and bound algorithm for the knapsack problem", *Management Science*, **13**, 723–735.

[68] D. Krass, S. P. Sethi and G. Sorger (1994), "Some complexity issues in a class of knapsack problems: What makes a Knapsack Problem 'hard' ", *INFOR*, **32**, 149–162.

[69] J. C. Lagarias, A. M. Odlyzko (1983), "Solving low-density subset sum problems", *Proc. 24 th Annual Symposium on Foundations of Computer Science*, Tucson, Arizona, 7–9 November 1983, 1–10.

[70] G. Laporte (1992), "The Vehicle Routing Problem: An overview of exact and approximate algorithms", *European Journal of Operational Research*, **59**, 345–358.

[71] G. S. Lueker (1975), "Two NP-complete problems in nonnegative integer programming", Report No. 178, Computer Science Laboratory, Priceton University, Princeton, NJ.

[72] S. Martello and P. Toth (1977), "An Upper Bound for the Zero-One Knapsack Problem and a Branch and Bound algorithm", *European Journal of Operational Research*, **1**, 169–175.

[73] S. Martello and P. Toth (1977), "Branch and bound algorithms for the solution of general unidimensional knapsack problems". In M. Roubens (ed.), *Advances in Operations Research*, North-Holland, Amsterdam, 295–301.

[74] S. Martello and P. Toth (1979), "The 0-1 knapsack problem". In A. Mingozzi, P. Toth, C. Sandi (ed.), *Combinatorial Optimization*, Wiley, Chichester, 237–279.

[75] S. Martello and P. Toth (1980), "Solution of the zero-one multiple knapsack problem", *European Journal of Operational Research*, **4**, 276–283.

[76] S. Martello and P. Toth (1981), "A bound and bound algorithm for the zero-one multiple knapsack problem", *Discrete Applied Mathematics*, **3**, 275–288.

[77] S. Martello and P. Toth (1984), "A mixture of dynamic programming and branch-and-bound for the subset-sum problem", *Management Science*, **30**, 765–771.

[78] S. Martello, and P. Toth (1987), "Algorithms for Knapsack Problems". In S. Martello, G. Laporte, M. Minoux and C. Ribeiro (Eds.), *Surveys in Combinatorial Optimization*, Ann. Discrete Math. **31**, North-Holland, Amsterdam, 1987, 213–257.

[79] S. Martello and P. Toth (1988), "A New Algorithm for the 0-1 Knapsack Problem", *Management Science*, **34**, 633–644.

[80] S. Martello and P. Toth (1990), *Knapsack Problems: Algorithms and Computer Implementations*, Wiley, Chichester, England.

[81] S. Martello and P. Toth (1990), "An exact algorithm for large unbounded knapsack problems", *Operations Research Letters*, **9**, 15–20.

[82] S. Martello and P. Toth (1995), "The bottleneck generalized assignment problem", *European Journal of Operational Research*, **83** 621–638.

[83] S. Martello and P. Toth (1997), "Upper Bounds and Algorithms for Hard 0-1 Knapsack Problems", *Operations Research*, **45**, 768–778.

[84] G. B. Mathews (1897), "On the Partition of Numbers", *Proc. of the London Mathematical Society*, **28**, 486-490.

[85] T. L. Morin and R. E. Marsten (1976), "An algorithm for nonlinear knapsack problems", *Management Science*, **22**, 1147–1158.

[86] H. Müller-Merbach (1979), "Improved upper bound for the zero-one knapsack problem. A note on the paper by Martello and Toth", *European Journal of Operational Research*, **2**, 212–213.

[87] J. I. Munro, R. J. Ramirez (1982), "Reducing Space Requirements for Shortest Path Problems", *Operations Research*, **30**, 1009–1013.

[88] R. M. Nauss (1976), "An Efficient Algorithm for the 0-1 Knapsack Problem", *Management Science*, **23**, 27–31.

[89] R. M. Nauss (1978), "The 0-1 knapsack problem with multiple choice constraint", *European Journal of Operational Research*, **2**, 125–131.

[90] A. Neebe and D. Dannenbring (1977), "Algorithms for a specialized segregated storage problem", *University of North Carolina*, Technical Report 77-5.

[91] G. L. Nemhauser and Z. Ullmann (1969), "Discrete dynamic programming and capital allocation", *Management Science*, **15**, 494–505.

[92] G.L. Nemhauser and L.A. Wolsey (1988), *Integer and Combinatorial Optimization*, Wiley, Chichester.

[93] M.W. Padberg (1975), "A note on zero-one programming", *Operations Research*, **23**, 833–837.

[94] C. H. Papadimitriou and K. Steiglitz (1982), *Combinatorial Optimization: Algorithms and Complexity*, Prentice Hall, Englewood Cliffs, New Jersey.

[95] U. Pferschy, D. Pisinger, G.J. Woeginger (1997), "Simple but Efficient Approaches for the Collapsing Knapsack Problem", *Discrete Applied Mathematics*, **77**, 271–280,

[96] D. Pisinger (1995), "An expanding-core algorithm for the exact 0-1 knapsack problem," *European Journal of Operational Research*, **87**, 175–187.

[97] D. Pisinger (1994), "Core Problems in Knapsack Algorithms", *DIKU, University of Copenhagen, Denmark*, Report 94/26. Submitted *Operations Research*, conditionally accepted.

[98] D. Pisinger (1994), "Dominance Relations in Unbounded Knapsack Problems", *DIKU, University of Copenhagen, Denmark*, Report 94/33. Submitted *European Journal of Operational Research*.

[99] D. Pisinger (1995), "An $O(nr)$ Algorithm for the Subset Sum Problem", *DIKU, University of Copenhagen, Denmark*, Report 95/6.

[100] D. Pisinger (1995), "A minimal algorithm for the Bounded Knapsack Problem", In: E. Balas, J. Clausen (eds.): *Integer Programming and Combinatorial Optimzation, Fourth IPCO conference*. Lecture Notes in Computer Science, **920**, 95–109.

[101] D. Pisinger (1995), "The Multiple Loading Problem", *Proc. NOAS'95*, University of Reykjavík, 18–19 August 1995. Submitted *European Journal of Operational Research*

[102] D. Pisinger (1995), "A minimal algorithm for the Multiple-choice Knapsack Problem," *European Journal of Operational Research*, **83**, 394–410.

[103] D. Pisinger (1996), "Strongly correlated knapsack problems are trivial to solve", *Proc. CO96*, Imperial College of Science, Technology and Medicine, London 27–29 March 1996. Submitted *Discrete Applied Mathematics*.

[104] D. Pisinger (1996), "The Bounded Multiple-choice Knapsack Problem", Proc. *AIRO'96* Perugia, 16–20 September 1996, 363–365.

[105] D. Pisinger (1997), "A minimal algorithm for the 0-1 knapsack problem", *Operations Research*, **45**, 758–767.

[106] G. Plateau and M. Elkihel (1985), "A hybrid method for the 0-1 knapsack problem", *Methods of Operations Research*, **49**, 277–293.

[107] S.A. Plotkin, D.B. Shmoys, E. Tardos (1991), "Fast approximation algorithms for fractional packing and covering problems", *Proc. 32nd Annual Symposium on Foundations of Computer Science*, San Juan, Puerto Rico, 1–4 October 1991, 495–504.

[108] T. J. van Roy and L. A. Wolsey (1987), "Solving mixed integer programming problems using automatic reformulation", *Operations Research*, **35**, 45–57.

[109] A. Sinha and A. A. Zoltners (1979), "The multiple-choice knapsack problem", *Operations Research*, **27**, 503–515.

[110] The Standard Performance Evaluation Corporation, http://www/specbench.org

[111] M. M. Syslo, N. Deo, J. S. Kowalik (1983), *Discrete Optimization Algorithms*, Prentice Hall, Englewood Cliffs, New Jersey.

[112] P. Toth (1980), "Dynamic programming algorithms for the zero-one knapsack problem", *Computing*, **25**, 29–45.

[113] R. Weismantel (1995), "Knapsack Problems, Test Sets and Polyhedra", Habilitationsschrift, TU Berlin, June 1995.

[114] C. Witzgal (1977), "On One-Row Linear Programs", Applied Mathematics Division, National Bureau of Standards.

[115] L.A. Wolsey (1975), "Facets of linear inequalities in 0-1 Variables", *Mathematical Programming*, **8**, 165–178.

[116] E. Zemel (1980), "The linear multiple choice knapsack problem", *Operations Research*, **28**, 1412–1423.

[117] E. Zemel (1984), "An $O(n)$ algorithm for the linear multiple choice knapsack problem and related problems", *Information Processing Letters*, **18**, 123–128.

[118] A.A. Zoltners (1978), "A direct descent binary knapsack algorithm", *Journal of ACM*, **25**, 304–311.

HANDBOOK OF COMBINATORIAL OPTIMIZATION (VOL.1)
D.-Z. Du and P.M. Pardalos (Eds.) pp. 429-478
©1998 Kluwer Academic Publishers

Fractional Combinatorial Optimization

Tomasz Radzik
Department of Computer Science
King's College London, London WC2R 2LS, UK
E-mail: radzik@dcs.kcl.ac.uk

Contents

1	Introduction	430
2	Fractional Combinatorial Optimization - the General Case	431
3	The Newton method	437
4	The Newton Method for the Linear Case	441
5	Megiddo's Parametric Search	448
6	Maximum Profit-to-Time Ratio Cycles	458
7	Maximum Mean Cycles	460
8	Maximum Mean-Weight Cuts	462
9	Concluding Remarks	472

References

1 Introduction

An instance of a *fractional combinatorial optimization* problem $\mathcal{F}$ consists of a specification of a set $\mathcal{X} \subseteq \{0,1\}^p$, and two functions $f : \mathcal{X} \longrightarrow \mathbf{R}$ and $g : \mathcal{X} \longrightarrow \mathbf{R}$. The task is to

$$\mathcal{F}: \quad \text{maximize} \quad \frac{f(\mathbf{x})}{g(\mathbf{x})}, \quad \text{for} \quad \mathbf{x} \in \mathcal{X}.$$

We assume that $f(\mathbf{x}) > 0$ for some $\mathbf{x} \in \mathcal{X}$, and $g(\mathbf{x}) > 0$ for all $\mathbf{x} \in \mathcal{X}$. The elements of the discrete domain $\mathcal{X}$ are often called *structures*, since in concrete fractional combinatorial optimization problems they represent combinatorial structures like cycles or spanning trees of a graph. Examples of fractional combinatorial optimization include the minimum-ratio spanning-tree problem [4], the maximum profit-to-time cycle problem [9, 13], the maximum-mean cycle problem [28], the maximum mean-weight cut problem [38], and the fractional 0–1 knapsack problem [24].

Fractional combinatorial optimization is a special case of (general) fractional optimization: maximize $f(\mathbf{x})/g(\mathbf{x})$ over a subset D of $\mathbf{R}^n$. Fractional optimization problems have been extensively studied and many algorithms have been designed and analyzed (see, for example, surveys [41, 42, 43, 44]). Computational methods designed for general fractional optimization can usually be adapted to fractional combinatorial optimization, but the analysis of a method for the combinatorial case may be considerably different from the analysis of the same method for the general case. For example, when general fractional optimization is considered, it is often assumed that the domain D is a compact subset of $\mathbf{R}^n$ and functions f and g are continuous. Such assumptions clearly do not have equivalent counterparts in fractional combinatorial optimization. Computational methods for general fractional optimization converge to the optimum objective value, but they cannot guarantee that this value is actually reached. In most cases, the corresponding methods for fractional combinatorial optimizations do produce optimal solutions. Note that the domain $\mathcal{X}$ is finite, so an optimal solution can be found in finite time by simply checking all structures $\mathbf{x} \in \mathcal{X}$.

This chapter is focused on two main methods for fractional combinatorial optimization, the *Newton method* and *Megiddo's parametric search method*. Both these methods, as well as most of the existing methods for general fractional optimization, are based on reduction from fractional optimization to non-fractional, parametric optimization. The Newton method was introduced for general fractional optimization by Dinkelbach [10], and there are a number of results concerning the fast convergence of this method and its variants (see for example [23, 35, 40]). The analyses of the Newton method for fractional combinatorial optimization presented in this chapter are based on [36] and [38]. Megiddo's parametric search method [31, 32] was designed for the case of fractional combinatorial optimization when both functions f and g are linear. This method yielded, for example, the first strongly polynomial algorithm for the maximum profit-to-time cycle problem. A strongly polynomial bound related to problem $\mathcal{F}$ is a bound which depends polynomially on p and does not depend on any other input parameters.

This chapter is organized in the following way. In Section 2 we describe basic properties of fractional combinatorial optimization, the reduction from fractional optimization to parametric optimization, and the binary search method. We also introduce in this section the maximum-ratio path problem for acyclic graphs, which we later use to illustrate both the Newton method and Megiddo's parametric search. In Section 3 we introduce the Newton method for fractional combinatorial optimization and present some general results about its convergence. In Section 4 we show that if both functions f and g of problem $\mathcal{F}$ are linear, then the Newton method finds an optimal solution in a strongly polynomial number of iterations, regardless of the structure of the domain $\mathcal{X}$. In Section 5 we discuss in depth Megiddo's parametric search method. Sections 6 and 8 contain two case studies. In Section 6 we show how Megiddo's parametric search yields fast algorithms for the maximum profit-to-time ratio cycle problem. In Section 8 we present an analysis of the Newton method for the maximum mean-weight cut problem, which gives the best known bound on the computational complexity of this problem. Most of the fastest known algorithms for fractional combinatorial optimization problems are based on either the Newton method, or Megiddo's parametric search, or the basic binary search method. The most notable exception to this rule is Karp's algorithm for the maximum-mean cycle problem [26]. We present this algorithm and its analysis in Section 7. Section 9 contains a few final comments and suggestions for further research.

2 Fractional Combinatorial Optimization - the General Case

For $\mathbf{x} \in \mathcal{X}$, numbers $f(\mathbf{x})$, $g(\mathbf{x})$, and $f(\mathbf{x})/g(\mathbf{x})$ are called *the cost*, the *weight*, and the *mean-weight cost* of structure $\mathbf{x}$. Using this terminology, problem $\mathcal{F}$ is to compute the maximum mean-weight cost of a structure in domain $\mathcal{X}$. We also want to find a structure which achieves this maximum mean-weight cost.

Consider the following problem.

$\mathcal{P}$: minimize $\delta \in \mathbf{R}$, subject to $f(\mathbf{x}) - \delta g(\mathbf{x}) \leq 0$, for all $\mathbf{x} \in \mathcal{X}$.

A pair $(\delta^*, \mathbf{x}^*) \in \mathbf{R} \times \mathcal{X}$ is an optimal solution of problem $\mathcal{P}$, if and only if,

$$f(\mathbf{x}) - \delta^* g(\mathbf{x}) \leq 0 = f(\mathbf{x}^*) - \delta^* g(\mathbf{x}^*), \quad \text{for each } \mathbf{x} \in \mathcal{X}.$$

This condition is equivalent to

$$f(\mathbf{x})/g(\mathbf{x}) \leq \delta^* = f(\mathbf{x}^*)/g(\mathbf{x}^*), \quad \text{for each } \mathbf{x} \in \mathcal{X},$$

which means that δ^* is the optimum objective value and $\mathbf{x}^*$ is an optimal solution of problem $\mathcal{F}$. Thus problems $\mathcal{F}$ and $\mathcal{P}$ are equivalent. Many iterative methods for solving problem $\mathcal{F}$ generate and solve a sequence of instances of the following problem, where $\delta \in \mathbf{R}$ is an additional input parameter.

$\mathcal{P}(\delta)$: maximize $f(\mathbf{x}) - \delta g(\mathbf{x})$, for $\mathbf{x} \in \mathcal{X}$.

Problem $\mathcal{P}(\delta)$ is called the *parametric problem* corresponding to problem $\mathcal{F}$, and sometimes also the *non-fractional version* of problem $\mathcal{F}$. Let $h(\delta)$ and $\mathbf{x}_\delta^*$ denote the optimum objective value and an optimal solution of problem $\mathcal{P}(\delta)$. We have $h(\delta) = 0$, if and only if, $(\delta, \mathbf{x}_\delta^*)$ is an optimal solution of problem $\mathcal{P}$, that is, if and only if, δ and $\mathbf{x}_\delta^*$ are the optimum objective value and an optimal solution of problem $\mathcal{F}$. Hence we have another equivalent formulation $\mathcal{Z}$ of problem $\mathcal{F}$.

$\mathcal{Z}$: find $\delta \in \mathbf{R}$ such that $h(\delta) = 0$, where

$$h(\delta) = \max\{f(\mathbf{x}) - \delta g(\mathbf{x}) \mid \mathbf{x} \in \mathcal{X}\}. \tag{1}$$

This formulation suggests that one can try to design algorithms for problem $\mathcal{F}$ by applying classical methods for finding a root of a function. The following properties of function h can be easily obtained from the fact that h is the maximum of a finite number of decreasing linear functions.

(i) function h is continuous on $(-\infty, +\infty)$ and strictly decreasing from $+\infty$ to $-\infty$.

(ii) $h(0) > 0$ (this follows from the assumption that $f(\mathbf{x}) > 0$, for some $\mathbf{x} \in \mathcal{X}$).

(iii) function h has exactly one root δ^*, and $\delta^* > 0$.

(iv) If $\delta_1 < \delta_2 < \cdots < \delta_q$ denote all values of δ for which two lines in $\{f(\mathbf{x}) - \delta g(\mathbf{x}) \mid \mathbf{x} \in \mathcal{X}\}$ intersect, then function h is linear on each interval $[-\infty, \delta_1], [\delta_i, \delta_{i+1}]$, for $i = 1, 2, \ldots, q-1$, and $[\delta_q, \infty]$.

(v) function h is convex.

Some methods for solving problem $\mathcal{F}$ require a subroutine only for the following weaker version of problem $\mathcal{P}(\delta)$.

$\mathcal{P}_0(\delta):$ find $\mathbf{y} \in \mathcal{X}$ such that
$$\operatorname{sign}(f(\mathbf{y}) - \delta g(\mathbf{y})) = \operatorname{sign}(\max\{f(\mathbf{x}) - \delta g(\mathbf{x}) : \mathbf{x} \in \mathcal{X}\}).$$

That is, problem $\mathcal{P}_0(\delta)$ is only to find the sign of the objective value of problem $\mathcal{P}(\delta)$. In our discussion of properties of computational methods for problem $\mathcal{F}$, we use the following three parameters:

$$\text{MAX}_f = \max\{f(\mathbf{x}) : \mathbf{x} \in \mathcal{X}\},$$
$$\text{MAX}_g = \max\{g(\mathbf{x}) : \mathbf{x} \in \mathcal{X}\},$$
$$\text{MIN}_g = \min\{g(\mathbf{x}) : \mathbf{x} \in \mathcal{X}\},$$
$$\text{GAP} = \min\left\{\left|\frac{f(\mathbf{x}')}{g(\mathbf{x}')} - \frac{f(\mathbf{x}'')}{g(\mathbf{x}'')}\right| : \mathbf{x}', \mathbf{x}'' \in \mathcal{X}, \text{ and } \frac{f(\mathbf{x}')}{g(\mathbf{x}')} \neq \frac{f(\mathbf{x}'')}{g(\mathbf{x}'')}\right\}.$$

Our assumptions on functions f and g imply that the optimum objective value of problem $\mathcal{F}$ is contained in interval $(0, \text{MAX}_f/\text{MIN}_g]$. Parameter GAP is the smallest difference between two different mean-weight costs.

If the weight function g counts the number of ones, then we have a *uniform* fractional combinatorial optimization problem:

$$\mathcal{F}_U: \text{ maximize } \frac{f(x_1, x_2, \ldots, x_p)}{x_1 + x_2 + \cdots + x_p}, \quad \text{for } (x_1, x_2, \ldots, x_p) \in \mathcal{X}.$$

If both the cost function f and the weight function g are linear, then we have a *linear* fractional combinatorial optimization problem:

$$\mathcal{F}_L: \text{ maximize } \frac{a_1 x_1 + a_2 x_2 + \cdots + a_p x_p}{b_1 x_1 + b_2 x_2 + \cdots + b_p x_p}$$
$$\text{subject to } (x_1, x_2, \ldots, x_p) \in \mathcal{X}.$$

An instance of problem $\mathcal{F}_L$ consists of a specification of a set of structures $\mathcal{X} \subseteq \{0,1\}^p$ and two real vectors $\mathbf{a} = (a_1, a_2, \ldots, a_p)$ and $\mathbf{b} = (b_1, b_2, \ldots, b_p)$. Throughout this paper, we denote the inner product $c_1 z_1 + c_2 z_2 + \cdots + c_p z_p$ of two vectors $\mathbf{c} = (c_1, c_2, \ldots, c_p)$ and $\mathbf{z} = (z_1, z_2, \ldots, z_p)$ by $\mathbf{cz}$. Thus in problem $\mathcal{F}_L$, the cost, the weight, and the mean-weight cost of a structure $\mathbf{x}$ are equal to $\mathbf{ax}$, $\mathbf{bx}$, and $(\mathbf{ax})/(\mathbf{bx})$, respectively.

Maximum-ratio paths in acyclic graphs

As an example of fractional combinatorial optimization, we consider the following MaxRatioPath problem. An input instance of this problem consists of an integer $n \geq 1$, a set of edges E of an acyclic directed graph with n vertices $1, 2, \ldots, n$, an edge-cost function $\mathbf{c} : E \longrightarrow \mathbf{R}$, and an edge-weight function $\mathbf{w} : E \longrightarrow \mathbf{R}$. To keep this example simple, we make the following assumptions: there are no multiple edges, the vertices are numbered according to a topological sort of the graph (that is, if $(v, u) \in E$, then $v < u$), and each vertex is reachable from vertex 1. For a path P, let $f(P) = \sum_{(v,u) \in P} \mathbf{c}(v, u)$, $g(P) = \sum_{(v,u) \in P} \mathbf{w}(v, u)$, and $f(P)/g(P)$ be the cost, the weight, and the mean-weight cost of this path. The task is to find a path from vertex 1 to vertex n which has the maximum mean-weight cost.

MaxRatioPath : maximize $\dfrac{f(P)}{g(P)}$,

over all paths $P \subseteq E$ from vertex 1 to vertex n.

This is a linear fractional combinatorial optimization problem. Using our terminology of general fractional combinatorial optimization, the set of structures $\mathcal{X}$ is here the set of characteristic vectors $\mathbf{x} \in \{0,1\}^E$ corresponding to paths from vertex 1 to vertex n.

The parametric problem corresponding to the MaxRatioPath problem is

MaxPath(δ) : maximize $f(P) - \delta g(P)$,

over all paths $P \subseteq E$ from vertex 1 to vertex n.

For a fixed $\delta \in \mathbf{R}$ and a path P, $f(P) - \delta g(P) = \sum_{e \in P} (\mathbf{c}(e) - \delta \mathbf{w}(e))$, so the optimum value of problem MaxPath(δ) is the maximum cost of a path from vertex 1 to vertex n according to the edge-cost function $\mathbf{c} - \delta \mathbf{w}$. An example of an input instance of the MaxRatioPath problem and the corresponding function h are shown in Figure 1.

In an acyclic graph, the shortest paths, as well as the longest ones, can be computed by considering the vertices according to a topological order [8,

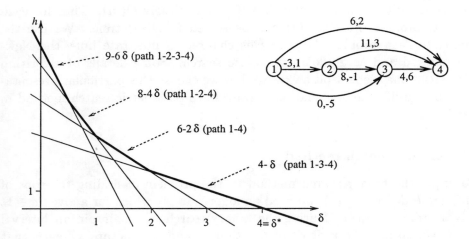

Figure 1: An input for the MaxRatioPath problem and the corresponding function $h(\delta)$. Lines $9 - 6\delta$, $8 - 4\delta$, $6 - 2\delta$, and $4 - \delta$ correspond to the four paths from vertex 1 to vertex 4. The optimal path is 1–3–4.

Chapter 25]. Algorithm MAXCOST described below computes the maximum cost of a path from vertex 1 to vertex n for a set of edges E and an arbitrary edge-cost function $\mathbf{a}$ by considering the vertices from 1 to $n-1$. While considering vertex v, the algorithm examines all edges outgoing from v. During the computation, the boolean variable $seen[u]$ indicates whether vertex u has already been encountered. At the end of the computation, number $d[v]$, for each $v = 1, 2, \ldots, n$, is equal to the maximum cost of a path from vertex 1 to vertex v. The *predecessor* pointers form a "longest-paths" tree rooted at vertex 1.

MAXCOST$(n, E, \mathbf{a})$
1) $d[1] \leftarrow 0$; $seen[1] \leftarrow$ **true**
2)   for $v \leftarrow 2$ to n do $seen[v] \leftarrow$ **false**
3)   for $v \leftarrow 1$ to $n - 1$ do
4)   for each u such that $(v, u) \in E$ do
5)   if (not $seen[u]$) or $(d[v] + \mathbf{a}(v, u) - d[u] > 0)$ then
6) $d[u] \leftarrow d[v] + \mathbf{a}(v, u)$; $predecessor[u] \leftarrow v$; $seen[u] \leftarrow$ **true**
7)   let $P = (v_1, v_2, \ldots, v_k)$ where $v_1 = 1$, $v_k = n$, and
      $v_i = predecessor[v_{i+1}]$, for $i = 1, 2, \ldots, k - 1$
8)   **return** $d[n]$ and path P

The running time of algorithm MAXCOST is clearly $O(m)$. Thus for each $\delta \in \mathbf{R}$, problem MaxPath(δ) can be solved in $O(m)$ time. We use the MaxRatioPath problem later in this chapter to illustrate both the Newton method and Megiddo's parametric search. Some details of algorithm MAXCOST may look a bit awkward, but we choose this particular presentation to simplify the explanation of Megiddo's parametric search method in Section 5.

The binary search method

To apply the binary search method to the problem of finding the root of function h defined in (1), we need an algorithm $\mathcal{A}_0$ which for a given $\delta \in \mathbf{R}$ solves problem $\mathcal{P}_0(\delta)$. During the binary search, we maintain an interval (α, β) containing the root δ^* of function h, and a structure $\mathbf{x}_\alpha$ such that $f(\mathbf{x}_\alpha) - \alpha g(\mathbf{x}_\alpha) > 0$. Structure $\mathbf{x}_\alpha$ is a "witness" that $\delta^* > \alpha$. During one iteration of the binary search, algorithm $\mathcal{A}_0$ is run for value $\delta = (\alpha + \beta)/2$. If it returns a structure $\mathbf{x}$ such that $f(\mathbf{x}) - \delta g(\mathbf{x}) = 0$, then $\mathbf{x}$ is an optimal solution of problem $\mathcal{F}$, and the computation terminates. If the returned structure $\mathbf{x}$ is such that $f(\mathbf{x}) - \delta g(\mathbf{x}) > 0$, then δ^* must be in interval (δ, β), so α is set to δ and $\mathbf{x}_\alpha$ is set to $\mathbf{x}$. Otherwise δ^* is in interval (α, δ), so β is set to δ.

The computation proceeds until an optimal solution has been found or the desired precision has been reached. At the end of the current iteration, $\alpha < f(\mathbf{x}_\alpha)/g(\mathbf{x}_\alpha) \le \delta^* < \beta$, so $\delta^* - f(\mathbf{x}_\alpha)/g(\mathbf{x}_\alpha) < \beta - \alpha$. Therefore, after $\lceil \log((\beta_0 - \alpha_0)/\epsilon) \rceil$ iterations, $\delta^* - f(\mathbf{x}_\alpha)/g(\mathbf{x}_\alpha) < \epsilon$, where (α_0, β_0) is the initial interval (α, β). Throughout this chapter, the base of logarithms is 2. If the computation is continued for a sufficient number of iterations, it actually finds an optimal solution of problem $\mathcal{F}$. Observe that when $\delta^* - f(\mathbf{x}_\alpha)/g(\mathbf{x}_\alpha)$, which is equal to $f(\mathbf{x}^*)/g(\mathbf{x}^*) - f(\mathbf{x})/g(\mathbf{x})$ for an optimal structure $\mathbf{x}^*$, becomes eventually less than GAP, then $f(\mathbf{x}_\alpha)/g(\mathbf{x}_\alpha) = \delta^*$, so $\mathbf{x}_\alpha$ is an optimal structure. This means that the binary search finds an optimal solution of problem $\mathcal{F}$ in at most $\lceil \log((\beta_0 - \alpha_0)/\text{GAP}) \rceil$ iterations. The initial interval (α_0, β_0) can be $(0, U/L)$, where U and L are an upper bound on MAX_f and a positive lower bound on MIN_g, respectively. For concrete fractional combinatorial optimization problems, such bounds are usually readily available. If algorithm $\mathcal{A}_0$ is somewhat weaker than we assumed and finds only a structure $\mathbf{x}$ with nonnegative $f(\mathbf{x}) - \delta g(\mathbf{x})$, provided that such a structure exists, then the binary search procedure described here can be modified in a straightforward way.

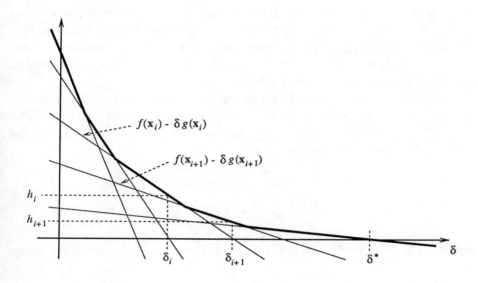

Figure 2: The Newton method for solving $h(\delta) = 0$

As an example, consider the special case of the MaxRatioPath problem when all edge costs and weights are integers not greater than U. For this problem we have $\text{MAX}_f \leq nU$, $\text{MIN}_g \geq 1$, and $\text{GAP} \geq 1/(nU)^2$, so the binary search method yields an $O(m \log(nU))$-time algorithm.

3 The Newton method

The Newton method for a fractional combinatorial optimization problem $\mathcal{F}$, called also the Dinkelbach method [10], is an application of the classical Newton method to the problem of finding the root of function h defined in (1). To use this method we need an algorithm $\mathcal{A}_{\max}$ which for a given $\delta \in \mathbf{R}$ computes $h(\delta)$ and a structure $\mathbf{x} \in \mathcal{X}$ such that $f(\mathbf{x}) - \delta g(\mathbf{x}) = h(\delta)$. That is, we need an algorithm $\mathcal{A}_{\max}$ which for a given $\delta \in \mathbf{R}$ solves problem $\mathcal{P}(\delta)$.

The Newton method for problem $\mathcal{F}$ is an iterative process which in each iteration generates a new, better lower estimate on the optimum objective value δ^*. During iteration i, the current estimate $\delta_i \leq \delta^*$ is considered and the following computation is performed. We first run algorithm $\mathcal{A}_{\max}$ for $\delta = \delta_i$ and obtain $h_i = h(\delta_i)$ and a structure $\mathbf{x}_i \in \mathcal{X}$ such that $f(\mathbf{x}_i) -$

$\delta_i g(\mathbf{x}_i) = h_i$. If $h_i = 0$, then $\delta_i = \delta^*$ and $\mathbf{x}_i$ is an optimal solution of problem $\mathcal{F}$, so the computation terminates. Otherwise we obtain the new estimate $\delta_{i+1} \leftarrow f(\mathbf{x}_i)/g(\mathbf{x}_i)$ (observe that $\delta_i < \delta_{i+1} \leq \delta^*$, since in this case $h_i > 0$) and we proceed to the next iteration. This process is illustrated in Figure 2. The computation begins with $\delta_1 = 0$.

Our aim is to derive bounds on the number of iterations. Let t be the index of the last iteration, or $+\infty$, if the computation does not terminate, and let $f_i = f(\mathbf{x}_i)$ and $g_i = g(\mathbf{x}_i)$, for each $1 \leq i < t+1$. From the description of the computation, we have

$$h_i = f_i - \delta_i g_i, \quad \text{for each } 1 \leq i < t+1, \tag{2}$$

$$\delta_{i+1} = \frac{f_i}{g_i}, \quad \text{for each } 1 \leq i < t. \tag{3}$$

Lemma 3.1 *The Newton method terminates in a finite number of iterations and*

(A) $h_1 > h_2 > \cdots > h_{t-1} > h_t = 0$,

(B) $0 = \delta_1 < \delta_2 < \cdots < \delta_{t-1} < \delta_t = \delta^*$,

(C) $g_1 > g_2 > \cdots > g_{t-1} \geq g_t$.

Proof. This lemma follows immediately from the following claim. For each $1 \leq i < t+1$,

(a) $h_1 > h_2 > \cdots > h_{i-1} > h_i \geq 0$,

(b) $0 = \delta_1 < \delta_2 < \cdots < \delta_{i-1} < \delta_i \leq \delta^*$,

(c) $g_1 > g_2 > \cdots > g_{i-1} \geq g_i$,

and the equalities at the end of these three chains are possible only if iteration i is the last iteration in the whole computation. We prove this claim by induction on i. For $i = 1$ the claim follows from the properties (ii) and (iii) of function h. Assume now that the claim is true for $i = j$, for some $1 \leq j < t$. Using (2) and (3), we have

$$\delta^* \geq \frac{f_j}{g_j} = \delta_{j+1} \quad \text{and} \quad \delta_{j+1} = \frac{f_j}{g_j} = \frac{h_j}{g_j} + \delta_j > \delta_j,$$

so (b) is true for $i = j + 1$. Since $\delta_j < \delta_{j+1} \leq \delta^*$, the monotonicity of function h implies that $h_{j+1} = h(\delta_{j+1}) < h(\delta_j) = h_j$, and $h_{j+1} = h(\delta_{j+1}) \geq$

$h(\delta^*) = 0$, so (a) is true for $i = j + 1$. To show that also (c) is true for $i = j + 1$, we use the following two inequalities

$$f_{j+1} - \delta_j g_{j+1} \leq \max_{\mathbf{x} \in \mathcal{X}} \{f(\mathbf{x}) - \delta_j g(\mathbf{x})\} = h_j = f_j - \delta_j g_j,$$

$$f_{j+1} - \delta_{j+1} g_{j+1} = h_{j+1} = \max_{\mathbf{x} \in \mathcal{X}} \{f(\mathbf{x}) - \delta_{j+1} g(\mathbf{x})\} \geq f_j - \delta_{j+1} g_j.$$

Subtracting these two inequalities we obtain $g_{j+1} \leq g_j$, and the equality is possible only if $h_{j+1} = f_j - \delta_{j+1} g_j$. Since $f_j - \delta_{j+1} g_j$ is by definition equal to 0 (see (3)), $g_{j+1} = g_j$ only if $h_{j+1} = 0$, that is, only if iteration $j+1$ is the last iteration of the whole computation. Hence (c) is true for $i = j + 1$. This concludes the proof of the claim.

The claim implies that all elements of sequence $(g_i)_{1 \leq i < t}$ are distinct, so t must be finite (numbers g_i are weights of structures from $\mathcal{X}$, so $t - 1 \leq |\mathcal{X}| < +\infty$). □

The following lemma indicates that sequences $(h_i)_{1 \leq i \leq t}$ and $(g_i)_{1 \leq i \leq t}$ should geometrically decrease to zero. This lemma is the basic tool in all analyses of the Newton method presented in this chapter.

Lemma 3.2 *For each* $i = 1, 2, \ldots, t - 1$,

$$\frac{h_{i+1}}{h_i} + \frac{g_{i+1}}{g_i} \leq 1. \tag{4}$$

Proof. Let $1 \leq i \leq t - 1$. Structure $\mathbf{x}_i$ maximizes $f(\mathbf{x}) - \delta_i g(\mathbf{x})$ over $\mathcal{X}$, so

$$\begin{aligned} h_i = f_i - \delta_i g_i &= f(\mathbf{x}_i) - \delta_i g(\mathbf{x}_i) \\ &\geq f(\mathbf{x}_{i+1}) - \delta_i g(\mathbf{x}_{i+1}) = f_{i+1} - \delta_i g_{i+1}. \end{aligned}$$

Therefore, using (2) and (3), we have

$$\begin{aligned} h_i &\geq f_{i+1} - \delta_i g_{i+1} \\ &= h_{i+1} + \delta_{i+1} g_{i+1} - \delta_i g_{i+1} \\ &= h_{i+1} + \frac{f_i}{g_i} g_{i+1} - \frac{h_i - f_i}{g_i} g_{i+1} \\ &= h_{i+1} + h_i \frac{g_{i+1}}{g_i}. \end{aligned}$$

This inequality immediately implies Inequality (4). □

We present now one general bound on the number of iterations of the Newton method, and two bounds for special classes of fractional combinatorial optimization problems.

Theorem 3.3 *The Newton method solves problem $\mathcal{F}$ in at most* $\log(\mathrm{MAX}_f) + \log(\mathrm{MAX}_g) - \log(\mathrm{MIN}_g) - \log(\mathrm{GAP}) + 2$ *iterations.*

Proof. Lemma 3.2 implies that for $i = 1, 2, \ldots, t-1$,

$$\frac{h_{i+1}g_{i+1}}{h_i g_i} \leq \frac{1}{4}. \tag{5}$$

Hence $h_{t-1}g_{t-1} \leq (1/4^{t-2}) h_1 g_1$ and

$$t \leq \log_4(h_1 g_1) - \log_4(h_{t-1}g_{t-1}) + 2. \tag{6}$$

Since $h_1 = \mathrm{MAX}_f$, so

$$h_1 g_1 \leq \mathrm{MAX}_f \cdot \mathrm{MAX}_g. \tag{7}$$

Equations (2) and (3) imply that for each $i = 1, 2, \ldots, t-1$, $\delta_{i+1} - \delta_i = h_i/g_i$. Hence

$$h_{t-1}g_{t-1} = (\delta_t - \delta_{t-1})g_{t-1}^2 = \left(\frac{f_{t-1}}{g_{t-1}} - \frac{f_{t-2}}{g_{t-2}}\right) g_{t-1}^2 \geq \mathrm{GAP} \cdot (\mathrm{MIN}_g)^2. \tag{8}$$

Inequalities (6), (7), and (8) imply the bound on the number of iterations stated in this theorem. □

The following two theorems have been frequently rediscovered in the context of many different fractional combinatorial optimization problems.

Theorem 3.4 *For a linear fractional combinatorial optimization problem $\mathcal{F}_L$ such that all input numbers a_i and b_i, $1 \leq i \leq p$, are integers not greater than U, the Newton method runs in $O(\log(pU))$ iterations.*

Proof. For such a problem $\mathcal{F}_L$, we have $\mathrm{MAX}_f \leq pU$, $\mathrm{MAX}_g \leq pU$, $\mathrm{MIN}_g \geq 1$, and $\mathrm{GAP} \geq 1/(pU)^2$, so Theorem 3.3 implies a $O(\log(pU))$ bound on the number of iterations. □

Theorem 3.5 *For a uniform fractional combinatorial optimization problem $\mathcal{F}_U$, the Newton method runs in at most $p+1$ iterations.*

Proof. Sequence $(g_i)_{1 \leq i \leq t-1}$ is strictly decreasing. For a uniform fractional optimization problem, each number g_i is a positive integer not greater than p, so $t \leq p+1$. □

4 The Newton Method for the Linear Case

In this section we show that for every linear fractional combinatorial optimization problem $\mathcal{F}_L$, the Newton method runs in a strongly polynomial number of iterations. We use the notation introduced in Section 3. Inequality (5), a direct consequence of Lemma 3.2, says that sequence $(h_i g_i)$ geometrically decreases to zero. Equations (2) and (3) imply that $h_1 g_1 = f_1 g_1$ and for $2 \leq i \leq t$,

$$h_i g_i = (f_1 - \delta_i g_i) g_i = \left(f_i - \frac{f_{i-1}}{g_{i-1}} g_i \right) g_i. \tag{9}$$

For each $1 \leq i \leq t$, numbers f_i and g_i are sums of some numbers drawn from $2p$ numbers $a_1, a_2, \ldots, a_p$, and $b_1, b_2 \ldots, b_p$. Thus the elements of sequence $(h_i g_i)$ are created from fixed $2p$ numbers using $O(p)$ additions/subtractions and up to three multiplications/divisions. Bounds on the numbers of iterations of the Newton method for linear fractional combinatorial optimization come from the fact that a geometric sequence whose elements are constructed in such a limited way cannot be long.

To establish intuition why such sequences cannot be long, let us assume for a moment that $b_1 \geq b_2 \geq \cdots \geq b_p \geq 0$, and that there exists a constant α such that for every $i = 1, 2, \ldots, t-1$, $g_{i+1}/g_i \leq \alpha < 1$. That is, we assume that sequence $(g_i)_{1 \leq i \leq t}$ on its own geometrically decreases to zero. Each number g_i is equal to the sum of distinct elements drawn from the multi-set $\{b_1, b_2, \ldots, b_p\}$. Obviously $g_1 \leq p b_1$. Since sequence (g_i) decreases geometrically, we must have $g_k < b_1$, for some $k = O(\log p)$. This means that number b_1 is not a term in the sum g_k, nor does it occur in any sum g_i, for $i \geq k$, so it can be excluded from further considerations. Since $g_k < b_1$, we must have $g_k \leq (p-1) b_2$, and after next $O(\log p)$ iterations we can exclude b_2, then b_3, and so on. Therefore, the length of sequence (g_i) is only $O(p \log p)$. (In fact, one can show that the length of such a sequence is $O(p)$. See Lemma 8.4.)

The actual analysis of the number of iterations is somewhat more involved than the previous paragraph may suggest, because numbers $h_i g_i$ have "more complicated structure" than numbers g_i (see Equation (9)). Moreover, we have to deal with sums of numbers which are not necessary nonnegative. Even if all numbers $a_1, a_2, \ldots, a_p, b_1, b_2 \ldots, b_p$ are positive, negative terms may appear since subtractions are used in the definition of numbers h_i. The following lemma, which gives a bound on the length of a geometrically decreasing sequence of sums of numbers, is our main tool in bounding the number of iterations of the Newton method. In the statement of this lemma, the elements of sequence $(\mathbf{y}_k \mathbf{c})_{k \geq 1}$ are constructed by adding and/or subtracting numbers drawn from the set of the components of vector $\mathbf{c}$.

Lemma 4.1 *[15] Let $\mathbf{c} = (c_1, c_2, \ldots, c_p)$ be a p dimensional vector with non-negative real components, and let $\mathbf{y}_1, \mathbf{y}_2, \ldots, \mathbf{y}_q$ be vectors from $\{-1, 0, 1\}^p$. If for all $i = 1, 2, \ldots, q-1$*

$$0 < \mathbf{y}_{i+1}\mathbf{c} \leq \frac{1}{2}\mathbf{y}_i \mathbf{c},$$

then $q = O(p \log p)$.

Proof. Consider the following polyhedron

$$\begin{aligned} P = \{ \mathbf{z} = (z_1, z_2, \ldots, z_p) \in \mathbf{R}^p \; : \; & \\ (\mathbf{y}_i - 2\mathbf{y}_{i+1})\mathbf{z} &\geq 0, \text{ for } i = 1, 2, \ldots, q-1, \\ \mathbf{y}_q \mathbf{z} &= 1, \\ z_i &\geq 0, \text{ for } i = 1, 2, \ldots, p \; \}. \end{aligned}$$

Let A and $\mathbf{b}$ denote the coefficient matrix and the right-hand side vector of the system defining polyhedron P. This polyhedron is not empty because it contains vector $\mathbf{c}/(\mathbf{y}_q \mathbf{c})$. From polyhedral theory we know that there exists a vector $\mathbf{c}^* = (c_1^*, c_2^*, \ldots, c_p^*) \in P$ such that $A'\mathbf{c}^* = \mathbf{b}'$ for some nonsingular $p \times p$ submatrix A' of matrix A and a subvector $\mathbf{b}'$ of vector $\mathbf{b}$ (the vertices of the polyhedron are such vectors). Cramer's rule says that for each $i = 1, 2, \ldots, p$,

$$c_i^* = \frac{\det A_i'}{\det A'},$$

where matrix A_i' is obtained from matrix A' by replacing the i-th column with vector $\mathbf{b}'$.

The determinant of a $p \times p$ matrix $M = (m_{ij})$ is equal to

$$\det M = \sum_\sigma \varepsilon_\sigma m_{1,\sigma(1)} m_{2,\sigma(2)} \cdots m_{p,\sigma(p)},$$

where the summation is over all permutations σ of the set of indices $\{1, 2, \ldots, p\}$, and each number ε_σ is equal to either 1 or -1. Thus $|\det M| \leq m^p p!$, where m is the maximum absolute value of an entry of matrix M. The entries of matrix A and vector $\mathbf{b}$, and consequently the entries of all matrices A'_i, are integers from interval $[-3,3]$. Thus $|\det A'_i| \leq 3^p p!$, for each $i = 1, 2, \ldots, p$. Since we also have $|\det A'| \geq 1$, then $c^*_i \leq 3^p p!$, for each $i = 1, 2, \ldots, p$, and

$$\mathbf{y}_j \mathbf{c}^* \leq \sum_{i=1}^p c^*_i \leq p\, 3^p p!,$$

for each $j = 1, 2, \ldots, q$. Finally we have

$$1 = \mathbf{y}_q \mathbf{c}^* \leq \frac{1}{2^{q-1}} \mathbf{y}_1 \mathbf{c}^* \leq \frac{1}{2^{q-1}} p\, 3^p p!,$$

so $q \leq \log(p3^p p!) + 1 = O(p \log p)$. □

In the analysis of the Newton method we also use the following rephrasing of Lemma 4.1.

Corollary 4.2 *Let* $\mathbf{c} = (c_1, c_2, \ldots, c_p) \in \mathbf{R}^p$, *and let* $\mathbf{y}_1, \mathbf{y}_2, \ldots, \mathbf{y}_q$ *be vectors from* $\{0,1\}^p$. *If for all* $i = 1, 2, \ldots, q-1$

$$0 < \mathbf{y}_{i+1} \mathbf{c} \leq \frac{1}{2} \mathbf{y}_i \mathbf{c},$$

then $q = O(p \log p)$.

We show two bounds on the number of iterations of the Newton method for linear fractional combinatorial optimization. First we show an $O(p^3 \log p)$ bound (Theorem 4.3), which comes from a direct analysis of sequence $(h_i g_i)$. This analysis uses Inequality (5), Equation (9), and Lemma 4.1. Then, by analyzing sequences (h_i) and (g_i) separately, we show an $O(p^2 \log^2 p)$ bound (Theorem 4.6).

Theorem 4.3 *The Newton method solves a linear fractional combinatorial optimization problem* $\mathcal{F}_L$ *in* $O(p^3 \log p)$ *iterations.*

Proof. For each $1 \leq i \leq t-1$, $h_i > 0$. This fact, Equation (9), and Inequality (5) imply that for each $i = 2, 3, \ldots, t-2$,

$$0 < \left(f_{i+1} - \frac{f_i}{g_i} g_{i+1}\right) g_{i+1} \leq \frac{1}{4}\left(f_i - \frac{f_{i-1}}{g_{i-1}} g_i\right) g_i,$$

and since $0 < g_i \leq g_{i-1}$, we further have

$$0 < f_{i+1} g_{i+1} g_i - f_i g_{i+1}^2 \leq \frac{1}{4} f_i g_i g_{i-1} - f_{i-1} g_i^2.$$

Hence, putting $s_i = f_i g_i g_{i-1} - f_{i-1} g_i^2$, we have for each $i = 2, 3, \ldots, t-2$,

$$0 < s_{i+1} \leq \frac{1}{4} s_i.$$

For a p^3-dimensional vector **c** whose set of components is $\{|a_j b_k b_l| : 1 \leq j, k, l \leq p\}$, there exist p^3-dimensional vectors $\mathbf{y}_2, \mathbf{y}_3, \ldots, \mathbf{y}_{t-1}$ with components from $\{-1, 0, 1\}$ such that $s_i = \mathbf{y}_i \mathbf{c}$, for each $i = 2, 3, \ldots, t-1$. Applying Lemma 4.1, we conclude that $t = O(p^3 \log p)$. □

The $O(p^2 \log^2 p)$ bound on the number of iterations is based on the following two lemmas.

Lemma 4.4 *There are at most $O(p \log p)$ iterations k such that $g_{k+1} \leq (2/3) g_k$.*

Proof. Let $k_1 < k_2 < \cdots < k_q$ be the indices k such that $g_{k+1} \leq (1/2) g_k$. Since sequence $(g_i)_{1 \leq i \leq t}$ is non-increasing, we have for each $i = 1, 2, \ldots, q-1$,

$$0 < \mathbf{x}_{k_{i+1}} \mathbf{b} = g_{k_{i+1}} \leq g_{k_i+1} \leq \frac{1}{2} g_{k_i} = \frac{1}{2} \mathbf{x}_{k_i} \mathbf{b}.$$

Applying Corollary 4.2 with $\mathbf{c} = \mathbf{b}$ and $\mathbf{y}_i = \mathbf{x}_{k_i}$, for $i = 1, 2, \ldots, q$, we conclude that $q = O(p \log p)$. Monotonicity of sequence $(g_i)_{1 \leq i \leq t}$ further implies that there are at most $2q = O(p \log p)$ iterations k such that $g_{k+1} \leq (2/3) g_k$. □

Lemma 4.5 *There are at most $O(p \log p)$ consecutive iterations k such that $g_{k+1} \geq (2/3) g_k$.*

Proof. Consider a sequence of q consecutive iterations $j+1, j+2, \ldots, j+q$ such that for each $k = j+1, j+2, \ldots, j+q-1$,

$$g_{k+1} \geq \frac{2}{3} g_k. \tag{10}$$

We show that Corollary 4.2 can be applied to vector $\mathbf{c} = \delta_{j+q} \mathbf{b} - \mathbf{a}$ and the sequence of $q - 2$ vectors $\mathbf{x}_k$, $k = j+1, \ldots, j+q-2$. Inequalities (10) and (4) imply that for $k = j+1, j+2, \ldots, j+q-1$,

$$h_{k+1} \leq \frac{1}{3} h_k. \tag{11}$$

Using Equations (2) and (3), and Inequalities (10) and (11), we obtain for each $k = j+1, j+2, \ldots, j+q-2$,

$$\delta_{k+2} - \delta_{k+1} = \frac{h_{k+1}}{g_{k+1}} \leq \frac{1}{2} \cdot \frac{h_k}{g_k} = \frac{1}{2}(\delta_{k+1} - \delta_k). \qquad (12)$$

Inequality (12) implies

$$\begin{aligned}
\delta_{j+q} - \delta_{k+1} &= (\delta_{j+q} - \delta_{j+q-1}) + (\delta_{j+q-1} - \delta_{j+q-2}) + \cdots + (\delta_{k+2} - \delta_{k+1}) \\
&\leq \frac{1}{2}(\delta_{j+q-1} - \delta_{j+q-2}) + \cdots + \frac{1}{2}(\delta_{k+1} - \delta_k) \\
&= \frac{1}{2}(\delta_{j+q-1} - \delta_k) \\
&< \frac{1}{2}(\delta_{j+q} - \delta_k).
\end{aligned} \qquad (13)$$

Using Equation (3), we obtain for $k = j+1, j+2, \ldots, j+q-2$,

$$\mathbf{x}_k(\delta_{j+q}\mathbf{b} - \mathbf{a}) = \delta_{j+q}\mathbf{b}\mathbf{x}_k - \mathbf{a}\mathbf{x}_k = \delta_{j+q}g_k - f_k = (\delta_{j+q} - \delta_{k+1})g_k > 0.$$

Thus, using Inequalities (13) and the monotonicity of sequence (g_i), we obtain for $k = j+1, j+2, \ldots, j+q-3$,

$$\begin{aligned}
0 < \mathbf{x}_{k+1}(\delta_{j+q}\mathbf{b} - \mathbf{a}) &= (\delta_{j+q} - \delta_{k+2})g_{k+1} \leq (\delta_{j+q} - \delta_{k+2})g_k \\
&\leq \frac{1}{2}(\delta_{j+q} - \delta_{k+1})g_k = \frac{1}{2}\mathbf{x}_k(\delta_{j+q}\mathbf{b} - \mathbf{a}).
\end{aligned}$$

Applying Corollary 4.2 to vector $\mathbf{c} = \delta_{j+q}\mathbf{b} - \mathbf{a}$ and the sequence of $q-2$ vectors $\mathbf{x}_k$, $k = j+1, \ldots, j+q-2$, we conclude that $q = O(p \log p)$. □

Lemmas 4.4 and 4.5 immediately imply the following bound on the number of iterations of the Newton method.

Theorem 4.6 *The Newton method solves a linear fractional combinatorial optimization problem $\mathcal{F}_L$ in $O(p^2 \log^2 p)$ iterations.*

Strongly polynomial bounds on the number of iterations of the Newton method for problem $\mathcal{F}_L$ are somewhat unexpected because, as the following example demonstrates, function $h(\delta)$ may consist of an exponential number of linear segments. Let $\mathbf{a} = (a_1, a_2, \ldots, a_{3p})$ and $\mathbf{b} = (b_1, b_2, \ldots, b_{3p})$ be vectors such that

$$a_i = \begin{cases} 2^{i-1}, & \text{for } i = 1, 2, \ldots, p, \\ 0, & \text{for } i = p+1, \ldots, 3p, \end{cases}$$

$$b_i = \begin{cases} 0, & \text{for } i = 1, 2, \ldots, p, \\ 2^{i-p-1}, & \text{for } i = p+1, \ldots, 3p. \end{cases}$$

Let $\mathcal{U} \subseteq \{0,1\}^{3p}$ (respectively, $\mathcal{W} \subseteq \{0,1\}^{3p}$) be the set such that a binary vector $\mathbf{x} = (x_1, \ldots, x_{3p})$ belongs to $\mathcal{U}$ (respectively, $\mathcal{W}$) if and only if $x_i = 0$ for each $p < i \leq 3p$ (respectively, for each $1 \leq i \leq p$). The 2^p numbers $\mathbf{au}$, for $\mathbf{u} \in \mathcal{U}$, are numbers $0, 1, \ldots, 2^p - 1$, and the 2^{2p} numbers $\mathbf{bw}$, for $\mathbf{w} \in \mathcal{W}$, are numbers $0, 1, \ldots, 2^{2p} - 1$. For $\mathbf{u} \in \mathcal{U}$, let $\mathbf{w_u}$ denote the vector for which $\mathbf{bw_u} = (\mathbf{au})^2$. Finally, let $\mathcal{X} = \{(\mathbf{u}, \mathbf{w_u}) : \mathbf{u} \in \mathcal{U}, \mathbf{u} \neq \mathbf{0}\}$. For this instance $(\mathcal{X}, \mathbf{a}, \mathbf{b})$ of linear fractional combinatorial optimization, function h is equal to

$$h(\delta) = \max\{k - \delta k^2 : k = 1, 2, \ldots, 2^p - 1\},$$

and consists of $2^p - 1$ linear segments.

The bound stated in Theorem 4.6 is the best known bound on the number of iterations of the Newton method for a general problem $\mathcal{F}_L$. However, when we consider a concrete problem of type $\mathcal{F}_L$, it often happens that special properties of this problem enable us to derive a better bound. For example, Theorem 4.6 implies that the Newton method solves the MaxRatioPath problem in $O(m^2 \log^2 m)$ iterations and, consequently, in $O(m^3 \log^2 m)$ total time. Using the properties of the MaxRatioPath problem, we show below that if the edge weights are nonnegative, then the Newton method requires actually only $O(m \log n)$ iterations. A more advanced example of such an analysis is presented in Section 8. We need one more technical lemma about the Newton method for the general case.

Lemma 4.7 *For each $1 \leq i \leq t - 2$ and $\delta \geq \delta_{i+2}$,*

$$f_i - \delta g_i \leq -h_{i+1}. \tag{14}$$

Proof. Since sequence (δ_i) monotonically increases and numbers g_i are positive, it is enough to show that Inequality (14) is true for each $1 \leq i \leq t - 2$ and $\delta = \delta_{i+2}$. Using Equations (2) and (3) and Lemma 3.1(C), we have for each $1 \leq i \leq t - 2$,

$$h_{i+1} = f_{i+1} - \delta_{i+1} g_{i+1} = (\delta_{i+2} - \delta_{i+1}) g_{i+1}$$
$$\leq (\delta_{i+2} - \delta_{i+1}) g_i = \delta_{i+2} g_i - f_i.$$

Fractional Combinatorial Optimization

Thus Inequality (14) is true for each $1 \leq i \leq t-2$ and $\delta = \delta_{i+2}$. □

Theorem 4.8 *The Newton method solves the MaxRatioPath problem in $O(m \log n)$ iterations and in $O(m^2 \log n)$ total time, if the weights of the edges are nonnegative.*

Proof. We use the notation from the description of the MaxRatioPath problem in Section 2. Consider an input instance $(n, E, \mathbf{c}, \mathbf{w})$ of the MaxRatioPath problem. We introduce the following definitions. For $\delta \in \mathbf{R}$, a vertex v, and an edge $(v, u) \in E$,

$$c_\delta(v, u) \stackrel{\text{def}}{=} (\mathbf{c} - \delta \mathbf{w})(v, u) = \mathbf{c}(v, u) - \delta \mathbf{w}(v, u),$$

$d_\delta(v) \stackrel{\text{def}}{=}$ the maximum length of a path from vertex 1 to vertex v according to the edge-cost function $\mathbf{c}_\delta$,

$$\overline{c}_\delta(v, u) \stackrel{\text{def}}{=} c_\delta(v, u) + d_\delta(v) - d_\delta(u).$$

For a subset of edges $Q \subseteq E$, $\mathbf{c}_\delta(Q)$ (respectively, $\overline{c}_\delta(Q)$) denotes the sum of $c_\delta(e)$ (respectively, $\overline{c}_\delta(e)$) over all edges $e \in Q$. The introduced definitions imply that for each path $P \subseteq E$,

$$\mathbf{c}_\delta(P) = f(P) - \delta g(P).$$

It is also easy to verify that for each edge $(v, u) \in E$,

$$\overline{c}_\delta(v, u) \leq 0, \tag{15}$$

and for each path P from vertex 1 to vertex n,

$$\overline{c}_\delta(P) = \mathbf{c}_\delta(P) - d_\delta(n). \tag{16}$$

Consider the computation of the Newton method on input $(n, E, \mathbf{c}, \mathbf{w})$. For $i = 1, 2, \ldots, t$, let P_i be the path from vertex 1 to vertex n computed during iteration i. We have for each $1 \leq i \leq t$,

$$h_i = f(P_i) - \delta_i g(P_i) = \mathbf{c}_{\delta_i}(P_i) = d_{\delta_i}(n). \tag{17}$$

We say that an edge e is *essential* at the beginning of iteration i, if it belongs to at least one of the paths $P_i, P_{i+1}, \ldots, P_t$. To prove the theorem, we show that a sequence of $k = \lceil \log n \rceil + 1$ consecutive iterations decreases the number of essential edges at least by one. Consider a sequence of iterations $i, i+1, \ldots, i+k$. Inequality (5) implies that

$$h_{i+k} g_{i+k} \leq \frac{1}{n^2} h_{i+1} g_{i+1}. \tag{18}$$

If $g_{i+k} \leq (1/n) g_i$, then let $e \in P_i$ be such an edge that

$$\mathbf{w}(e) \geq \frac{1}{|P_i|} \sum_{a \in P_i} \mathbf{w}(a) = \frac{1}{|P_i|} g(P_i) \geq \frac{1}{n-1} g_i > g_{i+k}.$$

Since sequence (g_j) is non-increasing, then for each $j \geq i+k$, $\mathbf{w}(e) > g_j = \sum_{a \in P_j} \mathbf{w}(a)$, so edge e does not belong to path P_j. Thus edge e is essential at the beginning of iteration i (since $e \in P_i$) but is not essential at the beginning of iteration $i+k$ (since $e \notin P_j$, for each $j \geq i+k$).

If $g_{i+k} > (1/n) g_i$, then also $g_{i+k} > (1/n) g_{i+1}$ and Inequality (18) implies that $h_{i+k} < (1/n) h_{i+1}$. This inequality and Lemma 4.7 imply that

$$\mathbf{c}_{\delta_{i+k}}(P_i) = f(P_i) - \delta_{i+k} g(P_i) = f_i - \delta_{i+k} g_i \leq -h_{i+1} < -n h_{i+k}.$$

Therefore, using (16) and (17), we have

$$\overline{\mathbf{c}}_{\delta_{i+k}}(P_i) = \mathbf{c}_{\delta_{i+k}}(P_i) - d_{\delta_{i+k}}(n) < -n h_{i+k} - h_{i+k} = -(n+1) h_{i+k}.$$

This means that there must be an edge $e \in P_i$ such that $\overline{\mathbf{c}}_{\delta_{i+k}}(e) < -h_{i+k}$. Let e be such an edge. For any path P from vertex 1 to vertex n which contains edge e, we have, using (15) and (16),

$$\mathbf{c}_{\delta_{i+k}}(P) = \overline{\mathbf{c}}_{\delta_{i+k}}(P) + d_{\delta_{i+k}}(n) \leq \overline{\mathbf{c}}_{\delta_{i+k}}(e) + h_{i+k} < 0.$$

Hence none of the paths P_j, for $j \geq i+k$, contains edge e, because $\delta_j \geq \delta_{i+k}$ and

$$\mathbf{c}_{\delta_{i+k}}(P_j) \geq \mathbf{c}_{\delta_j}(P_j) = h_j \geq 0.$$

Thus edge e is essential at the beginning of iteration i but is not essential at the beginning of iteration $i+k$. □

5 Megiddo's Parametric Search

In this section we present the parametric search method for fractional combinatorial optimization proposed by Megiddo [31, 32]. We introduce this method using the MaxRatioPath problem as an example. Let $(n, E, \mathbf{c}, \mathbf{w})$ be an input instance of this problem and let δ^* denote the optimum value for this input. See Section 2 for the notation concerning the MaxRatioPath problem. The maximum cost of a path from vertex 1 to vertex n according to the edge-cost function $\mathbf{c} - \delta^*\mathbf{w}$ is equal to 0. Thus if we know δ^* and run algorithm MAXCOST on input $(n, E, \mathbf{c} - \delta^*\mathbf{w})$, then it returns value 0 and a path P which is optimal for the instance $(n, E, \mathbf{c}, \mathbf{w})$ of the MaxRatioPath problem.

Now we try to run algorithm MAXCOST on input $(n, E, \mathbf{c} - \delta^*\mathbf{w})$ but *without* knowing the value of δ^*. We assume at first that each time we need the outcome of the comparison in line 5 of the algorithm, we get it from an oracle which is always correct. Having such an oracle, we can easily proceed with the computation, and the only change over the case with known δ^* is that at each point of the computation, each $d[v]$ is now not a number but a function of unknown δ^*. Observe that the way variables $d[v]$ are updated in line 6 implies that these functions are linear functions of δ^*. Therefore, the comparison

$$d[v] + (\mathbf{c}(v,u) - \delta^*\mathbf{w}(v,u)) - d[u] > 0 \;?$$

performed in line 5 is always of the form "$s - \delta^* t > 0$?" for some known numbers s and t. To find out the outcome of such a comparison, we only need to know the relation between numbers s/t and δ^*. The three possibilities, $s/t < \delta^*$, $s/t = \delta^*$, and $s/t > \delta^*$, are equivalent to the optimum value of problem MaxPath(s/t) being positive, equal to 0, or negative, respectively. Thus instead of the oracle, we can use algorithm MAXCOST itself. We obtain the following algorithm for the MaxRatioPath problem.

WEIGHTEDMAXCOST$(n, E, \mathbf{c}, \mathbf{w})$
1) $d[1] \leftarrow 0$; $seen[1] \leftarrow \mathbf{true}$
2)   **for** $v \leftarrow 2$ **to** n **do** $seen[v] \leftarrow \mathbf{false}$
3)   **for** $v \leftarrow 1$ **to** $n - 1$ **do**
4)   **for each** u such that $(v, u) \in E$ **do**
5)   let s and t be the numbers such that
               $s - \delta^* t = d[v] + \mathbf{c}(v,u) - \delta^*\mathbf{w}(v,u) - d[u]$
6)   **if** $t \neq 0$ **then**
7) $x \leftarrow$ the number returned by MAXCOST$(n, E, \mathbf{c} - (s/t)\mathbf{w})$
8)   **if** (not $seen[u]$) **or** ($t = 0$ and $s > 0$) **or** ($tx < 0$) **then**
9) $d[u] \leftarrow d[v] + \mathbf{c}(v,u) - \delta^*\mathbf{w}(v,u)$
10) $predecessor[u] \leftarrow v$; $seen[u] \leftarrow \mathbf{true}$
11)   let $P = (v_1, v_2, \ldots, v_k)$ where $v_1 = 1$, $v_k = n$, and

$$v_i = predecessor[v_{i+1}], \text{ for } i = 1, 2, \ldots, k-1$$
12) **return** path P

Theorem 5.1 *Algorithm* WEIGHTEDMAXCOST *solves the MaxRatioPath problem in* $O(m^2)$ *time.*

Proof. The computation MAXCOST$(n, E, \mathbf{c} - \delta^*\mathbf{w})$ returns an optimal path P for the instance $(n, E, \mathbf{c}, \mathbf{w})$ of the MaxRatioPath problem. If we trace, in parallel, the computations MAXCOST$(n, E, \mathbf{c} - \delta^*\mathbf{w})$ and WEIGHTEDMAXCOST$(n, E, \mathbf{c}, \mathbf{w})$, then we see that the *predecessor* pointers change in exactly the same way in both computations, because the outcomes of the corresponding conditions checked in line 5 of the first algorithm and in line 8 of the second one are always the same. Thus both computations return the same path, so the computation WEIGHTEDMAXCOST$(n, E, \mathbf{c}, \mathbf{w})$ returns an optimal path P for the instance $(n, E, \mathbf{c}, \mathbf{w})$ of the MaxRatioPath problem. The running time of algorithm WEIGHTEDMAXCOST is $O(m^2)$ because algorithm MAXCOST is executed at most m times. $\square$

To generalize the above idea, we need the notion of *linear algorithm*. Let $\mathcal{Q}(\delta)$ denote a problem whose input includes a parameter $\delta \in \mathbf{R}$. An algorithm $\mathcal{A}$ is a linear algorithm for problem $\mathcal{Q}(\delta)$, if it satisfies the following conditions.

1. For each fixed $\bar\delta \in \mathbf{R}$, algorithm $\mathcal{A}$ computes a correct solution of problem $\mathcal{Q}(\bar\delta)$.

2. The value of each arithmetic expression on each possible execution path of algorithm $\mathcal{A}$ is a linear function of parameter δ.

Observe that in our example we do not actually need an algorithm which computes a maximum cost path. Everything works equally well if we use an algorithm which only finds a positive-cost path, if such a path exists. In general terms, this means that instead of considering problem $\mathcal{P}(\delta)$, it is enough to consider problem $\mathcal{P}_0(\delta)$. The following theorem summarizes the basic form of Megiddo's parametric search.

Theorem 5.2 *If there exists a linear algorithm $\mathcal{A}$ for the parametric problem $\mathcal{P}_0(\delta)$ corresponding to a fractional combinatorial optimization problem $\mathcal{F}$ which runs in time T, then problem $\mathcal{F}$ can be solved in $O(T^2)$ time.*

Proof. Run algorithm $\mathcal{A}$ for problem $\mathcal{P}_0(\delta^*)$ with unknown δ^*. Since $\mathcal{A}$ is a linear algorithm, each comparison encountered during the computation is equivalent to a comparison "$s - \delta^* t > 0$?" for some numbers s and t, and can be evaluated by solving problem $\mathcal{P}_0(s/t)$. If each such problem is solved by running algorithm $\mathcal{A}$ with $\delta = s/t$, then the whole computation runs in $O(T^2)$ time. □

If we exploit possible independence of comparisons in algorithm $\mathcal{A}$, we may obtain a better bound than the bound stated in Theorem 5.2. Consider, for example, iteration v, $1 \leq v \leq n-1$, of the outer loop of the computation MAXCOST($n, E, \mathbf{c} - \delta^*\mathbf{w}$). Each comparison in line 5 performed during this iteration would give exactly the same outcome if it were performed right at the beginning of the iteration, because it does not depend on any updates done during this iteration. (Note that this statement would not be true if the graph had multiple edges.) Let $s_k - \delta^* t_k$, for $k = 1, 2, \ldots, \deg(v)$, be the expressions which have to be compared with 0 during this iteration, where $\deg(v)$ denotes the number of edges outgoing from vertex v. To know the outcome of these comparisons, we have to establish the relation between unknown number δ^* and all numbers s_k/t_k. We can either establish the relation between s_k/t_k and δ^* for each k independently, by computing MAXCOST($n, E, \mathbf{c} - \delta\mathbf{w}$) for each $\delta = s_k/t_k$, or we can first sort numbers s_k/t_k and then establish the position of δ^* with respect to the elements of the sorted sequence (s'_k/t'_k) using binary search. The latter method requires only $O(\log(\deg(v)))$ applications of algorithm MAXCOST, one application per one iteration of the binary search. The running time of sorting $\deg(v)$ numbers and $O(\log(\deg(v)))$ applications of algorithm MAXCOST is $O(\deg(v) \log(\deg(v))) + O(m \log(\deg(v))) = O(m \log(\deg(v)))$. Thus the running time of the whole computation is

$$O(m \sum_{v=1}^{n-1} \log(\deg(v))) = O(mn \log(m/n)), \qquad (19)$$

since $\sum_{v=1}^{n-1} \deg(v) = m$ and a logarithmic function is concave.

To generalize this idea, we introduce the notion of a *stage* of the computation. Consider a problem $\mathcal{Q}$ whose input includes a parameter $\delta \in \mathbf{R}$. A stage of the computation of an algorithm for problem $\mathcal{Q}$ is a (continual) part of the computation which has the following property. For each comparison performed during one stage, if this comparison is moved to the beginning of this stage, the outcome of the comparison does not change. We consider only comparisons whose outcomes may depend on the value of parameter δ. We say that an algorithm consists of r stages, if the computation of this algorithm can be partitioned into r stages, independently of the actual input values. In the theorem below, we distinguish between the main algorithm $\mathcal{A}_1$, which guides the whole computation, must be linear and should ideally consist of a small number of stages, and the "subordinate" algorithm $\mathcal{A}_2$, which is used to resolve individual comparisons performed by algorithm $\mathcal{A}_1$. Algorithms $\mathcal{A}_1$ and $\mathcal{A}_2$ may of course be the same, but they are usually different, if they are to provide the best possible running time of the whole computation (see Section 6).

Theorem 5.3 *If there exist*

(1) a linear algorithm $\mathcal{A}_1$ for the parametric problem $\mathcal{P}_0(\delta)$ corresponding to a fractional combinatorial optimization problem $\mathcal{F}$, which runs in time T_1 and consists of r stages with q_i comparisons during stage i, and

(2) an algorithm $\mathcal{A}_2$ for problem $\mathcal{P}_0(\delta)$ which runs in time T_2,

then problem $\mathcal{F}$ can be solved in $O(T_1 + T_2 \sum_{i=1}^{r} \log q_i)$ time.

Proof. Run algorithm $\mathcal{A}_1$ for problem $\mathcal{P}_0(\delta^*)$ with unknown δ^*. Consider stage i of this computation. During this stage we are to resolve q_i independent comparisons of the form $s - \delta^* t$. Computing the outcomes of all these comparisons can be reduced to computing the relation between δ^* and q_i numbers $x_1, x_2, \ldots, x_{q_i}$. If we sort these numbers and locate δ^* in the sorted sequence using binary search, as suggested in our example, then the running time of this stage is $O(q_i \log q_i + T_2 \log q_i)$. This gives, however, only an $O(T_1 \max_i \{\log q_i\} + T_2 \sum_{i=1}^{r} \log q_i)$ bound on the running time of the whole computation.

To obtain the bound claimed in this theorem, observe that sorting numbers $x_1, x_2, \ldots, x_{q_i}$ can be avoided. We can find the median of these numbers in $O(q_i)$ time [2] and compare this median with δ^* by running algorithm $\mathcal{A}_2$. The outcome of this comparison reveals the relation between δ^* and half of the numbers $x_1, x_2, \ldots, x_{q_i}$. We can continue this process by finding the median of the remaining half. This way we eventually partition numbers $x_1, x_2, \ldots, x_{q_i}$ into the numbers smaller than δ^* and the numbers greater than δ^*. The running time of this computation is

$$O\left(\sum_{k=0}^{\log q_i} \left(\frac{q_i}{2^k} + T_2\right)\right) = O(q_i + T_2 \log q_i).$$

Fractional Combinatorial Optimization

Summing this bound over all stages i, we obtain the bound claimed in this theorem. □

To minimize the bound of Theorem 5.3 for a given fractional combinatorial optimization problem, algorithm $\mathcal{A}_2$ should obviously be the fastest known algorithm for the non-fractional version of this problem. To find good candidates for algorithm $\mathcal{A}_1$, we should look among parallel algorithms, since such algorithm are especially designed to have a small number of stages. A parallel algorithm which runs in time T and uses P processors can be viewed as an algorithm which runs in sequential time TP and consists of T stages with at most P operations per stage. Thus the following theorem is an immediate corollary of Theorem 5.3.

Theorem 5.4 *If there exist*

(1) a linear parallel algorithm $\mathcal{A}_1$ for the parametric problem $\mathcal{P}_0(\delta)$ corresponding to a fractional combinatorial optimization problem $\mathcal{F}$, which runs in time T_1 and uses P processors, and

(2) a (sequential) algorithm $\mathcal{A}_2$ for problem $\mathcal{P}_0(\delta)$ which runs in time T_2,

then problem $\mathcal{F}$ can be solved in $O(T_1 P + T_2 T_1 \log P)$ time.

Consider another example of fractional combinatorial optimization, the maximum-ratio spanning-tree problem. An input instance of this problem consists of an undirected connected graph $G = (V, E)$, an edge-cost function $\mathbf{c} : E \longrightarrow \mathbf{R}$, and an edge-weight function $\mathbf{w} : E \longrightarrow \mathbf{R}$. The cost and the weight of a spanning tree in graph G is equal to the sum of the costs and the sum of the weights of the edges of this tree, respectively. The problem is to find a spanning tree in graph G with the maximum cost-to-weight ratio, or, using our terminology, a spanning tree with the maximum mean-weight cost. The corresponding parametric problem $\mathcal{P}(\delta)$ is the problem of computing a maximum spanning tree for the edge-cost function $\mathbf{c} - \delta \mathbf{w}$. The maximum-cost spanning-tree problem, which is equivalent to the minimum-cost spanning-tree problem, can be solved in $T_{\text{MST}} = O(\min\{m \log \log n, m + n \log n\})$ time [49, 14], where m is the number of edges and n is the number of vertices in the input graph.

A maximum-ratio spanning tree is a maximum-cost spanning-tree for the edge-cost function $\mathbf{c} - \delta^* \mathbf{w}$. Maximum-cost spanning trees depend only on the order of the edges by their costs. Thus to find a maximum-cost spanning-tree for the edge-cost function $\mathbf{c} - \delta^*\mathbf{w}$, it is enough to sort m numbers $\mathbf{c}(e) - \delta^*\mathbf{w}(e)$, $e \in E$. Sorting m numbers can be done in $O(\log m)$ parallel time using m processors [7]. Therefore, Theorem 5.4 implies that the maximum-ratio spanning-tree problem can be solved in $O(m \log m + T_{\text{MST}} \log^2 m) = O(T_{\text{MST}} \log^2 m)$ time.

We return now to the general case. Since we are interested in maximization problems, many sub-tasks appearing during the computation of algorithm $\mathcal{A}_1$ of Theorem 5.3 may be of the form

$$u(\delta^*) \leftarrow \max\{u_1(\delta^*), u_2(\delta^*), \ldots, u_q(\delta^*)\}, \tag{20}$$

where each u_i is a linear function of parameter δ. We discuss two different approaches to computing such a maximum. If the maximum (20) is calculated by comparing each u_i, for $i = 2, 3, \ldots, q$, with the already known maximum of u_1 through u_{i-1}, then this computation has $q - 1$ stages with one comparison per stage, and the maximum can be computed in $O(T_2 q)$ time. Here T_2 denotes, as in Theorem 5.3, the running time of an algorithm $\mathcal{A}_2$, which is used to resolve the comparisons performed during the computation of algorithm $\mathcal{A}_1$. Computation of the maximum (20) can be organized in $\lceil \log q \rceil$ stages by scheduling the comparisons in a tournament tree. Each stage consists of $O(q)$ comparisons, so Theorem 5.3 implies that the maximum (20) can be computed in $O(q + T_2 \log^2 q)$ time. The number of stages needed to compute the maximum can be further reduced to $O(\log \log q)$ in the following way. Follow the tournament for $O(\log \log \log q)$ stages to reduce the number of live elements to $k_1 \leq q/(\log \log q)$. Live elements are those which so far have won all their comparisons. Now perform a sequence of the following stages. During stage i, partition the set of k_i live elements into $k_i^2/(2k_1)$ equal-size groups, and perform all pairwise comparisons in each group. This stage requires $2k_1 = O(q/\log \log q)$ comparisons and reduces the number of live elements to $k_{i+1} = k_i^2/(2k_1)$. Thus $k_i = k_1/2^{2^{i-1}-1}$, so we have only $O(\log \log q)$ such stages. Therefore, this algorithm for computing the maximum of q numbers consists of $O(q)$ comparisons partitioned into $O(\log \log q)$ stages. Using this algorithm as algorithm $\mathcal{A}_1$, Theorem 5.3 implies that the maximum (20) can be computed in $O(q + T_2 \log q \log \log q)$ time. The above $O(\log \log q)$-stage algorithm for computing the maximum of q numbers is based on Valiant's parallel algorithm which runs in $O(\log \log q)$ time and uses q processors [47].

There is an alternative approach to computing the maximum (20). The maximum of q linear functions is a convex, piecewise linear function with

Fractional Combinatorial Optimization

at most q breakpoints. If we have two convex, piecewise linear functions π_1 and π_2 with q_1 and q_2 breakpoints, respectively, and the breakpoints of each function are sorted by their x-coordinates, then we can compute the sorted sequence of the breakpoints of function $\max\{\pi_1, \pi_2\}$ in $O(q_1+q_2)$ time by a straightforward merging process. Therefore, by following the merging scheme of the merge-sort algorithm, we can compute all breakpoints of the maximum of q linear functions u_i in $O(q \log q)$ time. Finding the linear segment of function u which contains δ^* can then be done by binary search over the breakpoints of function u. Using this approach, the maximum (20) can be computed in $O(q \log q + T_2 \log q)$ time.

Consider an algorithm $\mathcal{A}$ for a problem $\mathcal{Q}$ whose input includes a parameter $\delta \in \mathbf{R}$. A *maxima-computing phase* of an algorithm $\mathcal{A}$ is a (continual) part of the computation of $\mathcal{A}$ such that all comparisons performed during this part of the computation are (implicitly) involved in computing maxima, and all these maxima are independent, that is, they can be performed in arbitrary order without changing the outcome of the whole computation. As in the definition of a stage of an algorithm, we consider here only comparisons whose outcomes may depend on the value of parameter δ. The following theorem is based on the two approaches to computing the maximum (20) described above. The *size* of a maximum of q elements is equal to q.

Theorem 5.5 *If there exist*

(1)   a linear algorithm $\mathcal{A}_1$ for the parametric problem $\mathcal{P}_0(\delta)$ corresponding to a fractional combinatorial optimization problem $\mathcal{F}$, which runs in time T_1, consists of r maxima-computing phases, and the total size of the maxima in phase i is at most q_i, and

(2)   an algorithm $\mathcal{A}_2$ for problem $\mathcal{P}_0(\delta)$ which runs in time T_2,

then problem $\mathcal{F}$ can be solved in

(i) $O(T_1 + T_2 \sum_{i=1}^{r} \log q_i \log \log q_i)$ time, and

(ii) $O(T_1 + \sum_{i=1}^{r} (q_i + T_2) \log q_i)$ time.

Proof. Run algorithm $\mathcal{A}_1$ for problem $\mathcal{P}_0(\delta^*)$ with unknown δ^* and consider phase i of this computation. During this phase we have to compute independent maxima of the total size at most q_i. If we compute these maxima in parallel using the $O(\log \log q)$-stage algorithm described above, then the running time of this phase is $O(q_i + T_2 \log q_i \log \log q_i)$ and the running time of the whole computation is as in part (i) of the theorem. If we first compute the breakpoints of each maximum, sort together all breakpoints, and locate δ^* in the sorted list by binary search, then the running time of this phase is $O((q_i + T_2) \log q_i)$ and the running time of the whole computation is as in part (ii) of the theorem □

The best bound for the MaxRatioPath problem which we have shown so far is $O(mn \log(m/n))$ (see (19)). Using Theorem 5.5, we show now an $O(mn)$ bound. This bound, however, is based on other algorithm for computing a maximum-cost path than algorithm MAXCOST. Algorithm MAXCOST considers vertices $1, 2, \ldots, n-1$, and while considering vertex v, it examines all edges outgoing from v. The following algorithm MAXCOST2 considers vertices $2, 3 \ldots, n$ and while considering vertex v, it examines all edges incoming into v. This algorithm returns only the maximum cost of a path from vertex 1 to vertex n, but can be easily modified so that it returns also a maximum-cost path.

MAXCOST2$(n, E, \mathbf{a})$
1) $d[1] \leftarrow 0$
2) **for** $v \leftarrow 2$ **to** n **do** $d[v] \leftarrow -\infty$
3) **for** $v \leftarrow 2$ **to** n **do**
4) $d[v] \leftarrow \max\{ d[u] + \mathbf{a}(u, v) : (u, v) \in E \}$
5) **return** $d[n]$

Algorithm MAXCOST2 is still not quite what we need. If we apply Theorem 5.5 with algorithm MAXCOST2 as algorithm $\mathcal{A}_1$, then we obtain bound $O(mn \log(m/n) \log \log(m/n))$, from part (i) of the theorem, and bound $O(mn \log(m/n))$, from part (ii) of the theorem, so we do not improve bound (19).

By rearranging the computation of algorithm MAXCOST2, we obtain the recursive algorithm MAXCOSTRECURSIVE shown below. This algorithm considers first, recursively, the subgraph induced by vertices $u = 1, 2, \ldots, k = \lfloor (n+1)/2 \rfloor$, then considers the edges $(u, v) \in E$ such that $1 \leq u \leq k < v \leq n$ (observe that all maxima of this part of the computation — lines 7–8 below — are independent), and finishes up by considering, again recursively, the subgraph induced by vertices $v = k, k+1, \ldots, n$.

MAXCOSTRECURSIVE$(n, E, \mathbf{a})$
1) $d[1] \leftarrow 0$

2) **for** $v \leftarrow 2$ **to** n **do** $d[v] \leftarrow -\infty$
3) COMPUTEMAXIMA$(1, n)$
4) **return** $d[n]$

COMPUTEMAXIMA(p, r)
5) $k \leftarrow \lfloor (p + r)/2 \rfloor$
6) **if** $k > p$ **then** COMPUTEMAXIMA(p, k)
7) **for** $v \leftarrow k + 1$ **to** r **do**
8) $\quad d[v] \leftarrow \max\{ d[v], d[u] + \mathbf{a}(u, v) : p \leq u \leq k, (u, v) \in E \}$
9) **if** $k + 1 < r$ **then** COMPUTEMAXIMA$(k + 1, r)$

Algorithm MAXCOSTRECURSIVE runs in $O(m)$ time (each edge is considered only once) and, for each $k = 1, 2, \ldots, \log n$, it has $r_k = n/2^k$ maxima-computing phases (assume for convenience that n is a power of 2). Each of these r_k phases considers only edges $(u, v) \in E$ such that $a + 1 \leq u \leq a + 2^{k-1} < v \leq a + 2^k$, for some integer a, so the total size of the maxima in one such phase is at most $q_k = \min\{2^{2k}, m\}$. Therefore, part (ii) of Theorem 5.5 implies that the MaxRatioPath problem can be solved within the time bound:

$$O\left(m + \sum_{k=1}^{\log n} r_k (q_k + m) \log q_k \right) = O\left(m + \sum_{k=1}^{\log n} \frac{n}{2^k} \cdot m \cdot (2k) \right) = O(nm).$$

Bound (i) of Theorem 5.5 gives here the same time bound. We use Theorem 5.5 in next section to show low time bounds for computing a maximum profit-to-time ratio cycle.

We end this section with a comment about implementation of Megiddo's parametric search. Throughout the whole computation of algorithm $\mathcal{A}_1$, we should maintain an interval (α, β) containing the optimum value δ^*. Initially $(\alpha, \beta) = (0, +\infty)$. If during the computation we need to know the relation between a number x and unknown δ^*, we first check if x belongs to (α, β). If $x \notin (\alpha, \beta)$, then we know the relation between x and δ^*. If $x \in (\alpha, \beta)$, then we run algorithm $\mathcal{A}_2$ and find out whether $x = \delta^*$, $\delta^* \in (\alpha, x)$, or $\delta^* \in (x, \beta)$. If $x = \delta^*$, we should stop the entire computation since we have just found the optimum objective value we were looking for. In the other two cases we update interval (α, β) and proceed with the computation. Maintaining interval (α, β) does not improve the worst-case running times but it should significantly improve the average running times.

6 Maximum Profit-to-Time Ratio Cycles

In this section we consider the following classical fractional combinatorial optimization problem, which we call the MaxRatioCycle problem. An input instance of this problem consists of a directed graph $G = (V, E)$, an edge-cost function $\mathbf{c} : E \longrightarrow \mathbf{R}$, and an edge-weight function $\mathbf{w} : E \longrightarrow \mathbf{R}$. Let n and m denote the number of vertices and the number of edges in graph G, respectively. As in the MaxRatioPath problem, if P is a path in G, then $f(P) = \sum_{(i,j) \in P} \mathbf{c}(i,j)$, $g(P) = \sum_{(i,j) \in P} \mathbf{w}(i,j)$, and $f(P)/g(P)$ are the cost, the weight, and the mean-weight cost of this path. The task is to find a cycle in graph G with the maximum mean-weight cost.

MaxRatioCycle : maximize $\dfrac{f(\Gamma)}{g(\Gamma)}$ over all cycles Γ in graph G.

We assume that graph G is not acyclic, the weight of each cycle is positive, and there exists a cycle with positive cost. A cycle $\Gamma = (v_0, v_1, \ldots, v_j = v_0)$ is simple if the vertices $v_0, v_1, \ldots, v_{j-1}$ are distinct. There are infinitely many cycles in G, but to find a cycle with the maximum mean-weight cost, we can limit the search only to simple cycles. If a cycle Γ is not simple, then its set of edges can be partitioned into a number of simple cycles $\Gamma_1, \Gamma_2, \ldots, \Gamma_k$, and

$$\frac{f(\Gamma)}{g(\Gamma)} = \sum_{i=1}^{k} \frac{f(\Gamma_i)}{g(\Gamma)} = \sum_{i=1}^{k} \frac{g(\Gamma_i)}{g(\Gamma)} \cdot \frac{f(\Gamma_i)}{g(\Gamma_i)} \le \max_{1 \le i \le k} \left\{ \frac{f(\Gamma_i)}{g(\Gamma_i)} \right\}.$$

The inequality holds because numbers $g(\Gamma_i)/g(\Gamma)$ are positive and sum up to 1. Therefore, there must be a simple cycle which has the maximum mean-weight cost among all cycles. For convenience, however, instead of considering only simple cycles, we define the set of feasible solutions $\mathcal{X}_G$ as the set of the cycles in graph G which contain at most n edges.

The MaxRatioCycle problem models the following *tramp-steamer* problem [9, 29]. A trip from a port v to a port u takes $\mathbf{w}(v, u)$ time and gives profit of $\mathbf{c}(v, u)$ units. To maximize the mean daily profit, the steamer should follow a cycle which maximizes the ratio of the total profit to the total time. The MaxRatioCycle problem is often called the *maximum profit-to-time ratio cycle problem*.

The parametric problem corresponding to the MaxRatioCycle problem is

MaxCycle(δ) : maximize $f(\Gamma) - \delta g(\Gamma)$, over all cycles $\Gamma \in \mathcal{X}_G$.

An optimal solution of problem MaxCycle(δ) is a cycle from $\mathcal{X}_G$ with the maximum cost according to the edge-cost function $\mathbf{c} - \delta \mathbf{w}$. To be able to directly apply the Newton method to the MaxRatioCycle problem, we need an algorithm which for given edge costs finds a cycle $\Gamma \in \mathcal{X}_G$ with the maximum cost. (Observe that if we restricted ourselves only to simple cycles, this problem would be NP-hard.) A weaker algorithm, which finds a positive-cost cycle, if there exists one, is sufficient for Megiddo's parametric search.

Let $\mathbf{a} : E \longrightarrow \mathbf{R}$ be an arbitrary edge-cost function. Let $d_k(v, u)$ denote the maximum cost of a path from vertex v to vertex u containing at most k edges. If there is no path from v to u, then $d_k(v, u) = -\infty$. We have

$$d_1(v, u) = \begin{cases} \mathbf{a}(v, u), & \text{if } (v, u) \in E, \\ -\infty, & \text{if } (v, u) \notin E, \end{cases}$$

and

$$d_{2k}(v, u) = \max\{\, d_k(v, u),\, \max_{x \in V}\{d_k(v, x) + d_k(x, u)\}\,\}. \tag{21}$$

To abstract from technical details, we assume that n is a power of 2. For n other than a power of 2, we could simply change the set of feasible solutions to the set of all cycles containing at most $2^{\lceil \log n \rceil}$ edges. The recursive relation (21) gives immediately an $O(n^3 \log n)$ algorithm for computing numbers $d_n(v, u)$, for each pair of vertices v and u. If during the computation we store the information about the numbers which define the maxima in Equation (21), then we can also find a maximum-cost path from v to u, for each pair of vertices v and u. The maximum cost of a cycle $\Gamma \in \mathcal{X}_G$ is equal to $\max_{v \in V}\{d_n(v, v)\}$. Hence we have an $O(n^3 \log n)$ algorithm which for a given δ solves problem MaxCycle(δ). Using this algorithm and the Newton method, we obtain an algorithm for the MaxRatioCycle problem which runs in $O(n^3 m^2 \log^3 n)$ time for arbitrary edge costs and weights (see Theorem 4.6) and in $O(n^3 \log n \log(nU))$ time, if all edge costs and weights are integers not greater than U (see Theorem 3.4). Using a similar analysis to the analysis presented for the MaxRatioPath problem in the proof of Theorem 4.8, one can show that if the weights of the edges are nonnegative, then the Newton method solves the MaxRatioCycle problem in $O(m \log n)$ iterations and, consequently, in $O(n^3 m \log^2 n)$ time. This $O(m \log n)$ bound on the number of iterations for the MaxRatioCycle problem, as well as for the MaxRatioPath problem, can be further improved to $O(m)$ (an example of analysis which can give such a bound is presented in Section 8). However, we do not include here a proof of this bound, since this bound anyway yields a slower algorithm than algorithms which can be obtained by using Megiddo's parametric search.

Let $\text{MaxCycle}_0(\delta)$ denote the problem of detecting a positive-cost cycle according to the edge-cost function $\mathbf{c} - \delta\mathbf{w}$, if there exists such a cycle. To apply Megiddo's parametric search to the MaxRatioCycle problem, we have to combine two algorithms for the parametric problem $\text{MaxCycle}_0(\delta)$: a linear algorithm $\mathcal{A}_1$, which is to be run with unknown δ^* and should ideally have a small number of stages/phases, and a fast algorithm $\mathcal{A}_2$ used for resolving comparisons performed by the first algorithm (see Theorems 5.3 and 5.5). The problem of detecting a positive-cost cycle is equivalent to the problem of detecting a negative-cost cycle (simply negate the edge costs). All known algorithms for detecting a negative-cost cycle are actually based on single-source shortest-paths algorithms. The best known bound for shortest-paths algorithms on graphs with arbitrary edge costs is $O(nm)$. This bound is achieved, for example, by the Bellman-Ford algorithm [1, 12]. Golberg's shortest-paths algorithm [16] runs in $O(m\sqrt{n}\log U)$, if the edge costs are integers greater than $-U$, for a positive number U. Thus for arbitrary edge costs, the best known bound on the running time of algorithm $\mathcal{A}_2$ is $O(nm)$. It turns out that the best candidate for algorithm $\mathcal{A}_1$ is the algorithm which is based on the computation of numbers $d_k(v,u)$ using the recursive relation (21). This algorithm runs in $O(n^3 \log n)$ time, consists of $O(\log n)$ maxima-computing phases, and the total size of all maxima in one phase is $O(n^3)$. Therefore, Theorem 5.5 implies that the MaxRatioCycle problem can be solved in $O(n^3 \log n + nm \log^2 \log \log n)$ time (part (i) of the theorem) and in $O(n^3 \log^2 n)$ time (part (ii) of the theorem).

7 Maximum Mean Cycles

The maximum-mean cycle problem is the uniform MaxRatioCycle problem. An input instance of this problem consists of a directed graph $G = (V, E)$ and an edge-cost function $\mathbf{c} : E \longrightarrow \mathbf{R}$. We want to find a maximum-mean cycle in G, that is, a cycle which has the maximum mean cost. The mean cost of a cycle Γ is equal to $\left(\sum_{e \in \Gamma} \mathbf{c}(e)\right)/|\Gamma|$. In this section we describe the $O(nm)$ algorithm for the maximum-mean cycle problem designed by Karp [26]. This algorithm is an example that there are specialized algorithms for some fractional combinatorial optimization problems, which are faster than algorithms yielded by general methods such as the Newton method and Megiddo's parametric search.

Fractional Combinatorial Optimization

Let n and m denote the number of vertices and the number of edges in graph G. We assume that graph G is not acyclic. Let s be a vertex in graph G such that all other vertices are reachable from s. If such a vertex does not exist, we can add to graph G a new vertex s and edges (s,v) with arbitrary costs, for each $v \in V$. This modification does not change the cycles or their costs. For each vertex $v \in V$ and integer $k \geq 0$, let $d_k(v)$ be the maximum cost of a path of length k from vertex s to vertex v. If no such path exists, then $d_k(v) = -\infty$. For a subset of edges $A \subseteq E$, $\mathbf{c}(A)$ denotes the sum $\sum_{e \in A} \mathbf{c}(e)$. Let δ^* denote the maximum mean cost of a cycle in graph G.

Theorem 7.1

$$\delta^* = \max \left\{ \min_{0 \leq k \leq n-1} \left\{ \frac{d_n(v) - d_k(v)}{n - k} \right\} : v \in V \text{ and } d_n(v) > -\infty \right\}. \quad (22)$$

Proof. If we decrease the cost of each edge by the same amount α, then for each vertex $v \in V$ and each $k \geq 0$, $d_k(v)$ decreases by αk and, consequently, $(d_n(v) - d_k(v))/(n - k)$ decreases by α. Thus both sides of Equation (22) decrease by α. Therefore, it is enough to prove the theorem for the case when $\delta^* = 0$.

If $\delta^* = 0$, then Equation (22) is equivalent to

$$0 = \max \left\{ \min_{0 \leq k \leq n-1} \{d_n(v) - d_k(v)\} : v \in V \text{ and } d_n(v) > -\infty \right\},$$

which is equivalent to

$$0 = \max \left\{ d_n(v) - \max_{0 \leq k \leq n-1} \{d_k(v)\} : v \in V \text{ and } d_n(v) > -\infty \right\}. \quad (23)$$

Let v be a vertex such that $d_n(v) > -\infty$, and let P be a path of length n and cost $d_n(v)$ from s to v. Path P is not simple (that is, at least one vertex occurs on P more than once), so the edges of P can be partition into a cycle Γ and, possibly trivial, path R from s to v. Let $k = |R| \leq n - 1$. Since $\delta^* = 0$, then $\mathbf{c}(\Gamma) \leq 0$ and we have

$$d_n(v) = \mathbf{c}(P) = \mathbf{c}(R) + \mathbf{c}(\Gamma) \leq \mathbf{c}(R) \leq d_k(v) \leq \max_{0 \leq k \leq n-1} \{d_k(v)\}.$$

Thus the right-hand side of (23) is at most zero.

Now we show that the right-hand side of (23) is at least zero. Let Γ be a zero-cost cycle in graph G. Since $\delta^* = 0$, such a cycle must exist. Let v

be an arbitrary vertex on cycle Γ, and let P_1 be a maximum cost path from s to v. Such a path must exists because the graph does not have positive-cost cycles. Let P_2 be a path of length at least n which consists of path P_1 followed by some number of repetitions of cycle Γ. Since the cost of cycle Γ is equal to 0, the costs of paths P_1 and P_2 are the same, so path P_2 is a maximum cost path from s to v. Let P_3 be the path which consists of the first n edges of path P_2. Path P_3 is a maximum cost path from vertex s to some vertex w, because any initial part of a maximum cost path must be a maximum cost path. Thus we have

$$d_n(w) = \mathbf{c}(P_3) = \max_{0 \le k \le n-1} \{d_k(w)\}.$$

This means that the right-hand side of (23) is at least zero. □

Theorem 7.1 gives an algorithm for the maximum-mean cycle problem which runs in $O(nm)$ time. We have $d_0(s) = 0$ and $d_0(v) = -\infty$, for each $v \ne s$. Numbers $d_k(v)$, for each $v \in V$ and $k = 1, 2, \ldots n$, can be computed in $O(nm)$ time using the following relation:

$$d_k(v) = \max\{ d_{k-1}(u) + \mathbf{c}(u,v) : (u,v) \in E \}. \tag{24}$$

Knowing all numbers $d_k(v)$, for $v \in V$ and $0 \le k \le n$, we can compute δ^* in $O(n^2)$ time using Equation (22). Let v be a vertex which gives the maximum in (22), and let P be a maximum-cost path from s to v of length n. We can construct such a path, if during the computation of numbers $d_k(x)$ we store the information about the vertices which give the maxima in Equation (24). Each cycle extracted from path P, and there must be at least one, is a maximum-mean cycle.

8 Maximum Mean-Weight Cuts

In Section 6 we discussed a problem (the MaxRatioCycle problem) for which the algorithms obtained by applying Megiddo's parametric search are considerably better than the algorithms yielded by the Newton method, at least for the worst-case inputs. In this section we consider the MaxRatioCut problem, which is an example of the reverse situation. Megiddo's parametric search method does not yield fast algorithm for this problem, because there are no fast parallel algorithms for the maximum flow problem, which is de facto the non-fractional version of the MaxRatioCut problem. The main aim of this

Fractional Combinatorial Optimization

section is to show a linear bound on the number of iterations of the Newton method for the MaxRatioCut problem with non-negative edge weights. This linear bound gives the fastest known algorithm for this problem.

A *network* $G = (V, E, \mathbf{u}, \mathbf{d})$ is a directed, strongly connected graph with a set of vertices V, a set of edges E, a nonnegative edge-capacity function $\mathbf{u} : E \longrightarrow \mathbf{R}$, and a demand function $\mathbf{d} : V \longrightarrow \mathbf{R}$ such that $\sum_{v \in V} \mathbf{d}(v) = 0$. Without loss of generality, we assume that the set of edges is symmetric, that is, if $(x, y) \in E$, then also $(y, x) \in E$. If $\mathbf{d}(v)$ is negative, then v is a *source* — a node with supply. If $\mathbf{d}(v)$ is positive, then v is a *sink* — a node with demand. As before, n and m denote the number of vertices and the number of edges in network G, respectively.

For a subset of vertices $W \subseteq V$, $\mathbf{d}(W) \stackrel{\text{def}}{=} \sum_{v \in W} \mathbf{d}(v)$ is the net demand in W. If $S \subseteq V$, $T = V - S$, $S \neq \emptyset$ and $T \neq \emptyset$, then *cut* (S, T) is the set of edges (x, y) such that $x \in S$ and $y \in T$. The *capacity* and the *surplus* of a cut (S, T) are defined as, respectively,

$$\mathbf{u}(S, T) \stackrel{\text{def}}{=} \sum_{e \in (S,T)} \mathbf{u}(e),$$

$$surplus(S, T) \stackrel{\text{def}}{=} \mathbf{d}(T) - \mathbf{u}(S, T).$$

If network G is viewed as a model for designing shipment of the commodity from the sites with supply (the sources) to the sites with demand (the sinks), then a positive $surplus(S, T)$ means that not all demand in T can be satisfied. The amount of the demand in T which cannot be satisfied without violating the edge capacities is at least $surplus(S, T)$.

If we have an *edge-weight function* $\mathbf{w} : E \longrightarrow \mathbf{R}$, then $g(S, T) \stackrel{\text{def}}{=} \sum_{e \in (S,T)} \mathbf{w}(e)$ and $surplus(S, T)/g(S, T)$ are the *weight* and the *mean-weight surplus* of cut (S, T). The assumption that the underlying graph is strongly connected guarantees that each cut has at least one edge. We further assume, as in the general model of fractional combinatorial optimization, that the edge weights are such that the weight of each cut is positive. The MaxRatioCut problem is the problem of finding a cut with the maximum mean-weight surplus in a given network G and for a given edge-weight function $\mathbf{w}$. The corresponding parametric problem MaxSurplusCut(δ) is the problem of finding a maximum-surplus cut in network $G_\delta = (V, E, \mathbf{u} + \delta \mathbf{w}, \mathbf{d})$, that is, in network G with the edge-capacity function changed to function $\mathbf{u} + \delta \mathbf{w}$. The problem of finding a maximum-surplus cut in network G can be reduced in $O(m)$ time to the problem of computing a maximum flow in network G. All known algorithms for computing maximum-surplus cuts are based on maximum-flow algorithms. Let $T_{\text{MaxFlow}}(n,m)$ denote the worst-case time complexity of computing a maximum flow in a network with n vertices and m edges. The results from [17] and [27] imply that

$$T_{\text{MaxFlow}}(n,m) = O(\min\{nm\log(n^2/m), nm + n^{2+\epsilon}\}), \qquad (25)$$

where ϵ is an arbitrary positive constant.

The MaxMeanCut problem is the uniform MaxRatioCut problem, that is, the special case of the MaxRatioCut problem when all edge weights are equal to 1. The MaxMeanCut and the MaxRatioCut problems appear, for example, in the context of the classical *minimum-cost flow problem*. Goldberg and Tarjan [18] showed a simple strongly polynomial iterative method for solving the minimum-cost flow problem, which is based on computation of minimum-mean cycles. Ervolina and McCormick [11] (see also [39]) showed an analogous method for solving the dual problem of the minimum-cost flow problem, which is based on computation of maximum mean-surplus cuts. Wallacher [48] used minimum mean-weight cost cycles in a generalization of Goldberg and Tarjan's method. Analogously, maximum mean-weight surplus cuts can be used in a generalization of Ervolina and McCormick's method.

Theorem 5.2 implies that Megiddo's parametric search solves the MaxRatioCut problem in $O((T_{\text{MaxFlow}}(n,m))^2) = O(n^2m^2\log^2 n)$ time. The first parallel algorithm for the maximum flow problem is due to Schiloach and Vishkin [45]. This algorithm runs in $O(n^2\log n)$ time and uses n processors. Theorem 5.4 implies that Megiddo's parametric search based on Schiloach and Vishkin's algorithm solves the MaxRatioCut problem in $O(n^3m\log^3 n)$ time. So far no parallel maximum-flow algorithm has been designed, which would lead to a better bound for Megiddo's parametric search for the MaxRatioCut problem than $O^*(n^3m)$. Notation $O^*()$ hides a poly-logarithmic factor.

Theorem 4.6 gives an $O(m^2\log^2 n)$ bound on the number of iterations of the Newton method for the MaxRatioCut problem. In the remaining part of this section we show that special properties of the MaxRatioCut problem, notably the "maximum flow - minimum cut" duality, imply that the number of iterations is actually only $O(m)$, provided that the edge weights are nonnegative. This bound implies that the Newton method solves the MaxRatioCut problem with nonnegative edge weights in $O(mT_{\text{MaxFlow}}(n,m)) = O(nm^2\log n)$ time. Observe that in the case of the MaxMeanCut problem, an $O(m)$ bound on the number of iterations follows from Theorem 3.5 (in this case a stronger $O(n)$ bound can be shown, see [38]). From now on we assume that the edge weights are nonnegative.

A function $\mathbf{r} : E \longrightarrow \mathbf{R}$ is a *flow* in network G if for each edge $(x,y) \in E$,

$$\mathbf{r}(x,y) \leq \mathbf{u}(x,y),$$
$$\mathbf{r}(x,y) = -\mathbf{r}(y,x).$$

For a flow $\mathbf{r}$, if $A \subseteq E$, then $\mathbf{r}(A) \stackrel{\text{def}}{=} \sum_{e \in A} \mathbf{r}(e)$, and if $v \in V$, then $\mathbf{r}^{(\text{in})}(v)$ denotes the net flow into vertex v:

$$\mathbf{r}^{(\text{in})}(v) \stackrel{\text{def}}{=} \sum_{(v,x) \in E} \mathbf{r}(v,x).$$

We define $\mathbf{u_r} \stackrel{\text{def}}{=} \mathbf{u} - \mathbf{r}$, $\mathbf{d_r} \stackrel{\text{def}}{=} \mathbf{d} - \mathbf{r}^{(\text{in})}$, and for a cut (S,T),

$$surplus_{\mathbf{r}}(S,T) \stackrel{\text{def}}{=} \mathbf{d_r}(T) - \mathbf{u_r}(S,T).$$

Values $\mathbf{u_r}(e)$ and $\mathbf{d_r}(v)$ are commonly called the residual capacity of edge e and the residual demand at vertex v with respect to flow $\mathbf{r}$. A flow $\mathbf{r}$ is a *maximum flow* if it minimizes $\sum_{v \in V} |\mathbf{d_r}(v)|$. We need the following facts from the network-flows theory. Let $\mathbf{r}$, $\mathbf{r}_{\max}$, and $(S_{\max}, T_{\max})$ be a flow, a maximum flow, and a maximum surplus cut in network $G = (V, E, \mathbf{u}, \mathbf{d})$.

F1. For each cut (S,T), $surplus_{\mathbf{r}}(S,T) = surplus(S,T)$.

F2. $\mathbf{u}_{\mathbf{r}_{\max}}(S_{\max}, T_{\max}) = 0$.

F3. For each $T \subseteq V$, $\mathbf{d}_{\mathbf{r}_{\max}}(T) \leq \mathbf{d}_{\mathbf{r}_{\max}}(T_{\max})$.

F4. Let a new edge-capacity function $\mathbf{u}'$ be such that $\mathbf{u}'(e) \geq \mathbf{u}(e)$, for each edge $e \in E$. There exists a maximum flow $\mathbf{r}'_{\max}$ in network $G' = (V, E, \mathbf{u}', \mathbf{d})$ such that

(i) for each vertex $v \in V$, $d_{\mathbf{r}'_{\max}}(v)$ and $d_{\mathbf{r}_{\max}}(v)$ have the same sign,

(ii) for each edge $e \in E$, $|\mathbf{r}'_{\max}(e) - \mathbf{r}_{\max}(e)| \leq d_{\mathbf{r}_{\max}}(T_{\max})$, and

(iii) for each cut (S,T) in G, $|\mathbf{r}'_{\max}(S,T) - \mathbf{r}_{\max}(S,T)| \leq d_{\mathbf{r}}(T_{\max})$.

These facts can be derived from the properties of network flows described in [34] and [8, Chapter 27]. Fact F1 follows actually directly from the definitions we have introduced (for each flow $\mathbf{r}$ and cut (S,T), the differences $\mathbf{d}(T) - \mathbf{d_r}(T)$ and $\mathbf{u}(S,T) - \mathbf{u_r}(S,T)$ are the same). Facts F2 and F3 are closely related to the maximum-flow/minimum-cut theorem. Informally speaking, the conditions listed in Fact F4 are satisfied by a maximum flow $\mathbf{r}'_{\max}$ in network G' which "extends" the maximum flow $\mathbf{r}_{\max}$ in network G.

In our analysis of the Newton method for the MaxRatioCut problem we use the same notation as in Section 3, where this method is introduced. Thus

$$f_i = surplus(S_i, T_i),$$
$$g_i = g(S_i, T_i),$$
$$h_i = h(\delta_i) = surplus(S_i, T_i) - \delta_i g(S_i, T_i) = surplus_{\delta_i}(S_i, T_i), \quad (26)$$

where (S_i, T_i) is the maximum-surplus cut in network $G_{\delta_i} = (V, E, \mathbf{u} + \delta_i \mathbf{w}, \mathbf{d})$ computed in iteration i. Subscript δ will always indicate that the underlying network is network G_δ, that is, network G with the edge-capacity function changed to function $\mathbf{u} + \delta \mathbf{w}$. In particular,

$$surplus_\delta(S, T) \stackrel{def}{=} \mathbf{d}(T) - (\mathbf{u} + \delta \mathbf{w})(S, T)$$
$$= surplus(S, T) - \delta g(S, T), \quad (27)$$
$$surplus_{\delta, \mathbf{r}}(S, T) \stackrel{def}{=} \mathbf{d}_\mathbf{r}(T) - \mathbf{u}_{\delta, \mathbf{r}}(S, T), \quad (28)$$

where $\mathbf{r}$ is a flow in network G_δ. The analysis in this section is similar to the analysis of the Newton method for the MaxRatioPath problem presented in the proof of Theorem 4.8. We say that an edge e is *essential* at the beginning of iteration i, if it belongs to a cut (S_j, T_j), for some $j \geq i$. As the computation proceeds, the number of essential edges decreases and we want to analyze this decrement. As in the proofs of the other bounds on the number of iterations of the Newton method, we use Lemma 3.2 and its direct consequence expressed in Inequality (5) to measure the progress of the computation.

Let $\mathbf{r}_i$, for $i = 1, 2, \ldots, t$ (t is the index of the last iteration) be maximum flows in networks G_{δ_i} such that for each i, all conditions of Fact F4 above are satisfied with $G = G_{\delta_i}$, $G' = G_{\delta_{i+1}}$, $\mathbf{r}_{\max} = \mathbf{r}_i$, and $\mathbf{r}'_{\max} = \mathbf{r}_{i+1}$. Using (28), Facts F1 and F2, and then (26), we have

$$\mathbf{d}_{\mathbf{r}_i}(T_i) = surplus_{\delta_i, \mathbf{r}_i}(S_i, T_i) - \mathbf{u}_{\delta_i, \mathbf{r}_i}(S_i, T_i)$$
$$= surplus_{\delta_i}(S_i, T_i)$$
$$= h_i. \quad (29)$$

The following lemma gives a simple, sufficient condition for an edge not to be essential.

Fractional Combinatorial Optimization

Lemma 8.1 *If*
$$\mathbf{u}_{\delta_i,\mathbf{r}_i}(e) > h_i, \tag{30}$$
then edge e is not essential at the beginning of iteration i.

Proof. Let (S,T) be a cut containing an edge e which satisfies Inequality (30). Using Facts F1 and F3, and Equation (29), we obtain

$$\begin{aligned}
surplus_{\delta_i}(S,T) &= surplus_{\delta_i,\mathbf{r}_i}(S,T) = \mathbf{d}_{\mathbf{r}_i}(T) - \mathbf{u}_{\delta_i,\mathbf{r}_i}(S,T) \\
&\leq \mathbf{d}_{\mathbf{r}_i}(T_i) - \mathbf{u}_{\delta_i,\mathbf{r}_i}(S,T) = h_i - \mathbf{u}_{\delta_i,\mathbf{r}_i}(S,T) \\
&\leq h_i - \mathbf{u}_{\delta_i,\mathbf{r}_i}(e) < 0.
\end{aligned}$$

Therefore none of the cuts (S_j, T_j), for $j \geq i$, can contain edge e, because $\delta_j \geq \delta_i$ and

$$surplus_{\delta_i}(S_j, T_j) \geq surplus_{\delta_j}(S_j, T_j) = h_j \geq 0.$$

This means that edge e is not essential at the beginning of iteration i. □

The proof of the $O(m)$ bound on the number of iterations shown below in Theorem 8.3 involves quite a few technical details. Therefore, to highlight the main line of the argumentation, we first show a simpler $O(m \log m)$ bound.

Theorem 8.2 *The Newton method solves the MaxRatioCut problem in $O(m \log m)$ iterations, if the edge weights are nonnegative.*

Proof. We show that a sequence of $k = \lfloor \log m \rfloor + 2$ consecutive iterations decreases the number of essential edges at least by 1. This claim immediately implies the $O(m \log m)$ bound on the number of iterations. Consider iterations $i, i+1, \ldots, i+k$. Inequality (5) implies that

$$h_{i+k}g_{i+k} < \frac{1}{m^2}h_{i+1}g_{i+1}. \tag{31}$$

If $g_{i+k} < (1/m)g_i$, then, using the assumption that the edge weights are nonnegative, we know that for some edge $e \in (S_i, T_i)$,

$$\mathbf{w}(e) \geq \frac{1}{|(S_i, T_i)|} \sum_{a \in (S_i, T_i)} \mathbf{w}(a) \geq \frac{1}{m}g_i > g_{i+k}.$$

Such an edge e cannot belong to cut (S_{i+k}, T_{i+k}) or to any cut (S_j, T_j), for $j \geq i + k$ (since sequence (g_l) is nonincreasing), so e is essential at the beginning of iteration i but is not essential at the beginning of iteration $i+k$.

If $g_{i+k} \geq (1/m)g_i$, then also $g_{i+k} \geq (1/m)g_{i+1}$, and Inequality (31) implies that $h_{i+k} < (1/m)h_{i+1}$. Fact F3 implies that for each vertex $v \in T_i$, $d_{\mathbf{r}_i}(v)$ is nonnegative. Hence Fact F4(i) further implies that for each vertex $v \in T_i$, $d_{\mathbf{r}_{i+k}}(v)$ is nonnegative, so $d_{\mathbf{r}_{i+k}}(T_i)$ must be nonnegative as well. Therefore, using Lemma 4.7 and Fact F1, we obtain

$$\begin{aligned}
-mh_{i+k} &> -h_{i+1} \geq f_i - \delta_{i+k}g_i \\
&= \text{surplus}(S_i, T_i) - \delta_{i+k}g(S_i, T_i) \\
&= \text{surplus}_{\delta_{i+k}}(S_i, T_i) \\
&= \text{surplus}_{\delta_{i+k}, \mathbf{r}_{i+k}}(S_i, T_i) \\
&= \mathbf{d}_{\mathbf{r}_{i+k}}(T_i) - \mathbf{u}_{\delta_{i+k}, \mathbf{r}_{i+k}}(S_i, T_i) \\
&\geq -\mathbf{u}_{\delta_{i+k}, \mathbf{r}_{i+k}}(S_i, T_i).
\end{aligned} \qquad (32)$$

The above inequality implies that there is an edge $e \in (S_i, T_i)$ such that $\mathbf{u}_{\delta_{i+k}, \mathbf{r}_{i+k}}(e) > h_{i+k}$. Such an edge is essential at the beginning of iteration i, but Lemma 8.1 implies that it is not essential at the beginning of iteration $i + k$. □

Theorem 8.3 *The Newton method solves the MaxRatioCut problem in $O(m)$ iterations, if the edge weights are nonnegative.*

Proof. Lemma 3.2 implies that for each iteration i except the last one,

$$\frac{g_{i+1}}{g_i} \leq \frac{1}{2} \qquad (33)$$

or

$$\frac{h_{i+1}}{h_i} \leq \frac{1}{2}. \qquad (34)$$

We separately bound the number of iterations for which Inequality (33) holds and the number of iterations for which Inequality (34) holds. Let $i_1, i_2, \ldots, i_q$ be the indices of the iterations for which Inequality (33) holds. Let $\mu_j = g_{i_j}$, for $j = 1, 2, \ldots, q$, and let $(\alpha_k)_{k=1}^m$ be the sequence of the edge weights in nonincreasing order. Sequences $(\alpha_k)_{k=1}^m$ and $(\mu_j)_{i=j}^q$ satisfy the conditions of Lemma 8.4 below, so $q \leq m$.

Fractional Combinatorial Optimization

Now we bound the number of iterations for which Inequality (34) holds. Here the argument is more involved, because numbers h_i are not simply subsums of a fixed set of m numbers. From now on we consider only iterations for which Inequality (34) holds. To avoid towering subscripts, we number these iterations from 1 to $t' \leq t$. Thus "iteration i" and subscripts i refer now to the ith iteration for which Inequality (34) holds. We use the same notation and definitions as in the proof of Theorem 8.2, taking into account the new numbering of the iterations.

Consider cut $(S_i, T_i) = \{e_1, e_2, \ldots, e_p\}$ computed in iteration i. For technical reasons, we assume that there are at least $\log p + 5$ iterations following iteration i, that is, $t' \geq i + \log p + 5$. The proof of Theorem 8.2 was based on showing that at least one edge from cut (S_i, T_i) becomes unessential by the beginning of iteration $i + \lfloor \log m \rfloor + 2$. Here we show that there exists $l \geq 3$ such that at least $l - 2$ edges from cut (S_i, T_i) become unessential by the beginning of iteration $i + l + 2$. This claim immediately implies an $O(m)$ bound on the number of iterations.

Fact F4(ii), Equation (29), and Inequality (34) imply that for each edge $e \in E$ and index l such that $i + 3 \leq l \leq t'$,

$$\begin{aligned}
|\mathbf{u}_{\delta_{i+3}, \mathbf{r}_{l+1}}(e) - \mathbf{u}_{\delta_{i+3}, \mathbf{r}_l}(e)| &= |\mathbf{r}_{l+1}(e) - \mathbf{r}_l(e)| \\
&\leq \mathbf{d}_{\mathbf{r}_l}(T_l) = h_l \\
&\leq \frac{1}{2^{l-i-2}} h_{i+2}.
\end{aligned} \quad (35)$$

Analogously, using Fact F4(iii), we have for each cut (S, T) and index l such that $i + 3 \leq l \leq t'$,

$$|\mathbf{u}_{\delta_{i+3}, \mathbf{r}_{l+1}}(S, T) - \mathbf{u}_{\delta_{i+3}, \mathbf{r}_l}(S, T)| \leq \frac{1}{2^{l-i-2}} h_{i+2}. \quad (36)$$

The derivations (32) for $k = 3$ and Inequality (34) give

$$\mathbf{u}_{\delta_{i+3}, \mathbf{r}_{i+3}}(S_i, T_i) \geq h_{i+1} \geq 2 h_{i+2}. \quad (37)$$

Using this inequality and Inequality (36), we have for each index l such that $i + 4 \leq l \leq t'$,

$$\begin{aligned}
\mathbf{u}_{\delta_{i+3}, \mathbf{r}_l}(S_i, T_i) &= \mathbf{u}_{\delta_{i+3}, \mathbf{r}_{i+3}}(S_i, T_i) \\
&\quad + (\mathbf{u}_{\delta_{i+3}, \mathbf{r}_{i+4}}(S, T) - \mathbf{u}_{\delta_{i+3}, \mathbf{r}_{i+3}}(S, T)) \\
&\quad + \cdots + (\mathbf{u}_{\delta_{i+3}, \mathbf{r}_l}(S, T) - \mathbf{u}_{\delta_{i+3}, \mathbf{r}_{l-1}}(S, T)) \\
&\geq \mathbf{u}_{\delta_{i+3}, \mathbf{r}_{i+3}}(S_i, T_i) - h_{i+2} \left(\frac{1}{2} + \frac{1}{2^2} + \cdots + \frac{1}{2^{l-i-3}} \right) \\
&\geq h_{i+2}.
\end{aligned} \quad (38)$$

For each $l = 1, 2, \ldots, t' - i - 2$ and $j = 1, 2, \ldots, p$, let

$$\alpha_{l,j} = \frac{1}{h_{i+2}} \mathbf{u}_{\delta_{i+3}, r_{i+2+l}}(e_j).$$

Inequalities (37) and (38) imply that matrix $(\alpha_{l,j})$ satisfies Condition 2 of Lemma 8.5 below. (since we have $\sum_{j=1}^{p} \alpha_{l,j} = \mathbf{u}_{\delta_{i+3}, r_{i+2+l}}(S_i, T_i)/h_{i+2}$). Inequality (35) implies that Condition 3 is satisfied as well. Condition 1 holds because we assumed that $t' \geq i + \log p + 5$. Thus we can apply Lemma 8.5 to matrix $(\alpha_{l,j})$. According to the definition of "good entries" in the statement of Lemma 8.5, if $\alpha_{l,j}$ is a good entry, then there exists $1 \leq l' \leq l$ such that $\alpha_{l',j} > 1/2^{l'}$. If $\alpha_{l',j} > 1/2^{l'}$, then

$$\mathbf{u}_{\delta_{i+2+l'}, r_{i+2+l'}}(e_j) \geq \mathbf{u}_{\delta_{i+3}, r_{i+2+l'}}(e_j) = h_{i+2}\alpha_{l',j} > \frac{1}{2^{l'}} h_{i+2} \geq h_{i+l'+2},$$

so Lemma 8.1 implies that edge e_j is unessential at the beginning of iteration $i + l + 2$. Lemma 8.5 implies that there exists $l \geq 3$ such that there are at least $l - 2$ good entries among entries $\alpha_{l,1}, \alpha_{l,2}, \ldots, \alpha_{l,p}$. The $l - 2$ edges corresponding to these $l - 2$ entries are essential at the beginning of iteration i but become unessential by iteration $i + l + 2$. $\square$

It remains to prove the following two technical lemmas.

Lemma 8.4 *Let $\alpha_1 \geq \alpha_2 \geq \cdots \geq \alpha_m \geq 0$ and $\mu_1 > \mu_2 > \ldots > \mu_q > 0$ be such that*

1. $\mu_{j+1} \leq (1/2)\mu_j$, for $j = 1, 2, \ldots, q - 1$, and

2. $\mu_j \leq \sum \{\alpha_k \mid \alpha_k \leq \mu_j\}$, for $j = 1, 2, \ldots, q$.

Then $q \leq m$.

Proof. Let $\bar{\alpha}_k = \alpha_k + \alpha_{k+1} + \cdots + \alpha_m$. Condition 2 implies that $\mu_1 \leq \bar{\alpha}_1$, and if $\mu_j < \bar{\alpha}_m = \alpha_m$, then $\mu_j \leq 0$. Thus all numbers μ_j lie in the interval $[\bar{\alpha}_m, \bar{\alpha}_1]$. To prove that $q \leq m$, we show that each of the intervals $(\bar{\alpha}_m, \bar{\alpha}_{m-1}], (\bar{\alpha}_{m-1}, \bar{\alpha}_{m-2}], \ldots, (\bar{\alpha}_2, \bar{\alpha}_1]$ contains at most one element from sequence (μ_j). Let $\mu_j \in (\bar{\alpha}_{k+1}, \bar{\alpha}_k]$, for some $1 \leq j \leq q-1$ and $1 \leq k \leq m-1$. We show that $\mu_{j+1} \leq \bar{\alpha}_{k+1}$, so $\mu_{j+1} \notin (\bar{\alpha}_{k+1}, \bar{\alpha}_k]$. If $\bar{\alpha}_{k+1} \geq (1/2)\bar{\alpha}_k$, then

$$\mu_{j+1} \leq \frac{1}{2}\mu_j \leq \frac{1}{2}\bar{\alpha}_k \leq \bar{\alpha}_{k+1}.$$

If $\bar{\alpha}_{k+1} < (1/2)\bar{\alpha}_k$, then

$$\mu_{j+1} \leq \frac{1}{2}\mu_j \leq \frac{1}{2}\bar{\alpha}_k = \bar{\alpha}_k - \frac{1}{2}\bar{\alpha}_k = \bar{\alpha}_k + \bar{\alpha}_{k+1} - \frac{1}{2}\bar{\alpha}_k < \bar{\alpha}_k.$$

The above inequality and Condition 2 imply that $\mu_{j+1} \leq \bar{\alpha}_{k+1}$. □

Lemma 8.5 *Let $(\alpha_{l,j})$ be a $q \times p$ matrix such that*

1. $q \geq \log p + 3$,
2. $\sum_{j=1}^{p} \alpha_{l,j} \geq 1$, for each $1 \leq l \leq q$, and
3. $|\alpha_{l+1,j} - \alpha_{l,j}| \leq 1/2^l$, for each $1 \leq l \leq q-1$ and $1 \leq j \leq p$.

If $\alpha_{l,j} > 1/2^l$, then this and all subsequent entries in column j are "good entries." There exists l such that $3 \leq l \leq q$ and row l of the matrix contains at least $l - 2$ good entries.

Proof. Condition 3 implies that if $1 \leq l' < l'' \leq q$ and $1 \leq j \leq p$, then

$$\alpha_{l'',j} \leq \alpha_{l',j} + \frac{1}{2^{l'}} + \frac{1}{2^{l'+1}} + \cdots + \frac{1}{2^{l''-1}}$$
$$< \alpha_{l',j} + \frac{1}{2^{l'-1}}. \tag{39}$$

If $\alpha_{l,j}$ is the first good entry in column j and $l \geq 2$, then

$$\alpha_{l,j} \leq \alpha_{l-1,j} + \frac{1}{2^{l-1}} \leq \frac{1}{2^{l-1}} + \frac{1}{2^{l-1}} = \frac{1}{2^{l-2}}. \tag{40}$$

The first inequality above follows from Condition 3, and the second one holds because entry $\alpha_{l-1,j}$ is not a good entry.

Assume that for each $3 \leq l \leq q$, row l contains at most $l - 3$ good entries. (In particular, there are no good entries in rows 1, 2 and 3.) We show that this assumption implies that the sum of the entries in the last row is less than 1, which contradicts Condition 2 of the lemma. Let $j_1, j_2, \ldots, j_r$ be the indices of all columns which have at least one good entry. Let $\alpha_{l_1,j_1}, \alpha_{l_2,j_2}, \ldots, \alpha_{l_r,j_r}$ be the first good entries in these columns. Let $j_1, j_2, \ldots, j_r$ be ordered in such a way that $l_1 \leq l_2 \leq \cdots \leq l_r$. Hence for each $k = 1, 2, \ldots, r$, entries $\alpha_{l_k,j_1}, \alpha_{l_k,j_2}, \ldots, \alpha_{l_k,j_k}$ are good entries, so row l_k

contains at least k good entries. Our assumption says that row l_k contains at most $l_k - 3$ good entries, so

$$k \le l_k - 3. \tag{41}$$

The sum of the entries in the last row is at most

$$\begin{aligned}
&\frac{p-r}{2^q} + \alpha_{q,j_1} + \alpha_{q,j_2} + \cdots + \alpha_{q,j_r} \\
&\le \frac{p}{2^q} + \left(\alpha_{l_1,j_1} + \frac{1}{2^{l_1-1}}\right) + \cdots + \left(\alpha_{l_r,j_r} + \frac{1}{2^{l_r-1}}\right) \\
&\le \frac{1}{8} + \left(\frac{1}{2^{l_1-2}} + \frac{1}{2^{l_1-1}}\right) + \cdots + \left(\frac{1}{2^{l_r-2}} + \frac{1}{2^{l_r-1}}\right) \\
&= \frac{1}{8} + 6\left(\frac{1}{2^{l_1}} + \cdots + \frac{1}{2^{l_r}}\right) \\
&\le \frac{1}{8} + 6\left(\frac{1}{2^4} + \frac{1}{2^5} + \cdots\right) \\
&< 1.
\end{aligned}$$

The first inequality follows from Inequality (39), the second one from Condition 1 and Inequality (40), and the third one from Inequality (41). □

9 Concluding Remarks

The binary search, the Newton method, and Megiddo's parametric search form together a powerful set of methods for solving fractional combinatorial optimization problems. All three methods reduce a fractional problem to a sequence of instances of the corresponding non-fractional problem. Implementations of the binary search and the Newton method are straightforward once a procedure for solving the non-fractional problem is provided. Implementation of Megiddo's parametric search is a bit more complicated, because this method, unlike the other two, does not treat a procedure for the underlying non-fractional problem as a black box, but has to examine its structure and modify it appropriately.

Megiddo originally proposed his parametric search method in the context of fractional combinatorial optimization problems such as the minimum-ratio spanning-tree problem and the maximum profit-to-time cycle problem [31]. Since then, however, his method has been extended to far more general optimization problems; see for example [5, 6, 21, 33, 46]. If we use

Megiddo's parametric search method to obtain a fast algorithm for a given fractional combinatorial optimization problem, our main task is to find a parallel algorithm for the underlying non-fraction problem which exhibits a right trade-off between the parallel running time and the total amount of computation performed.

The Newton method for general fractional optimization has been extensively studied since Dinkelbach [10] introduced it in 1967, and a number of modifications and extensions of this method have been proposed and analyzed [23, 35, 40]. However, analyzing the Newton method for fractional combinatorial optimization problems is a relatively young research direction (the results presented in Sections 4 and 8 come from early 90's [37, 38]) and there are quite a few questions which remain to be answered. For example, how tight is the $O(p^2 \log^2 p)$ bound on the number of iterations of the Newton method for linear fractional combinatorial optimization shown in Theorem 4.6? The main tool used in the derivation of this bound is Lemma 4.1, which gives an $O(p \log p)$ bound on the length of a geometric sequence of subsums of a p-element set. We know that this $O(p \log p)$ bound is tight [19], but there may be a better way of using it in the analysis of the Newton method.

For problems such as MaxRatioPath, MaxRatioCycle, and MaxRatioCut, the Newton method requires only a linear number of iterations, provided that the edge weights are non-negative, and the proofs of these three bounds are remarkably similar to each other (compare the proofs of Theorems 4.8 and 8.2). However, a unified framework which would generalize these three examples has not been proposed yet. Such a framework would have to be based on the "primal-dual" structure of problems such as the three problems mentioned here. Another related task is to examine if the assumption that the edge weights are non-negative is really necessary to obtain a liner bound on the number of iterations.

The Newton method for fractional combinatorial optimization is an application of the classical Newton method to computing the root of a piecewise linear, convex function h defined in (1). The number of linear segments of function h is an obvious upper bound on the number of iterations of this method. The artificial example presented in Section 4 shows that function h may consist of an exponential number of linear segments, even for a linear fractional combinatorial optimization problem. The results of Gusfield [20] (upper bound) and Carstensen [3] (lower bound) imply that for the MaxRatioPath and the MaxRatioCycle problems function h consists, in the worst-case, of $n^{\Theta(\log n)}$ linear segments. Such tight bounds, however, are known only for relatively few problems. For example, no tight bounds on the complexity of function h for the MaxRatioCut problem and the maximum-ratio spanning-tree have been reported yet.

The Newton method constructs a sequence of improving lower bounds converging to the optimum objective value, but since it does not provide upper bounds, no bound on the error is available at any given iteration. There are methods for general fractional optimization, which are based on the Newton method but construct also improving upper bounds (see, for example, [23, 35, 40]). It is not clear yet if these methods can lead also to interesting results for fractional combinatorial optimization.

The methods for solving fractional combinatorial optimization problems rely on methods for solving the corresponding non-fractional optimization problems. The underlying non-fractional problem may however be difficult on its own, and a practical algorithm for solving this problem exactly may not exists. Hashizume et al [22] analyze Megiddo's parametric search method for maximizing $(a_0 + \mathbf{ax})/(b_0 + \mathbf{bx})$, when the algorithm for the underlying non-fractional problem of maximizing $(\mathbf{a} - \delta \mathbf{b})\mathbf{x}$ returns only approximate solutions. They show that the accuracy of the approximate solution obtained for the fractional problem is at least as good as the accuracy of the solutions computed for the non-fractional problem. As an example of their analysis, they present a fully polynomial approximation scheme for the fractional 0–1 knapsack problem. Examples of analogous analyses for the binary search method are shown in [25, 30]. However, a general approximation analysis of the Newton method has not been proposed yet.

References

[1] R. E. Bellman. On a routing problem. *Quarterly of Applied Mathematics*, 16:87–90, 1958.

[2] M. Blum, R. W. Floyd, V. Pratt, R. L. Rivest, and R. E. Tarjan. Time bounds for selection. *J. Comp. and Syst. Sci.*, 7(4):448–461, 1973.

[3] P. J. Carstensen. *The Complexity of Some Problems in Parametric, Linear, and Combinatorial Programming.* PhD thesis, Department of Mathematics, University of Michigan, Ann Arbor, Michigan, 1983. Doctoral Thesis.

[4] R. Chandrasekaran. Minimum ratio spanning trees. *Networks*, 7:335–342, 1977.

[5] E. Cohen and N. Megiddo. Maximizing concave functions in fixed dimension. In P. Pardalos, editor, *Complexity in Numerical Optimization*, pages 74–87. World Scientific, 1993.

[6] E. Cohen and N. Megiddo. Strongly polynomial time and NC algorithms for detecting cycles in dynamic graphs. *J. Assoc. Comput. Mach.*, 40(4):791–830, 1993.

[7] R. Cole. Parallel merge sort. *SIAM J. Comput.*, 17(4):770–785, August 1988.

[8] T. H. Cormen, C. E. Leiserson, and R. L. Rivest. *Introduction to Algorithms*. MIT Press, Cambridge, MA, 1990.

[9] G. B. Dantzig, W. O. Blattner, and M. R. Rao. Finding a cycle in a graph with minimu cost to time ratio with application to a ship routing problem. In P. Rosentiehl, editor, *Theory of Graphs*, pages 77–84. Gordon and Breach, New York, 1967.

[10] W. Dinkelbach. On nonlinear fractional programming. *Management Science*, 13:492–498, 1967.

[11] T. R. Ervolina and S. T. McCormick. Two strongly polynomial cut cancelling algorithms for minimum cost network flow. *Discrete Applied Math.*, 46:133–165, 1993.

[12] L. R. Ford, Jr. and D. R. Fulkerson. *Flows in Networks*. Princeton Univ. Press, Princeton, NJ, 1962.

[13] B. Fox. Finding minimal cost-time ratio circuits. *Oper. Res.*, 17:546–551, 1969.

[14] M. L. Fredman and R. E. Tarjan. Fibonacci heaps and their uses in improved network optimization algorithms. *J. Assoc. Comput. Mach.*, 34:596–615, 1987.

[15] M. Goemans. Personal communication, 1992.

[16] A. V. Goldberg. Scaling algorithms for the shortest paths problem. *SIAM J. Comput.*, 24(3):494–504, 1995.

[17] A. V. Goldberg and R. E. Tarjan. A new approach to the maximum flow problem. *J. Assoc. Comput. Mach.*, 35:921–940, 1988.

[18] A. V. Goldberg and R. E. Tarjan. Finding minimum-cost circulations by canceling negative cycles. *J. Assoc. Comput. Mach.*, 36:388–397, 1989.

[19] M. Goldmann. On a subset-sum problem. Personal Communication, 1994.

[20] D. Gusfield. Sensitivity analysis for combinatorila optimizaton. Technical Report UCB/ERL M90/22, Electronics Research Laboratory, University of California, Berkeley, May 1980.

[21] D. Gusfield. Parametric combinatorial computing and a problem of program module distribution. *J. Assoc. Comput. Mach.*, 30(3):551–563, 1983.

[22] S. Hashizume, M. Fukushima, N. Katoh, and T. Ibaraki. Approximation algorithms for combinatorial fractional programming problems. *Math. Programming*, 37:255–267, 1987.

[23] T. Ibaraki. Parametric approaches to fractional programs. *Math. Programming*, 26:345–362, 1983.

[24] H. Ishii, T. Ibaraki, and H. Mine. Fractional knapsack problems. *Math. Programming*, 13:255–271, 1976.

[25] K. Iwano, S. Misono, S. Tezuka, and S. Fujishige. A new scaling algorithm for the maximum mean cut problem. *Algorithmica*, 11(3):243–255, 1994.

[26] R. M. Karp. A Characterization of the minimum cycle mean in a digraph. *Discrete Math.*, 23:309–311, 1978.

[27] V. King, S. Rao, and R. Tarjan. A faster deterministic maximum flow algorithm. *J. Alg.*, 17(3):447–474, 1994.

[28] E. L. Lawler. Optimal cycles in doubly weighted directed linear graphs. In P. Rosentiehl, editor, *Theory of Graphs*, pages 209–213. Gordon and Breach, New York, 1967.

[29] E. L. Lawler. *Combinatorial Optimization: Networks and Matroids*. Holt, Reinhart, and Winston, New York, NY., 1976.

[30] S. T. McCormick. A note on approximate binary search algorithms for mean cuts and cycles. UBC Faculty of Commerce Working Paper 92-MSC-021, University of British Columbia, Vancouver, Canada, 1992.

[31] N. Megiddo. Combinatorial optimization with rational objective functions. *Math. Oper. Res.*, 4:414–424, 1979.

[32] N. Megiddo. Applying parallel computation algorithms in the design of serial algorithms. *J. Assoc. Comput. Mach.*, 30:852–865, 1983.

[33] C. Haibt Norton, S. A. Plotkin, and É. Tardos. Using separation algorithms in fixed dimension. *J. Alg.*, 13:79–98, 1992.

[34] C. H. Papadimitriou and K. Steiglitz. *Combinatorial Optimization: Algorithms and Complexity*. Prentice-Hall, Englewood Cliffs, NJ, 1982.

[35] P. M. Pardalos and A. T. Phillips. Global optimization of fractional programs. *J. Global Opt.*, 1:173–182, 1991.

[36] T. Radzik. Newton's method for fractional combinatorial optimization. In *Proc. 33rd IEEE Annual Symposium on Foundations of Computer Science*, pages 659–669, 1992.

[37] T. Radzik. Newton's method for fractional combinatorial optimization. Technical Report STAN-CS-92-1406, Department of Computer Science, Stanford University, January 1992.

[38] T. Radzik. Parametric flows, weighted means of cuts, and fractional combinatorial optimization. In P. Pardalos, editor, *Complexity in Numerical Optimization*, pages 351–386. World Scientific, 1993.

[39] T. Radzik and A. Goldberg. Tight bounds on the number of minimum-mean cycle cancellations and related results. *Algorithmica*, 11:226–242, 1994.

[40] S. Schaible. Fractional Programming 2. On Dinkelbach's algorithm. *Management Sci.*, 22:868–873, 1976.

[41] S. Schaible. A survey of fractional programming. In S. Schaible and W. T. Ziemba, editors, *Generalized Concavity in Optimization and Economics*, pages 417–440. Academic Press, New York, 1981.

[42] S. Schaible. Bibliography in fractional programming. *Zeitschrift für Operations Res.*, 26(7), 1982.

[43] S. Schaible. Fractional programming. *Zeitschrift für Operations Res.*, 27:39–54, 1983.

[44] S. Schaible and T. Ibaraki. Fractional programming. *Europ. J. of Operational Research*, 12, 1983.

[45] Y. Shiloach and U. Vishkin. An $O(n^2 \log n)$ Parallel Max-Flow Algorithm. *J. Algorithms*, 3:128–146, 1982.

[46] S. Toledo. Maximizing non-linear concave functions in fixed dimension. In P. Pardalos, editor, *Complexity in Numerical Optimization*, pages 429–447. World Scientific, 1993.

[47] L. G. Valiant. Parallelism in comparison problems. *SIAM J. Comput.*, 4:348–355, 1975.

[48] C. Wallacher. A Generalization of the Minimum-mean Cycle Selection Rule in Cycle Canceling Algorithms. Unpublished manuscript, Institut für Angewandte Mathematik, Technische Universität Carolo-Wilhelmina, Germany, November 1989.

[49] A. C. Yao. An $O(|E| \log \log |V|)$ algorithm for finding minimum spanning trees. *Information Processing Let.*, 4:21–23, 1975.

Reformulation-Linearization Techniques for Discrete Optimization Problems

Hanif D. Sherali
Department of Industrial and Systems Engineering
Virginia Polytechnic Institute and State University,
Blacksburg, VA 24061-0118
E-mail: hanifs@vt.edu

Warren P. Adams
Department of Math Sciences
Clemson University, Clemson, SC 29634-1907

1 Introduction

Discrete and continuous nonconvex programming problems arise in a host of practical applications in the context of production, location-allocation, distribution, economics and game theory, process design, and engineering design situations. Several recent advances have been made in the development of branch-and-cut algorithms for discrete optimization problems and in polyhedral outer-approximation methods for continuous nonconvex programming problems. At the heart of these approaches is a sequence of linear programming problems that drive the solution process. The success of such algorithms is strongly tied in with the strength or tightness of the linear programming representations employed.

This paper addresses the issue of generating tight linear programming (LP) representations via automatic reformulation techniques in solving discrete mixed-integer 0-1 linear (and polynomial) programming problems. The particular approach that we focus on in this paper is called the **Reformulation Linearization Technique (RLT)**, a procedure that can be used to generate tight linear (or sometimes convex) programming representation, for constructing not only exact solution algorithms, but also to design powerful heuristic procedures. Actually, this methodology can be extended to solve classes of continuous nonconvex programs as well, and we refer the interested reader to Sherali and Tuncbilek (1992, 1995, 1996), and Sherali (1996) for an introduction to this subject. (See Sherali and Adams, 1996, for a general survey on this topic.) The tight relaxations produced can often be used to derive good quality solutions (in polynomial time) to problems of practical sizes that arise in the aforementioned applications. Also, the RLT procedure is capable of generating representations of increasing degrees of strength, but with an accompanied increase in problem size. Coupled with the recent advances in LP technology, this permits one to incorporate tighter RLT based representations within the context of exact or heuristic methods.

The motivation for constructing "*good*" models, that is, models that have tight underlying linear programming representations, rather than simply "*mathematically correct*" models, has led to some crucial and critical research on the model formulation process as in Balas (1985), Jeroslow and Lowe (1984, 1985), Johnson (1989), Meyer (1981), Sherali and Adams (1989, 1990), Williams (1985), and Wolsey (1989). Much work has also been done in converting classes of separable or polynomial nonlinear integer programming problems into equivalent linear integer programs, and for generating tight, valid inequalities for such problems, as in Adams and Sherali (1986, 1987a,b), Balas and Mazzola (1984a,b), Glover (1975), and Sherali and Adams (1989, 1990). Sometimes, a special variable redefinition technique may be applicable depending on the problem structure as in Martin (1987). In this approach, a linear transformation is defined on the variables to yield an equivalent formulation that tightens the continuous relaxation by constructing a partial convex hull of a specially structured subset of constraints. A more generally applicable technique is to augment the formulation through the addition of valid or implied inequalities that typically provide some partial characterization for the convex hull of feasible solutions. Some cutting plane generation schemes in this vein include the ones described in Nemhauser and Wolsey (1990), Padberg (1980), Van Roy and Wolsey (1983), and Wolsey (1976). Automatic reformulation procedures utilizing such constraint generation schemes within a branch-and-bound or branch-and-cut framework are presented in Crowder et al. (1983), Hoffman and Padberg (1991), Johnson et al. (1985), Nemhauser et al. (1991), Oley and Sjouquist (1982), Padberg and Rinaldi (1991), and Van Roy and Wolsey (1987). Besides these studies, ample evidence is available in the literature on the efficacy of providing tight linear programming relaxations for pure and mixed zero-one programming problems as in Adams and Sherali (1986, 1987a,b), Crowder and Padberg (1986), Geoffrion and McBryde (1979), and Magnanti and Wong (1981), among many others.

In this paper, we shall be discussing in detail the Reformulation - Linearization Technique (RLT) of Sherali and Adams (1989, 1990, 1994) along with its several enhancements and extensions. This procedure is also an automatic reformulation technique that can be used to derive tight LP representations as well as to generate strong valid inequalities. Consider a mixed-integer zero-one linear programming problem whose feasible region X is defined in terms of some inequalities and equalities in binary variables $x = (x_1, \ldots, x_n)$ and a set of bounded continuous variables $y = (y_1, \ldots, y_m)$. Given a value of $d \in \{1, \ldots, n\}$, this RLT procedure constructs various polynomial factors of degree d comprised of the product of some d binary variables or their complements. These factors are then used to multiply each of the constraints defining X (including the variable bounding restrictions), to create a (nonlinear) polynomial mixed-integer zero-one programming problem. Using the relationship $x_j^2 = x_j$ for each binary variable $x_j, j = 1, \ldots, n$, substituting a variable w_J and v_{Jk}, respectively, in place of each nonlinear term of the type $\prod_{j \in J} x_j$, and $y_k \prod_{j \in J} x_j$, and relaxing integrality, the nonlinear polynomial problem is re-linearized into a higher dimensional polyhedral set X_d defined in terms of the original variables (x, y) and the new variables (w, v). For X_d to be equivalent to X, it is only necessary to enforce x to be binary valued and the remaining variables may be treated as continuous valued, since the binariness on the x-variables is shown to automatically enforce the required product relationships on the $w-$ and $v-$variables. Denoting the projection of X_d onto the space of the original (x, y)-variables as X_{Pd}, it can be shown that as d varies from 1 to n, we get,

$$X_{P0} \supseteq X_{P1} \supseteq X_{P2} \supseteq \ldots \supseteq X_{Pn} \equiv conv(X)$$

where X_{P0} is the ordinary linear programming relaxation, and $conv(X)$ represents the convex hull of X. The hierarchy of higher-dimensional representations produced in this manner markedly strengthens the usual continuous relaxation, as is evidenced not only by the fact that the convex hull representation is obtained at the highest level, but that in computational studies on many classes of problems, even the first level representation helps design algorithms that significantly dominate existing procedures in the literature (see Sherali and Adams, 1996). The theoretical implications of this hierarchy are noteworthy; the resulting representations subsume and unify many published linearization methods for nonlinear 0-1 programs, and the algebraic representation available at level n promotes new methods for identifying and characterizing facets and valid linear inequalities in the original variable space, as well as for providing information that directly bridges the gap between discrete and continuous sets. Indeed, since the level-n formulation characterizes the convex hull, all valid inequalities in the original variable space must be obtainable via a suitable projection; thus such a projection operation serves as an all-encompassing tool for generating valid inequalities.

Sherali and Adams (1989) also demonstrate the relationship between their hierarchy of relaxations and that which can be generated through disjunctive programming techniques. Balas (1985) has shown how a hierarchy spanning the spectrum from the linear programming relaxation to the convex hull of feasible solutions can be generated for linear mixed-integer zero-one programming problems by inductively representing the feasible region at each stage as a conjunction of disjunctions, and then taking its hull relaxation. This hull relaxation amounts to constructing the intersection of the individual convex hulls of the different disjunctive sets, and hence yields a relaxation. Sherali and Adams show that their hierarchy produces a different, stronger set of intermediate relaxations that lead to the underlying convex hull representation. To view their relaxations as Balas' hull relaxations requires manipulating the representation at any stage d to write it as a conjunction of a *non-standard* set of disjunctions. Moreover, by the nature of the RLT approach, Sherali and Adams also show (see Section 5) how one can readily construct a hierarchy of *linear* relaxations leading to the convex hull representation for mixed-integer zero-one *polynomial* programming problems having no cross-product terms among continuous variables.

In connection with the final comment above, we remark that Boros et al. (1989) have also independently developed a similar hierarchy for the special case of the unconstrained, quadratic pseudo-Boolean programming problem. For this case, they construct a standard linear programming relaxation that coincides with our relaxation at level $d = 1$, and then show in an *existential* fashion how a hierarchy of relaxations indexed by $d = 1, \ldots, n$ leading up to the convex hull representation at level n can be generated. This is done by including at level d, constraints corresponding to the extreme directions of the cone of nonnegative quadratic pseudo-Boolean functions that involve at most d of the n-variables. Each such relaxation can be viewed as the projection of one of our *explicitly* stated higher-order relaxations onto the variable space of the first level for this special case. Moreover, our approach also permits one to consider general pseudo-Boolean polynomials, constrained problems, as well as mixed-integer situations.

Lovasz and Shrijver (1989) have also independently proposed a similar hierarchy of relaxations for linear, *pure* 0-1 programming problems, which essentially amounts to deriving X_1 from X_0, finding the projection X_{P1}, and then repeating this step by replacing X_0 with X_{P1}. Continuing in this fashion, they show that in n steps, $conv(X)$ is obtained. However, from a practical viewpoint, while the relaxations $X_1, X_2, \ldots$ of Sherali and Adams are explicitly available and directly implementable, the projections required by Lovasz and Shrijver for level two and higher order relaxations are computationally burdensome, necessitating the potentially exponential task of vertex enumeration. Moreover, extensions to mixed-integer or to nonlinear zero-one problems are not evident using this development.

Another hierarchy along the same lines has been proposed by Balas et al. (1993). In this hierarchy, the set X_1 is generated as in Sherali and Adams (1989), but using factors involving only one of the binary variables, say, x_1. Projecting the resulting formulation onto the space of the original variables, produces the convex hull of feasible solutions to the original LP relaxation with the added restriction that x_1 is binary valued. This follows from Sherali and Adams (1989) since it is equivalent to treating only x_1 as binary valued and the remaining variables as continuous, and then generating the convex hull representation. Using the fact that x_1 is now binary valued at all vertices, this process is then repeated using another binary variable, say, x_2, in order to determine the convex hull of feasible vertices at which both x_1 and x_2 are binary valued. Continuing with the remaining binary variables $x_3, \ldots, x_n$ in this fashion, produces the convex hull representation at the final stage. Based on this construct, that amounts to a specialized application of Sherali and Adams' hierarchy, Balas *et al.* describe and test a cutting plane algorithm. Encouraging computational results are reported, despite the fact that this specialization produces weaker relaxations than the original Sherali-Adams' relaxations. Since this cutting plane generation scheme is based simply on the first level of Sherali and Adams' hierarchy in which binariness is enforced on only a single variable at a time, the prospect of enhancing computational performance by considering multiple binary variables at a time appears to be promising.

The remainder of this paper is organized as follows. We begin in Section 2 by presenting the basic RLT procedure of Sherali and Adams (1989, 1990, 1994). Its various properties are summarized in Section 3, and ideas for characterizing and generating facets of the convex hull of feasible solutions (or strong valid inequalities) are presented in Section 4. Extensions to multilinear programs and specializations for equality constrained situations are addressed in Section 5 and are illustrated using the popular quadratic assignment problem. Section 6 deals with a different RLT process proposed by Sherali, Adams and Driscoll (1996) that can be used to exploit special structures inherent in the problem, and Section 7 provides illustrations for some such structures, including generalized/variable upper bounding, covering, partitioning and parking constraints, and even problem sparsity. In any RLT approach, a significant additional tightening can be obtained by using conditional logic while generating the RLT constraints. This is discussed in Section 8 and is illustrated on the extensively-studied Traveling Salesman Problem. Section 9 concludes the paper by briefly discussing some special persistency results, extensions to general integer programs, and various computational guidelines. To lighten the reading, we omit proofs in this paper, and we refer the reader to the original papers for such details.

2 RLT Hierarchy for Mixed Integer Zero-One Problems

Consider a linear mixed integer 0-1 programming problem whose (nonempty) feasible region is given as follows:

$$X = \{(x,y) : \sum_{j=1}^{n} \alpha_{rj} x_j + \sum_{k=1}^{m} \gamma_{rk} y_k \geq \beta_r, \text{for } r = 1, \ldots, R,$$
$$0 \leq x \leq e_n, \; x \text{ integer}, \; 0 \leq y \leq e_m\}, \tag{1}$$

where e_n and e_m are, respectively, column vectors of n and m entries of 1, and where the continuous variables y_k are assumed to be bounded and appropriately scaled to lie in the interval $[0, 1]$ for $k = 1, \ldots, m$. (Upper bounds on the continuous variables are imposed here only for convenience in exposition, as we comment on later in the discussion.) Note that any equality constraints present in the formulation can be accommodated in a similar manner as are the inequalities in the following derivation, and we omit writing them explicitly in (1) only to simplify the presentation. However, we will show later that the equality constraints can be treated in a special manner which, in fact, might sometimes encourage the writing of the R inequalities in (1), as equalities by using slack variables.

In this section, we present the basic construction process of the *Reformulation Linearization Technique* (RLT). For the region described in (1), given any level $d \in \{0, \ldots, n\}$, this technique first converts the constraint set into a polynomial mixed-integer zero-one set of restrictions by multiplying the constraints with some suitable d-degree polynomial factors involving the n binary variables and their complements, and subsequently linearizes the resulting problem through appropriate variable transformations. For each level d, this produces a higher dimensional representation of the feasible region (1) in terms of the original variables x and y, and some new variables (w, v) that are defined to linearize the problem. Relaxing integrality, the projection, or "shadow," of this higher dimensional polyhedral set on the original variable space produces a tighter envelope for the convex hull of feasible solutions to (1), than does its ordinary linear programming (LP) or continuous relaxation. In fact, as d varies from zero to n, we obtain a hierarchy of such relaxations or shadows, each nested within the previous one, spanning the spectrum from the ordinary linear programming relaxation to the convex hull of feasible solutions. As mentioned earlier, the first level relaxation has itself proven to be sufficiently tight to benefit solution algorithms for several classes of problems. The RLT process for constructing a relaxation X_d of the region X defined in (1) corresponding to any level $d \in \{0, 1, \ldots, n\}$ proceeds as follows. For $d = 0$, the relaxation X_0 is simply the LP relaxation obtained by deleting the integrality restrictions on the x-variables. In order to construct the relaxation for any level $d \in \{1, \ldots, n\}$, let us consider the *bound-factors* $x_j \geq 0$ and $(1 - x_j) \geq 0$ for $j = 1, \ldots, n$ and let us compose *bound-factor products of degree (or order) d* by selecting some d distinct variables from the set $x_1, \ldots, x_n$, and by using either the bound-factor x_j or $(1 - x_j)$ for each selected variable in a product of these terms. Mathematically, for any $d \in \{1, \ldots, n\}$, and for each possible selection of d distinct variables, these (nonnegative polynomial) *bound factor products of degree (or order) d* are given by

$$F_d(J_1, J_2) = \left[\prod_{j \in J_1} x_j\right]\left[\prod_{j \in J_2}(1 - x_j)\right] \text{ for each } J_1, J_2, \subseteq N \equiv \{1, \ldots, n\}$$
$$\text{such that } J_1 \cap J_2 = \emptyset \text{ and } |J_1 \cup J_2| = d. \tag{2}$$

Any (J_1, J_2) satisfying the conditions in (2) will be said to be of *order d*. For example, for $n = 3$ and $d = 2$, these factors are $x_1 x_2$, $x_1 x_3$, $x_2 x_3$, $x_1(1 - x_2)$, $x_1(1 - x_3)$, $x_2(1 - x_1)$, $x_2(1 - x_3)$, $x_3(1 - x_1)$, $x_3(1 - x_2)$, $(1 - x_1)(1 - x_2)$, $(1 - x_1)(1 - x_3)$, and $(1 - x_2)(1 - x_3)$. In general, there are

$\binom{n}{d} 2^d$ such factors. For convenience, we will consider the single factor of degree zero to be $F_0(\emptyset, \emptyset) \equiv 1$, and accordingly assume products over null sets to be unity. Using these factors, let us construct a relaxation X_d of X, for any given $d \in \{0, \ldots, n\}$, using the following two steps that comprise the Reformulation-Linearization Technique (RLT).

Step 1 (Reformulation Phase): Multiply each of the inequalities in (1), including $0 \leq x \leq e_n$ and $0 \leq y \leq e_m$, by each of the factors $F_d(J_1, J_2)$ of degree d as defined in (2). Upon using the identity $x_j^2 \equiv x_j$ (and so $x_j(1 - x_j) = 0$) for each binary variable $x_j, j = 1, \ldots, n$, this gives the following set of additional, implied, nonlinear constraints:

$$\left[\sum_{j \in J_1} \alpha_{rj} - \beta_r\right] F_d(J_1, J_2) + \sum_{j \in N-(J_1 \cup J_2)} \alpha_{rj} F_{d+1}(J_1 + j, J_2)$$
$$+ \sum_{k=1}^{m} \gamma_{rk} y_k F_d(J_1, J_2) \geq 0$$

for $r = 1, \ldots, R$, and for each (J_1, J_2) of order d (3)
$F_D(J_1, J_2) \geq 0$ for each (J_1, J_2) of order $D \equiv \min\{d+1, n\}$, (4)
$F_d(J_1, J_2) \geq y_k F_d(J_1, J_2) \geq 0$
for $k = 1, \ldots, m$ and for each (J_1, J_2) of order d. (5)

Step 2 (Linearization Phase): Viewing the constraints in (3,4,5) in expanded form as a sum of monomials, linearize them by substituting the following variables for the corresponding nonlinear terms for each $J \subseteq N$:

$$w_J = \prod_{j \in J} x_j \text{ and } v_{Jk} \equiv y_k \prod_{j \in J} x_j, \text{ for } k = 1, \ldots, m. \quad (6)$$

We will assume the notation that

$$w_j \equiv x_j \text{ for } j = 1, \ldots, n, w_\emptyset \equiv 1, \text{ and } v_{\emptyset k} \equiv y_k \text{ for } k = 1, \ldots, m. \quad (7)$$

Denoting by $f_d(J_1, J_2)$ and $f_d^k(J_1, J_2)$ the respective linearized forms of the polynomial expressions $F_d(J_1, J_2)$ and $y_k F_d(J_1, J_2)$ under such a substitution, we obtain the following polyhedral set X_d whose projection onto the (x, y) space is claimed to yield a relaxation for X:

$$X_d = \{(x, y, w, v) :$$

$$\left[\sum_{j \in J_1} \alpha_{rj} - \beta_r\right] f_d(J_1, J_2) + \sum_{j \in N-(J_1 \cup J_2)} \alpha_{rj} f_{d+1}(J_1 + j, J_2)$$

$$+ \sum_{k=1}^{m} \gamma_{rk} f_d^k(J_1, J_2) \geq 0$$

for $r = 1, \ldots, R$, and for each (J_1, J_2) of order d \hfill (8)

$f_D(J_1, J_2) \geq 0$ for each (J_1, J_2) of order $D \equiv \min\{d+1, n\}$, \hfill (9)

$f_d(J_1, J_2) \geq f_d^k(J_1, J_2) \geq 0$

for $k = 1, \ldots, m$, and for each (J_1, J_2) of order $d\}$. \hfill (10)

Let us denote the projection of X_d onto the space of the original variables (x, y) by

$$X_{Pd} = \{(x, y) : (x, y, w, v) \in X_d\} \text{ for } d = 0, 1, \ldots, n. \tag{11}$$

Then, Sherali and Adams (1989, 1994) show that

$$X_{P0} \equiv X_0 \supseteq X_{p1} \supseteq X_{P2} \supseteq \ldots \supseteq X_{Pn} \equiv conv(X) \tag{12}$$

where $X_{P0} \equiv X_0$ (for $d = 0$) denotes the ordinary linear programming relaxation, and $conv(x)$ denotes the convex hull of X.

Notice that the nonlinear product constraints generated by this RLT process are implied by the original constraints, and were it not for the fact that we have explicitly imposed $x_j^2 = x_j$ (or $x_j(1-x_j) = 0$) for each binary variable, this process would have simply produced a relaxation that is equivalent to the ordinary LP relaxation. Hence, the key to the tightening lies in the recognition of binariness of x_j in replacing x_j^2 by x_j for each $j = 1, \ldots n$.

Example 2.1. Consider the following mixed-integer 0-1 constraint region.

$$X = \{(x, y) : x + y \leq 2, -x + y \leq 1, 2x - 2y \leq 1, x \text{ binary and } y \geq 0\}. \tag{13}$$

The corresponding LP relaxation of X is depicted in Figure 1. Note that no explicit upper bound on the continuous variable y is included in (13), although the other defining constraints of X imply the boundedness of y. The RLT process can be applied directly to X as above, without creating any explicit upper bound on y. Notice that for this instance, we have $n = 1$, and so by our foregoing discussion, the relaxation x_1 at level $d = 1$ should produce the convex hull representation. Let us verify this fact.

For $d = 1$, the bound-factor (products) of order 1 are simply x and $(1-x)$. Multiplying the constraints of X by x and by $(1-x)$, and using $x^2 = x$ along with the linearizing substitution $v = xy$ as given by (6), we get the following constraints in the higher dimensional space of the variables (x, y, v). This represents the set X_1.

$$x + y \le 2 \quad \begin{array}{l} \nearrow *x \Rightarrow \\ \searrow *(1-x) \Rightarrow \end{array} \quad \begin{array}{l} -x \le -v \\ 2x + y \le 2 + v \end{array}$$

$$-x + y \le 1 \quad \begin{array}{l} \nearrow *x \Rightarrow \\ \searrow *(1-x) \Rightarrow \end{array} \quad \begin{array}{l} -2x \le -v \\ x + y \le 1 + v \end{array}$$

$$2x - 2y \le 1 \quad \begin{array}{l} \nearrow *x \Rightarrow \\ \searrow *(1-x) \Rightarrow \end{array} \quad \begin{array}{l} x \le 2v \\ x - 2y \le 1 - 2v \end{array}$$

$$y \ge 0 \quad \begin{array}{l} \nearrow *x \Rightarrow \\ \searrow *(1-x) \Rightarrow \end{array} \quad \begin{array}{l} v \ge 0 \\ -y \le -v \end{array}$$

along with $0 \le x \le 1$. To examine the projection X_{P1} of X_1 onto the space of the (x, y) variables, let us rewrite X_1 as follows.

$$X_1 = \{(x, y, v) : v \ge 2x + y - 2, v \ge x + y - 1, v \ge x/2, v \ge 0,$$
$$v \le x, v \le 2x, v \le (1 - x + 2y)/2, v \le y\}.$$

This yields its projection (using Fourier-Motzkin elimination) as

$$X_{P1} = \{(x, y): \quad \max\{2x + y - 2, x + y - 1, x/2, 0\}$$
$$\le \min\{x, 2x, (1 - x + 2y)/2, y\}\}$$

Writing out the corresponding equivalent linear inequalities and dropping redundant constraints, we obtain

$$X_{P1} \equiv X_{Pn} = \{(x, y) : x \le 2y, 0 \le x \le 1, y \le 1\}$$

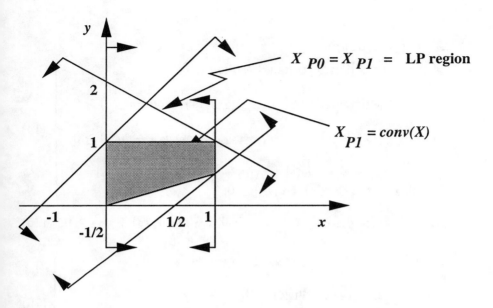

Figure 1: Illustration of RLT relaxations for Example 2.1.

which describes $conv(X)$ as seen in Figure 1.

Example 2.2. Consider the set

$$X = \{(x,y) : \alpha_1 x_1 + \alpha_2 x_2 + \gamma_1 y_1 + \gamma_2 y_2 \geq \beta, 0 \leq x \leq e_2, x \text{ integer},$$
$$0 \leq y \leq e_2\}$$

Hence, we have $n = m = 2$. Let us consider $d = 2$, so that $D = \min\{n+1, d\} = 2$ as well. The various sets (J_1, J_2) of order 2 and the corresponding factors are given below:

(J_1, J_2)	$(\{1,2\}, \emptyset)$	$(\{1\}, \{2\})$	$(\{2\}, \{1\})$	$(\emptyset, \{1,2\})$
$F_2(J_1, J_2)$	$x_1 x_2$	$x_1(1 - x_2)$	$x_2(1 - x_1)$	$(1 - x_1)(1 - x_2)$
$f_2(J_1, J_2)$	w_{12}	$x_1 - w_{12}$	$x_2 - w_{12}$	$1 - (x_1 + x_2) + w_{12}$
$f_2^k(J_1, J_2)$	v_{12k}	$v_{1k} - v_{12k}$	$v_{2k} - v_{12k}$	$y_k - (v_{1k} + v_{2k}) + v_{12k}$

Hence, we obtain the following constraints (14)-(16) corresponding to (8)-(10), respectively, where v_{Jk} has been written as $v_{J,k}$ for clarity:

$$X_2 = \{(x, y, w, v) :$$

$$(\alpha_1 + \alpha_2 - \beta)w_{12} + \gamma_1 v_{12,1} + \gamma_2 v_{12,2} \geq 0,$$

$$(\alpha_1 - \beta)[x_1 - w_{12}] + \gamma_1(v_{1,1} - v_{12,1}) + \gamma_1(v_{1,2} - v_{12,2}) \geq 0,$$

$$(\alpha_2 - \beta)[x_2 - w_{12}] + \gamma_1(v_{2,1} - v_{12,1}) + \gamma_2(v_{2,2} - v_{12,2}) \geq 0,$$

$$\beta[-1 + (x_1 + x_2) - w_{12}] + \gamma_1[y_1 - (v_{1,1} + v_{2,1}) + v_{12,1}]$$
$$+ \gamma_2[y_2 - (v_{1,2} + v_{2,2}) + v_{12,2}] \geq 0, \quad (14)$$

$$w_{12} \geq 0, x_1 - w_{12} \geq 0, x_2 - w_{12} \geq 0, 1 - (x_1 + x_2) + w_{12} \geq 0, \quad (15)$$

$$w_{12} \geq v_{12,k} \geq 0, (x_1 - w_{12}) \geq (v_{1,k} - v_{12,k}) \geq 0,$$

$$(x_2 - w_{12}) \geq (v_{2,k} - v_{12,k}) \geq 0,$$

$$\text{and } [1 - (x_1 + x_2) + w_{12}] \geq [y_k - (v_{1,k} + v_{2,k}) + v_{12,k}] \geq 0,$$
$$\text{for } k = 1, 2\}. \quad (16)$$

Some comments and illustrations are in order at this point. First, note that we could have computed inter-products of other defining constraints in constructing the relaxation at level d, so long as the degree of the intermediate polynomial program generated at the Reformulation step remains the same and no nonlinearities are created with respect to the y-variables. Hence, linearity would be preserved upon using the substitution (6,7). While the additional inequalities thus generated would possibly yield tighter relaxations for levels $1, \ldots, n - 1$, by our convex hull assertion in (12), these constraints would be implied by those defining X_n. Hence, our hierarchy results can be directly extended to include such additional constraints, but for simplicity, we omit these types of constraints. Nonetheless, one may include them in a computational scheme employing the sets $X_d, d < n$.

Second, as alluded to earlier, note that for the case $d = 0$, using the fact that $f_0(\emptyset, \emptyset) \equiv 1$, that $f_0^k(\emptyset, \emptyset) \equiv y_k$ for $k = 1, \ldots, m$, and that $f_1(j, \emptyset) \equiv x_j$ and $f_1(\emptyset, j) \equiv (1 - x_j)$ for $j = 1, \ldots, n$, it follows that X_0 given by (8)-(10) is precisely the continuous (LP) relaxation of X in which the integrality restrictions on the x-variables are dropped. Finally, for $d = n$, note that the inequalities (9) are implied by (10) and can therefore be omitted from the representation X_n.

3 Properties of the RLT Relaxation

Equation (12) asserts the main RLT result that for $d = 0, 1, \ldots, n$, the (conceptual) projections of the sets X_d represent a sequence of nested, valid relaxations leading up to the convex hull representation. In this section, we summarize some of the salient properties of the RLT relaxations that lead to this result.

Property 3.1. The constraints (9) and (10) for any level's relaxation can be surrogated in a particular fashion using nonnegative multipliers to produce the constraints of the lower level relaxations. Hence, the hierarchy among these relaxations is established via this property whereby lower level constraints are directly implied by higher level constraints.

Property 3.2. Consider the RLT constraints (10) that are generated by multiplying the bound factors of order d with the constraints $0 \leq y \leq e_m$. If $\hat{x}$ is any binary vector for which $(\hat{x}, \hat{y}, \hat{w}, \hat{v})$ is feasible to these constraints, then we must have $0 \leq \hat{y} \leq e_m$ and moreover, the identities (6) must hold true. Hence, these constraints serve to make the nonlinear relationship (6) hold true within the linear set X_d for any solution having binary values for x. In fact, for any binary value of $\hat{x}$, we have that $(\hat{x}, \hat{y}, \hat{w}, \hat{v}) \in X_d$ if and only if $(\hat{x}, \hat{y}) \in X$ and (2.4c) holds true.

Property 3.3. The highest level relaxation is equivalent via a nonsingular affine transformation to the polytope generated by taking the convex hull of all feasible solutions to X. Hence, this result along with Property 1 establishes (12). In fact, $(\hat{x}, \hat{y}, \hat{w}, \hat{v})$ is a vertex of X_n if and only if $\hat{x}$ is binary valued, (2.4c) holds true, and $\hat{y}$ is an extreme point of the set $\{y : (\hat{x}, y) \in X\}$.

A few remarks are pertinent at this point.

Remark 3.1. Observe that the upper bounds on the y-variables are not needed for (12) to hold true, and that the foregoing analysis holds by simply eliminating any RLT product constraints that correspond to nonexistent bounding restrictions. Given feasibility of the underlying mixed-integer program, we have that x is binary valued at an optimum to any linear program $\max\{cx + dy : (x, y) \in X_{Pn}\}$ for which there exists an optimal solution, that is, for which this LP is not unbounded. Hence, by the foregoing properties, we have that (12) holds true in this case as well.

Remark 3.2. A direct consequence of the convex hull result stated in (12) is that the convex hull representation over subsets of the x-variables can be computed in an identical fashion. In particular, suppose that the first $p \leq n$ variables are treated as binary in the set X of (1), with the remaining

$(n-p)x$-variables relaxed to be continuous. Then it directly follows that the pth level linearization of this relaxed problem produces $conv\{X_0 \cap \{(x,y) : x_j \text{ is binary } \forall j = 1, \ldots, p\}\}$. The special case in which $p = 1$ precisely recovers the main step in the convexification argument of Balas et al. (1993) mentioned in Section 1.

Remark 3.3. In concluding this section, let us comment on the situation in which there exist certain constraints from the first set of inequalities in (1) which involve only the x-variables. In this case, one can multiply such constraints with the factors y_k and $(1 - y_k)$ for $k = 1, \ldots, m$, and then linearize the resulting constraints using (6,7) as with the other constraints (3,4,5). By Property 3.2, these additional constraints are implied when x is binary in any feasible solution, but they might serve to tighten the continuous relaxations at the intermediate levels. Naturally, these constraints are implied by the other constraints defining the highest level relaxation X_n. In a likewise fashion, inequalities defining X that involve only the x-variables can be used in the same spirit as the factors $x_j \geq 0$ and $(1 - x_j) \geq 0$ to generate products of inequalities up to a given order level in order to further tighten intermediate level relaxations.

4 Generating Valid Inequalities and Facets Using RLT

Thus far, we have presented a hierarchy of relaxations leading up to the convex hull representation for zero-one mixed-integer programming problems. A key advantage of this development is that the RLT produces an algebraically explicit convex hull representation at the highest level. While it might not be computationally feasible to actually generate and solve the linear program based on this convex hull representation because of its potentially exponential size, there are other ways to exploit this information or facility to advantage.

One obvious tactic might be to identify a subset of variables along with either original or implied constraints involving these variables for which such a convex hull representation (or some higher order relaxation) would be manageable. Another approach might be to exploit any special structures in order to derive a simplified algebraic form of the convex hull representation for which the polyhedral cone that identifies all defining facets itself possesses a special structure. Exploiting this structure, it might then be possible to generate specific extreme directions of this cone, and hence identify certain (classes of) facets that could be used to strengthen the original problem formulation. Perhaps, this algebraic simplification could yield more compact, specially structured, representations of each level in the hierarchy, hence enabling an actual implementation of a full or partial application of RLT at some intermediate level in order to compose tight relaxations.

Each of these ideas have been explored in particular contexts. For example, for the Boolean Quadric Polytope, Sherali et al. (1995) have derived a characterization for all its defining facets via the extreme directions of a specially structured polyhedral cone. They then demonstrate how various known classes of facets including the clique, cut, and generalized cut inequality facets, emerge via the enumeration of different types of extreme directions. By examining a particular lower-dimensional projected restriction of this polyhedral cone, they discover a new class of facets. (This class can also be recovered via a combination of two types of facets for the related cut cone.) This type of an analysis was also used by Sherali and Lee (1995) for generalized upper bounding (GUB) constrained knapsack polytopes for which new classes of facets have been characterized through a polynomial-time sequential-simultaneous lifting procedure. Known lower bounds on the coefficients of lifted facets derived from minimal covers associated with the ordinary knapsack polytope have also been tightened using this framework. For the *set partitioning polytope*, Sherali and Lee (1996) show that a number of published valid inequalities along with constraint tightening procedures are automatically captured within the first- and second-level relaxations themselves. A variety of partial applications of the RLT scheme have also been developed in order to delete fractional linear programming solutions while tightening the relaxation in the vicinity of such solutions, and various simplified forms of RLT relaxations have been derived by exploiting its special structures. The polyhedral structure of many other specially structured combinatorial optimization problems can be studied using such constructs, and tight relaxations and strong valid inequalities can be generated for such problems. As recently shown by Sherali (1996), even convex envelopes of some special multilinear nonconvex functions can be generated using this approach.

We conclude this section by pointing out another strategy that can be used to derive strong valid inequalities in the space of the original variables through the intermediate higher dimensional RLT relaxations X_d, for $d \geq 1$. Note that by linear programming duality, the set of inequalities that define the projection X_{Pd} onto the (x,y) space of the higher dimensional set X_d are characterized via the set of surrogates of the constraints defining X_d that zero out the coefficients of the new variables (w, v) in the lifted space. While the enumeration of all such projected inequalities might be prohibitive, it is possible to formulate linear programming separation problems in this fashion to derive specific facets of X_{Pd} that delete a given infeasible continuous solution. The lift-and-project cutting plane algorithm of Balas *et al.* (1993) precisely does this for the special case of $d = 1$ and using RLT bound factor products involving only a single binary variable that turns out to be fractional at an optimum to the current LP relaxation. Stronger level one or higher relaxations can potentially lead to improved cuts within this framework.

5 Multilinear Programs and Equality Constraints with Applications to the Quadratic Assignment Problem

In this section, we present two important extensions of the RLT procedure described in Section 2. The first of these concerns multilinear mixed-integer zero-one *polynomial* programming problems in which the continuous variables $0 \leq y \leq e_m$ appear linearly in the constraints and the objective function. This is discussed below.

Extension 1. *Multilinear mixed-integer zero-one polynomial programming problems.*

Consider the set

$$X = \{(x, y) : \sum_{t \in T_{r0}} \alpha_{rt} p(J_{1t}, J_{2t}) + \sum_{k=1}^{m} y_k \sum_{t \in T_{rk}} \gamma_{rkt} p(J_{1t}, J_{2t}) \geq \beta_r,$$
$$r = 1, \ldots, R, 0 \leq x \leq e_n \text{ and integer}, 0 \leq y \leq e_m\}, \quad (17)$$

where for all t, $p(J_{1t}, J_{2t}) \equiv [\prod_{j \in J_{1t}} x_j][\prod_{j \in J_{2t}}(1-x_j)]$ are polynomial terms for the various sets (J_{1t}, J_{2t}), indexed by the sets T_{r0} and T_{r_k}, in (17). For $d = 0, 1, \ldots, n$, we can construct a polyhedral relaxation X_d for X by using the factors $F_d(J_1, J_2)$ to multiply the first set of constraints as before, where (J_1, J_2) are of order d. However, denoting δ_1 as the maximum degree of the polynomial terms in x not involving the y-variables, and δ_2 as the maximum degree of the polynomial terms in x that are associated with products involving y-variables, in lieu of (4), we now use $F_{D_1}(J_1, J_2) \geq 0$ for (J_1, J_2) of order $D_1 = \min\{d + \delta_1, n\}$, and in lieu of (5), we employ the constraints $F_{D_2}(J_1, J_2) \geq y_k F_{D_2}(J_1, J_2) \geq 0, k = 1, \ldots, m$, for all (J_1, J_2) of order $D_2 = \min\{d + \delta_2, n\}$. Of course, if $D_2 \geq D_1$, then the former restrictions are unnecessary, as they are implied by the latter. Note that in computing δ_1 and δ_2 in an optimization context, we consider the terms in the objective function as well, and that for the linear case, we have $\delta_1 = 1$ and $\delta_2 = 0$.

Now, linearizing the resulting constraints under the substitution (6,7) produces the desired set X_d. Because of Property 3.2, which again holds true here, when the integrality on the x-variables is enforced, each such set X_d is equivalent to the set X. Moreover, similar to Property 3.1, each constraint from the first set of inequalities in X_d for any $d < n$ is obtainable by surrogating two appropriate constraints from X_{d+1}. In fact, we again obtain the hierarchy of relaxations

$$conv(X) \equiv X_{Pn} \subseteq X_{P(n-1)} \subseteq \ldots \subseteq X_{P1} \subseteq X_0.$$

Extension 2: *Construction of Relaxations Using Equality Constraint Representations.*

Note that given any set X of the form (1), by adding slack variables to the first R constraints, determining upper bounds on these slacks as the sum of the positive constraint coefficients minus the right-hand side constant, and accordingly scaling these slack variables onto the unit interval, we may equivalently write the set X as

$$X = \{(x, y) : \sum_{j=1}^n \alpha_{rj} x_j + \sum_{k=1}^m \gamma_{rk} y_k + \gamma_{r(m+r)} y_{m+r} = \beta_r \text{ for } r = 1, \ldots, R,$$
$$0 \leq x \leq e_n, x \text{ integer}, 0 \leq y \leq e_{m+R}\}. \quad (18)$$

Now, for any $d \in \{1, \ldots, n\}$, observe that the factor $F_d(J_1, J_2)$ for any (J_1, J_2) of order d is a linear combination of the factors $F_p(J, \emptyset)$ for $J \subseteq N, p \equiv |J| = 0, 1, \ldots, d$. Hence, the constraint derived by multiplying an *equality* from (18) by $F_d(J_1, J_2)$ and then linearizing it via (6,7) is obtainable via an appropriate surrogate (with mixed-sign multipliers) of the constraints derived similarly, but using the factors $F_p(J, \emptyset)$ for $J \subseteq N, p \equiv |J| = 0, 1, \ldots, d$. Hence, these latter factors produce constraints that can generate the other constraints, and so X_d defined by (8)-(10) corresponding to X as in (18) is *equivalent* to the following, where $f_p(J, \emptyset) \equiv w_J$, and $f_p^k(J, \emptyset) \equiv v_{Jk}$ for all $p = 0, 1, \ldots, d+1$, as in (6,7):

$X_d = \{(x, y, w, v) :$ constraints of type (9) and (10) hold, and for $r = 1, \ldots, R$,

$$\left[\sum_{j \in J} \alpha_{rj} - \beta_r\right] w_J + \sum_{j \in \bar{J}} \alpha_{rj} w_{J+j} + \sum_{k=1}^{m} \gamma_{rk} v_{Jk} + \gamma_{r(m+r)} v_{J(m+r)} = 0$$

for all $J \subseteq N$ with $|J| = 0, 1, \ldots, d\}.$ \hfill (19)

Note that the savings in the number of constraints in (19) over that in (8)-(10) corresponding to the set X as in (18) is given by

$$R \left[2^d \binom{n}{d} - \sum_{i=0}^{d} \binom{n}{i} \right].$$

Also, observe that for $J = \emptyset$, the equalities in (19) are precisely the original equalities defining X in (18). Of course, because (19) is equivalent to the set of the type (8)-(10) which would have been derived using the factors $F_d(J_1, J_2)$ of degree d, all the foregoing results continue to hold true for (19).

While the approach in Section 2 developed for the inequality constrained case is convenient for theoretical purposes as it avoids the manipulation of surrogates that would be required for the equalities in (19), note that from a computational viewpoint, when $d < n$, the representation in (19) has fewer type (8) "complicating" constraints and variables (including slacks in (8)) than does (10) as given by the above savings expression, but has $R \times 2^d \binom{n}{d}$ additional constraints of the type (10), counting the nonnegativity restrictions on the slacks in (8), for the inequality constrained case. Hence, depending on the structure, either form of the representation of these relaxations may be employed, as convenient.

To summarize, when constructing the level d relaxation X_d in the presence of equality constraints, each equality constraint needs to be multiplied by the factors $F_p(J, \emptyset)$ for $J \subseteq N, p \equiv |J| = 0, 1, \ldots, d$ in the Reformulation step. Naturally, factors $F_p(J, \emptyset)$ that are known to be zeros, i.e., any such factor for which we know that there does not exist a feasible solution that has $x_j = 1, \forall j \in J$, need not be used in constructing these product constraints. The Linearization Phase is then applied as before.

Reformulation-Linearization for Discrete Optimization

Example 5.1. To illustrate both the foregoing extensions, let us consider the celebrated **quadratic assignment problem** defined as follows.

$$\textbf{QAP}: \text{Minimize} \quad \sum_{i=1}^{m}\sum_{j=1}^{m}\sum_{k>i}\sum_{l\neq j} c_{ijkl} x_{ij} x_{kl} \tag{20}$$

$$\text{subject to} \quad \sum_{i=1}^{m} x_{ij} = 1 \ \forall j = 1,\ldots,m \tag{21}$$

$$\sum_{j=1}^{m} x_{ij} = 1 \ \forall i = 1,\ldots,m \tag{22}$$

$$x \geq 0 \text{ and integer.} \tag{23}$$

First Level Relaxation:

To construct the relaxation at the first level, in addition to the constraints (21,22), and the nonnegativity restrictions on x in (23), we include the product constraints obtained by multiplying each constraint in (21) and (22) by each $x_{kl} \equiv F_1(\{x_{kl}\},\emptyset)$, and then apply the rest of the RLT procedure as usual, using the substitution $w_{ijkl} = x_{ij} x_{kl} \forall i,j,k,l$. Note that for any $j \in \{1,\ldots,m\}$, multiplying (21) by x_{kj} for each $k \in \{1,\ldots,m\}$, produces upon using $x_{kj}^2 = x_{kj}$ that

$$\sum_{i \neq k} w_{ijkj} = 0 \ \forall k = 1,\ldots,m. \tag{24}$$

Hence, since $w \geq 0$, this implies that $w_{ijkj} \equiv 0 \ \forall i \neq k$. Similarly, from (22), we get $w_{ijil} = 0 \ \forall j \neq l$. Furthermore, noting that $w_{ijkl} \equiv w_{klij}$, we only need to define $w_{ijkl} \ \forall i < k, j \neq l$. (For convenience in exposition below, however, we define $w_{ijkl} \ \forall i \neq k, j \neq l$, and explicitly impose $w_{ijkl} \equiv w_{klij}$.) Additionally, since we only need to impose $x_{ij} \geq 0$ in (23) for the continuous relaxation, since $x_{ij} \leq 1$ is then implied by the constraints (21,22), we only need to multiply the factors $x_{kl} \geq 0$ and $(1 - x_{kl}) \geq 0$ with each constraint $x_{ij} \geq 0$ in the reformulation phase. This produces the restrictions $w_{ijkl} \leq x_{ij}$ and $w_{ijkl} \leq x_{kl} \ \forall i,j,k,l$, along with $w \geq 0$. However, the former variable upper bounding constraints are implied by the constraints (28,29,30) below. Hence, the first level relaxation (that would yield a lower bound on QAP) is given as follows.

$$\text{Minimize} \quad \sum_{i}\sum_{j}\sum_{k>i}\sum_{l\neq j} c_{ijkl} w_{ijkl} \tag{25}$$

subject to $\quad \sum_i x_{ij} = 1 \ \forall j$ (26)

$$\sum_j x_{ij} = 1 \ \forall i \tag{27}$$

$$\sum_{i \neq k} w_{ijkl} = x_{kl} \ \forall j, k, l \neq j \tag{28}$$

$$\sum_{j \neq l} w_{ijkl} = x_{kl} \ \forall i, l, k \neq i \tag{29}$$

$$x \geq 0, w \geq 0, w_{ijkl} \equiv w_{klij} \ \forall i < k, j \neq l. \tag{30}$$

Second Level Relaxation:

Following the same procedure as above, and eliminating null variables as well as null and redundant constraints, we would obtain the same relaxation as in (25)-(30) with additional constraints (and variables) corresponding to multiplying each equality constraint in (21) by each factor $x_{kl}x_{pq}$ for $k < p, l \neq q \neq j$, and by multiplying each constraint in (22) by each factor $x_{kl}x_{pq}$ for $k < p, k \neq i \neq p, l \neq q$. This would produce the additional constraints

$$\sum_{i \neq k,p} w_{ijklpq} = w_{klpq} \ \forall j, k, l, p, q, \text{where } k < p, \text{ and } l \neq q \neq j \tag{31}$$

$$\sum_{j \neq l,q} w_{ijklpq} = w_{klpq} \ \forall j, k, l, p, q, \text{where } k < p, k \neq i \neq p, \tag{32}$$

and $l \neq q$

along with $w_{ij,kl,pq} \geq 0$ and that this is the same variable for each permutation of

$$ij, kl, pq, \forall \text{ distinct } i, k, p \text{ and } j, l, q. \tag{33}$$

Such relaxations have been computationally tested by Ramakrishnan et al. (1996), and Ramachandran and Pekny (1996) with promising results. Extensions of such specializations for general set partitioning problems are given by Sherali and Lee (1996). In fact, Adams and Johnson (1994) show that the lower bound produced by the first level relaxation itself subsumes a multitude of known lower bounding techniques in the literature, including a host of matrix reduction strategies. By designing a heuristic dual ascent procedure for the level-one relaxation, and by incorporating dual-based cutting planes within an enumerative algorithm, an exact solution technique has been developed and tested by Adams and Johnson (1996), that can competitively solve problems up to size 17. In an effort to make this algorithm generally applicable, no special exploitation of flow and/or distance symmetries was considered. As far as the strength of the RLT relaxation is concerned, on a set of standard test problems of sizes 8-20, the lower bounds produced by the dual ascent procedure uniformly dominated 12 other competing lower bounding schemes except for one problem of size 20, where our procedure yielded a lower bound of 2142, while an eigenvalue-based procedure produced a lower bound of 2229, the optimum value being 2570 for this problem. Recently, Resende *et al.* (1995) have been able to solve the first level RLT relaxation exactly for problems of size up to 30 using an interior-point method that employs a preconditioned conjugate gradient technique to solve the system of equations for computing the search directions. (For the aforementioned problem of size 20, the exact solution value of the lower bounding RLT relaxation turned out to be 2182, compared to our dual ascent value of 2142.) Sherali and Brown (1994) have also applied RLT to the problem of assigning aircraft to gates at an airport, with the objective of minimizing passenger walking distances. The problem is modeled as a variant of the quadratic assignment problem with partial assignment and set packing constraints. The quadratic problem is then equivalently linearized by applying the first-level of the RLT. In addition to simply linearizing the problem, the application of this technique generates additional constraints that provide a tighter linear programming representation. Since even the first-level relaxation can get quite large, we investigate several alternative relaxations that either delete or aggregate classes of RLT constraints. All these relaxations are embedded in a heuristic that solves a sequence of such relaxations, automatically selecting at each stage the tightest relaxation that can be solved with an acceptable estimated effort, and based on the solution obtained, it fixes a suitable subset of variables to 0-1 values. This process is repeated until a feasible solution is constructed. The procedure was computationally tested using realistic data obtained from *USAir* for problems having up to 7 gates and 36 flights. For all the test problems ranging from 4 gates and 36 flights to 7 gates and 14 flights, for which the size of the first-level relaxation was manageable (having 14, 494 and 4,084 constraints, respectively, for these two problem sizes), this initial relaxation itself always produced an optimal 0-1 solution. Finally, we mention that Adams and Sherali (1986) had earlier developed a technique for solving general 0-1 quadratic programming problems using a relaxation that turns out to be precisely the level-one RLT relaxation discussed above. This relaxation was shown to theoretically dominate other existing linearizations and was shown to computationally produce far tighter lower bounds. In these computations, we solved quadratic set covering problems having up to 70 variables and 40 constraints. For example, for this largest size problem, where the optimum objective value was 1312, our relaxation produced an initial lower bound of 1289 at the root node, and enumerated 14 nodes to solve the problem in 79 cpu seconds on an IBM 3081 Series D24 Group K computer. When the same algorithmic strategies were used on a relaxation that did not include the special RLT constraints, the initial lower bound obtained was only 398, and the algorithm enumerated 2130 nodes, consuming 197 cpu seconds.

6 Exploiting Special Structures to Generate Tighter RLT Representations

In Section 2, we discussed a technique for generating a hierarchy of relaxations that span the spectrum from the continuous LP relaxation to the convex hull of feasible solutions for linear mixed-integer 0-1 programming problems. The key construct was to compose a set of multiplication factors based on the bounding constraints $0 \leq x \leq e_n$ on the binary variables x, and to use these factors to generate implied nonlinear product constraints, then tighten these constraints using the fact that $x_j^2 \equiv x_j \ \forall j = 1, \ldots, n$, and subsequently linearize the resulting polynomial problem through a variable substitution process. This process yielded a family of tighter representations of the problem in higher-dimensional spaces.

It should seem intuitive that if one were to identify a set S of constraints involving the x-variables that imply the bounding restrictions $0 \leq x \leq e_n$, then it might be possible to generate a similar hierarchy by applying a set of multiplicative factors that are composed using the constraints defining S. Indeed, as we exemplify in this section, such generalized so-called S-*factors* can not only be used to construct a hierarchy of relaxations leading to the convex hull representation, but they also provide an opportunity to exploit inherent special structures. Through an artful application of this strategy, one can design relaxations that are more compact than the ones available through the RLT process of Section 2, while at the same time being tighter, as well as affording the opportunity to construct the convex hull representation at lower levels in the hierarchy in certain cases.

Consider the feasible region of a mixed-integer 0-1 programming problem stated in the form

$$X = \{(x,y) \in R^n \times R^m : Ax + Dy \geq b, x \in S, x \text{ binary}, y \geq 0\} \quad (34)$$

where

$$S \equiv \{x : g_i x - g_{0i} \geq 0 \text{ for } i = 1, \ldots, p\}. \quad (35)$$

Here, the constraints defining the set S have been specially composed to generate useful, tight relaxations as revealed in the sequel. For now, in theory, all that we require of the set S is that it implies the bounds $0 \leq x \leq e_n$ on the x-variables, where e_n is a vector of n-ones. Specifically, we assume that for each $t = 1, \ldots, n$,

$$\min\{x_t : x \in S\} = 0 \text{ and } \max\{x_t : x \in S\} = 1. \quad (36)$$

Note that if $\min(x_t) > 0$, we can fix $x_t = 1$, and if $\max(x_t) < 1$, we can fix $x_t = 0$, and if both these conditions hold for any t, then the problem is infeasible. Therefore, without loss of generality we will assume that the equalities of (36) hold for all t. We further note that (36) ensures that $p \geq n + 1$.

Now, define the sets P and $\overline{P}$ as follows, where $\overline{P}$ duplicates each index in P n times:

$$P = \{1, \ldots, p\}, \text{ and } \overline{P} = \{n \text{ copies of } P\} \quad (37)$$

The construction of the new hierarchy proceeds in a manner similar to that of Section 2. At any chosen level of relaxation $d \in \{1, \ldots, n\}$, we construct a higher dimensional relaxation $\overline{X}_d$ by considering the S-factors of order d defined as follows:

$$g(J) = \prod_{i \in J}(g_i x - g_{0i}) \text{ for each distinct } J \subseteq \overline{P}, |J| = d. \quad (38)$$

Note that to compose the S-factors of order d, we examine the collection of *constraint-factors* $g_i x - g_{0i} \geq 0, i = 1, \ldots, p$, and construct the product of some d of these constraint-factors, *including possible repetitions*. To permit such repetitions as d varies from 1 to n, we have defined $\overline{P}$ as in (37) for use in (38). Since the relaxation described below will recover the convex hull representation for $d = n$, we need not consider $d > n$. Using $(d-1)$ suitable dummy indices to represent duplications, it can be easily verified that there are a total of

$$\binom{p+d-1}{d} = (p+d-1)!/d!(p-1)!$$

distinct factors of this type at level d.

These factors are then used in a *Special Structures Reformulation Linearization Technique*, abbreviated **SSRLT**, as stated below, in order to generate the relaxation $\overline{X}_d$.

(a) **Reformulation Phase.** Multiply each inequality defining the feasible region (34) (including the constraints defining S) by each S-factor $g(J)$ of order d, and apply the identity $x_j^2 = x_j$ for all $j \in \{1, \ldots, n\}$.

(b) **Linearization Phase.** Linearize the resulting polynomial program by using the substitution defined in (39) below. This produces the dth level relaxation $\overline{X}_d$.

$$w_J = \prod_{j \in J} x_j \ \forall J \subseteq N, \ v_{Jk} = y_k \prod_{j \in J} x_j \ \forall J \subseteq N, \forall k. \tag{39}$$

For conceptual purposes, as in Section 2, define the projection of $\overline{X}_d$ onto the space of the original variables as

$$\overline{X}_{Pd} = \{(x,y) : (x,y,w,v) \in \overline{X}_d\} \ \forall d = 1, \ldots, n. \tag{40}$$

Additionally, as before, we will denote $\overline{X}_{P0} = \overline{X}_0$ (for $d = 0$) as the ordinary linear programming relaxation. Sherali, Adams and Driscoll (1996) then show that similar to (12),

$$\overline{X}_{P0} \equiv \overline{X}_0 \supseteq \overline{X}_{P1} \supseteq \overline{X}_{P2} \supseteq \ldots \supseteq \overline{X}_{Pn} = conv(X) \tag{41}$$

where $conv(\cdot)$ denotes the convex hull of feasible solutions. Moreover, this hierarchy dominates that of Section 2 in the sense that

$$\overline{X}_d \subseteq X_d \text{ and } \overline{X}_{Pd} \subseteq X_{Pd} \ \forall d = 0, 1, \ldots, n. \tag{42}$$

For convenience in notation, let us henceforth denote by $\{\cdot\}_L$ the process of linearizing a polynomial expression $\{\cdot\}$ in x and y via the substitution defined in (39), following the use of the identity $x_j^2 = x_j \forall j = 1, \ldots, n$.

Before proceeding further, let us highlight some important comments that pertain to the application of SSRLT. First, in an actual implementation, note that under the substitution $x_j^2 = x_j$ for all j, several terms defining the factors in (38) might be zeros, either by definition, or due to the restrictions of the set X in (34). For example, if $S \equiv \{x : 0 \leq x \leq e_n\}$, when $d = 2$, one such factor of type (38) is $x_j(1 - x_j)$ for $j \in \{1, \ldots, n\}$, which is clearly null when x_j^2 is equated with x_j. Additionally, if $S = \{x : 0 \leq x \leq e_n\}$, then multiplying the inequalities defining X in (3.1a) by any factor $F_d(J_1, J_2)$ for which the remaining constraints of $\overline{X}_d$ enforce $f_d(J_1, J_2) = 0$ will yield null constraints. Secondly, some of these factors might be implied in the sense that they can be reproduced as a nonnegative surrogate of other such factors that are generated in (38). For example, when $d = 3$ for $S \equiv \{x : 0 \leq x \leq e_n\}$, the factor $x_t^2 x_r \geq 0$ of order 3 is equivalent to $x_t x_r \geq 0$ of order 2, which is implied by other nonnegative factors of order 3 generated by the RLT constraints by Property 3.1. All such null and implied factors and terms should be eliminated in an actual application of SSRLT. Third, if any constraint in $\overline{X}_d$ of (34, 41) is implied by the remaining constraints, then we can simply remove this constraint from X without changing any resulting set $\overline{X}_d$. (This same logic hold relative to the foregoing RLT process and the sets X_d.) To illustrate, any single constraint in (21) or (22) can be removed from the QAP formulation of Example 5.1 while preserving the strengths of the prescribed relaxations at levels 1 and 2, saving $n(n-1)$ and $n(n-1)^2(n-2)$ constraints respectively.

As evident from the foregoing comments, the RLT process described in Section 2 is a special case of SSRLT when $S = \{x : 0 \leq x \leq e_n\}$. Note that one obvious scheme for generating tighter relaxations via SSRLT is to include in the latter set S certain suitable additional constraints depending on the problem structure, and hence generate S-factors at level d that include $f_d(J_1, J_2)$ of order d as defined in Section 2, along with any collection of additional S-factors as obtained via (38). Eliminating any null terms or implied factors thus generated, a hierarchy can be generated using SSRLT that would dominate the ordinary RLT hierarchy of relaxations at each level. Our focus here will largely reside on less obvious, and richer, instances where the set S possesses a special structure that implies the restrictions $0 \leq x \leq e_n$, without explicitly containing these bounding constraints.

Observe that we could conceptually think of SSRLT as being an inductive process, with the relaxation at level $(d+1)$ being produced by multiplying each of the constraints in the relaxation at level d with each S-factor defining S. Constraints produced by this process that effectively use null (zero) factor expressions $g(J)$ of order d are null constraints. Constraints produced by this process that effectively use factors $g(J)$ that are implied by other factors in (38), themselves implied by the constraints generated using the latter factors. Hence, the process of reducing the set of factors based on eliminating null or implied factors from use at the reformulation step, or that of eliminating the corresponding redundant constraints generated by such factors, are equivalent steps. It follows that an equivalent relaxation at level d would be produced by using only the non-null, non-implied factors, recognizing any zero variable terms in the resulting relaxation as identified by the S-factors. Furthermore, such non-redundant/non- null factors can be generated inductively through the levels, recognizing zero terms revealed at previous levels. This latter relaxation is what should actually be generated in practice. Sections 7 and 8 provide several examples. In a similar fashion to the RLT process of Section 2, a hierarchy leading to the convex hull representation can be generated in a piecewise manner by sequentially enforcing binariness on sets of variables, constructing the highest level relaxation for each considered set in turn.

Finally, let us comment on the treatment of equality constraints and equality factors. Note that whenever an equality constraint-factor defines an S-factor, any resulting product constraint is an equality restriction. Consequently, in the presence of equality constrained factors, in general, it is only necessary to multiply the corresponding equality constraint-factors simply with each x and y variable alone, as well as the constant 1, since the product with any other expression in x and y can be composed using these resulting products. Moreover, since the products with the x-variables are already being generated via other SSRLT constraints by virtue of the corresponding defining equality constraints of S already being included within X, and since $x \geq 0$ is implied by the inequality restrictions of $x \in S$, only products using y variables are necessary.

Furthermore, in this connection, note that if X contains equality structural constraints, in general, then these can be treated as in Section 5. That is, at level d, these equality constraints would simply need to be multiplied by the factors $F_p(J, \emptyset)$ for $J \subseteq N, p = |J| = 0, 1, \ldots, d$. Naturally, factors $F_p(J, \emptyset)$ that are known to be zeros, *i.e.*, any such factor for which we know that no feasible solution exists that has $x_j = 1 \ \forall j \in J$, need not be used in constructing these product constraints, and can be set equal to zero in the relaxation.

Example 6.1. For example, suppose that we have a set $S = \{x : e_n \cdot x = 1, x \geq 0\}$. Then the S-factors of order 1 are the expressions that define the restrictions

$$\{(1 - e_n \cdot x) = 0, \ x_1 \geq 0, \ldots, x_n \geq 0\}, \tag{43}$$

which include the equality constraint-factor along with the bound-factors

$x_1 \geq 0, \ldots, x_n \geq 0$. To compose the S- factors of order 2, note that

$$x_t(1 - e_n \cdot x) = 0 \text{ yields } \sum_{j \neq t} w_{(jt)} = 0 \; \forall t, \tag{44}$$

upon using $x_t^2 \equiv x_t$ and substituting $w_{(jt)}$ for $x_j x_t \; \forall j \neq t$ according to (39). (Note that we only need to define w_{jt} for $j < t$, and accordingly, we will denote $w_{(jt)} \equiv w_{jt}$ if $j < t$ and $w_{(jt)} \equiv w_{tj}$ if $t < j$.) Equation (44) along with

$$w_{(jt)} \equiv x_j x_t \geq 0 \; \forall j \neq t \tag{45}$$

produced by the other S-factors of order 2 imply that $w_{(jt)} \equiv 0 \; \forall j \neq t$, hence yielding null factors via (44) and (45). The only non-null S-factors of order 2 are therefore produced by pairwise self-products of the constraints defining S. But $(1 - e_n \cdot x)^2 = 0$ and $x_j^2 \geq 0, j = 1, \ldots, n$, respectively yield $(1 - e_n \cdot x) = 0$, and $x_j \geq 0 \; \forall j = 1, \ldots, n$, upon using $x_j^2 \equiv x_j$ and $x_j x_t = 0 \; \forall j \neq t$ as above. Hence, the reduced set of factors of order 2 are precisely the same as those of order 1, and this continues for all levels $2, \ldots, n$. Consequently, by (41), the convex hull representation would necessarily be produced at level 1 itself for this example.

To produce this level 1 representation, we would multiply all the constraints defining X (including the ones in S) by each factor $x_j \geq 0, j = 1, \ldots, n$, from (43). However, for the equality factor $(1 - e_n \cdot x) = 0$, by the foregoing discussion, we would only need to construct the RLT constraints $\{y_k(1 - e_n \cdot x)\}_L = 0$ and retain $e_n \cdot x = 1$. The resulting relaxation would produce the convex hull representation, as asserted above.

To further reinforce some of the preceding ideas before presenting additional specific details, we use another example that includes an equality constraint in S, but also explicitly includes the bound restrictions $0 \leq x \leq e_n$. As mentioned above, since the S-factors would now include the regular RLT bound-factors, any S-factors other than these bound-factor products are optional.

Example 6.2. Suppose that $n = 4$ and consider $S = \{x \in R^4 : x_1 + x_2 + x_3 + x_4 = 2, \; 0 \leq x \leq e_4\}$. The following factors are derived that can be applied in SSRLT, noting the equality constraint defining S.

(a) *Level 1 factors:* $x_j \geq 0$ and $(1 - x_j) \geq 0, j = 1, \ldots, 4$, and optionally, $(e_4 \cdot x - 2) = 0$ (to be multiplied by 1 and by each y variable alone as noted above).

(b) *Level 2 factors:* Bound factors of order 2 given by $\{x_i x_j, (1-x_i)x_j, x_i(1-x_j),$ and $(1-x_i)(1-x_j) \ \forall 1 \leq i < j \leq 4\}$, and optionally, any factors (to be applied to $y \geq 0$ alone) from the set $\{x_i - \sum_{j \neq i} x_i x_j = 0 \ \forall i = 1, \ldots, 4$ obtained by multiplying $e_4 \cdot x = 2$ by each $x_i, i = 1, \ldots, 4$, and $(e_4 \cdot x - 2) = 0$ itself, obtained from $(e_4 \cdot x - 2)^2 = 0$ upon using $\sum_{j \neq i} x_i x_j = x_i \ \forall i\}$.

(c) *Level d factors, $d = 3, 4$:* Bound factors $F_d(J_1, J_2) \geq 0$ of order d, with the additional restriction that all 3rd and 4th order terms are zeros, plus optionally, factors from the optional set at level 2. Note that the valid implication of polynomial terms of order 3 being zero, for example, is obtained through the RLT process by multiplying $x_i - \sum_{j \neq i} x_i x_j = 0$ with x_k, for each $i, k, i \neq k$. This gives $\sum_{j \neq i,k} x_i x_j x_k = 0$ which, by the nonnegativity of each triple product term, implies that $x_i x_j x_k = 0 \ \forall i \neq j \neq k$.

In a likewise fashion, for set partitioning problems, for example, any quadratic or higher order products of variables that involve a pair of variables that appear together in any constraint are zeros. More generally, any product term that contains variables or their complements that cannot simultaneously take on a value of 1 in any feasible solution can be restricted to zero. Sherali and Lee (1996) use this structure to present a specialization of RLT to derive explicit reduced level d representations in their analysis of set partitioning problems.

7 Applications of RLT for Some Particular Special Structures

We now demonstrate how some specific special structures can be exploited in designing an application of the general framework of SSRLT. This discussion will also illuminate the relationship between RLT and SSRLT, beyond the simple dominance result stated in (42). For this purpose, we employ various commonly occurring special structures such as generalized upper bounding (GUB) constraints, variable upper bounding (VUB) constraints, and to a lesser degree of structure, problem sparsity. These illustrations are by no means exhaustive; our motivation is to present the basic framework for this approach, and encourage the reader to design similar constructs for other applications on a case-by-case basis.

7.1 Generalized Upper Bounding (GUB) or Multiple Choice Constraints

Suppose that the set S of (35) is given as follows,

$$S = \{x : \sum_{j \in N_i} x_j \leq 1 \; \forall i \in Q = \{1,\ldots,q\}, x \geq 0\}, \qquad (46)$$

where $\bigcup_{i \in Q} N_i \equiv N \equiv \{1,\ldots,n\}$. Problems possessing this particular special structure arise in various settings including maximum cardinality node packing, set packing, capital budgeting, and menu planning problems among others (see Nemhauser and Wolsey, 1988).

First, let us suppose that $q \equiv 1$ in (46), so that

$$S \equiv \{x : e_n \cdot x \leq 1, x \geq 0\}$$

The S-factors of various orders for this particular set can be derived as follows:

(a) *S-factors at level 1.* These factors are directly obtained from the set S via the constraint factors $(1 - e_n \cdot x) \geq 0$, and $x_j \geq 0 \; \forall j = 1,\ldots,n$.

(b) *S-factors at level 2.* The linearization operation $[x_t(1 - e_n \cdot x)]_L \geq 0$ produces an expression

$$\sum_{j \neq t} w_{(jt)} \leq 0 \; \forall t = 1,\ldots,n$$

where $w_{(jt)} \equiv w_{tj}$ if $t < j$, and $w_{(jt)} \equiv w_{jt}$ if $j < t$. Moreover, via the pairwise products of x_j and $x_t, j \neq t$, we obtain factors of the type

$$(x_t x_j) \geq 0 \; \forall j \neq t, \text{ or } w_{(jt)} \geq 0 \; \forall j \neq t.$$

Similar to (44) and (45), the foregoing two sets of inequalities imply that

$$w_{(jt)} = 0 \; \forall j \neq t. \qquad (47)$$

Consequently, under (47), the only S-factors of order 2 that survive such a cancellation are self-product factors of the type $(x_t x_t) \geq 0 \; \forall t$, and $(1 - e_n \cdot x) \cdot (1 - e_n \cdot x) \geq 0$. These yield the same factors as at level 1, upon using $x_t^2 = x_t \; \forall t$ along with (47) as seen in Section 6. Hence, we only need to use the factors

$$(1 - e_n \cdot x) \geq 0 \text{ and } x_j \geq 0, j = 1,\ldots,n$$

to construct the equivalent set $\overline{X}_2$. Notice that $\overline{X}_2 \equiv \overline{X}_1$, and this equivalence relation continues through all levels of relaxations up to $\overline{X}_n$. Hence, the first level relaxation itself produces the convex hull representation in this case. There are two insightful points worthy of note in the context of this example. First, as illustrated next, although RLT recognizes that (47) holds true at each relaxation level, it may not produce the convex hull representation at the first level as does SSRLT.

Example 7.1. Let

$$X = \{(x_1, x_2) : 6x_1 + 3x_2 \geq 2, x_1 + x_2 \leq 1, x \text{ binary}\},$$

and consider the generation of the first level RLT relaxation. Note that the factors used in this context are x_j and $(1 - x_j)$ for $j = 1, 2$. Examining the product of $x_1 + x_2 \leq 1$ with x_1 yields $w_{12} \leq 0$, which together with $w_{12} \equiv [x_1 x_2]_L \geq 0$ yields $w_{12} \equiv 0$. Other products of the factors x_j and $(1 - x_j), j = 1, 2$, with $x_1 + x_2 \leq 1$ and $0 \leq x \leq e_2$ simply reproduce these same latter constraints.

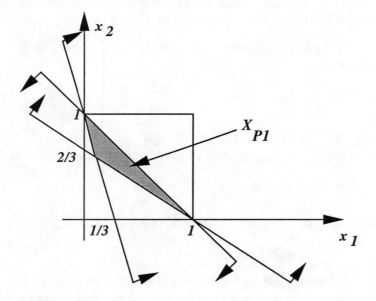

Figure 2: The first level relaxation using RLT.

Examining the products of $6x_1 + 3x_2 \geq 2$ with these first level factors yields nonredundant inequalities when the factors $(1-x_1)$ and $(1-x_2)$ are used, generating the constraints $2x_1+3x_2 \geq 2$ and $3x_1+x_2 \geq 1$, respectively. Hence, we obtain the first level relaxation (directly in projected form in this case) as

$$X_{P1} = \{(x_1, x_2) : 2x_1 + 3x_2 \geq 2, 3x_1 + x_2 \geq 1, x_1 + x_2 \leq 1, x \geq 0\}.$$

Figure 2 depicts the region X_{P1}. However, using SSRLT, by the above argument, we would obtain $\overline{X}_{P1} \equiv conv(X) \equiv \{x : x_1 + x_2 = 1, x \geq 0\}$ which is a strict subset of X_{P1}.

A second point to note is that we could have written the generalized upper bounding inequality $e_n \cdot x \leq 1$ as an equality $e_n \cdot x + x_{n+1} = 1$ by introducing a slack variable x_{n+1}, and then *recognizing the binariness of this slack variable*, we could have used this additional variable in composing the bound-factor products while applying RLT. Although this would have produced the same relaxation as with SSRLT, the process would have generated several more redundant constraints while applying the factors x_j and $(1-x_j)$ for $j = 1, \ldots, n+1$, to the constraints, as opposed to using the fewer factors $(1 - e_n \cdot x)$ and $x_j, j = 1, \ldots, n$, as needed by SSRLT. However, in more general cases of the set S, such a transformation that yields the same representation using RLT as obtained via SSRLT may not be accessible. (See Example 8.2 below, for instance.)

Example 7.2. Next, let us consider the case of $q = 2$ in (46). For the sake of illustration, suppose that

$$S = \{x \in R^5 : x_1 + x_2 + x_3 \leq 1, x_3 + x_4 + x_5 \leq 1, \text{ and } x \geq 0\} \qquad (48)$$

(a) *S-factors at level 1:* These are simply the constraints defining S.

(b) *S-factors at level 2:* As before, the pairwise products within each GUB set of variables will reproduce the same factors as at the first level, since (47) holds true within each GUB set. However, across the two GUB sets, we would produce the nonnegative quadratic bound factor products x_1x_4, x_1x_5, x_2x_4, and x_2x_5 along with the following factor products, recognizing that any quadratic product involving x_3 is zero, as this variable appears in both GUB sets.

$$x_4 \cdot (1 - x_1 - x_2 - x_3) \geq 0 \text{ yielding } x_4 - x_1x_4 - x_2x_4 \geq 0,$$
$$x_5 \cdot (1 - x_1 - x_2 - x_3) \geq 0 \text{ yielding } x_5 - x_1x_5 - x_2x_5 \geq 0,$$

$$x_1 \cdot (1 - x_3 - x_4 - x_5) \geq 0 \text{ yielding } x_1 - x_1 x_4 - x_1 x_5 \geq 0,$$
$$x_2 \cdot (1 - x_3 - x_4 - x_5) \geq 0 \text{ yielding } x_2 - x_2 x_4 - x_2 x_5 \geq 0,$$
and
$$(1 - x_1 - x_2 - x_3) \cdot (1 - x_3 - x_4 - x_5) \geq 0 \text{ yielding}$$
$$1 - x_1 - x_2 - x_3 - x_4 - x_5 + x_1 x_4 + x_1 x_5 + x_2 x_4 + x_2 x_5 \geq 0.$$

These can now be applied to the constraints defining X, recognizing the terms that have been identified to be zeros.

(c) *S-factors at levels* ≥ 3: Since there are only 2 GUB sets in this example, and since any triple product of distinct factors must involve a pair of factors coming from the defining constraints corresponding to the same GUB set, and the latter product is zero, all such products must vanish. Hence, all factors at level 3, and similarly at levels 4 and 5, coincide with those at level 2. In other words, the relaxation at level 2 (defined as $\overline{X}_2$) itself yields the convex hull representation.

In general, the *level equal to the independence number of the underlying intersection graph corresponding to the GUB constraints, which simply equals the maximum number of variables that can simultaneously be 1, is sufficient to generate the convex hull representation*. In the case of (46), the convex hull representation would be obtained at level q, or earlier.

An enlightening special case of (46) deals with the vertex packing problem. Given a graph $G = (V, E)$ with vertex set $V = \{v_1, v_2, \ldots, v_n\}$, an edge set E connecting pairs of vertices in V, and a weight c_j associated with each vertex v_j, the vertex packing problem is to select a maximum weighted subset of vertices such that no two vertices are connected by an edge. For each $j = 1, \ldots, n$ by denoting the binary variable x_j to equal 1 if vertex j is chosen and 0 otherwise, the vertex packing problem can be stated as maximize $\sum_j c_j x_j : x_i + x_j \leq 1 \ \forall (i,j) \in E, x$ binary}. The convex hull representation over any subset P of the variables can be obtained as above, by considering any clique cover of the subgraph induced by the corresponding vertices, with each set N_i corresponding to the variables defining some clique i. In fact, given a cover that has q cliques where each edge of E is included in some clique graph, the S-factors of level q themselves

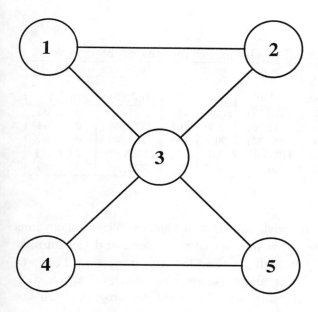

Figure 3: Vertex packing graph.

define the convex hull representation, since their products with the packing constraints, as well as with the nonnegativity restrictions on x, are implied by these factors. To illustrate, the inequalities of (48) can be considered as a (maximum cardinality) clique cover of the vertex packing problem on the graph in Figure 3, and so, the stated S-factors of level 2 themselves define the convex hull representation. This general packing observation may have widespread applicability since, as noted by Garfinkel and Nemhauser (1973), any finite integer linear program can be reformulated as a packing problem.

To illustrate the computational benefits of SSRLT over RLT in this particular context, we conducted the following experiment using pseudo-randomly generated set packing problems of the type maximize$\{\sum_{j=1}^{n} c_j x_j : \sum_{j \in N_i} x_j \geq 1 \; \forall i = 1, \ldots, q, x \text{ binary}\}$.

For several instances of such problems, we computed the optimal value of the 0-1 packing problem, that of its ordinary LP relaxation, as well as the optimal values of the first level relaxations produced by applying RLT and SSRLT, where the latter was generated by using all of the defining clique constraints, together with $x \geq 0$, to represent the set S. Table 1 gives the percentage gaps obtained for the latter three upper bounds with respect

Table 1: Computational results for set packing problems.

	Ordinary LP relaxation			Level-one via RLT			Level-one via SSRLT		
(q,n)	Density	% gap	Iter	(m',n')	% gap	Iter	(m',n')	% gap	Iter
(15,25)	66	42.0	22	(170,25)	16.8	160	(123,25)	0	35
(20,30)	56	26.7	36	(319,30)	7.2	267	(214,30)	0	32
(25,35)	52	57.3	43	(481,35)	20.2	152	(328,35)	0	113
(55,45)	27	37.0	103	(2256,63)	2.0	344	(1896,63)	0	171
(35,35)	56	36.5	48	(607,35)	10.7	292	(631,35)	0	60

to the optimal 0-1 value, along with the sizes of the respective relaxations (m' = number of constraints, n' = number of variables), and the number of simplex iterations (ITER) needed by the OSL solver version 2.001, to achieve optimality. Note that for all instances, the first level application of SSRLT *was sufficient to solve the underlying integer program*. On the other hand, although RLT appreciably improved the upper bound produced by the ordinary LP relaxation, it still left a significant gap that remains to be resolved in these problem instances. Moreover, the relatively simpler structure of SSRLT results in far fewer simplex iterations being required to solve this relaxation as compared with the effort required to solve RLT.

7.2 Variable Upper Bounding Constraints

This example points out that in the presence of variable upper bounding (VUB) types of restrictions, a further tightening of relaxations via SSRLT, beyond that of RLT, can be similarly produced.

Example 7.3. Consider a set S that is composed as follows in a particular problem instance:

$$S = \{x \in R^6 : 0 \leq x_1 \leq x_2 \leq x_3 \leq 1, 0 \leq x_4 \leq x_5 \leq 1, 0 \leq x_6 \leq 1\}.$$

The first level factors for this instance are given by $x_1 \geq 0, x_2 - x_1 \geq 0, x_3 - x_2 \geq 0, 1 - x_3 \geq 0, x_4 \geq 0, x_5 - x_4 \geq 0, 1 - x_5 \geq 0, x_6 \geq 0$, and $1 - x_6 \geq 0$. Compared with the RLT factors, these yield tighter constraints as they imply the RLT factors. For $d \in \{1, \ldots, 6\}$, taking these factors d at a time, including self-products, and simplifying these factors by eliminating null or implied factors, would produce the relaxation $\overline{X}_d$.

It is interesting to note in this connection that the VUB constraints of the type $0 \leq x_1 \leq x_2 \leq \ldots \leq x_k \leq 1$ used in this example can be equivalently transformed into a GUB constraint via the substitution $z_j = x_j - x_{j-1}$ for $j = 1, \ldots, k$, where $x_0 \equiv 0$. The inverse transformation yields $x_j = \sum_{t=1}^{j} z_t$ for $j = 1, \ldots, k$, thereby producing the equivalent representation $z_1 + z_2 + \ldots + z_k \leq 1, z \geq 0$. Under this transformation, and imposing binary restrictions on all the z-variables, the reformulation strategies described in Section 7.1 above can be employed. However, note that the process of applying RLT to the original or to the transformed problem can produce different representations. This is illustrated next.

Example 7.4. Consider the set

$$X = \{(x_1, x_2) : -6x_1 + 3x_2 \leq 1, 0 \leq x_1 \leq x_2 \leq 1, x \text{ binary}\}$$

The convex hull of feasible solutions is given by $0 \leq x_1 = x_2 \leq 1$ (see Figure 4(a)). This representation is produced by the level-1 SSRLT relaxation using the VUB constraints to define S, where the relevant constraint $x_1 \geq x_2$ which yields $x_1 = x_2$ is obtained by noting that the factor products $[x_1(x_2-x_1)]_L \geq 0$ and $[x_1(1-x_2)]_L \geq 0$ respectively give $w_{12} \geq x_1$ and $w_{12} \leq x_1$, or that $w_{12} = x_1$. This together with the constraint $[(x_2 - x_1)(1 + 6x_1 - 3x_2)]_L \geq 0$ yields $-2(x_2 - x_1) \geq 0$, or that $x_1 \geq x_2$.

On the other hand, constructing RLT at level 1 by applying the factors x_j and $(1 - x_j), j = 1, 2$, to the inequality restrictions of X, produces the relaxation (directly in projected form)

$$X_{P1} = \{(x_1, x_2) : 2x_1 - 3x_2 \geq -1, 3x_1 \geq x_2, 0 \leq x_1 \leq x_2 \leq 1\},$$

where $w_{12} = x_1$ is produced as with SSRLT, and where the first two constraints defining X_{P1} result from the product constraints $[(1+6x_1-3x_2)(1-x_1)]_L \geq 0$ and $[(1+6x_1-3x_2)x_2]_L \geq 0$, respectively. Figure 4(a) depicts the region defined by this relaxation.

However, if we were to apply the transformation $z_1 = x_1, z_2 = x_2 - x_1$ to X, where the inverse transformation is given by $x_1 = z_1$ and $x_2 = z_1 + z_2$, the problem representation in z-space becomes

$$Z = \{(z_1, z_2) : -3z_1 + 3z_2 \leq 1, z_1 + z_2 \leq 1, z \text{ binary}\},$$

where the *binariness on z_2 has been additionally recognized.* Figure 4(b) illustrates that the set $conv(Z)$ is given by the constraints $0 \leq z_1 \leq 1, z_2 = 0$.

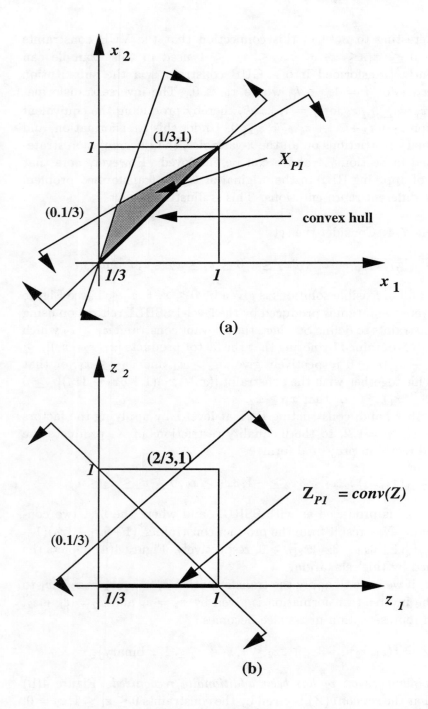

Figure 4: Depiction of the first level relaxations using RLT for Example 7.4.

Now, applying RLT to this transformed space, the relevant constraint $z_2 = 0$ is produced via $z_2 \geq 0$ and the first level product constraint $[(1 + 3z_1 - 3z_2)z_2]_L \geq 0$ which yields $-2z_2 \geq$, where $[z_1 z_2]_L \equiv 0$ from $[z_1 z_2]_L \geq 0$ and $[z_1(1 - z_1 \cdot z_2)]_L \geq 0$. Hence, for this transformed problem, RLT produces the first level relaxation $Z_{P1} = conv(Z)$, while we had $X_{P1} \subset conv(X)$ when applying RLT to the original problem. However, as with Example 7.1 for the case $q = 1$ (treating that as the transformed z- variable problem), we could have possibly obtained $Z_{P1} \supset conv(Z)$ as well, whereas SSRLT would necessarily produce the convex hull representation at level one in either case.

7.3 Sparse Constraints

In this example, we illustrate how one can exploit problem sparsity. Suppose that in some 0-1 mixed-integer problem, we have the knapsack constraint (either inherent in the problem or implied by it) given by $2x_1 + x_2 + 2x_3 \geq 3$. The facets of the convex hull of $\{(x_1, x_2, x_3) : 2x_1 + x_2 + 2x_3 \geq 3, x_i \text{ binary}, i = 1, 2, 3\}$, can be readily obtained as $\{x_1 + x_2 + x_3 \geq 2, x_1 \leq 1, x_2 \leq 1, x_3 \leq 1\}$. Similarly, another knapsack constraint might be of the type $x_4 + 2x_5 + 2x_6 \leq 2$, and the corresponding facets of the convex hull of feasible 0-1 solutions can be obtained as $\{x_4 + x_5 + x_6 \leq 1, x_4 \geq 0, x_5 \geq 0, x_6 \geq 0\}$. The set S can now be composed of these two sets of facets, along with other similar constraints involving the remaining variables on which binariness is being enforced, including perhaps, simple bound constraint factors. Note that in order to generate valid tighter relaxations, we can simply enforce binariness on variables that fractionate in the original linear programming relaxation in the present framework. Furthermore, entire convex hull representations of underlying knapsack polytopes are not necessary — simply, the condition (36) needs to be satisfied, perhaps by explicitly including simple bounding constraints. This extends the strategy of Crowder et al. (1983) in using facets obtained as liftings of minimal covers from knapsack constraints within this framework, in order to generate tighter relaxations.

8 Using Conditional Logic to Strengthen RLT Constraints: Application to the Traveling Salesman Problem

In all of the foregoing discussion, depending on the structure of the problem, there is another idea that we can exploit to even further tighten the RLT constraints that are generated. This trick deals with the use of conditional logic in the generation of RLT based constraints that can enable the tightening of a relaxation at any level, which would otherwise have been possible only at some higher level in the hierarchy. In effect, this captures within the RLT process the concepts of branching and logical preprocessing, features that are critical to the efficient solution of discrete optimization problems.

To introduce the basic concept involved, for simplicity, consider the following first level SSRLT constraint that is generated by multiplying a factor $(\alpha x - \beta) \geq 0$ with a constraint $(\gamma x - \delta) \geq 0$, where x is supposed to be binary valued, and where the data is all-integer. (Similar extensions can be developed for mixed-integer constraints, as well as for higher-order SSRLT constraints in which some factor is being applied to some other valid constraint-or-bound- factor product of order greater than one.)

$$\{(\alpha x - \beta)(\gamma x - \delta)\}_L \geq 0. \tag{49}$$

Observe that if $\alpha x = \beta$, then $(\alpha x - \beta)(\gamma x - \delta) \geq 0$ is valid regardless of the validity of $\gamma x \geq \delta$. Otherwise, we must have $\alpha x \geq \beta + 1$ (or possibly greater than $\beta + 1$, if the structure of $\alpha x \geq \beta$ so permits), and we can then perform standard logical preprocessing tests (zero-one fixing, coefficient reduction, etc. — see Nemhauser and Wolsey, 1988, for example) on the set of constraints $\alpha x \geq \beta + 1, \gamma x \geq \delta, x$ binary, along with possibly other constraints, to tighten the form of $\gamma x \geq \delta$ to the form $\gamma' x \geq \delta'$. For example, if $\alpha x \geq \beta$ is of the type $(1 - x_j) \geq 0$, for some $j \in \{1, \ldots, n\}$, then the restriction $\alpha x \geq \beta + 1, x$ binary, asserts that $x_j = 0$, and so, $\gamma x \geq \delta$ can be tightened under the condition that $x_j = 0$. (Similarly, in a higher-order constraint, if a factor $F_d(J_1, J_2)$ multiplies $\gamma x \geq \delta$, then the latter constraint can be tightened under conditional logical tests based on setting $x_j = 1 \ \forall j \in J_1$ and $x_j = 0 \ \forall j \in J_2$.)

Additionally, the resulting constraint $\gamma' x \geq \delta'$ can potentially be further tightened by finding the maximum $\theta \geq 0$ for which $\gamma' x \geq \delta' + \theta$ is valid when

Reformulation-Linearization for Discrete Optimization

$\alpha x \geq \beta + 1$ is imposed by considering the problem

$$\theta = -\delta' + \min\{\gamma' x : \gamma' x \geq \delta', \alpha x \geq \beta+1, \text{any other valid inequalities } x \text{ binary}\}, \quad (50)$$

and by increasing δ' by this quantity θ. Note that, of course, we can simply solve the continuous relaxation of (50) and use the resulting value after rounding it upwards (using $\theta = 0$ if this value is negative), in order to impose the following SSRLT constraint, in lieu of the weaker restriction (49), within the underlying problem:

$$\{(\alpha x - \beta)(\gamma' x - \delta' - \theta)\}_L \geq 0 \quad (51)$$

Observe that this also affords the opportunity to now tighten the factor $(\alpha x - \beta)$ in a similar fashion in (51), based on the valid constraint $\gamma' x \geq \delta' + \theta$. Interestingly, the sequential lifting process studied by Balas and Zemel (1978), for lifting a minimal cover inequality into a facet for a full-dimensional knapsack polytope can be viewed as a consequence of this RLT construct (see Sherali et al., 1996).

Example 8.1. To illustrate, consider the following knapsack constraint in binary variables x_1, x_2, and $x_3 : 2x_1 + 3x_2 + 3x_3 \geq 4$. Let us examine the RLT constraint

$$\{x_3[2x_1 + 3x_2 + 3x_3 - 4]\}_L \geq 0. \quad (52)$$

Applying the foregoing idea, we can tighten (52) under the restriction that $x_3 = 1$, knowing that it is always valid when $x_3 = 0$ regardless of the nonnegativity of any expression contained in $[\cdot]$. However, when $x_3 = 1$, the given knapsack constraint becomes $2x_1 + 3x_2 \geq 1$, which by coefficient reduction, can be tightened to $x_1 + x_2 \geq 1$. Hence, (52) can be replaced by the tighter restriction

$$\{x_3[x_1 + x_2 - 1]\}_L \geq 0. \quad (53)$$

Similarly, consider the RLT constraint

$$\{(1 - x_3)[2x_1 + 3x_2 + 3x_3 - 4]\}_L \geq 0. \quad (54)$$

This time, imposing $(1 - x_3) \geq 1$ or $x_3 = 0$, the knapsack constraint becomes $2x_1 + 3x_2 \geq 4$ which implies via standard logical tests that $x_1 = x_2 = 1$. Hence, we can impose the *equalities*

$$\{(1 - x_3)(x_1 - 1)\}_L = 0 \text{ and } \{(1 - x_3)(x_2 - 1)\}_L = 0 \quad (55)$$

in lieu of (54), which is now implied. Observe that in this example, the sum of the RLT constraints in (53) and (55) yield $x_1 + x_2 + x_3 \geq 2$ which happens to be a facet of the knapsack polytope $conv\{x : 2x_1 + 3x_2 + 3x_3 \geq 4, x \text{ binary}\}$. This facet can alternatively be obtained by lifting the minimal cover inequality $x_1 + x_2 \geq 1$ as in Balas and Zemel (1978).

Example 8.2. To illustrate the use of the foregoing conditional logic based tightening procedure in the context of solving an optimization problem using general S-factors, consider the following problem:

$$\text{Maximize } \{x_1 + x_2 : 6x_1 + 3x_2 \geq 2, x \in S, x \text{ binary}\} \quad (56)$$
$$\text{where } S = \{(x_1, x_2) : 2x_1 + x_2 \leq 2, x_1 + 2x_2 \leq 2, x \geq 0\}.$$

Figure 5 depicts this problem graphically. The integer problem has the optimal value $v(IP) = 1$, attained at the solution (1,0) or (0,1). The ordinary LP relaxation has the optimal value $v(LP) = 1/3$ attained at the solution (1/3,0).

Next, let us consider the first level relaxation (RLT-1) produced by RLT, using the factors x_j and $(1 - x_j), j = 1, 2$, to multiply all the problem constraints. Note that $[(2 - 2x_1 - x_2)x_1]_L$ yields $w_{12} \leq 0$, which together with $[x_1 x_2]_L \equiv w_{12} \geq 0$, gives $w_{12} = 0$. The other non-redundant constraints defining this relaxation are produced via the product constraints $[(1-x_1)(1-x_2)]_L \geq 0, [(6x_1 + 3x_2 - 2)(1 - x_1)]_L \geq 0$, and $[(6x_1 + 3x_2 - 2)(1 - x_2)]_L \geq 0$, which respectively yield (using $w_{12} = 0$), $x_1 + x_2 \leq 1, 2x_1 + 3x_2 \geq 2$, and $3x_1 + x_2 \geq 1$. This gives (directly in projected form)

$$X_{P1} = \{(x_1, x_2) : x_1 + x_2 \leq 1, 2x_1 + 3x_2 \geq 2, 3x_1 + x_2 \geq 1\}.$$

Figure 5 depicts this region. The optimal value using this relaxation is given by $v(\text{RLT-1}) = 5/7$, attained at the solution (1/7, 4/7).

On the other hand, to construct the first level relaxation (SSRLT-1) produced by SSRLT, we would employ the S-factors given by the constraints in (56). As in RLT, we obtain $w_{12} = 0$, and using this, the other nonredundant constraints defining this relaxation are produced via the product constraints

$$[(2 - 2x_1 - x_2)(2 - x_1 - 2x_2)]_L \geq 0,$$
$$[(6x_1 + 3x_2 - 2)(2 - 2x_1 - x_2)]_L \geq 0 \text{ and}$$
$$[(6x_1 + 3x_2 - 2)(2 - x_1 - 2x_2)]_L \geq 0,$$

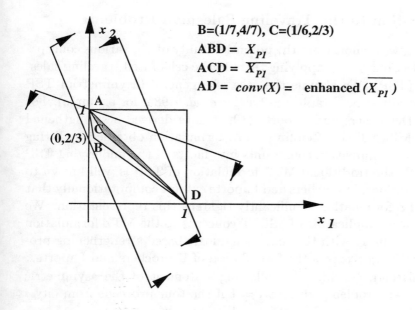

Figure 5: Depiction of the various first level RLT relaxations.

which respectively yield, $x_1 + x_2 \leq 1, 4x_1 + 5x_2 \geq 4$, and $2x_1 + x_2 \geq 1$. This gives (directly in projected form)

$$\overline{X}_{P1} = \{(x_1, x_2) : x_1 + x_2 \leq 1, 4x_1 + 5x_2 \geq 4, 2x_1 + x_2 \geq 1\}.$$

Figure 5 depicts this region. Note that $\overline{X}_{P1} \subset X_{P1}$ and that the optimal value of this relaxation is given by $v(\text{SSRLT-1}) = 5/6 > v(\text{RLT-1})$, and is attained at the solution $(1/6, 2/3)$.

Now, consider an enhancement of SSRLT-1 using conditional logic (a similar enhancement can be exhibited for RLT-1). Specifically, consider the product constraint $[(2 - 2x_1 - x_2)(6x_1 + 3x_2 - 2)]_L \geq 0$ of the form (49). Imposing $(2 - 2x_1 - x_2) \geq 1$ as for (49), i.e., $2x_1 + x_2 \leq 1$, yields $x_1 = 0$ by a standard logical test. This together with $6x_1 + 3x_2 \geq 2$ implies that $x_2 = 1$. Hence, the tightened form of this constraint is $(2 - 2x_1 - x_2)(x_1) = 0$ and $(2 - 2x_1 - x_2)(1 - x_2) = 0$. The first constraint yields $w_{12} = 0$ (as before), while the second constraint states that $x_1 + x_2 = 1$. This produces the convex hull representation, and so, this enhanced relaxation now recovers an optimal integer solution.

8.1 Application to the Traveling Salesman Problem

In this example, we demonstrate the potential utility of the various concepts developed in Sections 6-8 by applying them to the celebrated traveling salesman problem (TSP). We assume the case of a general (asymmetric) TSP defined on a totally dense graph (see Lawler et al., 1985, for example). For this problem, Desrochers and Laporte (1991) have derived a strengthened version of the Miller-Tucker-Zemlin (MTZ) formulation obtained by lifting the MTZ-subtour elimination constraints into facets of the underlying TSP polytope. While the traditional MTZ formulation of TSP is well known to yield weak relaxations, Desrochers and Laporte exhibit computationally that their lifted-MTZ formulation significantly tightens this representation. We show below that an application of SSRLT concepts to the MTZ formulation of TSP, used in concert with the conditional logic based strengthening procedure, *automatically recovers* the formulation of Desrochers and Laporte.

Toward this end, consider the following statement of the asymmetric traveling salesman problem, where $x_{ij} = 1$ if the tour proceeds from city i to city j, and is 0 otherwise, for all $i, j = 1, \ldots, n, i \neq j$.

ATSP: Minimize $\sum_i \sum_{j \neq i} c_{ij} x_{ij}$

$$\text{subject to:} \sum_{j \neq i} x_{ij} = 1 \quad \forall \quad i = 1, \ldots, n \tag{57}$$

$$\sum_{i \neq j} x_{ij} = 1 \quad \forall \quad j = 1, \ldots, n \tag{58}$$

$$u_j \geq (u_i + 1) - (n-1)(1 - x_{ij}) \quad \forall \quad i, j \geq 2, i \neq j \tag{59}$$

$$1 \leq u_j \leq (n-1) \quad \forall \quad j = 2, \ldots, n \tag{60}$$

$$x_{ij} \text{ binary} \quad \forall \quad i, j = 1, \ldots, n, i \neq j. \tag{61}$$

Note that (57, 58, and 61) represent the assignment constraints and (59) and (60) are the MTZ subtour elimination constraints. These latter constraints are derived based on letting u_j represent the rank order in which city j is visited, using $u_1 = 0$, and enforcing that $u_j = u_i + 1$ whenever $x_{ij} = 1$ in any binary feasible solution. Now, in order to construct a suitable reformulation using SSRLT, let us compose the set S as follows, and include this set of implied inequalities within the problem ATSP stated above.

$$\begin{aligned} S \equiv \quad & \{x : x_{ij} + x_{ji} \leq 1 \; \forall i, j \geq 2, i < j, x_{1j} + x_{j1} \leq 1 \; \forall j \geq 2, \\ & x_{ij} \geq 0 \; \forall i, j, i \neq j\}. \end{aligned} \tag{62}$$

Note that S is comprised of simple two-city subtour elimination constraints of the form proposed by Dantzig, Fulkerson, and Johnson (1954). Next, let us construct the following selected S-factor constraint products. First, consider the product constraint generated by multiplying (59) with the S-factor x_{ij}. Using conditional logic as with (49), and noting that we can impose $u_j = (u_i + 1)$ when $x_{ij} = 1$, this yields the constraint

$$x_{ij}u_j = x_{ij}(u_i + 1) \quad \forall i, j \geq 2, i \neq j. \tag{63}$$

Similarly, considering $[x_{1j}(u_j - 1)]_L \geq 0$ from (60), and enhancing this by the conditional logic that $u_j = 1$ when $x_{1j} = 1$, we get

$$x_{1j}u_j = x_{1j} \quad \forall j = 2, \ldots, n. \tag{64}$$

Repeating this with the upper bounding constraint in (60), we can enhance $[x_{j1}(n - 1 - u_j)]_L \geq 0$ to the following constraint, noting that $u_j = (n - 1)$ if $x_{j1} = 1$:

$$x_{j1}u_j = (n - 1)x_{j1} \quad \forall j = 2, \ldots, n. \tag{65}$$

Next, let us consider the product of (59) with the S-factor $(1 - x_{ij} - x_{ji}) \geq 0$. This gives the constraint $[(1 - x_{ij} - x_{ji})(u_j - u_i - 1 + (n-1)(1 - x_{ij}))]_L \geq 0$. Using $x_{ij}^2 = x_{ij}$ and $x_{ij}x_{ji} = 0$ (since $x_{ij} + x_{ji} \leq 1$), the foregoing constraint becomes

$$(u_j - u_i) \geq (u_j - u_i)(x_{ij} + x_{ji}) - (n - 2)(1 - x_{ij}) + (n - 2)x_{ji}.$$

From (63), we get $(u_j - u_i)x_{ij} = x_{ij}$, and upon interchanging i and j, this yields $(u_j - u_i)x_{ji} = -x_{ji}$. Substituting this into the foregoing SSRLT constraint, we obtain the valid inequality

$$u_j \geq (u_i + 1) - (n - 1)(1 - x_{ij}) + (n - 3)x_{ji} \quad \forall i, j \geq 2, i \neq j. \tag{66}$$

Similarly, multiplying (60) with the S-factor $(1 - x_{1j} - x_{j1}) \geq 0$ yields $[(1 - x_{1j} - x_{j1})(u_j - 1)]_L \geq 0$ and $[(1 - x_{1j} - x_{j1})(n - 1 - u_j)]_L \geq 0$. Using the conditional logic procedure, under $x_{1j} = x_{j1} = 0$, these constraints can be respectively tightened to $[(1 - x_{1j} - x_{j1})(u_j - 2)]_L \geq 0$ and $[(1 - x_{1j} - x_{j1})(n - 2 - u_j)]_L \geq 0$. Simplifying these products and using (64) and (65) yields the constraints

$$1 + (1 - x_{1j}) + (n - 3)x_{j1} \leq u_j \leq (n - 1) - (1 - x_{j1}) - (n - 3)x_{1j}. \tag{67}$$

Observe that (66) and (67) are tightened versions of (59) and (60), respectively, and are precisely the facet- defining, lifted-MTZ constraints derived by Desrochers and Laporte (1991). Hence, we have shown that a selected application of SSRLT used in concert with our conditional logic based strengthening procedure *automatically* generates this improved MTZ formulation. Sherali and Driscoll (1997) have developed further enhancements that tighten and subsume this formulation for both the ordinary as well as for the precedence constrained version of the asymmetric traveling salesman problem, using these SSRLT constructs along with conditional logic implications.

9 Conclusions and Extensions: Persistency, General Integer Programs, and Computational Expedients

The hierarchy of relaxations emerging from the RLT can be intuitively viewed as "stepping stones" between continuous and discrete sets, leading from the usual linear programming relaxation to the convex hull at level-n. By inductively progressing along these stepping-stone formulations, we have studied some novel *persistency* issues for certain constrained and unconstrained pseudo-Boolean programming problems. Given the tight linear programming relaxations afforded by RLT, a pertinent question that can be raised is that if we solve a particular d^{th} level representation in the RLT hierarchy, and some of the n variables turn out to be binary valued at optimality to the underlying linear program, then can we expect these binary values to *"persist"* at optimality to the original problem? Adams, Lassiter and Sherali (1993) derive sufficient conditions in terms of the dual solution that guarantee such a persistency result. For the unconstrained pseudo-Boolean program, we show that for $d = 1$ or for $d \geq n - 2$, persistency always holds. However, using an example with $d = 2$ and $n = 5$, we show that without the additional prescribed sufficient conditions, persistency will not hold in general. These results are also extended to constrained polynomial 0-1 programming problems. In particular, the analysis here reveals a class of 0-1 linear programs that possess the foregoing persistency property. Included within this class as a special case is the popular vertex packing problem, shown earlier in the literature to possess this property.

Thus far, we have been confining our attention to binary discrete variables. Conceptually, for a more general discrete integer program, one could employ a binary transformation to rewrite the problem as a 0-1 mixed-integer program, and then apply the foregoing RLT constructs. However, Adams and Sherali (1997) have shown that a more novel direct approach can lead to more insightful and compact representations, that are nonetheless equivalent to those obtained via the usual RLT process applied to the transformed binary problem.

To see the basic form of this approach, consider the following feasible region of a discrete mixed-integer program.

$$X = \{(x,y) \in R^n \times R^m : Ax + Dy \geq b \tag{68}$$
$$x_j \in S_j \equiv \{\theta_{jk}, k = 1,\ldots,k_j\},\ \forall j = 1,\ldots,n \tag{69}$$
$$y \geq 0\}, \tag{70}$$

where $\theta_{jk}, k = 1,\ldots,k_j, j = 1,\ldots,n$ are discrete real numbers (of either sign). As a generalization of the bound factors x_j and $(1 - x_j)$ when x_j is binary, let us define the *Lagrange Interpolation Polynomials* (**LIP**)

$$L_{jk} = \frac{\prod_{p \neq k}(x_j - \theta_{jp})}{\prod_{p \neq k}(\theta_{jk} - \theta_{jp})} \quad \forall k = 1,\ldots,k_j, j = 1,\ldots,n. \tag{71}$$

The RLT process employs *LIP factors or order d* composed by selecting some d distinct indices j, and for each selected index, choosing some LIP $L_{jk}, k \in \{1,\ldots,k_j\}$. The procedure then proceeds as follows for constructing the relaxation $\tilde{X}_d$, for any $d \in \{0,1,\ldots,n\}$.

Reformulation Phase:

1. Multiply (68, 70) by all possible LIP factors of order d.

2. Include nonnegativities on all possible LIP factors of order D, where $D = \min\{d+1, n\}$.

3. Use the following identity in the resulting nonlinear program

$$x_j L_{jk} = \theta_{jk} L_{jk} \ (\text{or } (x_j - \theta_{jk})L_{jk} = 0)\ \forall k = 1,\ldots,k_j, j = 1,\ldots,n. \tag{72}$$

Note that (72) corresponds to the step of letting $x_j^2 = x_j$, or setting $x_j(1 - x_j) = 0$, whenever x_j is binary as in the previous analysis.

4. Model (69) by including the following constraints via the introduction of binary variables $x_{jk}, k = 1,\ldots,k_j, j = 1,\ldots,n$. (Note that binariness on the x_{jk}-variables are relaxed below for defining the polyhedron

X_d.)

$$x_j = \sum_{k=1}^{k_j} \theta_{jk} x_{jk} \quad \forall j = 1, \ldots, n$$

$$\sum_{k=1}^{k_j} x_{jk} = 1, \quad \forall j = 1, \ldots, n$$

$$x_{jk} \geq 0 \quad \forall k = 1, \ldots, k_j, j = 1, \ldots, n.$$

Linearization Phase: Linearize the resulting problem obtained via Steps 1-4 of the Reformulation Phase by substituting a particular variable for each distinct product of the x-variables thus generated, and similarly, for each distinct product of y_t with the x-variables, $\forall t$. (Note that these products terms can include self-products of variables.) This produces a polyhedron $\tilde{X}_d$ in higher dimensions whose projection $\tilde{X}_{Pd}$ onto the (x, y) space can be shown to satisfy the usual RLT hierarchy

$$\tilde{X}_{P0} \supseteq \tilde{X}_{P1} \supseteq \ldots \supseteq \tilde{X}_{Pn} \equiv conv(X).$$

Moreover, the relationship between the binary transformed problem and the original discrete problem can be used to translate valid inequalities that are derived in the former space to those represented in the original (x, y) variable space. This construct can be useful in implementing a cutting plane or a branch-and-cut approach for general integer programs.

As far as the use of RLT as a practical computational aid for solving mixed-integer 0-1 problems is concerned, one may simply work with the relaxation X_1 itself, which has frequently proven to be beneficial. Using variations on the first-order implementation of RLT (a partial generation of X_1 enhanced by additional constraint products), Adams and Johnson (1994), Adams and Sherali (1984, 1986, 1990, 1991, 1993), Sherali and Brown (1994), Sherali, Krishnamurthy, and Al-Khayyal (1996), and Sherali, Ramachandran, and Kim (1994), have shown how highly effective algorithms can be constructed for various classes of discrete problems and applications. A study of special cases of this type has provided insights into useful implementation strategies based on first-order products. Additionally, techniques for generating tight valid inequalities that are implied by higher order relaxations may be devised, or explicit convex hull representations or facetial inequalities could be generated by applying the highest order RLT scheme to various subsets of sparse constraints that involve a manageable number of variables. The lattermost strategy would be a generalization of using facets of the knapsack polytope from minimal covers (see Balas and Zemel, 1978, and Zemel, 1987), so successfully implemented by Crowder et al. (1983) and Hoffman and Padberg (1991). An alternative would be to apply a higher order scheme on a subset of binary variables that turn out to be fractional in the initial linear programming solution. Letting $x_j, j \in J_f$, represent such a subset, it follows from Sherali and Adams (1989) that by deriving $X_{|J_f|}$ based on this subset, which is manageable if $|J_f|$ is relatively small, the resulting linear program will yield binary values for $x_j, \forall j \in J_f$. This would be akin to employing judicious partial convex hull representations. The development and testing of many such implementation strategies are under present study.

Finally, we address the task of solving RLT-based relaxations. While the RLT process leads to tight linear programming relaxations for the underlying discrete problem being solved as discussed above, one has to contend with the repeated solutions of such large-scale linear programs. By the nature of the RLT process, these linear programs possess a special structure induced by the replicated products of the original problem constraints (or its subset) with certain designated variables. At the same time, this process injects a high level of degeneracy in the problem since blocks of constraints automatically become active whenever the factor expression that generated them turns out to be zero at any feasible solution and the condition number of the bases can become quite large. As a result, simplex-based procedures and even interior-point methods experience difficulty in coping with such reformulated linear programs (see Adams and Sherali, 1993, for some related computational experience). On the other hand, a Lagrangian duality based scheme can not only exploit the inherent special structures, but can quickly provide near optimal primal and dual solutions that serve the purpose of obtaining tight lower and upper bounds. However, for a successful use of this technique, there are two critical issues. First, an appropriate formulation of the underlying Lagrangian dual must be constructed (see Fisher, 1981). Sherali and Myers (1989) also discuss and test various strategies and provide guidelines for composing suitable Lagrangian dual formulations. Second, an appropriate nondifferentiable optimization technique must be employed to solve the Lagrangian dual problem. For the size of problems that are encountered by us in the context of RLT, it appears imperative to use conjugate subgradient methods as in Camerini et al. (1975), Sherali and Ulular (1989), and Sherali et al. (1995), that employ higher-order information, but in a manner involving minor additional effort and storage over traditional subgradient algorithms. Since these types of algorithms are not usually dual adequate (Geoffrion, 1972), for algorithms that require primal solutions for partitioning purposes, some extra work becomes necessary. For this purpose, one can either use powerful LP solvers such as CPLEX (1990) or OB1 (Marsten, 1991) on suitable surrogate versions of the problem based on the derived dual solution, or apply primal solution recovery procedures as in Sherali and Choi (1995) along with primal penalty function techniques as in Sherali and Ulular (1989).

In closing, we anticipate that automatic reformulation techniques, such as the RLT scheme discussed in this paper will play a crucial role over the next decade in enhancing problem solving capability. Ongoing advances in solving large-scale linear programming problems will provide a further impetus to such techniques, which typically tend to derive tighter representations in higher dimensions and with additional constraints. Furthermore, newer relaxations based on semidefinite programming approaches as motivated by the work of Lovasz and Shrijver (1991), along with interior point approaches for solving such relaxations (see Overton and Wolkowicz, 1977), hold promise for future advancements.

Acknowledgement: This paper is based on excerpts from the book by the authors entitled, "A Reformulation-Linearization Technique for Solving Discrete and Continuous Nonconvex Problems," that is to be published by Kluwer Academic Publishers, 1997. Thanks are also due to the research support of the National Science Foundation under Grant Number DMI-9521398 and the Air Force Office of Scientific Researchunder Grant Number F49620-96-1-0274.

References

[1] Adams, W. P. and T. A. Johnson, *An Exact Solution Strategy for the Quadratic Assignment Problem Using RLT-Based Bounds,* Working paper, Department of Mathematical Sciences, Clemson University, Clemson, SC, 1996.

[2] Adams, W. P. and T. A. Johnson, *Improved Linear Programming-Based Lower Bounds for the Quadratic Assignment Problem,* DIMACS Series in Discrete Mathematics and Theoretical Computer Science, *Quadratic Assignment and Related Problems,* eds. P. M. Pardalos and H. Wolkowicz, 16, 43-75, 1994.

[3] Adams, W. P. and H. D. Sherali, *A Tight Linearization and an Algorithm for Zero-One Quadratic Programming Problems,* Management Science, 32(10), 1274-1290, 1986.

[4] Adams, W. P. and H. D. Sherali, *Linearization Strategies for a Class of Zero-One Mixed Integer Programming Problems,* Operations Research, 38(2), 217-226, 1990.

[5] Adams, W. P. and H. D. Sherali, *Mixed-Integer Bilinear Programming Problems,* Mathematical Programming, 59(3), 279-305, 1993.

[6] Adams, W. P., A. Billionnet, and A. Sutter, *Unconstrained 0-1 Optimization and Lagrangean Relaxation,* Discrete Applied Mathematics, 29(2-3), 131-142, 1990.

[7] Adams, W. P., J. B. Lassiter, and H. D. Sherali, *Persistency in 0-1 Optimization,* under revision for Mathematics of Operations Research, Manuscript, 1993.

[8] Balas, E., *Disjunctive Programming and a Hierarchy of Relaxations for Discrete Optimization Problems,* SIAM Journal on Algebraic and Discrete Methods, 6(3), 466-486, 1985.

[9] Balas, E. and J. B. Mazzola, *Nonlinear 0-1 Programming: I. Linearization Techniques,* Mathematical Programming, 30, 2-12, 1984a.

[10] Balas, E. and J. B. Mazzola, *Nonlinear 0-1 Programming: II. Dominance Relations and Algorithms,* Mathematical Programming, 30, 22-45, 1984b.

[11] Balas, E. and E. Zemel, *Facets of the Knapsack Polytope from Minimal Covers,* SIAM Journal of Applied Mathematics, 34, 119-148, 1978.

[12] Balas, E., S. Ceria, and G. Cornuéjols, *A Lift-and-Project Cutting Plane Algorithm for Mixed 0-1 Programs,* Mathematical Programming, 58(3), 295-324, 1993.

[13] Boros, E., Y. Crama and P. L. Hammer, *Upper Bounds for Quadratic 0-1 Maximization Problems,* RUTWR Report RRR # 14-89, Rutgers University, New Brunswick, NJ 08903, 1989.

[14] Camerini, P. M., L. Fratta, and F. Maffioli, *On Improving Relaxation Methods by Modified Gradient Techniques,* Mathematical Programming Study 3, North-Holland Publishing Co., New York, NY, 26-34, 1975.

[15] CPLEX, *Using the CPLEX Linear Optimizer,* CPLEX Optimization, Inc., Suite 279, 930 Tahoe Blvd., Bldg. 802, Incline Village, NV 89451, 1990.

[16] Crowder, H., E. L. Johnson, and M. W. Padberg, *Solving Large-Scale Zero-One Linear Programming Problems,* Operations Research, 31, 803-834, 1983.

[17] Desrochers, M. and G. Laporte, *Improvements and Extensions to the Miller-Tucker-Zemlin Subtour Elimination Constraints,* Operations Research Letters, 10(1), 27-36, 1991.

[18] Fisher, M. L., *The Lagrangian Relaxation Method for Solving Integer Programming Problems,* Management Science, 27(1), 1-18, 1981.

[19] Garfinkel, R. S. and G. L. Nemhauser, *A Survey of Integer Programming Emphasizing Computation and Relations Among Models,* In Mathematical Programming: Proceedings of an Advanced Seminar, T. C. Hu and S. Robinson (eds.), Academic Press, New York, NY, 77-155, 1973.

[20] Geoffrion, A. M., *Lagrangian Relaxation for Integer Programming,* Mathematical Programming Study 2, M. L. Balinski (ed.), North-Holland Publishing Co., Amsterdam, 82-114, 1974.

[21] Geoffrion, A. M. and R. McBryde, *Lagrangian Relaxation Applied to Facility Location Problems,* AIIE Transactions, 10, 40-47, 1979.

[22] Glover, F., *Improved Linear Integer Programming Formulations of Nonlinear Integer Problems,* Management Science, 22(4), 455-460, 1975.

[23] Hoffman, K. L. and M. Padberg, *Improving LP-Representations of Zero-One Linear Programs for Branch-and-Cut,* ORSA Journal on Computing, 3(2), 121-134, 1991.

[24] Jeroslow, R. G. and J. K. Lowe, *Modeling with Integer Variables,* Mathematical Programming Study 22, 167-184, 1984.

[25] Jeroslow, R. G. and J. K. Lowe, *Experimental Results on New Techniques for Integer Programming Formulations,* Journal of the Operational Research Society, 36, 393-403, 1985.

[26] Johnson, E. L., *Modeling and Strong Linear Programs for Mixed Integer Programming,* Algorithms and Model Formulations in Mathematical

Programming, NATO ASI 51, (ed.) S. Wallace, Springer-Verlag, 3-43, 1989.

[27] Johnson, E. L., M. M. Kostreva, and U. H. Suhl, *Solving 0-1 Integer Programming Problems Arising From Large Scale Planning Models*, Operations Research, 33(4), 803-819, 1985.

[28] Lovász, L. and A. Schrijver, *Cones of Matrices and Set Functions, and 0-1 Optimization*, SIAM J. Opt., 1, 166-190, 1991.

[29] Magnanti, T. L. and R. T. Wong, *Accelerating Benders Decomposition: Algorithmic Enhancement and Model Selection Criteria*, Operations Research, 29, 464-484, 1981.

[30] Martin, K. R., *Generating Alternative Mixed-Integer Programming Models Using Variable Redefinition*, Operations Research, 35, 820-831, 1987.

[31] Meyer, R. R., *A Theoretical and Computational Comparison of 'Equivalent' Mixed-Integer Formulations*, Naval Research Logistics Quarterly, 28, 115-131, 1981.

[32] Nemhauser, G. L. and L. A. Wolsey, *Integer and Combinatorial Optimization*, John Wiley & Sons, New York, 1988.

[33] Nemhauser, G. L. and L. A. Wolsey, *A Recursive Procedure for Generating all Cuts for Mixed-Integer Programs*, Mathematical Programming, 46, 379-390, 1990.

[34] Oley, L. A. and R. J. Sjouquist, *Automatic Reformulation of Mixed and Pure Integer Models to Reduce Solution Time in Apex IV,* Presented at the ORSA/TIMS Fall Meeting, San Diego, 1982.

[35] Overton, M. and H. Wolkowicz, "Semidefinite Programming," Mathematical Programming, 77(2), 105-110, 1997

[36] Padberg, M. W., *(1,k)-Configurations and Facets for Packing Problems*, Mathematical Programming, 18, 94-99, 1980.

[37] Padberg, M. and G. Rinaldi, *A Branch-and-Cut Algorithm for the Resolution of Large-Scale Symmetric Traveling Salesman Problems*, SIAM Review, 33, 60-100, 1991.

[38] Ramachandran, B. and J. F. Pekny, *Dynamic Factorization Methods for Using Formulations Derived from Higher Order Lifting Techniques in the Solution of the Quadratic Assignment Problem,* in State of the Art in Global Optimization, eds. C. A. Floudas and P. M. Pardalos, Kluwer Academic Publishers, 7, 75-92, 1996.

[39] Ramakrishnan, K. G., M. G. C. Resende, and P. M. Pardalos *A Branch and Bound Algorithm for the Quadratic Assignment Problem Using a Lower Bound Based on Linear Programming,* in State of the Art in Global Optimization, eds. C. A. Floudas and P. M. Pardalos, Kluwer Academic Publishers, 7, 57-74, 1996.

[40] Sherali, H. D., *On the Derivation of Convex Envelopes for Multlinear Functions,* Working Paper, Department of Industrial and Systems Engineering, Virginia Polytechnic Institute and State University, Blacksburg, VA 24061-0118, 1996.

[41] Sherali, H. D. and W. P. Adams, *A Decomposition Algorithm for a Discrete Location-Allocation Problem,* Operations Research, 32(4), 878-900, 1984.

[42] Sherali, H. D. and W. P. Adams, *A Hierarchy of Relaxations Between the Continuous and Convex Hull Representations for Zero-One Programming Problems,* SIAM Journal on Discrete Mathematics, 3(3), 411-430, 1990.

[43] Sherali, H. D. and W. P. Adams, *A Hierarchy of Relaxations and Convex Hull Characterizations for Mixed- Integer Zero-One Programming Problems,* Discrete Applied Mathematics, 52, 83-106, 1994. (Manuscript, 1989).

[44] Sherali, H. D. and E. L. Brown, *A Quadratic Partial Assignment and Packing Model and Algorithm for the Airline Gate Assignment Problem,* DIMACS Series in Discrete Mathematics and Theoretical Computer Science, *Quadratic Assignment and Related Problems,* eds. P. M. Pardalos and H. Wolkowicz, 16, 343-364, 1994.

[45] Sherali, H. D. and G. Choi, *Recovery of Primal Solutions When Using Subgradient Optimization Methods to Solve Lagrangian Duals of Linear Programs,* Operations Research Letters, 19(3), 105-113, 1996.

[46] Sherali, H. D. and P. J. Driscoll, *On Tightening the Relaxations of Miller-Tucker-Zemlin Formulations for Asymmetric Traveling Salesman Problems*, Working Paper, Department of Industrial and Systems Engineering, Virginia Polytechnic Institute and State University, Blacksburg, Virginia, 1996.

[47] Sherali, H. D. and Y. Lee, *Sequential and Simultaneous Liftings of Minimal Cover Inequalities for GUB Constrained Knapsack Polytopes*, SIAM Journal on Discrete Mathematics, 8(1), 133-153, 1995.

[48] Sherali, H. D. and Y. Lee, *Tighter Representations for Set Partitioning Problems*, Discrete Applied Mathematics, 68, 153-167, 1996.

[49] Sherali, H. D. and D. C. Myers, *Dual Formulations and Subgradient Optimization Strategies for Linear Programming Relaxations of Mixed-Integer Programs*, Discrete Applied Mathematics, 20(S-16), 51-68, 1989.

[50] Sherali, H. D. and O. Ulular, *A Primal-Dual Conjugate Subgradient Algorithm for Specially Structured Linear and Convex Programming Problems*, Applied Mathematics and Optimization, 20, 193-221, 1989.

[51] Sherali, H. D. and C. H. Tuncbilek, *A Global Optimization Algorithm for Polynomial Programming Problems Using a Reformulation- Linearization Technique*, Journal of Global Optimization, 2, 101-112, 1992.

[52] Sherali, H. D. and C. H. Tuncbilek, *A Reformulation-Convexification Approach for Solving Nonconvex Quadratic Programming Problems*, Journal of Global Optimization, 7, 1-31, 1995.

[53] Sherali, H. D. and C. H. Tuncbilek, *New Reformulation-Linearization Technique Based Relaxations for Univariate and Multivariate Polynomial Programming Problems*, Operation Research Letters, to appear, 1996.

[54] Sherali, H. D., W. P. Adams, and P. Driscoll, *Exploiting Special Structures in Constructing a Hierarchy of Relaxations for 0-1 Mixed Integer Problems*, Operations Research, to appear, 1996.

[55] Sherali, H. D., G. Choi, and C. H. Tuncbilek, *A Variable Target Value Method*, under revision for Mathematical Programming, 1995.

[56] Sherali, H. D., R. Krishnamurthy, and F. A. Al-Khayyal, *A Reformulation -Linearization Approach for the General Linear Complementarity Problem,* Presented at the Joint National ORSA/TIMS Meeting, Phoenix, Arizona, 1993.

[57] Sherali, H. D., R. S. Krishnamurthy, and F. A. Al-Khayyal, *An Enhanced Intersection Cutting Plane Approach for Linear Complementarity Problems,* Journal of Optimization Theory and Applications, to appear, 1995.

[58] Sherali, H. D., Y. Lee, and W. P. Adams, *A Simultaneous Lifting Strategy for Identifying New Classes of Facets for the Boolean Quadric Polytope,* Operations Research Letters, 17(1), 19-26, 1995.

[59] Van Roy, T. J. and L. A. Wolsey, *Solving Mixed Integer Programs by Automatic Reformulation,* Operations Research, 35, 45-57, 1987.

[60] Van Roy, T. J. and L. A. Wolsey, *Valid Inequalities for Mixed 0-1 Programs,* CORE Discussion Paper No. 8316, Center for Operations Research and Econometrics, Universite Catholique de Louvain, Belgium, 1983.

[61] Williams, H. P., *Model Building in Mathematical Programming.* John Wiley and Sons (Wiley Interscience), Second Edition, New York, NY, 1985.

[62] Wolsey, L. A. *Facets and Strong Valid Inequalities for Integer Programs,* Operations Research, 24, 367-373, 1976.

[63] Wolsey, L. A., *Strong Formulations for Mixed Integer Programming: A Survey,* Mathematical Programming, 45, 173-191, 1989.

[64] Wolsey, L. A., *Valid Inequalities for 0-1 Knapsacks and MIPs with Generalized Upper Bound Constraints,* Discrete Applied Mathematics, 29, 251-262, 1990.

Gröbner Bases in Integer Programming

Rekha R. Thomas
Department of Mathematics
Texas A&M University, College Station, TX 77843
E-mail: rekha@math.tamu.edu

Contents

1 Introduction   534

2 Parametric linear programming and
   regular triangulations   537

3 Parametric integer programming and
   Gröbner bases   542
   3.1 Algebraic fundamentals 542
   3.2 The Conti-Traverso algorithm 543
   3.3 Test sets in integer programming 546

4 Universal test sets for linear and
   integer programming   550

5 Variation of cost functions in linear and
   integer programming   558

6 Unimodular matrices   565

7 Implementation Issues   568

   References

1 Introduction

Recently, application of the theory of Gröbner bases to integer programming has given rise to new tools and results in this field. Here we present this algebraic theory as the natural integer analog of the simplex approach to linear programming. Although couched in algebra, the theory of Gröbner bases and its consequences for integer programming are intimately intertwined with polyhedral geometry and lattice arithmetic which are staples of the traditional approach to this subject.

Throughout this paper we fix a d by n integral matrix A of full row rank. Let $ker(A)$ denote the $n-d$ dimensional subspace $\{u \in \mathbf{R}^n : Au = 0\}$ and $ker_\mathbf{Z}(A)$ the saturated lattice $\{u \in \mathbf{Z}^n : Au = 0\}$ of rank $n-d$. For technical simplicity we assume that $ker(A)$ and hence $ker_\mathbf{Z}(A)$ have no non-trivial intersection with the non-negative orthant of $\mathbf{R}^n$. We often identify A with the point configuration $\mathcal{A} = \{a_1, \ldots, a_n\} \subset \mathbf{Z}^d$ where a_j is the jth column of A. Let $cone(\mathcal{A}) = \{Au : u \in \mathbf{R}^n_{\geq 0}\}$ be the d-dimensional polyhedral cone in $\mathbf{R}^d$ generated by elements of $\mathcal{A}$ and $cone_\mathbf{N}(\mathcal{A}) = \{Au : u \in \mathbf{N}^n\}$ be the additive monoid in $\mathbf{Z}^d$ generated by $\mathcal{A}$. Throughout this article, $\mathbf{N}$ denotes the non-negative integers and $\mathbf{R}_{\geq 0}$ the non-negative real numbers. For $x \in \mathbf{R}^n$, we let $supp(x) = \{i : x_i \neq 0\}$ denote the support of x.

We study $LP_{A,c}(b) := minimize \, \{c \cdot x \,:\, Ax = b, \, x \geq 0\}$, the linear program in which the *coefficient matrix* A is as above, the *right hand side vector* $b \in \mathbf{R}^d$, and the *cost vector* $c \in \mathbf{R}^n$. The set of *feasible solutions* to $LP_{A,c}(b)$ is the polyhedron $P_b = \{x \in \mathbf{R}^n_{\geq 0} : Ax = b\}$ which is non-empty if and only if b lies in $cone(\mathcal{A})$. Notice that P_b is the *fiber* of b under the linear map $\pi_A : \mathbf{R}^n_{\geq 0} \to \mathbf{R}^d$, such that $x \mapsto Ax$, i.e., $P_b = \pi_A^{-1}(b)$. We call P_b the *b-fiber* of π_A. The assumption $ker(A) \cap \mathbf{R}^n_{\geq 0} = \{0\}$ implies that P_b is a polytope (bounded polyhedron) for all $b \in cone(\mathcal{A})$. We let $LP_{A,c}$ denote the family of programs $LP_{A,c}(b)$ obtained by varying $b \in cone(\mathcal{A})$ while keeping A and c fixed, and LP_A the family obtained by keeping only A fixed. The cost vector c is *generic* with respect to LP_A if, $LP_{A,c}(b)$ has a unique optimum (which is automatically a vertex of P_b) for each $b \in cone(\mathcal{A})$.

Let $IP_{A,c}(b) := minimize \, \{c \cdot x \,:\, Ax = b, \, x \in \mathbf{N}^n\}$ be the integer program where A and c are as above. We may assume that b is integral since $IP_{A,c}(b)$ is feasible if and only if $b \in cone_\mathbf{N}(\mathcal{A})$. Let $P_b^I = conv\{x \in \mathbf{N}^n : Ax = b\}$ where $conv$ stands for "convex hull". Since $P_b^I \subseteq P_b$ and P_b is a polytope, P_b^I is again a polytope for each $b \in cone_\mathbf{N}(\mathcal{A})$. By an abuse of nomenclature, we call P_b^I the *b-fiber* of π_A^I where $\pi_A^I : \mathbf{N}^n \to \mathbf{Z}^d$, $x \mapsto Ax$. As above, we let $IP_{A,c}$ denote the family of programs obtained by varying

$b \in cone_{\mathbf{N}}(\mathcal{A})$ while keeping A and c fixed and IP_A the family obtained by keeping only A fixed. The cost vector c is generic with respect to IP_A if $IP_{A,c}(b)$ has a unique optimum (which is automatically a vertex of P_b^I) for each $b \in cone_{\mathbf{N}}(\mathcal{A})$. In Section 5 we will see that if c is generic for IP_A then it is also generic for LP_A but not conversely, in general.

We now recall some polyhedral facts that will be needed. The reader is refered to the books [24] and [33] for details. A *polyhedral complex* Δ is a finite collection of polyhedra such that (i) if $P \in \Delta$, then every face of P is in Δ and (ii) if $P_1, P_2 \in \Delta$, then $P_1 \cap P_2$ is a face of P_1 and P_2. Elements of Δ are called *cells* or *faces* of Δ and the *support* of Δ is the union of all faces in Δ. A polyhedral complex is completely specified by its maximal (with respect to inclusion) cells. If all cells in Δ are cones, then Δ is called a *polyhedral fan*. A fan in $\mathbf{R}^n$ is *complete* if its support is all of $\mathbf{R}^n$.

For a polyhedron $P \subseteq \mathbf{R}^n$ and cost vector $c \in \mathbf{R}^n$ we write $face_c(P)$ for the face of P at which c gets minimized, i.e., $face_c(P) = \{v \in P : c \cdot v \leq c \cdot u, \forall u \in P\}$. If F is any face of P, then $\mathcal{N}(F;P)$ denotes the cone of (inner) normals, called the *inner normal cone* of P at F. In symbols, $\mathcal{N}(F;P) = \{c \in \mathbf{R}^n : c \cdot x \leq c \cdot y \text{ for all } x \in F, y \in P\}$. The collection of cones $\mathcal{N}(F;P)$ is denoted $\mathcal{N}(P)$ and called the (inner) *normal fan* of P. The normal fan of P is a polyhedral fan in $\mathbf{R}^n$ that is complete if and only if P is a polytope.

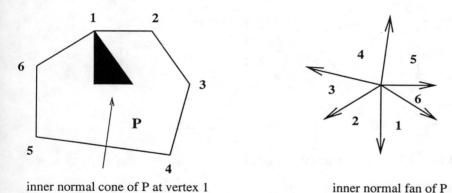

inner normal cone of P at vertex 1   inner normal fan of P

Figure 1: Inner normal cone and fan.

We say that two polytopes are *normally equivalent* if they have the same normal fan. Given two polytopes P and Q in $\mathbf{R}^n$, their *Minkowski sum* is the polytope $P + Q = \{p + q : p \in P, q \in Q\} \subset \mathbf{R}^n$ and P and Q are called *Minkowski summands* of $P + Q$. For all $c \in \mathbf{R}^n$, $face_c(P + Q) =$

$face_c(P) + face_c(Q)$. This implies that every vertex of $P + Q$ is a sum of vertices of the summands and that every edge of $P+Q$ is a parallel translate of an edge of some summand. The definition of Minkowski sum extends to the case of finitely many summands and as in the usual extension of addition to integration, the operation of taking Minkowski sums of finitely many polytopes extends naturally to the operation of taking *Minkowski integrals* of infinitely many polytopes. See [5] for details. The common *refinement* of two fans $\mathcal{F}$ and $\mathcal{G}$ in $\mathbf{R}^n$, denoted $\mathcal{F} \cap \mathcal{G}$, is the fan of all intersections of cones from $\mathcal{F}$ and $\mathcal{G}$. We say that $\mathcal{F} \cap \mathcal{G}$ is a refinement of $\mathcal{F}$ (respectively $\mathcal{G}$). The following are two useful facts in this context: (i) for polytopes P and Q in $\mathbf{R}^n$, the fan $\mathcal{N}(P + Q) = \mathcal{N}(P) \cap \mathcal{N}(Q)$ and (ii) the fan $\mathcal{N}(P)$ is a refinement of $\mathcal{N}(Q)$ if and only if λQ is a Minkowski summand of P for some positive real number λ. For a hyperplane $H = \{x \in \mathbf{R}^n : ax = 0\}$ in $\mathbf{R}^n$, let H^+ denote the closed half space $\{x \in \mathbf{R}^n : ax \geq 0\}$ and H^- denote $\{x \in \mathbf{R}^n : ax \leq 0\}$. A *hyperplane arrangement* in $\mathbf{R}^n$ is the common refinement of finitely many fans of the form $\{H^+, H^-\}$. The arrangement is usually specified by listing the associated hyperplanes. The Minkowski sum of finitely many line segments is called a *zonotope* and by (i), its normal fan is a hyperplane arrangement.

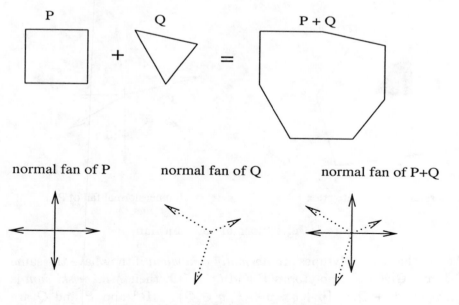

Figure 2: Minkowski sum of polytopes.

Gröbner Bases in Integer Programming 537

This article is organized as follows: In Section 2 we introduce the *regular triangulation* Δ_c of $\mathcal{A}$ with respect to c and show that Δ_c determines the optimal solutions of programs in $LP_{A,c}$. This is a reformulation of the usual duality theory for linear programming. We also introduce *test sets* for linear programming which give rise to a generalization of the simplex algorithm for programs in $LP_{A,c}$. In Section 3 we develop the Gröbner basis algorithm for $IP_{A,c}$ and examine the underlying geometry. Section 4 constructs and studies *universal test sets* for both LP_A and IP_A. In Section 5 we introduce the *secondary* and *state* polytopes associated with A. The normal fans of these polytopes model the effect of varying cost functions in linear and integer programming. The lay out of Sections 2 to 5 is strongly guided by the intention to reaffirm the philosophy that integer programming is an arithmetic refinement of linear programming. Section 6 examines unimodular matrices and we conclude in Section 7 with a brief discussion of implementation issues associated with solving integer programs using Gröbner bases.

2 Parametric linear programming and regular triangulations

In this section we study the family of linear programs $LP_{A,c}$ from the point of view of regular triangulations ([3], [4] and [16]) and test sets [24]. We identify the point configuration $\mathcal{A} = \{a_1, \ldots, a_n\}$ with the index set $\{1, \ldots, n\}$. A *subdivision* of $\mathcal{A}$ is a collection of subsets σ of $\{1, \ldots, n\}$ such that the cones $cone(\sigma) = cone(\{a_i : i \in \sigma\})$ form a *polyhedral fan* with support $cone(\mathcal{A})$. If $cone(\sigma)$ is k dimensional, then σ is called a k-cell of the subdivision. A subdivision is a *triangulation* if for each cell σ, $cone(\sigma)$ is simplicial.

Every cost vector $c \in \mathbf{R}^n$ induces a subdivision Δ_c of $\mathcal{A}$ as follows: $\{i_1, \ldots, i_k\}$ is a cell of Δ_c if and only if there exists a row vector $y \in \mathbf{R}^d$ such that $y \cdot a_j = c_j$ if $j \in \{i_1, \ldots, i_k\}$ and $y \cdot a_j < c_j$ if $j \in \{1, \ldots, n\} \setminus \{i_1, \ldots, i_k\}$. Subdivisions obtained in this way are called *regular* (or *coherent*).

The regular subdivision Δ_c can also be constructed geometrically:

1. Let $\hat{\mathcal{A}} = \{(a_1, c_1), (a_2, c_2), \ldots, (a_n, c_n)\} \subset \mathbf{R}^{d+1}$ be the point configuration obtained by lifting each point a_i of $\mathcal{A}$ to height c_i in one higher dimensional space.

2. The set of "lower" faces of $cone(\hat{\mathcal{A}})$ form a d-dimensional polyhedral complex on $\{1, \ldots, n\}$: a lower face is represented by the set of indices of points in $\hat{\mathcal{A}}$ that lie on it. The set of these faces is Δ_c. Equivalently, Δ_c is the projection onto $cone(\mathcal{A})$ of the lower faces of $cone(\hat{\mathcal{A}})$.

A face is "lower" if it has a normal vector with negative last coordinate. One should verify that the above construction gives Δ_c. The algebraic theory of Gröbner bases for integer programming is especially suited for configurations $\mathcal{A}$ whose elements span an affine hyperplane in $\mathbf{R}^d$: the toric ideal I_A constructed in Section 3 is *homogeneous* in this case. It then suffices to construct $conv(\hat{\mathcal{A}})$ instead of $cone(\hat{\mathcal{A}})$ and project the lower faces of the $(d+1)$-polytope $conv(\hat{\mathcal{A}})$ onto $conv(\mathcal{A})$, to obtain Δ_c.

Example 2.1 Let $\mathcal{A} = \{1, 4, 6, 10\} \subset \mathbf{R}$ and $c = (5, 1, 3, 5)$. Then $cone(\mathcal{A})$ is the non-negative real line and the unique lower face of $cone(\hat{\mathcal{A}})$ is the extreme ray generated by (a_2, c_2). Hence $\Delta_c = \{\{2\}\}$.

Example 2.2 Let $\mathcal{A} = \{(1, 0, 0), (1, 1, 0), (1, 2, 1), (1, 1, 2), (1, 0, 1)\}$, be the three dimensional configuration whose affine span is the hyperplane $x_1 = 1$ in $\mathbf{R}^3$. Then $conv(\mathcal{A})$ is a pentagon with five distinct regular triangulations that may be represented as $\Delta_{(2,3,5,0,0)}$, $\Delta_{(3,2,1,0,0)}$, $\Delta_{(2,3,1,0,0)}$, $\Delta_{(0,1,0,1,1)}$ and $\Delta_{(0,1,2,0,1)}$. Notice that the cost vector $(0, 0, 0, 0, 1)$ is not generic since $\Delta_{(0,0,0,0,1)} = \{\{1, 4, 5\}, \{1, 2, 3, 4\}\}$ is not a triangulation. It may be refined to either $\Delta_{(0,1,0,1,1)} = \{\{1, 4, 5\}, \{1, 3, 4\}, \{1, 2, 3\}\}$ or $\Delta_{(0,1,2,0,1)} = \{\{1, 4, 5\}, \{1, 2, 4\}, \{2, 3, 4\}\}$. Figure 3 shows the five triangulations and the above sudivision of $\mathcal{A}$.

We now wish to relate the regular subdivision Δ_c to the optimal solutions of programs in the family $LP_{A,c}$. Let $LP_{A,c}(b)^{dual}$ denote the linear program dual to $LP_{A,c}(b)$:

$$\text{maximize}\{y \cdot b : y \cdot A \leq c, \ y \in \mathbf{R}^d\},$$

with feasible region $Q_c = \{y \in \mathbf{R}^d : yA \leq c\}$.

Theorem 2.3 *For each $b \in cone(\mathcal{A})$, a feasible solution x to $LP_{A,c}(b)$ is optimal if and only if the support of x, $supp(x)$, is a subset of a cell of Δ_c.*

Proof. Let x be an optimal solution of $LP_{A,c}(b)$ and y an optimal solution of $LP_{A,c}(b)^{dual}$. By complementary slackness, $x_j > 0$ implies $y \cdot a_j = c_j$, which means that $supp(x)$ lies in a face of Δ_c. Conversely, let x be any feasible solution to $LP_{A,c}(b)$ such that $supp(x)$ is a subset of a face of Δ_c. Then there exists $y \in \mathbf{R}^d$ with $supp(x) \subseteq \{j : y \cdot a_j = c_j\}$. This implies $c \cdot x = y \cdot A \cdot x = y \cdot b$ and hence, x is an optimal solution of $LP_{A,c}(b)$. $\square$

Gröbner Bases in Integer Programming

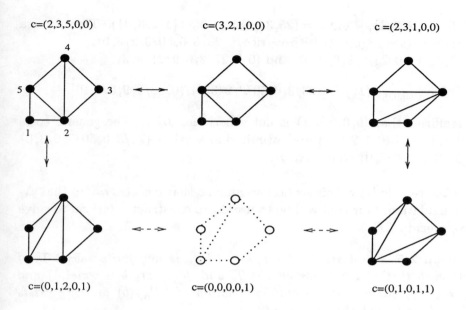

Figure 3: Regular triangulations and subdivision of $conv(\mathcal{A})$.

Corollary 2.4 *For a cost vector $c \in \mathbf{R}^n$, the following are equivalent:*
(a) c is generic with respect to LP_A
(b) Δ_c is a triangulation
(c) Q_c is a simple polyhedron.

Proof. (a) $\Leftrightarrow$ (b): The subdivision Δ_c is a triangulation if and only if for each $b \in cone(\mathcal{A})$, the smallest cone in Δ_c containing b is simplicial. The latter condition is equivalent to the existence, for each $b \in cone(\mathcal{A})$, of a unique point in P_b with support in a cell of Δ_c. By Theorem 2.3 this is the unique optimal solution to $LP_{A,c}(b)$. Therefore Δ_c is a triangulation if and only if c is generic with respect to LP_A.
(b) $\Leftrightarrow$ (c): By definition, $F \subseteq \{1, \ldots, n\}$ is a face of Δ_c if and only if $\{y \in \mathbf{R}^d : a_j \cdot y = c_j \, \forall j \in F, a_j \cdot y < c_j \, \forall j \notin F\}$ is non-empty. But every set of this form is a, possibly empty, face of Q_c. Using the relationship between the dimension of the face given by F and the cardinality of F, it can be seen that Δ_c is polar to the *boundary complex* of Q_c, and Δ_c is a triangulation if and only if Q_c is a simple polyhedron. (The boundary complex of a polyhedron P is the collection (complex) of all proper faces of P.) $\square$

Example 2.2 continued. Consider the regular subdivision $\Delta_{(0,0,0,0,1)} =$

$\{\{1,4,5\},\{1,2,3,4\},\}$ and $b = (25, 34, 18) \in cone(\{1, 2, 3, 4\})$. Then P_b is a quadrilateral (see Figure 4) with vertices: $(23/3, 0, 50/3, 2/3, 0)$, $(7, 0, 17, 0, 1)$, $(0, 23/2, 9, 9/2, 0)$, and $(0, 7, 27/2, 0, 9/2)$ with

$$face_{(0,0,0,0,1)}(P_b) = [(23/3, 0, 50/3, 2/3, 0), (0, 23/2, 9, 9/2, 0)].$$

This reaffirms that $(0, 0, 0, 0, 1)$ is not generic for LP_A. The generic costs $(0, 1, 0, 1, 1)$ and $(0, 1, 2, 0, 1)$ are optimized at vertices $(23/3, 0, 50/3, 2/3, 0)$ and $(0, 23/2, 9, 9/2, 0)$ respectively.

For the remainder of this section we assume that c is generic, so that Δ_c is a triangulation. Our goal will be to use Δ_c to construct a *test set* to solve all programs in $LP_{A,c}$.

Definition 2.5 *A test set for the family $LP_{A,c}$ is any finite subset $\mathcal{T}_c$ of $ker(A)$ such that $c \cdot t > 0$ for all $t \in \mathcal{T}_c$ and, for every $b \in cone(\mathcal{A})$ and every $x \in P_b$, either x is the optimal solution of $LP_{A,c}(b)$ or there exists $t \in \mathcal{T}_c$ and $\epsilon > 0$ such that $x - \epsilon t \geq 0$.*

A test set is *minimal* if it has minimal cardinality. We say that $I \subset \{1, \ldots, n\}$ is a *minimal non-face* of Δ_c if I is not a face of Δ_c but every proper subset of I is a face of Δ_c.

Proposition 2.6 *A finite subset $\mathcal{T}_c \subset ker(A)$ is a minimal test set for $LP_{A,c}$ if and only if for every minimal non-face I of Δ_c there is a unique vector $t \in \mathcal{T}_c$ such that $I = \{i : t_i > 0\}$.*

Proof. By Theorem 2.3, a feasible solution x of $LP_{A,c}(b)$ is non-optimal if and only if $supp(x)$ contains a minimal non-face of Δ_c. Hence $\mathcal{T}_c \subset ker(A)$ is a minimal test set for $LP_{A,c}$ if and only if for every minimal non-face I of Δ_c there is a unique vector $t \in \mathcal{T}_c$ such that $I = \{i : t_i > 0\}$. This property of $\mathcal{T}_c$ is necessary and sufficient to guarantee the existence of an improving direction in $\mathcal{T}_c$ for every non-optimal solution to a program in $LP_{A,c}$. □

Example 2.2 continued. For the generic cost vector $c = (0, 1, 0, 1, 1)$, we had $\Delta_c = \{\{1, 4, 5\}, \{1, 2, 3\}, \{1, 3, 4\}\}$. The minimal non-faces of Δ_c are $\{2, 5\}$, $\{2, 4\}$ and $\{3, 5\}$. A minimal test set for $LP_{A,c}$ is therefore, $\mathcal{T}_c = \{2e_2 + e_5 - 2e_1 - e_3,\ 3e_2 + e_4 - 2e_1 - 2e_3,\ e_3 + 3e_5 - 2e_1 - 2e_4\} \subset ker(A)$ where e_i denotes the ith unit vector in $\mathbf{R}^n$.

Gröbner Bases in Integer Programming 541

We may assume without loss of generality that $T_c \subset ker_{\mathbf{Z}}(A)$. It is sometimes convenient to view $u \in T_c$ as the line segment $[u^+, u^-]$ where u^+ and u^- are the unique vectors in $\mathbf{N}^n$ of disjoint supports such that $u = u^+ - u^-$. Since $c \cdot u^+ > c \cdot u^-$, we may assume that $[u^+, u^-]$ is directed from u^+ to u^-. Then solving $LP_{A,c}(b)$ using T_c is a generalization of the simplex algorithm for this linear program since in this method, the directed segments of T_c trace a monotone path from every non-optimal solution of the program to the optimum that stays entirely within P_b. Unlike in the simplex method, this path is not required to be along the edge-skeleton of P_b. Figure 4 is an illustration of this method on $LP_{A,c}(b)$ where A is as in Example 2.2, $c = (0, 1, 0, 1, 1)$, $b = (25, 34, 18)$ and T_c as above. All objects have been projected onto the x_4, x_5 plane. The dotted line shows a monotone path from the initial solution $(0, 7, 27/2, 0, 9/2)$ to the optimum $(23/3, 0, 50/3, 2/3, 0)$ traced by the directed segments of T_c.

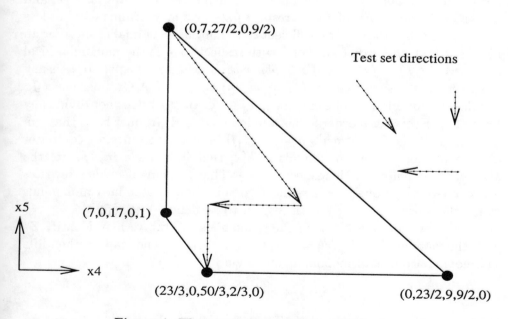

Figure 4: The generalized simplex method.

3 Parametric integer programming and Gröbner bases

3.1 Algebraic fundamentals

Our goal in this section is to present the Gröbner basis algorithm for integer programs in the family $IP_{A,c}$, due to Conti and Traverso [8]. (See also Section 2.8 in [1].) We use this subsection to introduce the essential algebraic notions that will be needed. We refer the reader to [1] and [11] for the theory of Gröbner bases and to [27] for an algebraic development of the connections between Gröbner bases and convex polytopes in general.

Let $k[\mathbf{x}] = k[x_1, \ldots, x_n]$ be the polynomial ring in the variables $x_1, \ldots, x_n$ where k is any field. We identify a monomial $x^\alpha = x_1^{\alpha_1} x_2^{\alpha_2} \cdots x_n^{\alpha_n} \in k[\mathbf{x}]$ with its exponent vector $\alpha = (\alpha_1, \ldots, \alpha_n) \in \mathbf{N}^n$. A *term order* $\succ$ on $\mathbf{N}^n$ is a total order on $\mathbf{N}^n$ such that $0 \prec \alpha$ for all $\alpha \in \mathbf{N}^n \backslash \{0\}$ and $\alpha + \delta \succ \beta + \delta$ for all $\delta \in \mathbf{N}^n$ whenever $\alpha \succ \beta$. For a polynomial $f \in k[\mathbf{x}]$, there exists a unique term $c_\alpha x^\alpha$ in f such that x^α has the most expensive exponent vector with respect to $\succ$, among all monomials in f. Then $c_\alpha x^\alpha$ is the *initial term* and x^α the *initial monomial* of f with respect to $\succ$. We may assume without loss of generality that the scalar coefficient of the initial monomial is one. For an ideal $I \subseteq k[\mathbf{x}]$, the *initial ideal* of I with respect to $\succ$ is the monomial ideal $in_\succ(I) = \langle in_\succ(f) : f \in I\rangle$ and a *Gröbner basis* for I with respect to $\succ$ is any finite set $\mathcal{G}_\succ \subset I$ such that $in_\succ(I) = \langle in_\succ(g) : g \in \mathcal{G}_\succ\rangle$. A Gröbner basis $\mathcal{G}_\succ$ is *reduced* if for any pair of elements $g_i, g_j \in \mathcal{G}_\succ$, $in_\succ(g_j)$ does not divide any term of g_i. Every term order $\succ$ has a unique reduced Gröbner basis that can be computed via *Buchberger's algorithm* [7] implemented in most computer algebra packages. The monomials in $k[\mathbf{x}]$ that lie outside $in_\succ(I)$ are the *standard monomials* with respect to $\succ$. The *division algorithm* rewrites every $f \in k[\mathbf{x}]$ uniquely as a k-linear combination of standard monomials called the *normal form* of f with respect to $\succ$, denoted $nf_\succ(f)$.

Just as we identify $\mathbf{N}^n$ with the monomials in $k[\mathbf{x}]$ we may identify $\mathbf{Z}^d$ with the monomials in $k[\mathbf{t}^{\pm 1}] = k[t_1, t_1^{-1}, \ldots, t_d, t_d^{-1}]$. The map π_A^I then lifts to a homomorphism of monoid algebras via

$$\hat{\pi}_A^I : k[\mathbf{x}] \to k[t^{\pm 1}], \quad x_j \mapsto t^{a_j},$$

where a_j is the jth column of A. The kernel of $\hat{\pi}_A^I$, denoted I_A, is a prime ideal in $k[\mathbf{x}]$ called the *toric ideal* of A, a name that follows from the fact that the *affine variety* $V(I_A) \subset k^n$ is a (not necessarily *normal*) toric variety. For more details of this connection see [15] and [27].

Clearly, if $u \in ker_{\mathbf{Z}}(A)$ then the *binomial* $x^{u^+} - x^{u^-} \in I_A$. Conversely, every binomial in I_A is of the form $x^{u^+} - x^{u^-}$ for some $u \in ker_{\mathbf{Z}}(A)$.

Lemma 3.1 *The toric ideal* $I_A = \langle x^{u_i^+} - x^{u_i^-} : u_i \in ker_{\mathbf{Z}}(A), i = 1, \ldots, t \rangle.$

Sketch of proof. The ideal I_A is generated as a k-vector space and hence as an ideal by the infinite set of binomials $\{x^{u^+} - x^{u^-} : u \in ker_{\mathbf{Z}}(A)\}$. Therefore, by the Hilbert basis theorem, I_A has a finite generating set of the form $\{x^{u_i^+} - x^{u_i^-} : u \in ker_{\mathbf{Z}}(A), i = 1, \ldots, t\}$ for some $t \in \mathbf{N}\setminus\{0\}$. □

The mechanics of the Buchberger algorithm assures that if a set of binomials is input to the algorithm then, regardless of the term order, every intermediate polynomial created during the course of the algorithm as well as the Gröbner basis that is output, again consists of binomials. For the toric ideal I_A, the following stronger result holds.

Corollary 3.2 *Every reduced Gröbner basis of* I_A *consists of a finite set of binomials of the form* $x^{u^+} - x^{u^-}$, *where* $u \in ker_{\mathbf{Z}}(A)$.

Corollary 3.3 *While dividing (reducing) a monomial* $x^p \in k[\mathbf{x}]$ *by a reduced Gröbner basis of* I_A, *every remainder formed (and in particular the normal form of* x^p*) is a monomial whose exponent vector lies in* P_{Ap}^I.

Proof. An element $x^{u^+} - x^{u^-} \in \mathcal{G}_\succ$ divides a monomial x^p if and only if the initial monomial x^{u^+} divides x^p. The remainder is $(x^p/x^{u^+}) \cdot x^{u^-} = x^{p-u}$ which is again a monomial in $k[\mathbf{x}]$ since $p \geq u^+$ and hence $p - u \in \mathbf{N}^n$. Further, since $Au = 0$, it follows that $Ap = A(p - u)$. □

3.2 The Conti-Traverso algorithm

The Conti-Traverso algorithm requires that the cost vector c in $IP_{A,c}$ is generic with respect to IP_A. Every vector $c \in \mathbf{R}^n$ orders the points in $\mathbf{N}^n$ via the inner product $c \cdot x$. Whenever this is a partial order, we fix a term order $\succ$ to break ties and denote by $\succ_c$, the refinement of c by $\succ$. The total order $\succ_c$ is such that $\alpha \succ_c \beta$ if either $c \cdot \alpha > c \cdot \beta$ or $c \cdot \alpha = c \cdot \beta$ and $\alpha \succ \beta$. The optimal cost value of both $IP_{A,c}(b)$ and $IP_{A,\succ_c}(b)$ are the same, but we now have the advantage that $\succ_c$ is generic with respect to IP_A. Since every non-generic $c \in \mathbf{R}^n$ can be made generic as above without affecting the optimal value of the programs in $IP_{A,c}$, we assume in the rest of this subsection that c is generic with respect to IP_A.

In the first step of the algorithm, we need to compute the reduced Gröbner basis $\mathcal{G}_c$ of the toric ideal I_A. In usual Gröbner basis computations one requires a term order as input to the Buchberger algorithm. However, the order induced by c is often not a term order since it can fail to satisfy the condition $c \cdot \alpha > c \cdot 0 = 0$ for all non-zero $\alpha \in \mathbf{N}^n$. It is this property that usually guarantees that the Buchberger algorithm will terminate after finitely many steps. In our situation, the termination of the Buchberger algorithm follows from the assumption that P_b^I is a polytope for all $b \in cone_{\mathbf{N}}(\mathcal{A})$.

Algorithm 3.4 The Conti-Traverso algorithm for $IP_{A,c}$.
(i) Compute the reduced Gröbner basis $\mathcal{G}_c$ of I_A.
(ii) To solve $IP_{A,c}(b)$ for $b \in cone_{\mathbf{N}}(\mathcal{A})$, find any solution u of the program and compute $x^{u'} = nf_c(x^u)$. Then u' is the unique optimum of $IP_{A,c}(b)$.

Proof of Algorithm. Let $\mathcal{G}_c = \{x^{\alpha_i^+} - x^{\alpha_i^-}, i = 1, \ldots, p\}$ where the underlined term is the initial monomial of each binomial. By Corollary 3.3 u' is a solution to $IP_{A,c}(b)$. Suppose there exists $v \in P_b^I \cap \mathbf{N}^n$ such that $c \cdot u' > c \cdot v$. Then $x^{u'}$ is the initial term of $\underline{x}^{u'} - x^v \in I_A$ with respect to c and hence there exists a binomial in $\mathcal{G}_c$ whose initial monomial divides $x^{u'}$. This contradicts that $x^{u'}$ is the normal form of x^u with respect to $\mathcal{G}_c$. □

Example 2.2 continued. For $A = \begin{pmatrix} 1 & 1 & 1 & 1 & 1 \\ 0 & 1 & 2 & 1 & 0 \\ 0 & 0 & 1 & 2 & 1 \end{pmatrix}$ and the cost vector $c = (0, 1, 0, 1, 1)$ which is generic with respect to IP_A,

$$\mathcal{G}_{(0,1,0,1,1)} = \{\underline{x_3 x_5^3} - x_1^2 x_4^2, \underline{x_2 x_5^2} - x_1^2 x_4, \underline{x_2 x_4} - x_3 x_5, \underline{x_2^2 x_5} - x_1^2 x_3\}$$

where the underlined terms are the initial terms. For $b = (25, 34, 18)$, the point $(1, 10, 10, 4, 0)$ is a solution of $IP_{A,c}(b)$. The normal form of the monomial $x_1 x_2^{10} x_3^{10} x_4^4$ with respect to $\mathcal{G}_c$ is $x_1^7 x_3^{17} x_5$ and hence the optimal solution of $IP_{A,c}(b)$ is $(7, 0, 17, 0, 1)$. A possible reduction path of $x_1 x_2^{10} x_3^{10} x_4^4$ is given by the following chain of monomials:

$$x_1 x_2^{10} x_3^{10} x_4^4 \xrightarrow[\text{4 times}]{x_2 x_4 - x_3 x_5} x_1 x_2^6 x_3^{14} x_4^4 x_5^4 \xrightarrow[\text{3 times}]{x_2^2 x_5 - x_1^2 x_3} x_1^7 x_3^{17} x_5. \quad \square$$

There are a number of issues that have to be dealt with while implementing the above version of the Conti-Traverso algorithm. In particular, finding a generating set for I_A to be used as input to the Buchberger algorithm in Step (i) and finding an initial solution u to $IP_{A,c}(b)$ are both non-trivial.

(We will briefly address the former problem in Section 7.) Once these two issues are taken care of, the reduced Gröbner basis $\mathcal{G}_c$ and the normal form of x^u can be computed using a software package that does Gröbner basis computations like Macaulay 2 [18] or GRIN [21]. We note that Algorithm 3.4 is a condensed version of the original Conti-Traverso algorithm in [8], useful for highlighting the main computational steps involved. The original algorithm uses a single Gröbner basis computation on a larger ideal to bypass the computation of generators for I_A. Then a reduction phase takes care of finding both an initial solution u and the optimal solution u' of $IP_{A,c}(b)$.

Algorithm 3.5 The Original Conti-Traverso Algorithm.
Consider the ideal $J = \langle x_j t^{a_j^-} - t^{a_j^+}, j = 1,\ldots,n, t_0 t_1 \cdots t_d - 1 \rangle$ in the polynomial ring $k[x_1,\ldots,x_n,t_0,t_1,\ldots,t_d]$. Let $t_{\bar{0}} = \{t_1,\ldots,t_d\}$.
(i) Compute the reduced Gröbner basis $\mathcal{G}_{\succ'}$ of J with respect to any *elimination* term order $\succ'$ such that $\{t_0,t_1,\ldots,t_d\} \succ' \{x_1,\ldots,x_n\}$ and $\succ'$ restricted to $k[\mathbf{x}]$ induces the same total order as c.
(ii) In order to solve $IP_{A,c}(b)$, form the monomial $t^b = t_0^\beta t_{\bar{0}}^{b+\beta(e_1+\cdots+e_d)}$ where $\beta = max\{|b_j| : b_j < 0\}$ and e_i is the i-th unit vector in $\mathbf{R}^d$. Compute $nf_{\succ'}(t^b) = t^\gamma x^{u'}$. If $\gamma = 0$ then $IP_{A,c}(b)$ is feasible with optimal solution u'. Else $IP_{A,c}(b)$ is infeasible.

Proof. The ideal $J = \langle x_j t^{a_j^-} - t^{a_j^+}, j = 1,\ldots,n, t_0 t_1 \cdots t_d - 1 \rangle$ has the property that $J \cap k[\mathbf{x}] = I_A$, which in turn implies that $\mathcal{G}_{\succ'} \cap k[\mathbf{x}] = \mathcal{G}_c$. For a proof of this fact see Theorem 2 in Section 2.2 of [11]. Therefore, Step (i) of Algorithm 3.5 indirectly achieves Step (i) of Algorithm 3.4. Since $\succ'$ is an elimination order with $t \succ' x$, once a monomial in $k[\mathbf{x}]$ has been encountered during the reduction of t^b, all subsequent monomials also lie in $k[\mathbf{x}]$ and their exponent vectors are all solutions to $IP_{A,c}(b)$. The algorithm will reduce t^b to a monomial $x^u \in k[\mathbf{x}]$ if and only if $IP_{A,c}(b)$ is feasible. The *only if* direction is clear. To see the *if* direction, suppose the normal form $t^\gamma x^{u'}$ of t^b has $\gamma \neq 0$ and $IP_{A,c}(b)$ has a solution v. Then since $\succ'$ was an elimination order with $\{t_0,t_1,\ldots,t_d\} \succ' \{x_1,\ldots,x_n\}$, the binomial $t^\gamma x^{u'} - x^v \in J$ has $t^\gamma x^{u'}$ as initial term. This contradicts that $t^\gamma x^{u'}$ is the normal form of t^b with respect to $\mathcal{G}_{\succ'}$. Hence, if $\gamma \neq 0$ we may conclude that $IP_{A,c}(b)$ is infeasible. However, if $\gamma = 0$, by the same argument as in the proof of Algorithm 3.4, u' is the optimal solution to $IP_{A,c}(b)$. □

If all entries of the matrix A are non-negative, we do not need the variable t_0 and the binomial $t_0 t_1 \cdots t_d - 1$ in the above computation. We refer the reader to [8] for computational tests of the above algorithm. Algorithm 3.5 is like a "Phase 1 - Phase 2" procedure for $IP_{A,c}(b)$ while Algorithm 3.4 is just "Phase 2". Building the ideal J is equivalent to considering the "extended" integer program :

$$minimize\{M \cdot t + c \cdot x : -\mathbf{1} \cdot t_0 + I \cdot t_{\overline{0}} + A \cdot x = b,\, t, x \geq 0,\, integer\}$$

for which $(t_0, t_{\overline{0}}) = (\beta, b + \beta(e_1 + \cdots + e_d)), x = 0$ is a solution. Here M is a vector with sufficiently large entries, I is the $d \times d$ identity matrix and $-\mathbf{1}$ is a column vector of minus ones. In the reduction step, we start at the above "artificial solution" of $IP_{A,c}(b)$ and reduce this whenever $IP_{A,c}(b)$ is feasible, to an initial solution of $IP_{A,c}(b)$ and then to the optimum of $IP_{A,c}(b)$.

3.3 Test sets in integer programming

We now examine the Conti-Traverso algorithm and establish the integer analogs of the results in Section 2. For a general $\omega \in \mathbf{R}^n$ and a polynomial $f \in k[\mathbf{x}]$, we define the *initial form* of f with respect to ω, denoted $in_\omega(f)$, to be the sum of all terms in f whose monomials are of maximal ω-weight. For an ideal $I \subset k[\mathbf{x}]$ the initial ideal of I with respect to ω is $in_\omega(I) = \langle in_\omega(f) : f \in I \rangle$ which in general may not be a monomial ideal (ideal generated by monomials). The reduced Gröbner basis $\mathcal{G}_w$ may be defined as before and all monomials that lie outside $in_\omega(I)$ are said to be standard with respect to ω. The following theorem is the integer analog to Theorem 2.3.

Theorem 3.6 *For each $b \in cone_\mathbf{N}(\mathcal{A})$, a feasible solution u to $IP_{A,c}(b)$ is optimal if and only if x^u is standard with respect to c.*

Proof. A solution u to $IP_{A,c}(b)$ is non-optimal if and only if $nf_c(x^u) \neq x^u$ which happens if and only if $x^u \in in_c(I_A)$. □

For a generic cost vector c with reduced Gröner basis $\mathcal{G}_c = \{\underline{x}^{\alpha_i^+} - x^{\alpha_i^-} : i = 1, \ldots, p\}$, the initial ideal $in_c(I_A) = \langle x^{\alpha_i^+} : i = 1, \ldots, p \rangle$ is a monomial ideal. The monomials in $in_c(I_A)$ are in bijection with the lattice points in the "staircase" like subset $\mathcal{S}_c = \cup_{i=1}^p (\alpha_i^+ + \mathbf{N}^n)$ of $\mathbf{N}^n$. The optimal solutions to the programs in $IP_{A,c}$ are precisely the lattice points of $\mathbf{N}^n$ that do not lie in $\mathcal{S}_c$ and there is precisely one optimal solution per program.

Corollary 3.7 *A cost vector $c \in \mathbf{R}^n$ is generic with respect to IP_A if and only if $in_c(I_A)$ is a monomial ideal.*

Proof. A cost vector c is not generic with respect to IP_A if and only if there exists a minimal generator of the form $x^{u^+} - x^{u^-}$ for $in_c(I_A)$. Such a generator would have $c \cdot u^+ = c \cdot u^-$ and would allow multiple optimal solutions in certain fibers of π_A^I. □

Example 2.2 continued. For the non-generic cost vector $c = (0,0,0,0,1)$, $in_c(I_A) = \langle x_3 x_5, x_2 x_5^2, x_2^2 x_5, x_2^3 x_4 - x_1^2 x_3^2 \rangle$. For $b = (25, 34, 18)$, P_b^I is a quadrilateral with vertices: $(7,0,17,0,1)$, $(7,1,6,1,0)$, $(1,10,10,4,0)$ and $(1,6,14,0,4)$ with $face_{(0,0,0,0,1)}(P_b) = [(7,1,6,1,0), (1,10,10,4,0)]$. □

For the rest of this section we assume that c is generic with respect to IP_A.

Definition 3.8 *A set $\mathcal{R}_c \subseteq ker_{\mathbf{Z}}(A)$ is called a test set for the family $IP_{A,c}$ if $c \cdot r > 0$ for all $r \in \mathcal{R}_c$ and, for each non-optimal solution u to a program $IP_{A,c}(b)$, there exists $r \in \mathcal{R}_c$ such that $u - r \in P_b^I$.*

Existence of a finite test set $\mathcal{R}_c$ for $IP_{A,c}$ implies a trivial solution method for all programs in $IP_{A,c}$. Starting at a solution u of $IP_{A,c}(b)$, we can successively move to improved solutions of the program by subtracting appropriate elements of $\mathcal{R}_c$. A solution u' will be optimal for $IP_{A,c}(b)$ if and only if there does not exist $r \in \mathcal{R}_c$ such that $u' - r$ is feasible for $IP_{A,c}(b)$.

Proposition 3.9 *A finite subset $\mathcal{R}_c \subseteq ker_{\mathbf{Z}}(A)$ is a minimal test set for $IP_{A,c}$ if and only if for every minimal generator α_i^+ of $\mathcal{S}_c$, there is a unique vector r in $\mathcal{R}_c$ such that $r^+ = \alpha_i^+$.*

Proof. By Theorem 3.6, the non-optimal solutions to the programs in $IP_{A,c}$ are precisely the lattice points in $\mathcal{S}_c$. □

As before we may interpret each element $\underline{x^{\alpha_i^+} - x^{\alpha_i^-}}$ of the reduced Gröbner basis $\mathcal{G}_c = \{x^{\alpha_i^+} - x^{\alpha_i^-}, i = 1, \ldots, p\}$ uniquely, as either the line segment $[\alpha_i^+, \alpha_i^-]$ in $P_b^I(A\alpha_i^+)$ directed from α_i^+ to α_i^- or as the vector $\alpha_i = \alpha_i^+ - \alpha_i^- \in ker_{\mathbf{Z}}(A)$.

Corollary 3.10 *The reduced Gröbner basis $\mathcal{G}_c$ of I_A is a uniquely defined minimal test set for $IP_{A,c}$.*

Proof. Since $\mathcal{G}_c$ is reduced, for every minimal generator of $in_c(I_A)$, there is a unique binomial in $\mathcal{G}_c$ with initial monomial equal to this generator. Hence $\mathcal{G}_c$ is a minimal test set by Proposition 3.9. For a fixed initial monomial $x^{\alpha_i^+}$, the trailing monomial $x^{\alpha_i^-}$ is such that α_i^- is the unique optimal solution to $IP_{A,c}(A\alpha_i^+)$. Else, the trailing monomial would be divisible by some initial monomial of $\mathcal{G}_c$ contradicting that $\mathcal{G}_c$ is reduced. □

The above theorems show that $in_c(I_A)$ is to $IP_{A,c}$ what Δ_c is to $LP_{A,c}$. As might be expected, these two entities are related as shown in the following theorem which first appeared in [25].

Theorem 3.11 *A subset $F \subseteq \{1, \ldots, n\}$ is a face of Δ_c if and only if there does not exist a minimal generator $x^{\alpha_i^+}$ of the initial ideal $in_c(I_A)$ such that $supp(\alpha_i^+)$ is contained in F.*

Example 2.2 continued. For $\Delta_{(0,1,0,1,1)} = \{\{1,4,5\}, \{1,3,4\}, \{1,2,3\}\}$ and $\mathcal{G}_{(0,1,0,1,1)} = \{\underline{x_3 x_5^3} - x_1^2 x_4^2, x_2 x_5^2 - x_1^2 x_4, \underline{x_2 x_4} - x_3 x_5, x_2^2 x_5 - x_1^2 x_3\}$, no minimal generator of $in_{(0,1,0,1,1)}(I_A)$ is supported on a face of $\Delta_{(0,1,0,1,1)}$.

Proof of Theorem 3.11. (*taken from [27]*). For a subset $F \subseteq \{1, \ldots, n\}$ the following are equivalent:

F is a face of Δ_c
$\Leftrightarrow \exists\ y \in \mathbf{Z}^d$ feasible for $LP_{A,c}(b)^{dual}$ such that $F = \{j : a_j \cdot y = c_j\}$
$\Leftrightarrow \exists\ b \in \mathbf{Z}^d$: an optimal solution y of $LP_{A,c}(b)^{dual}$
   satisfies $F = \{j : a_j \cdot y = c_j\}$
$\Leftrightarrow \exists\ b \in \mathbf{Z}^d$: an optimal solution u of $LP_{A,c}(b)$ has $supp(u) = F$
$\Leftrightarrow \exists\ b \in \mathbf{Z}^d$: an optimal solution u of $LP_{A,c}(b)$ has $supp(u) = F$
   and is integral
$\Leftrightarrow \exists$ a monomial x^u such that $F = supp(u)$ and no power of x^u is in $in_c(I_A)$
$\Leftrightarrow F$ does not contain the support of any minimal generator of $in_c(I_A)$.

The first equivalence follows from the definition of Δ_c. Since every solution of $LP_{A,c}(b)^{dual}$ is optimal for some cost vector, we get the second equivalence. The third equivalence follows from complimentary slackness and the fourth from the fact that scaling b by a suitable integer multiple until the optimal solution to $LP_{A,c}(b)$ is integral does not change any of the preceding arguments. However, then u is optimal for $IP_{A,c}(b)$ and all its integer multiples are optimal for the integer programs $IP_{A,c}(\cdot)$ in which they are solutions. This condition is equivalent to the fact that x^u and all its powers are standard with respect to c which is equivalent to saying that no minimal generator of $in_c(I_A)$ can have support $F = supp(u)$. □

In other words, $in_c(I_A)$ is determined by Δ_c at the level of its *radical ideal* although typically this correspondence is many-one, reflecting the philosophy that integer programming is an arithmetic refinement of linear programming.

Example 2.2 continued. The triangulation $\Delta_{(0,1,0,1,1)} = \Delta_{(2,7,8,0,0)}$ supports both the initial ideals, $in_{(0,1,0,1,1)}(I_A) = \langle x_3 x_5^3, x_2 x_5^2, x_2 x_4, x_2^2 x_5 \rangle$ and $in_{(2,7,8,0,0)} = \langle x_3 x_5, x_2 x_5^2, x_2^2 x_5, x_2^3 x_4 \rangle$, in the sense of Theorem 3.11. □

If the elements of $\mathcal{G}_c$ are thought of as directed line segments, then the above results imply that there exists a directed path in P_b^I (possibly more than one) from *every* non-optimal solution of $IP_{A,c}(b)$ to the unique optimum, comprised of translates of elements in $\mathcal{G}_c$. In fact, the elements of $\mathcal{G}_c$ build a connected directed graph in each fiber P_b^I wherein the nodes are the lattice points in that fiber and the edges are elements of $\mathcal{G}_c$. The graph in a fiber P_b^I is connected since every non-optimal solution to $IP_{A,c}(b)$ has out degree at least one and the graph has a unique sink at the optimal solution to $IP_{A,c}(b)$. If the directions on all edges of this graph are reversed, then starting at the unique optimum of $IP_{A,c}(b)$, one can trace a directed path to every lattice point in P_b^I. This allows all lattice points in P_b^I to be enumerated. These ideas have been applied to a class of stochastic integer programs from manufacturing in [29] to statistical sampling in [13].

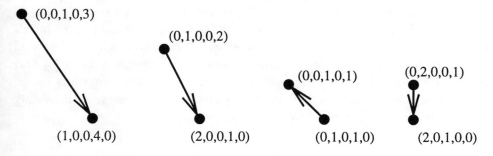

Figure 5: Elements of $\mathcal{G}_c$

Example 2.2 continued. Figure 5 shows the projections onto the x_4, x_5 plane, of the elements in $\mathcal{G}_{(0,1,0,1,1)}$, interpreted as directed line segments. Figure 6 shows the directed path corresponding to the particular reduction discussed earlier, of $x_1 x_2^{10} x_3^{10} x_4^4$ to its normal form $x_1^7 x_3^{17} x_5$. The black points in the figure are the projections onto the x_4, x_5 coordinates of the lattice points in the polytope $P^I_{(25,34,18)}$. The two types of directed segments in this path correspond to the two distinct binomials used in the reduction. □

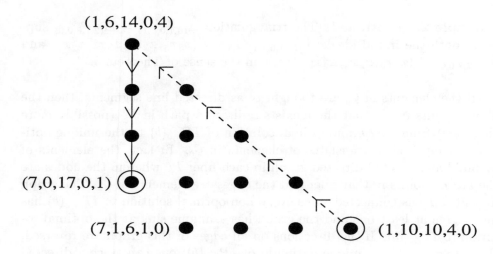

Figure 6: A reduction path.

Given that the idea of solving integer programs via test sets is rather natural, it is not surprising that a number of test sets can be found in the integer programming literature. In 1975, Graver [17] showed the existence of a finite set of vectors that solve integer programs of the form $IP_{A,c}(b)$ for all b and all c. We call this test set the *Graver basis of A* and it will be discussed further in Section 4. Variants of the Graver basis appear in both [6] and [9]. In 1981, Scarf [23] introduced a test set for integer programs called the *neighbors of the origin*. Relationships among these test sets (including Gröbner bases) are discussed in [30].

4 Universal test sets for linear and integer programming

Thus far we examined test sets for linear and integer programs in which the cost vector c was fixed. A *universal test set* for LP_A (respectively IP_A) is a finite subset of $ker(A)$ (respectively $ker_{\mathbf{Z}}(A)$) that contains a test set for $LP_{A,c}$ (respectively $IP_{A,c}$) for all $c \in \mathbf{R}^n$ that are generic with respect to

Gröbner Bases in Integer Programming

LP_A (respectively IP_A). In this section we construct universal test sets for LP_A and IP_A, examine their geometry and establish their relationships. We say that a lattice point is *primitive* if the g.c.d. of its coordinates is one.

Definition 4.1 *A* circuit *of A is a primitive non-zero vector u in $ker_{\mathbb{Z}}(A)$ such that its support is minimal with respect to inclusion.*

For the purposes of linear programming it suffices to define a circuit of A as any non-zero vector of minimal support in $ker(A)$ where two circuits t and t' are equivalent if one is a real multiple of the other. The above definition fixes a representative from each equivalence class (up to sign) and is more precise for integer programming. We denote the set of circuits of A as $\mathcal{C}_A$ and as before we may interpret $u \in \mathcal{C}_A$ as the line segment $[u^+, u^-] \subset P_{Au^+} = P_{Au^-}$. The polytope P_b is called a *circuit fiber* of π_A (respectively *circuit fiber* of π_A^I) if $b = Au^+$ for some $u \in \mathcal{C}_A$.

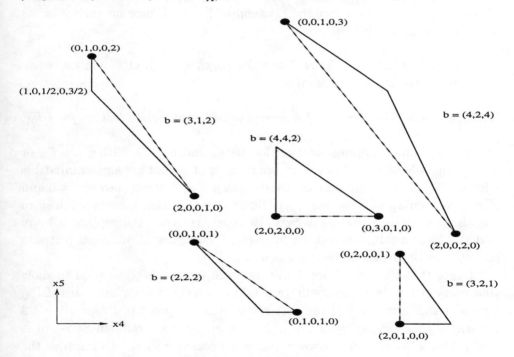

Figure 7: Circuit fibers of π_A.

Example 2.2 continued. The circuits of A are $2e_2 + e_5 - 2e_1 - e_3$, $3e_2 + e_4 - 2e_1 - 2e_3$, $e_3 + 3e_5 - 2e_1 - 2e_4$, $e_2 + 2e_5 - 2e_1 - e_4$, and $e_2 + e_4 - e_3 - e_5$.

Figure 7 shows the circuit fibers of π_A each accompanied by the right hand side vector it corresponds to. It may be observed that every circuit is a primitive edge in its fiber. Proposition 4.2 confirms that this is true in general.

Proposition 4.2 *If u is a circuit of A then, $[u^+, u^-]$ is a primitive edge of the circuit fiber, P_{Au^+}.*

Proof. We need only show that $[u^+, u^-]$ is an edge of P_{Au^+}. Let $c \in \mathbf{R}^n$ be such that $c_j = 0$ for all $j \in supp(u)$ and $c_j = 1$ otherwise. Then, $[u^+, u^-]$ lies in $face_c(P_{Au^+})$. Suppose $v \in face_c(P_{Au^+}) \backslash [u^+, u^-]$. Then $u^+ - v$ is a non-zero vector in $ker(A)$ whose support is contained in $supp(u)$. Since $u \in \mathcal{C}_A$, there exists $\lambda \in \mathbf{R}\backslash\{0\}$ such that $u^+ - v = \lambda u$. Therefore, $v = (1 - \lambda)u^+ + \lambda u^-$ which shows that v lies in the affine span of u^+ and u^-. Since $v \notin [u^+, u^-]$ either $u^+ \in]v, u^-[$ or $u^- \in]u^+, v[$. However, neither of these can occur since $supp(u^+) \cap supp(u^-) = \emptyset$. Hence no such v exists and $face_c(P_{Au^+}) = [u^+, u^-]$. □

The dashed lines in Figure 7 are the edges in each circuit fiber corresponding to the circuit it contains.

Theorem 4.3 *The circuits of A form a minimal universal test set for LP_A.*

Proof. Let $\mathcal{T}_c$ be a minimal test set for $LP_{A,c}$ and $u \in \mathcal{T}_c$. Either $u \in \mathcal{C}_A$ or $u = \sum \lambda_i p_i$ where $\lambda_i \in \mathbf{R}_{\geq 0}$ and p_i are circuits of A that are sign compatible with u. We say that x is sign compatible to y if $x_i = 0$ whenever $y_i = 0$ and if $x_i \neq 0$ then $sign(x_i) = sign(y_i)$. Since $c \cdot u > 0$ there exists some p_i in the above sum such that $c \cdot p_i > 0$. However, then we can replace u by p_i and still have a minimal test set for $LP_{A,c}$. Therefore all minimal test sets of LP_A can be chosen to be subsets of $\mathcal{C}_A$.

To see that $\mathcal{C}_A$ is a minimal universal test set for LP_A we need to show that every circuit is necessary in some minimal test set associated with LP_A. Consider the circuit u and the cost vector c in the proof of Proposition 4.2 for which $face_c(P_{Au^+}) = [u^+, u^-]$. For an appropriate refinement $\succ_c$ of c, $face_{\succ_c}(P_{Au^+}) = u^-$ which shows that u is necessary in $\mathcal{T}_{\succ_c}$ to improve the non-optimal solution u^+. □

Corollary 4.4 *For every generic cost vector $c \in \mathbf{R}^n$, there exists a minimal test set for $LP_{A,c}$ that consists only of edges of certain fibers of π_A.*

We now describe a universal test set for IP_A due to Graver [17]. For each $\sigma \in \{+,-\}^n$, consider the monoid $S_\sigma = ker_{\mathbf{Z}}(A) \cap \mathbf{R}^n_\sigma$, where $\mathbf{R}^n_\sigma$ is the orthant with sign pattern σ. Let H_σ be the unique minimal *Hilbert basis* of the pointed polyhedral $(n-d)$-cone $cone(S_\sigma)$ in $\mathbf{R}^n$. The Hilbert basis of a polyhedral cone K in $\mathbf{R}^n$ is a minimal subset of $K \cap \mathbf{Z}^n$ such that every integral vector in K can be written as a non-negative integral combination of the elements in the basis. Pointed cones have unique Hilbert bases (see Chapter 16, [24]). We call $Gr_A := \cup_\sigma H_\sigma \backslash \{0\}$ the *Graver basis* of A.

Theorem 4.5 *[17] The Graver basis of A is a universal test set for IP_A.*

Proof. Consider an arbitrary generic cost vector $c \in \mathbf{R}^n$ and a right hand side vector $b \in cone_{\mathbf{N}}(\mathcal{A})$. Let u be a non-optimal solution to $IP_{A,c}(b)$ for which u' is the optimal solution. If $u - u' \in S_\sigma$ then, there exists $h_i \in H_\sigma$ and $n_i \in \mathbf{N} \backslash \{0\}$ such that $u - u' = \sum n_i h_i$. Since $c \cdot (u - u') > 0$ there exists some h_i in the above sum such that $c \cdot h_i > 0$. Subtracting this h_i from u, we get an improved solution to $IP_{A,c}(b)$. $\square$

The elements of Gr_A are precisely all the elements $u \in ker_{\mathbf{Z}}(A)$ for which there does not exist $v \in ker_{\mathbf{Z}}(A)$ such that $v^+ \leq u^+$ and $v^- \leq u^-$. However, one can verify that the elements in a reduced Gröbner basis of I_A also have this property, giving rise to the following result.

Theorem 4.6 *Every reduced Gröbner basis of I_A is a subset of the Graver basis of A.*

Corollary 4.7 *There exists only finitely many distinct reduced Gröbner bases for I_A as the cost function is varied.*

Let UGB_A denote the union of all the finitely many distinct reduced Gröbner bases of I_A. Then UGB_A is a subset of Gr_A and is a universal test set for IP_A. We call UGB_A the *universal Gröbner basis* of A. If $g \in UGB_A$ (respectively Gr_A), we call the fiber $P^I_{Ag^+}$ a *Gröbner fiber* (respectively *Graver fiber*) of π^I_A.

Both Gr_A and UGB_A can be computed using Gröbner basis calculations as described in [28]. In order to compute Gr_A we consider the *Lawrence lifting* of A which is the enlarged matrix $\Lambda(A) = \begin{bmatrix} A & 0 \\ I & I \end{bmatrix} \in \mathbf{Z}^{(n+d) \times 2n}$ where 0 is a $d \times n$ matrix of all zeros and I is an $n \times n$ identity matrix. The matrices A and $\Lambda(A)$ have isomorphic kernels: $ker_{\mathbf{Z}}(\Lambda(A)) = \{(u, -u) : u \in ker_{\mathbf{Z}}(A)\}$. The toric ideal $I_{\Lambda(A)}$ is the homogeneous prime ideal

$$I_{\Lambda(A)} = \langle x^{\alpha}y^{\beta} - x^{\beta}y^{\alpha} : \alpha, \beta \in \mathbf{N}^n, A\alpha = A\beta \rangle$$

in the polynomial ring $k[x_1, \ldots, x_n, y_1, \ldots, y_n]$.

Theorem 4.8 *For the matrix $\Lambda(A)$, the following sets coincide:*
(i) the Graver basis of $\Lambda(A)$,
(ii) the universal Gröbner basis of $\Lambda(A)$,
(iii) any reduced Gröbner basis of $I_{\Lambda(A)}$,
(iv) any minimal generating set of $I_{\Lambda(A)}$ (up to scalar multiples), and
(v) the set of binomials $x^{\alpha}y^{\beta} - x^{\beta}y^{\alpha}$ supported on primitive one-dimensional fibers $[(\alpha, \beta), (\beta, \alpha)]$.

Proof. The Graver bases Gr_A and $Gr_{\Lambda(A)}$ are related as follows: $Gr_{\Lambda(A)} = \{x^{\alpha}y^{\beta} - x^{\beta}y^{\alpha} : \alpha, \beta \in \mathbf{N}^n, x^{\alpha} - x^{\beta} \in Gr_A\}$. Since $Gr_{\Lambda(A)}$ is the Graver basis of $\Lambda(A)$, it is a generating set of $I_{\Lambda(A)}$ and by Theorem 4.6, it is a Gröbner basis of $I_{\Lambda(A)}$ (not necessarily reduced), with respect to every generic cost vector. Notice that it suffices to show that $Gr_{\Lambda(A)}$ is the unique minimal generating set of $I_{\Lambda(A)}$ in order to prove the equality of the sets in (i),(ii),(iii) and (iv). This is because of Theorem 4.6, the definition of $UGB_{\Lambda(A)}$, and the fact that every reduced Gröbner basis of $I_{\Lambda(A)}$ contains a minimal generating set for $I_{\Lambda(A)}$. We show below that the sets in (i) and (iv) coincide. Choose any element $g := x^{\alpha}y^{\beta} - x^{\beta}y^{\alpha}$ of $Gr_{\Lambda(A)}$, and fix $\sigma \in \{-, +\}^n$ such that $\alpha - \beta$ lies in $S_{\sigma} = ker_{\mathbf{Z}}(A) \cap \mathbf{R}^n_{\sigma}$. Let $\mathcal{B}$ be the set of all binomials $x^{\gamma}y^{\delta} - x^{\delta}y^{\gamma}$ in $I_{\Lambda(A)}$ except g. Suppose that $\mathcal{B}$ generates $I_{\Lambda(A)}$. Then $x^{\alpha}y^{\beta} - x^{\beta}y^{\alpha}$ can be written as a linear combination of elements in $\mathcal{B}$. But this is only possible if there exists a binomial $x^{\gamma}y^{\delta} - x^{\delta}y^{\gamma}$ in $\mathcal{B}$ such that $x^{\gamma}y^{\delta}$ divides $x^{\alpha}y^{\beta}$. This implies that $\gamma - \delta$ lies in the semigroup S_{σ}. Moreover, since $\gamma \leq \alpha$ and $\delta \leq \beta$, the non-zero vector $(\alpha - \beta) - (\gamma - \delta)$ lies in S_{σ} as well. Therefore $\alpha - \beta$ cannot be an element in the Hilbert basis of $cone(S_{\sigma})$. This is a contradiction, and we conclude that every minimal generating set of $I_{\Lambda(A)}$ requires (a scalar multiple of) the binomial g.

For the equality of (i) and (v) we shall prove that every Graver fiber contains precisely two lattice points. Let $g \in Gr_{\Lambda(A)}$ as above. Suppose that the common fiber of (α, β) and (β, α) contains a third point $(\gamma, \delta) \in \mathbf{N}^{2n}$. Then $\alpha, \beta, \gamma, \delta \in \mathbf{N}^n$ all lie in the same fiber of π_A^I and $\alpha + \beta = \gamma + \delta$. This implies that $\alpha - \beta = (\gamma - \beta) + (\delta - \beta)$. We will show that the non-zero vectors

$\gamma-\beta$ and $\delta-\beta$ are sign compatible with $\alpha-\beta$. This contradicts $\alpha-\beta \in Gr_A$ and thus completes the proof. Let $j \in \{1,\ldots,n\}$. If $\alpha_j > 0$ then $\beta_j = 0$, and this implies $(\alpha-\beta)_j = \alpha_j > 0$, $(\gamma-\beta)_j = \gamma_j \geq 0$, and $(\delta-\beta)_j = \delta_j \geq 0$. If $\alpha_j = 0$ then $(\alpha-\beta)_j = -\beta_j \leq 0$, $(\gamma-\beta)_j = \gamma_j - \beta_j = -\delta_j \leq 0$, and $(\delta-\beta)_j = \delta_j - \beta_j = -\gamma_j \leq 0$. □

Algorithm 4.9 How to compute the Graver basis of A.
(i) Compute the reduced Gröbner basis $\mathcal{G}$ of $I_{\Lambda(A)}$ with respect to any term order.
(ii) The Graver basis Gr_A consists of all elements $\alpha-\beta$ such that $x^\alpha y^\beta - x^\beta y^\alpha$ appears in $\mathcal{G}$.

Proof. By Theorem 4.8, any reduced Gröbner basis of $I_{\Lambda(A)}$ is also the Graver basis of $\Lambda(A)$. The bijection between the kernels of A and $\Lambda(A)$ implies that a reduced Gröbner basis of $I_{\Lambda(A)}$ with the variables y_j set to one, is the Graver basis of A. □

Since $UGB_A \subseteq Gr_A$, all we need now for the computation of UGB_A is a way to identify the elements of UGB_A from among the elements of Gr_A. To do this, we establish the integer analog of Proposition 4.2. A second test for whether an element in the Graver basis lies in UGB_A can be found in [28].

Proposition 4.10 *If a vector u lies in the universal Gröbner basis UGB_A then $[u^+, u^-]$ is a primitive edge of the Gröbner fiber $P^I_{Au^+}$.*

Lemma 4.11 *Let x^α be a minimal generator of the initial monomial ideal $in_c(I_A)$, and let δ be any lattice point in $P^I_{A\alpha}$ such that $c \cdot \alpha \geq c \cdot \delta$. Then $supp(\delta) \cap supp(\alpha) = \emptyset$.*

Proof. Suppose $k \in supp(\alpha) \cap supp(\delta)$ for a lattice point δ in the $A\alpha$-fiber of π^I_A for which $c \cdot \alpha \geq c \cdot \delta$. Then $\alpha - e_k$ and $\delta - e_k$ are lattice points in the same fiber of π^I_A and $c \cdot (\alpha - e_k) \geq c \cdot (\delta - e_k)$. This implies that x^α/x_k lies in the initial monomial ideal $in_c(I_A)$, which is a contradiction to x^α being a minimal generator. □

Lemma 4.12 *For an element u of UGB_A, both u^+ and u^- are vertices in the Au^+-fiber of π^I_A.*

Proof. By definition, u^- is the optimal vertex with respect to some cost function c in the Au^--fiber of π_A^I. Recall our assumption that the integer programs $IP_{A,c}(b)$ are bounded. This implies the existence of an integral vector M with all coordinates positive in the row space of A. After replacing M by a multiple if necessary, we may assume that $M - c$ has all coordinates positive. Clearly, the cost function $\omega := M - c$ attains its **maximum** over $P_{Au^+}^I$ at u^-. Let v denote the restriction of ω to the support of u^+ (i.e, $v_i = w_i$ if $u_i^+ > 0$ and $v_i = 0$ if $u_i^+ = 0$). We claim that v attains a unique maximum over $P_{Au^+}^I$ at u^+. If not, then there exists another lattice point δ in $P_{Au^+}^I$ with $v \cdot \delta \geq v \cdot u^+$. Since $v \cdot u^+ > 0$, the set $supp(v) \cap supp(\delta) = supp(u^+) \cap supp(\delta)$ is not empty. By Lemma 4.11, this implies $c \cdot u^+ < c \cdot \delta$. In view of $M \cdot u^+ = M \cdot \delta$, we conclude that $v \cdot u^+ = w \cdot u^+ > w \cdot \delta \geq v \cdot \delta$, as desired. □

Proof of Proposition 4.10. Let $g \in UGB_A$ and choose $w, v \in \mathbf{N}^n$ as in the proof of Lemma 4.12. Consider the cost vector $u := (v \cdot g) w + (w \cdot -g) v \in \mathbf{N}^n$. We have $u \cdot g^+ = u \cdot g^-$. It suffices to show that $u \cdot g^+ > u \cdot \gamma$ for all lattice points γ other than g^+ and g^- in the Ag^+-fiber of π_A^I. If $supp(\gamma) \cap supp(g^+) = \emptyset$, then $supp(\gamma) \cap supp(v) = \emptyset$ which implies that $u \cdot g^- = (v \cdot g)(w \cdot g^-) > (v \cdot g)(w \cdot \gamma)$. If $supp(\gamma) \cap supp(g^+) \neq \emptyset$, then by Lemma 4.11, $w \cdot g^+ > w \cdot \gamma$. This implies that $u \cdot g^+ = (v \cdot g)(w \cdot g^+) + (w \cdot -g))(v \cdot g^+) > (v \cdot g)(w \cdot \gamma) + (w \cdot -g)(v \cdot \gamma) = u \cdot \gamma$. Therefore, $[g^+, g^-]$ is an edge of the Ag^+-fiber of π_A^I with outer normal vector u. □

Corollary 4.13 *For every generic cost vector $c \in \mathbf{R}^n$, the reduced Gröbner basis $\mathcal{G}_c$ consists only of edges of certain fibers of π_A^I.*

Algorithm 4.14 How to compute UGB_A.
1. Compute the Graver basis Gr_A using Algorithm 4.9.
2. For each element $x^\alpha - x^\beta$ of Gr_A decide whether $[\alpha, \beta]$ is an edge of $P_{A\alpha}^I$.

Example 2.2 continued. Figure 8 shows the Gröbner fibers of π_A^I. As before, the dashed lines indicate the edge defined by the elements of UGB_A in each Gröbner fiber. In general, it is possible for two or more elements of UGB_A to come from a given Gröbner fiber. □

The following proposition ties together the three universal test sets introduced in this section to give the result : $\mathcal{C}_A \subseteq UGB_A \subseteq Gr_A$. However, as Example 4.16 shows any of these containments may or may not be strict.

Gröbner Bases in Integer Programming

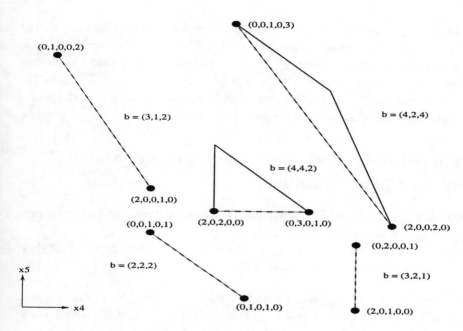

Figure 8: Gröbner fibers of π_A^I.

Proposition 4.15 *The circuits of A are contained in UGB_A.*

Proof. Let $u = u^+ - u^-$ be a circuit of A. Consider the cost function $c := \sum\{e_i : i \notin supp(u)\}$. After refining c to be generic, we may suppose that $c \cdot u^+ > c \cdot u^-$. Then the monomial x^{u^+} lies in $in_c(I_A)$, and there exists a binomial $x^{\alpha_i^+} - x^{\alpha_i^-}$ in $\mathcal{G}_c$ such that $x^{\alpha_i^+}$ divides x^{u^+}. Since $c \cdot \alpha_i^+ \geq c \cdot \alpha_i^-$ and $supp(\alpha_i^+) \subseteq supp(u^+) \subseteq supp(u)$, we conclude that $supp(\alpha_i^-) \subseteq supp(u)$. Since u is a circuit, these facts imply $u = \alpha \in UGB_A$. □

Example 4.16 (i) For $A = \begin{pmatrix} 1 & 1 & 1 & 1 \\ 0 & 1 & 2 & 3 \end{pmatrix}$, $\mathcal{C}_A = \{x_2^2 - x_1x_3, x_3^2 - x_2x_4,$ $x_3^3 - x_1x_4^2, x_2^3 - x_1^2x_4\}$ while $UGB_A = Gr_A = \mathcal{C}_A \cup \{x_1x_4 - x_2x_3\}$.

(ii) For $A = \begin{pmatrix} 1 & 1 & 1 & 1 \\ 0 & 5 & 6 & 10 \end{pmatrix}$, $\mathcal{C}_A = UGB_A = \{x_3^5 - x_2^4x_4, x_3^5 - x_1^2x_4^3, x_2^2 - x_1x_4, x_2^6 - x_1x_3^5\}$ while $Gr_A = \mathcal{C}_A \cup \{x_3^5 - x_1x_2^2x_4^2\}$.

(iii) For our running example, $\mathcal{C}_A = UGB_A = Gr_A = \{x_3x_5^3 - x_1^2x_4^2, x_2x_5^2 - x_1^2x_4, x_2x_4 - x_3x_5, x_2^2x_5 - x_1^2x_3, x_1^2x_3^2 - x_2^3x_4\}$. □

It is an important problem to bound the *degree* of the elements in the three test sets given above. The degree of a binomial $x^{u^+} - x^{u^-} \in I_A$ is the 1-norm of the vector $u \in \ker_{\mathbf{Z}}(A)$. It was shown in [25] that the degree of an element in Gr_A is at most $(n-d)(d+1)\mathcal{D}(A)$ where $\mathcal{D}(A)$ is the largest maximal minor of A. However, this bound is not sharp, and further discussions on degree issues can be found in [20].

5 Variation of cost functions in linear and integer programming

We now study the effect of varying cost vectors in linear and integer programming and study their relationships.

There is a natural equivalence relation on the space of all (not just generic) cost vectors with respect to IP_A.

Definition 5.1 *Two cost vectors c and c' in $\mathbf{R}^n$ are equivalent (with respect to IP_A) if the integer programs $IP_{A,c}(b)$ and $IP_{A,c'}(b)$ have the same set of optimal solutions for all b in $\text{cone}_{\mathbf{N}}(\mathcal{A})$.*

We establish a structure theorem for these equivalence classes (Theorem 5.10). This theorem can also be derived from more general results of Mora-Robbiano [22] and Bayer-Morrison [2] on *Gröbner fans* and *state polytopes* for *graded polynomial ideals*. What we present here is an alternate construction for toric ideals which is self-contained and provides more precise information for integer programming.

Recall that a cost vector c is *generic* for IP_A if the optimal solution with respect to c in every fiber P_b^I of π_A^I is a unique vertex. Generic equivalence classes are characterized as follows:

Proposition 5.2 *Given two generic cost vectors c and c' in $\mathbf{R}^n$, the following are equivalent:*
(i) For every $b \in \text{cone}_{\mathbf{N}}(\mathcal{A})$, the programs $IP_{A,c}(b)$ and $IP_{A,c'}(b)$ have the same optimal solution.
(ii) The cost vectors c and c' support the same optimal vertex in each fiber P_b^I of π_A^I.
(iii) The reduced Gröbner bases $\mathcal{G}_c$ and $\mathcal{G}_{c'}$ of I_A are equal.

Proof. Conditions (i) and (ii) are equivalent since the optimal solution of $IP_{A,c}(b)$ is the vertex of P_b^I supported by c. The set of all non-optimal

Gröbner Bases in Integer Programming

solutions to the programs $IP_{A,c}(\cdot)$ and $IP_{A,c'}(\cdot)$ are the monomial ideals $in_c(I_A)$ and $in_{c'}(I_A)$ respectively. Then (i) holds if and only if $in_c(I_A) = in_{c'}(I_A)$. This is equivalent to (iii) since an initial ideal uniquely determines the reduced Gröbner basis associated with it. □

Lemma 5.3 *For the reduced Gröbner basis* $\mathcal{G}_c \subset \mathbf{Z}^n$, $span_{\mathbf{Z}}(\mathcal{G}_c) = ker_{\mathbf{Z}}(A)$.

Proof. Every vector $g_i = g_i^+ - g_i^-$ in $\mathcal{G}_c$ lies in $ker_{\mathbf{Z}}(A)$. Hence $span_{\mathbf{Z}}(\mathcal{G}_c) \subseteq ker_{\mathbf{Z}}(A)$. Let $\alpha \in ker_{\mathbf{Z}}(A)$. We can write α uniquely as $\alpha^+ - \alpha^-$ where α^+, α^- are vectors in $\mathbf{N}^n$ with disjoint supports. Further, $A\alpha^+ = A\alpha^-$, and hence α^+ and α^- lie in the same fiber of π_A^I. Let β be the unique optimum in this fiber with respect to c. Since $\mathcal{G}_c$ is a test set for $IP_{A,c}$, there exist non-negative integral multipliers n_i and n'_i such that $\alpha^+ - \beta = \sum_{g_i \in \mathcal{G}_c} n_i g_i$ and $\alpha^- - \beta = \sum_{g_i \in \mathcal{G}_c} n'_i g_i$. Hence $\alpha = \sum_{g_i \in \mathcal{G}_c} (n_i - n'_i) g_i$ which implies that $span_{\mathbf{Z}}(\mathcal{G}_c) = ker_{\mathbf{Z}}(A) \simeq \mathbf{Z}^{n-d}$. □

Let $St(A)$ denote the Minkowski sum of all Gröbner fibers of π_A^I. This is a well-defined polytope in $\mathbf{R}^n$ which we call the *state polytope* of A. Lemma 5.3 implies $dim(St(A)) = n - d$. The complete polyhedral fan $\mathcal{N}(St(A))$ is called the *Gröbner fan* of A.

Lemma 5.4 *Every fiber of π_A^I is a Minkowski summand of $St(A)$.*

Proof. It suffices to show that $\mathcal{N}(St(A))$ is a refinement of $\mathcal{N}(P_b^I)$ for all $b \in cone_{\mathbf{N}}(\mathcal{A})$. Let c be a generic cost vector and let $w \neq c$ belong to the interior of the cone $\mathcal{N}(face_c(St(A)); St(A))$. Then w lies in $\mathcal{N}(\beta_i; P_{A\beta_i}^I)$ for each element $\alpha_i - \beta_i$ in the reduced Gröbner basis $\mathcal{G}_c$. This implies $w \cdot \alpha_i > w \cdot \beta_i$ for all i, and therefore $\mathcal{G}_w = \mathcal{G}_c$.

Now consider an arbitrary $b \in cone_{\mathbf{N}}(\mathcal{A})$. Let u be the unique optimum of $IP_{A,c}(b)$. The equality of test sets $\mathcal{G}_w = \mathcal{G}_c$ implies that u is also the unique optimum of $IP_{A,w}(b)$. Hence w lies in the interior of $\mathcal{N}(u; P_b^I)$. Therefore, $\mathcal{N}(face_c(St(A)); St(A)) \subseteq \mathcal{N}(u; P_b^I)$, as desired. □

Proposition 5.5 *Let db denote any probability measure with support $cone_{\mathbf{N}}(\mathcal{A})$ such that $\int_b b\, db$ is finite. Then the Minkowski integral $\int_b P_b^I\, db$ is a polytope normally equivalent to $St(A)$.*

Proof. The hypothesis $\int_b b\, db < \infty$ guarantees that $\int_b P_b^I\, db$ is bounded. By Lemma 5.4, $\int_b P_b^I\, db$ is a summand of $St(A)$ and is hence a polytope. However, each Gröbner fiber is a summand of $\int_b P_b^I\, db$ and hence $\int_b P_b^I\, db$ is an $(n-d)$-polytope in $\mathbf{R}^n$ that has the same normal fan as $St(A)$. □

Corollary 5.6 *There exists only finitely many facet directions among the fibers of of π_A^I.*

¿From now on we shall use the term *state polytope* for any polytope normally equivalent to $\int_b P_b^I db$. We define the *Gröbner cone* associated with $\mathcal{G}_c$ to be the closed convex polyhedral cone

$$\mathcal{K}_c := \{ x \in \mathbf{R}^n : g_i \cdot x \geq 0,\ g_i \in \mathcal{G}_c \}$$

Observation 5.7 *The Gröbner cone $\mathcal{K}_c$ has full dimension n. Its lineality space $\mathcal{K}_c \cap -\mathcal{K}_c$ equals $rowspan(A) \simeq \mathbf{R}^d$.*

Proof. We have $dim(\mathcal{K}_c) = n$ because c lies in the interior of $\mathcal{K}_c$. The lineality space $\mathcal{K}_c \cap -\mathcal{K}_c$ equals the orthogonal complement of $\mathcal{G}_c$ in $\mathbf{R}^n$, which coincides with the row span of A by Lemma 5.3. □

Proposition 5.8 *The Gröbner fan of A is the collection of all Gröbner cones $\mathcal{K}_c$ together with their faces, as c varies over all generic cost vectors.*

Proof. The argument in the proof of Lemma 5.4 shows that, for c generic, the Gröbner cone $\mathcal{K}_c$ equals $\mathcal{N}(face_c(St(A)); St(A))$. □

We remark that each cone in the Gröbner fan has the same lineality space $rowspan(A) \simeq \mathbf{R}^d$ and it is often more convenient to work with its image in $ker(A) \simeq \mathbf{R}^n/rowspan(A) \simeq \mathbf{R}^{n-d}$. We call this image of the Gröbner fan in $\mathbf{R}^{n-d}$, the *pointed Gröbner fan* of A.

Corollary 5.9 *The equivalence classes of cost functions with respect to IP_A (cf. Definition 5.1) are precisely the cells of the Gröbner fan.*

Proof. By Proposition 5.2, two cost vectors c and c' are equivalent if and only if they support the same optimal face in each fiber of π_A^I. Using Propositions 5.5 and 5.8, it follows that c and c' are equivalent if and only if they lie in the relative interior of the same cell in $\mathcal{N}(St(A))$. □

Example 2.2 continued. Adding the five Gröbner fibers of π_A^I we see that $St(A)$ is an octagon on $\mathbf{R}^5$. Figure 9 shows the pointed Gröbner fan of A in $\mathbf{R}^2$. The numbers of the Gröbner cones correspond to the numbers indexing the eight distinct Gröbner bases of I_A which are given below along with a representative cost vector. □

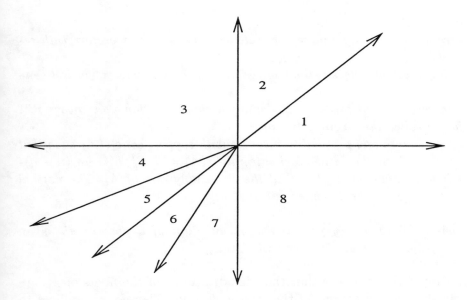

Figure 9: The pointed Gröbner fan of A.

1.	$c = (2, 3, 5, 0, 0)$ $\{x_3x_5 - x_2x_4,$ $x_1^2x_4 - x_2x_5^2,$ $x_1^2x_3 - x_2^2x_5\}$	2.	$c = (3, 2, 1, 0, 0)$ $\{x_2x_4 - x_3x_5,$ $x_1^2x_4 - x_2x_5^2,$ $x_1^2x_3 - x_2^2x_5\}$	3.	$c = (2, 3, 1, 0, 0)$ $\{x_2x_4 - x_3x_5,$ $x_1^2x_4 - x_2x_5^2,$ $x_2^2x_5 - x_1^2x_3\}$
4.	$c = (2, 5, 3, 0, 0)$ $\{x_2x_4 - x_3x_5,$ $x_2^2x_5 - x_1^2x_3,$ $x_2x_5^2 - x_1^2x_4,$ $x_1^2x_4^2 - x_3x_5^3\}$	5.	$c = (0, 1, 0, 1, 1)$ $\{x_2x_4 - x_3x_5,$ $x_2^2x_5 - x_1^2x_3,$ $x_2x_5^2 - x_1^2x_4,$ $x_3x_5^3 - x_1^2x_4^2\}$	6.	$c = (2, 7, 8, 0, 0)$ $\{x_3x_5 - x_2x_4,$ $x_2x_5^2 - x_1^2x_4,$ $x_2^2x_5 - x_1^2x_3,$ $x_2^3x_4 - x_1^2x_3^2\}$
7.	$c = (0, 1, 2, 0, 1)$ $\{x_3x_5 - x_2x_4,$ $x_2x_5^2 - x_1^2x_4,$ $x_2^2x_5 - x_1^2x_3,$ $x_1^2x_3^2 - x_2^3x_4\}$	8.	$c = (0, 2, 5, 1, 0)$ $\{x_3x_5 - x_2x_4,$ $x_2x_5^2 - x_1^2x_4,$ $x_1^2x_3 - x_2^2x_5\}$		

A cost vector w lies in the interior of a Gröbner cone $\mathcal{K}_c$ if and only if w is generic and equivalent to c. Hence the interiors of the top-dimensional cells in the Gröbner fan are precisely the equivalence classes of generic cost vectors. The following theorem summarizes the above discussion.

Theorem 5.10
(i) There are only finitely many equivalence classes of cost vectors with respect to IP_A.
(ii) Each equivalence class is the relative interior of a convex polyhedral cone in $\mathbf{R}^n$.
(iii) The collection of these cones defines a polyhedral fan that covers $\mathbf{R}^n$. This fan is called the Gröbner fan *of A.*
(iv) Let db denote any probability measure with support $cone_\mathbf{N}(\mathcal{A})$ such that $\int_b b\,db < \infty$. Then the Minkowski integral $St(A) = \int_b P_b^I db$ is an $(n-d)$-dimensional convex polytope, called the state polytope *of A. The normal fan of $St(A)$ equals the Gröbner fan of A.*

Proposition 5.11 *Every primitive edge direction of a fiber of π_A^I is an element of the universal Gröbner basis UGB_A.*

Proof. Proposition 5.5 says that the edge directions of the fibers of π_A^I are precisely the edge directions of the state polytope. Therefore it suffices to show that every primitive edge direction of $St(A)$ is an element of UGB_A.

Suppose u is primitive and defines an edge direction of the state polytope $St(A)$. Then u is the normal vector to a facet of a maximal cone $\mathcal{K}_c$ in the Gröbner fan $\mathcal{N}(St(A))$. Therefore u appears in the inequality presentation of $\mathcal{K}_c$. In other words, u is equal to one of the elements u_i of the reduced Gröbner basis $\mathcal{G}_c$. □

Theorem 5.12 *The universal Gröbner basis of A consists precisely of the primitive edge directions in the fibers of π_A^I.*

Proof. This follows from Propositions 4.10 and 5.11. □

Corollary 5.13 *For an element u in UGB_A, there exists two cost vectors c and c' in $\mathbf{R}^n$ such that $u \in \mathcal{G}_c$ and $-u \in \mathcal{G}_{c'}$.*

Proof. Every element in UGB_A appears as a facet normal of some cell in the Gröbner fan. Take as $\mathcal{G}_c$ and $\mathcal{G}_{c'}$ the Gröbner bases associated with the two Gröbner cones that share this facet. □

Direct translations of the work of Gel'fand-Kapranov-Zelevinsky ([16], Chapter 7) and Billera-Gel'fand-Sturmfels [4] give an analogous theory for linear programming. Their work predates the integer results above and had different motivations. However, their results also follow from the natural integer analog of the above line of reasoning.

Congruent to Definition 5.1 we have the following definition for LP_A.

Definition 5.14 *Two cost vectors c and c' in $\mathbf{R}^n$ are equivalent (with respect to LP_A) if the linear programs $LP_{A,c}(b)$ and $LP_{A,c'}(b)$ have the same set of optimal solutions for all b in $cone(\mathcal{A})$.*

Theorem 5.15 *For two generic (with respect to LP_A) cost vectors $c, c' \in \mathbf{R}^n$, the following conditions are equivalent:*
(i) c and c' are equivalent with respect to LP_A.
(ii) c and c' support the same optimal vertex in each fiber P_b of π_A.
(iii) The vectors c and c' define the same regular triangulation $\Delta_c = \Delta_{c'}$.

The above theorem follows from Theorem 2.3 and says that the equivalence class of a generic $c \in \mathbf{R}^n$ is

$$\mathcal{L}_c = \{\omega \in \mathbf{R}^n : \Delta_\omega = \Delta_c\}$$

which is a full dimensional open polyhedral cone in $\mathbf{R}^n$ whose closure $\overline{\mathcal{L}}_c$ is called the *secondary cone* of c. The set of all secondary cones together with their faces is the *secondary fan* of A, a term that is justified in the following theorem from [16].

Theorem 5.16 *There exists an $(n-d)$-dimensional polytope called the secondary polytope of A, denoted $\Sigma(A)$, whose inner normal fan $\mathcal{N}(\Sigma(A))$ is the secondary fan of A.*

This result then establishes the following results for linear programming.

Corollary 5.17 *The equivalence classes of cost functions with respect to LP_A are precisely the cells of the secondary fan.*

Theorem 5.18
(i) There are only finitely many equivalence classes of cost vectors with respect to LP_A.
(ii) Each equivalence class is the relative interior of a convex polyhedral cone in $\mathbf{R}^n$.
(iii) The collection of these cones defines a polyhedral fan that covers $\mathbf{R}^n$. This fan is called the secondary fan *of A.*

(iv) Let db denote any probability measure with support cone($\mathcal{A}$) such that $\int_b bdb < \infty$. Then the Minkowski integral $\Sigma(A) = \int_b P_b db$ is an $(n-d)$-dimensional convex polytope, called the secondary polytope *of A. The normal fan of $\Sigma(A)$ equals the secondary fan of A.*

We call any $(n-d)$-polytope that is normally equivalent to $\Sigma(A)$, the secondary polytope of A. A specific secondary polytope of A is obtained by taking the Minkowski sum of all circuit fibers of π_A which we shall call $\Sigma(A)$ in the rest of this section. Using this secondary polytope we can prove Theorem 5.18 by establishing the following three linear analogs of results stated earlier.

Lemma 5.19 *Every fiber of π_A is a Minkowski summand of $\Sigma(A)$.*

Proposition 5.20 *Let db denote any probability measure with support cone($\mathcal{A}$) such that $\int_b bdb$ is finite. Then the Minkowski integral $\int_b P_b db$ is a polytope normally equivalent to $\Sigma(A)$.*

Proposition 5.21 *The fan $\mathcal{N}(\Sigma(A))$, is the collection of all cones $\overline{\mathcal{L}}_c$ together with their faces, as c varies over all generic cost vectors.*

As for Gröbner cones, each secondary cone has the same lineality space $rowspan(A)$ and one can work with its image in $\mathbf{R}^{n-d}$. We call this image the *pointed secondary fan* of A.

Example 2.2 continued. We saw that $\mathcal{A}$ in our running example had five distinct regular triangulations. Figure 10 shows the pointed secondary fan of A. The Minkowski sum of the five circuit fibers of π_A from Figure 7 is a pentagon in $\mathbf{R}^5$. □

Proposition 4.2 and Proposition 5.20 together imply the following theorem which is the linear counterpart of Theorem 5.12.

Theorem 5.22 *The circuits of A consists precisely of the primitive edge directions in the fibers of π_A.*

We conclude this section by establishing the relationship between the secondary and state polytope associated with a matrix A and hence between the secondary and Gröbner fan of A.

Proposition 5.23 *(i) The Gröbner fan of A is a refinement of the secondary fan of A.*
(ii) The secondary polytope of A is a summand of the state polytope of A.

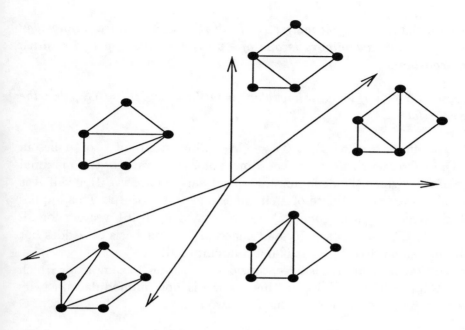

Figure 10: The pointed secondary fan of A.

Proof. From the general facts about Minkowski sums in the introduction, it suffices to prove (i). Let c, c' be generic cost vectors with respect to IP_A that belong to the same Gröbner cone $\mathcal{K}_c$. Then $in_c(I_A) = in_{c'}(I_A)$ and hence by Theorem 3.11, $\Delta_c = \Delta_{c'}$. This shows that every top dimensional cell of the Gröbner fan lies entirely within a top dimensional cell of the secondary fan and hence the Gröbner fan is a refinement of the secondary fan. □

The above proposition proves our claim in the introduction that cost functions that are generic with respect to IP_A are also generic with respect to LP_A since it proves that the interior of every Gröbner cone is contained in the interior of a secondary cone. The converse is true if and only if the secondary fan and Gröbner fan coincides.

6 Unimodular matrices

Unimodular matrices occupy a very special place among all matrices, in the context of the theory developed above.

Definition 6.1 *An integral matrix A of full row rank is called unimodular if each of its maximal minors is one of $-p$, 0 or p, where p is a positive integral constant.*

Theorem 6.2 [26] *If A is unimodular, then $C_A = UGB_A = Gr_A$ but the converse is false.*

Proof. It suffices to show that $C_A = Gr_A$. Since a circuit u of A lies in $ker_\mathbf{Z}(A)$, by Cramer's rule, every coordinate of u is the ratio of two maximal minors of A of which the denominator is necessarily non-zero. Hence if A is unimodular, every coordinate of a circuit is one of $0, 1$ or -1. This implies that the extreme rays of $cone(S_\sigma)$ are generated by $0, \pm 1$ vectors for all $\sigma \in \{+,-\}^n$. Therefore, each H_σ and hence the Graver basis Gr_A does not contain any integral vector that is not a circuit of A.

To see that the converse is false, consider our running example for which $C_A = UGB_A = Gr_A$. However this matrix is not unimodular since its maximal minors have absolute value one, two and three. $\square$

Theorem 6.3 [26] *If A is unimodular, then the secondary polytope of A coincides with the state polytope of A, but the converse is false.*

Proof. If A is unimodular then the integer programming fiber P_b^I coincides with the linear programming fiber P_b for all $b \in cone_\mathbf{N}(\mathcal{A})$ ([24], Theorem 19.2). Moreover, if $b \in cone(\mathcal{A}) \setminus cone_\mathbf{N}(\mathcal{A})$, then there exists $b' \in cone_\mathbf{N}(\mathcal{A})$ such that P_b and $P_{b'}$ are normally equivalent. Therefore the Minkowski integrals in Theorems 5.18 (iv) and 5.10 (iv) coincide, which proves that $St(A) = \Sigma(A)$.

To see that the converse is false, consider the matrix

$$A = \begin{bmatrix} 1 & -2 & 1 & 0 & 0 & 0 & 0 & 0 \\ 1 & -1 & 0 & 1 & 0 & 0 & 0 & 0 \\ 1 & 0 & 0 & 0 & 1 & 0 & 0 & 0 \\ 0 & 1 & 0 & 0 & 0 & 1 & 0 & 0 \\ 0 & 0 & 1 & 0 & 0 & 0 & 1 & 0 \\ 0 & 0 & 0 & 1 & 0 & 0 & 0 & 1 \end{bmatrix}.$$

This matrix is the Lawrence lifting of the two by four matrix in its top left corner. Its Graver basis consists precisely of the four circuits. There are eight distinct reduced Gröbner bases associated with this matrix each of which corresponds to a distinct triangulation. This implies that the state and secondary polytopes coincide. However, A is not unimodular since it has maximal minors of absolute value zero, one and two. □

We now examine unimodular matrices that are of *Lawrence type*. A matrix is of Lawrence type if it is the Lawrence lifting of another matrix.

Definition 6.4 *The circuit arrangement of A is the hyperplane arrangement consisting of the hyperplanes in $\mathbf{R}^n$ which are orthogonal to the circuits of A.*

The following result was stated in Lemma 5.2 of [3].

Proposition 6.5 *(i) The circuit arrangement of A is a refinement of the secondary fan of A.*
(ii) If A is of Lawrence type, then the circuit arrangement equals the secondary fan.

Definition 6.6 *The* Graver arrangement *of A is the arrangement consisting of the hyperplanes in $\mathbf{R}^n$ which are orthogonal to the elements in the Graver basis Gr_A.*

The following proposition is then a direct consequence of Theorem 4.6 and Proposition 4.15.

Proposition 6.7 *(i) The Graver arrangement of A is a refinement of the Gröbner fan of A.*
(ii) The Graver arrangement of A is a refinement of the circuit arrangement of A.

Invoking Theorem 4.8 we get the following easy corollary.

Corollary 6.8 *If A is of Lawrence type, then the state polytope of A is a zonotope (i.e., a Minkowski sum of line segments). The Gröbner fan of A coincides with the Graver arrangement of A.*

However, if a matrix A is unimodular and of Lawrence type, then the above results imply the following.

Corollary 6.9 *If the matrix A is unimodular and of Lawrence type, then its secondary fan, circuit arrangement, Gröbner fan and Graver arrangement all coincide.*

Example 6.10 Let A_G be the vertex-edge incidence matrix of a directed graph $G = (V, E)$, and consider the *capacitated transshipment problem:*

minimize $c \cdot x$ subject to $A_G \cdot x = b$ and $0 \leq x \leq b'$, $x \in \mathbf{Z}^E$.

When rewriting this integer program in the form (2.1), we get the enlarged coefficient matrix $\Lambda(A_G) = \begin{pmatrix} A_G & 0 \\ 1 & 1 \end{pmatrix}$. This matrix has format $(|E|+|V|) \times 2|V|$ and it is unimodular and of Lawrence type. Hence, for the family of flow problems $IP_{\Lambda(A_G)}$, the secondary fan, the circuit arrangement, the Gröbner fan and the Graver arrangement all coincide.

7 Implementation Issues

We conclude in this section with a brief discussion of issues relating to the implementation of the Gröbner basis method for integer programming.

In Section 3 we saw some of the problems that have to be faced while solving an integer program using the Conti-Traverso algorithm. The first issue is finding a generating set for the toric ideal, I_A, which can be a stumbling block for large integer programs. By Theorem 4.8, finding a generating set for a toric ideal is in the worst case as hard as finding a universal Gröbner basis of the ideal. In theory, generators for I_A can be found using the original Conti-Traverso algorithm via a Gröbner basis computation on a larger ideal in at most $d + 1$ extra variables. In practice, this method works only when the integer programs at hand are reasonably small. Newer algorithms given in [14] and [21] run entirely within the polynomial ring $k[\mathbf{x}]$ and have been found to be faster (as one might expect). Both these algorithms use many relatively short Gröbner basis computations.

Once a generating set for I_A has been found, the reduced Gröbner basis $\mathcal{G}_c$ has to be calculated. This can be done in principle using any software package that computes Gröbner bases like Macaulay 2. As the problem size increases this computation may become difficult or sometimes impossible. In many situations, one can push the computations further by exploiting the specific structure of the integer program at hand. One such methodology was adopted in [29] where it was possible to decompose the Gröbner basis computation into smaller sub-computations that could be carried out efficiently. The perfect f-matching problem was studied in [12] where it was shown that the Gröbner basis elements can be interpreted graph theoretically. Recently Hayer and Hochstättler [19] investigated Gröbner bases of *vertex cover* problems where once again graphical interpretations of the Gröbner basis is possible. In these situations combinatorial arguments aid both in the computation and understanding of $\mathcal{G}_c$ and UGB_A. Many other special cases can be found in [27]. The recent paper [10] introduces certain decomposition techniques that aid in the computation of test sets.

The software GRIN (GRöbner bases for INteger programming) exploits the combinatorial interpretation of the Buchberger algorithm possible in the case of toric ideals, along with some of the usual criteria used to speed up Gröbner bases computations. A detailed discussion of the special features implemented in GRIN can be found in [21]. This paper also reports experiments with randomly generated integer matrices of sizes ranging from 4×8 to 8×16 with non-negative entries in the range 0 to 20 and comparisons with CPLEX. According to their results GRIN becomes competitive with CPLEX when the matrix A is dense and randomly generated with entries that are not all 0/1. In these situations, the optimal solution of the integer program is typically far from the optimal solution of its linear relaxation. The expensive part of the work done by GRIN is in the computation of $\mathcal{G}_c$, after which reducing a non-optimal solution in a given fiber to the optimum was found to be extremely fast. The computation of $\mathcal{G}_c$ has to be thought of as a preprocessing procedure given the matrix A and cost vector c after which any $IP_{A,c}(b)$ for $b \in cone_\mathbf{N}(\mathcal{A})$ can be solved very quickly. Traditional algorithms for integer programming typically need to re-start calculations if the right hand side vector is changed.

Often in practice, one is interested in solving $IP_{A,c}(b)$ for a fixed b. In this situation, typically only a small fraction of $\mathcal{G}_c$ is required. So a natural question to ask is if the Buchberger algorithm can be truncated with respect to the right hand side vector b to output a subset of $\mathcal{G}_c$ that will be sufficient to solve $IP_{A,c}(b)$. A combinatorial algorithm in this direction was given in [32] which was generalized and placed in an algebraic setting in [31]. The idea involves imposing a multivariate grading on I_A with respect to $cone_\mathbf{N}(\mathcal{A})$ and a partial order on the elements of $cone_\mathbf{N}(\mathcal{A})$. The Buchberger algorithm is then truncated with respect to a given b to produce a test set sufficient for $IP_{A,c}(b')$ where b' is any right hand side vector less than or equal to b in the above partial order.

References

[1] W.W. Adams and P. Loustaunau, *An Introduction to Gröbner Bases*,

American Mathematical Society, Graduate Studies in Math., Vol. III, 1994.

[2] D. Bayer and I. Morrison, Gröbner bases and geometric invariant theory I, *Journal of Symbolic Computation* Vol.6 (1988) pp. 209-217.

[3] L.J. Billera, P. Filliman and B. Sturmfels, Constructions and complexity of secondary polytopes, *Advances in Mathematics* Vol. 83 (1990) pp. 155-179.

[4] L.J. Billera, I.M. Gel'fand and B. Sturmfels, Duality and minors of secondary polyhedra, *Journal of Combinatorial Theory* **B** Vol. 57 (1993) pp. 258-268.

[5] L.J. Billera and B. Sturmfels, Fiber polytopes, *Annals of Mathematics* Vol. 135 (1992) pp. 527–549.

[6] C.E. Blair and R.G. Jeroslow, The value function of an integer program, *Mathematical Programming* Vol.23 (1982) pp. 237-273.

[7] B. Buchberger, *On Finding a Vector Space Basis of the Residue Class Ring Modulo a Zero Dimensional Polynomial Ideal* (German), Ph.D. Thesis, Univ of Innsbruck, Austria, 1965.

[8] P. Conti and C. Traverso, Gröbner bases and integer programming, *Proceedings AAECC-9* (New Orleans), Springer Verlag, LNCS Vol.539 (1991) pp. 130-139.

[9] W. Cook, A.M.H. Gerards, A. Schrijver and É. Tardos, Sensitivity theorems in integer linear programming, *Mathematical Programming* Vol.34 (1986) pp. 251-264.

[10] G. Cornuejols, R. Urbaniak, R. Weismantel and L. Wolsey, Decomposition of integer programs and of generating sets, Fifth Annual European Symposium on Algorithms (ESA'97), Graz, Austria, 1997. To appear in *LNCS*, Springer-Verlag.

[11] D. Cox, J. Little and D. O'Shea, *Ideals, Varieties, and Algorithms*, Second edition, Springer-Verlag, New York, 1996.

[12] J. de Loera, B. Sturmfels and R.R.Thomas, Gröbner bases and triangulations of the second hypersimplex, *Combinatorica*, Vol.15 (1995) pp. 409-424.

[13] P. Diaconis and B. Sturmfels, Algebraic algorithms for sampling from conditional distributions, *Annals of Statistics*, to appear.

[14] F. Di Biase and R. Urbanke, An algorithm to calculate the kernel of certain polynomial ring homomorphisms, *Experimental Mathematics*, Vol.4 (1995) pp. 227-234.

[15] W. Fulton, *Introduction to Toric Varieties*, Princeton University Press, Princeton, New Jersey, 1993.

[16] I.M. Gel'fand, M. Kapranov and A. Zelevinsky, *Multidimensional Determinants, Discriminants and Resultants*, Birkhäuser, Boston, 1994.

[17] J.E. Graver, On the foundations of linear and integer programming I, *Mathematical Programming* Vol.8 (1975) pp. 207-226.

[18] D. Grayson and M. Stillman, *Macaulay 2:* a computer algebra system, available from http://www.math.uiuc.edu/~dan/.

[19] M.Hayer and W.Hochstättler, personal communication.

[20] S. Hosten, *Degrees of Gröbner Bases of Integer Programs*, Ph. D. Thesis, Cornell University, 1997.

[21] S. Hosten and B. Sturmfels, GRIN : An implementation of Gröbner bases for integer programming, in *Integer Programming and Combinatorial Optimization* (E. Balas and J. Clausen eds.), LNCS Vol. 920 (1995) pp. 267-276.

[22] T. Mora and L. Robbiano, The Gröbner fan of an ideal, *Journal of Symbolic Computation* Vol.6 (1988) pp. 183-208.

[23] H.E. Scarf, Neighborhood systems for production sets with indivisibilities, *Econometrica* Vol. 54 (1986) pp. 507-532.

[24] A. Schrijver, *Theory of Linear and Integer Programming*, Wiley-Interscience Series in Discrete Mathematics and Optimization, New York, 1986.

[25] B. Sturmfels, Gröbner bases of toric varieties, *Tôhoku Math. Journal* Vol. 43 (1991) pp. 249-261.

[26] B. Sturmfels, Asymptotic analysis of toric ideals, *Memoirs of the Faculty of Science*, Kyushu University Ser.A, Vol. 46 (1992) pp. 217-228.

[27] B. Sturmfels, *Gröbner Bases and Convex Polytopes*, American Mathematical Society, Providence, RI, 1995.

[28] B. Sturmfels and R.R.Thomas, Variation of cost functions in integer programming, *Mathematical Programming* Vol. 77 (1997) pp. 357-387.

[29] S.R. Tayur, R.R. Thomas and N.R. Natraj, An algebraic geometry algorithm for scheduling in the presence of setups and correlated demands, *Mathematical Programming*, Vol. 69 (1995) pp. 369-401.

[30] R.R. Thomas, A geometric Buchberger algorithm for integer programming, *Mathematics of Operations Research* Vol. 20 (1995) pp. 864-884.

[31] R.R.Thomas and R.Weismantel, Truncated Gröbner bases for integer programming, *Applicable Algebra in Engineering, Communication and Computing*, to appear.

[32] R.Urbaniak, R.Weismantel and G.Ziegler, A variant of Buchberger's algorithm for integer programming, *SIAM J. on Discrete Mathematics*, Vol. 1 (1997) pp. 96-108.

[33] G. Ziegler, *Lectures on Polytopes*, Graduate Texts in Mathematics, Springer Verlag, New York, 1995.

HANDBOOK OF COMBINATORIAL OPTIMIZATION (VOL. 1)
D.-Z. Du and P.M. Pardalos (Eds.) pp. 573-746
©1998 Kluwer Academic Publishers

Applications of Set Covering, Set Packing and Set Partitioning Models: A Survey

R.R. Vemuganti
Merrick School of Business
University of Baltimore
1420 North Charles Street
Baltimore, Maryland 21201
E-mail: rvemuganti@ubmail.ubalt.edu

Contents

1 Introduction   575

2 SP and SPT Models and Their Variants   576

3 Transformation of the Models   578
   3.1 SPT to SC 579
   3.2 SPT to SP 579
   3.3 SC to GSP and SP to GSC 580
   3.4 Zero-one MDK to GSP 580
   3.5 GSP to SP 582
   3.6 Zero-one LIP to Zero-one MD Knapsack 585
   3.7 Mixed SPT and SP to SP 586

4 Graphs and Networks   586
   4.1 Vertex (Node) Packing Problem 588
   4.2 Maximum Matching 589
   4.3 Minimum Covering Problem 592
   4.4 Chromatic Index and Chromatic Number 592
   4.5 Multi-Commodity Disconnecting Set Problem 594
   4.6 Steiner Problem in Graphs 596

5	**Personnel Scheduling Models**	**598**
	5.1 Days Off or Cyclical Scheduling	599
	5.2 Shift Scheduling	601
	5.3 Tour Scheduling	602
6	**Crew Scheduling**	**603**
7	**Manufacturing**	**609**
	7.1 Assembly Line Balancing Problem	609
	7.2 Discrete Lot Sizing and Scheduling Problem	610
	7.3 Ingot Size Selection	611
	7.4 Spare Parts Allocation	613
	7.5 Pattern Sequencing in Cutting Stock Operations	613
	7.6 Constrained Network Scheduling Problem	615
	7.7 Cellular Manufacturing	617
8	**Miscellaneous Operations**	**618**
	8.1 Frequency Planning	618
	8.2 Timetable Scheduling	619
	8.3 Testing and Diagnosis	620
	8.4 Political Districting	622
	8.5 Information Retrieval and Editing	623
	8.6 Check Clearing	625
	8.7 Capital Budgeting Problem	626
	8.8 Fixed Charge Problem	627
	8.9 Mathematical Problems	628
9	**Routing**	**628**
	9.1 Traveling Salesman Problem	629
	9.2 Single Depot Vehicle Routing	631
	9.3 Multiple Depots and Extensions	633
10	**Location**	**633**
	10.1 Plant Location Problem	634
	10.2 Lock Box Location Problem	636
	10.3 P-Center Problem	637
	10.4 P-Median Problem	638
	10.5 Service Facility Location Problem	638
11	**Review of Bibliography**	**640**
	11.1 Theory	641
	11.2 Transformations	642
	11.3 Graphs	642
	11.4 Personnel Scheduling	643
	11.5 Crew Scheduling	644

11.6 Manufacturing 645
11.7 Miscellaneous Operations 646
11.8 Routing 647
11.9 Location 652

12 Conclusions   **658**

References

Abstract

Set covering, set packing and set partitioning models are a special class of linear integer programs. These models and their variants have been used to formulate a variety of practical problems in such areas as capital budgeting, crew scheduling, cutting stock, facilities location, graphs and networks, manufacturing, personnel scheduling, vehicle routing and timetable scheduling among others. Based on the special structure of these models, efficient computational techniques have been developed to solve large size problems making it possible to solve many real world applications. This paper is a survey of the applications of the set covering, set packing, set partitioning models and their variants, including generalizations.

1 Introduction

Set covering (SC), set packing (SP) and set partitioning (SPT) problems are a very useful and important class of linear integer programming models. A variety of practical problems have been formulated as one of these models or their variants. Because of their special structure, it has been possible to develop efficient techniques (see the list of references under the theory) to solve large size problems. Due to the facility to solve large problems and the flexibility to model a variety of systems coupled with the advances in computer technology, these models have been used to solve many real world problems. This paper is a survey of SC, SP and SPT formulations (including their variants) of capital budgeting, crew scheduling, cutting stock, facilities location, graphs and networks, manufacturing, personnel scheduling, vehicle routing and time table scheduling problems among others. In addition, relationships among these models and the conversion of finite linear integer programs (LIP) to one of these models are presented.

Another objective of this paper is to provide an extensive bibliography on both theory and applications of SC, SP and SPT models. For convenience, the reference list is divided into nine groups namely, general, theory, graphs, personnel scheduling, crew scheduling, manufacturing, miscellaneous operations, routing and location. The general list consists of useful references related to general integer programming techniques and concepts including Lagrangian relaxation, surrogate constraints, subgradient methods, tabu search, disjunctive programming, branch and bound and cutting plane methods. These references are selected from the papers listed in the other eight groups which deal with both application and theory of SC, SP and SPT models. The theory list contains articles exclusively devoted to algorithms and mathematical properties of the SC, SP and SPT models. The remaining seven groups deal with papers related to the specific application area the title suggests except the group miscellaneous operations. The miscellaneous operations group contains papers related to a variety of applications including time table scheduling, information retrieval, political redistricting, diagnostic systems, distribution of broadcasting frequencies among others. It should be noted that many papers listed in the application group contain theoretical contributions also. Also some papers in the application group especially in location and routing, may be concerned with nonlinear programming models and general integer programming models. The purpose of including such articles is to provide as comprehensive a bibliography as possible. The list of references included in this paper is by no means complete. The interested reader may review the list of the papers included in the bibliography for additional references. The author apologizes, if some useful and relevant papers are not included.

The next section deals with the basic SC, SP, and SPT models and their variants. The subsequent sections address relationships among these models including LIP and each application group. Several numerical examples are provided throughout the paper to illustrate various applications.

2 SP and SPT Models and Their Variants

Consider a finite set $M = 1, 2, ..., m$ and $M_j, j \in N$, a collection of subsets of the set M where $N = 1, 2, ..., n$. A subset $F \subseteq N$ is called a *cover* of M if $\cup_{j \in F} M_j = M$. The subset $F \subseteq N$ is called a *packing* of M if $M_j \cap M_k = \emptyset$

Set Covering, Set Packing and Set Partitioning

for all $j, k \in F$ and $j \neq k$. If $F \subseteq N$ is both a cover and packing then it is called a *partitioning*.

Suppose c_j is the cost associated with M_j. Then the set covering problem is to find a minimum cost cover. If c_j is the value or weight of M_j, then the set packing problem is to find a maximum weight or value packing. Similarly the set partitioning problem is to find a partitioning with minimum cost. These problems can be formulated as zero-one linear integer programs as shown below. For all $i \in M$ and $j \in N$ let

$$a_{ij} = \begin{cases} 1 & \text{if } i \in M_j \\ 0 & \text{otherwise} \end{cases}$$

and

$$x_j = \begin{cases} 1 & \text{if } j \in F \\ 0 & \text{otherwise} \end{cases}$$

Then the set covering (1), set packing (2) and set partitioning (3) formulations are given by

$$\begin{aligned} \min \quad & \sum_{j=1}^{n} c_j x_j \\ \text{s.t.} \quad & \sum_{j=1}^{n} a_{ij} x_j \geq 1 \quad i = 1, 2, ..., m \\ \text{and} \quad & x_j = 0, 1 \quad j = 1, 2, ..., n \end{aligned} \quad (1)$$

$$\begin{aligned} \max \quad & \sum_{j=1}^{n} c_j x_j \\ \text{s.t.} \quad & \sum_{j=1}^{n} a_{ij} x_j \leq 1 \quad i = 1, 2, ..., m \\ \text{and} \quad & x_j = 0, 1 \quad j = 1, 2, ..., n \end{aligned} \quad (2)$$

$$\begin{aligned} \max \quad & \sum_{j=1}^{n} c_j x_j \\ \text{s.t.} \quad & \sum_{j=1}^{n} a_{ij} x_j = 1 \quad i = 1, 2, ..., m \\ \text{and} \quad & x_j = 0, 1 \quad j = 1, 2, ..., n \end{aligned} \quad (3)$$

The following numerical example illustrates the above models.

Example 1 Suppose $M = \{1, 2, 3, 4, 5, 6\}$, $M_1 = \{1, 2\}$, $M_2 = \{1, 3, 4\}$, $M_3 = \{2, 4, 5\}$, $M_4 = \{3, 5, 6\}$, $M_5 = \{4, 5, 6\}$, $c_1 = 5$, $c_2 = 4$, $c_3 = 6$, $c_4 = 2$, and $c_5 = 4$. The formulation of the set covering model is given by

$$\begin{aligned} \min \quad & 5x_1 + 4x_2 + 6x_3 + 2x_4 + 4x_5 \\ \text{s.t.} \quad & x_1 + x_2 \geq 1 \end{aligned}$$

$$x_1 + x_3 \geq 1$$
$$x_2 + x_4 \geq 1$$
$$x_2 + x_3 + x_5 \geq 1$$
$$x_3 + x_4 + x_5 \geq 1$$
$$x_4 + x_5 \geq 1$$
$$\text{and} \quad x_1, x_2, x_3, x_4, x_5 \in \{0,1\}$$

The formulation of the corresponding set packing and set partitioning models is straight forward.

Clearly, the three models are special structured zero-one linear integer programs since all the elements of the constraint coefficient matrix $A = (a_{ij})$ are 0 or 1 and the right hand side (RHS) of the constraints are all unity. For some applications the RHS of the constraints may be not all be unity but positive integers. The corresponding models are called general set covering (GSC), general set packing (GSP) and general set partitioning (GSPT). For these general models, while the variables are required to be non-negative integers, they need not be constrained to zero or one. For some other applications the constraint set may include two types of inequalities or all three types of inequalities. Such models are called mixed models. For example, a model with both less than or equal to and greater than or equal to constraints with all RHS equal to unity is called mixed SC and SP model.

It should be noted that the integer programming formulations of the SC, SP and SPT models including their generalizations are NP-complete, except for a few special cases. The relationships among these models and the transformation of the linear integer programs, including the zero-one multi-dimensional knapsack problem into a SP problem, are presented next.

3 Transformation of the Models

Transformation of the models is useful in comparing various computational techniques to generate the optimal solutions. Several transformations to convert one model to another model including the conversion of a zero-one LIP to a zero-one multi-dimensional knapsack (MDK) problem and the conversion of a MDK problem to a SP problem are explored in this section.

3.1 SPT to SC

Subtracting artificial surplus variables y_i from the ith constraint of the SPT model formulation (3), the constraints of the SPT model can be written as:

$$\sum_{j=1}^{n} a_{ij} x_j - y_i = 1.$$

To insure all artificial surplus variables remain at zero level in any optimal solution (when SPT is feasible), the objective function is changed to

$$\sum_{j=1}^{n} c_j x_j + \theta \sum_{i=1}^{m} y_i$$

where θ is any number greater than $\sum_j c_j$. Substituting for y_i in the objective function and eliminating y_i from the constraints, the following SC formulation yields an optimal solution to the SPT problem.

$$\begin{aligned}
\min \quad & \sum_{j=1}^{n} c'_j x_j - m\theta \\
\text{s.t.} \quad & \sum_{j=1}^{n} a_{ij} x_j \geq 1 \quad i = 1, 2, ..., m \\
\text{and} \quad & x_j = 0, 1 \quad j = 1, 2, ..., n
\end{aligned} \quad (4)$$

where $c'_j = c_j + \theta \sum_{j=1}^{m} a_{ij}$.

3.2 SPT to SP

Adding artificial slack variables y_i to the ith constraint of the SPT model formulation (3), the constraints of the SPT model can be written as:

$$\sum_{j=1}^{n} a_{ij} x_j + y_i = 1.$$

As noted earlier, the modified objective function

$$\sum_{j=1}^{n} c_j x_j + \theta \sum_{i=1}^{m} y_i$$

guarantees that all artificial slack variables remain at zero level in any optimal solution. Substituting for y_i in the objective function and eliminating

y_i from the constraints, the following formulation yields an optimal solution to the SPT problem.

$$\begin{aligned}
\min \quad & \sum_{j=1}^n c'_j x_j + m\theta \\
\text{s.t.} \quad & \sum_{j=1}^n a_{ij} x_j \leq 1 \quad i=1,2,...,m \\
\text{and} \quad & x_j = 0,1 \quad j=1,2,...,n
\end{aligned} \quad (5)$$

where $c'_j = c_j - \theta \sum_{i=1}^m a_{ij}$. Noting the fact c'_j are negative numbers and changing the objective function to

$$\max \sum_{j=1}^n (-c'_j) x_j$$

results in a SP formulation.

3.3 SC to GSP and SP to GSC

Substituting $x_j = 1 - y_j$, the formulation (1) of the SC problem is equivalent to

$$\begin{aligned}
\max \quad & \sum_{j=1}^n c_j y_j - \sum_{j=1}^n c_j \\
\text{s.t.} \quad & \sum_{j=1}^n a_{ij} y_j \leq b_i \quad i=1,2,...,m \\
\text{and} \quad & y_j = 0,1 \quad j=1,2,...,n
\end{aligned} \quad (6)$$

where $b_i = \sum_{j=1}^n a_{ij}$. Noting the fact that all b_i are nonnegative integers (when SC is feasible), the above is a GSP formulation of the SC problem. A similar transformation can be used to convert a SP problem into a GSC problem.

3.4 Zero-one MDK to GSP

The LIP formulation of a zero-one MDK problem is

$$\begin{aligned}
\max \quad & \sum_{j=1}^n c_j x_j \\
\text{s.t.} \quad & \sum_{j=1}^n a_{ij} x_j \leq b_i \quad i=1,2,...,m \\
\text{and} \quad & x_j = 0,1 \quad j=1,2,...,n
\end{aligned} \quad (7)$$

where a_{ij}, c_j, and b_i are all nonnegative numbers. When all numbers involved are rational they can be converted to integers by multiplying them with an appropriate positive integer. In transforming this problem to a GSP problem it is assumed that all a_{ij}, c_j, and b_i are nonnegative integers. Let

$$a_j = \max_{i=1}^{m} a_{ij} \quad j = 1, 2, ..., n, \text{ and}$$

$$x_{jk} = 0, 1 \quad \begin{array}{l} j = 1, 2, ..., n \\ k = 1, 2, ..., a_j \end{array}$$

If all x_{jk} are guaranteed to be equal for a given j, clearly (when $a_{ij} > 0$)

$$a_{ij} x_j = \sum_{k=1}^{a_{ij}} x_{jk}.$$

Then an equivalent formulation of the MDK problem is given by

$$\begin{array}{ll} \max & \sum_{j=1}^{n} \sum_{k=1}^{a_j} c_{jk} x_{jk} \\ \text{s.t.} & \sum_{j=1}^{n} \sum_{k=1}^{a_{ij}} x_{jk} \leq b_i \quad i = 1, 2, ..., m \\ & x_{jk} + z_j = 1 \quad \begin{array}{l} j = 1, 2, ..., n \\ k = 1, 2, ..., a_j \end{array} \\ & x_{jk} = 0, 1 \quad \begin{array}{l} j = 1, 2, ..., n \\ k = 1, 2, ..., a_j \end{array} \\ \text{and} & z_j = 0, 1 \quad j = 1, 2, ..., n \end{array} \quad (8)$$

where $c_{jk} = c_j/a_j$ for $k = 1, 2, ..., a_j$. It should be noted that the second set of constraints guarantee that all x_{jk} are equal for a given j. Add artificial slack variables to the equality constraints with a very large negative coefficient in the objective function as in the conversion of the SPT problem to SP problem and eliminate the artificial slack variables from both the objective function and the constraints to obtain the required formulation.

Example 2. Consider the following zero-one two dimensional knapsack problem.

$$\begin{array}{ll} \max & 10x_1 + 4x_2 + 12x_3 + x_4 \\ \text{s.t.} & 3x_1 + x_2 + 4x_3 + x_4 \leq 4 \\ & x_1 + 2x_2 + 2x_3 + 2x_4 < 3 \\ \text{and} & x_j = 0, 1 \quad j = 1, 2, 3, 4. \end{array}$$

Since $a_1 = \max(3,1) = 3$, $a_2 = \max(1,2) = 2$, $a_3 = \max(4,2) = 4$, and $a_4 = \max(1,2) = 2$, by defining eleven $x_{jk}(x_{11}, x_{12}, x_{13}, x_{21}, x_{22}, x_{31}, x_{32}, x_{33}, x_{34},$

x_{41}, and x_{42}) and four $z_j(z_1, z_2, z_3,$ and $z_4)$ zero-one variables, the problem can be reformulated as

max $(10/3)(x_{11} + x_{12} + x_{13}) + 2(x_{21} + x_{22}) + 3(x_{31} + x_{32} + x_{33} + x_{34})$
$\qquad +(1/2)(x_{41} + x_{42})$

s.t. $\quad x_{11} + x_{12} + x_{13} + x_{21} + x_{31} + x_{32} + x_{33} + x_{34} + x_{41} \leq 4$

$\qquad x_{11} + x_{21} + x_{22} + x_{31} + x_{32} + x_{41} + x_{42} \leq 3$

$\qquad x_{1k} + z_1 = 1 \qquad k = 1, 2, 3$

$\qquad x_{2k} + z_2 = 1 \qquad k = 1, 2$

$\qquad x_{3k} + z_3 = 1 \qquad k = 1, 2, 3, 4$

$\qquad x_{4k} + z_4 = 1 \qquad k = 1, 2$

$\qquad z_j = 0, 1 \qquad j = 1, 2, 3, 4$

$\qquad x_{1k} = 0, 1 \qquad k = 1, 2, 3$

$\qquad x_{2k} = 0, 1 \qquad k = 1, 2$

$\qquad x_{3k} = 0, 1 \qquad k = 1, 2, 3, 4$

and $\quad x_{4k} = 0, 1 \qquad k = 1, 2.$

The equality constraints can be changed to less than or equal to constraints by subtracting the following expression

$$\theta(\sum_{k=1}^{3} x_{1k} + \sum_{k=1}^{2} x_{2k} + \sum_{k=1}^{4} x_{3k} + \sum_{k=1}^{2} x_{4k} + 3z_1 + 2z_2 + 4z_3 + 2z_4) - 11\theta$$

from the objective function where $\theta > (10 + 4 + 12 + 1) = 27$.

3.5 GSP to SP

The LIP formulation of a GSP problem when the variables are restricted to binary is

$$\begin{aligned} \max \quad & \sum_{j=1}^{n} c_j x_j \\ \text{s.t.} \quad & \sum_{j=1}^{n} a_{ij} x_j \leq b_i \quad i = 1, 2, ..., m \\ \text{and} \quad & x_j = 0, 1 \quad j = 1, 2, ..., n \end{aligned} \qquad (9)$$

where b_i are positive integers. In order to transform this to a SP problem, the ith constraint is replaced by b_i inequality constraints with all right hand sides equal to one. Let

$$y_{ijk} = 0, 1 \quad \text{for} \quad j = 1, 2, ..., n$$
$$i = 1, 2, ..., m$$
$$k = 1, 2, ..., b_i$$
$$z_j = 0, 1 \quad \text{for} \quad j = 1, 2, ..., n$$
$$\text{and} \quad b = \max_{i=1}^{m} b_i$$

Rearrange the constraints if necessary so that $b_1 = b$ and consider the following mixed SP and SPT formulation.

$$\max \quad \sum_{j=1}^{n} c_j \sum_{k=1}^{b_i} y_{1jk}$$
$$\text{s.t.} \quad \sum_{j=1}^{n} a_{ij} y_{ijk} \leq 1 \quad i = 1, 2, ..., m$$
$$k = 1, 2, ..., b_i$$
$$\sum_{k=1}^{b_i} y_{ijk} + z_j = 1 \quad i = 1, 2, ..., m$$
$$j = 1, 2, ..., n$$
$$y_{ijk} = 0, 1 \quad j = 1, 2, ..., n$$
$$i = 1, 2, ..., m$$
$$k = 1, 2, ..., b_i$$
$$\text{and} \quad z_j = 0, 1 \quad j = 1, 2, ..., n \qquad (10)$$

To see the equivalence between these two formulations summing up the inequalities over k for a given i yields

$$\sum_{j=1}^{n} a_{ij} \sum_{k=1}^{b_i} y_{ijk} \leq b_i.$$

From the equality constraints it is clear that

$$\sum_{k=1}^{b_i} y_{ijk} = x_j$$

is zero-one for all $i = 1, 2, ..., m$. It is straight forward to verify that x_j constructed from a feasible solution to the mixed SP and SPT formulation yields a feasible solution to the GSP problem. Now suppose x_j is a feasible solution to the GSP. If $a_{ij} = 0$ and $x_j = 1$ then set any one of the variables

$y_{ijk}, k = 1, 2, ..., b_i$ equal to 1 and the remaining to zero. If $x_j(t)$ and $a_{ij}(t)$ are equal to one for $t = 1, 2, ..., h$ (note that $h \leq b_i$), then set

$$y_{ij(t)t} = 1 \text{ for all } t = 1, 2, ..., h$$

and the rest of the variables to zero. This provides a feasible solution to the mixed SP and SPT formulation with objective values of both formulations the same. Since there is one to one correspondence, the mixed SP and SPT formulation yields an optimal solution to the GSP problem. The equalities can be replaced by inequalities to obtain the SP formulation using the procedure described in coverting SPT to SP.

When x_j are not binary, they can be replaced by binary variables $y_{jk}, k = 1, 2, ..., t_j$ using the transformation

$$x_j = \sum_{k=1}^{t_j} 2^k y_{jk}$$

where t_j is suitably chosen to insure all possible values of x_j are included. Since all x_j are bounded above ($\leq b$) it is possible to select such t_j. Now substituting for x_j, in the GSP formulation yields a zero-one MD knapsack problem which can be converted to a binary GSP problem.

Example 3. Consider the following binary GSP problem.

$$\begin{aligned} \max \quad & 4x_1 + 6x_2 + 7x_3 + 8x_4 + 10x_5 \\ \text{s.t.} \quad & x_2 + x_3 + x_4 + x_5 \leq 3 \\ & x_1 + x_2 + x_3 \leq 2 \\ & x_1 + x_3 + x_4 \leq 2 \\ & x_j = 0, 1 \quad j = 1, 2, 3, 4 \end{aligned}$$

Using several zero-one variables y_{ijk}, an equivalent mixed SP and SPT formulation is

$$\begin{aligned} \max \quad & \sum_{k=1}^{3}(4y_{11k} + 6y_{12k} + 7y_{13k} + 8y_{14k} + 10y_{15k}) \\ \text{s.t.} \quad & y_{12k} + y_{13k} + y_{14k} + y_{15k} \leq 1 & k = 1, 2, 3 \\ & y_{21k} + y_{22k} + y_{23k} \leq 1 & k = 1, 2 \\ & y_{31k} + y_{33k} + y_{34k} \leq 1 & j = 1, 2 \\ & \sum_{k=1}^{3} y_{1jk} + z_j = 1 & j = 1, 2, ..., 5 \\ & \sum_{k=1}^{2} y_{2jk} + z_j = 1 & j = 1, 2, ..., 5 \end{aligned}$$

$$\sum_{k=1}^{2} y_{3jk} + z_j = 1 \qquad j = 1, 2, ..., 5$$
$$y_{1jk} = 0, 1 \qquad \begin{array}{l} j = 1, 3, ..., 5 \\ k = 1, 2, 3 \end{array}$$
$$y_{2jk} = 0, 1 \qquad \begin{array}{l} j = 1, 2, ..., 5 \\ k = 1, 2 \end{array}$$
$$y_{3jk} = 0, 1 \qquad \begin{array}{l} j = 1, 2, ..., 5 \\ k = 1, 2 \end{array}$$
and
$$z_j = 0, 1 \qquad j = 1, 2, ..., 5$$

3.6 Zero-one LIP to Zero-one MD Knapsack

When the variables are bounded above, any LIP can be converted to a Zero-one LIP using the standard binary transformation. Consider the Zero-one LIP

$$\begin{array}{ll} \max & \sum_{j=1}^{n} c_j x_j \\ \text{s.t.} & \sum_{j=1}^{n} a_{ij} x_j \leq b_i \quad i = 1, 2, ..., m \\ \text{and} & x_j = 0, 1 \quad j = 1, 2, ..., n \end{array} \qquad (11)$$

where all $b_i \geq 0$, and integers. When all $a_{ij} > 0$, no changes are necessary to convert the problem to a Zero-one MD Knapsack problem. When some $a_{ij} \leq 0$, let

$$a_{ij1} = \max(0, a_{ij}) \geq 0$$
$$\text{and} \quad a_{ij2} = \min(0, a_{ij}) \leq 0$$

for $i = 1, 2, ..., m$ and $j = 1, 2, ..., n$. For each variable x_j (only for j for which some $a_{ij} < 0$)

$$a_{ij} x_j = (a_{ij1} x_{j1} + a_{ij2} x_{j2})$$

when the binary variables x_{j1} and x_{j2} are equal. Replacing x_{j2} with $1 - y_{j2}$, an equivalent formulation of the Zero-one LIP is

$$\begin{array}{ll} \max & \sum_{j=1}^{n} c_j x_{j1} \\ \text{s.t.} & \sum_{j=1}^{n} (a_{ij1} x_{j1} + (-a_{ij2}) y_{j2}) \leq b_i + \sum_{j=1}^{n} (-a_{ij2}) \quad i = 1, 2, ..., m \\ & x_{j1} + y_{j2} = 1 \quad j = 1, 2, ..., n \\ & x_{j1} = 0, 1 \\ & y_{j2} = 0, 1 \quad j = 1, 2, ..., n \end{array}$$

$$(12)$$

Since all $a_{ij2} \leq 0$, all the coefficients involved are nonnegative including the right hand sides of the inequalities. Using the standard procedure of adding artificial slack variables, the equality constraints can be converted to inequality constraints which yields the equivalent Zero-one Knapsack formulation.

3.7 Mixed SPT and SP to SP

The equality constraints can be converted to less than or equal to constraints, by adding the artificial slack variables and eliminating them from both the constraints and the objective function. Similarly other mixed formula transformations presented in this section. Even though it is possible to convert one model to another it should be noted that some conversions require a large number of additional variables and constraints. In the next section, models related to Graphs and Networks are presented.

4 Graphs and Networks

Let N be a finite set of points and A be a finite set of ordered pair of points, (i,j), from the set N. The pair N and A is called a directed graph and is denoted by $G = (N, A)$. The elements of A are called nodes (or vertices) and the elements of A are called arcs. For any arc $(i,j) \in A$, i is called the beginning node and j is called the ending node. If the elements of A are unordered, they are called edges and the corresponding graph denoted by $G = (N, E)$, is called an undirected graph. Usually no distinction is made between graphs and networks. However, when a subset of nodes are singled out for a specific purpose such as sources and sinks to transport some commodity, the corresponding graph is called a network. A graph is called a bipartite graph if the nodes can be partitioned into two sets such that the beginning node of every arc belongs to one set and the ending node of every arc belongs to the other set. The cardinality of these sets N and A (or E) are denoted by n and m which represent the number of nodes and arcs (edges) respectively.

Two arcs (or two edges) are called adjacent if they have at least one node in common. Similarly two nodes are adjacent to each other if they are connected by an edge or arc. A chain is a sequences of arcs $(A_1, A_2, ... A_r)$ such that each arc has one node in common with its successor and predecessor with the exception of A_1 and A_r which have a common node with the successor and predecessor respectively. If i is the beginning node of A_1 and

j is the ending of A_r, then it is a chain from node i and to node j. If all the nodes encountered are distinct than it is called an elementary chain. If the beginning and end points of an elementary chain are the same then it is called a cycle. A path is a sequence of arcs $(A_1, A_2, ...A_r)$ such that the ending node of every arc in the sequence is the beginning node of the next arc. If i is the beginning node of A_1 and j is the ending node of A_r, then it is a path from node i to j. The path is elementary if all nodes encountered are distinct. If the beginning and ending nodes of a path are same then it is called a circuit. An undirected graph $G = (N, E)$ is called a tree if it has exactly $(n-1)$ arcs and has no cycles. A directed graph $G = (N, A)$, is called a tree if it contains exactly $m = n - 1$ arcs, has no circuits and every node is the ending node of exactly one arc except one node which is the beginning node of one or many arcs but not the ending node of any arc. The following examples are is used to illustrate the concepts.

Example 4. Suppose $N = \{1,2,3,4,5\}$ and $A = \{(1,2), (1,3), (4,1), (1,5), (2,3), (4,2), (3,4), (3,5), (4,5)\}$. In the above graph the arc set $A_1 =$

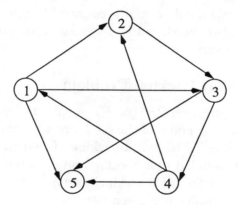

Figure 1: A Directed Graph

$\{(1,2), (2,3), (3,4)\}$ is an elementary path from node 1 to node 5, the arc set $A_2 = \{(1,2), (2,3), (3,4), (4,1)\}$ is a circuit, the arc set $A_3 = \{(1,5), (3,5)\}$ is an elementary chain, and the arc set $A_4 = \{(1,5), (4,5), (4,1)\}$ is a cycle. Clearly in undirected graphs there are only chains and cycles.

Example 5. Suppose $N = \{1,2,3,4,5,6\}$ and $A = \{(1,2), (1,3), (3,4), (1,5), (4,6)\}$. Clearly the above graph is a tree (in fact it is a called rooted tree with root 1).

There are many problems related to graphs and networks such as vertex packing (stability number), maximum matching, minimum covering, chro-

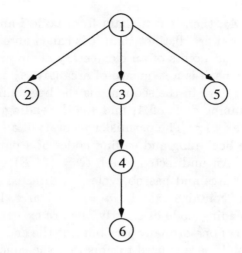

Figure 2: A Directed Tree

matic index, chromatic number, multi- commodity minimum disconnecting set and Steiner problem which can be formulated as one of the SC, SP and SPT models or their variants.

4.1 Vertex (Node) Packing Problem

Consider an undirected graph $G = (N, E)$. A subset of nodes P of N is called a *vertex packing* if no two nodes of the set P are adjacent to each other. The Vertex Packing problem is to find a packing of maximum cardinality. Let $x_i = 1$, if node i is included in a packing and $x_i = 0$ otherwise. Since two nodes connected by an edge cannot be included in a packing a SP formulation of the vertex packing problem is given by

$$\begin{aligned}
\max \quad & \sum_{i=1}^{n} x_i \\
\text{s.t.} \quad & x_i + x_j \leq 1 \quad (i,j) \in E \\
\text{and} \quad & x_j = 0, 1 \quad i = 1, 2, ..., n
\end{aligned} \quad (13)$$

The maximum value of the objective function is also called the stability number of a graph. To illustrate the usefulness of this model, consider a franchise business whose objective is to maximize the number of profitable franchises in a given area. Certain locations are so close to each other, if franchises are open in both neither will make a profit. Represent each location by a node and connect any two nodes by an edge, if the corresponding locations are unprofitable, when franchises are open in both. To provide another example (even though esoteric) consider placing eight queens on a chessboard so that no queen can capture another queen. In order to determine the feasibility of placing eight queens construct a graph with 64 nodes, each node representing a square on the chessboard and connect two nodes by an edge if the corresponding squares are in the same row or in the same column or in the same diagonal. If the corresponding stability number is eight or more it is possible to place eight queens on a chessboard without one capturing another.

When each node i is assigned a nonnegative weight of c_i and the coefficient of x_i is c_i in the objective function of the formulation (13), then it is called a weighted Vertex Packing Problem. An interesting application of this model is the transformation of a SP problem to a weighted Vertex Packing Problem. To see the connection between these two models, construct an undirected graph with n nodes, each node representing a variable in the formulation (2) and connect two nodes j and k by an edge if there is a constraint i such that $a_{ij} = a_{ik} = 1$. The equivalence between the SP problem and the weighted Vertex Packing Problem generated by the corresponding graph is illustrated below.

Example 6.

$$\max \quad \sum_{j=1}^{5} c_i x_j$$
$$\text{s.t.} \quad x_1 + x_2 + x_3 \leq 1$$
$$x_2 + x_3 + x_4 \leq 1$$
$$x_3 + x_4 + x_5 \leq 1$$
$$x_1 + x_3 + x_5 \leq 1$$
$$\text{and} \quad x_j = 0, 1 \quad j = 1, 2, ..., 5$$

The undirected graph with 5 nodes and 8 edges corresponding to the above SP problem is shown below. From the first constraint, it is clear that among the variables x_1, x_2, x_3 no two variables can be found equal to 1 in any feasible solution. This is equivalent to Vertex Packing constraints on nodes (1,2), (2,3) and (1,3).

4.2 Maximum Matching

Consider an undirected graph $G = (N, E)$. Two edges (or arcs) are said to be adjacent to each other if they have a node in common. A subset of

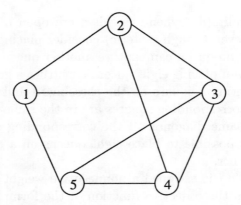

Figure 3: A Set Packing Graph

edges D is called a matching if no two edges in D are adjacent to each other. The maximum matching problem is to determine a matching of maximum cardinality. Let $x_j = 1$, if edge j is included in the matching and $x_j = 0$ otherwise. Also for each edge (i, j), let $a_{ij} = a_{ji} = 1$. Then the SP formulation of the maximum matching problem is given by

$$\begin{aligned}
\max \quad & \sum_{j=1}^{m} x_j \\
\text{s.t.} \quad & \sum_{j=1}^{m} a_{ij} x_j \leq 1, \quad i = 1, 2, ..., n \\
\text{and} \quad & x_j = 0, 1 \quad j = 1, 2, ..., m
\end{aligned} \qquad (14)$$

Example 7. To illustrate the above model consider the following graph with numbers on each edge representing the number assigned to each edge. The corresponding maximum matching problem is

$$\begin{aligned}
\max \quad & \sum_{j=1}^{7} x_j \\
\text{s.t.} \quad & x_1 + x_2 \leq 1 \\
& x_1 + x_3 + x_4 \leq 1 \\
& x_3 + x_5 \leq 1 \\
& x_4 + x_5 + x_6 \leq 1 \\
& x_6 + x_7 \leq 1 \\
& x_2 + x_7 \leq 1 \\
\text{and} \quad & x_j = 0, 1 \quad j = 1, 2, ..., 7
\end{aligned}$$

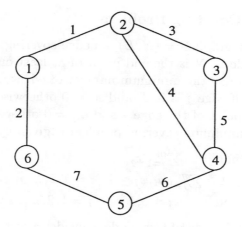

Figure 4: An Undirected Graph

To illustrate the application of this model, suppose there are p workers and q jobs and each worker is trained to perform at least one job. The problem is to determine whether each worker can be assigned to a job for which the individual is qualified. Consider a bipartite graph with p nodes corresponding to the workers in one group and q nodes corresponding to the jobs in the second group. Connect the node in the first group with an edge to a node in the second group if the individual is qualified to perform the corresponding job. If the value of the maximum matching for this graph is p then all workers can be assigned to jobs for which they are qualified.

A closely related problem is the standard assignment problem. Suppose there are n workers and n jobs. It costs c_{ij} if worker i is assigned to job j. The problem is to assign each worker to one job and each job to one worker so that the total cost is a minimum. Let $x_{ij} = 1$, if worker i is assigned to job j and $x_{ij} = 0$ otherwise. The SPT formulation of this problem is given by

$$
\begin{aligned}
\min \quad & \sum_{i=1}^{n} \sum_{j=1}^{n} c_{ij} x_{ij} \\
\text{s.t.} \quad & \sum_{j=1}^{n} x_{ij} = 1 \quad i = 1, 2, ..., n \\
& \sum_{i=1}^{n} x_{ij} = 1 \quad j = 1, 2, ..., n \\
\text{and} \quad & x_{ij} = 0, 1 \quad i = 1, 2, ..., n \\
& \phantom{x_{ij} = 0, 1} \quad j = 1, 2, ..., n
\end{aligned}
\tag{15}
$$

Efficient techniques have been developed to solve this problem which require polynomial computational time to determine the optimum solution.

4.3 Minimum Covering Problem

Given an undirected graph $G = (N, E)$, an edge covering F is a subset of E such that every node in N is the end point of at least one edge in F. The problem is to determine the minimum number of edges needed to cover all nodes. Let $x_j = 1$, if edge j is in F and $x_j = 0$ otherwise. Also let $a_{ij} = 1$ if node i is an end point of the edge j and $a_{ij} = 0$ otherwise. Then the SC formulation of the minimum covering problem is given by

$$\begin{aligned}\min \quad & \sum_{j=1}^m x_j \\ \text{s.t.} \quad & \sum_{j=1}^m a_{ij} x_j \leq 1 \quad i = 1, 2, ..., n \\ \text{and} \quad & x_j = 0, 1 \quad j = 1, 2, ..., m\end{aligned} \quad (16)$$

For a simple application of this model consider a fort with towers at the endpoints of each wall. A guard stationed at a wall can watch both towers at the end of the wall. The problem is to determine the minimum number of guards needed to watch all towers. Define a node for each tower and connect any two nodes by an edge if they are connected by a wall. Clearly, the minimum number of edges to cover all nodes yields the minimum number of guards needed to watch all the towers. Many more important and useful applications of this model are included in the section related to location problems.

Another related model is to cover all edges by nodes. That is find a subset of nodes P of N such that at least one node of every edge belongs to P. The problem is to determine the minimum number of nodes needed to cover all edges. Let $x_i = 1$ if node i is included in the cover and $x_i = 0$ otherwise. Then the SC formulation of the node covering problem is given by

$$\begin{aligned}\min \quad & \sum_{i=1}^n x_i \\ \text{s.t.} \quad & x_i + x_j \geq 1 \quad (i, j) \in E \\ \text{and} \quad & x_i = 0, 1 \quad i = 1, 2, ..., n\end{aligned} \quad (17)$$

It is easy to see the vertex packing problem (13) and the node covering problem (17) are closely related [set $x_i = 1 - y_i$ in formulation (17) to obtain the formulation (13)].

4.4 Chromatic Index and Chromatic Number

The chromatic index of an undirected graph is the minimum number of colors needed to color all edges of the graph so that no two adjacent edges receive the same color. It is clear that no more than m colors are needed since m represents the number of edges. Let $y_k = 1$, if color k is used and $y_k = 0$ otherwise, for $k = 1, 2, ..., m$. Also let $x_{jk} = 1$, if edge j is given color k and $x_{jk} = 0$ otherwise, for $j = 1, 2, ..., m$ and $k = 1, 2, ..., m$. Finally let $a_{ij} = 1$, if node i is an endpoint of edge j and $a_{ij} = 0$ otherwise for $i = 1, 2, ..., n$ and $j = 1, 2, ..., m$. Then an integer programming formulation is given by

$$\begin{aligned}
\min \quad & \sum_{k=1}^{m} y_k \\
\text{s.t.} \quad & \sum_{k=1}^{m} x_{jk} = 1 && j = 1, 2, ..., m \\
& \sum_{j=1}^{m} a_{ij} x_{jk} \leq y_k && k = 1, 2, ..., m \\
& && i = 1, 2, ..., n \\
& y_k = 0, 1 && k = 1, 2, ..., m \\
\text{and} \quad & x_{jk} = 0, 1 && k = 1, 2, ..., m \\
& && j = 1, 2, ..., m
\end{aligned} \quad (18)$$

The first set of constraints ensures that every arc is assigned a color and the second set of constraints guarantees that all edges adjacent to a given node receive at most one color if it is used to color some edge. Substituting $z_k = 1 - y_k$, the problem can be transformed to a mixed SP and SPT model.

A related problem is to find the minimum number of colors needed to color all nodes of an undirected graph so that no two adjacent nodes receive the same color. The minimum number of colors needed is called the chromatic number. Let $y_k = 1$, if color k is used and $y_k = 0$ otherwise for $k = 1, 2, ..., n$. Also let $x_{ik} = 1$, if node i given color k and $x_{ik} = 0$ otherwise for $i = 1, 2, ..., n$ and $k = 1, 2, ..., n$. In addition, let $a_{ij} = a_{ji} = 1$, if $(i, j) \in E$ and $a_{ij} = 0$ otherwise and $a_{ii} = 1$ for all $i = 1 2, ..., n$. Then a linear integer programming formulation of the problem is given by

$$\begin{aligned}
\min \quad & \sum_{k=1}^{n} y_k \\
\text{s.t.} \quad & \sum_{k=1}^{n} x_{ik} = 1 && k = 1, 2, ..., n \\
& \sum_{j=1}^{n} a_{ij} x_{jk} \leq y_k && i = 1, 2, ..., n \\
& && k = 1, 2, ..., n \\
\text{and} \quad & y_k = 0, 1 && k = 1, 2, ..., n \\
& x_{ik} = 0, 1 && k = 1, 2, ..., n \\
& && i = 1, 2, ..., n.
\end{aligned} \quad (19)$$

The first set of constraints guarantees that every node receives exactly one color and the second set of constraints ensures that none of the nodes adjacent to a given node receive the same color. Substituting $z_k = 1 - y_k$, this problem also can be transferred to a mixed SP and SPT model.

To illustrate a simple application of this model suppose at the end of an academic year several students must take oral exams from several professors. The problem is to determine the minimum number of periods needed to schedule the oral examinations. During an oral exam only one student can be examined by a professor during any period. To model this problem construct a bipartite graph with N_1 nodes representing the students and N_2 nodes representing the professors. Connect a node $i \in N_1$ and $j \in N_2$ with an edge if student i must be examined by professor j. If the edges are colored so that no two adjacent edges receive the same color, each color can correspond to a period. Clearly, the chromatic index of the graph yields the minimum number of time periods needed to complete all oral examinations. Other models of time table scheduling problems are discussed in the miscellaneous operations section.

4.5 Multi-Commodity Disconnecting Set Problem

Consider a directed network $G = (N, A)$ and let $S = \{s_1, s_2, ... s_k\} \subseteq N$ and $T = \{t_1, t_2, ..., t_k\} \subseteq N$ be the source set and sink set. A set of arcs $D \subseteq A$ is called a disconnecting set which when removed from the network would block all paths from s_i to t_i for $i = 1, 2, ..., k$. To disrupt communications from each s_i to t_i, all arcs in a disconnecting set from the network must be removed. Suppose it costs cj to remove (destroy) the arc c_j, for $j = 1, 2, ..., m$. The problem of interest is to find a disconnecting D which costs the least. Such a disconnecting set is called a multi-commodity minimum disconnecting and is useful in attacking an enemy network to disrupt all communications between the sources and the corresponding sinks. Suppose $P_1, P_2, ..., P_r$ represent all elementary paths from every point in S to the corresponding point in T. Let $x_j = 1$, if the jth edge is selected for removal from the network and $x_j = 0$ otherwise. Also let $a_{ij} = 1$, if path P_i contains the arc j and $a_{ij} = 0$ otherwise for $i = 1, 2, ..., r$ and $j = 1, 2, ..., m$. A SC formulation of the multi-commodity minimum disconnecting problem is given by

$$\begin{aligned} \min \quad & \sum_{j=1}^{m} c_j x_j \\ \text{s.t.} \quad & \sum_{j=1}^{m} a_{ij} x_j \geq 1 \quad i = 1, 2, ..., r \\ \text{and} \quad & x_j = 0, 1 \quad j = 1, 2, ..., m \end{aligned} \quad (20)$$

Set Covering, Set Packing and Set Partitioning

Even for a network of moderate size the number of paths could be prohibitively large. A method called row generation scheme may be used (for $k \geq 3$) to solve this problem which does not require the explicit knowledge of all the constraints. Efficient computational techniques are available when the number of sources k is equal to 1 or 2. The following numerical example is used to illustrate the model.

Example 8. Consider the following network with source set and sink set consisting of three nodes with numbers on each arc representing the arc number assigned to it. There is only one path (elementary) from each source to the corresponding sink and these are listed below.

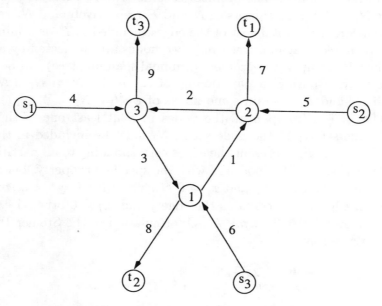

Figure 5: A Multi-Commodity Network

$$P_1 = \{(s_1, 3), (3, 1), (1, 2), (2, t_1)\}$$
$$P_2 = \{(s_2, 2), (2, 3), (3, 1), (1, t_2)\}$$
$$P_3 = \{(s_3, 1), (1, 2), (2, 3), (3, t_3)\}$$

The corresponding SC formulation is given below.

$$\min \sum_{j=1}^{7} c_j x_j$$

s.t.
$$x_4 + x_3 + x_1 + x_7 \geq 1$$
$$x_5 + x_2 + x_3 + x_8 \geq 1$$
$$x_6 + x_1 + x_2 + x_9 \geq 1$$
and $x_j = 0, 1 \quad j = 1, 2, ..., 7$

4.6 Steiner Problem in Graphs

Consider an undirected graph $G = (N, E)$ and cost c_j associated with edge j, $j = 1, 2..., m$. Suppose $N = S \cup P$ and the node set P is designated as the set of Steiner points. The problem is to determine a tree $T = (N_1, E_1)$ such that N_1 and E_1 are subsets of N and E respectively and N_1 contains all nodes in S and the total cost of the edges included in E_1 is a minimum. It should be noted that in a tree every two nodes are connected by a single path. When P is empty, the problem (minimal spanning tree) can be solved very efficiently. Consider a partitioning of the nodes $N = N_1 \cup N_2$ such that both N_1 and N_2 contain some nodes of $S(N_1 \cap N_2 = \emptyset, N_1 \cap S \neq \emptyset$ and $N_2 \cap S \neq \emptyset)$. To span all the nodes in S at least one of the edges from the node set N_1 to the node set in N_2 must be included in the tree. Suppose $E_1, E_2, ..., E_r$ represent the edges corresponding to all partitionings (also called cut sets) of the nodes N with the specified property. Let $x_j = 1$, if jth edge is included in E_1 and $x_j = 0$ otherwise for $j = 1, 2, ..., m$. Also let $a_{ij} = 1$ if edge set E_i contains the edge j and $a_{ij} = 0$ otherwise, for all $i = 1, 2, ..., r$ and $j = 1, 2, ..., m$. A SC formulation of the Steiner Problem in graphs is given by

$$\min \quad \sum_{j=1}^{m} c_j x_j$$
s.t. $\sum_{j=1}^{m} a_{ij} x_j \geq 1 \quad i = 1, 2, ..., r$
and $x_j = 0, 1 \quad j = 1, 2, ..., m.$ \hfill (21)

This problem also can be solved using the row generation scheme which does not require the explicit knowledge of all constraints similar to the multi-disconnecting set problem. Models of this type can be used to determine the minimum cost needed to determine communication links between several locations so that communication is possible between any two pair of locations.

Example 9. Consider the following undirected graph with 5 nodes and 8 edges with numbers on each edge representing the number assigned to it and $S = \{1, 2, 3\}$. The list of all possible partitionings of the nodes and the edges corresponding to each partitioning are given in Table 1. The SC

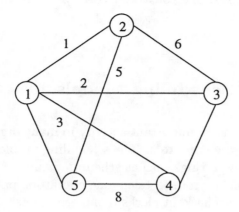

Figure 6: An Undirected Graph for the Steiner Problem

Table 1:

N_1	N_2	Edges
(1)	(2,3,4,5)	$E_1 = (1,2,3,4)$
(1,4)	(2,3,5)	$E_2 = (1,2,4,7,8)$
(1,5)	(2,3,4)	$E_3 = (1,2,3,5,8)$
(1,4,5)	(2,3)	$E_4 = (1,2,5,7)$
(2)	(1,3,4,5)	$E_5 = (1,5,6)$
(2,4)	(1,3,5)	$E_6 = (1,3,5,6,7,8)$
(2,5)	(1,3,4)	$E_7 = (1,4,6,8)$
(2,4,5)	(1,3)	$E_8 = (1,4,6)$
(3)	(1,2,4,5)	$E_9 = (2,6,7)$
(3,4)	(1,2,5)	$E_{10} = (2,3,6,8)$
(3,5)	(1,2,4)	$E_{11} = (2,4,6,7,8)$
(3,4,5)	(1,2)	$E_{12} = (2,3,4,6)$

formulation consisting of 8 variables, one variable corresponding to each edge and 12 constraints, one constraint corresponding to each partitioning is straight forward.

5 Personnel Scheduling Models

Scheduling personnel is an important activity in many organizations in both manufacturing and service sectors. The scheduling problems arise in a variety of service delivery settings such as scheduling nurses in hospitals, airline and hotel reservations personnel, telephone operators, patrol officers, workers in a postal facility, checkout clerks in supermarkets, ambulance and fire service personnel, toll collectors, check encoders in banks, train and bus crew, personnel in fast food restaurants, and others. The scheduling problems also arise in manufacturing activities especially those requiring continuity in the production process such as chemicals and steel. Operational performance of both service and manufacturing operations such as quality and level of service, labor cost and productivity are effected by employee scheduling decisions. Tardiness, turnover and absenteeism complicates the situation due to unsatisfactory work schedules. Providing satisfactory work schedules to employees, maintaining the required service levels, insuring the human needs such as breaks for lunch and rest, meeting the governmental and union legal requirements, achieving the production goals and controlling the labor costs are some of the major issues in modeling the employee scheduling problem.

The solution to the scheduling problem is relatively simple in organizations which operate five days a week and one standard shift a day, since all employees are required to follow one schedule except possibly lunch breaks. This section deals with scheduling problems which arise in organizations which operate six or more days a week, with one or more shifts per day. There are three basic models called cyclical or days off scheduling, shift scheduling and tour scheduling to structure a variety of personnel scheduling problems. Many criteria such as the total number of employees, total number of hours of labor, total cost of labor, unscheduled labor costs, overstaffing, understaffing, number of schedules with consecutive days off, number of different work schedules utilized may be used in conjunction with the three basic models to suit a particular application.

5.1 Days Off or Cyclical Scheduling

Consider an organization which operates seven days a week with one shift per day. The number of employees required on each day of the week may vary but is stable from week to week. There is only one employee category and the number of employees required on any given day is determined (or estimated) on the basis of the required level of service. Every employee must be given two consecutive days off in a week. The problem of interest is to determine the minimum number of employees required to meet the daily demand for their services. Suppose the number of employees needed on the ith day of the week is $b_i, i = 1, 2, ..., 7$. Clearly there are 7 possible schedules which satisfy the two consecutive days off requirement. Let x_i represent the number of employees who start work on ith day and continue work for a total of 5 consecutive days. For example an employee who begins work on the 7th day will also work on the first four days of the week and is off on the 5th and 6th days of the week. This problem can be formulated as a GSC and is given below.

$$\min \quad \sum_{j=1}^{7} x_j$$

$$\text{s.t.} \quad x_1 + x_4 + x_5 + x_6 + x_7 \geq b_1$$
$$x_1 + x_2 + x_5 + x_6 + x_7 \geq b_2$$
$$x_1 + x_2 + x_3 + x_6 + x_7 \geq b_3$$
$$x_1 + x_2 + x_3 + x_4 + x_7 \geq b_4$$
$$x_1 + x_2 + x_3 + x_4 + x_5 \geq b_5$$
$$x_2 + x_3 + x_4 + x_5 + x_6 \geq b_6$$
$$x_3 + x_4 + x_5 + x_6 + x_7 \geq b_7$$

and x_i is a nonnegative integer for $i = 1, 2, ..., 7$ \hfill (22)

Since the unused number of man days is given by

$$5\sum_{i=1}^{7} x_i - \sum_{i=1}^{7} b_i$$

minimizing the total employees will also minimize the overstaffing. If the labor costs vary depending upon the days off, it is possible to obtain minimum cost solution by changing the objective function to

$$\sum_{i=1}^{7} c_i x_i$$

where c_i is the cost of the employee who starts work on the ith day of the week. If nonconsecutive days off are permitted, there are a total of 21 possible schedules. The corresponding model can be formulated with 21 variables.

If the optimal solution of the model (22) is implemented, it is possible that some employees may never get a weekend off. Such solutions can be avoided by extending the planning horizon to several weeks and incorporating only those schedules which satisfy the required minimum number of weekends off. For example if the planning horizon consists of four weeks and the schedules are restricted to consecutive days off in each week, there are a total of $(7)4 = 2401$ possible schedules. Many of these schedules may not include even one weekend off. In addition, some schedules may require a long work stretch. For example, a work schedule with days 1 and 2 off in the first week, days 6 and 7 off in the next three weeks requires an individual to work 10 consecutive days without a break. This schedule also provides 4 consecutive days off if repeated once in four weeks. Such undesirable schedules may be eliminated in formulating the problem. For a given planning horizon consisting of m days, suppose the total number of schedules which meet the requirements is n. Let $a_{ij} = 1$, if the day i in the planning horizon is a work day in the schedule j and $a_{ij} = 0$ otherwise. Also let x_j be the number employees with work schedule j. A GSC formulation of this model is given by

$$\begin{aligned} \min \quad & \sum_{j=1}^{n} x_j \\ \text{s.t.} \quad & \sum_{j=1}^{n} a_{ij} x_j \geq b_i \qquad i = 1, 2, ..., m \\ \text{and} \quad & x_j \text{ is a nonnegative integer} \quad j = 1, 2, ..., n \end{aligned} \qquad (23)$$

where b_i is the required number of employees on ith day of the planning horizon. When there are several categories of tasks to be performed and each employee can perform only one task, the scheduling problem can be solved by treating each category of tasks separately. The interesting case is when employees can perform multiple tasks. Models (22) and (23) can easily be extended to incorporate the ability of the employees to perform multiple tasks by defining $a_{ijk} = 1$, if day i of the planning horizon is a work day in schedule j and the employee is required to perform task k. Obviously, the number of schedules, consequently the number of variables in model (23) increase substantially with the length of the planning horizon and the number of tasks.

5.2 Shift Scheduling

Many service facilities and manufacturing companies operate more than one shift a day. For example hospitals operate twenty four hours a day and seven days a week. Shift scheduling deals with problems related to start time, work span, lunch breaks and rest periods in assigning shifts to employees. The work day is divided into several periods of equal duration such as an hour. Based on shift length, constraints on work span (number of periods of continuous work), lunch breaks, rest and start time, several feasible schedules can be generated. For example, if the work day consists of 14 hours from 8 am through 10 pm and is divided into 28 half hour periods, each work schedule can be represented by a sequence of ones and zeroes with one corresponding to work period and zero corresponding to nonwork period. The sequence (0,0,0,0,0,0,0,0,1,1,1,1,1,1,1,1,0,0,1,1,1,1,1,1,0,0, 0,0) corresponds to start work at 12 pm, take one hour break at 4 pm after working for 4 hours, resume work again at 5 pm till 8 pm and quit for the day. Assuming that all employees must start at the beginning of a period and work eight hours at a stretch (including breaks), the number of possible schedules or shifts is 11. However, if there is flexibility in the work stretch many more schedules are possible. If part time employees are permitted, additional schedules can be added to the list of feasible schedules. An important factor in determining the length of the period is the availability of reliable estimates of the personnel requirement during each period of the work day. Suppose b_i represents the number of personnel needed during the ith period and $a_{ij} = 1$, if ith period is a work period in the jth schedule. A GSC formulation of the shift scheduling problem is given by

$$\begin{aligned} \min \quad & \sum_{j=1}^{n} c_j x_j \\ \text{s.t.} \quad & \sum_{j=1}^{n} a_{ij} x_j \geq b_i \quad && i = 1, 2, ..., m \\ \text{and} \quad & x_j \text{ is a nonnegative integer} \quad && j = 1, 2, ..., n \end{aligned} \quad (24)$$

where c_j is the cost of schedule j, n is the total number of feasible schedules and m is the number periods in a work day.

While this model provides the number of personnel required to work on each shift during each work day to minimize the total costs, it does not give individual employee schedules for a week including the days off. A model which integrates the shift scheduling and days off scheduling models is more useful.

5.3 Tour Scheduling

The combined model of days-off and shift scheduling is known as tour scheduling. Suppose a planning horizon such as a week or a month is chosen. Each working day of the planning horizon is divided into several periods of equal length. Several feasible tours are selected taking into account the days off, work stretch, starting time of a shift, lunch breaks and rest, legal constraints, part-time employees, and other restrictions. Suppose there are several tasks to be performed during each period of the planning horizon and bik is the number personnel required during period i for task k. Also let $a_{ijk} = 1$, if task k is performed during period i of tour j and $a_{ijk} = 0$ otherwise. Further, suppose m is the total number of periods in the planning horizon, n is the total number of feasible tours and r is the number of tasks. The GSC model of the tour scheduling is problem is given by

$$\begin{aligned}
\min \quad & \sum_{j=1}^{n} c_j x_j \\
\text{s.t.} \quad & \sum_{j=1}^{n} a_{ijk} x_j \geq b_{ik} \quad i = 1, 2, ..., m \\
& \qquad\qquad\qquad\qquad k = 1, 2, ..., r \\
\text{and} \quad & x_j \text{ is a positive integer} \quad j = 1, 2, ..., n
\end{aligned} \quad (25)$$

where c_j is the cost of tour j. The number of variables (feasible schedules) increase significantly with the length of the planning horizon.

All the models described so far attempt to determine the number of schedules of each type to minimize some chosen criteria. Individual employees or their preferences are not incorporated. Suppose x_{j*} is the optimal solution to problem (25), w_{ij} is the preference index of the employee i to tour j and the total number of employees is p. Also let $y_{ij} = 1$, if employee is assigned to tour j and $y_{ij} = 0$ otherwise. The GSPT model (also called generalized assignment or transportation problem) to determine the optimal assignment of tours to employees is given by

$$\begin{aligned}
\max \quad & \sum_{i=1}^{p} \sum_{j=1}^{n} w_{ij} y_{ij} \\
\text{s.t.} \quad & \sum_{j=1}^{n} y_{ij} = 1 \quad i = 1, 2, ..., k \\
& \sum_{i=1}^{k} y_{ij} = x_{j*} \quad j = 1, 2, ..., n \\
\text{and} \quad & y_{ij} = 0, 1 \quad i = 1, 2, ..., p \\
& \qquad\qquad\qquad j = 1, 2, ..., n
\end{aligned} \quad (26)$$

It is possible to explicitly incorporate each employee in model (25). Let $x_{tj} = 1$, if employee t is assigned to tour j and $x_{tj} = 0$, otherwise and c_{tj} is the corresponding cost. A mixed GSC and SPT formulation is given by

$$\begin{aligned}
\min \quad & \sum_{t=1}^{p} \sum_{j=1}^{n} c_{tj} x_{tj} \\
\text{s.t.} \quad & \sum_{j=1}^{n} x_{tj} = 1 & t = 1, 2, ..., p \\
& \sum_{t=1}^{p} \sum_{j=1}^{n} a_{ijk} x_{tj} \geq b_{ik} & i = 1, 2, ..., m \\
& & k = 1, 2, ..., r \\
\text{and} \quad & x_{tj} = 0, 1 & t = 1, 2, ..., p \\
& & j = 1, 2, ..., n \qquad (27)
\end{aligned}$$

Clearly the size of problem (27) is p times the size of the problem (25) with respect to the number of variables.

Incorporating additional constraints, such as a limit on the ratio of part time employees to the total number of employees, productivity of the employees, requires general integer programming formulation. The following numerical example is used to illustrate tour generation.

Example 10. Consider a facility which operates Monday through Saturday of every week with two shifts per day. Every employee is off on Sunday and must work one shift per day for five days a week. In addition, at least four shifts during a week must be either the first or the second shift. Suppose an employee is off on Saturday. There are two tours which correspond to either all or first shift all are second shift. There are five tours with four first shifts and one second shift. Similarly, there are five tours with four second shifts and one first shift, therefore, the number of possible tours with Saturday off is 12. Since any one of the six days can be selected for day off, the total number of tours is 72. If each shift is treated as a time period, each week consists of twelve time periods. A twelve dimension vector may be used to represent each tour. For example, the vector T = (0,1,0,1,0,1,0,1,1,0,0,0) represents a tour working on the first shift Monday through Thursday, second shift on Friday and is off on Saturday.

6 Crew Scheduling

Crew scheduling is an important problem in many transportation systems such as airlines, cargo and package carriers, mass transit systems, buslines and trains since a significant portion of the cost of operations is due to the payments to the crews which include salaries, benefits and expenses. The primary objective in crew scheduling is to sequence the movements of the crew in time and locations so as to staff the desired vehicle movements at a minimum cost. As in the employee scheduling problem generating a set of feasible schedules is necessary to formulate the corresponding SC and SPT (or their variants) for the crew scheduling problem. However, generating a set of feasible schedules for the crew problem is much more complex, since the crew has to be paired with a specified number of sequence of legs of trips or flights over time and space, in addition to incorporating overnight stay away from home base, return to the home base at least once in a specified number of days, constraints on rest periods, restrictions on work stretch and many other legal and safety rules and regulations. The terminology related to scheduling airline crews is used to describe the crew scheduling problem which can be easily interpreted in the context of train, busline, ship, mass transit and other crew scheduling problems.

A flight leg or flight segment is an airborne trip between an origin and destination city pair. Each flight segment has a specified departure time from the city of origin and a scheduled arrival time at the city of the destination. The duration of the flight segment is the difference between the arrival time and the departure time. Each flight segment must be assigned a crew consisting of a specified number of pilots and flight attendants. Crews reside in various cities called bases. The number and location of the bases depend upon the size of the operation. The union work rules and government regulations for assigning crews vary depending upon the crew type (pilot or flight attendant), crew size, aircraft type and type of operations (international or domestic). A duty period consists of one or several flight segments with a limit on both the duration of the flight and the number of flight segments. A duty period is similar to shift in the employees scheduling problem. During a duty period a crew might be attending to their duties or traveling as passengers to reposition themselves for other assignments which is called deadheading. During deadheading the crew may be assigned flights operated by another carrier. Two duty periods must be separated by rest periods which are called overnights. There are minimum and maximum limits on the duration of the overnights and the minimum limit depends upon the duration of the duty period. Time away from the base which is the elapsed time from departure to the return of the crew base cannot exceed a specified number of days as mandated by the work rules. A pairing is a sequence of flight segments which may be grouped into duty and rest periods, the first flight segment beginning and the las flight segment ending at the crew base.

Calculating the cost of pairing may vary from one organization to another. One may use the salary paid to the crew, plus hotel, per-diem, ground transportation and deadheading fare paid during the rest periods. One may also use the opportunity cost which is obtained by calculating the difference between salary of the crew and the actual salary earned during the flying time plus expenses incurred during rest periods. An adjustment has to be made to account for carry over flights not covered in the current planning horizon. Clearly the planning horizon has a significant impact on the number of pairings, since the number of flight segments included in calculating the pairings increases with the length of the planning horizon.

Determining all pairings is fairly time consuming. All legs are linked together to form resolved legs which must be flown as a unit without changing the crew. Resolved legs are linked together to form trips which can be completed in one duty period. In the third level the trips are linked into pairings. After determining the optimal pairings, they are grouped into bid-line to form monthly schedules for the crew. Suppose the total number of pairings is n and c_j is the cost of jth pairing. Also suppose there are m flight segments during the planning horizon. Let $a_{ij} = 1$, if jth pairing includes the flight segment i and $a_{ij} = 0$ otherwise. Let $x_j = 1$, if jth pairing is elected and $x_j = 0$ otherwise. Then a SC model formulation of the crew scheduling problem is given by

$$\begin{aligned}&\min &&\sum_{j=1}^{n} c_j x_j \\ &\text{s.t.} &&\sum_{j=1}^{n} a_{ij} x_j \geq 1 &&i = 1, 2, ..., m \\ &\text{and} &&x_j = 0, 1 &&j = 1, 2, ..., n\end{aligned} \quad (28)$$

Inequalities are used in the constraint set to allow deadheading which could be used to move a given crew from one place to another (from one base to another) to reduce the costs. If deadheading is not permitted, an SPT formulation can be obtained by converting the inequalities into equalities. A major problem in solving the crew pairing problem is the enormous number of pairings generated in real life applications.

The crew base concept is not meaningful in a mass-transport system. As in the employee scheduling problem days off is relevant and a crew pairing problem may be viewed as all feasible weekly or monthly schedules depending on the length of the planning horizon.

Example 11. To illustrate the construction of crew pairings consider a small domestic airline operating between three cities with two morning and two afternoon flights between each pair of cities during each day. Suppose the flight time between the pairs 1 and 2 is 5 hours, 1 and 3 is 4 hours and 2 and 3 is 3 hours. The departure and arrival schedules of the daily flights

Table 2:

From City	Departure Time	To City	Arrival Time
1	08	2	13
1	15	2	20
1	10	3	14
1	16	3	20
2	09	1	14
2	14	1	19
2	07	3	10
2	15	3	18
3	08	1	12
3	16	1	20
3	09	2	12
3	18	2	21

are given in Table 2.

The planning horizon is three days which consists of 36 flight legs. Each duty period cannot exceed 12 hours. The minimum duration of the overnight is 8 hours. Time away from the base is limited to three days. Since all flights reach destinations on the same day and considerable time is available for rest (overnight) before returning to duty the following day, duty periods can be constructed treating each day separately. There are eight activities including four departures and four arrivals at each city. The times of these activities during each day for city 1, city 2, and city 3 are (08, 10, 12,14, 15, 16, 19, 20), (07, 09, 12, 13, 14, 15, 20, 21) and (08, 09, 10, 14, 16, 18, 20) respectively. To generate all possible duty periods, first construct a network consisting of 23 nodes representing the departure and arrivals times of each flight and connect the departure node and the corresponding arrival node by an arc. Also, join two consecutive nodes corresponding to each city by an arc as shown in the network below where the first number associated with each node represents the city and the second number the departure or arrival time. Set the length of each arc equal to the difference between the time of the ending node and the beginning node. Clearly the length of each arc represents either the duration of a flight or wait time at an airport to catch another flight. Starting with any departure node enumerate all paths

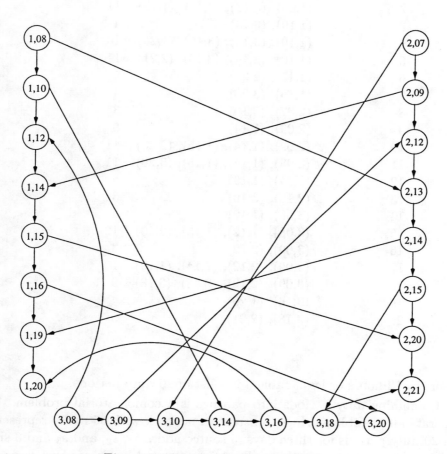

Figure 7: Graph of a Flight Schedule

Table 3:

Duty Period Number	List of Nodes in the Path	Duration of the Path
1	(1,08), (2,13)	5
2	(1,08), (2,13), (2,14), (1,19)	11
3	(1,10), (3,14)	4
4	(1,10), (3,14), (3,16), (1,20)	10
5	(1,10), (3,14), (3,18), (2,21)	11
6	(1,15), (2,20)	5
7	(1,16), (3,20)	4
8	(2,07), (3,10)	3
9	(2,09), (1,14)	5
10	(2,09), (1,14), (1,15), (2,20)	11
11	(2,09), (1,14), (1,15), (3,20)	11
12	(2,14), (1,19)	5
13	(2,15), (3,18)	3
14	(3,08), (1,12)	4
15	(3,08), (1,12), (1,15), (2,20)	12
16	(3,09), (2,12)	3
17	(3,09), (2,12), (2,14), (1,19)	10
18	(3,09), (2,12), (2,15), (3,18)	9
19	(3,16), (1,20)	4
20	(3,18), (2,21)	3

of length 12 hours or less. Table 3 is a list of 20 duty periods.

Combining duty periods into pairings is a combinatorial problem. To generate all possible pairings, construct a network with 66 nodes representing 20 duty periods for three days, 3 source nodes s_1, s_2, and s_3 and 3 sink nodes t_1, t_2, and t_3. A node on day 1 is connected by an arc to a node on day 2 provided the city of destination on day 1 and the city of origin on day 2 are the same. Similarly the nodes on day 2 and day 3 are connected. Finally connect source node s_i with an arc to all nodes on day 1, with city of origin i and connect all nodes on day 3 with city of destination i to t_i. All paths from (s_i, t_i) for $i = 1, 2, 3$ yield the required pairings. Duty period 5 on day 1, duty period 11 on day 2, and duty period 14 on day 3 is an example of a pairing.

7 Manufacturing

A variety of problems related to manufacturing activity such as assembly line balancing, discrete lot size and scheduling, ingot size selection, spare parts allocation and cutting stock which can be formulated as SC, SP and SPT models or their variants are presented in this section.

7.1 Assembly Line Balancing Problem

In an assembly line balancing problem there are a set of tasks to be performed in a specified order determined by a set of precedence relations. Given the time required for processing each activity, the problem is to determine the minimum number of stations, so that the total time required for processing all activities assigned to a station does not exceed a specified number called cycle time without violating the precedence relations. Suppose the number of tasks is n with processing times $t_i, i = 1, 2, ..., n$. The set of precedence relations P is specified by ordered pairs (i, j) which implies that the task i must be completed before the task j. Also suppose c is the cycle length. Obviously the number of stations required is no more than n. To implement the precedence relations, if $(i, j) \in P$, and tasks i and j are assigned to stations $s(i)$ and $s(j)$, then $s(i) \leq s(j)$.

Let $x_{ik} = 1$, if task i is assigned to station k and $x_{ik} = 0$ otherwise. Also let $y_k = 1$, if station k is open for assigning activities and $y_k = 0$ otherwise. An integer programming formulation of the problem is

$$
\begin{aligned}
\min \quad & \sum_{k=1}^{n} y_k \\
\text{s.t.} \quad & \sum_{k=1}^{n} x_{ik} = 1 && i = 1, 2, ..., n \\
& \sum_{t=1}^{n} t_i x_{ik} \leq c y_k && k = 1, 2, ..., n \\
& \sum_{k=1}^{h} x_{ik} \geq x_{jh} && (i, j) \in P \\
& && \text{and } h = 1, 2, ..., n \\
& y_k = 0, 1 && k = 1, 2, ..., n \\
\text{and} \quad & x_{ik} = 0, 1 && i = 1, 2, ..., n \\
& && k = 1, 2, ..., n
\end{aligned}
\tag{29}
$$

The first set of constraints insure that every task is assigned exactly to one station. The second set guarantees tasks are assigned only to a station if it is open and the total time of the activities does not exceed the cycle length. The third set maintains the precedence relations. Defining complimentary variables v_k and z_{ik} such that

$$v_k + y_k = 1$$
$$\text{and} \quad x_{ik} + z_{ik} = 1.$$

The second set of constraints can be converted to binary knapsack constraints and the third set of constraints can be transformed to SC constraints. Converting the binary knapsack constraints to SP constraints as noted earlier results in a mixed SP, SC and SPT model. Other formulations of this model are available.

7.2 Discrete Lot Sizing and Scheduling Problem

Consider a production scheduling problem where several items are manufactured on a single machine over a finite planning horizon consisting of several time periods. During each time period either the machine is idle or the entire duration is devoted to the production of a single item. Each item may require a set up time of zero, one or several time periods before production can start if this item is not produced in the previous period. The setup cost and time are item dependent but independent of the item produced in the previous period. Given that an entire duration of a period is devoted to a single item, the demand for each item can be measured in terms of the number of time periods of production needed to satisfy the demand. Inventory cost is incurred when excess production is carried from one time period to the next period. Shortages are not permitted and the setup cost, inventory cost and production cost may vary from period to period. The problem is to determine a production schedule for each item which satisfies the demands at a minimum cost.

Without loss of generality the demand for any item can be assumed to be 0 or 1 during any time period. Clearly, the demand must be an integer since it is measured in terms of the number of time periods of production needed to meet the demand. Suppose the demand is 3 units in time period 6. Since only 1 unit of demand can be met from the production of time period 6, the demand for the remaining two units must be met from the units produced in the first two periods. Make the demand equal to one in period 6 and increase the demands in periods 4 and 5 by 1. Examine the total demand in period 5 and it is more than 1, continue the procedure. Finally in period 1 if the demand is more than 1 and cannot be met by the initial inventory, the problem is infeasible.

Set Covering, Set Packing and Set Partitioning

Suppose the planning horizon consists of m time periods and the number of items is n. Let k_i be the number of feasible schedules for product i, $x_{ij} = 1$, if jth schedule is selected for product i and $x_{ij} = 0$ otherwise. Also let c_{ij} be the cost of jth schedule of product i, and $a_{ijt} = 1$, if period t is being used either to setup or produce product i in the jth schedule and $a_{ijt} = 0$ otherwise. A mixed SP and SPT formulation of this problem is given by

$$\begin{aligned}
\min \quad & \sum_{i=1}^{n} \sum_{j=1}^{n} c_{ij} x_{ij} \\
\text{s.t.} \quad & \sum_{i=1}^{m} \sum_{j=1}^{k_i} a_{ijt} x_{ij} \leq 1 \quad t = 1, 2, ..., m \\
& \sum_{j=1}^{k_i} x_{ij} = 1 \quad i = 1, 2, ..., n \\
\text{and} \quad & x_{ij} = 0, 1 \quad i = 1, 2, ..., n \\
& \qquad\qquad\qquad j = 1, 2, ..., k_i
\end{aligned} \qquad (30)$$

The first set of constraints insures no more than one product is made during any time period and the second set of constraints guarantees that exactly one schedule is selected for each product i. The number of variables required to formulate this model grows exponentially with the number of time periods.

7.3 Ingot Size Selection

In the steel industry, ingot is an intermediate product which is a mass of metal shaped in a bar or a block. Ingot size or dimensions is an important factor in producing a finished product to customer specifications. In order to make the finished product an ingot must be processed to scale the dimensions resulting in scrap metal. Clearly, producing an optimal ingot size within the technological constraints to manufacture a specific finished product, will reduce the scrap and waste. However, when each finished product requires a different ingot size, producing too many sizes of ingots may result in significant increase in inventory and material handling costs and in logistical problems. One way to deal with the problem is to produce the minimum number of ingots sizes necessary to produce all the finished products. If every finished product can be made from all ingot sizes the problem is trivial; however, this is not usually the case. Suppose the total number of all ingots sizes is n and the number of finished products to be made is m. Let $a_{ij} = 1$, if the finished product i can be produced from ingot of size j and $a_{ij} = 0$ otherwise. Also, let $y_j = 1$, if ingot size j is used to manufacture some finished product and $y_j = 0$ otherwise. Finally let $x_{ij} = 1$ if ingot size j is used to manufacture the finished product i and $x_{ij} = 0$ otherwise. The following formulation provides the minimum number of ingot sizes needed to make all finished products.

$$\begin{align}
\min \quad & \sum_{j=1}^{n} y_i \\
\text{s.t.} \quad & \sum_{j=1}^{n} a_{ij} x_{ij} = 1 & i = 1, 2, ..., m \\
& x_{ij} leq y_j & i = 1, 2, ..., m \\
& & j = 1, 2, ..., n \\
& y_j = 0, 1 & j = 1, 2, ..., n \\
\text{and} \quad & x_{ij} = 0, 1 & i = 1, 2, ..., m \\
& & j = 1, 2, ..., n
\end{align} \tag{31}$$

Substituting $z_j = 1 - y_j$, yields the mixed SP and SPT model. The objective function can be replaced by

$$\sum_{i=1}^{m} \sum_{j=1}^{n} c_{ij} x_{ij}$$

to minimize the total scrap where c_{ij} is the scrap generated when finished product i is made from an ingot of size j. When $a_{ij} = 0$, the corresponding c_{ij} value can be considered to be infinite (a large number).

A similar formulation may be used to minimize the number of various metallurgical grades of ingots needed to make a variety of steel plates used in production of railroad cars, ships, pipes and boilers. When customer orders are received for steel plates, they are categorized by sales grade based on the required chemical and mechanical properties. The ingots produced in a basic oxygen furnace are also assigned metallurgical grade called met grade based on the chemical or metallurgical composition. Assigning one met grade to each sales grade and producing all customers orders within a sales grade category using only one met grade may require a substantial number of met grades. Producing a variety of met grades can reduce productivity due to change over time required in switching from one met grade to another, in addition to maintaining a large inventory of different metallurgical grades. When more than one met grade can be used to satisfy a customer order, minimizing the number of met grades required to satisfy all customers orders is useful in increasing the productivity and reducing the inventory. If the size is replaced by grade, the formulation of this problem is identical to the above problem.

Set Covering, Set Packing and Set Partitioning

7.4 Spare Parts Allocation

Consider a repair shop where several types of engines are paired. Each engine may require one or several types of modules to repair. Given the available number of spare modules of each type, the problem is to determine the optimal allocation of modules in order to maximize the number of engines repaired. Suppose m is the number of various types of modules, bi is the number of spares of module type i, n is the number of engines requiring repair and $a_{ij} = 1$, if engine j requires module i and $a_{ij} = 0$ otherwise. Let $x_j = 1$, if engine j is repaired and $x_j = 0$ otherwise. A GSP formulation of this model is

$$\begin{align} \max \quad & \sum_{j=1}^n x_j \\ \text{s.t.} \quad & \sum_{j=1}^n a_{ij} x_j \leq b_i \quad i = 1, 2, ..., m \\ \text{and} \quad & x_j = 0, 1 \quad j = 1, 2, ..., m \end{align} \quad (32)$$

When more than one module of the same type is required for repair, the corresponding zero-one MD knapsack formulation can be converted to a GSP problem.

7.5 Pattern Sequencing in Cutting Stock Operations

Certain types of products such as rolls of paper are produced in a variety of large dimensions due to the economies of scale and technological restrictions. These rolls must be cut to specifications using various patterns to meet customer demands. With respect to a given pattern, there is a certain amount of trim loss due to lack of demand for the left over roll because of its small width. Given the demands for various width of rolls, the cutting stock problem is to determine the number of rolls to be cut using a specific pattern to minimize the total trim loss. This problem can be formulated as a linear integer program. Having determined the optimal patterns and the number of rolls to be cut using a specific pattern, determining the sequence in which to cut the different patterns is an important problem.

Suppose there are p optimal patters to meet the demands for m different types of widths. Let d_{ki} equal the number of rolls of width k to be cut in pattern i and $P_i = (d_{1i}, d_{2i}, ..., d_{mi})$. In order to cut a pattern, slitting knives are to be set at appropriate locations. The number of slitter knife settings required to cut pattern i is equal to $b_i = \sum_k d_{ki}$ the number of rolls to be cut. When one pattern is followed by another, the settings of some slitter knives have to be changed to match the required pattern. The objective in sequencing the patterns is to minimize the total number of slitter knife settings required for all patterns combined. The slitter knives can be arranged beginning from the left of the role called single-ended slitting plan where the trim loss occurs at the right end of the roll. One may also use a double-ended plan in which widths may be matched from either or both ends of the roll.

Obviously the maximum number of settings required is $\sum b_i$ which corresponds to slitting each pattern separately. Suppose a single-ended slitting plan is used and consider a subsequence S of patterns $s = (j_1, j_2, ..., j_t)$. Let $h_k(s)$ represent the number of rolls of width k common to all patterns in the subset S which is given by

$$h_k(s) = \min_{j \in s} d_{kj}$$

Also let $c(s)$ represent the total number of widths of all sizes common to all patterns in the subset. Clearly

$$c(s) = \sum_{k=1}^{m} h_k(s)$$

and only subsets for which $c(s) > 0$ contribute to the reduction of slitter knife settings. From these subsets an optimal sequence can be obtained by selecting a combination of subsets in such a way that each pattern appears in exactly one subset. Suppose the number of subsets for which $c(s) > 0$ is n and f_j is the optimal number of slitter knife settings required for the subset j. Also let $a_{ij} = 1$, if the subset j contains the pattern i and $a_{ij} = 0$ otherwise. For feasibility each individual pattern is included in the list of the sets. A SPT formulation of the pattern sequencing problem is

$$\begin{align} \min \quad & \sum_{j=1}^{n} f_j x_j \\ \text{s.t.} \quad & \sum_{j=1}^{n} a_{ij} x_j = 1 \quad i = 1, 2, ..., p \\ \text{and} \quad & x_j = 0, 1 \quad j = 1, 2, ..., n \end{align} \quad (33)$$

where $x_j = 1$, if the subset j is selected and $x_j = 0$ otherwise.

Example 12. Suppose rolls of 215" width are cut to satisfy the demand for 8 rolls of 64" width, 4 rolls of 60" width, 3 rolls of 48" width, 3 rolls of 45" width, 7 rolls of 33" width, 6 rolls of 32" width and 1 roll of 16" width. Cutting one roll for each one of the 7 patterns listed in Table 4 will satisfy the demand.

Table 4: PATTERNS

Width	No. of Rolls	1 2 3 4 5 6 7
64"	8	2 0 0 1 2 0 3
60"	4	0 3 0 1 0 0 0
48"	3	1 0 1 0 0 1 0
45"	3	0 0 0 2 1 0 0
33"	7	1 0 5 0 1 0 0
32"	6	0 1 0 0 0 5 0
16"	1	0 0 0 0 0 0 1

Consider the subset $s = (1,3)$ of patterns. The individual number of slitter knives required for patterns 1 and 3 are 4 and 6 respectively. Since the set s has two settings in common the number of settings required for the subset is 8. The subsets, the corresponding optimal arrangement and the number of setting required are listed in Table 5.

The SPT model corresponding to this example requires 26 variables and 7 constraints.

7.6 Constrained Network Scheduling Problem

Consider a scheduling problem involving m jobs and job i consisting of several tasks. Further suppose that p_{ij}, a nonnegative integer is the time required for processing task j of job i and no preemption of a task is allowed. A set of K resources (or machines) are available to process the tasks and a specified amount of resource of each type is required to process each task of each job. For each job i, there are a set of precedence relations which require that a certain task be completed before processing can begin on another task. The amount of resource k available in period t is R_{kt}, during a planning horizon consisting of T time periods. The cost of completing the job i in f units of time is $g_i(f)$. The problem is to determine the time at which the processing of each task should begin in order to minimize the total cost of completing all jobs in T units of time or less without violating the precedence relations and the resource constraints.

Suppose the number of schedules to complete job i, in T units of time or less without violating the precedence relations is n_i, f_{ih} is the completion time of job i under schedule h and $c_{ih} = g_i(f_{ih})$. Further suppose that a_{ihkt}

Table 5:

Number of Subset	Patterns in the Subset	Optimal Arrangement	Number of Settings (f_j)
1	(1, 3)	(1, 3)	8
2	(1, 4)	(1, 4)	7
3	(1, 5)	(1, 5)	5
4	(1, 6)	(1, 6)	9
5	(1, 7)	(1, 7)	6
6	(2, 4)	(2, 4)	7
7	(2, 6)	(2, 6)	9
8	(3, 5)	(3, 5)	9
9	(3, 6)	(3, 6)	11
10	(4, 5)	(4, 5)	6
11	(4, 7)	(4, 7)	7
12	(5, 7)	(5, 7)	6
13	(1, 3, 5)	(1, 5, 3)	10
14	(1, 3, 6)	(1, 3, 6)	13
15	(1, 4, 5)	(1, 5, 4)	8
16	(1, 4, 7)	(1, 7, 4)	4
17	(1, 5, 7)	(1, 5, 7)	7
18	(4, 5, 7)	(4, 5, 7)	9
19	(1, 4, 5, 7)	(1, 5, 7, 4)	10
20	(1)	(1)	4
21	(2)	(2)	4
22	(3)	(3)	6
23	(4)	(4)	4
24	(5)	(5)	3
25	(6)	(6)	6
26	(7)	(7)	4

is the amount of resource k required for all tasks of job i in process at time t in schedule h. Let $x_{ih} = 1$, if schedule h is selected for job i and $x_{ih} = 0$ otherwise. An integer programming formulation of the resource constrained network scheduling problem is given by

$$\begin{aligned}
\min \quad & \sum_{i=1}^{m} \sum_{h=1}^{n_i} c_{ih} x_{ij} \\
\text{s.t.} \quad & \sum_{h=1}^{n_i} x_{ih} = 1 & i = 1, 2, ..., m \\
& \sum_{i=1}^{m} \sum_{h=1}^{n_i} a_{ihkt} x_{ih} \leq R_{kt} & k = 1, 2, ..., K \\
& & t = 1, 2, ..., T \\
\text{and} \quad & x_{ih} = 0, 1 & i = 1, 2, ..., m \\
& & h = 1, 2, ..., n_i
\end{aligned} \quad (34)$$

When each task requires the use of a single machine and only one machine of each type is available during each period of the planning horizon, clearly $a_{ihkt} = 0$ or 1 and $R_{kt} = 1$ for all k and t. For this special case (a version of the job-shop problem), the formulation corresponds to a mixed SP and SPT model.

7.7 Cellular Manufacturing

Consider a manufacturing operation where several parts are produced and each part requires processing on a specified set of machines. The machines are grouped into cells and several cells may contain the same machine type. The problem of interest is to determine the minimum number of cells that each part must visit to complete processing on all the required machines. Since there is no interaction between parts, each part can be treated independently. Let $M(i)$, be the set of machines required for processing part i and $a_{jk} = 1$ if cell k contains the machine j and $a_{jk} = 0$ otherwise. Also let $y_{ik} = 1$ if part i visits cell k and $y_{ik} = 0$ otherwise. A SC model to determine the minimum number of cells for part i is

$$\begin{aligned}
\min \quad & \sum_{k=1}^{n} y_{ik} \\
\text{s.t.} \quad & \sum_{k=1}^{n} a_{jk} y_{ik} \geq 1 & j \in M(i) \\
\text{and} \quad & y_{ik} = 0, 1 & k = 1, 2, ..., n
\end{aligned} \quad (35)$$

where n is the total number of cells.

More realistic problems with machine capacity limitations to process the parts can be formulated as linear integer programs including the optimal allocation of machines to cells to minimize the total number of cells visited by all parts combined. The above formulation may be used as a subproblem in developing efficient techniques to solve optimal allocation machines to cells.

8 Miscellaneous Operations

As the title of this section suggests, a variety of unique and unrelated problems are discussed in this section.

8.1 Frequency Planning

Transmit and receive sites, links to connect the sites and frequency bands available for transmission are important components of any satellite communication system. Each ground terminal in a communication system can transmit and receive communications. Because of restrictions on the availability of channels in the region where the transmitter and receiver are located, constraints due to interference and technological limitations of the satellite, the number channels available to a system are limited. A frequency plan is an assignment of a separate frequency interval within the available channels to each link in a communication system.

Suppose r is the number of ground terminals which can be both a transmitter and receiver. Each link j is an ordered pair of stations. Obviously the maximum number of links required is $n = r(r-1)$. If x_j is the frequency assigned to the transmitter of link j then the corresponding frequency of the receiver must be x_j+s where s is a specified number. Due to highly nonlinear form of link inference function, the available range of the frequencies for the entire satellite system is divided into m intervals of equal but small bandwidths. In addition, the link interference constraint requires if a frequency interval i assigned to a link j, then none of the intervals, $i, i+1, ..., i+mj-1$ can be assigned to any other link. This is equivalent to assigning all frequency intervals $(i, i+1, ..., i+mj-1)$ to link j. Suppose p_{ij} is a measure of link interference representing the transmitter and link interference if the interval i is assigned to link j. Let $x_{ij} = 1$, if interval i is assigned to link j and $x_{ij} = 0$ otherwise. A SPT formulation of this model is given by

$$\min \quad \sum_{j=1}^{n} \sum_{i=1}^{m-m_j+1} p_{ij} x_{ij}$$
$$\text{s.t.} \quad \sum_{i=1}^{m-m_j+1} x_{ij} = 1 \quad j = 1, 2, ..., n$$
$$\sum_{j=1}^{n} \sum_{i=k}^{m-m_j+1} x_{ij} + s_k = 1 \quad k = 1, 2, ..., m$$

Set Covering, Set Packing and Set Partitioning

$$x_{ij} = 0, 1 \quad j = 1, 2, ..., n$$
$$\text{and} \quad s_k = 0, 1 \quad k = 1, 2, ..., m \quad (36)$$

where s_k is the slack variable. The first set of constraints insures exactly one bandwidth is assigned to link j and the second set of constraints guarantees that the interference constraints are satisfied. Note that $x_{ij} = 0$, if $j \geq (m - m_j)$ in the above formulation.

8.2 Timetable Scheduling

Scheduling classes and examinations to avoid conflicts, without violating the resources available such as number of class rooms, room capacity and other constraints such as no consecutive examinations are computationally difficult problems. One simple model for each scheduling problem is discussed below.

Suppose there are m classrooms and n classes to be assigned to classrooms each day. Further, suppose each day is divided into t periods and $a_{ik} = 1$ if class i is required to be scheduled during period k and $a_{ik} = 0$ otherwise. Let $x_{ij} = 1$, if class i is scheduled in room j and $x_{ij} = 0$ otherwise. A mixed SP and SPT model of this problem is given by

$$\begin{array}{ll} \min & \sum_{i=1}^{n} \sum_{j=1}^{m} c_{ij} x_{ij} \\ \text{s.t.} & \sum_{j=1}^{m} x_{ij} = 1 \quad i = 1, 2, ..., n \\ & \sum_{i=1}^{n} a_{ik} x_{ij} \leq 1 \quad j = 1, 2, ..., m \\ & \quad k = 1, 2, ..., t \\ \text{and} & x_{ij} = 0, 1 \quad i = 1, 2, ..., n \\ & \quad j = 1, 2, ..., m \end{array} \quad (37)$$

where c_{ij} is the cost of assigning class i to room j. The first set of constraints ensures that every class is assigned a classroom and the second set guarantees that no more than one class is scheduled during any period in any classroom.

Scheduling examinations is a difficult problem in large universities because of the number of students and the number of courses involved. Simple versions of this problem can be typically formulated in one of the two ways.

Given a set of examinations, determine the minimum number of time periods necessary to schedule all examinations with no conflict. A conflict occurs when two examinations are scheduled concurrently and one or more students must take both examinations. A model of this problem is discussed in the section graphs and networks (see Chromatic Number).

A second formulation of the problem is more detailed, having determined the groups of examinations to be scheduled simultaneously called examination block, assign at most one block to a time slot on a given day. Suppose there are m examination blocks and t time periods during a given day. If all examinations are to be completed in D days then $Dt \geq m$. Adding dummy examination blocks if necessary it can be assumed that $Dt = m$. Clearly there are $n = \binom{m}{t}$ possible combinations ($m \geq t$) of examinations schedules on any day. Suppose c_j is the total number of students having two or more examinations in the jth combination. Let $x_j = 1$, if jth combination is selected and $x_j = 0$ otherwise. Also let $a_{ij} = 1$, if examination block i is in jth combination and $a_{ij} = 0$ otherwise. A GSPT formulation of this problem is given by

$$\begin{aligned} \min \quad & \sum_{j=1}^{n} c_j x_j \\ \text{s.t.} \quad & \sum_{j=1}^{n} a_{ij} x_j = 1 \quad i = 1, 2, ..., m \\ & \sum_{j=1}^{n} x_j = D \\ \text{and} \quad & x_j = 0, 1 \quad j = 1, 2, .., N \end{aligned} \qquad (38)$$

The first set of constraints insures that every exam block is scheduled and the second constraint guarantees exactly D days are scheduled for all examination blocks.

8.3 Testing and Diagnosis

Testing and diagnosis are very important problems in medicine, repair service, software reliability and others. In this section two models are presented one related blood analysis and the second related to diagnostic inference.

Blood Analysis Several tests have to be performed on a blood specimen. To execute tests several cuvetts are filled with a blood specimen and any necessary testing agents are added. Because of the testing equipment configuration, the tests must be partitioned into r clusters of m tests each with the maximum number of tests $n = rm$. Clearly the total number of possible clusters is $s = \binom{n}{m}$. A priori it is not possible to determine a combination of tests to be performed for a blood specimen. If the clusters are grouped improperly a specimen requiring three tests may require three clusters which is time consuming when compared to all three tests are performed in the same cluster. Given p historical data sets of tests, the problem is to determine the cluster configuration which minimizes the expected number clusters required per specimen.

Suppose the vector $z = (z_1, z_2, ...z_n)$ denotes the test composition of an arbitrary blood specimen where $z_k = 1$, if test k is required and $z_k = 0$ otherwise. Further suppose $z_{ik} = 1$, if test k is performed in sample i and $z_{ik} = 0$ otherwise. Since at least one test is performed for any specimen, the minimum number of clusters required for any specimen is 1. Let $a_{kj} = 1$, if test k is in cluster j and $a_{kj} = 0$ otherwise. The number of tests performed for sample i in cluster j is

$$\sum_{k=1}^{n} z_{ik} a_{kj}$$

and therefore, the fraction of samples using cluster j is equal to

$$c_j = \frac{1}{p} \sum_{k=1}^{p} \min(1, \sum_{k=1}^{n} z_{ik} a_{kj}).$$

Let $x_j = 1$. If cluster j is selected and $x_j = 0$ otherwise. A SPT formulation of the model is given by

$$\begin{aligned} \min \quad & \sum_{j=1}^{n} c_j x_j \\ \text{s.t.} \quad & \sum_{j=1}^{n} a_{kj} x_j = 1 \quad k = 1, 2, ..., n \\ \text{and} \quad & x_j = 0, 1 \quad j = 1, 2, ..., s \end{aligned} \quad (39)$$

Diagnostic Expert System Diagnostic systems deal with identifying causes or reasons for various symptoms, examination findings on laboratory test results. Such problems of interest in determining the diseases based on various tests or symptoms, repairs needed to correct automobile problems and others.

Suppose a set of disorders or diseases and the manifestations caused by each disorder is specified by experts. Given a set of manifestations the problem is to determine the minimum possible number of disorders causing the manifestations. Let n be number of disorders, m be the number of manifestations and $a_{ij} = 1$, if disorder j causes manifestation i and $a_{ij} = 0$ otherwise. Also let $x_j = 1$, if disorder j is selected and $x_j = 0$, otherwise. An SC formulation of this problem is given by

$$\begin{aligned} \min \quad & \sum_{j=1}^{n} x_j \\ \text{s.t.} \quad & \sum_{j=1}^{n} a_{ij} x_j \geq 1 \quad i = 1, 2, ..., m \\ \text{and} \quad & x_j = 0, 1 \quad j = 1, 2, ..., n \end{aligned} \quad (40)$$

8.4 Political Districting

Dividing a region such as a state into small areas such as a district to elect political representatives is called political districting. Suppose the region consists of m population units such as counties (or census tracks) and the population units must be grouped together to form r districts. Due to court rulings and regulations, the deviation of the population per district cannot exceed a certain proportion of the average population. In addition, each district must be contiguous and compact. A district is contiguous if it is possible to reach any two places of the district without crossing another district. Compactness essentially means, the district is somewhat circular or a square in shape rather than a long and thin strip. Such shapes reduce the distance of the population units to the center of the district or between two population centers of a district.

Suppose $p_i, i = 1, 2, ..., m$ is the population of the unit i. The mean population is

$$\bar{p} = (\sum_{i=1}^{m} p_i)/m.$$

Then for feasibility every district j must satisfy

$$|p(j) - \bar{p}| \leq a\bar{p}$$

where $p(j)$ is the population of district j and $0 < a < 1$.

Suppose $a_{ij} = 1$, if the unit i is included in district j and $a_{ij} = 0$ otherwise. Clearly $p(j)$ is given by

$$p(j) = \sum_{i=1}^{m} a_{ij}.$$

To test for the contiguity of a district, construct an undirected graph whose nodes a re the units of the district and connect two nodes by an edge if they have a common border. The district is contiguous if there is a path between any two nodes. For compactness, consider any two populations units i and k of a district. If population units i and k are included in a district j, $a_{ij} = a_{kj} = 1$ and

$$d_j = \max_{i,k}(d_{ik} a_{ij} a_{kj})$$

is the distance between two units which are farthest apart where d_{ik} is the distance between units i and k. If A_j is the area of the district j, d_j^2/A_j may be used as a measure of the compactness. Suppose n represents the number of all feasible districts. Suppose c_j a measure of the deviation of population of district j is

$$c_j = |p(j) - \bar{p}|/a\bar{p}.$$

A GSPT formulation of the political districting problem is given by

$$\begin{aligned}
\min \quad & \sum_{j=1}^n c_j x_j \\
\text{s.t.} \quad & \sum_{j=1}^n a_{ij} x_j = 1, \quad i = 1, 2, ..., m \\
& \sum_{j=1}^n x_j = r \\
\text{and} \quad & x_j = 0, 1 \quad j = 1, 2, ..., n
\end{aligned} \qquad (41)$$

where $x_j = 1$, if the district j is selected and $x_j = 0$ otherwise. If the objective function is changed to

$$\min \max_{j=1}^n c_j x_j$$

the problem is called a "bottleneck" problem.

8.5 Information Retrieval and Editing

Consider a multiple file data storage system with a distinct file for each supercategory of information. Each file contains several records with items corresponding to more detailed categories. Multiple files or super categories of files are typically overlapping. A record may contain information relevant to various supercategories. Because of the overlapping nature of information stored, a request for certain specified items of information related to a category can be obtained by interrogating any of the several different files. Given the time required to search a file (depends on the number of records stored) and several requests for information related to various categories the problem is to select the files which provide information in the least amount of time.

Suppose there are n files and f_j is the length of the file $j, j = 1, 2, ..., n$. Also suppose there are m requests and $a_{ij} = 1$, if request i can be met from file j and $a_{ij} = 0$ otherwise. A SC formulation of the model is

$$\begin{aligned}
\min \quad & \sum_{j=1}^n f_j x_j \\
\text{s.t.} \quad & \sum_{j=1}^n a_{ij} x_j \geq 1 \quad i = 1, 2, ..., m \\
\text{and} \quad & x_j = 0, 1 \quad j = 1, 2, ..., n
\end{aligned} \qquad (42)$$

The constraints insure that at least one file is selected for each request.

Table 6:

Edit	Fields					
	1	2	3	4	5	6
1	-	(0,1)	(0)	-	(0,1)	-
2	(1)	-	(1)	(0,1)	-	(2,3)
3	-	(1,2)	-	(1,2,3)	-	-
4	-	(0,2)	-	-	-	(0,1)
5	(1)	-	-	(0)	(1,2)	-

Another application related to data is consider a database generated from surveys or questionnaires. Usually the responses to surveys o r questionnaires contain categorical data meaning that the magnitude of the coded information such as 1 = single, 2 = married, has no intrinsic value. Responses to surveys contain large amounts of incorrect data. To check for the accuracy of data, all data sets are examined through a set of edits or tests. Failed data sets are corrected by making as few changes as possible.

Suppose each questionnaire requires n fields $(y_1, y_2, ..., y_n)$ to represent the data and R_j represents all possible entries in field j. An edit E_i consists of a set of logically unacceptable values. $R_{ij} \subseteq R_j$ for some $j \in F_i \subseteq \{1, 2, ..., n\}$. A data set y fails or inaccurate if $y \in E_i$. Given a set of m edits E_i with corresponding R_{ij} and F_i and $y \in E_i$, the problem is to determine the minimum number of fields to be corrected to obtain a meaningful data.

Let $x_j = 1$, if field j is selected and $x_j = 0$, otherwise. Also let $a_{ij} = 1$, if $j \in F_i$ and $a_{ij} = 0$ otherwise. An SC formulation of the problem is

$$\begin{aligned} \min \quad & \sum_{j=1}^{n} x_j \\ \text{s.t.} \quad & a_{ij} x_j \geq 1 \quad i = 1, 2, ..., m \\ & x_j = 0, 1 \quad j = 1, 2, ..., n \end{aligned} \qquad (43)$$

Solution to this problem may not generate a feasible record. In generating the constraints in addition to the m edits, all implied edits must be included.

Example 8. Suppose a questionnaire contains 6 fields and the possible values for each field are $R_1 = (0, 1)$, $R_2 = (0, 1, 2)$, $R_3 = (0, 1)$, $R_4 = (0, 1, 2, 3)$, $R_5 = (0, 1, 2, 3)$. The five edits selected are listed in Table 6.

Now consider $y = (1, 0, 0, 0, 1, 0)$. Clearly this data set fails E_1, E_4 and E_5.

Without adding the implied edits the set covering formulation is

$$\begin{aligned}
\min \quad & x_1 + x_2 + x_3 + x_4 + x_5 + x_6 \\
\text{s.t.} \quad & x_2 + x_3 + x_5 \geq 1 \\
& x_2 + x_6 \geq 1 \\
& x_1 + x_4 + x_5 \geq 1 \\
\text{and} \quad & x_j = 0, 1 \quad j = 1, 2, ..., 6
\end{aligned}$$

Generating implied edits requires considerable effort due to the combinatorial nature of the problem.

8.6 Check Clearing

Clearing checks is an important activity in commercial banks since cleared checks guarantee the availability of funds to the bank in which the check is deposited. Checks drawn on out-of-town banks (transit) require considerably longer duration for clearance in comparison with checks drawn on local banks or checks drawn on the bank itself. A check not cleared in time costs the bank since funds must be made available to the customer. Various methods are available for check clearance such as clearing the checks through the Federal Reserve System or shipping checks directly to the bank using various transportation modes. Deciding which method to use for clearing is complicated by additional factors such as the time and day of the week the check is deposited and bank availability schedule. Each bank has an availability schedule which outlines the number of days required to clear checks in each region of the country. In addition, checks are grouped based on the drawee check classification and each group or type must be treated separately. The selection of the time period during which a check is sent for clearance is also an important factor.

Suppose the number time periods, the numbers modes for clearance and the number of types of checks are t, m and n respectively. Let $x_{ijk} = 1$, if check type i is sent for clearance by mode j in period k and $x_{ijk} = 0$ otherwise.

Also let $y_{jk} = 1$, if clearing mode j is used in period k and $y_{ik} = 0$ otherwise. Suppose c_{ijk} is the opportunity cost of check type i cleared by mode j in period k, v_j is the variable cost and f_{jk} is the fixed cost for clearing method j in period k and d_{ik} is the number checks of type i available for clearance in period k. Let $a_{ij} = 1$, if mode j can be used to clear check type i and $a_{ij} = 0$ otherwise.

A mixed SP and SPT model of the check clearing problem is given by (substitute $z_{jk} = 1 - y_{jk}$)

$$\min \sum_{i=1}^{n}\sum_{j=1}^{m}\sum_{k=1}^{t}(c_{ijk}+v_{jdjk})x_{ijk} + \sum_{j=1}^{m}\sum_{k=1}^{t} f_{jk}y_{jk}$$

s.t. $\sum_{j=1}^{m} a_{ij}x_{ijk} = 1$ $i = 1,2,...,n$

$x_{ijk} \leq y_{jk}$ $i = 1,2,...,n$

$x_{ijk} = 0,1$ $i = 1,2,...,n$

$j = 1,2,...,m$

$k = 1,2,...,t$

and $y_{jk} = 0,1$ $j = 1,2,...,m$

$k = 1,2,...,t$ (44)

8.7 Capital Budgeting Problem

Suppose there are n investment projects and c_j is the net-present value of the project j, for $j = 1, 2, ..., n$. Let a_{ij} be the cash-outlay or capital expenditure required during period i for $i = 1, 2, ..., m$. Given a budget b_i, for period i, the problem is to determine a subset of projects which maximizes the total net-present value without violating budget restrictions in each period. Let $x_j = 1$, if project j is selected and $x_j = 0$ otherwise. A LIP formulation of the problem is

$$\max \quad \sum_{j=1}^{n} v_j x_j$$
s.t. $\sum_{j=1}^{n} a_{ij}x_j \leq b_i$ $i = 1,2,...,m$

and $x_j = 0,1$ $j = 1,2,...,n$ (45)

where v_j is the net-present value of project j.

In this formulation both nonnegative and negative values of a_{ij} are permitted. A positive value implies that the project requires capital expenditure and a negative value corresponds to the situation where the income generated is greated than the capital expenditure. As noted earlier, this problem can be converted to a Zero-one MDK problem which in turn can be transformed to a GSP problem. It is possible to incorporate SP constraints such as

$$x_k + x_j < 1$$

which implies that at most one of the projects k or j may be selected.

8.8 Fixed Charge Problem

A fixed charge bounded linear programming problem may be formulated as

$$\min \sum_{j=1}^n c_j x_j + \sum_{j=1}^n f_j y_j$$
$$\text{s.t.} \quad \sum_{j=1}^n a_{ij} x_j \geq b_i \quad i = 1, 2, ..., n$$
$$x_j \leq u_j y_j \quad j = 1, 2, ..., n$$
$$x_j \geq 0 \quad j = 1, 2, ..., n$$
$$\text{and} \quad y_j = 0, 1 \quad j = 1, 2, ..., n \quad (46)$$

where u_j is an upper bound on the values of x_j and $f_j > 0$, is the fixed cost which is incurred only when the corresponding $x_j > 0$. In this formulation, it is assumed that all b_i are nonnegative. Consider the following related SC problem.

$$\min \sum_{j=1}^n f_j y_j$$
$$\text{s.t.} \quad \sum_{j=1}^n b_{ij} y_j \geq 1 \quad i = i_1, i_2, ..., i_t$$
$$\text{and} \quad y_j = 0, 1 \quad j = 1, 2, ..., n \quad (47)$$

where $b_{ij} = 1$ if $a_{ij} > 0$ and $b_{ij} = 0$ otherwise (corresponding to the constraints $i = (i_1, i_2, ..., i_t)$ for which $b_i > 0$). Clearly any feasible y_j for problem (46) is also a feasible solution to problem (47) and the converse is not true. When an optimal solution from (47) substituted in (46) yields an optimal x, the corresponding value of the objective function yields an upper bound for the fixed charge problem. When the optimal solution to (47) does not lead to a feasible solution to (46), suppose J^* is the set of indices j for which $y_j = 1$ in the current optimal solution and $t_j = 0$ if j in J^* and $t_j = 1$ otherwise. The SC constraint

$$\sum_{j=1}^n t_j y_j \geq 1$$

can be added to formulation (47) to eliminate the current optimal solution. The procedure can be continued to generate a good feasible solution. In addition, the formulation (47) can also be embedded in branch and bound algorithms to solve the fixed charge linear programs including the fixed charge transportation problem.

8.9 Mathematical Problems

Suppose $A = (a_{ij})$ is a mxn matrix with all $a_{ij} = 0$ or 1. The a-width of such a matrix is the minimum number of columns of matrix A necessary so that the sum of each row of the resulting submatrix is at least equal to an integer a. Clearly, a cannot exceed

$$a^* = \min_{i=1}^{m} \sum_{j=1}^{n} a_{ij}$$

and GSC formulation of this problem is straight forward.

Suppose S is a (finite) set of integers and $s_i \subseteq s$ for $i = 1, 2, ..., n$ and each s_i is an arithmetic progression. For an arithmetic progression the difference between two consecutive numbers is same. For example (3,5,7,9) is an arithmetic progression which can be expressed as $(2i + 1)$, $i = 1, 2, 3$ and 4. If all elements of S can be covered by the sets $s_i, i = 1, 2, ..., n$ it is called n-cover by arithmetic progressions. Formulation of this problem is also straight forward. Two dimensional version of this problem is useful in production operations of VLSI chips.

9 Routing

In distribution management, strategic decisions regarding the location of plants, warehouses and depots, tactical decisions concerning the fleet size and mix, operational decisions dealing with routing and scheduling of vehicles have a significant impact on the cost of delivery of goods and services to customers and maintaining a satisfactory level of service. Because of considerable capital requirements, it is not possible to relocate facilities and to some extent change the fleet size and mix frequently. Consequently selection of routes and scheduling vehicles is an important problem in adapting to changing market conditions in many operations such as supermarkets, department stores, package delivery, cargo pickup and delivery, newspaper delivery, preventive maintenance tours and others.

Consider a distribution system with one or several depots delivering a product to customers located over a network using several vehicles. Each customer requires a specified amount of the product to be delivered and each vehicle has a capacity which limits the amount of the product that can be delivered in one trip. Usually a vehicle starts from a given depot and must return to the same depot. Given the distances between customers, and the distance between the depots and customers, the routing and scheduling problem is to determine the number of vehicles needed and the assignment of customers to each vehicle without violating the capacity constraints which minimizes the total distance traveled by all vehicles. If the list of customers assigned to each vehicle is known, minimizing the total distance traveled by each vehicle separately and combining the results for all vehicles provides the desired solution. Given a list of customers, each route corresponds to the order in which the customers are visited. Finding the optimal order of visiting customers is the well known Traveling Salesman Problem which is presented next.

9.1 Traveling Salesman Problem

Consider a directed graph $G = (N, A)$. Let c_{ij} be the distance (length or cost) of the arc $(i,j) \in A$. A tour (Hamiltonian cycle) is an elementary circuit which is also equivalent to starting at any given node, visiting every other node exactly once and returning to the starting node. The sum of the distances of the arcs in the circuit is the length of the tour. The objective of the Traveling Salesman Problem (TSP) is to determine a tour of shortest length. Assuming all possible arcs are included in the arc set A, the total number of all possible tours is $n!$ The TSP is a difficult combinatorial problem because of the enormous number of possible tours for a large n. The following mixed SC and SPT formulation of the TSP is useful in developing models for a variety of routing problems. Let $x_{ij} = 1$, if arc (i,j) is in the tour and $x_{ij} = 0$ otherwise.

$$\begin{aligned}
\min \quad & \sum_{i=1}^{n} \sum_{j=1}^{n} c_{ij} x_{ij} \\
\text{s.t.} \quad & \sum_{i=1}^{n} x_{ij} = 1 && j = 1, 2, ..., n \\
& \sum_{j=1}^{n} x_{ij} = 1 && i = 1, 2, ..., n \\
& \sum_{i \in q} \sum_{j \in q} x_{ij} \geq 1 && \text{for all nonempty } q \subseteq N \\
\text{and} \quad & x_{ij} = 0, 1 && (i,j) \in A
\end{aligned} \quad (48)$$

The first two sets of constraints ensure that each node is visited exactly once and the third set of constraints eliminates subtours. When $c_{ij} = c_{ji}$, the problem is called a symmetric TSP. The distance matrix is Euclidean if the distances satisfy the triangle inequality $c_{ij} \leq c_{ik} + c_{kj}$ for all $(i,j), (i,k), (k,j) \in A$. Other formulations of this problem are available. An extension of the problem is called M Traveling Salesman Problem (MTSP) where M salesman are to visit the nodes in such a way so that the total distance traveled by all salesman is a minimum. Each node must be visited by exactly one salesman except the common node. Each salesman must travel along a subtour of the nodes which includes a node common to all salesmen. The MTSP can be formulated as TSP by creating M copies of the common node and connecting each copy of the node with the rest of the nodes as the original node. The M copies of the node are either not connected or connected by an arc with a distance exceeding

$$\sum_{i=1}^{n}\sum_{j=1}^{n} c_{ij}.$$

The nodes connecting two copies of the node is a subtour which can be assigned to a salesman. A direct integer programming formulation of the MTSP is given by

$$\begin{aligned}
\min \quad & \sum_{i=1}^{n}\sum_{j=1}^{n} c_{ij} x_{ij} \\
\text{s.t.} \quad & \sum_{j=1}^{n} x_{1j} = M \\
& \sum_{j=1}^{n} x_{ij} = 1 \quad i = 2,...,n \\
& \sum_{i=1}^{n} x_{i1} = M \\
& \sum_{i=1}^{n} x_{ij} = 1 \quad j = 2,...,n \\
& \sum_{i \in q}\sum_{j \in q} x_{ij} \geq 1 \quad \text{for all nonempty } q \subseteq (N-1) \\
\text{and} \quad & x_{ij} = 0,1 \quad (i,j) \in A
\end{aligned} \quad (49)$$

In this formulation node 1 is assumed to be the common starting node for all salesmen. This formulation is also a mixed GSPT and SC model.

Because of the number of variables and constraints involved the computational effort grows exponentially with the number of nodes to determine the optimal tour using integer programming formulations. Several techniques based on the relaxations of the TSP such as assignment, 2-matching, 1-tree, 1-arborescence and n-path have been successful in generating optimal tours for moderate size problems. Heuristic methods which generate good tours (not necessarily optimal) are useful in practical applications. One approach called the k-opt method seems to work well in generating good tours. Starting with any tour, the k-opt method is a systematic search for better tours by deleting and adding a specified number of arcs. The minimum number of arcs to be replaced to generate a new tour is two. Select any two arcs of the tour and replace them with two new arcs not in the tour if the new tour is better than the current one. After examining all combinations of two arcs, combinations of three arcs can be examined. In a 3-opt method, the search procedure is stopped after examining all combinations of three arcs. Clearly for large values of k, the k-opt method also requires substantial computational effort.

9.2 Single Depot Vehicle Routing

Suppose $G = (N, A)$ is a directed graph and all customers are located at the nodes of the graph. Without loss of generality suppose node 1 is a single depot. If the capacity of a single vehicle is sufficient to satisfy the demand of all customers the problem can be formulated as a TSP. When several vehicles are needed to satisfy the demand and the capacity of the vehicles are different, each individual vehicle must be treated explicitly in developing a model. To get a feel for the size of the problem, a direct integer programming formulation is presented below.

Let $x_{ijk} = 1$, if vehicle k is used to visit node j directly after visiting node i and $x_{ijk} = 0$ otherwise. Let q_i be the demand at node i, Q_k be the capacity of vehicle k and v be the number of vehicles used.

$$
\begin{aligned}
\min \quad & \sum_{i=1}^{n} \sum_{j=1}^{n} \sum_{k=1}^{v} c_{ijk} x_{ijk} \\
\text{s.t.} \quad & \sum_{i=1}^{n} \sum_{k=1}^{v} x_{ijk} = 1 && j = 1, 2, ..., n \\
& \sum_{j=1}^{n} \sum_{k=1}^{v} x_{ijk} = 1 && i = 1, 2, ..., n \\
& \sum_{j=1}^{n} x_{ijk} \leq 1 && k = 1, 2, ..., v \\
& \sum_{i=1}^{n} x_{ijk} - \sum_{i=1}^{n} x_{jik} = 0 && k = 1, 2, ..., v \\
& && j = 2, ..., n \\
& \sum_{i=1}^{n} q_i \sum_{j=1}^{n} x_{ijk} \leq Q_k && k = 1, 2, ..., v \\
& \sum_{i \in s} \sum_{j \in s} x_{ijk} \leq |s| - 1 && \text{for all nonempty} \\
& && s \in \{2, ..., n\} \\
& && k = 1, 2, ..., v \\
\text{and} \quad & x_{ijk} = 0, 1 && i = 1, 2, ..., n \\
& && j = 2, 2, ..., n \\
& && k = 1, 2, ..., v && (50)
\end{aligned}
$$

The first and second set of constraints ensure that a vehicle enters and departs each node. The third set guarantees that each vehicle is used at most once. The fourth set ensures that if a vehicle enters a node it must also depart from that node. The fifth set guarantees that the total demand of the nodes visited by a vehicle is no more than the capacity of the vehicle. The last set corresponds to the usual subtour elimination constraints for each vehicle. Other formulations of this problem including additional constraints on total travel time are available.

It is possible to formulate the above problem as a mixed SP and SPT problem. Suppose r is the total number of feasible tours. For feasibility the combined demand of the nodes in a tour (demand of the tour) must not exceed the maximum capacity of the vehicles. Suppose $a_{ij} = 1$, if node i is included in tour j and $a_{ij} = 0$ otherwise. Let $b_{kj} = 1$, if the vehicle k can carry the demand of tour j and $b_{kj} = 0$ otherwise. Also let c_j be the minimum cost of tour j, $x_j = 1$ if tour is selected and $x_j = 0$ otherwise. Then the problem is

$$\begin{align} \min \quad & \sum_{j=1}^{r} c_j x_j \\ \text{s.t.} \quad & \sum_{j=1}^{r} a_{ij} x_j = 1 \quad i = 2, 3, ..., n \\ & \sum_{j=1}^{r} b_{kj} x_j \leq 1 \quad k = 1, 2, ..., v \\ \text{and} \quad & x_j = 0, 1 \quad j = 1, 2, ..., r \end{align} \tag{51}$$

The first set of constraints ensures that every node is included in exactly one tour and the second set guarantees that each vehicle is assigned no more than one tour.

Generating all feasible tours is a very time consuming process. In addition, one TSP has to be solved for each tour to determine the cost associated with each tour. A variety of heuristics have been developed to generate solutions to the routing problem which can be grouped into four categories namely cluster first-route second, route first-cluster second, savings or insertion and improvement or exchange. In cluster first-route second approach the nodes are first assigned to vehicles and one TSP is solved for each vehicle to determine the optimal tour. In the route first-cluster second approach first a long tour is constructed through all nodes, then the tour is partitioned into pieces or segments which can be assigned to vehicles. In the savings or insertion approach, unassigned nodes are inserted into the existing route or routes based on the least cost or maximum savings, taking into account the vehicle capacity constraints. Having generated a set of feasible routes for each vehicle, improvement or exchange procedures examine exchanging arcs in a given route or exchanging nodes between two routes and leads to an improved feasible solution. Exact procedures to solve the routing problems include specialized branch and bound methods, dynamic programming and

9.3 Multiple Depots and Extensions

When there are multiple depots a vehicle starting at a specific depot may be required to return to the same depot or permitted to visit another depot. In some applications the delivery of goods at each node must be completed within the time window defined by the earliest and latest time between which the product must be delivered. When the time windows are full days and the delivery must occur on a specified number of days of the planning horizon, the problem is called multi-period vehicle or periodic routing problem. The integer programming formulation (50) can be modified to incorporate multiple depots, and time window constraints. In formulation (51), the additional constraints are taken into account when generating the tours. However, it is no longer feasible to determine the minimum cost of each tour using TSP when time window constraints are imposed. One may have to resort to heuristics to determine the best possible arrangement of each tour. Other extensions include the integrated inventory and routing model and the integrated depot location and routing model which can be formulated as mixed integer and integer programs. The formulation (51) may also be used in scheduling a fleet of ships to pick-up and deliver cargo as well as pick-up passengers and deliver them to their destinations (dial-a-ride problem).

10 Location

The selection of sites for locating facilities has a major impact on both public and private sector operations. Examples of facilities in manufacturing activity include plants, warehouses and retail outlets for producing, distributing and selling products. The location of centers for processing checks (lock box) has significant impact on banking operations. The locations of emergency medical services, fire stations, social service centers, day care centers, post offices, bus stops, shopping centers and hospitals are a major concern of most regional and urban planners. Distance measure, cost structure, time to travel to service centers, supply and demand for services and products are some of the important factors in modeling location problems. When the location of the facilities is unrestricted (can be located on a two dimensional plane) both Euclidean metric and Rectilinear metric can be used to calculate the distance between two points. Given two points $P_1 = (x_1, y_1)$ and

$P_2 = (x_2, y_2)$, the Euclidean distance between the two points is

$$d_{12} = \{(x_1 - x_2)^2 + (y_1 - y_2)^2\}^{1/2}$$

and the rectilinear distance is

$$d_{12} = |x_1 - x_2| + |y_1 - y_2|.$$

If the location of facilities are restricted to the nodes of a graph, the distance between two nodes is the shortest distance between them using the arcs or edges of the graph. When the location of facilities is permitted on the edges (or arcs) of a graph the shortest distance between two points can be calculated using the following approach. Suppose x is a point on the edge (a, b), y is a point on the edge (g, h) and d_{ij} is the shortest distance between nodes i and j. Then the shortest distance between the points x and

$$\begin{aligned}d_{xy} = &\min(d(x,a) + dag + d(g,y), d(x,a) + dah + d(h,y),\\ &d(x,b) + dbg + d(g,y), d(x,b) + dbh + d(h,y))\end{aligned}$$

where $d(x, a), d(x, b), d(g, y)$ and $d(h, y)$ are the lengths of the edge segments.

Depending on the selection criteria of the locations, distance measure used and the restrictions on the location of the facilities a variety of models have been developed to determine the optimum location of facilities. Models based on the Euclidean distance measure are not included since they require nonlinear programming formulations. In majority of the models, the location of facilities is restricted to the nodes or edges of a graph (which may be used to represent transportation networks) due to the physical travel involved in delivering the goods and services. The location models are divided into five categories, namely plant location problem, lock box location problem, p-center problem, p-median problem and service facilities location problem which are presented next.

10.1 Plant Location Problem

Consider a manufacturing operation with m potential sites for plants to produce a single commodity and ship a specified number of units of the product to each one of the n customers. Suppose fi is the fixed cost of opening a plant at location i, c_{ij} is the unit shipping cost from plant i to customer j, b_j is the demand at customer j and the capacity of the plant is unlimited. The problem is to determine the location of the plants which minimizes the total cost. Since the capacity of each plant is unlimited it is optimal to ship all the quantity to a customer from one location. Let $x_{ij} = 1$, if customer j is supplied from plant i and $x_{ij} = 0$ otherwise. Also let $y_i = 1$, if plant i is open and $y_i = 0$ otherwise. An integer programming formulation of the plant location problem is given by

$$\begin{aligned}
\min \quad & \sum_{i=1}^{m}\sum_{j=1}^{n} c_{ij}b_j x_{ij} + \sum_{i=1}^{m} f_i y_i \\
\text{s.t.} \quad & \sum_{i=1}^{m} x_{ij} = 1 & j = 1, 2, ..., n \\
& x_{ij} \leq y_i & i = 1, 2, ..., m \\
& & j = 1, 2, ..., n \\
& x_{ij} = 0, 1 & j = 1, 2, ..., n \\
& & i = 1, 2, ..., m \\
\text{and} \quad & y_i = 0, 1 & i = 1, 2, ..., m \quad (52)
\end{aligned}$$

To transform this problem into a SP problem define $z_i = 1 - y_i$ and select a large number M such that

$$M > \sum_{i=1}^{m}\sum_{j=1}^{n} c_{ij}b_j.$$

Then problem (52) is equivalent to

$$\begin{aligned}
\max \quad & \sum_{i=1}^{m}\sum_{j=1}^{n}(M - c_{ij}b_j)x_{ij} \\
& + \sum_{i=1}^{m} f_i z_i - \sum_{i=1}^{m} f_i - mM \\
\text{s.t.} \quad & \sum_{i=1}^{m} x_{ij} \leq 1 & j = 1, 2, ..., n \\
& x_{ij} + z_i \leq 1 & i = 1, 2, ..., m \\
& & j = 1, 2, ..., n \\
& x_{ij} = 0, 1 & i = 1, 2, ..., m \\
& & j = 1, 2, ..., n \\
\text{and} \quad & z_i = 0, 1 & i = 1, 2, ..., m \quad (53)
\end{aligned}$$

It is also possible to transform problem (53) into a SC problem. Set $c_{ij}b_j = M$, a large number if plant i cannot supply customer j and arrange $c_{ij}b_j$ for each customer in increasing order. Suppose r_{kj} for $k = 1, 2, ..., j*$ are the distinct values of the sequence and $r_{kj} \leq r_{k+1,j}$. Let $z_{kj} = 1$, if customer j is not served by a facility with transportation cost less than or equal to

r_{kj} and $z_{kj} = 0$ otherwise. Also let $a_{ijk} = 1$, if $c_{ij}b_j < r_{kj}$ and $a_{ijk} = 0$ otherwise. A SC formulation is given by

$$\begin{aligned}
\min \quad & \sum_{i=1}^m f_i y_i + \sum_{j=1}^n \sum_{k=1}^{j^*} (r_{k+1,j} - r_{kj}) x_{kj} \\
\text{s.t.} \quad & \sum_{i=1}^m a_{ijk} y_i + z_{kj} \geq 1 \quad && k = 1, 2, ..., j^* \\
& && j = 1, 2, .., n \\
& y_i = 0, 1 \quad && i = 1, 2, ..., m \\
\text{and} \quad & z_{kj} = 0, 1 \quad && k = 1, 2, ..., j* \\
& && j = 1, 2, ..., n
\end{aligned} \quad (54)$$

10.2 Lock Box Location Problem

The number of days required to clear a check drawn on a bank in city i depends upon the city j in which the check is cashed. For a company which pays bills to many clients, it is profitable to maintain accounts at various strategically located banks and pay the clients from checks drawn on one of the banks so that large clearing times can be achieved. It costs the company to maintain an account (lock box) in a bank. Suppose there are n potential lock box locations and m client locations. Suppose c_{ij} is the monetary value per dollar of a check issued in city j and cashed in city i. Suppose b_i is the dollar volume of checks paid in city i. Let $x_{ij} = 1$, if customer in city i is paid from an account in city j. Also let $y_j = 1$, if an account is maintained in city j and $y_j = 0$ otherwise. An integer programming formulation of the lock box location problem is given by

$$\begin{aligned}
\max \quad & \sum_{i=1}^m \sum_{j=1}^n c_{ij} b_i x_{ij} - \sum_{j=1}^n f_j y_j \\
\text{s.t.} \quad & \sum_{j=1}^n x_{ij} = 1 \quad && i = 1, 2, ..., m \\
& x_{ij} \leq y_j \quad && i = 1, 2, ..., m \\
& && j = 1, 2, ..., n \\
& x_{ij} = 0, 1 \quad && i = 1, 2, ..., m \\
& && j = 1, 2, ..., n \\
\text{and} \quad & y_j = 0, 1 \quad && j = 1, 2, ..., n
\end{aligned} \quad (55)$$

where f_j is the fixed cost of maintaining an account at location j. This model is similar to the plant location model (47) which can be transformed into a SP model.

10.3 P-Center Problem

Suppose the demand for a service is located at the points $x_i, i = 1, 2, ..., m$ and locations $y = (y_1, y_2, ..., y_p)$ for p service centers are to be selected from a set of points s. For any two points x and y, suppose $d(x, y)$ is the shortest distance between them. For each demand point xi the distance to the closest service center is given by

$$d(x_i, y) = \min_{y_j \in y} d(x_i, y_j).$$

The maximum closest distance between demand points and service centers is

$$d(y) = \max_{x_i} d(x_i, y).$$

The P-Center problem is to determine $y^* \in s$ which minimizes $d(y)$. The problem of locating p emergency service facilities which can be reached from demand points in the shortest possible time is the P-Center problem.

When the number of points in the set S is finite such as the nodes of a graph the optimal locations of the centers can be obtained by solving a series of SC problems. Suppose the number of points in S is $n (n \geq p)$ and h_{ij} is the shortest distance between demand point i and location j. Select any p points from the set s and calculate the corresponding value of the objective function $d_1 = d(y)$. To determine if the objective function can be improved set $a_{ij} = 1$, if $h_{ij} < d_1$ and $a_{ij} = 0$ otherwise and solve the following SC problem

$$\begin{aligned}
\min \quad & \sum_{j=1}^{n} x_j \\
\text{s.t.} \quad & \sum_{j=1}^{n} a_{ij} x_j \geq 1 \quad i = 1, 2, ..., m \\
\text{and} \quad & x_j = 0, 1 \quad j = 1, 2, ..., n
\end{aligned} \quad (56)$$

If the value of the objective function of SC is greater than p then the existing solution is optimal. Otherwise calculate d_2 corresponding to the new centers, generate the corresponding SC problem and continue the procedure.

Clearly, when the service centers are restricted to the nodes of a graph, the set of possible locations is finite. Even when the service centers are permitted along the edges, it is possible to reduce the number of possible locations to a finite set. When the locations of service centers are permitted along the edges of a graph the problem is called absolute P-center problem.

10.4 P-Median Problem

Suppose there are m demand points with a user population of a_i at demand point i and there are n possible locations for service centers. Also, suppose d_{ij} is the distance between demand point i and location j. Given a set of permissible locations for each demand point, the problem is to assign one or more service centers to each demand point so that the sum of the population weighted distance for all demands points from the respective service centers is a minimum. To formulate this problem suppose $a_{ij} = 1$ if the demand point i can receive service from location j and $a_{ij} = 0$ otherwise. Also let x_{ij} be the fraction of the population at node i receiving service from location j. When the number of service centers required is p, a mixed integer programming formulation is given by

$$\begin{aligned}
\min \quad & \sum_{i=1}^{m} \sum_{j=1}^{n} a_i d_{ij} x_{ij} \\
\text{s.t.} \quad & \sum_{j=1}^{n} a_{ij} x_{ij} = 1 & i = 1, 2, ..., m \\
& x_{ij} \leq y_j & i = 1, 2, ..., m \\
& & j = 1, 2, ..., n \\
& \sum_{j=1}^{n} y_j = p \\
& x_{ij} \geq 0 & i = 1, 2, ..., m \\
& & j = 1, 2, ..., n \\
\text{and} \quad & y_j = 0, 1 & j = 1, 2, ..., n
\end{aligned} \tag{57}$$

where $y_j = 1$, if a service center is open in location j and $y_j = 0$ otherwise.

Since service centers have no capacity restrictions, assigning the nearest service center among the selected sites to each demand point is feasible, an optimal solution to problem (57) can always be found with all $x_{ij} = 0, 1$. The model (57) is called generalized p-median problem. When the location of service centers and demand points are restricted to the nodes of a graph, the problem is called a p-median problem. Even when the location of service center is permitted along the edges of the graph, an optimal solution can always be found by restricting the locations to the nodes.

10.5 Service Facility Location Problem

In all the applications discussed in this section the demand points and the service facility locations are restricted to the nodes of a graph. Suppose d_{ij} and t_{ij} represent the shortest distance and time required to travel from node i to node j. A set covering location problem is to find the minimum number

Set Covering, Set Packing and Set Partitioning

facilities required so that every demand point has at least one facility which can be reached with in a specified distance or time or both. Let $a_{ij} = 1$, if the facility located at node j can be reached from the demand point i with in the specified distance or time and $a_{ij} = 0$ otherwise. Also, let $x_j = 1$, if a service facility is located at node j and $x_j = 0$ otherwise. The SC formulation of the this model is straight forward.

A related problem called, maximal covering problem is to maximize the coverage when the number of service centers is restricted to a specified number p. An integer programming formulation of this model is

$$
\begin{aligned}
\max \quad & \sum_{i=1}^{m} y_i \\
\text{s.t.} \quad & a_{ij} x_j \geq y_i \quad i = 1, 2, ..., m \\
& \sum_{j=1}^{n} x_j = p \\
& x_j = 0, 1 \quad j = 1, 2, ..., n \\
& y_i = 0, 1 \quad i = 1, 2, ..., m
\end{aligned} \quad (58)
$$

Substituting $y_i = 1 - z_i$, the problem can be transformed to a mixed SC and GSPT model.

Another useful model is called hierarchical objective set covering model which has multiple objectives. One is to minimize the number of facility locations to cover all demand points and the second is to maximize excess coverage. A model to maximize excess coverage is given by

$$
\begin{aligned}
\max \quad & \sum_{i=1}^{m} y_i \\
\text{s.t.} \quad & \sum_{j=1}^{n} a_{ij} x_j - y_i \geq 1 \quad i = 1, 2, ..., m \\
\text{and} \quad & x_j = 0, 1 \quad j = 1, 2, ..., n
\end{aligned} \quad (59)
$$

Both objectives can be incorporated by changing the objective function to

$$
\min w \sum_{j=1}^{n} x_j - \sum_{i=1}^{m} y_i
$$

where $w > 0$ is some chosen weight. Substituting $y_i = (1 - z_i)$, this model can be transformed to a mixed SPT and GSC model.

Another extension of the set covering location problem is to incorporate backup coverage. Suppose a backup coverage must be located with in a specified distance or time. let $b_{ij} = 1$, if node j is with in the distance or time to provide backup coverage. A SC formulation of this model is

$$
\max \quad \sum_{i=1}^{m} y_i
$$

$$\begin{aligned}
\text{s.t.} \quad & \sum_{j=1}^{n} a_{ij}x_j \geq 1 & i = 1, 2, ..., m \\
& \sum_{j=1}^{n} b_{ij}x_j - y_i \geq 1 & i = 1, 2, ..., m \\
& \sum_{j=1}^{n} x_j = p^* \\
\text{and} \quad & x_j = 0, 1 & j = 1, 2, ..., n \\
& y_j = 0, 1 & i = 1, 2, ..., m
\end{aligned} \quad (60)$$

where p^* is the optimal number of service centers needed for the primary coverage. The first set of constraints ensures the primary coverage and the second set represents the backup coverage.

The most general extension is the multiple response units model. Suppose multiple response units or several facilities can be located at a given node. Also suppose the demand node i requires r_i response units and the response unit k must be located with in a distance (or time) of h_{ik}. Let x_j be the number response units located at node j and $a_{ijk} = 1$, if location j, can provide response unit k to demand point j and $a_{ijk} = 0$ otherwise. Note that it is reasonable to assume that the maximum distance for the first response unit is less than or equal to the maximum distance for the second response unit, the maximum distance for the second response unit is less than or equal to the maximum distance for the third response unit and so on. This assumption implies that if a location can provide a first response unit, it can also provide second response unit, third response unit and so on. A GSC formulation is given by

$$\begin{aligned}
\min \quad & \sum_{j=1}^{n} x_j \\
\text{s.t.} \quad & \sum_{j=1}^{n} a_{ijk}x_j \geq k & k = 1, 2, ..., r_i \\
& & i = 1, 2, ..., m \\
\text{and} \quad & x_j \text{ is a nonnegative integer} & j = 1, 2, ..., n
\end{aligned} \quad (61)$$

11 Review of Bibliography

The list of references in each category except the general list are divided into subgroups related to the topics discussed in each section. An attempt is made to include articles containing information relevant to more than one subgroup in all appropriate subgroups. A few articles may not have been included in any subgroup due to the inability of the author to secure copies of the articles. Even though the papers have been reviewed carefully, it is very possible that some of them may have been listed in the wrong category. The

Set Covering, Set Packing and Set Partitioning

list of references included in this paper is by no means complete. However, it is the hope of the author that all significant papers have been included. The subgroups and the references related to each subgroup for each category are presented next.

11.1 Theory

Set Covering: Avis (80) Baker (81), Balas and Ng (89), Balas (84), Balas and Ho (80), Balas (80), Balas and Padberg (76, 75b), Beasley (87), Bellmore and Ratliff (71), Benvensite (82) Chang and Nemhauser (85), Chaudary, Moon and McMormick (87), Christofides and Paixo (86), Christofides and Korman (75), Chvatal (79), Comforti, Corneil and Mahjoub (86), Cornuejols and Sassano (89), Crama, Hammer and Ibaraki (90) El-Darzi (88), El-Darzi and Mitra (88a, 88b), Ectheberry (77) Fisher and Kedia (90), Fisher and Wolsey (82), Fowler, Paterson and Tanimoto (81) Garfinkel and Nemhauser (72) Hammer and Simeone (87), Hammer, Johnson and Peled (79), Ho (82), Hochbaum (80) John and Kochenberger (88), Johnson (74) Lawler (66), Leigh, Ali, Ezell and Noemi (88), Lemke, Salkin and Speilberg (71), Lavasz (75) Murty (73) Padberg (79), Peled and Simeone (85), Pierce and Lasky (75), Pierce (68) Roth (69), Roy (72) Salkin (75), Salkin and Koncal (73) Vasko and Wolfe (88), and Vasko and Wilson (86, 84a, 84b)

Set Packing: Balas and Padberg (76, 75b) Chang and Nemhauser (85), Coffman and Leuker (91), Crama, Hammer and Ibaraki (90) Fisher and Wolsey (82), Fowler, Paterson and Tanimoto (81), Fox and Scudder (86) Padberg (79, 73), Pierce (68)

Set Partitioning: Albers (80), Anily and Federgruen (91) Balas and Padberg (76, 75a, 75b) Chan and Yano (92), Coffman and Leuker (91), Crama, Hammer and Ibaraki (90) El-Darzi (88), El-Darzi and Mitra (88a, 88b) Fisher and Kedia (90, 86) Garfinkel and Nemhauser (69) Hammer and Simeone (87), Hwang, Sum and Yao (85) Marsten (74), Michaud (72) Nemhauser, Trotter and Nauss (74) Padberg (79, 73), Pierce and Lasky (75), Pierce (68) Ryan and Falkner (88) Trubin (69)

The constraint co-efficient matrix for all the three models SC, SP and SPT including their generalizations and the mixed models is a matrix of zeroes and ones. Properties of the zero-one matrices in developing solution strategies for linear models with zero-one constraint co-efficient matrix have been explored by Balas and Padberg (76), Balas (72), Berge (72), Fulkerson, Hoffman and Oppenheim (74), Padberg (74a, 74b) and Ryan and Falkner (88).

11.2 Transformations

The transformations and model conversions presented in this section can be found in Balas and Padberg (76), Hammer, Johnson and Peled (79), Lemke, Salkin and Spielberg (71), and Padberg (79) of the list of references on theory and Garfinkel and Nemhauser (75, 73) and Karp (72) of the general list of references.

11.3 Graphs

Vertex Packing: Berge (73) Chang and Nemhauser (85), Chvatal (77) Edmonds (62) Houck and Vemuganti (77) Nemhauser and Trotter (75)

Maximum Matching: Balanski (70), Berge (73) Edmonds (62) Norman and Rabin (59)

Minimum Cover: Balanski (70), Berge (73) Edmonds (62) Norman and Rabin (59) Weinberger (76)

Chromatic Index and Chromatic Number: Berge (73), Brelaz (79), Brown (72) Corneil and Graham (73) Leighton (79) Mehta (81) Salazar and Oakford (74) Wang (74), Wood (69)

Multi-Commodity Disconnecting Set: Aneja and Vemuganti (77) Bellmore and Ratliff (71)

Steiner Problem On Graphs: Aneja (80) Beasley (89, 84) Chopra (92), Cockane and Melzak (69) Dreyfus and Wagner (71) Gilbert and Pollack (68) Hakimi (71), Hanan (66), Hwang and Richards (92) Khoury, Pardalos and Hearn (93), Khoury, Pardalos and Du (93) Maculan (87) Winter (87), Wong (84), Wu, Widmayer and Wong (86)

The formulations and applications of the Vertex Packing, Maximum Matching, Minimum Cover, Chromatic Index and Chromatic Number models can be found in Berge (73), Balanski (70), Edmonds (62), Houck and Vemuganti (77), Nemhauser and Trotter (75) and Weinberger (76). The book by Nemhauser and Wolsey (88) from the general list of references is a good source for additional information on these four models. The Multi-Commodity Disconnecting Set Problem and the Steiner Problem on graphs formulations are due to Aneja and Vemuganti (77) and Aneja (80). Chopra (92) and Khoury, Pardalos and Hearn (93) present many formulations of the Steiner Problem on graphs. Implementation of the Examination Scheduling Problem at Cedar Crest College, Allentown, Pennsylvania is reported in Mehta (81).

11.4 Personnel Scheduling

Days Off (Or Cyclical) Scheduling: Abernathy, Baloff and Hershey (74) Bailey (85), Bailey and Field (85), Baker, Burns and Carter (79), Baker and Magazine (77), Baker (76,74), Baltholdi (81), Bartholdi, Orlin and Ratliff (80), Bartholdi and Ratliff (78), Bechtold (88, 81), Bechtold and Showalter (87, 85), Bennett and Potts (68), Bodin (73), Brown and Tibrewala (75), Brownell and Lowerre (75), Burns and Koop (87), Burns and Carter (85), Burns (78) Emmons and Burns (91), Emmons (85) Howell (66) Koop (86), Krajewski and Ritzman (77) Miller, Pierskalla and Rath (76), Morris and Showalter (83) Rothstein (73, 72) Tibrewala, Phillippe and Browne (72) Vohra (88)

Shift Scheduling: Abernathy, Baloff and Hershey (74), Altman, Beltrami and Rappaport (71) Bailey (85), Bailey and Field (85), Baker (76), Baker, Crabil and Magazine (73), Bartholdi (81), Bechtold and Jacobs (90), Bechtold and Showalter (87, 85), Bodin (73), Browne (79), Byrne and Potts (73). Dantazig (54) Gaballa and Pearce (79) Henderson and Berry (77, 76) Ignall, Kolesar and Walker (72) Keith (79), Koop (88), Krajewski, Ritzman and McKenzie (80), Krajewski and Ritzman (77) Lessard, Rousseu and DuPuis (81), Lowerre (79, 77) Mabert (79), Mabert and Raedels (77), Maier-Rothe and Wolfe (73), Moondra (76), Morris and Showalter (83) Paixo and Pato (89) Segal (74), Shepardson and Marsten (80) Vohra (88)

Tour Scheduling: Abernathy, Baloff and Hershey (74), Abernathy, Baloff, Hershey and Wandell (73) Bailey (85), Bechtold, Brusco and Showalter (91), Bechtold (88), Bechtold and Showalter (87, 85), Bodin (73), Buffa, Cosagrave and Luce (76) Easton and Rossin (91a, 91b) Francis (66) Glover and McMillan (86), Glover, McMillan and Glover (84), Guha and Browne (75) Hagberg (85), Holloran and Byrn (86), Hung and Emmons (90) Krajewski and Ritzman (77) Li, Robinson and Mabert (91) Mabert and Watts (82), Mabert and McKenzie (80), McGinnis, Culver and Deane (78), Megeath (78), Monroe (70), Morris and Showalter (83), Morrish and O'Connor (70) Ozkarahan and Bailey (88), Ozkarahan (87) Papas (67) Ritzman, Krajewski and Showalter (76) Showalter and Mabert (88), Smith (76), Smith and Wiggins (77), Stern and Hersh (80) Taylor and Huxley (89), Tien and Kamiyama (82) Warner (76), Warner and Prawda (72)

Miscellaneous: Abernathy, Baloff and Hershey (71), Ahuja and Sheppard (75) Bechtold (91, 79), Bechtold and Sumners (88), Bechtold, Janaro and Sumners (84) Chelst (81, 78), Chen (78), Church (73) Eilon (64) Gentzler, Khalil and Sivazlian (77), Green and Kolesar (84) Hershey, Abernathy and Baloff (74) Klasskin (73) Linder (69), Loucks and Jacobs (91) McGrath (80) Price (70) Showalter, Krajewski and Ritzman (78) Wolfe and Young (65a, 65b).

The SC formulation of the personnel scheduling problem is due to Dantazig (54). Scheduling models of telephone operators at the General Telephone Company of California and the Illinois Bell Telephone Company are described in Buffa, Cosgrove and Luce (76) and Keith (79). Applications of scheduling models to encode and process checks at the Ohio National Bank, Chemical Bank and the Purdue National Bank are reported in Krajewski, Ritzman and McKenzie (80), Mabert (79), and Mabert and Readels (77). Applications of scheduling models to staffing nursing personnel at the Pediatrics ward of the Colorado General Hospital, Harper Hospital (Detroit) are reported in Megeath (78), and Morrish and O'Connor (70). Models of scheduling patrol officers in San Francisco, aircraft cleaning crews for an international airline, sanitary workers (household refuse collection) in New York and bus drivers in Quebec city are described in Taylor and Huxley (89), Stern and Hersh (80), Altman, Beltrami and Rappaport (71) and Lessard (85). Scheduling of Sales Personnel and Clerical Employees at Qantas Airlines and United Airlines are reported in Gabella and Pearce (79) and Holloran and Byrn (86).

11.5 Crew Scheduling

Airline Crew Scheduling: Anbil, Gelman, Patty and Tanga (91), Arabeyre, Fearnley, Steiger and Teather (69), Arabeyre (66) Baker and Fisher (81), Baker and Frey (80), Baker, Bodin, Finnegan and Ponder (79), Ball and Roberts (85), Barnhart, Johnson, Anbil and Hatay (91), Bronemann (70) Darby-Dowman and Mitra (85) Evers (56) Gerbract (78), Gershkoff (90, 89, 87) Jones (89) Kabbani and Patty (93), Kolner (66) Lavoie, Minoux and Odier (88) Marsten and Shepardson (81), Marsten, Muller and Killion (79), McCloskey and Hansman (57), Minoux (84) Niederer (66) Rannou (86), Rubin (73) Spitzer (87, 61), Steiger (65) Bodin, Golden, Assad and Ball (83) of the Routing references.

Mass Transit Crew Scheduling: Amar (85) Ball, Bodin and Dial (85, 83, 81, 80), Belletti and Davani (85), Bodin, Ball, Duguid and Mitchell (85), Bodin, Rosenfield and Kydes (81), Bodin and Dial (80), Booler (75), Borrett and Roes (81) Carraresi and Gallo (84), Cedar (85) Edwards (80) Falkner and Ryan (87) Hartley (85, 81), Henderson (75), Heurgon and Hervillard (75), Hoffstadt (81), Howard and Moser (85) Keaveny and Burbeck (81),

Koutsopoulos (85) Leprince and Mertens (85), Lessard, Rousseau and DuPuis (81), Leudtke (85) Marsten and Shepardson (81), Mitchell (85), Mitra and Darby-Dowman (85), Mitra and Welsh (81) Paixo, Branco, Captivo, Pato, Eusebio and Amado (86), Parker and Smith (81), Piccione, Cherici, Bielli and LaBella (81) Rousseau and Lessard (85), Ryan and Foster (81) Scott (85), Shepardson (85), Stern and Cedar (81), Stern (80) Tykulsker, O'Neil, Cedar and Scheffi (85) Ward, Durant and Hallman (81), Wren, Smith and Miller (85), Wren (81) Bodin, Golden, Assad and Ball (83) of the Routing references.

Application of the Airline Crew Scheduling models at American Airlines, Flying Tiger, Continental Airlines and Swiss Air are reported in Anbil, Gelman, Patty and Tanga (91), Gershkoff (89, 87), Kabbani and Patty (93), Marsten and Shepardson (81), Marsten, Muller and Killion (79), and Steiger (65).

Modelling and implementation of the Mass Transit Crew Scheduling systems in various metropoliton areas (Amsterdam, Christchurch in New Zealand, Hamburg, New York, Helsinki, Los Angeles, Dublin and Rome) are described in Borrett and Roes (81), Falkner and Ryan (87), Hoffstadt (81), Howard and Moser (85), Marsten and Shepardson (81), Mitchell (85), Mitra and Darby-Dowman (85), Piccione, Cherici, Bielli and LaBella (81) and Ryan and Foster (81).

11.6 Manufacturing

Assembly Line Balancing: Baybars (86a, 86b), Bowman (60) Freeman and Jucker (67) Gutjahr and Nemhauser (64) Hackman, Magazine and Wee (89), Hoffman (92) Ignall (65) Johnson (83, 81) Kilridge (62) Patterson and Albracht (75) Salveson (55) Talbot, Patterson and Gehrlein (86), Talbot and Patterson (84) White (61)

Discrete Lot Sizing and Scheduling Problem: Cattrysse, Saloman, Kirk and Van Wassenhove (93), Cattrysse, Maes and Van Wassenhove (90, 88) Dizelenski and Gomory (65) Lasdon and Terjung (71) Manne (58)

Ingot Size Selection: Vasko, Wolfe and Scott (89, 87)

Spare Parts Allocation: Scudder (84)

Pattern Sequencing In Cutting Stock Operations: Pierce (70)

Resource Constrained Network Scheduling Problem: Fisher (73)

Cellular Manufacturing: Stanfel (89)

The Assembly Line Balancing problem formulation (29) is based upon the formulations of Bowman (60) and White (61). Other formulations of this model can be found in the rest of the references listed under this topic. The Discrete Lot Sizing and scheduling model is due to Cattrysse, Saloman, Kirk and Van Wassenhove (93). Generalizations of this model are described in Cattrysse, Maes and Van Wassenhove (90, 88), Dizelenski and Gomory (65), Lasdon and Terjung (71), and Manne (58). Models and formulations of Ingot Size Selection, Spare Parts Allocation, Pattern Sequencing In Cutting Stock Operations, Resource Constrained Network Scheduling Problem and Cellular Manufacturing are described in Vasko, Wolfe and Scott (89, 87), Scudder (84), Pierce (70), Fisher (73), and Stanfell (89). Implementation of the Ingot Size Selection models at the Bethlehem Steel Corporation are reported in Vasko, Wolfe and Scott (89, 87).

11.7 Miscellaneous Operations

Frequency Planning: Thuve (81)

Timetable Scheduling: Almond (69, 66), Arani and Lofti (89), Aubin (89), Aust (76) Barham and Westwood (78), Broder (64) Carter and Tovey (92), Carter (86), Csima and Gotleib (64) Dempster (71), Dewerra (78, 75) Even, Itai and Shamir (76) Ferland and Roy (85) Gans (81), Glassey and Mizrach (86), Gosselin and Trouchon (86), Grimes (70) Hall and Action (67), Hertz (92) Knauer (74) LaPorte and Desroches (84), Lions (67) Mehta (81), Mulvey (82) Tripathy (84, 80) White and Chan (79), Wood (69)

Testing and Diagnosis: Nawijn (88) Reggia, Naw and Wang (83)

Political Districting: Garfinkel and Nemhauser (70)

Information Retrieval and Editing: Day (65) Garfinkel, Kunnathur and Liepins (86)

Check Clearing: Markland and Nauss (83) Nauss and Markland (85)

Capital Budgeting: Valenta (69)

Fixed Charge Problem: Aneja and Vemuganti (74) Frank (72) McKeown (81)

Mathematical Problems: Fulkerson, Nemhauser and Trotter (74) Heath (90)

The Models presented in this section on Frequency Planning, Testing and Diagnosis, Political Districting, Information Retrieval and Editing, Check Clearing, Capital Budgeting, Fixed Charge Problem and Mathematical Problems are based upon Thuve (81), Nawijn (88), Reggia, Naw and Wang (83), Garfinkel and Nemhauser (70), Day (65), Garfinkel, Kunnathur and Liepins (86), Markland and Nauss (83), Valenta (69), McKeown (81) and Fulkerson, Nemhauser and Trotter (74) and Heath (90).

Applications of Timetable Scheduling modles at SUNY, Buffalo, University of Waterloo, Ontario School System and Cedar Crest College are reported in Arani and Lofti (89), Carter (89), Lions (67) and Mehta (81). Implementation of Check Clearing model at the Maryland National Bank is presented in Markland and Nauss (83).

11.8 Routing

Travelling Salesman Problem: Bellmore and Hong (74), Bodin, Golden, Assad and Ball (83) Christofides (85b), Christofides, Mingozzi and Toth (81b), Christofides and Eilon (73) Eilon, Watson-Gandy and Christofides (71) Gavish and Shlifer (78), Golden, Levy and Dahl (81), Golden, Magnanti and Nguyen (77) Held and Karp (70) LaPorte (92b), Lenstra and Rinnooy Kan (75), Lin and Kernighan (73), Lin (65) Magnanti (81), Malandraki and Daskin (89) Russell (77) Solomon and Desrosiers (88)

Single And Multiple Depots: Agarwal, Mathur and Salkin (89), Agin (75), Altinkemer and Gavish (91, 90, 87), Anily and Federgruen (90), Averbakh and Berman (92) Baker (92), Balinski and Quandt (64), Ball, Golden, Assad, Bodin (83), Bartholdi, Platzman, Collins and Warden (83), Beasley (84, 83, 81), Bell, Dalberto, Fisher, Greenfield, Jaikumar, Kedia, Mack and Prutzman (83), Bellman (58), Beltrami and Bodin (74), Bertsimas (92, 88), Bertsimas and Ryzin (91), Bodin, Golden, Assad and Ball (83), Bodin and Golden (81), Bodin and Kursh (79, 78), Bodin (75), Bonder, Cassell and Andros (70), Bramel and Simchi-Levi (93), Bramel, Coffman, Shor and Simchi-Levi (92), Brown and Graves (81), Butt and Cavalier (91) Chard (68), Cheshire, Melleson and Naccache (82), Christofides (85a, 85b), Christofides, Mingozzi and Toth (81a, 81b, 79), Christofides (76, 71), Christofides and Eilon (69), Clark and Wright (64), Crawford and Sinclair (77), Cullen, Jarvis and Ratliff (81), Cunto (78) Daganzo (84), Dantazig and Ramser (59), Doll (80), Dror and Trudeau (90) Eilon, Watson-Gandy and Christofides (71), Eilon and Christofides (69), Etezadi and Beasley (83), Evans and Norbeck (85) Ferebee (74), Ferguson and Dantazig (56), Fisher, Greenfield, Jaikumar and Lester (82), Fisher and Jaikumar (81), Fletcher (63), Fleuren (88), Foster and Ryan (76), Frederickson, Hecht and Kim (78) Garvin, Crandall, John and Spellman (57), Gaskell (67), Gavish and Shlifer (78), Gavish, Schweitzer and Shlifer (78), Gendreau, Hertz and LaPorte (92), Gheysens, Golden and Assad (84), Gillett and Johnson (76), Gillett and Miller (74), Golden and Assad (88, 86a), Golden, Bodin and Goodwin (86), Golden and Baker (85), Golden, Gheysens and Assad (84), Golden and Wong

(81), Golden, Magnanti and Nguyen (77), Golden (77) Haimovich, Rinnooy Kan and Stouge (88), Haimovich and Rinnooy Kan (85), Hauer (71), Holmes and Parker (76), Hyman and Gordon (68) Kirby and McDonald (72), Kirby and Potts (69), Krolak, Felts and Nelson (72) Labbe, LaPorte and Mercure (91), Lam (70), LaPorte (92a), LaPorte, Nobert and Taillefer (88, 87), LaPorte and Nobert (87, 84), LaPorte, Mercure and Nobert (86), LaPorte and Nobert and Derochers (85), LaPorte, Derochers and Nobert (84), Lenstra and Rinnooy Kan (81, 76, 75), Levary (81), Levy, Golden and Assad (80), Li and Simchi-Levi (93, 90), Lecena (86) Magnanti (81), Malandraki and Daskin (89), Male, Liebman and Orloff (77), Marquez Diez- Canedo and Escalante (77), Minas and Mitten (58), Minieka (79), Mole (83, 79), Mole, Johnson and Wells (83), Mole and Jameson (76) Nelson, Nygard, Griffin and Shreve (85), Norbeck and Evans (84) Orloff (76a, 76b, 74a, 74b), Orloff and Caprera (76) Passens (88), Psarafits (89, 88, 83c), Pullen and Webb (67) Robertson (69), Ronen (92) Salvelsbergh (90), Scharge (83), Solomon and Desrosiers (88), Stern and Dror (79), Stewart and Golden (84), Stricker (70), Sumichrast and Markham (93), Sutcliffe and Board (91) Tillman and Cain (72), Tillman and Hering (71), Tillman (69), Tillman and Cochran (68), Turner, Ghare and Foulds (76), Turner and Hougland (75), Tyagi (68) Unwin (68) Van Leeuwen (83) Watson-Gandy and Foulds (72), Webb (72), Williams (82), Wren and Holliday (72) Yellow (70)

Routing With Time Windows: Baker and Schaffer (86), Bodin, Golden, Assad, Ball (83), Bodin and Golden (81) Cassidy and Bennett (72) Desrochers, Desrosiers and Soloman (92), Desrochers, Lenstra, Savelsbergh and Soumis (88), Desrochers, Soumis, Desrosiers and Sauve (85), Desrochers, Soumis, Desrosiers (84), Dumas, Desrosiers and Soumis (91) El-Azm (85) Fleuren (88) Gertsbach and Gurevich (77), Golden and Assad (88, 86a, 86b), Golden and Wasssel (87), Golden and Baker (85), Golden, Magnanti and Nguyen (77) Jaw, Odoni, Psaraftis and Wilson (86) Knight and Hofer (68), Kolen, Rinnooy Kan and Trienekens (87), Koskosidis, Powell and Soloman (92) Malandraki and Daskin (89) Potvin and Rousseau (93), Potvin, Kervahut and Rousseau (92) Salvelsbergh (85), Scharge (83), Sexton and Choi (72), Soloman and Desrosiers (88), Solomon (87, 86)

Periodic Routing Problem: Bodin, Golden, Assad, Ball (83) Cheshire, Melleson and Naccache (82), Christofides (85b, 84) Foster and Ryan (76) Gaudioso and Paletta (92), Golden and Assad (88) Hausman and Gilmour (67) Russel and Igo (79), Raft (82) Sexton and Bodin (85a, 85b), Solomon and Desrosiers (88) Tan and Beasley (84)

Integrated Inventory And Routing: Anily and Federgruen (93, 92, 90a, 90b), Arisawa and Elmaghraby (77a, 77b) Bramel and Simchi-Levi (93) Chien, Balakrishnan and Wong (89), Christofides (85b) Dror and Ball (87), Dror and Levy (86), Dror, Ball and Golden (86) Farvolden, LaPorte and Xu (93), Federgruen and Simchi-Levi (92), Federgruen and Zipkin (84) Golden and Assad (86a), Golden and Baker (85), Golden, Assad and Dahl (84) Hall (91)

Dial-A-Ride Problem: Angel, Caudle, Noonan and Whinston (72) Bennett and Gazis (72), Bodin and Sexton (86), Bodin and Berman (79) Cullin, Jarvis and Ratliff (81) Daganzo (78), Desrosiers, Dumas and Soumis (86), Dulac, Ferland and Forgues (80), Dumas, Desrosiers and Soumis (91) Fleuren (88), Foulds, Read and Robinson (77) Golden and Assad (88) Jaw, Odoni, Psarafitis and Wilson (86) McDonald (72) Newton and Thomas (74, 69) Psaraftis (86, 83a, 83b, 80) Solomon and Desrosiers (88), Stein (78), Stewart and Golden (81, 80)

Location And Routing: Jacobson and Madsen (80) LaPorte, Nobert and Taillefer (88), LaPorte, Nobert and Arpin (86), LaPorte and Nobert (81)

Scheduling A Fleet Of Ships, Aircrafts, Trains and Buses: Applegreen (71, 69), Assad (81, 80) Bartlett (57), Bartlett and Charnes (57), Barton and Gumer (68), Bodin, Golden, Schuster and Romig (80), Brown, Graves and Ronen (87) Ceder and Stern (81), Charnes and Miller (56) Fisher and Rosenwein (89), Florian, Guerin and Bushell (76) Laderman, Gleiberman and Egan (66), Levin (71) Martin-Lof (70), McKay and Hartley (74) Nemhauser (69) Peterson and Fullerton (73), Pierce (69), Pollak (77) Rao and Zionts (68), Richardson (76), Ronen (86, 83) Saha (70), Salzborn (74, 72a, 72b, 70, 69), Simpson (69), Smith and Wren (81), Soumis, Ferland and Rousseau (80), Spaccamela, Rinnooy Kan and Stougie (84), Szpiegel (72) White and Bomberault (69), Wolters (79), Wren (81) Young (70)

The routing problem is introduced by Dantazig and Ramser (59). The SPT formulation of routing problem is due to Balanski and Quandt (64). A variety of routing applications are listed in Table 7.

Table 7: APPLICATIONS

Bartholdi, Platzman, Collins and Warden (3)	Meals-on-Wheels Senior Citizens Inc., Atlanta
Bell, Dalberto, Fisher, Greenfield, Jaikumar, Kedia, Mack and Prutzman (83)	Distribution of Oxygen, Hydrogen etc., at Air Products and Chemicals, Inc.
Bodin and Berman (79)	School Bus Routing at Brentwood School District, Long Island, New York
Bodin and Kursh (78)	Routing and Scheduling of street sweepers in New York City and Washington, D.C.
Brown, Graves and Ronen (87)	Scheduling of Crude Oil Tankers for a major oil company
Brown and Graves (81)	Routing Petroleum tank trucks at Chevron, USA
Cassidy and Bennett (72)	Catering of meals to the schools of the Inner London Education Authority
Ceder and Stern (81)	Scheduling bus trips at Egged, the Israel National Bus Carrier
Crawford and Sinclair (72)	Scheduling beer tankers at WAIKATO Brewers Ltd., Hamilton, Nwe Zealand
Cunto (78)	Routing of boats to sample oil wells at Lake Maracaibo, Venezuela
Evans and Norbeck (85)	Food Distribution at KRAFT
Fisher and Rosenwein (89)	Military Sealift Command of the U.S. Navy
Fisher, Greenfield, Jaikumar and Lester (82)	Distribution of a major product at DUPONT
Golden, Magnanti and Nguyen (77)	Distributing newspaper with large circulation

Golden and Wassil (87)	Distribution of soft drinks at Joyce Beverages, Baltimore Division of Mid-Atlantic Coca-Cola, Pepsi-Cola Bottling Group of Purchase, New York and others
Gavish, Schweitzer and Shlifer (78)	Scheduling buses for large bus company
Jacobsen and Madsen (80)	Designing transfer points and routes for distributing newspaper for a company in Denmark
Jaw, Odoni, Psarafitis and Wilson (86)	Dial-A-Ride Model application at Rufbus GmbhBodenseekreis, Friedrichshafen, Germany
Knight and Hofer (68)	Routing vehicles to collect and deliver small consignments for a contract transport undertaking in London
McDonald (72)	Transporting specimens from a hospital to laboratories
McKay and Hartley (74)	Distribution of bulk petroleum products at the Defence Fuel Supply Center (DFSC) and the Military Sealift Command (MSC)
Salzborn (70)	Scheduling trains at the Adelaide Metropolitan Passenger Service of South Australian Railways
Smith and Wren (81)	Bus scheduling at the West Yorkshire Passenger Transport System
Stern and Dror (79)	Reading Electric Meters in the City of Beersheva, Israel

11.9 Location

Plant (Warehouse) Location and Allocation: Akinc and Khumawala (77), Atkins and Shriver (68) Baker (74), Ballou (68), Barcelo and Casanovas (84), Baumol and Wolfe (58), Bilde and Krarup (77), Brown and Gibson (72), Burstall, Leaver and Sussams (62) Cabot, Francis and Stary (70), Cerveny (80), Cho, Johnson, Padberg and Rao (83), Cho, Padberg and Rao (83), Cohon, ReVelle, Current, Eagles, Eberhart and Church (80), Cooper (64, 63), Cornuejols, Nemhauser and Wolsey (90) Davis and Ray (69), Dearing (85), Drysdale and Sandiford (69), Dutton, Hinman and Millham (74) Efroymson and Ray (66), Ellwein and Gray (71), El-Shaieb (73), Elson (72), Erlenkotter (78, 73), Feldman, Lehrer and Ray (66) Geoffrion and McBride (78), Gelders, Printelon and Van Wassenhove (87), Guignard (80), Guignard and Spielberg (79, 77) Hammer (68), Hoover (67), Hormozi and Khumawala (92) Khumawala, Neebe and Dannenbring (74), Khumawala (73a, 72), Khumawala and Whybark (71), Kolen (83), Koopmans and Beckman (74), Krarup and Pruzan (83), Kuehn and Hamburger (63), Kuhn and Kuenne (62) LaPorte, Nobert and Arpin (86), Levy (67), Louveaux and Peeters (92) Manne (64), Maranzana (64), Marks, ReVelle and Liebman (70) Nambiar, Gelders and Van Wassenhove (89, 81) Perl and Daskin (85, 84), Polopolus (65) ReVelle, Marks and Liebman (70) Sa (69), Saedat (81), Scott (70), Shannon and Ignizio (70), Spielberg (70, 69a, 69b), Smith, Mangtelsdorf, Luna and Reid (89), Swain (74) Tapiero (71) Van Roy and ErlenKotter (82), Vergin and Rogers (67) Wendell and Hurter (73)

Lock Box Location: Cornuejols, Fisher and Nemhauser (77a, 77b) Kramer (66), Kraus, Janssen and McAdams (70) Maier and Vanderwede (76, 74), Malczewski (90), Mavrides (79), McAdams (68) Nauss and Markland (81, 79) Shankar and Zoltners (72), Stancil (68)

P-Center Problem: Aneja, Chandrasekaran and Nair (88) Chandrasekaran and Daugherty (81), Chhajed and Lowe (92), Christofides and Viola (71) Dearing (85), Drenzner (86, 84), Dyer and Frieze (85) Eisemann (62) Garfinkel, Neebe and Rao (77), Goldman (72a, 69) Hakimi, Schmeichel and Pierce (78), Hakimi and Maheshwari (72), Hakimi (64), Halfin (74), Halpern (76), Handler (73), Hansen, Labbe, Peters and Thisse (87), Hooker, Garfinkel and Chen (91) Kariv and Hakimi (79a), Kolen (85) Lin (75) Musuyama, Ibaraki and Hasegawa (81), Minieka (77, 70), Moon and Chaudary (84) Richard, Beguin and Peeters (90) Tansel, Francis and Lowe (83a, 83b) Vijay (85)

P-Median Problem: Chhajed and Lowe (92), Church and Weaver

(86), Church and Meadows (77), Church and ReVelle (76) Dearing (85) Erkut, Francis and Lowe (88) Goldman (72b, 71), Goldman and Witzgall (70) Hakimi (65, 64), Halpern (76), Hansen, Labbe, Peters and Thisse (87), Hooker, Garfinkel and Chen (91) Jarvinen, Rajala and Sinerro (72) Kariv and Hakimi (79a), Khumawala (73b) Mavrides (79), Minieka (77), Mirchandani (79), Moon and Chaudary (84) Narula, Ogbu and Samuelsson (77), Neebe (78) ReVelle and Elzinga (89), ReVelle and Hogan (89b), Richard, Beguin and Peeters (90), Rydell (71, 67) Snyder (71b) Tansel, Francis and Lowe (83a, 83b), Teitz and Bart (68), Toregas, Swain, ReVelle and Bergman (71) Weaver and Church (85) **Service Facilities Location** Ball and Lin (93), Berlin and Liebman (74) Chrissis, Davis and Miller (82), Church and Meadows (79), Church and ReVelle (76), Current and Storbeck (88) Dee and Liebman (72), Deighton (71), Drezner (86) Erlenkotter (73) Foster and Vohra (92), Francis, Lowe and Ratliff (78) Goodchild and Lee (89), Gunawardane (82) Holmes, Williams and Brown (72) Kolen (85), Kolesar and Walker (74) Marks, ReVelle and Liebman (70), Moon and Chaudary (84), Mukundan and Daskin (91) Neebe (88) Orloff (77) Patel (79) Rao (74), Ratick and White (88), ReVelle (89), ReVelle and Hogan (89a), ReVelle, Toregas and Falkson (76), ReVelle, Marks and Liebman (70), ReVelle and Swain (70), Richard, Beguin and Peeters (90), Rojeski and ReVelle (70) Saatcioglue (82), Saydam and McKnew (85), Schilling, Jayaraman and Barkhi (93), Schilling (82), Schilling, ReVelle, Cohen and Elzinga (80), Schilling (80), Schreuder (81), Slater (81), Storbeck (82) Toregas and ReVelle (73), Toregas, Swain, ReVelle and Bergman (71), Toregas and ReVelle (70) Valinsky (55) Wagner and Falkson (75), Walker (74), White and Case (74), Weaver and Church (83)

Maximal Covering Problem: Balakrishnan and Storbeck (91), Balas (83), Batta, Dolan and Krishnamurty (89), Bennett, Eaton and Church (82) Church and Weaver (86), Church and Roberts (83), Church and Meadows (79), Church and ReVelle (76), Current and O'Kelly (92), Current and Schilling (90), Current and Storbeck (88) Daskin, Haghani, Khanal and Malandraki (89), Daskin (83, 82) Eaton, Daskin, Simmons, Bulloch and Jansma (85) Fuziwara, Makjamroen and Gupta (87) Kalstorin (79) Medgiddo, Zemel and Hakimi (83), Mehrez and Stulman (84, 82), Mehrez (83), Meyer and Brill (88) Pirkul and Schilling (91) ReVelle and Hogan (88) Schilling, Jayaraman and Barkhi (93), Storbeck and Vohra (88)

Hierarchical Objective Set Covering Model: Charnes and Storbeck (80), Church, Current and Storbeck (91), Church and Eaton (87) Daskin, Hogan and ReVelle (88), Daskin and Stern (81) Flynn and Ratick (88) Moore and ReVelle (82), Mukundan and Daskin (91) Plane and Hendrick (77)

Backup Coverage Model: Church and Weaver (86) Daskin, Hogan and ReVelle (88) Hogan and ReVelle (86, 83) Pirkul and Schilling (89) Storbeck and Vohra (88)

Multiple Response Unit Model: Batta and Mannur (80) Marianov and ReVelle (91) Schilling, Elzinga, Cohen, Church and ReVelle (79)

Miscellaneous: Alao (71), Armour and Buffa (63) Beckman (63), Bell and Church (85), Bellman (65), Bertsimas (88), Bindschedler and Moore (61), Bouliane and LaPorte (92) Chan and Francis (76), Chaudary, McCormick and Moon (86), Chaiken (78), Church and Garfinkel (78), Conway and Maxwell (61), Cooper (72, 68, 67), Current and Schilling (89), Current and Storbeck (87), Current, ReVelle and Cohen (85) Dearing and Francis (74a, 74b) Eislet (92), Elzinga, Hearn and Randolph (76), Elzinga and Hearn (73, 72a, 72b), Erkut, Francis, Lowe and Tamir (89), Eyster, White and Wierwille (73) Fitzsimmons and Allen (83), Fitzsimmons (69), Francis and Mirchandani (89), Francis, McGinnis and White (83), Francis and Goldstein (74), Francis and White (74), Francis and Cabot (72), Francis (72, 67a, 67b, 64, 63), Frank (66) Gavett and Plyter (66), Ghosh and Craig (86), Gleason (75), Goldberg and Paz (91), Goldberg, Dietrich, Chen and Mitwasi, Valenzuela and Criss (90) Handler and Mirchandani (79), Hansen, Thisse and Wendell (86), Hitchings (69), Hodgson (90, 81), Hogg (68), Hopmans (86), Hsu and Nemhauser (79), Hurter, Schaeffer and Wendell (75)

Keeny (72), Kimes and Fitzsimmons (90), Kirca and Erkip (88) Larson (75, 72), Lawrence and Pengilly (69), Leamer (68), Love, Morriss and Wesolowsky (88), Love, Wesolowsky and Kraemer (73), Love and Morris (72), Love (72, 69, 67) McKinnon and Barber (72), McHose (61), Mirchandani (80, 79), Mole (73), Mycielski and Trzechiakowske (63) Nair and Chandrasekaran (71) Osleeb, Ratick, Buckley, Lee and Kuby (86) Palermo (61), Picard and Ratliff (78), Price and Turcotte (86), Pritsker (73), Pritsker and Ghare (70) Rand (76), ReVelle and Serra (91), ReVelle (86), Roodman and Schwartz (77, 75), Rosing (92), Ross and Soland (77), Rushton (89) Schaefer and Hurter (74), Schneider (71), Schniederjans, Kwak and Helmer (82), Simmons (71, 69), Snyder (71a), Storbeck (90, 88) Tansel and Yesilkokeen (93), Tansel, Francis and Lowe (80), Taylor (69), Teitz (68), Tewari and Jena (87), Tideman (62) Volz (71) Watson-Gandy (82, 72), Watson-Gandy and Eilon (72), Wesolowsky (73a, 73b, 72), Wesolowsky and Love (72, 71a, 71b), Weston (82), Wirasinghe and Waters (83), Weaver and Church (83) Young (63)

Table 8: APPLICATIONS

Bennett, Eaton and Church (82)	Selecting Sites for Rural Health Workers; Valle del Cauca, Columbai
Cerveny (80)	Location of Bloodmobile Operations
Cohon, ReVelle, Currnt, Eagles, Eberhart and Church	Power plant locations in a six-state region of U.S
Current and O'Kelly (92)	Locating emergency warning sirens in a midwestern city of U.S.
Daskin (82)	Emergency medical vehicles located in Austin, Texas
Daskin and Stern (81)	Emergency medical vehicles location in Austin, Texas
Drysdale and Sandiford (69)	Locating warehouses for R + CA Victor Company Ltd., Canada
Dutton, Hinman and Millham (74)	Locating electrical power generating plants in the Pacific Northwest
Eaton, Daskin, Simmons, Bulloch and Jansma (85)	Emergency medical service vehicle location in Texas
Fitzsimmons and Allen (83)	Selection of out-of-state audit offices
Flynn and Ratic (88)	Air Service locations for small communities in North and South Dakota
Fujiwara, Makjamroen and Gupta (87)	Ambulance deployment - A case study in Bangkok

Goldberg, Dietrich, Chen and Mitwasi, Valenzuela and Criss (90)	Locating emergency medical services in Tucson, Arizona
Hogg (68)	Siting of fire stations in Briston County Borough, England
Holmes, Williams and Brown (72)	Location of public day care facilities in Columbus, Ohio
Hopmans (86)	Locating bank branches in Netherlands
Kimes and Fitzsimmons	Selecting profitable sites at La Quinta Inns
Kirca and Erkip (88)	Selecting solid waste transfer points in Turkey
Kolesar and Walker (74)	Dynamic relocation of fire companies in New York City
Nambiar, Gelders and Van Wassenhove (89, 81)	Location of rubber processing factories in Malaysia
Patel (79)	Locating rural social service centers in India
Plane and Hendrick (77)	Location of fire companies for Denver Fire Department
Price and Turcotte (86)	Location of blood bank in Canada
Saedat (81)	Location of grass drying plants in the Netherlands
Schniederjans, Kwak and Helmer (82)	Locating a trucking terminal
Schreuder (81)	Locating fire stations in Rotterdam

Smith, Mangelsdorf, Luna and Reid (89)	Supply centers for Ecuador's health workers just-in-time
Tewari and Jena (87)	Location of high schools in rural India
Volz (71)	Ambulance location in semi-rural areas of Washtenaw County, Michigan
Walker (74)	Location of fire stations in New York City
Weston (82)	Telephone answering sites in a service industry (U.S.)
Wirasinghe and Waters (83)	Location of solid waste transfer points (Canada)

The models and formulations addressed in this section can be found in Dearing (85), Lourveaux and Peeters and Manne (64) (Plant Location), Cornuejols, Fisher and Nemhauser (77a), Kraus, Janseen and McAdams (70), Nauss and Markland (81) (Lock Box), Garfinkel, Neebe and Rao (77) and Minieka (70) (P-Center), Hakimi (64), Narula, Ogbu and Samuelsson (77) and ReVelle and Swain (70) (P-Median), Toregas and ReVelle (70), Toregas, Swain, ReVelle and Bergman (71) (Service Facilities), Church and ReVelle (74) (Maximal Covering), Daskin and Stern (81) (Hierarchical Objective Set Covering), Hogan and ReVelle (86) (Backup Coverage) and Batta and Mannur (90) (Multiple Response Unit).
91

12 Conclusions

In this paper a variety of applications of set covering, set packing, and set partitioning models including their variants and generalizations are presented. In addition, transformations to convert one model to another including the transformation of the MD Knapsack problem and the LIP to one of these models are discussed. It should be noted that some applications such as Travelling Salesman Problem, Multicommodity Disconnecting Problem and Steiner Problem in Graphs require enormous number of constraints where as the formulation of the Personnel Scheduling, Crew Scheduling, and the Routing models require a very large number of variables. The transformations of the MD Knapsack problem and the LIP require both a very large number of variables and constraints.

Moderate size SC, SP and SPT models can be solved efficiently with the existing algorithms and techniques and have been used in many real life situations. Efficient solution techniques to solve very large SC, SP and SPT models will enhance the application of these mpodels to solve many real life lgoistics problems. Clearly, these special structured models are a very useful and important class of linear integer programs and deserve the effort devoted by many researchers.

References

Theory

Albers, S., "Implicit Enumeration Algorithms for the Set Partitioning Problem", OR Spektrum, Vol. 2, pp. 23-32 (1980).

Anily, S., and Federguren, A., "Structured Partitioning Problems", Operations Research, Vol. 39, pp. 130- 149 (1991).

Avis, D., "A Note on Some Computationally Difficult Set Covering Problems", Mathematical Programming, Vol. 18, pp. 138-145 (1980).

Baker, E., "Efficient Heuristic Algorithms for the Weighted Set Covering Problem", Computers and Operations Research, Vol. 8, pp. 303-310 (1981).

Balas, E., and Ng, S.M., "On the Set Covering Polytope: 1. All the Facets With Coefficients in 0, 1,2", Mathematical Programming, Vol. 43, pp. 57-69 (1989).

Balas, E., "A Sharp Bound on the Ratio Between Optimal Integer and Fractional Covers", Mathematics of Operations Research, Vol. 9, pp. 1-5 (1984).

Balas, E., and Ho, A., "Set Covering Algorithms Using Cutting Planes Heuristics and Subgradient Optimization: A Computational Study", Mathematical Programming Study, Vol. 12, pp. 37-60 (1980).

Balas, E., "Cutting Planes from Conditional Bounds: A New Approach to Set Covering" Mathematical Programming Study, Vol. 12, pp. 19-36 (1980).

Balas, E., and Padberg, M.W., "Set Partitioning: A Survey", Combinatorial Optimization (Edited by N. Christofides, A. Mingozzi, P. Toth and C. Sandi), John Wiley and Sons, New York, pp. 151 - 210 (1979).

Balas, E., "Set Covering with Cutting Planes for Conditional Bounds", Ninth International Symposium on Mathematical Programming (1976).

Balas, E., and Padberg, M.W., "Set Partitioning: A Survey", SIAM Review, Vol. 18, pp. 710-760 (1976).

Balas, E., and Padberg, M.W., "On Set Covering Problem II: An Algorithm for Set Partitioning", Operations Research, Vol. 22, 1, pp. 74-90 (1975a).

Balas, E., and Padberg, M.W., "Set Partitioning", Combinatorial Programming: Methods and Applications (Edited by B. Roy), Reidel Publishing Co., pp. 205-258 (1975b).

Balas, E., and Padberg, M.W., "On the Set-Covering Problem", Operations Research, Vol. 20, pp. 1152- 1161 (1972).

Beasley, J.E., "An Algorithm for Set Covering Problem", European Journal of Operational Research, Vol. 31, pp. 85-93 (1987).

Bellmore, M., and Ratliff, H.D., "Set Covering and Involutory Basis", Management Science, Vol. 18, pp. 194-206 (1971).

Benvensite, R., "A Note on the Set Covering Problem", Journal of the Operational Research Society, Vol. 33, pp. 261-265 (1982).

Berge, C., "Balanced Matrices", Mathematical Programming, Vol. 2, pp. 19-31 (1972).

Chan, T.J., and Yano, C.A., "A Multiplier Adjustment Approach for the Set Partitioning Problem", Operations Research, Vol. 40, pp. S40-S47 (1992).

Chaudry, S.S., Moon, I.D., and McCormick, S.T., "Conditional Covering: Greedy Heuristics and Computational Results", Computers and Operations Research, Vol. 14, pp. 11-18 (1987).

Christofides, N., and Paixo, J., "State Space Relaxation for the Set Covering Problem" Faculdade, de ciencias Lisboa, Portugal (1986).

Christofides, N., and Korman, S., "A Computational Survey of Methods for the Set Covering Problem", Management Science, Vol. 21, pp. 591-599 (1975).

Chvatal, V., "A Greedy-Heuristic for the Set Covering Problem", Mathematics of Operations Research, Vol. 4, pp. 233-235 (1979).

Coffman, E.G., Jr., and Lueker, G.S., "Probabilistic Analysis of Packing and Partitioning Algorithms", John Wiley (1991).

Conforti, M., Corneil, D.G., and Mahjoub, A.R., "K-Covers 1: Complexity and Polytope", Discrete Mathematics, Vol. 58, pp. 121-142 (1986).

Cornvejols, G., and Sassano, A., "On the (0,1) Facets of the Set Covering Polytope", Mathematical Programming, Vol. 43, pp. 45-55 (1989).

Crama, Y., Hammer, P., and Ibaraki, T., "Packing, Covering and Partitioning Problems with Strongly Unimodular Constraint Matrices", Mathematics of Operaions Research, Vol. 15, pp. 258-269 (1990).

El-Darzi, E., "Methods for Solving the Set Covering and Set Partitioning Problems using Graph Theoretic (Relaxations) Algorithms", PhD Thesis, Brunel University, Uxbridge (UK) (1988).

El-Darzi, E., and Mitra, G., "A Tree Search for the Solution of Set Problems Using Alternative Relaxations", TR/03/88, Department of Mathematics and Statistics, Brunel University, Uxbridge (UK) (1988a).

El-Darzi, E., and Mitra, G., "Set Covering and Set Partitioning: A Collection of Test Problems", TR/01/88, Department of Mathematics and Statistics, Brunel University, Uxbridge (UK) (1988b).

Etcheberry, J., "The Set Covering Problem: A New Implicit Enumeration Algorithm", Operations Research, Vol. 25, pp. 760-772 (1977).

Fisher, M.L., and Kedia, P., "Optimal Solution of Set Covering/Partitioning Problems Using Dual Heuristics", Management Science, Vol. 36, pp. 674-688 (1990).

Fisher, M.L., and Kedia, P., "A Dual Algorithm for Large Scale Set Partitioning", Paper No. 894, Purdue University, West Lafayette, Indiana (1986).

Fisher, M.L., and Wolsey, L., "On the Greedy Heuristic for Continuous Covering and Packing Problems", SIAM Journal On Algebraic and Discrete Methods, Vol. 3, pp. 584-591 (1982).

Fowler, R.J., Paterson, M.S., and Tanimoto, S.L., "Optimal Packing and Covering in the Plane are NP- Complete", Information Processing Letters, Vol. 12, pp. 133-137 (1981).

Fox, G.W., and Scudder, G.D., "A Simple Strategy for Solving a Class of 0-1 Integer Programming Models", Computers and Operations Research, Vol. 13, pp. 707-712 (1986).

Fulkerson, D.R., Hoffman, A.J., and Oppenheim, R., "On Balanced Matrices", Mathematical Programming Study, Vol. 1, pp. 120-132 (1974).

Garfinkel, R., and Nemhauser, G.L., "Optimal Set Covering: A Survey", Perspectives on Optimization., (Edited by A.M. Geoffrion), Addison-Wesley, pp. 164-183 (1972).

Garfinkel, R.S., and Nemhauser, G.L., "The Set Partitioning Problem: Set Covering with Equality Constraints", Operations Research, Vol. 17, pp. 848-856 (1969).

Hammer, P.L., and Simeone, B., "Order Relations of Variables in 0-1 Programming", Annals of Discrete Mathematics, Vol. 31, pp. 83-112 (1987).

Hammer, P.L., Johnson, F.L., and Peled, U.N., "Regular 0-1 Programs", Cahiers du Centre d'Etudes de Recherche Operationnelle, Vol. 16, pp. 267-276 (1974).

Ho, A., "Worst Case of a Class of Set Covering Heuristic", Mathematical Programming, Vol. 23, pp. 170- 181 (1982).

Hochbaum, D., "Approximation Algorithms for the Weighted Set Covering and Node Cover Problems", GSIA, Carnegie-Mellon University (1980).

Hwang, F.K., Sum, J., and Yao, E.Y., "Optimal Set Partitioning", SIAM Journal On Algebraic Discrete Methods, Vol. 6, pp. 163-170 (1985).

John, D.G., and Kochenberger, G.A., "Using Surrogate Constraints in a Lagrangian Relaxation Approach to Set Covering Problems" Journal of Operational Research Society, Vol. 39, pp. 681-685 (1988).

Johnson, D., "Approximation Algorithms for Combinatorial Problems", Journal of Computer and System Scinces, Vol. 9, pp. 256-278 (1974).

Lawler, E.L., "Covering Problems: Duality Relations and a New Method of Solutions", SIAM Journal of Applied Mathematics, Vol. 14, pp. 1115-1132 (1966).

Leigh, W., Ali, D., Ezell, C., and Noemi, P., "A Branch and Bound Algorithm for Implementing Set Covering Model Expert System", Computers and Operations Research, Vol. 11, pp. 464-467 (1988).

Lemke, C.E., Salkin, H.M., and Spielberg, K., "Set Covering by Single Branch Enumeration with Linear Programming Subproblems", Operations Research, Vol. 19, pp. 998-1022 (1971).

Lovasz, L., "On the Ratio of Optimal Integer and Fractional Covers", Discrete Mathematics, Vol. 13, pp. 383-390 (1975).

Marsten, R.E., "An Algorithm for Large Set Partitioning Problems", Management Science, Vol. 20, pp. 774- 787 (1974).

Michaud, P., "Exact Implicit Enumeration Method for Solving the Set Partitioning Problem", IBM Journal of Research and Development, Vol. 16, pp. 573-578 (1972).

Murty, K., "On the Set Representation and Set Covering Problem", Symposium on the Theory of Scheduling and Its Applications, (Edited by S.E., Elmaghraby), Springer-Verlag (1973).

Nemhauser, G.L., Trotter, L.E., and Nauss, R.M., "Set Partitioning and Chain Decomposition", Management Science, Vol. 20, pp. 1413-1423 (1974).

Padberg, M., "Covering, Packing and Knapsack Problems", Annals of Discrete Mathematics, Vol. 4, pp. 265-287 (1979).

Padberg, M., "Perfect Zero-One Matrices", Mathematical Programming, Vol. 6, pp. 180-196 (1974a).

Padberg, M., "Characterization of Totally Unimodular, Balanced and Perfect Matrices", Combinatorial Programming: Methods and Applications, (Edited by B. Roy), D. Reidal Publishing, pp. 275-284 (1974b).

Padberg, M., "On the Facial Structure of Set Packing Polyhedra", Mathematical Programming, Vol. 5, pp. 199-215 (1973).

Peled, U.N., and Simeone, B., "Polynomial-Time Algorithms for Regular Set-Covering and Threshold Synthesis", Discrete Applied Mathematics, Vol. 12,, pp. 57-69 (1985).

Pierce, J.F., and Lasky, J.S., "Improved Combinatorial Programming Algorithms for a Class of All Zero-One Integer Programming Problems", Management Science, Vol. 19, pp. 528-543 (1975).

Pierce, J.F., "Applications of Combinatorial Programming to a Class of All-Zero-One Integer Programming Problems", Management Science, Vol. 15, pp. 191-209 (1968).

Roth, R., "Computer Solutions to Minimum Cover Problems", Operations Research, Vol. 17, pp. 455-465 (1969).

Roy, B. "An Algorithm for a General Constrained Set Covering Problem", Computing and Graph Theory, Academic Press (1972).

Ryan, D.M. and Falkner, J.C., "On the Integer Properties of Scheduling Set Partitioning Models", European Journal of Operational Research, Vol. 35, pp. 442-456 (1988).

Ryzhkov, A.P., "On Certain Covering Problems", Engineering Cybernetics, Vol. 2, pp. 543-548 (1973).

Salkin, H.M., "The Set Covering Problem", Integer Programming, Addison Wesley, pp. 439-481, (1975).

Salkin, H.M., and Koncal, R.D., "Set Covering by an All-Integer Algorithm: Computational Experience", ACM Journal, Vol. 20, pp. 189-193, (1973).

Trubin, V.A., "On a Method of Integer Linear Programming of a Special Kind", Soviet Math. Dokl., Vol. 10, pp. 1544-1546 (1969).

Vasko, F.J., and Wolfe F.E., "Solving Large Set Covering Problems on a Personal Computer", Computers and Operations Research, Vol. 15, pp. 115-121 (1988).

Vasko, F.J., and Wilson, G.R., "Hybrid Heuristics for Minimum Cardinality Set Covering Problems", Naval Research Logistic Quarterly, Vol. 33, pp. 241-250 (1986).

Vasko, F.J., and Wilson, G.R., "Using Facility Location Algorithms to Solve Large Set Covering Problems", Operations Research Letters, Vol. 3, pp. 85-90 (1984a).

Vasko, F.J., and Wilson, G.R., "An Efficient Heuristic for Large Set Covering Problems", Naval Research Logistics Quarterly, Vol. 31, pp. 163-171 (1984b).

Graphs

Aneja, Y.P., "An Integer Programming Approach to the Steiner Problem in Graphs", Networks, Vol. 10, pp. 167-178 (1980).

Aneja, Y.P., and Vemuganti, R.R., "A Row Generation Scheme for Finding Multi-Commodity Minimum Disconnecting Set", Management Science, Vol. 23, pp. 652-659 (1977).

Balanski, M.L., "On Maximum Matching, Minimum Covering and Their Connections", Proceedings of the International Symposium on Mathematical Programming, (Edited by H.W.Kuhn), Princeton University Press, pp. 303-311, (1970).

Beasley, J.E., "An SST-Based Algorithm for the Steiner Problem on Graphs", Networks, Vol. 19, pp. 1-16 (1989).

Beasley, J.E., "An Algorithm for the Steiner Problem in Graphs", Networks, Vol. 14, pp. 147-159 (1984).

Bellmore, M., and Ratliff, H.D., "Optimal Defence of Multi-Commodity Networks", Management Science, Vol. 18, pp. 174 - 185 (1971).

Berge, C., "Graphs and Hypergraphs", Translated by E. Minieka, North-Holland Publishing Company, Amsterdam-London (1973).

Brelaz, D., "New Method to Color the Vertices of a Graph", Communications of the ACM, Vol.22, pp. 251- 256 (1979).

Brown, J.R., "Chromatic Scheduling and the Chromatic Number Problem", Management Science, Vol. 19, pp. 456-663 (1972).

Chang, G.J., and Nemhauser, G.L., "Covering, Packing and Generalized Perfection", SIAM Journal on Algebraic and Discrete Methods, Vol. 6, pp. 109-132 (1985).

Chopra, S., "Comparison of Formulations and Heuristics for Packing Steiner Trees on a Graph", Technical Report, J.L. Kellogg Graduate School of Management, Northwestern University, Illinois (1992).

Chvatal, V., "Determining the Stability Number of a Graph", SIAM Journal on Computing, Vol. 6, pp. 643-662 (1977).

Cockayne, E.J., and Melzak, Z.A., "Steiner's Problem for Set Terminals", Quarterly of Applied Mathematics, Vol. 26, pp. 213-218 (1969).

Corneil, D.G., and Graham, B., "An Algorithm for Determining the Chromatic Number of a Graph", SIAM Journal of Computing, Vol. 2, pp. 311-318 (1973).

Dreyfus, S.E., and Wagner, R.A., "The Steiner Problem in Graphs", Networks, Vol. 1, pp. 195-207 (1971).

Edmonds, J., "Maximum Matching and a Polyhedron with 0,1 Vertices", Journal of Research of National Bureau of Standards, Vol. 69b, pp. 125-130 (1965a)

Edmonds, J., "Paths, Trees and Flowers", Canadian Journal of Mathematics, Vol. 17, pp. 449-467 (1965b).

Edmonds, J., "Covers and Packing in a Family of Sets", Bulletin of the American Mathematical Society, Vol. 68, pp. 494-499 (1962).

Gilbert, E.N., and Pollak, H.O., "Steiner Minimal Trees", SIAM Journal of Applied Mathematics, Vol. 16, pp. 1-29 (1968).

Hakimi, S.L, "Steiner's Problem in Graphs and Its Implications", Networks, Vol. 1, pp. 113-133 (1971).

Hanan, M., "On Steiner's Problem with Rectilinear Distance", SIAM Journal of Applied Mathematics, Vol. 14, pp. 255-265 (1966).

Houck, D.J., and Vemuganti, R.R., "An Algorithm for the Vertex Packing Problem", Operations Research, Vol. 25, pp. 773-787 (1977).

Hwang, F.K., and Richards, D.S., "Steiner Tree Problems", Networks, Vol. 22, pp. 55-89 (1992).

Khoury, B.N., Pardalos, P.M., and Hearn, D.W., "Equivalent Formulations for the Steiner Problem on Graphs", Network Optimization Problems (Edited by D.-Z, Du and P.M. Pardos), World Scientific Publishing Co., pp. 111-124 (1993).

Khoury, B.N., Pardalos, P.M., and Du, D.-Z., "A Test Problem Generator for the Steiner Problem in Graphs", Department of Industrial and Systems Engineering Working Paper, University of Florida, Gainsville, Florida (1993).

Leighton, F., "A Graph-Coloring Algorithm for Large Scheduling Problems", Journal of Research of the National Bureau of Standards, Vol. 84, pp. 489-506 (1979).

Maculan, N., "The Steiner Problem in Graphs", Annals of Discrete Mathematics, Vol. 31, pp. 185-212 (1987).

Mehta, N.K., "The Application of Graph Coloring Model to an Examination Scheduling Problem", Interfaces, Vol. 11, pp. 57 - 64 (1981).

Nemhauser, G.L., and Trotter, L.E., "Vertex Packing: Structural Properties and Algorithm", Mathematical Programming, Vol. 8, pp. 232-248 (1975).

Nemhauser, G.L., and Trotter, L.E., "Properties of Vertex Packing and Independent Systems Polyhedra", Mathematical Programming, Vol. 6, pp. 48-61 (1974).

Norman, R.Z., and Rabin, M.O., "An Algorithm for a Minimum Cover of a Graph", Proceedings of the American Mathematical Society, Vol. 10, pp. 315-319 (1959).

Salazar, A., and Oakford, R.V., "A Graph Formulation of a School Scheduling Algorithm", Communications of the ACM, Vol. 17, pp. 696-698 (1974).

Wang, C.C., "An Algorithm for the Chromatic Number of a Graph", Journal of ACM, Vol. 21, pp.385-391 (1974).

Weinberger, D.B., "Network Flows, Minimum Coverings, and the Four Color Conjecture", Operations Research, Vol. 24, pp. 272-290 (1976).

Winter, P., "Steiner Problem in Networks: A Survey", Networks, Vol. 17, pp. 129-167 (1987).

Wong, R.T., "A Dual Ascent Approach to Steiner Tree Problems on a Directed Graph", Mathematical Programming, Vol. 28, pp. 271-287 (1984).

Wood, D.C., "A Technique for Coloring a Graph Applicable to Large Scale Time-Tableing Problems", Computing Journal, Vol. 12, pp. 317-319 (1969).

Wu, Y.F., Widmayer, P., and Wong, C.K., "A Faster Approximation Algorithm for the Steiner Problem in Graphs", Acta Informatica, Vol. 23, pp. 223-229 (1986).

Personnel Scheduling

Abernathy, W., Baloff, N., and Hershey, J., "A Variable Nurse Staffing Model", Decision Sciences, Vol. 5, pp. 58-72 (1974).

Abernathy, W., Baloff, N., Hershey, J., and Wandel, S., "A Three Stage Manpower Planning and Scheduling Model - A Service Sector Example", Operations Research, Vol. 21, pp. 693-711 (1973).

Abernathy, W., Baloff, N., and Hershey, J., "The Nurse Staffing Problem: Issues and Prospects", Sloan Management Review, Vol. 12, pp. 87 - 99 (1971).

Ahuja, H., and Sheppard, R., "Computerized Nurse Scheduling" Industrial Engineering, Vol. 7, pp. 24-29 (1975).

Altman, S., Beltrami, E.J., Rappoport, S.S., and Schoepfle, G.K., "Nonlinear Programming Model of Crew Assignments for Household Refuse Collection" , IEEE Transactions on Systems, Man and Cybernetics, Vol. 1, pp. 289-291 (1971).

Bailey, J., "Integrated Days Off and Shift Personnel Scheduling", Computers and Industrial Engineering, Vol. 9, pp. 395-404 (1985).

Bailey, J., and Field, J., "Personnel Scheduling with Flexshift Models", Journal of Operations Management, Vol. 5, pp. 327-338 (1985).

Baker, K.R., Burns, R.N., and Carter, M., "Staff Scheduling with Days-Off and Work Stretch Constraints", AIIE Transactions, Vol. 11, pp. 286-292 (1979).

Baker, K.R., and Magazine, M.J., "Workforce Scheduling with Cyclic Demands and Days-Off Constraints", Management Science, Vol. 24, pp. 161-167 (1977).

Baker, K.R., "Workforce Allocation in Cyclic Scheduling Problem: A Survey", Operational Research Quarterly, Vol. 27, pp. 155-167 (1976).

Baker, K.R., "Scheduling a Full-Time Workforce to Meet Cyclic Staffing Requirements", Management Science, Vol. 20, pp. 1561-1568 (1974).

Baker, K.R., Crabil, T.B., and Magazine, M.J., "An Optimal Procedure for Allocating Manpower with Cyclic Requirements", AIIE Transactions, Vol. 5, pp. 119-126 (1973).

Bartholdi, J.J., "A Guaranteed-Accuracy Round-off Algorithm for Cyclic Scheduling and Set Covering", Operations Research, Vol. 29, pp. 501-510 (1981).

Bartholdi, J.J., Orlin, J.B., and Ratliff, H.D., "Cyclic Scheduling via Integer Programs with Circular Ones," Operations Research, Vol. 29, pp.1074-1085 (1980).

Bartholdi, J.J., and Ratliff, H.D., "Unnetworks, With Applications to Idle Time Scheduling", Management Science, Vol. 24, pp. 850-858 (1978).

Bechtold, S., Brusco, M., and Showalter, M., "A Comparative Evaluation of Labor Tour Scheduling Methods", Decision Sciences, Vol. 22, pp. 683-699 (1991).

Bechtold, S.E., "Optimal Work-Rest Schedules with a Set of Fixed-Duration Rest Periods", Decision Sciences, Vol. 22, pp. 157-170 (1991).

Bechtold, S.E., and Jacobs, L.W., "Implicit Modelling of Flexible Break Assignments in Optimal Shift Scheduling", Management Science, Vol. 36, pp. 1339-1351 (1990).

Bechtold, S.E., and Sumners, D.L., "Optimal Work-Rest Scheduling with Exponential Work Rate Decay", Management Science, Vol. 34, pp. 547-552 (1988).

Bechtold, S.E., "Implicit Optimal and Heuristic Labor Staffing in a Multi-objective Multilocation Environment", Decision Sciences, Vol. 19, pp. 353 - 372 (1988).

Bechtold, S.E., and Showalter, M., "A Methodology for Labor Scheduling in a Service Operating System", Decision Sciences, Vol. 18, pp. 89-107 (1987).

Bechtold, S.E., and Showalter, M., "Simple Manpower Scheduling Methods for Managers", Production and Inventory Management, Vol. 26, pp. 116-132 (1985).

Bechtold, S.E., Janaro, R.E., and Sumners, D.L., "Maximization of Labor Productivity through Optimal Rest-Break Schedules", Management Science, Vol. 30, pp. 1442-1458 (1984).

Bechtold, S.E., "Work Force Scheduling for Arbitrary Cyclic Demand", Journal of Operations Management, Vol. 1, pp. 205-214 (1981).

Bechtold, S.E., "Quantitative Models for Optimal Rest Period Scheduling: A Note", OMEGA, The International Journal of Management Science, Vol. 7, pp. 565-566 (1979).

Bennett, B.T., and Potts, R.B., "Rotating Roster for a Transit System", Transportation Science, Vol. 2, pp. 14-34 (1968).

Bodin, L.D., "Toward a General Model for Manpower Scheduling: Parts 1 and 2", Journal of Urban Analysis, Vol. 1, pp. 191-245 (1973).

Browne, J.J., "Simplified Scheduling of Routine Work Hours and Days Off", Industrial Engineering, Vol. 11, pp. 27 - 29 (1979).

Browne, J.J. and Tibrewala, R.K., "Manpower Scheduling", Industrial Engineering, Vol. 7, pp. 22-23 (1975).

Brownell, W.S., and Lowerre, J.M., "Scheduling of Workforce Required in Continuous Operations Under Alternate Labor Policies", Management Science, Vol. 22, pp. 597-605 (1976).

Buffa, E.S., Cosagrove, M.J., and Luce, B.J., "An Integrated Work Shift Scheduling System", Decision Sciences, Vol. 7, pp. 620-630 (1976).

Burns, R.N., and Koop, G.J., "A Modular Approach to Optimal Shift Manpower Scheduling", Operations Research, Vol. 35, pp. 100-110 (1987).

Burns, R.N., and Carter, M.W., "Work Force Size and Single Shift Schedules with Variable Demands", Management Science, Vol. 31, pp. 599-607 (1985).

Burns, R.N., "Manpower Scheduling with Variable Demands and Alternate Weekends Off", INFOR, Canadian Journal of Operations Research and Information Processing, Vol. 16, pp. 101-111 (1978).

Byrne, J.L., and Potts, R.B., "Scheduling of Toll Collectors", Transportation Science, Vol. 30, pp. 224-245 (1973).

Chaiken, J., and Dormont, P., "A Patrol Car Allocation Model: Background and A Patrol Car Allocation Model: Capabilities and Algorithms", Management Science, Vol. 24, pp. 1280-1300 (1978)

Chelst, K., "Deployment of One vs. Two-Officer Patrol Units: A Comparison of Travel Times", Management Science, Vol. 27, pp. 213-230 (1981).

Chelst, K., "Algorithm for Deploying a Crime Directed Patrol Force", Management Science, Vol. 24, pp. 1314-1327 (1978).

Chen, D., "A Simple Algorithm for a Workforce Scheduling Model", AIIE Transactions, Vol. 10, pp. 244 - 251 (1978).

Church, J.G., "Sure Staff: A Computerized Staff Scheduling System for Telephone Business Offices" Management Science, Vol. 20, pp. 708 - 720 (1973).

Dantazig, G.W., "A Comment on Edie's Traffic Delays at Toll Booths", Operations Research, Vol. 2, pp. 339 - 341 (1954).

Easton, F.F., and Rossin, D.F., "Sufficient Working Subsets for the Tour Scheduling Problem", Management Science, Vol. 37, pp. 1441-1451 (1991a).

Easton, F.F., and Rossin, D.F., "Equivalent Alternate Solutions for the Tour Scheduling Problems", Decision Sciences, Vol. 22, pp. 985-1007 (1991b).

Eilon, S., "On a Mechanistic Approach to Fatigue and Rest Periods", International Journal of Production Research, Vol. 3, pp. 327-332 (1964).

Emmons, H., and Burns, R.N., "Off-Day Scheduling with Hierarchical Worker Categories", Operations Research, Vol. 39, pp. 484-495 (1991).

Emmons, H., "Workforce Scheduling with Cyclic Requirements and Constraints on Days Off, Weekends Off and Workstretch", IIE Transactions, Vol. 17, pp. 8-16 (1985).

Frances, M.A., "Implementing a Program of Cyclical Scheduling of Nursing Personnel", Hospitals, Vol. 40, pp. 108 - 123 (1966).

Gaballa, A., and Pearce, W., "Telephone Sales Manpower Planning at Qantas", Interfaces, Vol. 9, pp. 1-9 (1979).

Gentzler, G.L., Khalil, T.M., and Sivazlian, B.B., "Quantitative Methods for Optimal Rest Period Scheduling", OMEGA, The International Journal of Management Science, Vol. 5, pp. 215-220 (1977).

Glover, F., and McMillan, C., "The General Employee Scheduling Problem: An Integration of Management Science and Artificial Intelligence", Computers and Operations Research, Vol. 13, pp. 563-573 (1986).

Glover, F., McMillan, C. and Glover, R., "A Heuristic Programming Approach to the Employee Scheduling Problem and Some Thoughts on 'Managerial Robots'", Journal of Operations Management, Vol. 4, pp. 113- 128 (1984).

Green, L. and Kolesar, P., "The Feasibility of One Officer Patrol in New York City", Management Science, Vol. 30, pp. 964-981 (1984).

Guha, D., and Browne, J., "Optimal Scheduling of Tours and Days Off", Preprints, Workshop on Automated Techniques for Scheduling of Vehicle Operators for Urban Public Transportation Services (Edited by Bodin and Bergman), Chicago, Illinois (1975).

Hagberg, B., "An Assignment Approach to the Rostering Problem: An Application to Taxi Vehicles", Computer Scheduling of Public Transport 2 (Edited by J-M Rousseau), pp. 313 - 318, North-Holland (1985).

Henderson, W.B., and Berry, W.L., "Determining Optimal Shift Schedules for Telephone Traffic Exchange Operators", Decision Sciences, Vol. 8, pp. 239-255 (1977)

Henderson, W.B., and Berry, W.L., "Heuristic Methods for Telephone Operator Shift Scheduling: An Experimental Analysis", Management Science, Vol. 22, pp.1372-1380 (1976).

Hershy, J.C., Albernathy, W.J., and Baloff, N., "Comparison of Nurse Allocation Policies - A Monte Carlo Model", Decision Sciences, Vol. 5, pp. 58 - 72 (1974).

Holloran, T.J., and Byrn, J.E., "United Airline Station Manpower Planning System", Interfaces, Vol. 16, pp. 39-50 (1986).

Howell, J.P., "Cyclical Scheduling of Nursing Personnel," Hospitals, J.A.H.A., Vol. 40, pp. 77 - 85 (1966).

Hung, R., and Emmons, H., "Multiple-Shift Workforce Scheduling Under the 3-4 Compressed Workweek With a Hierarchical Workforce", Department of Operations Research Working Paper, Case Western Reserve University, Cleveland, Ohio (1990).

Ignall, E. Kolesar, P., and Walker, W., "Linear Programming Models of Crew Assignments for Refuse Collection", IEEE Transactions on Systems, Man and Cybernetics, Vol. 2, pp. 664-666 (1972).

Keith, E.G., "Operator Scheduling", AIIE Transactions, Vol. 11, pp. 37-41 (1979).

Klasskin, P.M., "Operating to Schedule Operators", Telephony, Vol. pp. 29 - 31 (1973).

Koop, G.J., "Multiple Shift Workforce Lower Bounds", Management Science, Vol. 34, pp. 1221-1230 (1988).

Koop, G.J., "Cyclic Scheduling of Weekends", Operations Research Letters, Vol. 4, pp. 259-263 (1986).

Krajewski, L.J., Ritzman, L.P., and McKenzie, P., "Shift Scheduling in Banking Operations: A Case Application", Interfaces, Vol. 10, pp. 1-8 (1980).

Krajewski, L.J., and Ritzman, L.P., "Disaggregation in Manufacturing and Service Organizations: Survey of Problems and Research", Decision Sciences, Vol. 8, pp. 1 - 18, (1977).

Lessard, R., Rousseau, J.M., and DuPuis, D., "Hatus I: A Mathematical Programming Approach to the Bus Driver Scheduling Problem", Computer Scheduling of Public Transport, (Edited by A. Wren), North-Holland Publishing Company, pp. 255-267 (1981).

Li, C., Robinson, E.P., and Mabert, V.A., "An Evaluation of Tour Scheduling Heuristics with Differences in Employee Productivity and Cost", Decision Sciences, Vol. 22, pp. 700-718 (1991).

Linder, R.W., "The Development of Manpower and Facilities Planning Methods for Airline Telephone Reservation Offices", Operational Research Quarterly, Vol. 20, No. 1, pp. (1969).

Loucks, J.S., and Jacobs, F.R., "Tour Scheduling and Task Assignment of a Heterogeneous Work Force", Decision Sciences, Vol. 22, pp. 719-738 (1991).

Lowerre, J.M., "On Personnel Budgeting on Continuous Operations (With Emphasis on Hospitals)", Decision Sciences, Vol, 10, pp. 126-135 (1979).

Lowerre, J.M., "Work Stretch Properties for the Scheduling of Continuous Operations Under Alternative Labor Policies", Management Science, Vol. 23, pp. 963-971 (1977).

Mabert, V.A., and Watts C.A., "A Simulation Analysis of Tour Shift Construction Procedures", Management Science, Vol. 28, pp. 520-532 (1982).

Mabert, V.A., and McKenzie, J.P., "Improving Bank Operations: A Case Study at Bank Ohio/Ohio National Bank", OMEGA, The International Journal of Management Science, Vol. 8, pp. 345-354 (1980).

Mabert, V.A., "A Case Study of Encoder Shift Scheduling Under Uncertainty", Management Science, Vol. 25, pp. 623-631 (1979).

Mabert, V.A., and Raedels, A., "The Detail Scheduling of A Part-Time Work Force: A Case Study of Tellar Staffing", Decision Sciences, Vol. 8, pp. 109-120 (1977).

Maier-Rothe, C., and Wolf, H.B., "Cylical Scheduling and Allocation of Nursing Staff", Socio-Economic Planning Sciences, Vol. 7, pp. 471-487 (1973).

McGinnis, L.F., Culver, W.D., and Deane, R.H., "One and Two-Phase Heuristics for Workforce Scheduling", Computers and Industrial Engineering, Vol. 2, pp. 7-15 (1978).

McGrath, D., "Flextime Scheduling: A Survey", Industrial Management, Vol. 22 , pp. 1-4 (1980).

Megeath, J.D., "Successful Hospital Personnel Scheduling", Interfaces, Vol. 8, pp. 55 - 59 (1978).

Miller, H.E., Pierskalla, W.P., and Rath, G.J., "Nurse Scheduling, Using Mathematical Programming", Operations Research, Vol. 24, pp. 857-870 (1976).

Monroe, G., "Scheduling Manpower for Service Operations", Industrial Engineering, Vol. 2, pp. 10-17 (1970).

Moondra, S.L., "An L.P. Model for Work Force Scheduling for Banks", Journal of Bank Research, Vol. 6, pp. 299-301 (1976).

Morris, J.G., and Showalter, M.J., "Simple Approaches to Shift, Days-Off and Tour Scheduling Programs", Management Science, Vol. 29, pp. 942-950 (1983).

Morrish, A.R., and O'Connor, A.R., "Cyclic Scheduling" Hospitals J.A.H.A., Vol. 14, pp. 66-71 (1970).

Ozkarahan, I., and Bailey, J.E., "Goal Programming Model Subsystem of Flexible Nurse Scheduling Support System", IIE Transactions, Vol. 20, No. 3, pp. 306-316 (1988).

Ozkarahan, I., "A Flexible Nurse Scheduling Support System", Ph.D. Dissertation, Arizona State University (1987).

Paixao, J., and Pato, M., "A Structural Lagrangean Relaxation for Two-Duty Period Bus Drive Scheduling Problems", European Journal of Operational Research, Vol. 39, No. 2, pp. 213-222 (1989).

Pappas, I.A., "Dynamic Job Assignment for Railway Personnel", Management Science, Vol. 13, pp. B809- B816 (1967).

Price, E., "Techniques to Improve Staffing", American Journal of Nursing, Vol. 70, pp. 2112 - 2115 (1970).

Ritzman, L.P., Krajewski, L.J., and Showalter, M.J., "The Disaggregation of Aggregate Manpower Plans", Management Science, Vol. 22, pp. 1204-1214 (1976).

Rothstein, M., "Hospital Manpower Shift Scheduling by Mathematical Programming", Health Service Research, Vol. 8, pp. 60 - 66 (1973).

Rothstein, M., "Scheduling Manpower by Mathematical Programming", Industrial Engineering, Vol. 4, pp. 29-33 (1972).

Segal, M., "The Operator-Scheduling Problem: A Network Flow Approach", Operations Research, Vol. 22, pp. 808-823 (1974).

Shepardson, F., and Marsten, R.E., "A Lagrangean Relaxation Algorithm for the Two Duty Period Scheduling Problem", Management Science, Vol. 26, pp. 274-281 (1980).

Showalter, M.J., and Mabert, V.A., "An Evaluation of A Full-Part Time Tour Scheduling Methodology", International Journal of Operations and Production Management, Vol. 8, pp. 54-71 (1988).

Showalter, M.J., Krajewski, L. J., and Ritzman, L.P., "Manpower Allocation in U.S. Postal Facilities: A Heuristic Approach", Computers and Operations Research, Vol. 2, pp. 141-13 (1978).

Smith, D.L., "The Application of an Interactive Algorithm to Develop Cyclical Schedules for Nursing Personnel", INFOR, Canadian Journal of Operations Research and Information Processing,, Vol. 14, pp. 53-70 (1976).

Smith, H.L., Mangelsdorf, K.R., Luna, J.C., and Reid, R.A., "Supplying Ecuador's Health Workers Just in Time", Interfaces, Vol. 19, pp. 1-12 (1989).

Smith, L., and Wiggins, A., "A Computer-Based Nursing Scheduling System", Computers and Operations Research, Vol. 4, pp. 195-212 (1977).

Stern, H.I., and Hersh, M., "Scheduling Aircraft Cleaning Crews", Transportation Science, Vol. 14, pp. 277 - 291 (1980).

Taylor, P.E., and Huxlery, S.J., "A Break from Tradition for the San Francisco Police: Patrol Officer Scheduling Using An Optimization-Based Decision Support System", Interfaces, Vol. 19, pp. 4-24 (1989).

Tibrewala, R., Phillippe, D., and Browne, J., "Optimal Scheduling of Two Consecutive Idle Periods", Management Science, Vol. 19, pp. 71-75 (1972).

Tien, J.M., and Kamiyama, A., "On Manpower Scheduling Algorithms", SIAM Review, Vol. 24, pp. 275- 287 (1982).

Vohra, R.V., "A Quick Heuristic for Some Cyclic Staffing Problems with Breaks", Journal of Operations Research Society, Vol. 39, pp. 1057-1061 (1988).

Warner, D.M., "Scheduling Nursing Personnel According to Nursing Preference: A Mathematical Programming Approach", Operations Research, Vol. 24, pp. 842-856 (1976).

Warner, D.M., and Prawda, J., "A Mathematical Programming Model for Scheduling Nursing Personnel in Hospitals", Management Science, Vol. 19, pp. 411-422 (1972).

Wolfe, H., and Young, J.P., "Staffing the Nursing Unit Part I Controlled Variable Staffing", Nursing Research, Vol. 14, pp. 236-243 (1965a).

Wolfe, H., and Young, J.P., "Staffing the Nursing Unit Part II The Multiple Assignment Technique", Nursing Research, Vol. 14, pp. 299-303 (1965b).

Crew Scheduling

Amar, G., "New Bus Scheduling Methods at RATP", Computer Scheduling of Public Transport, (Edited by J.H. Roussean), Elsevier Science Publishers, North Holland, pp. 415-426 (1985).

Anbil, R., Gelman,E., Patty, B., and Tanga, R., "Recent Advances in Crew-Paring Optimization at American Airlines", Interfaces, Vol. 21, pp. 62-74 (1991).

Arabeyre, J.P., Fearnley, J., Steiger, F.C., and Teather, W., "The Airline Crew Scheduling Problem: A Survey", Transportation Science, Vol. 3, pp. 140-163 (1969).

Arabeyre, J.P., "Methods of Crew Scheduling", Proceedings, 6th AGIFORS (Airline Group of International Federation of Operations Research) Symposium, Air France (1966).

Baker, E., and Fisher, M., "Computational Results for Very Large Air Crew Scheduling Problems", OMEGA, The International Journal of Management Science, Vol. 9, pp. 613-618 (1981).

Baker, E.K., and Frey, K., "A Heuristic Set Covering Based System for Scheduling Air Crews", Proceedings, SE AIDS (1980).

Baker, E.K., and Bodin, L.D., Finnegan, W.F., and Ponder, R., "Efficient Heuristic Solutions to an Airline Crew Scheduling Problem", AIIE Transactions, Vol. 11, pp. 79-85 (1979).

Ball, M.O., Bodin, D.L., and Greenberg, J., "Enhancement to the RUCUS - II Crew Scheduling System", Computer Scheduling of Public Transport 2 (Edited by J.-M. Rousseau), Elsevier Science Publishers, North- Holland Publishing Company, pp. 279-293 (1985).

Ball, M., and Roberts, A., "A Graph Partitioning Approach to Airline Crew Scheduling", Transportation Science, Vol. 19, pp. 106-126 (1985).

Ball, M., Bodin, L., and Dial, R., "A Matching Based Heuristic for Scheduling Mass Transit Crews and Vehicles" Transportation Science, Vol. 17, pp. 4-31 (1983).

Ball, M.O., Bodin, D.L., and Dial, R., "Experimentation with Computerized System for Scheduling Mass Transit Vehicles and Crews", Computer Scheduling of Public Transport (Edited by A. Wren), North-Holland Publishing Company, pp. 313-334, (1981).

Ball, M., Bodin, L., and Dial, R., "Scheduling of Drivers for Mass Transit Systems Using Interactive Optimization", World Conference on Transportation Research, London, England (April 1980).

Barnhart, C., Johnson, E., Anbil, R., and Hatay, L., "A Column Generation Technique for the Long-haul Crew Assignment Problem", ORSA/TIMS (1991).

Belletti, R., and Davani, A., "BDROP: A Package for the Bus Drivers' Rostering Problem", Computer Scheduling of Public Transport 2 (Edited by J. -M. Rousseau), Elsevier Science Publishers, North-Holland Publishing Company, pp. 319-324 (1985).

Bodin, L., Ball, M., Duguid, R., and Mitchell, M., "The Vehicle Scheduling Problem with Interlining", Computer Scheduling of Public Transport 2, (Edited by J. -M. Rousseau), Elsevier Science Publishers, North-Holland, (1985).

Bodin, L., Rosenfield and Kydes, A., "Scheduling and Estimation Techniques for Transportation Planning", Computers and Operations Research, Vol. 8, pp. 25-38 (1981).

Bodin, L., and Dial, R., "Hierarchical Procedures for Determining Vehicle and Crew Requirements for Mass Transit Systems", Transportation Research Record, 746, pp. 58-64 (1980).

Booler, J.M., "A Method for Solving Crew Scheduling Problems", Operational Research Quarterly, Vol. 26, pp. 55-62 (1975).

Borret, J.M.J., and Roes, A.W., "Crew Scheduling by Computer: A Test on the Possibility of Designing Duties for a Certain Bus Line", Computer Scheduling Public Transport, (Edited by A. Wren), North- Holland Publishing Company, pp. 237-253 (1981).

Bronemann, D.R., "A Crew Planning and Scheduling System", Proceedings, 10th AGIFORS (Airline Group of International Federation of Operations Research) Symposium , (1970).

Carraresi, P., and Gallo, G., "Network Models for Vehicle and Crew Scheduling", European Journal of Operational Research, Vol. 16, pp. 139-151 (1984).

Ceder, A., "The Variable Trip Procedure Used in the Automobile Vehicle Scheduler", Computer Scheduling of Public Transport 2 (Edited by J.-W. Rousseau), Elsevier Science Publishers, North-Holland Publishing Company, pp. 371-390, (1985).

Darby-Dowman, K., and Mitra, G., "An Extension of Set Partitioning with Applications to Scheduling Problems", European Journal of Operational Research, Vol. 21, pp. 200-205 (1985).

Edwards, G.R., "An Approach to the Crew Scheduling Problem", New Zealand Operational Research, Vol. 8, pp. 153-171 (1980).

Evers, G.H.E., "Relevant Factors Around Crew-Utilization", AGIFORS (Airline Group of International Federation of Operations Research) Symposium, KLM (1956).

Falkner, J.C., and Ryan, D.M., "A Bus Crew Scheduling System Using a Set Partitioning Model", Asia- Pacific Journal of Operations Research, Vol. 4, pp. 39-56 (1987).

Gerbract, R., "A New Algorithm for Very Large Crew Pairing Problems", Proceedings, 18th AGIFORS (Airline Group of the International Federation of Operations Research) Symposium (1978).

Gershkoff, I., "Overview of the Crew Scheduling Problem", ORSA/TIMS National Conference (1990).

Gershkoff, I., "Optimizing Flight Crew Schedules", Interfaces, Vol. 19, pp. 29-43 (1989).

Gershkoff, I., "American's System for Building Crew Pairings", Airline Executive, Vol. 11, pp. 20-22 (1987).

Hartley, T., "Two Complementary Bus Scheduling Programs", Computer Scheduling of Public Transport 2 (Edited by J. - M. Rousseau), Elsevier Science Publishers, pp. 345 - 367 (1985).

Hartley, T., "A Glossary of Terms in Bus and Crew Scheduling", Computer Scheduling of Public Transport (Edited by A. Wren), North-Holland Publishing Company, pp. 353-359 (1981).

Henderson, W., "Relationships Between the Scheduling of Telephone Operators and Public Transportation Vehicle Drivers", Preprint, Workshop on Automated Techniques for Schedules of Vehicle Operators for Urban Public Transportation Services (Edited by L. Bodin, and D. Bergmann), Chicago, Illinois (1975).

Heurgon, E., and Hervillard, R., "Preparing Duty Rosters for Bus by Computers", UITP Revue, Vol. 24, pp. 33 - 37 (1975).

Hoffstadt, J., "Computerized Vehicle and Driver Scheduling for the Hamburger Hochbahn Aktiengesellschaft", Computer Scheduling of Public Transport: Urban Passenger and Crew Scheduling (Edited by A. Wren), North-Holland Publishing Company, pp. 35 - 52 (1981).

Howard, S.M., and Moser, P.I., "Impacs: A Hybrid Interactive Approach to Computerized Crew Scheduling", Computer Scheduling of Public Transport 2 (Edited by J.-M.Rousseau), Elsevier Science Publishers, North-Holland Publishing Company, pp. 211-220, (1985).

Jones, R.D., "Development of an Automated Airline Crew Bid Generations Systems", Interfaces, Vol. 19, pp. 44-51 (1989).

Kabbani, N.M., and Patty, B.W., "Aircraft Routing at American Airlines", ORSA/TIMS, Joint National Meeting (1993).

Keaveny, I.T., and Burbeck, S., "Automatic Trip Scheduling and Optimal Vehicle Assignments", Computer Scheduling of Public Transport (Edited by A.Wren), North-Holland Publishing Company, pp. 125 - 145, (1981).

Kolner, T.K., "Some Highlights of a Scheduling Matrix Generator System", Proceedings, 6th AGIFORS (Airline Group of the International Federation of Operations Research) Symposium (1966).

Koutsopoulos, H.N., Odoni, A.R., and Wilson, N.H.M., "Determination of Headways as a Function of Time Varying Characteristics on a Transient Network", Computer Scheduling of Public Transport 2 (Edited by J. M., Rousseau), Elsevier Science Publishers, North-Holland, pp. 391-413 (1985).

Lavoie, S., Minoux, M., and Odier, E., "A New Approach for Crew Pairing Problems by Column Generation Scheme with An Application to Air Transportation", European Journal of Operational Research, Vol. 35, pp. 45-58 (1988).

LePrince, M., and Mertens, W., "Vehicle and Crew Scheduling at the Societe Des Transports Intercommunaux De Bruxelles", Computer Scheduling of Public Transport 2 (Edited by J.-M. Rousseau), Elsevier Science Publishers, North-Holland, pp. 149-178, (1985).

Lessard, R., Rouseau, J.-M., and DuPuis, D., "HASTUS I: A Mathematical Programming Approach to the Bus Driver Scheduling Problem", Computer Scheduling of Public Transport (Edited by A. Wren), North-Holland Publishing Company, pp. 255-267, (1980).

Leudtke, L.K., "RUCUS II: A Review of System Capabilities", Computer Scheduling of Public Transport 2 (Edited by J.-M. Rousseau), Elsevier Science Publishers, North-Holland Publishing Company, pp. 61-116 (1985).

Marsten, R.E., and Muller, M.R., and Killion, D.L., "Crew Planning at Flying Tiger: A Successful Application of Integer Programming", Management Science, Vol. 25, pp. 1175-1183 (1989).

Marsten, R.E., and Shepardson, F., "Exact Solution of Crew Scheduling Problems Using the Set Partitioning Model: Recent Successful Applications", Networks, Vol. 11, pp. 165-177 (1981).

McCloskey, J.F., and Hanssman, F., "An Analysis of Stewardess Requirements and Scheduling for a Major Airline", Naval Research Logistic Quarterly, Vol. 4, pp. 183-192 (1957).

Minoux, M., "Column Generation Techniques in Combinatorial Optimization, A new Application to Crew- Pairing Problems", Proceedings, 24th AGIFORS (Airline Group of the International Federation of Operations Research) Symposium, Strasbourg, France (1984).

Mitchell, R., "Results and Experience of Calibrating HASTAUS-MARCO for Work Rule Cost at the Southern California Rapid Transit District, Los Angeles", Computer Scheduling of Public Transport 2 (Edited by J.-M. Rousseau), Elsevier Science Publishers, North-Holland, (1985).

Mitra, G., and Darby-Dowman K., "CRU-SCHED: A Computer Based Bus Crew Scheduling System Using Integer Programming" Computer Scheduling in Public Transport (Edited by J-M. Rousseau), Elsevier Publishers, North-Holland Publishing Company, pp. 223-232 (1985).

Mitra, G., and Welsh, A., "A Computer Based Crew Scheduling System Using a Mathematical Programming Approach", Computer Scheduling Public Transport: Urban Passenger Vehicle and Crew Scheduling (Edited by A.Wren) North Holland Publishing Company, pp. 281-296 (1981).

Niederer, M., "Optimization of Swissair's Crew Scheduling by Heuristic Methods Using Integer Linear Programming Models", Proceedings, 6th AGIFORS (Airline Group of the International Federation of Operations Research), Symposium (1966).

Paixao, J.P., Branco, M.I., Captivo, M.E., Pato, M.V., Eusebio, R., and Amado, L., "Bus and Crew Scheduling on a Microcomputer", (Edited by J. D. Coelho and L. V. Tavers), North-Holland Publishing Company (1986).

Parker, M.E., and Smith, B.M., "Two Approaches to Computer Crew Scheduling", Computer Scheduling of Public Transport, (Edited by A. Wren), North-Holland Publishing Company, pp. 193-221 (1981).

Piccione, C., Cherici, A., Bielli, M., and LaBella, A., "Practical Aspects in Automatic Crew Scheduling", Computer Scheduling of Public Transport: Urban Passenger Vehicle and Crew Scheduling (Edited by A. Wren), North-Holland Publishing Company, pp. 223-236 (1981).

Rannou, B., "A New Approach to Crew Pairing Optimization", Proceedings, 26th AGIFORS (Airline Group of the International Federation of Operations Research) Symposium, England (1986).

Rousseau, J.-M., (Ed)., Computer Scheduling of Public Transport 2, Elsevier Publishers, North Holland (1985).

Rousseau, J. -M, and Lessard, R., "Enhancements to the HASTUS Crew Scheduling Algorithm", Computer Scheduling of the Public Transport 2 (Edited by J. -M. Rousseau), Elsevier Science Publishers, North-Holland, pp. 295 - 310, (1985).

Rubin, J., "A Technique for the Solution of Massive Set Covering Problem with Application to Airline Crew Scheduling", Transportation Science, Vol. 7, No. 1, pp. 34-48 (1973).

Ryan, D.M., and Foster, B.A., "An Integer Programming Approach to Scheduling", Computer Scheduling of Public Transport: Urban Passenger Vehicle and Crew Scheduling (Edited by A. Wren), North-Holland Publishing Company, pp. 269-280 (1981).

Scott, D., "A Large Scale Linear Programming Approach to the Public Transport Scheduling and Costing Problem", Computer Scheduling of Public Transport 2 (Edited by J. - M. Rousseau), Elsevier Science Publishers, North-Holland Publishing Company, pp. 473-491, (1985).

Shepardson, F., "Modelling the Bus Crew Scheduling Problem", Computer Scheduling of Public Transport 2 (Edited by J. -M. Rousseau), Elsevier Science Publishers, North-Holland Publishing Company, pp. 247-261 (1985).

Stern, H.I., and Ceder, A., "A Deficit Funciton Approach for Bus Scheduling", Computer Scheduling of Public Transport (Edited by A. Wren), North-Holland Publishing Company, pp. 85 - 96, (1981).

Spitzer, M., "Crew Scheduling with Personal Computer", Airline Executive, Vol. 11, pp. 24-27 (1987).

Spitzer, M., "Solution to the Crew Scheduling Problem", Proceedings, 1st AGIFORS (Airline Group of International Federation of Operations Research) Symposium (1961).

Steiger, F., "Optimization of Swissair's Crew Scheduling by an Integer Linear Programming Model", Swissair, O.R. SDK 3.3.911 (1965).

Stern, H., "Bus and Crew Scheduling (Note)", Transportation Research, Vol. 14A, pp. 154- (1980).

Tykulsker, R.J., O'Niel, K.K., Ceder, A., and Scheffi, Y., "A Commuter Rail Crew Assignment/Work Rules Model", Computer Scheduling of Transport 2, (Edited by J-M. Rousseau), Elsevier Publishers, North-Holland, pp. 232-246 (1985).

Ward, R.E., Durant, P.A., and Hallman,A.B., "A Problem Decomposition Approach to Scheduling the Drivers and Crews of Mass Transit Systems", Computer Scheduling of Public Transport, (Edited by A. Wren), North-Holland Publishing Company, pp. 297-316 (1981).

Wren, A., Smith, B.M., and Miller, A.J., "Complimentary Approaches to Crew Scheduling", Computer Scheduling of Transport 2, (Edited by J.H. Rousseau), North-Holland Publishing Company, pp. 263-278 (1985).

Wren, A., "General Review of the Use of Computers in Scheduling Buses and Their Crews", Computer Scheduling of Public Transport: Urban Passenger Vehicle and Crew Scheduling, (Edited by A. Wren), North-Holland Publishing Company, pp. 3-16 (1981).

Manufacturing

Baybars, I., "A Survey of Exact Algorithms for the Simple Assembly Line Balancing Problem", Management Science, Vol. 32, pp. 909-932 (1986a).

Baybars, I., "An Efficient Heuristic Method for the Simple Assembly Line Balancing Problem", International Journal of Production Research, Vol. 24, pp. 149-166 (1986b).

Bowman, E.H., "Assembly-Line Balancing by Linear Programming", Operations Research, Vol. 8, pp. 385- 389 (1960).

Cattrysse, D., Saloman, M., Kuik, R., and Van Wassenhove, L.N., "A Dual Ascent and Column Generation Heuristic for the Discrete Lotsizing an Scheduling Problem with Setup Times", Management Science, Vol. 39, pp. 477-486 (1993).

Cattrysse, D., Maes, J. and Van Wassenhove, L.N., "Set Partitioning and Column Generation Heuristics for Capacitated Dynamic Lotsizing", European Journal of Operational Research, Vol. 46, pp. 38-47 (1990)

Cattrysse, J., Maes, J., and Van Wassenhove, L.N., "Set Partitioning Heuristic for Capacitated Lotsizing", Working Paper 88-12, Division of Industrial Management, Katholieke Universiteit Leuven, Belgium. (1988).

Dzielinski, B.P., and Gomory, R.E., "Optimal Programming of Lot Sizes Inventory and Labor Allocations", Management Science, Vol. 11, pp. 874 - 890 (1965).

Fisher, M.L., "Optimal Solution of Scheduling Problems Using Lagrange Multipliers - Part I", Operations Research, Vol. 21, pp. 1114-1127 (1973).

Freeman, D.R., and Jucker, J.V., "The Line Balancing Problem", Journal of Industrial Engineering, Vol. 18, pp. 361-364 (1967).

Gutjahr, A.L., and Nemhauser,G.L., "An Algorithm for the Line Balancing Problem", Management Science,Vol. 11, pp. 308-315 (1964).

Hackman, S.T., Magazine, M.J., and Wee, T.S., "Fast Effective Algorithms for Simple Assembly Line Balancing Problems", Operations Research, Vol. 37, pp. 916-924 (1989).

Hoffmann, T.R., "Eureka: A Hybrid System for Assembly Line Balancing", Management Science, Vol. 38, pp. 39-47 (1992).

Ignall, E.J., "A Review of Assembly Line Balancing", The Journal of Industrial Engineering, Vol. 16, pp. 244-254 (1965).

Johnson, R.V., "A Branch and Bound Algorithm for Assembly Line Balancing Problems with Formulation Irregularities", Management Science, Vol. 29, pp. 1309-1324 (1983).

Johnson, R.V., "Assembly Line Balancing Algorithms: Computational Comparisons:, International Journal of Production Research, Vol. 19, pp. 277-287 (1981).

Kilbridge, M.D., and Webster, L., "A Review of Analytical Systems of Line Balancing", Operations Research, Vol. 10, pp. 626-638 (1962).

Lasdon, L.S., and Terjung, R.C., "An Efficient Algorithm for Multi-Item Scheduling", Operations Research, Vol. 19, pp. 946 - 969 (1971).

Manne, A.S., "Programming of Economic Lot Sizes", Management Science, Vol. 4, pp. 115-135 (1958).

Patterson, J.H., and Albracht, J.J., "Assembly-Line Balancing: Zero-One Programming with Fibonacci Search", Operations Research, Vol. 23, pp. 166 - 174 (1975).

Pierce, J.F., "Pattern Sequencing and Matching a Stock Cutting Operations", Tappi, Vol. 53, pp. 668-678 (1970).

Salveson, M.E., "The Assembly Line Balancing Problem", Journal of Industrial Engineering, Vol. 6, pp. 18- 25 (1955).

Scudder, G.D., "Priority Scheduling and Spare Parts Stocking Policies for a Repair Shop: The Multiple Failure Case", Management Science, Vol. 30, pp. 739-749 (1984).

Stanfell, L.E., "Successive Approximation Procedures for a Cellular Manufacturing Problem with Machine Loading Constraints", Engineering Costs and Production Economics, Vol. 17, pp. 135-147 (1989).

Talbot, F.B., and Gehrlein, W.V., "A Comparative Evaluation of Heuristic Line Balancing Techniques", Management Science, Vol. 32, pp. 430-454 (1986).

Talbot, F.B., and Patterson, J.H., "An Integer Programming Algorithm with Network Cuts Solving the Assembly Line Balancing Problem", Management Science, Vol. 30, pp. 85-99 (1984).

Vasko, F.J., Wolf, F.E., and Scott, K.L., Jr., "A Set Covering Approach to Metallurgical Grade Assignment", European Journal of Operations Research, Vol. 38, pp. 27 - 34 (1989).

Vasko, F.J., Wolf, F.E., and Scott, K.L., "Optimal Selection of IGNOT Sizes Via Set Covering", Operations Research, Vol. 35, pp. 346-353 (1987).

White, W.W., "Comments on a Paper by Bowman", Operations Research, Vol. 9, pp. 274 - 276 (1961).

Miscellaneous Operations

Almond, M., "A University Faculty Time Table", Computer Journal, Vol. 12, pp. 215-217 (1969).

Almond, M., "An Algorithm for Constructing University Time-Table", Computer Journal, Vol. 8, pp. 331 - 340 (1966).

Aneja, Y.P., and Vemuganti, R.R., "Set Covering and Fixed Charge Transportation Problem", Technical Report, University of Baltimore, Maryland (1974).

Arani, T., and Lotfi, V., "A Three Phased Approach to Final Exam Scheduling", IIE Transactions, Vol. 21, pp. 86 - 96 (1989).

Aubin, J., and Ferland, J.A., "A Large Scale Timetabling Problem", Computers and Operations Research, Vol. 16, pp. 67 - 77 (1989).

Aust, R.J., "An Improvement Algorithm for School Timetablikng", Computer Journal, Vol. 19, pp. 339-345 (1976).

Barham, A.M., and Westwood, J.B., "A Simple Heuristic to Facilitate Course Timetabling", Journal of the Operational Research Society, Vol. 29, pp. 1055-1060 (1978).

Broder, S., "Final Examination Scheduling", Communications of ACM, Vol. 7, pp. 494-498 (1964).

Carter, M.W., and Tovey, C.A., "When is the Classroom Assignment Problem Hard?", Operations Research, Vol. 40, pp. S28-S39 (1992).

Carter, M.W., "A Lagrangian Relaxation Approach to Classroom Assignment Problem", INFOR, Canadian Journal of Operations Research and Information Processing, Vol. 27, pp. 230-246 (1989).

Carter, M.W., "A Survey of Practical Applications of Examination Timetable Scheduling", Operations Research, Vol. 34, pp. 193-202 (1986).

Csima, J., and Gotleib, G.C., "Tests on a Computer Method for Constructing Timetables", Communications of the ACM, Vol. 7, pp. 160 - 163 (1964).

Day, R.H., "On Optimal Extracting from a Multiple Data Storage System: An Application of Integer Programming", Operations Research, Vol. 13, pp. 482-494 (1965).

Dempster, M.A.H., "Two Algorithms for the Timetabling Problem", Combinatorial Mathematics and Applications (Edited by D.J.A. Welsh), Academic Press, pp. 63 - 85 (1971).

DeWerra, D., "Some Comments on a Note About Timetabling", INFOR, Canadian Journal of Operations Research and Information Processing, Vol. 16, pp. 90-92 (1978).

DeWerra, D., "On a Particular Conference Scheduling Problem", INFOR, Canadian Journal of Operations Research and Information Processing, Vol. 13, pp. 308 - 315 (1975).

Even, S., Itai, A., and Shamir, A., "On the Complexity of Timetable and Multicommodity Flow Problems", SIAM Journal on Computing, Vol. 5, pp. 691 - 703 (1976).

Ferland, P.C., and Roy, S., "Timetabling Problem for University as Assignment of Activities to Resources", Computers and Operations Research, Vol. 12, pp. 207-218 (1985).

Frank, R.S., "On the Fixed Charge Hitchcock Transportation Problem", Ph.D. Dissertation, The Johns Hopkins University, Baltimore, Maryland (1972).

Fulkerson, D.R., Nemhauser, G.L., and Trotter, I.E., "Two Computationally Difficult Set Covering Problems that Arises in Computing the 1-With of Incidence Matrices of Steiner Triple Systems", Mathematical Programming Study, Vol. 2, pp. 72-81 (1974).

Gans, O.B. de., "A Computer Timetabling System for Secondary Schools in Netherlands", European Journal of Operational Research, Vol. 7, pp. 175 - 182 (1981).

Garfinkel, R.S., Kunnathur, A.S., and Liepins, G.E., "Optimal Imputation of Erroneous Data: Categorial Data, General Edits", Operations Research, Vol. 34, pp. 744-751 (1986).

Garfinkel, R.S., and Nemhauser, G.L., "Optimal Political Distracting by Implicitly Enumeration Techniques", Management Science, Vol. 16, pp. B495-B508 (1970).

Glassey, C.R., and Mizrach, M., "A Decision Support System for Assigning Classes to Rooms", Interfaces, Vol. 16, pp. 92 - 100 (1986).

Gosselin, K., and Trouchon, M., "Allocation of Classrooms by Linear Programming", Journal of Operational Research Society, Vol. 37, pp. 561 - 569 (1986).

Grimes, J., "Scheduling to Reduce Conflict in Meetings", Communications of the ACM, Vol. 13, pp. 351- 352 (1970).

Hall, A., and Action, F., "Scheduling University Course Examination by Computer", Communications of the ACM, Vol. 10, pp. 235-238 (1967).

Heath, L.S., "Covering a Set with Arithmetic Progressions is NP-Complete", Information Processing Letters", Vol. 34, pp. 293-298 (1990).

Hertz, A., "Find a Feasible Course Schedule Using Tabu Search", Discrete Applied Mathematics, Vol. 35, pp. 255-270 (1992).

Knauer, B.A., "Solution of Timetable Problem", Computers and Operations Research, Vol. 1, pp. 363 - 375 (1974).

LaPorte, G., and Desroches, S., "Examination Timetabling by Computer", Computers and Operations Research, Vol. 11, pp. 351-360 (1984).

Lions, J., "The Ontario School Scheduling Problem", Computer Journal, Vol. 10, pp. 14 - 21 (1967).

Markland, R.E., and Nauss, R.M., "Improving Transit Clearing Operations at Maryland National Bank", Interfaces, Vol. 13, pp. 1-9 (1983).

McKeown, P.G., "A Branch-and-Bound Algorithm for Solving Fixed-Charge Problems", Naval Research Logistics Quarterly, Vol. 28, pp. 607-617 (1981).

Mehta, N.K., "The Application of a Graph Coloring Method to an Examination Scheduling Problem", Interfaces, Vol. 11, pp. 57-64 (1981).

Mulvey, J.M., "A Classroom/Time Assignment Model", European Journal of Operational Research, Vol. 9, pp. 64-70 (1982).

Nawijn, W.M., "Optimizing the Performance of a Blood Analyzer: Application of the Set Partitioning Problem", European Journal of Operational Research, Vol. 36, pp. 167-173 (1988).

Reggia, J.A., Naw, D.S., and Wang, P.Y., "Diagnostic Expert Systems Based on Set Covering Model", International Journal of Man-Machine Studies, Vol. 19, pp. 437-460, (1983).

Thuve, H., "Frequency Planning as a Set on Partitioning Problem", European Journal of Operational Research, Vol. 6, pp. 29-37 (1981).

Tripathy, A., "School Timetabling - A Case in Large Binary Linear Integer Programming", Management Science, Vol. 30, pp. 1473-1489 (1984).

Tripathy, A., "A Lagrangian Relaxation Approach to Course Timetabling", Journal of the Operational Research Society, Vol. 31, pp. 599-603 (1980).

Valenta, J.R., "Capital Equipment Decision: A Model for Optimal Systems Interfacing", M.S. Thesis, Massachusetts Institute of Technology (1969).

Van Slyke, R., "Redundant Set Covering in Telecommunications Networks", Proceedings of the 1982 IEEE Large Scale Systems Symposium, pp. 217-222 (1982).

White, G.M., and Chan, P.W., "Towards the Construction of Optimal Examination Schedules", INFOR, Canadian Journal of Operations Research and Information Processing, Vol. 17, pp. 219-229 (1979).

Wood, D.C., "A Technique for Coloring a Graph Applicable to Large Scale Time-Tabling Problems", Computer Journal, Vol. 12, pp. 317-319 (1969).

Woodbury, M., Ciftan, E., and Amos, D., "HLA Serum Screening Based on an Heuristic Solution of the Set Cover Problem", Computer Programs in Biomedicine, Vol. 9, pp. 263-273 (1979).

Routing

Agarwal, Y., Mathur, K., and Salkin, H.M., "A Set-Partitioning Based Algorithm for the Vehicle Routing Problem", Networks, Vol. 19, pp. 731-750 (1989).

Agin, N., and Cullen, D., "An Algorithm for Transportation Routing and Vehicle Loading", Logistics (Edited by M. Geisler), North Holland, Amsterdam, pp. 1-20 (1975).

Altinkemer, K., and Gavish, B., "Parallel Savings Based Heuristics for the Delivery Problem", Operations Research, Vol. 39, pp. 456-469 (1991).

Altkinkemer, K., and Gavish, B., "Heuristics for the Delivery Problem with Constant Error Guarantees", Transportation Science, Vol. 24, pp. 294-297 (1990).

Altkinkemer, K., and Gavish, B., "Heuristics for Unequal Weight Delivery Problem with a Fixed Error Guarantee", Operations Research Letters, Vol. 6, pp. 149-158 (1987).

Angel, R., Caudle, W., Noonan, R., and Whinston, A., "A Computer Assisted School Bus Scheduling", Management Science, Vol. 18, pp. 279-288 (1972).

Anily, S., and Federgruen, A., "Two-Echelon Distribution Systems with Vehicle Routing Costs and Central Inventories", Operations Research, Vol. 41, pp. 37-47 (1993).

Anily, S., and Federgruen, A., "Rejoinder to Comments on One-Warehouse Multiple Retailer Systems with Vehicle Routing Costs", Management Science, Vol. 37, pp. 1497-1499 (1992).

Anily, S., and Federgruen, A., "A Class of Euclidean Routing Problems with General Route Costs Functions", Mathematics of Operations Research, Vol. 15, pp. 268-285 (1990a).

Anily S., and Federgruen, A., "One-Warehouse Multiple Repair Systems with Vehicle Routing Costs", Management Science, Vol. 36, pp. 92-114 (1990b).

Appelgren, L., "Integer Programming Methods for a Vessel Scheduling Problem", Transportation Science, Vol. 5, pp. 64-78 (1971).

Appelgren, K., "A Column Generation Algorithm for a Ship Scheduling Problem", Transportation Science, Vol. 3, pp. 53-68 (1969).

Arisawa, S., and Elmaghraby, S.E., "The 'Hub' and 'Wheel' Scheduling Problems; I. The Hub Scheduling Problem: The Myopic Case", Transportation Science, Vol. 11, pp. 124 - 146 (1977a).

Arisawa, S., and Elmaghraby S.E., "The 'Hub' and 'Wheel' Scheduling Problems, II. The Hub Operation Scheduling Problem (HOSP): Multi-Period and Infinite Horizon and the Wheel Operation Scheduling Problem (WOSP)" Transportation Science, Vol. 11, pp. 147-165 (1977b).

Assad, A., "Analytic Models in Rail Transportation: An Annotated Bibliography", INFOR, Canadian Journal of Operations Research and Information Processing, Vol. 19, pp. 59-80 (1981).

Assad, A., "Models for Rail Transportation", Transportation Research, Vol. 14A, pp. 205-220 (1980).

Averabakh, I., and Berman, O., "Sales-Delivery Man Problems on Tree-Like Networks", Working Paper, Faculty of Management, University of Toronto, Canada (1992).

Baker, B.M., "Further Improvements to Vehicle Routing Heuristics", Journal of the Operational Research Society, Vol. 43, pp. 1009-1012 (1992).

Baker, E., and Schaffer, J., "Solution Improvement Heuristics for the Vehicle Routing and Scheduling Problem with Time Window Constraints", American Journal of Mathematical and Management Sciences, Vol. 6, pp. 261-300 (1986).

Balinksi, M., and Quandt, R., "On an Integer Program for the Delivery Problem", Operations Research, Vol. 12, pp. 300 -304 (1964).

Ball, M., Golden, B., Assad, A., and Bodin, L., "Planning for Truck Fleet Size in the Presence of a Common Carrier Option", Decision Sciences, Vol. 14, pp. 103-130 (1983).

Bartholdi, J., Platzman, L., Collins, R., and Warden, W., "A Minimal Technology Routing System for Meals on Wheels", Interfaces, Vol. 13, pp. 1-8 (1983).

Bartlett, T., "An Algorithm for the Minimum Number of Transport Units to Maintain a Fixed Schedule", Naval Research Logistics Quarterly, Vol. 4, pp. 139-149 (1957).

Bartlett, T., and Charnes, A., "Cyclic Scheduling and Combinatorial Topology: Assignment of Routing and Motive Power to Meet Scheduling Maintenance Requirements: Part II, Generalizations and Analysis.", Naval Research Logistics Quarterly, Vol. 4, pp. 207-220 (1957).

Barton, R., and Gumaer, R., "The Optimum Routing for an Air Cargo Carrier's Mixed Fleet", Transportation - A Series, pp. 549-561. New York Academy of Science, New York (1968).

Beasley, J., "Fixed Routes", Journal of the Operational Research Society, Vol. 35, pp. 49-55 (1984).

Beasley, J.E., "Route First-Cluster Second Methods for Vehicle Routing", OMEGA, The International Journal of Management Science, Vol. 11, pp. 403-408 (1983).

Beasley, J., "Adapting the Savings Algorithm for Varying Inter-Customer Travel Times", OMEGA, The International Journal of Management Science, Vol. 9, pp. 658-659 (1981).

Bell, W., Dalberto, L., Fisher, M., Greenfield, A., Jaikumar, R., Keedia, P., Mack, R., and Prutzman, P., "Improving the Distribution of Industrial Gases with an On-Line Computerized Routing and Scheduling Optimizer", Interfaces, Vol. 13, pp. 4-23 (1983).

Bellman R., "On a Routing Problem", Quarterly Journal of Applied Mathematics, Vol. 16, pp. 87 - 90 (1958).

Bellmore, M., and Hong, S., "Transformation of the Multisalesman Problem to the Standard Traveling Salesman Problem", Journal of the ACM, Vol. 21, pp. 500 - 504 (1974).

Beltrami, E., and Bodin, L., "Networks, and Vehicle Routing for Municipal Waste Collection", Networks, Vol. 4, pp. 65-94 (1974).

Bennett, B., and Gazis, D., "School Bus Routing by Computer", Transportation Research, Vol. 6, pp. 317- 325 (1972).

Bertsimas, D.J., "A Vehicle Routing Problem with Stochastic Demand", Operations Research, Vol. 40, pp. 574-585 (1992).

Bertismas, D.J., and Van Ryzin G., "A Stochastic and Dynamic Vehicle Routing Problem in the Euclidean Plane", Operations Research, Vol. 39, pp. 601-615 (1991).

Bertsimas, D., "The Probabilistic Vehicle Routing Problem", Sloan Working Paper No. 2067-88, Massachusetts Institute of Technology, Cambridge, Massachusetts (1988).

Bodin, L, and Sexton, T., "The Multiple - Vehicle Subscriber Dial -A-Ride Problems", TIMS Studies in the Management Sciences, Vol. 22, pp. 73-76 (1986).

Bodin, L.D., Golden, L.B., Assad, A.A., and Ball, M.O., "Routing and Scheduling of Vehicles and Crews: The State of Art", Computers and Operations Research, Vol. 10, pp. 65-211 (1983).

Bodin, L., and Golden, B., "Classification in Vehicle Routing and Scheduling", Networks, Vol. 11, pp. 97- 108 (1981).

Bodin, L., and Golden, B., Schuster A.D., and Romig, W., "A Model for the Blocking of Trains", Transportation Research, Vol. 14B, pp. 115-120 (1980).

Bodin, L., and Berman, L., "Routing and Scheduling of School Busses by Computer", Transportation Science, Vol. 13, pp. 113-129 (1979).

Bodin, L., and Kursh, S., "A Detailed Description of a Computer System for the Routing and Scheduling of Street Sweepers", Computers and Operations Research, Vol. 16, pp. 181-198 (1979).

Bodin, L., and Kursh, S., "A Computer-Assisted System for the Routing and Scheduling of Street Sweepers", Operations Research, Vol. 26, pp. 527-637 (1978).

Bodin, L., "A Taxonomic Structure for Vehicle Routing and Scheduling Problems", Computers and Urban Society, Vol. 1, pp. 11-29 (1975).

Bodner, R., Cassell, E., and Andros, P., "Optimal Routing of Refuse Collection Vehicles", Journal of Stationary Engineering Division, ASCE, 96(SA4) Proceedings Paper 7451, pp. 893-904 (1970).

Bramel, J., and Simchi-Levi, D., "A Location Based Heuristic for General Routing Problems", Technical Report, Graduate School of Business, Columbia University (1993).

Bramel, J., Coffman, E.G., Shor, P.W., and Simchi-Levi, D., "Probabilistic Analysis of the Capacitated Vehicle Routing Problem With Unsplit Demands", Operations Research, Vol. 40, pp. 1095-1106 (1992).

Brown, G.B., Graves, G.W., and Ronen, D., "Scheduling Ocean Transportation of Crude Oil", Management Science, Vol. 33, pp. 335-346 (1987).

Brown, G., and Graves, G., "Real Time Dispatch of Petroleum Tank Trucks", Management Science, Vol. 27, pp. 19-32 (1981).

Butt, S., and Cavalier, T.M., "A Heuristic for the Multiple Tour Maximum Collection Problem", Department of Industrial and Managerial Systems Working Paper, The Pennsylvania State University, University Park, Pennsylvania (1991).

Cassidy, P.J., and Bennett, H.S., "Tramp - A Multiple Depot Vehicle Scheduling System", Operational Research Quarterly, Vol. 23, pp. 151-163 (1972).

Ceder, A., and Stern, H., "Deficit Function Bus Scheduling and Deadheading Trip Insertions for Fleet Size Reduction", Transportation Science, Vol. 15, pp. 338-363 (1981).

Chard, R., "An Example of An Integrated Man-machine System for Truck Scheduling", Operational Research Quarterly, Vol. 19, pp. 108 (1968).

Charnes, A., and Miller, M.H., "A Model for the Optimal Programming of Railway Freight Train Movements", Management Science, Vol. 3, pp. 74-92 (1956).

Cheshire, I.M., Malleson, A.M., and Naccache, P.F., "A Dual Heuristic for Vehicle Scheduling", Journal of the Operational Research Society, Vol. 33, pp. 51-61 (1982).

Chien, T.W., Balakrishnan, A., and Wong, R.T., "An Integrated Inventory Allocation and Vehicle Routing Problem", Transportation Science, Vol. 23, pp. 67-76 (1989).

Christofides, N., "Vehicle Routing", Traveling Salesman Problem, (Edited by E.L. Lawler, J.K. Lenstra, A.H.A. Rinooy Kan and D.S. Shmoys), John Wiley & Sons, pp. 431-448 (1985a).

Christofides, N., "Vehicle Routing", Combinatorial Optimization: Annotated Bibliographics (Edited by M. O'Heigeartaigh, J.K. Lenstra and A.H.G Kinnooy Kan), pp. 148-163, Centre for Mathematics and Computer Science, Amsterdam (1985b).

Christofides, N., and Beasley, J., "The Period Routing Problem", Networks, Vol. 14, pp. 237-256 (1984).

Christofides, N., Mingozzi, A., and Toth, P., "Exact Algorithms for the Vehicle Routing Problem, Based on Spanning Tree and Shortest Path Relaxations", Mathematical Programming, Vol. 20, pp. 255-282 (1981a).

Christofides, N., Mingozzi, A., and Toth, P., "State Space Relaxation Procedures for the Computation of Bounds to Routing Problems", Networks, Vol. 11, pp. 145-164 (1981b).

Christofides, N., Mingozzi, A., and Toth, P., "The Vehicle Routing Problem", In Christofides et al., Combinatorial Optimization, Wiley and Sons, New York, pp. 315-338 (1979).

Christofides, N., "The Vehicle Routing Problem", Revue Francoise d'Automatique, Informatique et Recherche Operationnelle (RAIRO), Vol. 10, pp. 55-70 (1976).

Christofides, N., and Eilon, S., "Algorithms for Large Scale Traveling Salesman Problem", Operational Research Quarterly, Vol. 23, pp. 511-518 (1973).

Christofides, N., "Fixed Routes and Areas of Delivery Operations", International Journal of Physical Distribution, Vol. 1, pp. 87-92 (1971).

Christofides, N., and Eilon, S., "An Algorithm for the Vehicle-Dispatching Problem", Operational Research Quarterly, Vol. 20, pp. 309-318 (1969).

Clarke, G., and Wright, J.W., "Scheduling of Vehicles from a Central Depot to a Number of Delivery Points", Operations Research, Vol. 12, pp. 568-681 (1964).

Crawford, J.L., and Sinclair, G.B., " Computer Scheduling of Beer Tanker Deliveries", International Journal of Physical Distribution, Vol. 7, pp. 294-304 (1977).

Cullen, F.H., Jarvis, J.J., and Ratliff, H.D., "Set Partitioning Based Heuristic for Interactive Routing", Networks, Vol. 11, pp. 125-143 (1981).

Cunto, E., "Scheduling Boats to Sample Oil Wells in Lake Maracaibo", Operations Research, Vol. 26, pp. 183-196 (1978).

Daganzo, C., "The Distance Traveled to Visit N Points With A Maximum of C Stops Per Vehicle: An Analytical Model and An Application", Transportation Science, Vol. 18, pp. 331-350 (1984).

Daganzo, C., "An Approximate Analytic Model of Many-to-Many Demand Responsive Transportation Systems", Transportation Research, Vol. 12, pp. 325-333 (1978).

Dantazig, G.G., and Ramser, J.H., "The Truck Dispatching Problem", Management Science, Vol. 6, pp. 80- 91 (1959).

Desrochers, M., Desrosiers, J., and Solomon, M., "A New Optimization Algorithm for the Vehicle Routing with Time Windows", Operations Research, Vol. 40, pp. 342-354 (1992).

Desrochers, M., Lenstra, J.K., Savelsbergh, M.M.P, and Soumis, F., "Vehicle Routing with Time Windows: Optimization and Approximation", Vehicle Routing: Methods and Studies, (Edited by B.L. Golden and A.A. Assad), North-Holland, Amsterdam, pp. 65-84 (1988).

Desrosiers, J., Dumas, Y., and Soumis, F., "A Dynamic Programming Solution of the Large-Scale Single- Vehicle Dial-A-Ride Problem with Time Windows", The American Journal of Mathematical and Management Sciences, Vol. 6, pp. 301 - 325 (1986).

Desrosiers, J., Soumis, F., Desrochers, M., and Sauve, M., "Routing and Scheduling Time Windows Solved by Network Relaxation and Branch-and-Bound on Time Variables", Computer Scheduling of Public Transport 2 (Edited by J.-M. Rousseau), pp. 451-471, Elsevier Science Publishers (1985).

Desrosiers, J., Soumis, F., and Desrochers, M., "Routing with Time Windows by Column Generation", Networks, Vol. 14, pp. 545-565 (1984).

Doll, L.L., "Quick and Dirty Vehicle Routing Procedures", Interfaces, Vol. 10, pp. 84-85 (1980).

Dror, M., and Trudeau, P., "Split-Delivery Routing", Naval Research Logistics Quarterly, Vol. 37, pp. 383- 402 (1990).

Dror, M., and Ball, M., "Inventory/Routing: Reduction from an Annual to a Short-Period Problem", Naval Research Logistics Quarterly, Vol. 34, pp. 891-905 (1987).

Dror, M., and Levy, L., "A Vehicle Routing Improvement Algorithm Comparison of a Greedy and Matching Implementation of Inventory Routing", Computers and Operations Research, Vol. 13, pp. 33-45 (1986).

Dror, M., Ball, M., and Golden, B., "A Computational Comparison of Algorithms for the Inventory Routing Problem", Annals of Operations Research, Vol. 4, pp. 3-23 (1986).

Dulac, G., Ferland, J., and Forgues, P., "School Bus Routes Generator in Urban Surroundings", Computers and Operations Research, Vol. 1, pp. 199-213 (1980).

Dumas, Y., Desrosiers, J., and Soumis, F., "The Pick-up and Delivery Problem with Time Windows", European Journal of Operational Research, Vol. 54, pp. 7-22 (1991).

Eilon, S., Watson-Gandy, G., and Christofides, N., "Distribution Management", Griffin, London (1971).

Eilon, S., and Christofides, N., "An Algorithm for the Vehicle Dispatching Problem", Operational Research Quarterly, Vol. 20, pp. 309-318 (1969).

El-Azm, A., "The Minimum Fleet Size Problem and Its Applications to Bus Scheduling", Computer Scheduling of Public Transport 2 (Edited by J.-M. Rousseau), pp. 493-512, Elsevier Science Publishers (1985).

Etezadi, T., and Beasley, J., "Vehicle Fleet Composition", Journal of the Operational Research Society, Vol. 34, pp. 87-91 (1983).

Evans, S.R., and Norback, J.P., "The Impact of a Decision Support System for a Vehicle Routing in a Food Service Supply Situation", Journal of the Operational Research Society, Vol. 36, pp. 467-472 (1985).

Farvolden, J.M., Laporte, G., and Xu, J., "Solving an Inventory Allocation and Routing Problems Arising in Grocery Distribution", CRT-886, Centre De Recherche Sur Les Transports, Universite De Montreal, Montreal, Canada (1993).

Federgruen, A., and Simchi-Levi, D., "Analytical Analysis of Vehicle Routing and Inventory Routing Problems", Handbooks in Operations Research and Management Science, the volume on "Networks and Distribution", (Edited by M. Ball, C. Magnanti, C. Monma and G. Nemhauser) (1992).

Federgruen, A., and Zipkin, P., "A Combined Vehicle Routing and Inventory Allocation Problem", Operations Research, Vol. 32, pp. 1019-1037 (1984).

Ferebee, J., "Controlling Fixed-Route Operations", Industrial Engineering, Vol. 6, pp. 28-31 (1974).

Ferguson, A., and Dantzig, G., "The Allocation of Aircraft to Routes - An Example of Linear Programming Under Uncertain Demand", Management Science, Vol. 3, pp. 45-73 (1956).

Fisher, M.L., and Rosenwein, M.B., "An Interactive Optimization System for Bulk Cargo Ship Scheduling", Naval Research Logistic Quarterly, Vol. 36, pp. 27-42 (1989).

Fisher, M.L., Greenfield, A.J., Jaikumar, R., and Lester, J.T. III, " A Computerized Vehicle Routing Application", Interfaces, Vol. 12, No. 4, pp. 42-52 (1982).

Fisher, M.L., and Jaikumar, R., "A Generalized Assignment Heuristic for Vehicle Routing", Networks, Vol. 11, pp. 109-124 (1981).

Fletcher, A., "Alternative Routes Round the Delivery Problem", Data and Control, Vol. 1, pp. 20-22 (1963).

Fleuren, H.A., "A Computational Study of the Set Partitioning Approach for Vehicle Routing and Scheduling Problems", Ph.D Dissertation, Universiteit Twente, Netherlands (1988).

Florian, M., Guerin, G., and Bushel, G., "The Engine Scheduling Problem in a Railway Network", INFOR, Canadian Journal of Operations Research and Information Processing, Vol. 15, pp. 121-138 (1976).

Foster, B.A., and Ryan, D.M., "An Integer Programming Approach to the Vehicle Scheduling Problem", Operational Research Quarterly, Vol. 27, pp. 367-384 (1976).

Foulds, L.R., Read, E.G., and Robinson, D.F., "A Manual Solution Procedure for the School Bus Scheduling Problem", Australian Road Research, Vol. 7, pp. 1-35 (1977).

Frederickson, G., Hecht, M., and Kim, C., "Approximation Algorithms for Some Routing Problems", SIAM Journal on Computing, Vol. 7 pp. 178-193 (1978).

Garvin, W., Crandall, H., John, J., Spellman, R., "Applications of Vehicle Routing in the Oil Industry", Management Science, Vol. 3, pp. 407-430 (1957).

Gaskell, T., "Bases for Vehicle Fleet Scheduling", Operational Research Quarterly, Vol. 18, pp. 281-295 (1967).

Gaudioso, M., and Paletta, G., "A Heuristic for the Periodic Vehicle Routing Problem", Transportation Science, Vol. 26, pp. 86-92 (1992).

Gavish B., and Shlifer, E., "An Approach for Solving a Class of Transportation Scheduling Problem", European Journal of Operational Research, Vol. 3, pp. 122-134 (1978).

Gavish, B., and Schweitzer, P., and Shlifer, E., "Assigning Buses to Schedules in a Metropolitan Area", Computers and Operations Research, Vol. 5, pp. 129-138 (1978).

Gendreau, M., Hertz, A., and LaPorte, G., "A Tabu Search Heuristic for the Vehicle Routing Problem", CRT 777, Centre De Recherche Sur Les Transports, Universite De Montreal, Montreal, Canada (1992).

Gertsbach, I., and Gurevich, Y., "Constructing an Optimal Fleet for a Transportation Schedule", Transportation Science, Vol. 11, pp. 20-36 (1977).

Gheysens, F., Golden, B., and Assad, A., "A Comparison of Techniques for Solving the Fleet Size and Mix Vehicle Routing Problem", OR Spektrum, Vol. 6, pp. 207-216 (1984).

Gillett, B.E., and Johnson, J., "Multi-Terminal Vehicle-Dispatch Algorithm", OMEGA, The International Journal of Management Science, Vol. 4, pp. 711-718 (1976).

Gillett, B.E., and Miller, L.R., "A Heuristic Algorithm for the Vehicle-Dispatch Problem", Operations Research, Vol. 22, pp. 340-349 (1974).

Golden, B.L., and Assad, A.A., "Vehicle Routing: Methods and Studies", North-Holland, Amsterdam (1988).

Golden, B., and Wasil, E., "Computerized Vehicle Routing in the Soft Drink Industry", Operations Research, Vol. 35, pp. 6-17 (1987).

Golden, B., and Assad, A.A., "Prospectives of Vehicle Routing: Exciting New Developments", Operations Research, Vol. 34, pp. 803-810 (1986a).

Golden, B., and Assad, A., "Vehicle Routing with Time Window Constraints", American Journal of Mathematical and Management Sciences, Vol. 15, pp. . (1986b).

Golden, B., Bodin, L., and Goodwin, T., "Micro Computer-Based Vehicle Routing and Scheduling Software", Computers and Operations Research, Vol. 13, pp. 277-285 (1986).

Golden, B., and Baker, E., "Future Directions in Logistics Research", Transportation Research, Vol. 19A, pp. 405-409 (1985).

Golden, B., Gheysens, F., and Assad, A., "On Solving the Vehicle Fleet Size and Mix Problem", Operations Research, (Edited by J.P. Barnes), North Holland (1984).

Golden, B., Assad, A., and Dahl, R., "Analysis of a Large-Scale Vehicle Routing Problem with An Inventory Component", Large Scale Systems, Vol. 7, pp. 181-190 (1984).

Golden, B., Levy, L., and Dahl, R., "Two Generalizations of the Traveling Salesman Problem", OMEGA, The International Journal of Management Science, Vol. 9, pp. 439-441 (1981).

Golden, B., and Wong, R., "Capacitated Arc Routing Problems", Networks, Vol. 11, pp. 305-315 (1981).

Golden, B., Magnanti, T.L., and Nguyen, H.Q., "Implementing Vehicle Routing Algorithms", Networks, Vol. 7, pp. 113-148 (1977).

Golden, B., "Evaluating a Sequential Vehicle Routing Algorithm", AIIE Transactions, Vol. 9, pp. 204-208 (1977).

Haimovich, M., Rinnooy Kan, H.G., and Stouge, L., "Analysis of Heuristics for Vehicle Routing Problem", Vehicle Routing: Methods and Studies, (Edited by B.L. Golden, A.A. Assad), Elsevier Science Publishers, pp. 47-61 (1988).

Haimovich, M., and Rinnooy Kan, A., "Bounds and Heuristics for Capacitated Routing Problems", Mathematics for Operations Research, Vol. 10, pp. 527-542 (1985).

Hall, R.W., "Comments on One-Warehouse Multiple Retailer Systems with Vehicle Routing Costs", Management Science, Vol. 37, pp. 1496-1497 (1991).

Hauer, E., "Fleet Selection for Public Transportation Routes", Transportation Science, Vol. 5, pp. 1-21 (1971).

Hausman, W., and Gilmour, P., "A Multi Period Truck Delivery Problem", Transportation Research, Vol. 1, pp. 349-357 (1967).

Held, M., and Karp, R.M., "The Traveling Salesman Problem and Minimum Spanning Trees", Operations Research, Vol. 18, pp. 1138-1162 (1970).

Holmes, R.A., and Parker, R.G., "A Vehicle Scheduling Procedure Based Upon Savings and A Solution Perturbation Scheme", Operational Research Quarterly, Vol. 27, pp. 83-92 (1971).

Hyman, W., and Gordon, L., "Commercial Airline Scheduling Technique", Transportation Research, Vol. 2, pp. 23-30 (1968).

Jacobsen, S.K., and Madsen, O., "A Comparative Study of Heuristics for a Two-Level Routing Location- Problem", European Journal of Operational Research, Vol. 5, pp. 378-387 (1980).

Jaw, J., Odoni, H., Psaraftis, H., and Wilson, N., "A Heuristic Algorithm for the Multi-Vehicle Advance- Request Dial-A-Ride Problem with Time Windows", Transportation Research, Vol. 20B, pp. 243-257 (1986).

Kirby, R.F., and McDonald, J.J., "The Savings Method for the Vehicle Scheduling", Operational Research Quarterly, Vol. 24, pp. 305-306 (1972).

Kirby, R.F., and Potts, R.B., "The Minimization Route Problem with Turn Penalties and Prohibitions", Transportation Research, Vol. 3, pp. 397-408 (1969).

Knight, K.W., and Hofer, J.P., "Vehicle Scheduling with Timed and Connected Calls: A Case Study", Operational Research Quarterly, Vol. 19, pp. 299-309 (1968).

Kolen, A., Rinnooy Kan, A., and Trienekens, H., "Vehicle Routing with Time Windows", Operations Research, Vol. 35, pp. 266-273, (1987).

Koskosidis, Y.A., Powell, W.B., and Solomon, M.M., "An Optimization-Based Heuristic for Vehicle Routing and Scheduling with Soft Time Window Constraints", Transportation Science, Vol. 26, pp. 69-85 (1992).

Krolak, P., Felts, W., and Nelson, J., "A Man-Machine Approach Toward Solving the Generalized Truck Dispatching Problem", Transportation Science, Vol. 6, pp. 149-169 (1972).

Labbe, M., Laporte, G., and Mercure, H., "Capacitated Vehicle Routing on Trees", Operations Research, Vol. 39, pp. 616-622 (1991).

Laderman, J.L., Gleiberman, L., and Egan, J.F., "Vessal Allocation by Linear Programming", Naval Research Logistics Quarterly, Vol. 13, pp. 315-320 (1966).

Lam, T., "Comments on a Heuristic Algorithm for the Multiple Terminal Delivery Problem", Transportation Science, Vol. 4, pp. 403-405 (1970).

LaPorte, G., "The Vehicle Routing Problem: An Overview of Exact and Approximate Algorithms", European Journal of Operational Research, Vol. 59, pp. 345-358 (1992a).

LaPorte, G., "The Traveling Salesman Problem: An Overview of Exact and Approximate Algorithms", European Journal of Operational Research, Vol. 59, pp. 231-248 (1992b).

LaPorte, G., Nobert, Y., and Taillefer, S., "Solving a Family of Multi-Depot Vehicle Routing and Location Routing Problems", Transportation Science, Vol. 22, pp. 161-172 (1988).

LaPorte, G., Nobert, Y., and Taillefer, S., "A Branch-and-Bound Algorithm for the Assymetrical Distance Constrained Vehicle Routing Problem", Mathematical Modelling, Vol. 9, pp. 857-868 (1987).

LaPorte, G., and Nobert, Y., "Exact Algorithms for the Vehicle Routing Problem", Annals of Discrete Mathematics, Vol. 31, pp. 147-184 (1987).

Laporte, G., Mercure and Nobert, Y., "An Exact Algorithm for the Asymmetrical Capacitated Vehicle Routing Problem", Networks, Vol. 16, pp. 33-46 (1986).

LaPorte, G., Nobert, Y., and Arpin, D., "An Exact Algorithm for Solving Capacitated Location-Routing Problem", Annals of Operations Research, Vol. 6, pp. 239 - 310 (1986).

LaPorte, G., Nobert, Y., and Derochers, M., "Optimal Routing Under Capacity and Distance Restrictions", Operations Research, Vol. 33, pp. 1050-1073 (1985).

LaPorte, G., Desrochers, M., and Nobert, Y., "Two Exact Algorithms for the Distance - Constrained Vehicle Routing Problem", Networks, Vol. 14, pp. 161-172 (1984).

LaPorte, G., and Nobert, Y., "Comb Inequalities for the Vehicle Routing Problem", Methods of Operations Research, Vol. 51, pp. 271-276 (1984).

LaPorte, G., and Nobert, Y., "An Exact Algorithm for Minimizing Routing and Operating Costs in Depot Locations", European Journal of Operational Research, Vol. 6, pp. 224-226 (1981).

Lenstra, J.K., and Rinnooy Kan, A.H.G., "Complexity of Vehicle Routing and Scheduling Problems", Networks, Vol. 11, pp. 221-227 (1981).

Lenstra, J.K,. and Rinnooy Kan, A.H.G., "On General Routing Problems", Networks, Vol. 6, pp. 273-280 (1976).

Lenstra, J.K., and Rinnooy Kan, A.H.G., "Some Simple Applications of the Traveling Salesman Problem", Operational Research Quarterly, Vol. 26, pp. 717-733 (1975).

Levary, R., "Heuristic Vehicle Scheduling" OMEGA, The International Journal of Management Science, Vol. 9, pp. 660-663 (1981).

Levin, A., "Scheduling and Fleet Routing Models for Transportation Systems", Transportation Science, Vol. 5, pp. 232-255 (1971).

Levy, L., Golden B., and Assad, A., "The Fleet Size and Mix Vehicle Routing Problem", Management Science and Statistics Working Paper 80-011, College of Business and Management, University of Maryland College Park, Maryland (1980).

Li, Chung-Lun, Simchi-Levi, D., and Desrochers, S.M., "On the Distance Constrained Vehicle Routing Problem", Operations Research, Vol. 40, pp. 790-799 (1993).

Li, Chung-Lun, and Simchi-Levi, D., "Worst Case Analysis of Heuristics for Multidepot Capacitated Vehicle Routing Problems", ORSA Journal on Computing, Vol. 2, pp. 64-73 (1990).

Lin, S., and Kernighan, B., "An Effective Heuristic Algorithm for the Traveling Salesman Problem", Operations Research, Vol. 21, pp. 498-516 (1973).

Lin, S., "Computer Solutions of the Traveling Salesman Problem", Bell Systems Technical Journal, Vol. 44, pp. 2245-2269 (1965).

Lucena Filho, A.P., "Exact Solution Approaches for the Vehicle Routing Problem", Ph.D. Thesis, Imperial College, London, (1986).

Magnanti, T.L., "Combinatorial Optimization and Vehicle Fleet Planning: Perspectives and Prospects", Networks, Vol. 11, pp. 179-213 (1981).

Malandraki, C., an Daskin, M.S., "Time Dependent Vehicle Routing Problems: Formulations and Heuristic Algorithms", Technical Report, Department of Civil Engineering, Northwestern University (1989).

Male, J., Liebman, J., and Orloff, C., "An Improvement of Orloff's General Routing Problem", Networks, Vol. 7, pp. 89-92 (1977).

Marquez Diez-Canedo, J., and Escalante, O., "A Network Solution to a General Vehicle Scheduling Problem", European Journal of Operational Research, Vol. 1, pp. 255-261 (1977).

Martin-Lof, A., "A Branch-And-Bound Algorithm for Determining the Minimal Fleet Size of a Transportation System", Transportation Science, Vol. 4, pp. 159-163 (1970).

McDonald, J.J., "Vehicle Scheduling: A Case Study", Operational Research Quarterly, Vol. 23, pp. 433-444 (1972).

McKay, M.D., and Harley, H.O., "Computerized Scheduling of Seagoing Tankers", Naval Research Logistics Quarterly, Vol. 21, pp. 255-264 (1974).

Minas, J.G., and Mitten, L.G., "The Hub Operation Scheduling Problem", Operations Research, Vol. 6, pp. 329-345 (1958).

Minieka, E., "The Chinese Postman Problem for Mixed Networks", Management Science, Vol. 25, 643-648 (1979).

Mole, R., "The Curse of Unintended Rounding Error, A Case from the Vehicle Scheduling Literature", Journal of the Operational Research Society, Vol. 34, pp. 607-613 (1983).

Mole, R., Johnson, D.G., and Wells, R., "Combinatorial Analysis for the Route First-Cluster Second Vehicle Routing", OMEGA, The International Journal of Management Science, Vol. 11, pp. 507-512 (1983).

Mole, R., "A Survey of Local Delivery Vehicle Routing Methodology", Journal of the Operational Research Society, Vol. 30, pp. 245-252 (1979).

Mole, R., and Jameson, S., "A Sequential Route-Building Algorithm Employing a Generalized Savings Criterion", Operational Research Quarterly, Vol. 27, pp. 503-511 (1976).

Nelson, M.D., Nygard, K.E., Griffin, J.H., and Shreve, W.E., "Implementation Techniques for the Vehicle Routing Problem", Computers and Operations Research, Vol. 12, pp. 273-283 (1985).

Nemhauser, G., "Scheduling Local and Express Trains", Transportation Science, Vol. 3, pp. 164-175 (1969).

Newton, R., and Thomas, W., "Bus Routing in a Multi-School System", Computers and Operations Research, Vol. 1, pp. 213-222 (1974).

Newton, R., and Thomas, W., "Design of School Bus Routes by Computer", Socio-Economic Planning Sciences, Vol. 3, pp. 75-85 (1969).

Norback, J.P., and Evans, S.R., "An Heuristic Method for Solving Time-Sensitive Routing Problems", Journal of the Operational Research Society, Vol. 35, pp. 407-414 (1984).

Orloff, C., "On General Routing Problems Comments", Networks, Vol. 6, pp. 281-284 (1976a).

Orloff, C., "Route Constrained Fleet Scheduling", Transportation Science, Vol. 10, pp. 149-168 (1976b).

Orloff, C., and Caprera, D., "Reduction and Solution of Large Scale Vehicle Routing Problems", Transportation Science, Vol. 10, pp. 361-373 (1976).

Orloff, C., "A Fundamental Problem in Vehicle Routing", Networks, Vol. 4, pp. 35-64 (1974a).

Orloff, C., "Routing A Fleet of M Vehicles to/from a Central Facility", Networks, Vol. 4, pp. 147-162 (1974b).

Olson, C.A., Sorenson, E.E., and Sullivan, W.J., "Medium Range Scheduling for Freighter Fleet", Operations Research, Vol. 17, pp. 255-264 (1969).

Paessens, H., "The Savings Algorithms for the Vehicle Routing Problem", European Journal of Operational Research, Vol. 34, pp. 336-344 (1988).

Peterson, F.R., and Fullerton, H.V., "An Optimizing Network Model for the Canadian Railways", Rail Int., Vol. 4, pp. 1187-1192 (1973).

Pierce, J.F., "Direct Search Algorithms for Truck Dispatching Problems - Part I", Transportation Science, Vol. 3, pp. 1-42 (1969).

Pollack, M., "Some Elements of the Airline Fleet Planning Problem", Transport Research, Vol. 11, pp. 301- 310 (1977).

Potvin, J-Y., and Rousseau, J-M., "A Parallel Route Building Algorithm for the Vehicle Routing and Scheduling Problem with Time Windows", European Journal of Operational Research, Vol. 66, pp. 331-340 (1993).

Potvin, J-Y., Kervahut, T., and Rousseau, J-M., "A Tabu Heuristic for the Vehicle Routing Problem with Time Windows", CRT-855, Centre De Recherche Sur Les Transports, Universite De Montreal, Montreal, Canada (1992).

Psaraftis, H.N., "Dynamic Vehicle Routing: Is It A Simple Extension of Static Routing", CORS/TIMS/ORSA, Vancouver, Working Paper - MIT-OR-89-1 (1989).

Psaraftis, H.N., "Dynamic Vehicle Routing Problems", Vehicle Routing: Methods and Studies, (Edited by B. Golden and A. Assad), North Holland (1988).

Psaraftis, H.N., "Scheduling Large Scale Advance Request Dial-A-Ride Systems", American Journal of Mathematical and Management Science, Vol. 6, pp. (1986).

Psaraftis, H.N., "An Exact Algorithm for the single Vehicle Many-to-Many Dial-A-Ride Problem with Time Windows", Transportation Science, Vol. 17, pp. 351-357 (1983a).

Psaraftis, H.N., "Analysis of an $O(N2)$ Heuristi for the Single Vehicle Many-to-Many Euclidean Dial-a-Ride Problem", Transportation Research, Vol. 17B, pp. 133-145 (1983b).

Psaraftis, H.N., "K-Interchange Procedures for Local Search in a Precedence-Constrained Routing Problem", European Journal of Operational Research, Vol. 13, pp. 391-402 (1983c).

Psaraftis, H.N., "A Dynamic Programming Solution to the Single Vehicle Many-to-Many Immediate Request Dial-A-Ride Problem", Transportation Science, Vol. 14, pp. 130-154 (1980).

Pullen, H.G.M., and Webb, M.H.J., "A Computer Application to a Transport Scheduling Problem", Computer Journal, Vol. 10, pp. 10-13 (1967).

Raft, O.M., "A Modular Algorithm for an Extended Vehicle Scheduling Problem", European Journal of Operational Research, Vol. 11, pp. 67 - 76 (1982).

Rao, M.R., and Zionts, S., "Allocation of Transportation Units to Alternative Trips-A Column Generation Scheme with out-of-killer Subproblems", Operations Research, Vol. 16, pp. 52-63 (1968).

Richardson, R., "An Optimization Approach to Routing Aircraft", Transportation Science, Vol. 10, pp. 52-71 (1976).

Robertson, W.C., "Route and Van Scheduling in the Newspaper Industry", Operational Research Quarterly, Special Conference Issue, Vol. 20, pp. 99 - (1969).

Ronen, D., "Allocation of Trips to Trucks Operating from a Single Terminal", Computers and Operations Research, Vol. 19, pp. 445-451 (1992).

Ronen, D., "Short-Term Scheduling of Vessals for Shipment of Bulk and Semi-Bulk Commodities Originating in a Single Area", Operations Research, Vol. 34, pp. 164-173 (1986).

Ronen, D., "Cargo Ships Routing and Scheduling: A Survey of Models and Problems", European Journal of Operational Research, Vol. 12, pp. 119-126 (1983).

Russell, R., and Igo, W., "An Assignment Routing Problem", Networks, Vol. 9, pp. 1-17 (1979).

Russell, R.A., "An Effective Heuristic for the M-Tour Traveling Salesman Problem with Some Side Constraints", Operations Research, Vol. 25, pp. 517-524 (1977).

Saha, J.L., "An Algorithm for Bus Scheduling Problems", Operational Research Quarterly, Vol. 21, pp. 463- 474 (1970).

Salzborn, F., "Minimum Fleet Size Models for Transportation Systems", Transportation and Traffic Theory (Edited by D. Buckley), Reed, Sydney, pp. 607-624 (1974).

Salzborn, F., "A Note on Fleet Routing". Transportation Research, Vol. 7, pp. 335-355 (1972a).

Salzborn, F., "A Note on Fleet Routing Models for Transportation Systems", Transportation Science, Vol. 6, pp. 335-337 (1972b).

Salzborn, F., "The Minimum Fleet Size for a Suburban Railways System", Transportation Science, Vol. 4, pp. 383-402 (1970).

Salzborn, F., "Timetables for a Suburban Rail Transit System", Transportation Science, Vol. 3, pp. 297-316 (1969).

Salvelsbergh, M.W.P., "An Efficient Implementation of Local Search Algorithms for Constrained Routing Problems", European Journal of Operational Research", Vol. 47, pp. 75-85 (1990).

Salvelsbergh, M.W.P., "Local Search in Routing Problems with Time Windows", Annals of Operations Research, Vol. 4, pp. 285-305 (1985).

Schrage, L., "Formulation and Structure of More Complex/Realistic Routing and Scheduling Problem", Networks, Vol. 11, pp. 229-232 (1981).

Schultz, H., "A Practical Method for Vehicle Scheduling", Interfaces, Vol. 9, pp. 13-19 (1979).

Sexton, T., and Choi, Y., "Pick-up and Delivery of Partial Loads with Time Windows", The American Journal of Mathematical and Management Sciences, Vol. 6, pp. 369-398 (1986).

Sexton, T., and Bodin, L., "Optimizing Single Vehicle Many-to-Many Operations with Desired Delivery Times: II Routing", Transportation Science, Vol. 19, pp. 411-435 (1985a).

Sexton, T., and Bodin, L., "Optimizing Single Vehicle Many-to-Many Operations with Desired Delivery Times: I, Scheduling", Transportation Science, Vol. 19, pp. 378-410 (1985b).

Sexton, T., and Choi, Y., "Routing and Scheduling Problems with time Windows, Partial Loads and Dwell Times", American Journal of Mathematical and Management Sciences, Vol. pp. (1972).

Simpson, R., "A Review of Scheduling and Routing Models for Airline Scheduling", Proceedings, Ninth AGIFORS Symposium, Operations Research Division, American Airlines, New York (1969).

Smith, B., and Wren, A., "VAMPIRES and TASC: Two Successfully Applied Bus Scheduling Programs", Computer Scheduling of Public Transport: Urban Passenger Vehicle and Crew Scheduling (Edited by A. Wren), North-Holland Publishing Company, Amsterdam, pp. 97-124. (1981).

Solomon, M., and Desrosiers, J., "Time Window Constrained Routing and Scheduling Problems", Transportation Science, Vol. 22, pp. 1-13 (1988).

Solomon, M., "Algorithms for the Vehicle Routing and Scheduling Problems with Time Windows Constraints", Operations Research, Vol. 35, pp. 254-265 (1987).

Solomon, M.M., "On the Worst-Case Performance of Some Heuristics for the Vehicle Routing and Scheduling Problems with Time Window Constraints", Networks, Vol. 16, pp. 161-174 (1986).

Soumis, F. Ferland, J., Rousseau, J., "A Model for Large Scale Aircraft Routing and Scheduling Problems", Transportation Research, Vol. 14B, pp. 191-201 (1980).

Spaccamela, M.A., Rinnooy Kan, A., and Stougie, L., "Hierarchical Vehicle Routing Problems", Networks, Vol. 14, pp. 571-586 (1984).

Stein, D., "Scheduling Dial-A-Ride Transportation Systems", Transportation Science, Vol. 12, pp. 232-249 (1978).

Stein, D.M., "An Asymptotic, Probabilistic Analysis of a Routing Problem", Mathematics of Operations Research, Vol. 3, pp. 89-101 (1978).

Stern, H., and Dror, M., "Routing Electric Meter Readers", Computers and Operations Research, Vol. 6, pp. 209-233 (1979).

Stewart, W.R., and Golden, B., "A Lagrangian Relaxation Heuristic for Vehicle Routing", European Journal of Operational Research, Vol. 15, pp. 84-88 (1984).

Stewart, W., and Golden, B., "Computing Effective Subscriber Bus Routes", Proceedings, 1980 SE TIMS Conference (Edited by P. Dearing, G. Worm), Virginia Beach, pp. 170-178 (1981).

Stewart, W., and Golden, B., "The Subscriber Bus Routing Problem", Proceedings, IEEE International Conference on Circuits and Computers, Port Chester, New York, pp. 153-156 (1980).

Stricker, R., "Public Sector Vehicle Routing: The Chinese Postman Problem", M.S. Thesis, Massachusetts Institute of Technology, Cambridge, Massachusetts (1970).

Sumichrast, R.T., and Markham, I.S., "Routing Delivery Vehicles with Multiple Sources, Destinations and Depots", TIMS/ORSA Joint National Meeting (1993).

Sutcliffe, C., and Board, J., "The Ex-Ante Benefits of Solving Vehicle-Routing Problems", Journal of the Operational Research Society, Vol. 42, pp. 135-143 (1991).

Szpigel, V., "Optimal Train Scheduling on a Single Track Railway", Operations Research, Vol. 20, pp. 343- 352 (1972).

Tan, C., and Beasley, J., "A Heuristic Algorithm for the Period Vehicle Routing Problem", OMEGA,The International Journal of Management Science, Vol. 12, pp. 497-504 (1984).

Tillman, F., and Cain, T., "An Upper Bounding Algorithm for the Single and Multiple Terminal Delivery Problem", Management Science, Vol. 18, pp. 664-682 (1972).

Tillman, F., and Hering, R., "A Study of a Look-Ahead Procedure for Solving the Multiterminal Delivery Problem", Transportation Research, Vol. 5, pp. 225-229 (1971).

Tillman, F.A., "The Multiple Terminal Delivery Problem with Probabilistic Demands", Transportation Science, Vol. 3, pp. 192-204 (1969).

Tillman, F.A., and Cochran, H., "A Heuristic Approach for Solving the Delivery Problem", Journal of Industrial Engineering, Vol. 19, pp. 354-358 (1968).

Turner, W., Ghare, P., and Foulds, L., "Transportation Routing Problem - A Survey", AIIE Transactions, Vol. 6, pp. 288-301 (1976).

Turner, W., and Hougland, E., "The Optimal Routing of Solid Waste Collection Vehicles", AIIE Transactions, Vol 7, pp. 427-431 (1975).

Tyagi, M., "A Practical Method for the Truck Dispatching Problem", Journal Operational Research Society of Japan, Vol. 10, pp. 76-92 (1968).

Unwin, E., "Bases for Vehicle Fleet Scheduling", Operational Research Quarterly, Vol. 19, pp. 201-202 (1968).

Van Leeuwen P., and Volgenant, A., "Solving Symmetric Vehicle Routing Problems Asymetrically", European Journal of Operational Research, Vol. 12, pp. 388-393 (1983).

Watson-Gandy, C.D.T., and Foulds, L.R., "The Vehicle Scheduling Problem: A Survey", New Zealand Operational Research Quarterly, Vol. 23, pp. 361-372 (1972).

Webb, M.H.J., "Relative Performance of Some Sequential Methods of Planning Multiple Delivery Journeys", Operational Research Quarterly, Vol. 23, pp. 361-372 (1972).

White, W., and Bomberault, A., "A Network Algorithm for Empty Freight Car Allocation", IBM Systems Journal, Vol. 9, pp. 147-169 (1969).

Williams, B., "Vehicle Scheduling: Proximity Searching", Journal of Operational Research Society, Vol. 33, pp. 961-966 (1982).

Wolters, J., "Minimizing the Number of Aircraft for a Transportation Network", European Journal of Operational Research, Vol. 3, pp. 394-402 (1979).

Wren, A., and Holliday, A., "Computer Scheduling of Vehicle from One or More Depots to a Number of Delivery Points", Operational Research Quarterly, Vol. 23, pp. 333-344 (1972).

Wren, A., (Ed.), Computer Scheduling of Public Transportation: Urban Passenger Vehicle and Crew Scheduling", North-Holland Publishing Company, Amsterdam (1981).

Yellow, P., "A Computational Modification to the Savings Method of Vehicle Scheduling", Operational Research Quarterly, Vol. 21, pp. 281-283 (1970).

Young, D., "Scheduling a Fixed Schedule, Common Carrier Passenger Transportation System", Transportation Science, Vol. 4, pp. 243-269 (1970).

Location

Akinc, V., and Khumawala, B.M., "An Efficient Branch and Bound Algorithm for the Capacitated Warehouse Location Problem", Management Science, Vol. 23, pp. 585-594 (1977).

Alao, N., "Two Classes of Distance Minimization Problems", A Review, Some Interpretations and Extensions", Geographical Analysis, Vol. 3, pp. 299-319 (1971).

Aneja, Y.P., and Chandra Sekaran, R., and Nair, R.P.K., "A Note on the M-Center Problem with Rectilinear Distances", European Journal of Operational Research, Vol. 35, pp. 118-123 (1988).

Armour, G.C., and Buffa, E.S., "A Heuristic Algorithm and Simulation Approach to the Relative Location of Facilities", Management Science, Vol. 9, pp. 294-309 (1963).

Atkins, R.J., and R.H. Shriver, "New Approaches to Facilities Location," Harvard Business Review, pp. 70-79 (1968).

Baker, K.R., "A Heuristic Approach to Locating a Fixed Number of Facilities", Logistics and Transportation Review, Vol. 10, pp. 195-205 (1974).

Balakrishnan, P.V., and Storbeck, J.E., "MCTHRESH: Modeling Maximum Coverage with Threshold Constraints", Environment and Planning B: Planning and Design, Vol. 18, pp. 459-472 (1991).

Balas, E., "A Class of Location, Distribution and Scheduling Problems: Modelling and Solution Methods", Revue Belge de Statistique, d'Informatique et de Recherche Operationnelle, Vol. 22, pp. 36-57 (1983).

Ball, M.O., and Lin, F.L., "A Reliability Model Applied to Emergency Service Vehicle Location", Operations Research, Vol. 41, pp. 18-36 (1993).

Ballou, R., "Locating Warehouses in a Logistics System", The Logistics Review, Vol. 4, pp. 23-40 (1968).

Barcelo, J., and Casanovas, J., "A Heuristic Lagrangian Algorithm for the Capacitated Plant Location Problem", European Journal of Operational Research, Vol. 15, pp. 212-226 (1984).

Batta, R., and Mannur, N. R., "Covering-Location Models for Emergency Situations that Require Multiple Response Units", Management Science, Vol. 36, pp. 16-23 (1990).

Batta, R., Dolan, J.M., and Krishnamurty, N.N., "The Maximal Expected Covering Location Problem: Revisited", Transportation Science, Vol. 23, pp. 277-287 (1989).

Baumol, W.J., and Wolfe, P., "A Warehouse-Location Problem", Operations Research, Vol. 6, pp. 252-263 (1958).

Beckmann, M., "Principles of Optimum Location for Transportation Networks", Quantitative Geography, Atherton Press, NY (1963).

Bell, T., and Church, R., "Location and Allocation Modelling in Archeolgoical Settlement Pattern Search: Some Preliminary Applications", World Archeology, Vol. 16, pp. 354-371 (1985).

Bellman, R., "An Application of Dynamic Programming to Location-Allocation Problems," SIAM Review, Vol. 7, pp. 126-128 (1965).

Bennett, V.L., Eaton, D.J., and Church, R.L., "Selecting Sites for Rural Health Workers", Social Sciences of Medicine, Vol. 16, pp. 63-72 (1982).

Berlin, G.N., and Liebman, J.C., "Mathematical Analysis of Emergency Ambulance Location", Socio- Economical Planning Science, Vol. 8, pp. 323-328 (1971).

Bertsimas, D., "Two Traveling Salesman Facility Location Problems", Sloan Working Paper No. 2068-88, Massachusetts Institute of Technology, Cambridge, Massachusetts (1988).

Bilde, O., and Krarup, J., "Sharp Lower Bounds and Efficient Algorithms for the Simple Plant Location Problem", Annals of Discrete Mathematics, Vol. 1, pp. 79-97 (1977).

Bindschedler, A.E., and Moore, J.M., "Optimal Location of New Machines in Existing Plant Layouts", The Journal of Industrial Engineering, Vol. 12, pp. 41-47 (1961).

Bouliane, J., "Locating Postal Relay Boxes Using a Set Covering Alogorithm", American Journal of Mathematical and Management Sciences, Vol. 12, pp. 65-74 (1992).

Brandeau, M., and Chiu, S., "An Overview of Representative Problems in Location Research", Management Science, Vol. 35, pp. 645-673 (1989).

Brown, P.A., and Gibson, D.F., "A Quantified Model for Facility Site Selection-Application to a Multiplant Location Problem", AIIE Transactions, Vol. 4, pp. 1-10 (1972).

Burstall, R.M., Leeaver, R.A., and Sussmas, J.E., "Evaluation of Transport Costs for Alternative Factory Sites - A Case Study," Operational Research Quarterly, Vol. 13, pp. 345-354 (1962).

Cabot, A.V., and Francis, R.L., and Stary, M.A., "A Network Flow Solution to a Rectilinear Distance Facility Location Problem", AIIE Transactions, Vol. 2, pp. 132-141 (1970).

Cerveny, R.P., "An Application of Warehouse Location Technique to Bloodmobile Operations", Interfaces, Vol. 10, pp. 88-94 (1980).

Chaiken, J.M., "Transfer of Emergency Service Deployment Models to Operating Agencies", Management Science, Vol. 24, pp. 719-731 (1978).

Chan, A.W., and Francis, R.L., "A Round-Trip Location Problem on a Tree Graph", Transportation Science, Vol. 10, pp. 35-51 (1976).

Chandrasekaran, R., and Daughety, A., "Location on Tree Networks: P-Center and N-Dispersion Problems", Mathematics of Operations Research, Vol. 6, pp. 50-57 (1981).

Charnes, A., and Storbeck, J., "A Goal Programming Model for the Siting of Multilevel EMS Systems", Socio-Economic Planning Sciences", Vol. 14, pp. 155-161 (1980).

Chaudry, S.S., McCormick, S., and Moon, D., "Locating Independent Facilities with Maximum Weight: Greedy Heuristics." OMEGA", The International Journal of Management Science, Vol. 14, pp. 383-389 (1986).

Chhajad, D., and Lowe, T.J., "M-Median and M-Center Problems with Mutual Communication: Solvable Special Cases", Operations Research, Vol. 40, pp. S56-S66 (1992).

Cho, D.C., Johnson, E.L., Padberg, M., and Rao, M.R., "On the Uncapacitated Plant Location Problem I: Valid Inequalities and Facets", Mathematics of Operations Research, Vol. 8, pp. 579-589 (1983).

Cho, D.C., Padberg, M., and Rao, M.R., "On the Uncapacitated Plant Location Problem II: Facets and Lifting Theorems", Mathematics of Operations Research, Vol. 8, pp. 590-612 (1983).

Chrissis, J.W., Davis, R.P., and Miller, D.M., "The Dynamic Set Covering Problem", Applied Mathematics Modelling, Vol. 6, pp. 2-6 (1982).

Christofides, N., and Viola, P., "The Optimum Location of Multi-Centers on a Graph", Operational Research Quarterly, Vol. 22, pp. 145-154 (1971).

Church, R., Current, J. and Storbeck, J., "A Bicriterion Maximal Covering Location Formulation Which Considers the Satisfaction of Uncovered Demand", Decision Sciences, Vol. 22, pp. 38-52 (1991).

Church, R.L., and Eaton, D.J., "Hierarchical Location Analysisd Using Covering Objectives", Spatial Analysis and Location - Allocation Models (Edited by A. Ghosh and G. Ruston), Van Nostrand Reinhold Company, Inc., New York, pp. 163-185 (1987).

Church, R.L., and Weaver, J.R. "Theoretical Links Between Median and Coverage Location Problems", Annals of Operations Research, Vol. 6, pp. 1-19 (1986).

Church, R.L., and Roberts, K.L., "Generalized Coverage Models and Public Facility Location", Papers of the Regional Science Association, Vol. 53, pp. 117-135 (1983).

Church, R.L., and Meadows, M.E., "Location Modeling Utilizing Maximum Service Distance Criteria", Geographical Analysis, Vol. 11, pp. 358-373 (1979).

Church, R.L., and Garfinkel, R.S., "Locating an Obnoxious Facility on a Network", Transportation Science, Vol. 12, pp. 107-118 (1979).

Church, R.L., and Meadows, M.E., "Results of a New Approach to Solving the P-median Problem with Maximum Distance Constraints", Geographical Analysis, Vol. 9, pp. 364-378 (1977).

Church, R.L., and Revelle, C., "Theoretical and Computational Links Between the p-median, Location Set-Covering, and the Maximal Covering Location Problem", Geographical Analysis, Vol. 8, pp. 406- 415 (1976).

Church, R.L., and Revelle, C. "The Maximum Covering Location Problems", Papers of the Regional Science Association, Vol. 32, pp. 101-118 (1974).

Cohon, J.L., ReVelle, C.S., Current, J., Eagles, T., Eberhart, R., and Church, R., "Application of Multiobjective Facility Location Model to Power Plant Siting in a Six-State Region of the U.S.", Computers and Operations Research, Vol. 7, pp. 107-123 (1980).

Conway, R.W., and Maxwell, W.L., "A Note on the Assignment of Facility Locations", The Journal of Industrial Engineering, Vol. 12, pp. 34-36 (1961).

Cooper, L., "The Transportation-Location Problem", Operations Research, Vol. 20, pp. 94-108 (1972).

Cooper, L., "An Extension of the Generalized Weber Problem", The Journal of Regional Science, Vol. 8, pp. 181-197 (1968).

Cooper, L., "Solutions of Generalized Location Equilibrium Models," Journal of Regional Science, Vol. 7, pp. 1-18 (1967).

Cooper, L., "Heuristic Methods for Location and Allocation Problems", SIAM Review, Vol. 6, pp. 37- 53 (1964).

Cooper, L., "Location - Allocation Problems", Operations Research, Vol. 11, pp. 331-343 (1963).

Cornuejols, G., Nemhauser, G.L., and Wolsey, L.A., "The Uncapacitated Facility Location Problem", In Discrete Location Theory, R.L., (Edited by Francis and P. Mirchandani), Wiley Interscience (1990).

Cornuejols, G., Fisher, M.L., and Nemhauser, G.L., "Location of Bank Accounts to Optimize Float: An Analytic Study of Exact and Approximate Algorithms", Management Science, Vol. 23, pp. 789-810 (1977a).

Cornuejols, G., Fisher, M.L., and Nemhauser, G.L., "On the Uncapacitated Location Problem", Annals of Discrete Mathematics, Vol. 1, pp. 163-177 (1977b).

Current, J., and O'Kelly, M., "Locating Emergency Warning Sirens", Decision Sciences, Vol. 23, pp. 221-234 (1992).

Current, J.R., and Schilling D.A. "Analysis of Errors Due to Demand Data Aggregation in the Set Covering and Maximal Covering Location Problems", Geographical Analysis, Vol. 22, pp. 116-126 (1990).

Current, J.R., and Schilling, D.A., "The Covering Salesman Problem", Transportation Science, Vol. 23, pp. 208-213 (1989).

Current, J.R., and Storbeck, J.E., "Capacitated Covering Models", Environment and Planning B: Planning and Design, Vol. 15, pp. 153-163 (1988).

Current, J.R., and Storbeck, J.E., "Satisfying Solutions to Infeasible Set Partitions", Environment and Planning B: Planning and Design, Vol. 14, pp. 182-192 (1987).

Current, J.R., ReVelle, C.S., and Cohon, J.L., "The Maximum/Shortest Path Problem: A Multiobjective Network Design and Routing Formulation", European Journal of Operational Research Vol. 21, ppp. 189-199 (1985).

Daskin, M., Haghani, A.E., Khanal, M., and Malandraki, C., "Aggregation Effects in Maximum Covering Models", Annals of Operations Research, Vol. 18, 115-140 (1989).

Daskin, M., Hogan, K., and Revelle, C., "Integration of Multiple, Excess, Backup and Expected Covering Models", Environment and Planning B: Planning and Design, Vol. 15, pp. 15-35 (1988).

Daskin, M.S., "A Maximum Expected Covering Location Model: Formulation, Properties, and Heuristic Solution", Transportation Science, Vol. 17, pp. 48-70 (1983).

Daskin, M.S., "Application of an Expected Covering Model to Emergency Medical Service Systems Design", Decision Sciences, Vol. 13, pp. 416-439 (1982).

Daskin, M., and Stern, E., "A Hierarchical Objective Set Covering Model for Emergency Medical Service Vehicle Deployment", Transportation Science, Vol. 15, pp 137-152 (1981).

Davis, P.S., and Ray, T.L., "A Branch and Bound Algorithm for the Capacitated Facilities Location Problem", Naval Research Logistics Quarterly, Vol. 16, pp. 331-344 (1969).

Dearing, P.M., "Location Problems", Operations Research Letters, Vol. 4, pp. 95-98 (1985).

Dearing, P.M., and Francis, R.L., "A Minimax Location Problem on a Network", Transportation Science, Vol. 8, pp. 333-343 (1974a).

Dearing, P.M., and Francis, R.L., "A Network Flow Solution to a Multi-Facility Minimax Location Problem Involving Rectilinear Distances", Transportation Science, Vol. 8, pp. 126-141 (1974b).

Dee, N., and Liebman, J.C., "Optimal Location of Public Facilities", Naval Research Logistics Quarterly, Vol. 19, pp. 753-760 (1972).

Deighton, D., "A Comment on Location Models", Management Science, Vol. 18, pp. 113-115 (1971).

Drezner, Z., "The P-Cover Problem", European Journal of Operational Research, Vol. 26, pp. 312-313 (1985).

Drezner, Z., "The P-Center Problem-Heuristic and Optimal Algorithms", Journal of the Operational Research Society, Vol. 35, pp. 741-748 (1984).

Drysdale, J.K., and Sandiford, P.J., "Heuristic Warehouse Location - A Case History Using a New Method", Canadian Operations Research Society, Vol. 7, pp. 45-61 (1969).

Dutton, R., Hinman, G., and Millham, C.B., "The Optimal Location of Nuclear Power Facilities in the Pacific Northwest", Operations Research, Vol. 22, pp. 478-487 (1974).

Dyer, M.E., and Frieze, A.M., "A Simple Heuristic for the P-Center Problem", Operations Research Letters, Vol. 3, pp. 285-288 (1985).

Eaton, D., Daskin, M., Simmons, D., Bulloch, B., and Jansma, G., "Determining Emergency Medical Service Deployment in Austin, Texas", Interfaces, Vol. 15, pp. 96-108 (1985).

Efroymson, M., and Ray, T., "A Branch and Bound Algorithm for Plant Location", Operations Research, Vol. 14, pp. 361-368 (1966).

Eiselt, H.A., "Location Modeling in Practice", American Journal of Mathematical and Management Sciences, Vol. 12, pp. 3-18 (1992).

Eiselt, H.A., and Pederzoli, G., "A Location Problem in Graphs", New Zealand Journal of Operations Research, Vol. 12, pp. 49-53 (1984).

Eisemann, K., "The Optimum Location of a Center", SIAM Review, Vol. 4, pp. 394-401 (1962).

Ellwein, L.B., and Gray, P., "Solving Fixed Charge Location - Allocation Problems with Capacity and Configuration Constraints", AIIE Transactions, Vol. 3, pp. 290-298 (1971).

El-Shaieb, A.M., "A New Algorithm for Locating Sources Among Destinations", Management Science, Vol. 20, pp. 221-231 (1973).

Elson, D.G., "Site Location via Mixed Integer Programming", Operational Research Quarterly, Vol. 23, pp. 31-43 (1972).

Elzinga, J., Hearn, D., and Randolph, W.D., "Minimax Multifacility Location with Euclidean Distances", Transportation Science, Vol. 10, pp. 321-336 (1976).

Elzinga, J., and Hearn, D., "A Note on a Minimax Location Problem", Transportation Science, Vol. 7, pp. 100-103 (1973).

Elzinga, J., and Hearn, D., "Geometrical Solutions for Some Minimax Location Problems", Transportation Science, Vol. 6, pp. 379-394 (1972a).

Elzinga, J., and Hearn, D., "The Minimum Covering Sphere Problem," Management Science, Vol. 19, pp. 96-104 (1972b).

Erkut, E., Francis, R.L., Lowe, T.J., and Tamir, A., "Equivalent Mathematical Programming Formulations of Montonic Tree Network Location Problems", Operations Research, Vol. 37, pp. 447- 461 (1989).

Erkut, E., Francis, R.L., and Lowe, T.J., "A Multimedian Problem with Interdistance Constraints", Environment and Planning B: Planning and Design, Vol. 15, pp. 181-190 (1988).

Erlenkotter, D., "A Dual-Based Procedure for Uncapacitated Facility Location", Operations Research, Vol. 26, pp. 992-1009 (1978).

Erlenkotter, D., "Facility Location with Price-Sensitive Demand, Private, Public and Quasi-Public", Management Science, Vol. 24, pp. 378-386 (1977).

Erlenkotter, D., "A New Algorithm for Locating Sources Among Destinations", Management Science, Vol. 20, pp. 221-231 (1973).

Eyster, J.W., White, J.A., and Wierwille, W.W., "On Solving Multifacility Location Problems Using a Hyperboloid Approximation Procedure," AIIE Transactions, Vol. 5, pp. 1-6 (1973).

Feldmann, E., Lehrer, F., and Ray, T., "Warehouse Location Under Continuous Economics of Scale", Management Science, Vol. 12, pp. 670-684 (1966).

Fitzsimmons, J.A., and Allen, L. A., "A Warehouse Location Model Helps Texas Comptroller Select Out-of-State Audit Offices", Interfaces, Vol. 13, pp. 40 - 46 (1983).

Fitzsimmons, J.A., "A Methodology for Emergency Ambulance Deployment", Management Science, Vol. 15, pp. 627-636 (1969).

Flynn, J., and Ratick, S., "A Multiobjective Hierarchical Covering Model for the Essential Air Services Program", Transportation Science, Vol. 22, pp. 139-147 (1988).

Foster, D.P., and Vohra, R.K., "A Probabilistic Analysis of the K-Location Problem", American Journal of Mathematical and Management Sciences, Vol. 12, pp. 75 - 87 (1992).

Francis, R.L., and Mirchandani, P.B., (Eds.), "Discrete Location Theory", John Wiley & Sons (1989).

Francis, R.L., McGinnis, L.F., and White J.A., "Locational Analysis", European Journal of Operational Research, Vol. 12, pp. 220-252 (1983).

Francis, R.L., Lowe, T.J., and Ratliff, H.D., "Distance Constraints for Tree Network Multifacility Location Problem", Operations Research, Vol. 26, pp. 570-596 (1978).

Francis, R.L., and Goldestein, J.M., "Location Theory: A Selective Bibliography", Operations Research, Vol. 22, pp. 400-410 (1974).

Francis, R.L., and White, J.A., "Facilities Layout and Location", Prentice Hall, Inc., (1974).

Francis, R.L., and Cabot, A.V., "Properties of a Multifacility Location Problem Involving Euclidean Distances", Naval Research Logistics Quarterly, Vol. 19, pp. 335-353 (1972).

Francis, R.L., "A Geometric Solution Procedure for a Rectilinear Distance Minimax Location", AIIE Transactions, Vol. 4, pp. 328-332 (1972).

Francis, R.L., "Some Aspects of Minimax Location Problem", Operations Research, Vol. 15, pp. 1163-1168 (1967a).

Francis, R.L., "Sufficient Conditions for Some Optimum - Property Facility Design", Operations Research, Vol. 15, pp. 448-466 (1967b).

Francis, R.L., "On the Location of Multiple New Facilities With Respect to Existing Facilities", The Journal of Industrial Engineering, Vol. 15, pp. 106-107 (1964).

Francis, R.L., "A Note on the Optimum Location of New Machines in Existing Plant Layouts", The Journal of Industrial Engineering, Vol. 14, pp. 57-59 (1963).

Frank, H., "Optimum Location on a Graph with Probabilistic Demand," Operations Research, Vol. 14, pp. 409-421 (1966).

Fujiwara, O., Makjamroen, T., and Gruta, K.K., "Ambulance Deployment Analysis: A Case Study of Bangkok", European Journal of Operational Research, 31, pp. 9-18 (1987).

Garfinkel, R.S., and Noebe, A.W., and Rao, M.R., "The M-Center Problem: Minimax Facility Location", Management Science, Vol. 23, pp. 1133-1142 (1977).

Gavett, J.W., and Plyter, N.V., "The Optimal Assignments of Facilities to Locations by Branch and Bound", Operations Research, Vol. 14, pp. 210-231 (1966).

Gelders, L. F., Pintelow, L.M., and Van Wassenhove, L.N., "A Location-Allocation Problem in a Large Belgian Brewery", European Journal of Operational Research, Vol. 28, pp. 196-206 (1987).

Geoffrion, A.M., and McBride, R., "Lagrangian Relaxation Applied to Capacitated Facility Location Problem", AIIE Transactions, Vol. 10, pp. 40-47 (1978).

Ghosh, A., and Craig, C.S., "An Approach to Determining Optimal Location of New Services", Journal of Marketing Research, Vol. 23, pp. 354-362 (1986).

Gleason, J., "A Set Covering Approach to Bus Stop Location", OMEGA, The International Journal of Management Science, Vol. 3, pp. 605-608 (1975).

Goldberg, J., and Paz, L., "Locating Emergency Vehicle Bases When Service Time Depends On Call Location", Transportation Science, Vol. 25, pp. 264-280 (1991).

Goldberg, J. R., Dietrich, R., Chen, J.M., Mitwasi, M.G., Valenzuela, T., and Criss, E., "Validating and Applying a Model for Locating Emergency Medical Vehicles in Tucson, AZ (Case Study)", European Journal of Operational Research, Vol. 49, pp. 308-324 (1990).

Goldman, A.J., "Minimax Location of a Facility on a Network", Transportation Science, Vol. 6, pp. 407-418 (1972a).

Goldman, A.J., "Approximate Localization Theorems for Optimal Facility Placement," Transportation Science, Vol. 6, pp. 195-201 (1972b).

Goldman, A.J., "Optimal Center Locations in Simple Networks", Transportation Science, Vol. 5, pp. 212-221 (1971).

Goldman, A.J., and Witzgall, C.J., "A Localization Theorem for Optimal Facility Placement," Transportation Science, Vol. 4, pp. 406-408 (1970).

Goldman, A.J., "Optimum Location for Centers in a Network", Transportation Science, Vol. 3, pp. 352-360 (1969).

Goodchild, M., and Lee, J., "Coverage Problems and Visibility Regions on Topographic Surfaces", Annals of Operations Research, Vol. 18, pp. 175-186 (1989).

Guignard, M., "Fractional Vertices, Cuts, Facets of the Simple Plant Location Problem", Mathematical Programming Study, Vol. 12, pp. 150-162 (1980).

Guignard M., and Spielberg, K., "A Direct Dual Method for the Mixed Plant Location Problem with Some Side Constraints", Mathematical Programming, Vol. 17, pp. 198-228 (1979).

Guignard, M., and Spielberg, K., "Algorithms for Exploiting the Structure of Simple Plant Location Problems", Annals of Discrete Mathematics, Vol. 1, pp. 247-271 (1977).

Gunawardane, G., "Dynamic Version of Set Covering Type Public Facility Location Problems", European Journal of Operational Research, Vol. 10, pp. 190-195 (1982).

Hakimi, S., Schmeichel, E., and Pierce, J., "On P-Centers in Networks", Transportation Science, Vol. 12, pp. 1-15 (1978).

Hakimi, S., and Maheshwari, S.N., "Optimum Locations of Centers in Networks", Operations Research, Vol. 20, pp. 967-973 (1972).

Hakimi, S.L., "Optimum Distribution of Switching Centers in a Communication Network and Some Related Graph Theoretic Problems", Operations Research, Vol. 13, pp. 462-475 (1965).

Hakimi, S., "Optimum Locations of Switching Centers and the Absolute Centers and Medians of a Graph", Operations Research, Vol. 12, pp. 450-459 (1964).

Halfin, S., "On Finding the Absolute and Vertex Centers of a Tree with Distances", Transportation Science, Vol. 8, pp. 75-77 (1974).

Halpern, J., "The Location of a Center-Median Convex Combination on an Undirected Tree", Journal of Regional Science, Vol. 16, pp. 237-245 (1976).

Hammer, P.L., "Plant Location - A Pseudo-Boolean Approach", Israel Journal of Technology, Vol. 6, pp. 330-332 (1968).

Handler, G.Y., and Mirchandani, P.B., "Location on Networks, Theory and Algorithms", MIT Press, Cambridge (1979).

Handler, G., "Minimax Location Facility in an Undirected Tree Graph", Transportation Science, Vol. 7, pp. 287-293 (1973).

Hansen, P., Labbe, M., Peters, D., and Thisse, J-F., "Single Facility Location on Networks", Annals of Discrete Mathematics, Vol. 31, pp. 113-146 (1987).

Hansen, P., Thisse, J.F., and Wendell, R.E., "Equivalence of Solutions to Network Location Problems", Mathematics of Operations Research, Vol. 11, pp. 672-678 (1986).

Hitchings, G.F., "Analogue Techniques for the Optimal Location of a Main Facility in Relation to Ancillary Facilities", International Journal Production Research, Vol. 7, pp. 189-197 (1969).

Hodgson, M.J., "A Flow Capturing Location and Allocation Model", Geographical Analysis, Vol. 22, pp. 270 - 279 (1990).

Hodgson, M.J., "The Location of Public Facilities Intermediate to the Journey to Work", European Journal of Operational Research, Vol. 6, pp. 199-204 (1981).

Hogan, K., and Revelle, C., "Concepts and Applications of Backup Coverage", Management Science, Vol. 32, pp. 1434-1444 (1986).

Hogan, K., and Revelle, C.S., "Backup Coverage Concepts in the Locaiton of Emergency Services", Modleing and Simulation, Vol. 14, pp. 1423 (1983).

Hogg, J., "The Siting of Fire Stations", Operational Research Quarterly, Vol. 19, pp. 275-287 (1968).

Holmes, J., Williams, F.B., and Brown, L.A., "Facility Location Under Maximum Travel Restriction: An Example Using Day Care Facilities", Geographical Analysis, Vol. 4, pp. 258-266 (1972).

Hooker, J.N., Garfinkel, R.S., and Chen, C.K., "Finite Dominating Sets for Network Location Problems", Operations Research, Vol. 39, pp 100-118 (1991).

Hoover, E.M., "Some Programmed Models of Industry Location", Land Economics Vol. 43, pp. 303- 311 (1967).

Hopmans, A.C.M., "A Spatial Interaction Model for Branch Bank Accounts", European Journal of Operations Research, Vol. 27, pp. 242 - 250 (1986).

Hormozi, A.M., and Khumawala, B.M., "An Improved Multi-Period Facility Location Model", CBA Working Paper Series - 252, University of Houston (1992).

Hsu, W.L., and Nemhauser, G.L., "Easy and Hard Bottleneck Location Problems", Discrete Applied Mathematics., Vol. 1, pp. 209-215 (1979).

Hurter, A.P., Schaeffer, M.K., and Wendell, R.E., "Solution of Constrained Location Problems", Management Science, Vol. 22, pp. 51-56 (1975).

Jarvinen, P., Rajala, J., and Sinerro, J., "A Branch-and-Bound Algorithm for Seeking P-Median", Operations Research, Vol. 20, pp. 173-178 (1972).

Kalstorin, T.D., "On the Maximal Covering Location Problem and the General Assignment Problem", Management Science, Vol. 25, pp. 106-112 (1979).

Kariv, O., and Hakimi, S.L., "An Algorithmic Approach to Network Location Problem I: The P- Centers", SIAM Journal of Applied Mathematics, Vol. 37, pp 513-538 (1979a).

Kariv, O., and Hakimi, S.L., "An Algorithmic Approach to Network Location Problems II: The P- Medians", SIAM Journal of Applied Mathematics, Vol. 37, pp. 539-560 (1979b).

Keeney, R.L., "A Method for Districting Among Facilities", Operations Research, Vol. 20, pp. 613- 618 (1972).

Khumawala, B.M., Neebe, A.,and Dannenbring, D.G., "A Note on El-Shaieb's New Algorithm for Locating Sources Among Destinations", Management Science, Vol. 21, pp. 230-233 (1974).

Khumawala, B.M., "An Efficient Heuristic Procedure for the Uncapacitated Location Problem", Naval Research Logistics Quarterly, Vol. 20, pp. 109-121 (1973a).

Khumawala, B.M., "An Efficient Algorithm for the P-Median Problem with Maximum Distance Constraints", Geographical Analysis, Vol. 5, pp. 309-321 (1973b).

Khumawala, B.M., "An Efficient Branch and Bound Algorithm for the Warehouse Location Problem", Management Science, Vol. 18, pp. 718-731 (1972).

Khumawala, B.M., and Whybark, W.E., "A Comparison of Some Recent Warehouse Location Techniques", Logistics Review, Vol. 7, pp. 3-19 (1971).

Kimes, S.E., and Fitzsimmons, J.A., "Selecting Profitable Sites ast La Quinta Inns", Interfaces, Vol. 20, pp. 12 - 20 (1990).

Kirca, O., and Eurkip, N., "Selecting Transfer Station Locations for Large Solid Waste Systems", European Journal of Operational Research, Vol. 38, pp. 339 - 349 (1988).

Kolen, A., "The Round-Trip P-Center and Covering Problem on a Tree", Transportation Science, Vol. 19, pp. 222-234 (1985).

Kolen, A., "Solving Problems and the Uncapacitated Plant Location Problems on Trees", European Journal of Operational Research, Vol. 12, pp. 266-278 (1983).

Kolesar, P., and Walker W.E., "An Algorithm for the Dynamic Relocation of Fire Companies", Operations Research, Vol. 11, pp. 244-274 (1974).

Koopmans, Tjalling, C., and Beckmann, M., "Assignment Problems and the Location of Economic Activities," Econometrica, Vol. 25, pp. 53-76 (1957).

Kramer, R.L., "Analysis of Lock-Box Locations", Bankers Monthly Magazine, Vol. pp. 50-55 (1966).

Krarup, J., and Pruzan, P.M., "The Simple Plant Location Problem: Survey and Synthesis", European Journal of Operational Research, Vol. 12, pp. 36-81 (1983).

Kraus, A., Janssen, C., and McAdams, A.K., "The Lock-Box Location Problem: A Class of Fixed Charge Transportation Problem", Journal of Bank Research, Vol. 1, pp. 51-58 (1970).

Kuehn, A., and Hamburger, N., "A Heuristic Program for Locating Warehouses", Management Science, Vol. 9, pp. 643-666 (1963).

Kuhn, H.W., and Kuenne, R.E., "An Efficient Algorithm for the Numerical Solution of the Generalized Weber Problem in Spatial Economics," Journal of Regional Science, Vol. 4, pp. 21-33 (1962).

Laporte, G., Nobert, Y., and Arpin, D., "An Exact Algorithm for Solving Capacitated Location - Routing Problems", In Location Decisions: Methodology and Applications, (Edited by J.C. Baltzer), Scientific Publishing Company, pp. 246-257 (1986).

Larson, R.C., "Approximating the Performance of Urban Emergency Service Systems", Operations Research, Vol. 23, pp. 845-869 (1975).

Larson, R.C., and Stevenson, K.A., "On Insensitivities in Urban Redistricting and Facility Location", Operations Research, Vol. 20, pp. 595-612 (1972).

Lawrence, R.M., and Pengilly, P.J., "The Number and Location of Depots Required for Handling Products for Distribution to Retail Stores in South-East England", Operational Research Quarterly, Vol. 20, pp. 23-32 (1969).

Leamer, E.E., "Locational Equilibria," Journal of Regional Science, Vol. 8, pp. 229-242 (1968).

Levy, J., "An Extended Theorem for Location on a Network", Operational Research Quarterly, Vol. 18, pp. 433-442 (1967).

Lin, C.C., "On Vertex Addends in Minimax Location Problems", Transportation Science, Vol. 9, pp. 165-168, (1975).

Louveaux, F.V., and Peters, D., "A Dual-Based Procedure for Stochastic Facility Location", Operations Research, Vol. 40, pp. 564-573 (1992).

Love, R., F., Morris, J.G., and Weslowsky, G.O., "Facilities Location: Models and Methods", North Holland Publishing Company, New York (1988).

Love, R.F., Wesolowsky, G.O. , and Kraemer, S.A., "A Multi-Facility Minimax Location Method for Euclidean Distances", International Journal of Production Research, Vol. 11, pp. 37-46 (1973).

Love, R.F., and Morris, J.G., "Modelling Inter-City Road Distances by Mathematical Functions", Operational Research Quarterly, Vol. 23, pp. 61-71 (1972).

Love, R.F., "A Computational Procedure for Optimally Locating a Facility with Respect to Several Rectangular Regions," Journal of Regional Science, Vol. 12, pp. 233-243 (1972).

Love, R.F., "Locating Facilities in Three-Dimensional Space by Convex Programming," Naval Research Logistics Quarterly., Vol. 16, pp. 503-516 (1969).

Love, R.F., "A Note on the Convexity of the Problem of Siting Depots", International Journal Production Research, Vol. 6, pp. 153-154 (1967).

MacKinnon, Ross, G., and Barber, G.M., "A New Approach to Network Generation and Map Representation: The Linear Case of the Location-Allocation Problem," Geographical Analysis, Vol. 4, pp. 156-158 (1972).

Maier, S.F., and Vanderweide, J.H., "A Unified Location Model for Cash Disbursements and Lock- Box Collections", Journal of Bank Research, Vol. 7, pp. 166-172 (1976).

Maier, S.F., and Vanderweide, J.H., "The Lock-Box Location Problem: A Practical Reformulation", Journal of Bank Research, Vol. 5, pp. 92-95 (1974).

Malczewski, J., and Ogyczak, W., "An Interactive Approach to the Centeral Facility Locaiton Problem", Geographical Analysis, Vol. 22, pp. 244 - 258 (1990).

Manne, A.S., "Plant Location Under Economics-of-Scale-Decentralization and Computation", Management Science, Vol. 11, pp. 213-235 (1964).

Maranzana, F., "On the Location of Supply Points to Minimize Transport Costs", Operational Research Quarterly, Vol. 15, pp. 261-270 (1964).

Marks, D., Revelle, C.S., and Liebman, J.C., "Mathematical Models of Location: A Review", Journal of Urban Planning and Development Division, Vol. 96, pp. 81-93 (1970).

Marianov, V., and Revelle, C., "The Standard Response Fire Protection Siting Problem", Information Systems and Operations Research, Vol. 29, pp. 116-129 (1991).

Masuyama, S., Ibaraki, T., Hasegawa, T., "The Computational Complexity of the M-Center Problem in the Plane" Transactions of IECE Japan, Vol. E 64, pp. 57-64 (1981).

Mavrides, L.P., "An Indirect Method for the Generalized K-Median Problem Applied to Lock-Box Location", Management Science, Vol. 25, pp. 990-996 (1979).

McAdams, A.K., "Critique of: A Lock-Box Location Model", Management Science, Vol. 15, pp. 888- 890 (1968).

McHose, A.H., "A Quadratic Formulation of the Activity Location Problem," The Journal of Industrial Engineering, Vol. 12, pp. 334 (1961).

Medgiddo, N., Zemel, E., and Hakimi, L., "The maximum coverage location problem", SIAM Journal of Algebra and Discrete Methods, Vol. 4, pp 253-261 (1983).

Mehrez, A., and Stulman, A., "An Extended Continuous Maximal Covering Location Problem With Facility Placement", Computers and Operations Research, Vol. 11, pp. 19-23 (1984).

Mehrez, A., "A Note on the Linear Integer Formulation of the Maximal Covering Location Problem with Facility Placement on the Entire Plane", Journal of Regional Science, Vol. 23, pp. 553-555 (1983).

Mehrez, A., and Stulman, A., "The Maximal Covering Location Problem With Facility Placement on the Entire Plane", Journal of Regional Science, Vol. 22, pp. 361-365 (1982).

Meyer, P.D., and Brill, E.D., Jr., " A Method for Locating Wells in a Groundwater Monitoring Networking Under Conditions of Uncertainty", Water Resources Research, Vol. 24, pp. 1277-1282 (1988).

Minieka, E., "The Centers and Medians of a Graph", Operations Research, Vol. 25, pp. 641-650 (1977).

Minieka, E., "The M-Center Problem", SIAM Review, Vol. 12, pp. 138-139 (1970).

Mirchandani, P.B., "Locational Decisions on Stochastic Networks", Geographical Analysis, Vol. 12, pp. 172-183 (1980).

Mirchandani, P.B., and Odoni, A.R., "Locating New Passenger Facilities on a Transportation Network", Transportation Research, Vol. 13-B, pp. 113-122 (1979a).

Mirchandani, P.B., and Odoni, A.R., "Location of Medians on Stochastic Networks", Transportation Science, Vol. 13, pp. 85-97 (1979b).

Mole, R.H., "Comments on the Location of Depots", Management Science, Vol. 19, pp 832-833 (1973).

Moon, I.D., and Chaudhry, S.S., "An Analysis of Network Location Problem with Distance Constraints", Management Science, Vol. 30, pp. 290-307 (1984).

Moore, G., and Revelle, C., "The Hierarchical Service Location Problem", Management Science, Vol. 28, pp. 775-780 (1982).

Mukundan, S., and Dakin, M., "Joint Location/Sizing Maximum Profit Covering Models", Information Systems and Operations Research, Vol. 29, pp. 139-152 (1991).

Mycielske, J., and Trzechiakowske, W., "Optimization of the Size and Location of Service Stations", Journal of Regional Science, Vol. 5, pp. 59-68 (1963).

Nair, K.P.K., and R. Chandrasekaran, "Optimal Location of a Single Service Center of Certain Types," Naval Research Logistics Quarterly, Vol. 18, pp. 503-510 (1971).

Nambiar, J.M., Gelders, L.F., and Van Wassenhove, L.N., "Plant Location and Vehicle Routing in the Malaysian Rubber Smallholder Sector: A Case Study", European Journal of Operational REsearch, Vol. 38, pp. 14 - 26 (1989).

Nambiar, J., M., Gelders, L. F.,, and Van Wassenhove, L.N., "A Large-Scale Location-Allocation Problem in a Natural Rubber Industry", European Journal of Operational Research, Vol. 6, pp. 181- 189 (1981).

Narula, S.C., Ogbu, U.I., and Samuelsson, H.M., "An Algorithm for the P-Median Problems", Operations Research, Vol. 25, pp. 709-713 (1977).

Nauss, R.M., and Markland,R.E., "Theory and Application of an Optimizing Procedure for Lock-Box Location Analyses", Management Science, Vol. 27, pp. 855-865 (1981).

Nauss, R.M., and Markland, R.E., "Solving Lock-Box Location Problems", Financial Management, Vol. pp. 21-31 (1979).

Neebe, A.W., "A Procedure for Locating Emergency-Service Facilities For All Possible Response Distances", Journal of Operation Research Society, Vol. 39, 743-748 (1988).

Neebe, A.W., "A Branch and Bound Algorithm for the P-Median Transportation Problem", Journal of the Operational Research Society, Vol. 29, pp. 989 (1978).

Orloff, C.S., "A Theoretical Model of Net Accessibility in Public Facility Location", Geographical Analysis, Vol. 9, pp. 244-256 (1977).

Osleeb, J.P., Ratick, S.J., Buckley, P., Lee, K., and Kuby, M., "Evaluation of Dredging and Offshore Loading Locations for U.S. Coal Exports Using Local Logistics System", Annals of Operations Research, Vol. 6, pp. 163 - 180 (1986).

Palermo, F.P., "A Network Minimization Problem", IBM Journal, Vol. pp. 335-337 (1961).

Patel, N., "Locating Rural Social Service Centers in India", Management Science, Vol. 28, pp. 775-780 (1979).

Perl, J., and Daskin, M., "A Warehouse Location-Routing Problem", Transportation Research, Vol. 19B, pp. 381-396 (1985).

Perl, J., and Daskin, M.S., "A Unified Warehouse Location-Routing Methodology", Journal of Business Logistics, Vol. 5, pp. 92-111 (1984).

Picard, J.C., and Ratliff, H.D., "A Cut Approach to the Rectilinear Distance Facility Location Problem", Operations Research, Vol. 28, pp. 422-433 (1978).

Pirkul, H., and Schilling, D., "The Maximal Covering Location Problem With Capacities on Total Workload", Management Science, Vol. 37, pp. 233-248 (1991)

Pirkul, H., and Schilling, D., "The Capacitated Maximal Covering Location Problem With Backup Service", Annals of Operations Research, Vol. 18, pp. 141-154 (1989).

Plane, D.R., and Hendrick, T.E., "Mathematical Programming and the Location of Fire Companies for the Denver Fire Department", Operations Research, Vol. 25, pp. 563-578 (1977).

Polopolus, L., "Optimum Plant Numbers and Locations for Multiple Produce Processing", Journal of Farm Economics, Vol. 47, pp. 287- 295 (1965).

Price, W.L., and Turcotte, M., "Locating a Blood Bank", Interfaces, Vol. 16, pp. 17 - 26 (1986).

Pritsker, A.A.B., "A Note to Correct the Procedure of Pritsker and Ghare for Locating Facilities with Respect to Existing Facilities", AIIE Transactions, Vol. 5, pp. 84-86 (1973).

Pritsker, A.A., Ghare, P.M., "Locating New Facilities with Respect to Existing Facilities," AIIE Transactions, Vol. 2, pp. 290-297 (1970).

Rand, G.K., "Methodological Choices in Depot Location Studies", Operational Research Quarterly, Vol. 27, pp. 241-249 (1976).

Rao, A., "Counterexamples for the Location of Emergency Service Facilities", Operations Research, Vol. 22, pp. 1259-1261 (1974).

Ratick, S.J., and White, A.L., "A Risk-Sharing Model for Locating Noxious Facilities", Environment and Planning B: Planning and Design, Vol. 15, pp. 165-179 (1988).

ReVelle, C., and Serra, D., "The Maximum Capture Problem Including Relocation", Information Systems and Operations Research, Vol. 29, pp. 130-138 (1991).

ReVelle, C., "Review, Extension and Prediction in Emergency Service Siting Models", European Journal of Operational Research, Vol. 40, pp. 58-69 (1989).

ReVelle, C., and Elzinga, D.J., "An Algorithm for Facility Location in A Districted Region." Environment and Planning B: Planning and Design, Vol. 16, pp. 41 -50 (1989).

ReVelle, C., and Hogan, K., "The Maximum Availability Location Problem", Transportation Science, Vol. 23, pp. 192-200 (1989).

ReVelle, C., and Hogan, R., "A Reliability - Constrained Siting Model with Local Estimates of Busy Fractions", Environment and Planning B: Planning and Design, Vol. 15, pp. 143-152 (1988).

ReVelle, C., "The Maximum Capture or 'Sphere of Influence' Location Problem: Hotelling Revisited on a Network", Journal of Regional Science, Vol. 26, pp. 343-358 (1989b).

ReVelle, C., and Hogan, K., "The Maximum Reliability Location Problem and Alpha-Reliable P- Center Problem: Derivatives of the Probabilistic Location Set Covering Problem", Annals of Operations Research, Vol. 18, pp. 155-174 (1986).

ReVelle, C., Toregas, C., and Falkson, L., "Applications of the Location Set-Covering Problem", Geographical Analysis, Vol. 8, pp. 65-76 (1976).

ReVelle, C., Marks, D., and Liebman, J.C., "An Analysis of Private and Public Sector Location Models", Management Science, Vol. 16, pp. 692-707 (1970).

ReVelle, C., and Swain, R., "Central Facilities Location", Geographical Analysis, Vol. 2, pp. 30-42 (1970).

Richard, D., Beguin, H., and Peeters, D., "The Location of Fire Stations in a Rural Environment: A Case Study", Environment and Planning A, Vol. 22, pp. 39-52 (1990).

Rojeski, P., and Revelle, C., "Central Facilities Location Under an Investment Constraint", Geographical Analysis, Vol. 2, pp. 343-360 (1970).

Roodman, G. M., and Schwarz, L.B., "Extensions of the Multi-period Facility Phase-Out Model New Procedures and Application to a Phase-in/Phase-out Problem", AIIE Transactions, Vol. 9, pp. 103-107 (1977).

Roodman, G.M., and Schwarz, L.B., "Optimal and Heuristic Facility Phase-Out Strategies", AIIE Transactions, Vol. 7, pp. 177-184 (1975).

Rosing, K.E., "The Optimal Location of Steam Generators in Large Heavy Oil Fields", American Journal of Mathematical and Management Sciences, Vol. 12, pp. 19-42 (1992).

Ross, G.T., and Soland, R.M., "Modeling Facility Location Problems as Generalized Assignment Problem", Management Science, Vol. 24, pp. 345-357 (1977).

Rushton, G., "Applications of Location Models", Annals of Operations Research", Vol. 18, pp. 25 - 42 (1989).

Rydell, P.C., "A Note on the Principle of Median Location: Comments," Journal of Regional Science, Vol. 11, pp. 395-395-396 (1971).

Rydell, P.C., "A Note on a Location Principle: Between the Median and the Mode," Journal of Regional Science, Vol. 7, pp. 185-192 (1967).

Sa, G., "Branch-and-Bound and Approximate Solutions to the Capacitated Plant-Location Problem", Operations Research, Vol. 17, pp. 1007-1016 (1969).

Saatcioglu, O., "Mathematical Programming Models for Airport Site Selection", Transportation Research-B, Vol. 16B, pp. 435-447 (1982).

Saedt, A.H.P., "The Siting of Green Forage Drying Plants for Use by a Large Number of Farms - A Location-Allocation Case Study", European Journal of Operational Research, Vol. 6, pp. 190-194, (1981).

Saydam, C., and McKnew, M., "A Separable Programming Approach to Expected Coverage: An Application to Ambulance Location", Decision Sciences, Vol. 16, pp. 381-397 (1985).

Schaefer, M.R., and Hurtur, A.P., "An Algorithm for the Solution of a Location Problem", Naval Research Logistics Quarterly, Vol. 4, pp. 625-636 (1974).

Schilling, D.A., Jayaraman, V., and Barkhi, R., "A Review of Covering Problems in Facility Location", Location Science, Vol. 1, pp. 25 - 55 (1993).

Schilling, D., "Strategic Facility Planning: The Analysis of Options", Decision Sciences, Vol. 13, pp. 1-14 (1982).

Schilling, D., ReVelle, C., Cohen, J., and Elzinga, D.J., "Some Models for Fire Protection Location Decisions", European Journal of Operations Research, Vol. 5, pp. 1-7 (1980).

Schilling, D., "Dynamic Location Modeling for Public Sector Facilities: A Multicriteria Approach", Decision Sciences, Vol. 11, pp. 714-724 (1980).

Schilling, D., Elzinga, D.J., Cohen, J., Church, R.,and ReVelle, C., "The Team/Fleet Models for Simultaneous Facility and Equipment Siting", Transportation Science, Vol. 13, pp. 163-175 (1979).

Schneider, J.B., "Solving Urban Location Problems: Human Intuition Versus the Computer", American Institute of Planners, Vol. 37, pp. 95-99 (1971).

Schneiderjans, M.J., Kwak, N.K, and Helmer, M.C., "An Application of Good Programming to Resolve A Site Locaiton Problem", Interfaces, Vol. 12, pp. 65 - 72 (1982).

Schreuder, J.A.M., "Application of a Location Model for Fire Stations in Rotterdam", European Journal of Operational Research, Vol. 6, pp. 212-219 (1981).

Scott, A.J., "Location-Allocation Systems: A Review", Geographical Analysis, Vol. 2, pp. 95-119 (1970).

Shanker, R.J., and Zoltners, A.A., "An Extension of the Lock-Box Problem", Journal of Bank Research , Vol. 2, pp. 62 - 62 (1972).

Shannon, R.D., and Ignizio, J.P., "A Heuristic Programming Algorithm for Warehouse Location," AIIE Transactions, Vol. 2, pp. 334-339 (1970).

Simmons, D.M., "A Further Note on One-Dimensional Space Allocation", Operations Research, Vol. 19, pp. 249-249 (1971).

Simmons, D.M., "One Dimensional Space Allocation: An Ordering Algorithm", Operations Research, Vol. 17, pp. 812-826 (1969).

Slater, P.J., "On Locating a Facility to Service Areas Within a Network", Operations Research, Vol. 29, pp. 523-531 (1981).

Smith, H.L., Mangelsdorf, K.R., Luna, J.S., and Reid, R.A., "Supplying Ecuador's Health Workers Just in Time", Interfaces, Vol. 19, pp. 1-12 (1989).

Snyder, R.D., "A Note on the Location of Depots", Management Science, Vol. 18, pp. 97 - 97 (1971a).

Snyder, R.D., "A Note on the Principles of Median Location", Journal of Regional Science, Vol. 11, pp. 391-394 (1971b).

Spielberg, K., "On Solving Plant Location Problems", In Applications of Mathematical Programming Techniques, (Edited by E. M. Beale), American Elsevier (1970).

Spielberg, K., "Plant Location with Generalized Search Origin", Management Science, Vol. 16, pp. 165-178 (1969a).

Spielberg, K., "Algorithms for the Simple Plant Location Problem with Some Side Constraints", Operations Research, Vol. 17, pp. 85 - 111 (1969b).

Stancill, J.M., "A Decision Rule Model for Establishment of a Lock-Box", Management Science, Vol. 15, pp. 884-887 (1968).

Storbeck, J.E., "Classical Central Places As Protected Thresholds", Geographical Analysis, Vol. 22, pp. 4-21 (1990).

Storbeck, J.E., "The Spatial Structuring of Central Places", Geographical Analysis, Vol. 20, pp. 93- 110 (1988).

Storbeck, J.E., and Vohra, V. "A Simple Trade-Off Model for Maximal and Multiple Coverage", Geographical Analysis, Vol. 20, pp. 220-230 (1988).

Storbeck, J.E., "Slack, Natural Slack, and Location Covering", Socio-Economic Planning Sciences, Vol. 16, pp. 99-105 (1982).

Swain, R.W., "A Parametric Decomposition Approach for the Solution of Uncapacitated Location Problems", Management Science, Vol. 21, pp. 189-98 (1974).

Tansel, B. C., and Yesilkokcen, G., "Composite Regions of Feasibility for Certain Classes of Distance Constrained Network Location Problems", IEOR - 9313, Department of Industrial Engineering, Bilkent University, Ankara, Turkey (1993).

Tansel, B.C., Francil, R.L., and Lowe, T.J., "Location on Networks, Part I, the P-Center and P- Median Problems", Management Science, Vol. 29, pp. 482-497 (1983a).

Tansel, B.C., Francis, R.L., and Lowe, T.J., "Location on Networks, Part II, Exploiting Tree Network Structure", Management Science, Vol. 29, pp. 498-511 (1983b).

Tansel, B.C., Francis, and R.L., Lowe, T.J., "Binding Inequalities for Tree Network Location Problems with Distance Constraints", Transportation Science, Vol. 14, pp. 107-124 (1980).

Tapiero, C.S., "Transportation-Location-Allocation Problems Over Time", Journal of Regional Science, Vol. 11, pp. 377-386 (1971).

Taylor, P.J., "The Location Variable in Taxonomy," Geographical Analysis, Vol. 1, pp. 181-195 (1969).

Teitz, M., and Bart, P., "Heuristic Methods for Estimating Generalized Vertex Median of a Weighted Graph", Operations Research, Vol. 16, pp. 955-961 (1968).

Teitz, M., "Towards a Theory of Urban Public Facility Location", Papers of the Regional Science Association, Vol. 11, pp. 35-51 (1968).

Tewari, V.K., and Jena, S., "High School Location Decision Making in Rural India an Location - Allocation Models", Spatial Analysis and Location and Allocation Models (Editors, A. Ghosh and G. Rushton), Van Nostrand Reinhold Company, Inc., New York, pp. 137-162 (1987).

Tideman, M., "Comment on a Network Minimization Problem,'" IBM Journal, pp. 259-259 (1962).

Toregas, C., and ReVelle, C., "Binary Logic Solutions to a Class of Location Problems", Geographical Analysis, Vol. 5, pp. 145-155 (1973).

Toregas, C., ReVelle C., Swain, R., Bergman, L., "The Location of Emergency Service Facilities", Operations Research, Vol. 19, pp. 1363-1373 (1971).

Toregas, C., and ReVelle, C., "Optimal Location Under Time or Distance Constraints", Papers of the Regional Science Association, Vol. 28, pp. 133-143 (1970).

Valinsky, D., "A Determination of the Optimum Location of the Firefighting Units in New York City", Operations Research, Vol. 3, pp. 494-512 (1955).

Van Roy, T.J., and Erlenkotter, D., "A Dual Based Procedure for Dynamic Facility Location", Management Science, Vol. 28, pp 1091-1105 (1982).

Vergin, R.C., and Rogers, J.D., "An Algorithm and Computational Procedure for Locating Economic Facilities", Management Science, Vol. 13, pp. 240-254 (1967).

Vijay, J., "An Algorithm for the P-Center Problem in the Plane", Transportation Science, Vol. 19, pp. 235-245 (1985).

Volz, R.A., "Optimum Ambulance Location in Semi-Rural Areas", Transportation Science, Vol. 5, pp. 193-203 (1971).

Wagner, J.L., and Falkson, L.M., "The Optimal Nodal Location of Public Facilities with Price- Sensitive Demand", Geographical Analysis, Vol. 7, pp. 69-83 (1975).

Walker, W., "Using the Set-Covering Problem to Assign Fire Companies to Fire Houses", Operations Research, Vol. 22, pp. 275-277 (1974).

Watson-Gandy, C.D.T., "Heuristic Procedures for the M-Partial Cover Problem on a Plane", European Journal of Operations Research, Vol. 11, pp. 149-157 (1982).

Watson-Gandy, C.D.T., "A Note on the Centre of Gravity in Depot Location", Management Science, Vol. 18, pp. B478-B481 (1972).

Watson-Gandy, C.D.T., and Eilon, S., "The Depot Siting Problem with Discontinuous Delivery Cost", Operational Research Quarterly, Vol. 23, pp. 277 - 287 (1972).

Weaver, J. -R., and Church, R.L., "A Median Locaiton Model with Nonclosest Facility Service", Transportation Science, Vol. 19, pp. 58 (1985).

Weaver, J. - R, Church, R.L., "A Comparison of Solution Procedures for Covering Location Problems", Modeling and Simulation, Vol. 14, pp. 147 (1983).

Weaver, J. -R., and Church, R.L., "Computational Procedures for Location Problems on Stochastica Networks", Transportation Science, Vol. 17, pp. 168 (1983).

Wendell, R.E., and Hurter, Jr., A.P., "Optimal Location on a Network", Transportation Science, Vol. 7, pp. 18-33 (1973).

Wesolowsky, G.O., "Dynamic Facility Location," Management Science, Vol. 19, pp. 1241-1248 (1973).

Wesolowsky, G.O., "Location in Continuous Space", Geographical Analysis, Vol. 5, pp. 95-112 (1973).

Wesolowsky, G.O., and Love, R.F., "A Nonlinear Approximation for Solving a Generalized Rectangular Distance Weber Problem," Management Science, Vol. 18, pp. 656-663 (1972).

Wesolowsky, G.O., "Rectangular Distance Location Under the Minimax Optimality Criterion", Transportation Science, Vol. 6, pp. 103-113 (1972).

Wesolowsky, G.O., and Love, R.F., "Location of Facilities with Rectangular Distances Among Point and Area Destinations", Naval Research Logistics Quarterly, Vol. 18, pp. 83-90 (1971).

Wesolowsky, G.O., and Love, R.F., "The Optimal Location of New Facilities Using Rectangular Distances", Operations Research, Vol. 19, pp. 124-129 (1971).

Weston, Jr., F.C., "Optimal Configuration of Telephone Answering Sites in a Service Industry", European Journal of Operational Research, Vol. 10, pp. 395 - 405 (1982).

White, J.A., and Case, K.E. "On Covering Problems and the Central Facilities Location Problem", Geographical Analysis, Vol. 6, pp. 281-293 (1974).

Wirasinghe, S.C., and Waters, N.M., "An Approximate Procedure for Determing the Number, Capacities and Locations of Solid Waste Transfer Stations in an Urban Region", European Journal of Operational Research, Vol. 12, pp. 105-111 (1983).

Young, H.A., "On the Optimum Location of Checking Stations", Operations Research, Vol. 11, pp. 721-731 (1963).

General

Balas, E., "Disjunctive Programming", Annals of Discrete Mathematics, Vol. 5, pp. 3-51 (1979).

Balas, E., "Machine Sequencing via Disjunctive Graphs: An Implicit Enumeration Algorithm" Operations Research, Vol. 17, pp. 941-957 (1969).

Bland, R. G., "New Finite Rules for the Simplex Method", Mathematics of Operations Research, Vol. 2, pp. 103-107 (1977).

Balanski, M., and Spielberg, K., "Methods for Integer Programming: Algebraic, Combinatorial and Enumerative", Progress in Operations Research, Vol. III, Wiley (1969).

Balanski, M., "Integer Programming: Methods, Uses, Computation", Management Science, Vol. 12, pp. 253-313 (1965).

Balanski, M.L., "Fixed Cost Transportation Problem", Naval Research Logistics Quarterly, Vol. 8, pp. 41 - 54 (1961).

Barr, R.S., Glover, F. and Klingman, D., "A New Optimization Method for Large-Scale Fixed-Charge Transportation Problems", Operations Research, Vol. 29, pp. 448-463 (1981).

Breu, R., and Burdet, C.A., "Branch and Bound Experiments in 0-1 Programming", Mathematical Programming Study, Vol. 2, pp. 1-50 (1974).

Camerini, P., Fratta, L. and Maffioli, F., "On Improving Relaxation Methods by Modified Gradient Techniques", Mathematical Programming Study, Vol. 3, pp. 26-34 (1975).

Christofides, N., "Zero-one Programming Using Non-Binary Tree Search", The Computer Journal, Vol. 14, pp. 418-421 (1971).

Dyer, M.E., "Calculating Surrogate Constraints", Mathematical Programming, Vol. 19, pp. 255-278 (1980).

Fisher, M.L., "An Applications Oriented Guide to Lagrangian Relaxation", Interfaces, Vol. 15, pp. 10- 21 (1985).

Fisher, M.L., "The Langrangean Relaxation Method for Solving Integer Programming Problems", Management Science, Vol. 27, pp. 1-18 (1981).

Fisher, M.L., "Worst-Case Analysis of Integer Programming Heuristic Algorithms", Management Science, Vol. 26, pp. 1-17 (1980).

Fisher, M. L., Nemhauser, G.L., Wolsey, L., "An Analysis of Approximations for Finding a Maximum Weight Hamiltonian Circuit", Operations Research, Vol. 27, pp. 799-809 (1979).

Fisher, M.L., Northup, W.D., and Shapiro, J.F., "Using Duality to Solve Discrete Optimization Problems: Theory and Computational Experience", Mathematical Programming Study, Vol. 3, pp. 56-94 (1975).

Fisher, M.L., and Shapiro, J. "Constructive Duality in Integer Programming". SIAM Journal of Applied Mathematics, Vol. 27, pp. 31-52 (1974).

Fitzpatrick, D.W., "Scheduling on Disjunctive Graphs", Ph.D Dissertation, The Johns Hopkins University, Baltimore, Maryland (1976).

Garfinkel, R.S., and Neuhauser, G.L., "Integer Programming", Wiley, New York (1975).

Garfinkel, R.S., and Nemhauser, G.L. , "A Survey of Integer Programming Emphasizing Computation an Relations Among Models", Mathematical Programming, (Edited by T.C. Hu and S.M. Robinson), Academic Press (1973).

Gavish, B., "On Obtaining the 'Best' Multipliers for a Lagrangean Relaxation For Integer Programming". Computers and Operations Research, Vol. 5, pp. 55-71 (1978).

Geoffrion, A.M., "Lagrangean Relaxation for Integer Programming", Mathematical Programming Study, Vol. 2, pp. 82 - 114 (1974).

Geoffrion, A.M., and Marsten, R.E., "Integer Programming Algorithms: A Framework and State-of- the-Art Survey", Management Science, Vol. 18, pp. 465-491 (1972).

Glover, F., "Tabu Search - Part II", ORSA Journal on Computing, Vol. 2, pp. 4-32 (1990).

Glover, F., "Tabu Search - Part I", ORSA Journal on Computing", Vol. 1, pp. 190-206 (1989).

Glover, F., "Heuristics for Integer Programming Using Surrogate Constraints", Decision Sciences, Vol. 8, pp. 156-166 (1977).

Glover, F. "Surrogate Constraint Duality in Mathematical Programming", Operations Research, Vol. 23, pp. 434-451 (1975).

Goffin, J.L., "On the Convergence Rates of Subgradient Optimization Methods", Mathematical Programming, Vol. 13, pp. 329-348 (1977).

Harrison, T.P., "Micro Versus Mainframe Performance for a Selected Class of Mathematical Programming Problems", Interfaces, Vol. 15, pp. 14-19 (1985).

Held, M., Wolfe, P., and Crowder, H.P., "Validation of Subgradient Optimization", Mathematical Programming, Vol. 6, pp. 62-88 (1974).

Jeroslow, R.G., "Cutting Plane Theory: Disjunctive Methods", Annals of Discrete Mathematics, Vol. 1, pp. 293-330, (1977).

Johnson, D.S., Demers, A., Ullman, J.D., Garey, M.R., and Graham, R.L., "Worst-Case Performance Bounds for Simple One-Dimensional Packing Problems", SIAM Journal on Computing, Vol. 3, pp. 299 - 326 (1974).

Karp, R.M., "On the Computational Complexity of Combinatorial Problems", Networks, Vol. 5, pp. 45 - 68 (1975).

Karp, R.M., "Reducibility Among Combinatorial Problems", Complexity of Computer Computations, (Edited by R.E. Miller and J.W. Thatcher), Plenum Press, New York, pp. 85-103 (1972).

Karwan, M.H., and Rardin, R., "Surrogate Dual Multiplier Search Procedures in Integer Programming", Operations Research, Vol. 32, pp. 52-69 (1984).

Karwan, M.H., and Rardin, R.L., "A Lagrangian Surrogate Duality in a Branch and Bound Procedure", Naval Research Logistics Quarterly, Vol. 28, pp. 93 - 101 (1973).

Klee, V., "Combinatorial Optimization: What is the State of Art?" Mathematics of Operations Research, Vol. 5, pp. 1-26 (1980).

Lawler, E.L., Lenstra, J.K., Rinooy Khan, A.H.G., and Shmoys, D.S., "Traveling Salesman Problem", John Wiley and Sons (1985).

Lawler, E.L., "Combinatorial Optimization", Holt, Reinhart and Winston (1976).

McKeown, P.G, "A Branch-and-Bound Algorithm for Solving Fixed-Charge Problems", Naval Research Logistics Quarterly, Vol. 28, pp. 607-617 (1981).

Mitra, G., "Investigation of some Branch and Bound Strategies for the Solution of Mixed Integer Linear Programs", Mathematical Programming, Vol. 4, pp. 150-170 (1973).

Nemhauser, G.L., and Wolsey, L.A., "Integer Programming", Handbooks in Operations Research and Manangement Science, Vol. 1, Optimization, (Edited by G.L. Nemhauser, A.H.G. Rinnooy Kan, M.J. Tood), North-Holland, Amsterdam (1989).

Nemhauser, G.L., and Wolsey, L.A., "Integer and Combinatorial Optimization", John Wiley and Sons, Inc. (1988).

Owen, J., "Cutting Planes for Programs with Disjunctive Constraints", Journal of Optimization Theory and Applications, Vol. 11, pp. 49-55 (1973).

Padberg, M., "Perfect Zero-One Matrices", Mathematical Programming, Vol. 6, pp. 180-196 (1974).

Ragsdale, C.T., and McKeewn, P.G., "An Algorithm for Solving Fixed-Charge Problem with Surrogate Constraints", Computers and Operations Research, Vol. 18, pp. 87-96 (1991).

Raimond, J.F., "Minimaximal Paths in Disjunctive Graphs by Direct Search", IMB Journal of Research and Development, Vol. 13, pp. 391-399 (1969).

Shapiro, J., "A Survey of Lagrangian Techniques for Discrete Optimization", Annals of Discrete Mathematics, Vol. 5, pp. 113-138 (1979).

Singhal, J., Marsten, R.E., and Morin T.L., "Fixed Order Branch and Bound Methods for Mixed- Integer Programming: The Zoom System", ORSA Journal on Computing, Vol. 1, pp. 44-51 (1989).

Author Index

Abernathy, W., 643, 668
Abramson, D., 234-235, 284
Achlioptas, D., 138, 140
Action, F., 646, 689
Adams, W.P., 179, 187-188, 236, 282, 479-484, 498-499, 517, 524-527 530-532
Adams, W.W., 542, 569
Adjiman, C.S., 1, 46-48, 50, 67, 71-72
Adler, I., 209, 212, 217, 282, 292
Agarwal, Y., 647, 691
Aggarwal, A., 8, 56, 60, 72-73
Agin, N., 647, 691
Agrawal, V., 86, 142
Ahrens, J.H., 355, 418
Ahuja, H., 643, 668
Ahuja, N.K., 215, 217, 221, 233, 282
Aittoniemi, L, 388, 418
Akinc, V., 652, 714
Al-Khayyal, F.A., 41, 72, 524-525, 532
Alao, N., 654, 714
Albers, S., 641, 659
Albracht, J.J., 645, 686
Ali, D., 641, 663
Alimonti, P., 122, 140
Alizadeh, F., 151, 179, 182, 276-277, 282
Allen, L.A., 654-655, 722
Almond, M., 646, 687
Altinkemer, K., 647, 691
Altman, S., 644, 668
Alvernathy, W.J., 644, 673
Amado, L., 645, 682

Amar, G., 644, 677
Amos, D., 646, 690
Anbil, R., 644-645, 677-678
Andersen, E.D., 212, 282
Andros, P., 647, 694
Androulakis, I.P., 46-48, 50, 67, 71-72
Aneja, Y.P., 642, 646, 652, 665, 687, 714
Angel, R., 649, 691
Anily, S., 641, 647, 649, 659, 691
Anthony Brooke, 51, 73
Apkarian, P., 166, 184
Appelgren, K., 649, 692
Appelgren, L., 649, 691
Arabeyre, J.P., 644, 677
Arani, T., 647, 687
Arisawa, S., 649, 692
Armour, G.C., 654, 714
Arora, S., 116, 140-141
Arpin, D., 649, 652, 704, 729
Asano, T., 116, 141
Asirelli, P., 79, 141
Assad, A., 647-648, 649, 692, 700, 705
Assad, A.A., 647-649, 694, 701
Atkins, R.J., 652, 714
Atkinson, D.S., 258, 283
Aubin, J., 646, 687
Aurenhammer, F., 174, 183
Ausiello, G., 79, 90, 117, 141
Aust, R.J., 646, 687
Averabakh, I., 647, 692
Avis, D., 174, 183, 310, 418, 641, 659

Bahn, O., 258, 283
Bailey, J., 643, 668
Bailey, J.E., 643, 675
Baker, B.M., 647, 692
Baker, E. 641, 644, 647-649, 659, 677, 692, 701
Baker, E.K., 644, 678
Baker, K.R., 643, 652, 668-669, 714
Balakrishnan, A., 649, 695
Balakrishnan, P.V., 653, 714
Balanski, M., 742
Balanski, M.L., 642, 665
Balas, E., 39, 72, 246, 283, 311-313, 321, 324-325, 338-340, 342, 418-419, 480, 482-483, 492, 494, 517-518, 525, 527, 641, 653, 659-660, 714, 742
Balinksi, M., 647, 649, 692
Ball, M., 647, 649, 692, 698
Ball, M.O., 644, 647-648, 653, 678, 694, 714
Ballou, R., 652, 714
Baloff, N., 643-644, 668, 673
Barber, G.M., 654, 730
Barcelo, J., 652, 714
Barcia, P., 366, 419
Barham, A.M., 646, 687
Barkhi, R., 653, 737
Barnes, E.R., 217, 283
Barnhart, C., 644, 678
Barr, R.S., 742
Bart, P., 653, 739
Bartholdi, J., 647, 650, 692
Bartholdi, J.J., 643, 669
Bartlett, T., 649, 693
Barton, R., 649, 693
Bastos, F., 220-223, 293
Batta, R., 653, 654, 658, 715
Battiti, R., 77, 133-136, 141

Baumol, W.J., 652, 715
Baybars, I., 645, 684
Bayer D., 558, 570
Bayer, D.A., 212, 283
Bazaraa, M.S., 217, 283
Beale, E.M.L., 8, 30, 72
Beasley, J., 647-648, 696, 698, 712
Beasley, J.E., 641-642, 642, 647, 660, 665, 693
Beaumont, N., 39, 72
Bechtold, S., 643, 669-670
Beckmann, M., 652, 654, 715, 728
Beguin, H., 652-653, 735
Bell, T., 654, 715
Bell, W., 647, 650, 693
Belletti, R., 644, 678
Bellman, R., 647, 654, 693, 715
Bellman, R.E., 336, 388, 419, 460, 474
Bellmore, M., 641-642, 647, 660, 665, 693
Beltrami, E., 647, 693
Beltrami, E.J., 644, 668
Benders, J.F., 9, 72
Benichou, M., 33, 72
Bennett, B., 649, 693
Bennett, B.T., 643, 670
Bennett, H.S., 648, 650, 695
Bennett, V.L., 653, 655, 715
Benson, H.P., 156, 176, 182, 183
Benson, S.J., 280, 283
Benvensite, R., 641, 660
Berge, C., 641-642, 660, 665
Bergman, L., 653, 658, 740
Berlin, G.N., 653, 715
Berman, L., 649-650, 694
Berman, O., 647, 692
Berry, W.L., 643, 672
Bertsimas, D., 654, 714

Bertsimas, D.J., 647, 694
Bhattacharya, B.K., 174, 183
Bielli, M., 645, 683
Bilde, O., 652, 715
Billera, L.J., 536-537, 562, 567, 570
Billionnet, A., 527
Bindschedler, A.E., 654, 715
Blair, C.E., 84, 141, 550, 570
Bland, R.G., 742
Blattner, W.O., 430, 458, 475
Blum, M., 452, 474
Board, J., 648, 712
Bodin, D.L., 644, 678-679
Bodin, L., 647-650, 692-694, 701, 710
Bodin, L.D., 643-644, 670, 678
Bodner, R., 647, 694
Boehm, M., 83, 142
Bomberault, A., 649, 713
Booler, J.M., 644, 679
Borchers, B., 8, 30, 33, 50, 73, 234-235, 247, 257, 279-280, 283-284, 291-292
Boros, E., 482, 527
Borret, J.M.J., 645, 679
Boufkhad, Y., 140, 143
Bouliane, J., 654, 715
Bowman, E.H., 645, 684
Boyd, S., 277, 296
Bramel, J., 647, 649, 695
Branco, M.I., 645, 682
Brandeau, M., 654, 716
Brelaz, D., 642, 665
Breu, R., 742
Brill, E.D., 653, 731
Broder, A., 140, 142
Broder, S., 646, 687
Bronemann, D.R., 644, 679
Brown, E.L., 499, 524, 530

Brown, G.B., 647, 649-650, 695
Brown, J.R., 642, 665
Brown, L.A., 653, 656, 726
Brown, P.A., 652, 716
Browne, J., 643, 672, 676
Browne, J.J., 643, 670
Brownell, W.S., 643, 670
Brusco, M., 643, 669
Buchberger, B., 542, 570
Buckley, P., 654, 733
Buffa, E.S., 644, 654, 670, 714
Bulfin, R.L, 388, 419
Bulloch, B., 653, 655, 721
Burbeck, S., 644, 681
Burdet, C.A., 742
Burkard, R.E., 309, 392, 419
Burns, R.N., 643, 668, 670-671
Buro, M., 83, 142
Burstall, R.M., 652, 716
Bushel, G., 649, 699
Bushnell, M., 86, 142
Butt, S., 647, 695
Byrn, J.E., 644, 673
Byrne, J.L., 643, 671

CPLEX, 526, 528
Cabot, A.V., 652, 654, 716, 723
Cabot, V., 183
Cain, T., 648, 712
Camerini, P., 743
Camerini, P.M., 525, 527
Caprara, A., 309, 419
Caprera, D., 648, 707
Captivo, M.E., 645, 682
Carraresi, P., 180, 183, 644, 679
Carstensen, P.J., 473, 474
Carter, M., 643, 668
Carter, M.W., 643, 646-647, 670, 687

Casanovas, J., 652, 714
Case, K.E., 653, 742
Cassell, E., 647, 694
Cassidy, P.J., 648, 650, 695
Cattrysse, D., 646, 685
Cattrysse, J., 646, 685
Caudle, W., 649, 691
Cavalier, T.M., 647, 695
Ceder, A., 644-645, 649-650, 679, 683-684, 695
Ceria, S., 246, 283, 483, 492, 494, 527
Cerveny, R.P., 652, 655, 716
Chaiken, J., 643, 671
Chaiken, J.M., 654, 716
Chakradar, S., 86, 142
Chan, A.W., 654, 716
Chan, P.W., 646, 690
Chan, T.J., 641, 660
Chandra Sekaran, R., 652, 714
Chandra, A.K., 373, 408, 419
Chandrasekaran, R., 430, 474, 652, 654, 716, 732
Chang, G.J., 642, 665
Chao, M.-T., 140, 142
Chard, R., 647, 695
Charnes, A., 649, 653, 693, 695, 716
Chaudhry, S.S., 652-653, 732
Chaudry, S.S., 641, 654, 660, 716
Chelst, K., 643, 671
Chen, C.K., 652-653, 727
Chen, D., 643, 671
Chen, J., 92, 94, 96-97, 142
Chen, J.M., 654, 656, 724
Cherici, A., 645, 683
Cheriyan, J., 85, 142
Cheshire, I.M., 647, 695
Chevalley, C., 204, 284

Chhajad, D., 652, 716
Chien, T.W., 649, 695
Chilali, M., 166, 184
Chiu, S., 654, 716
Cho, D.C., 652, 717
Choi, G., 517, 525-526, 530
Choi, Y., 648, 710
Chopra, S., 642, 665
Chrissis, J.W., 653, 717
Christof, T., 246, 284
Christofides, N., 403, 419, 641, 652, 647-649, 660, 696, 698, 717, 743
Church, J.G., 643, 671
Church, R., 654, 715, 737
Church, R.L., 652-655, 658, 715, 717-718, 741
Chvátal, V., 310, 362, 419
Chv:atal, V., 138-139, 142, 310, 362, 419, 641-642, 660, 666
Ciftan, E., 646, 690
Ciric, A.R., 8, 73
Clarke, G., 647, 696
Clarke, M.R.B., 308, 421
Cochran, H., 648, 712
Cockayne, E.J., 642, 666
Coffman, E.G., 641, 647, 660, 695
Cohen, B., 129, 147
Cohen, E., 472, 475
Cohen, J., 653-654, 737
Cohon, J.L., 652, 654-655, 718-719
Cole, R., 454, 475
Collins, R., 647, 650, 692
Conforti, M., 179, 183, 641, 661
Conti, P., 542, 545-546, 570
Conway, R.W., 654, 718
Cook, S.A., 79, 138, 142
Cook, W., 178, 183, 550, 570
Cooper, L., 652, 654, 718

Cormen, T.H., 419, 434, 465, 475
Corneil, D.G., 641, 661
Corneil, D.G., 642, 666
Cornuéjols, G., 179, 183, 246, 283, 483, 492, 494, 527, 569-570, 641, 652, 658, 661, 718-719
Cosagrove, M.J., 644, 670
Cox, D., 542, 570
Crabil, T.B., 643, 669
Craig, C.S., 654, 724
Crama, Y., 482, 527, 641, 661
Crandall, H., 647, 700
Crawford, J.L., 647, 650, 697
Crescenzi, P., 79, 90, 117-118, 122, 141, 143
Criss, E., 654, 656, 724
Crowder, H., 302, 419, 480, 515, 525, 528
Crowder, H.P., 744
Csima, J., 646, 688
Cullen, D., 647, 691
Cullen, F.H., 647, 649, 697
Culver, W.D., 643, 674
Cunningham, W.H., 85, 142
Cunto, E., 647, 650, 697
Current, J., 652, 653, 655, 717-719

d'Atri, A., 117, 141
d'Atri, G., 352, 418
Daganzo, C., 647, 649, 697
Dahl, R., , 647, 649, 701
Dakin, M., 654, 732
Dalberto, L., 647, 650, 693
Dallwig, S., 46-47, 50, 72
Dannenbring, D., 403, 425
Dannenbring, D.G., 652, 727
Danninger, G., 180, 183
Dantazig, G.G., 647, 649, 697

Dantazig, G.W., 644, 671
Dantzig, G., 647, 699
Dantzig, G.B., 208, 216, 284., 311, 320, 419, 430, 458, 475
Darby-Dowman, K., 644-645, 679, 682
Daskin, M., 653-655, , 658, 719-721
Daskin, M.S., 647-648, 652, 705, 733
Daughety, A., 652, 716
Davani, A., 644, 678
Davis, M., 81, 143, 200, 284
Davis, P.S., 652, 720
Davis, R.P., 653, 717
Day, R.H., 646, 688
de Loera, J., 568, 570
de Santis, M., 79, 141
de Silva, A., 234-235, 284
de Simone, C., 202, 254, 284
DeWerra, D., 646, 688
Deane, R.H., 643, 674
Dearing, P.M., 652-654, 658, 720
Dee, N., 653, 720
Deighton, D., 653, 720
Dembo, R.S., 331, 387, 408, 419
Demers, A., 744
Dempster, M.A.H., 646, 688
Deo, N., 318, 427
Derochers, M., 648, 704
Desrochers, M., 520, 522, 528, 648, 697
Desroches, S., 646, 689
Desrosiers, J., 133, 145, 647-649, 697-698, 711
Deza, M., 245, 284
Di Biase, F., 568, 571
Diaconis, P., 549, 571
Dial, R., 644, 678-679

Diehl, M., 202, 254, 284
Dietrich, R., 654, 656, 724
Dietrich, B.L., 303, 352, 420
Diffe, W., 305, 420
Dikin, I.I., 208, 217, 285
Dinkelbach, W., 430, 437, 473, 475
Dolan, J.M., 653, 715
Doll, L.L., 647, 698
Dormont, P., 643, 671
Dreyfus, S.E., 642, 666
Drezner, Z., 196, 294, 652, 720
Driscoll, P.J., 517, 531
Dror, M., 647-648, 649, 651, 698, 711
Drysdale, J.K., 652, 655, 720
Du Merle, O., 258, 283
Du, D., 137, 143
Du, D.-Z., 86, 144, 170-171, 183-184, 198, 213, 217, 285, 642, 667
DuPuis, D., 643, 645, 673, 681
Dubois, O., 140, 143
Dudziński, 309, 318, 324, 373, 374, 376, 399, 420
Duguid, R., 644, 678
Dulac, G., 649, 698
Dumas, Y., 648-649, 697-698
Dunford, N., 156, 184
Duran, M.A., 9, 44, 73
Durant, P.A., 645, 684
Dutton, R., 652, 655, 720
Dyer, M.E., 372-373, 420, 652, 720, 743
Dzielinski, B.P., 646, 685

Eagles, T., 652, 655, 718
Easton, F.F., 643, 671
Eaton, D., 653, 655, 721
Eaton, D.J., 653, 655, 715, 717

Eberhart, R., 652, 655, 718
Edmonds, J., 167, 184, 245, 285, 642, 666
Edwards, G.R., 644, 679
Efroymson, M., 652, 721
Egan, J.F., 649, 703
Eilon, S., 643, 647, 671, 696, 698
Eiselt, H.A., 654, 721
Eisemann, K., 652, 721
El-Azm, A., 648, 698
El-Bakry, A.S., 226, 235, 285
El-Darzi, E., 641, 661
El-Halwagi, M., 42-44, 75
El-Shaieb, A.M., 652, 721
Elkihel, M., 359, 427
Ellwein, L.B., 652, 721
Elmaghraby, S.E., 649, 692
Elson, D.G., 652, 721
Elzinga, D.J., 653-654, 735, 737
Elzinga, J., 654, 721
Emmons, H., 643, 671-673
Erkut, E., 173, 184, 653, 654, 721-722
Erlenkotter, D., 652-654, 722, 740
Ervolina, T.R., 464, 475
Escalante, O., 648, 706
Escudero, L.F., 303, 352, 418, 420
Etcheberry, J., 641, 661
Etezadi, T., 647, 698
Eurkip, N., 654, 656, 728
Eusebio, R., 645, 682
Evans, S.R., 647-648, 650, 698, 707
Even, S., 646, 688
Evers, G.H.E., 644, 679
Eyster, J.W., 654, 722
Ezell, C., 641, 663

Falk, J.E., 72
Falkner, J.C., 641, 644, 664, 679

Falkson, L., 653, 735
Falkson, L.M., 653, 740
Farinaccio, F., 180, 183
Farvolden, J.M., 649, 699
Favati, P., 178, 184
Fayard, D., 309, 321, 323, 342-343, 420-421
Fearnley, J., 644, 677
Federgruen, A., 641, 647, 649, 659, 691, 699
Feige, U., 116, 143
Feldman, I., 397, 422
Feldmann, E., 652, 722
Felts, W., 648, 703
Feo, T.A., 130, 147
Ferebee, J., 647, 699
Ferguson, A., 647, 699
Ferland, J., 649, 698, 711
Ferland, J.A., 646, 687
Ferland, P.C., 646, 688
Ferreira, C.E., 305, 372, 407, 421
Fiacco, A.V., 209, 285
Field, J., 643, 668
Finke, G., 355, 418
Finnegan, W.F., 644, 678
Fischetti, M., 400, 409, 421
Fisher, M., 644, 647, 650, 677, 693
Fisher, M.L., 366, 421, 525, 528, 641, 646-647, 649-650, 652, 655, 661, 685, 699, 718-719, 743
Fisk, J.C., 403, 405, 410, 423
Fitzpatrick, D.W., 743
Fitzsimmons, J.A., 654-656, 722, 728
Fletcher, A., 647, 699
Fletcher, R., 9, 23, 73
Fleuren, H.A., 647-649, 699
Florian, M., 649, 699

Floudas, C., 281, 285
Floudas, C.A., 1, 8-9, 46-48, 50-52, 60, 67, 71-75
Floyd, R.W., 452, 474
Flynn, J., 653, 655, 722
Ford, L.R., 217, 285, 460, 475
Forgues, P., 649, 698
Foster, B.A., 645, 647, 683, 700
Foster, D.P., 653, 722
Foulds, L., 648, 712
Foulds, L.R., 648-649, 700, 713
Fowler, R.J., 641, 661
Fox, B., 430, 475
Fox, G.W., 641, 662
Frances, M.A., 643, 672
Francis, R.L., 175-176, 187, 652-654, 716, 720-723, 739
Franco, J., 83, 137, 139-140, 142-144
Frank, A., 184
Frank, H., 654, 723
Frank, R.S., 646, 688
Fratta, L., 525, 527, 743
Frederickson, G., 647, 700
Fredman, M.L., 453, 475
Freedman, B.A., 217, 295
Freeman, D.R., 645, 685
Fréville, A., 308, 421
Frey, K., 644, 678
Friesen, D., 92, 94, 96-97, 142
Frieze, A., 140, 142-143
Frieze, A.M., 308, 421, 652, 720
Fujisawa, K., 279, 285
Fujishige, S., 167, 184, 474, 476
Fujiwara, O., 653, 655, 723
Fukushima, M., 474, 476
Fulkerson, D.R., 217, 285, 460, 475, 641, 646, 662, 688
Fullerton, H.V., 649, 707

Fulton, W., 542, 571

Gaballa, A., 643, 672
Gafinkel, R.S., 654, 717
Gahinet, P., 166, 184
Gallaire, H., 79, 143
Gallo, G., 308, 421, 644, 679
Gans, O.B., de., 646, 688
Garey, M.R., 408, 421, 744
Garfinkel, R., 641, 662
Garfinkel, R.S., 528, 642, 646, 652-653, 658, 689, 723, 727, 743
Garvin, W., 647, 700
Gaskell, T., 647, 700
Gaudioso, M., 647, 700
Gauthier, J.M., 33, 72
Gavett, J.W., 654, 724
Gavish, B., 308, 422, 647, 691, 700, 744
Gazis, D., 649, 693
Gehrlein, W.V., 645, 686
Gel'fand, I.M., 537, 562-563, 570-571
Gelatt Jr., C.D., 132, 145
Gelders, L.F., 652, 656, 724, 732-733
Gelman, E., 645, 677
Gendreau, M., 647, 700
Gens, G., 352, 422
Gent, I.P., 129, 143
Gentzler, G.L., 643, 672
Geoffrion, A.M., 8-9, 73, 526, 528, 652, 724, 744
George, A., 223, 286
Gerards, A.M., 178, 183
Gerards, A.M.H., 550, 570
Gerbract, R., 644, 680
Gershkoff, I., 644, 648, 680, 700

Ghare, P., 648, 712
Ghare, P.M., 654, 734
Gheysens, F., 647, 700-701
Ghosh, A., 654, 724
Giannessi, F., 149, 152, 176, 184
Gibson, D.F., 652, 716
Gilbert, E.N., 642, 666
Gillett, B.E., 647, 701
Gilmore, P.C., 305, 388, 397, 422
Gilmour, P., 648, 702
Girlich, E., 167, 185
Girodet, P., 33, 72
Glassey, C.R., 646, 689
Gleason, J., 654, 724
Gleiberman, L., 649, 703
Glover, F., 46, 73, 132, 134, 143, 319, 382, 422, 643, 672, 742
Glover, G., 480, 528
Glover, R., 643, 672
Glover, f., 744
Goemans, M., 442, 475
Goemans, M.X., 85, 116, 143-144
Goemans, Michel X., 276-277, 286
Goerdt, A., 115, 144
Goffin, J.L., 258, 283, 286, 744
Goldberg, A., 464, 477
Goldberg, A.V., 342, 422, 460, 464, 475-476
Goldberg, J., 654, 656, 724
Golden, B., 647-649, 692, 698, 700, 705, 711-712
Golden, B.L., 647-649, 650, 651, 701-702
Goldern, L.B., 647-649, 694
Goldestein, J.M., 654, 723
Goldman, A.J., 652-653, 724-725
Goldmann, M., 473, 476
Goldstein, A.S., 221, 223, 288

Golub, G.H., 221, 286
Gomory, R.E., 245, 286, 305, 388, 397, 422, 646, 685
Gondzio, J., 212, 258, 282, 286
Goodchild, M., 653, 725
Goodwin, T., 647, 701
Gordon, L., 648, 702
Gosselin, K., 646, 689
Gotleib, G.C., 646, 688
Graham, B., 642, 666
Graham, R.L., 744
Granot, F., 178, 185
Graver, J.E., 550, 553, 571
Graves, G.W., 647, 649-650, 695
Gray, P., 652, 721
Grayson, D., 540, 571
Green, L., 643, 672
Greenberg, H., 397, 422
Greenberg, J., 644, 678
Greenfield, A., 647, 650, 693
Greenfield, A.J., 647, 650, 699
Griffin, J.H., 648, 706
Grimes, J., 646, 689
Grossmann, I.E., 6, 9, 18, 20-21, 30, 32-34, 37-39, 44, 51, 60, 70, 73, 75-76
Gr:otschel, M., 167, 185, 245-248, 286, 305, 421
Gruta, K.K., 653, 655, 723
Gu, J., 79, 83, 86, 119, 137, 143-144, 147, 198, 285
Gu, Q.-P., 86, 144
Guerin, G., 649, 699
Guha, D., 643, 672
Guignard, M., 652, 725
Guisewite, G.M., 213, 287
Gumaer, R., 649, 693
Gunawardane, G., 653, 725
Gupta, O.K., 8, 30, 32-33, 50, 73

Gurevich, Y., 648, 700
Gusfield, D., 472-473, 476
Gutjahr, A.L., 645, 685

Ha'cijan, L.G., 186
Hackman, S.T., 645, 685
Hagberg, B., 643, 672
Hager, W.W., 164, 185
Haghani, A.E., 653, 719
Haimovich, M., 648, 702
Hakimi, L., 653, 731
Hakimi, S., 652-653, 658, 725
Hakimi, S.L., 176, 185., 642, 652-654, 666, 727
Halfin, S., 652, 725
Hall, A., 646, 689
Hall, R.W., 649, 702
Hallman, A.B., 645, 684
Halpern, J., 652-653, 725
Hamburger, N., 652, 729
Hammer, P., 641, 661
Hammer, P.L., 85, 144, 182, 186, 308, 324, 331, 387, 408, 419, 421, 422, 482, 527, 641-642, 652, 662, 726
Hanan, M., 642, 666
Handler, G.Y., 175, 185, 652, 654, 726
Hansen, P., 84-85, 120, 129, 132-133, 144, 166, 185, 652-654, 726
Hanssman, F., 644, 682
Hanthikumar, J.G., 179, 186
Hara, S., 277, 290
Harjunkoski, I., 44, 75
Harlety, H.O., 649, 651, 706
Harrison, T.P., 744
Hartley, T., 644, 680
Hasegawa, T., 654, 731

Hashizume, S., 474, 476
Hastad, J., 116, 145
Hatay, L., 644, 678
Hauer, E., 648, 702
Haurie, A., 258, 286
Hausman, W., 648, 702
Hayer, M., 568, 571
Hearn, D., 654, 721
Hearn, D.W., 642, 666
Heath, L.S., 646, 689
Heath, M., 223, 286
Hecht, M., 647, 700
Held, M., 647, 702, 744
Helgason, R.V., 217, 290
Hellman, M.E., 305, 420
Helmberg, C., 277-280, 287
Helmer, M.C., 654, 656, 737
Henderson, W., 644, 680
Henderson, W.B., 643, 672
Hendrick, T.E., 654, 656, 734
Hentges, G., 33, 72
Hering, R., 648, 712
Hersh, M., 644, 676
Hershey, J., 643, 668
Hershy, J.C., 644, 673
Hertz, A., 646-647, 689, 700
Hervillard, R., 644, 680
Heurgon, E., 644, 680
Hinman, G., 652, 655, 720
Hirata, T., 116, 141
Hiriart-Urruty, J.-B., 180, 185
Hirsch, W.M., 162, 186
Hirschberg, D.S., 373, 394, 408, 419, 422
Hitchcock, F.L., 216, 287
Hitchings, G.F., 654, 726
Ho, A., 641, 659, 662
Hochbaum, D., 641, 662

Hochbaum, D.S., 168, 179, 186, 309, 422
Hochst:attler, W., 568, 571
Hodgson, M.J., 654, 726
Hofer, J.P., 648, 651, 703
Hoffman, A.J., 162, 186, 641, 662
Hoffman, K.L., 246, 287, 305, 422, 480, 525, 528
Hoffmann, T.R., 645, 685
Hoffstadt, J., 645, 680
Hogan, K., 653-654, 656, 658, 719, 726, 735
Hogg, J., 654, 656, 726
Holland, O., 245, 247, 286
Holliday, A., 648, 713
Holloran, T.J., 644, 673
Holmberg, K., 9, 26, 73
Holmes, J., 653, 656, 726
Holmes, R.A., 648, 702
Hong, S., 647, 693
Hooker, J.N., 79, 84-85, 144, 652-653, 727
Hoover, E.M., 652, 727
Hopmans, A.C.M., 654, 656, 727
Hormozi, A.M., 652, 727
Horowitz, E., 313, 333, 337, 376, 423
Horst, R., 156, 176, 182, 186
Hosten, S., 545, 558, 568-569, 571
Houck, D.J., 642, 666
Hougland, E., 648, 712
Housos, E., 196, 288
Howard, S.M., 645, 680
Howell, J.P., 643, 673
Hsu, W.L., 654, 727
Huang, C., 196, 288
Hung, M.S., 403, 405, 410, 423
Hung, R., 643, 673
Hurter, A.P., 654, 727

Hurter, Jr., A.P., 652, 741
Hurtur, A.P., 654, 736
Huxlery, S.J., 644, 676
Hwang, F.K., 171, 183, 186, 641-642, 662, 666
Hyman, W., 648, 702

Ibaraki, S., 654, 731
Ibaraki, T., 168, 186, 318, 336, 346, 423, 430-431, 473-474, 476, 478, 641, 661
Ibarra, O.H., 346-347, 397, 423
Ignall, E., 643, 673
Ignall, E.J., 645, 685
Ignizio, J.P., 652, 737
Igo, W., 648, 709
Ingargiola, G.P., 331-332, 388, 408, 423
Ishii, H., 430, 476
Itai, A., 646, 688
Iwano, K., 474, 476

J:ornsten, 366, 419
Jacobs, F.R., 644, 674
Jacobs, L.W., 643, 669
Jacobsen, S.K., 649, 651, 702
Jaikumar, R., 647, 650, 693, 699
Jameson, S., 648, 706
Janaro, R.E., 643, 670
Janssen, C., 652, 658, 729
Jansen, B., 270, 272, 296
Jansma, G., 653, 655, 721
Jarvinen, P., 653, 727
Jarvis, J.J., 217, 283, 647, 649, 697
Jaumard, B., 84, 120, 129, 132-133, 144-145, 166, 185
Jaw, J., 648-649, 651, 702
Jayaraman, V., 653, 737
Jena, S., 654, 657, 739

Jeroslow, R.G., 84, 141, 362, 423, 480, 528, 550, 570, 744
John, D.G., 641, 662
John, J., 647, 700
Johnson, D., 641, 662
Johnson, D.G., 648, 706
Johnson, D.S., 91-92, 94, 136, 145, 198, 288, 408, 421, 744
Johnson, E., 644, 678
Johnson, E.L., 302, 324, 365, 372, 419, 422-423, 480, 515, 525, 528-529, 641, 652, 662, 717
Johnson, F.L., 641, 662
Johnson, J., 647, 701
Johnson, J.L., 86, 145
Johnson, R.V., 645, 685-686
Johnson, T.A., 236, 282, 498-499, 524, 526
Johnston, N.D., 119, 146
Jones, R.D., 644, 680
Joshi, A., 221, 223, 288
Joy, S., 279, 283
Jucker, J.V., 645, 685
Judice, J., 220-223, 293
Junger, M., 202, 245, 248, 254, 284-286, 288

Kabbani, N.M., 645, 681
Kalantari, B., 186
Kaliski, J.A., 221, 288
Kalstorin, T.D., 653, 727
Kamath, A., 140, 145
Kamath, A.P., 79, 85, 145, 198, 269, 272, 276, 288-289
Kamiyama, A., 643, 677
Kannan, R., 394, 423
Kapoor, A., 179, 183
Karisch, S.E., 277, 279, 280, 297
Kariv, O., 652-654, 727

Karloff, H., 116, 145
Karmarkar, N., 212, 217, 282
Karmarkar, N.K., 79, 85, 145, 196,
    198, 201, 208, 212, 225,
    237, 259, 261-262, 269, 272,
    276, 281, 288-289, 292, 294
Karp, R.M., 431, 460, 476, 647,
    702, 744-745
Karwan, M.H., 745
Katoh, N., 168, 186, 474, 476
Kautz, H., 126, 129, 133, 147
Kayal, N., 372-372, 420
Keaveny, I.T., 644, 681
Kedia, P., 641, 661
Keedia, P., 647, 650, 693
Keeney, R.L., 654, 727
Keith, E.G., 644, 673
Kelley, J.E., 34, 73
Kendrick, D., 51, 73
Kennington, J.L., 217, 290
Kernighan, B., 647, 705
Kervahut, T., 648, 708
Khachiyan, L.G., 203, 208, 290
Khalil, T.M., 643, 672
Khanal, M., 653, 719
Khanna, S., 122, 145
Khoury, B.N., 642, 666-667
Khumawala, B.M., 652-653, 714,
    727-728
Kiefl, S., 305, 421
Kilbridge, M.D., 645, 686
Killion, D.L., 645, 681
Kim, C., 647, 700
Kim, C.E., 346-347, 397, 423
Kimes, S.E., 654, 656, 728
Kindervater, G.A.P., 314, 423
King, V., 464, 476
Kirby, R.F., 648, 703
Kirca, O., 654, 656, 728

Kirkpatrick, S., 132, 138, 145-146
Kirousis, L.M., 138-140, 146
Klasskin, P.M., 644, 673
Klee, V., 174, 186, 745
Kleine Buening, H., 83, 142
Klingman, D., 742
Knauer, B.A., 646, 689
Knight, K.W., 648, 651, 703
Knuth, D.E., 355, 423
Kochenberger, G.A., 641, 662
Kocis, G.R., 6, 9, 18, 37, 73
Kojima, M., 277, 279, 285, 290
Kolen, A., 648, 652-653, 703, 728
Kolesar, P., 643, 653, 656, 672-
    673, 728
Kolesar, P.J., 332, 424
Kolner, T.K., 644, 681
Koncal, R.D., 641, 664
Koop, G.J., 643, 670, 673
Koopmans, T.C., 652, 728
Korman, S., 641, 660
Korsh, J.F., 331-332, 388, 408, 423
Koskosidis, Y.A., 648, 703
Kostreva, M.M., 480, 529
Koutsopoulos, H.N., 645, 681
Koutsoupias, E., 125, 146
Kovalev, M., 167, 185
Kowalik, J.S., 318, 427
Kozolov, M.K., 186
Kraemer, S.A., 654, 730
Krajewski, L.J., 643-644, 673, 675-
    676
Kramer, R.L., 652, 728
Kranakis, E., 138-140, 146
Krarup, J., 652, 715 728
Krass, D., 310, 424
Kraus, A., 652, 658, 729
Kravanja, Z., 44, 75
Krinzac, D., 138, 140

Krishnamurthy, R.S., 524-525, 532
Krishnamurty, N.N., 653, 715
Krispenz, C., 305, 421
Krizanc, D., 138-139, 146
Krolak, P., 648, 703
Krushal, J.B., 216, 290
Kuby, M., 654, 733
Kuehn, A., 652, 729
Kuenne, R.E., 652, 729
Kuhn, H.W., 652, 729
Kuik, R., 646, 685
Kullmann, O., 83, 146
Kunnathur, A.S., 646, 689
Kursh, S., 647, 650, 694
Kwak, N.K., 654, 656, 737
Kydes, A., 644, 679

LaBella, A., 645, 683
LaPorte, G., 646-649, 652, 689, 699, 700, 703-704, 729
Labbé, M., 175-176, 186
Labbe, M., 648, 652-653, 703, 726
Laderman, J.L., 649, 703
Laffrey, T.J., 79, 146
Lagarias, 212, 289
Lagarias, J.C., 212, 282, 310, 424
Laird, P., 119, 146
Lam, T., 648, 703
Laporte, G., 305, 424, 520, 522, 528
Larsen, C., 182, 186
Larson, R.C., 654, 729
Lasdon, L.S., 646, 686
Lasky, J.S., 641, 664
Lassiter, J.B., 522, 527
Laurent, M., 245, 284
Lavoie, S., 644, 681
Lawler, E.L., 32, 74, 430, 458, 476, 641, 662, 745

Lawler, Eugene, 217, 290
Lawrence, R.M., 654, 729
LePrince, M., 645, 681
Leamer, E.E., 654, 729
Lee, E.K., 290
Lee, J., 653, 725
Lee, K., 654, 733
Lee, Y., 493, 498, 506, 517, 525, 531-532
Leeaver, R.A., 652, 716
Lehrer, F., 652, 722
Leigh, W., 641, 663
Leighton, F., 642, 667
Leiserson, C.E., 419, 434, 465, 475
Lemarechal, C., 180, 185
Lemke, C.E., 641-642, 663
Lenstra, J.K., 314, 423, 647-648, 697, 704-705, 745
Lessard, R., 643, 645, 673, 681, 683
Lester, J.T., 647, 650, 699
Leudtke, L.K., 645, 681
Levary, R., 648, 705
Levenberg, K., 290
Levesque, H., 119, 129, 138, 146-147
Levin, A., 649, 705
Levner, E., 352, 422
Levy, J., 652, 729
Levy, L., 647-649, 698, 701, 705
Leyffer, S., 9, 23, 73
Li, C., 643, 674
Li, Chung-Lun, 648, 705
Li, Y., 237, 239, 290-294
Liebman, J., 648, 705
Liebman, J.C., 652-653, 715, 720, 731, 735
Liepins, G.E., 646, 689
Lin, C.C., 652, 729

Lin, F.L., 653, 714
Lin, S., 647, 705
Linder, R.W., 644, 674
Lions, J., 647, 689
Little, J., 542, 570
Liu, L., 196, 288
Logemann, G., 81, 143
Lotfi, V., 646, 687
Loucks, J.S., 644, 674
Loustaunau, P., 542, 569
Louveaux, F.V., 652, 729
Lovász, L., 167, 178, 185, 187, 246, 277, 286, 290, 483, 526, 529, 641, 663
Love, R., 654, 729-730
Love, R.F., 654, 741
Loveland, D., 81, 143
Loveland, D.W., 85, 146
Lowe, J.K., 84, 141, 480, 528
Lowe, T.J., 652-654, 716, 721-723, 739
Lowerre, J.M., 643, 670, 674
Luce, B.J., 644, 670
Lucena Filho, A.P., 648, 705
Luckhardt, H., 83, 146
Lueker, G.S., 305, 394, 424, 641, 660
Luna, J.C., 643, 676
Luna, J.S., 652, 657, 738
Lund, C., 116, 140
Luo, Z.Q., 258, 286
Lustig, I.J., 208, 210, 253, 290

Mabert, V.A., 643, 674, 676
MacKinnon, 654, 730
Mack, R., 647, 650, 693
Maculan, N., 642, 667
Madsen, O., 649, 651, 702
Maes, J., 646, 685

Maffioli, F., 525, 527, 743
Magazine, M.J., 643, 645, 668-669, 685
Magnanti, T.L., 481, 529, 647-648, 650, 702, 705
Magnanti, Thomas L., 215, 217, 221, 233, 282
Maheshwari, S.N., 652, 725
Mahjoub, A.R., 641, 661
Maier, S.F., 652, 730
Maier-Rothe, C., 643, 674
Makjamroen, T., 653, 655, 723
Malandradi, C., 647-648, 653, 719, 705
Malczewski, J., 652, 730
Male, J., 648, 705
Malleson, A.M., 647, 695
Malucelli, F., 180, 183
Mangelsdorf, K.R., 643, 652, 657, 676, 738
Manne, A.S., 646, 652, 686, 730
Mannila, H., 117, 146
Mannur, N.R., 654, 658, 715
Maranas, C.D., 46-47, 50, 72
Maranzana, F., 652, 730
Marchetti-Spaccamela, A., 342, 422
Marianov, V., 654, 731
Markham, I.S., 648, 712
Markland, R.E., 647, 652, 658, 689, 733
Marks, D., 652-653, 731, 735
Marquardt, D., 291
Marquez Diez-Canedo, J., 648, 706
Marsten, R.E., 208, 210, 253, 309, 425, 641, 643, 645, 663, 676, 681-682, 744, 746
Martelli, A., 79, 141
Martello, S., 309, 314, 318, 323, 325-329, 332-334, 340, 342-

343, 345, 347-348,352, 359,
361-362, 382, 384, 387, 391,
394, 396-400, 403-404, 406,
408-410, 415, 418, 420-421,
424-425
Martin, A., 305, 372, 407, 421
Martin, K.R., 480, 529
Martin-Lof, A., 649, 706
Masuyama, S., 654, 731
Mathews, G.B., 305, 425
Mathon, V., 166, 185
Mathur, K., 647, 691
Mavrides, L.P., 652-653, 731
Mawengkang, H., 9, 36, 74
Maxwell, W.L., 654, 718
Mazzola, J.B., , 480, 527
McAdams, A.K., 652, 658, 729, 731
McBryde, R., 481, 528, 652, 724
McCloskey, J.F., 644, 682
McCormick, G.P., 41, 74, 209, 285
McCormick, S., 654, 716
McCormick, S.T., 464, 474-475, 477, 641, 660
McDonald, J.J., 648-649, 651, 703, 706
McGinnis, L.F., 643, 654, 674, 722
McGrath, D., 644, 675
McHose, A.H., 654, 731
McKay, M.D., 649, 651, 706
McKeewn, P.G., 746
McKenzie, J.P., 643, 674
McKenzie, P., 644, 673
McKeown, P.G., 646, 689, 745
McKnew, M., 653, 736
McMillan, C., 643, 672
Meadows, M.E., 653, 717
Medgiddo, N., 653, 731
Megeath, J.D., 644, 675
Megiddo, N., 431, 448, 472, 475, 477
Mehrez, A., 653, 731
Mehrotra, S., 208, 212, 221, 225, 291
Mehta, N.K., 642, 647, 667, 690
Meketon, M.S., 217, 296
Melzak, Z.A., 642, 666
Mercure, H., 648, 703-704
Mertens, W., 645, 681
Meszaros, C., 212, 282
Meyer, P.D., 653, 731
Meyer, R.R., 480, 529
Michaud, P., 641, 663
Mikhailow, G.W., 8, 30, 32, 50, 74
Miller, A.J., 645, 684
Miller, D.M., 653, 717
Miller, H.E., 643, 675
Miller, L.R., 647, 701
Miller, M.H., 649, 695
Millham, C.B., 652, 655, 720
Minas, J.G., 648, 706
Mine, H., 430, 476
Mingozzi, A., 403, 419, 647, 696
Minieka, E., 648, 652-653, 658, 706, 732
Minker, J., 79, 143
Minoux, M., 644, 681-682
Minton, S., 119, 146
Mirchandani, P.B., 175-176, 185, 187, 654, 722, 726, 732
Misono, S., 474, 476
Mitchell, D., 119, 129, 138, 146-147
Mitchell, D.G., 138, 142
Mitchell, E., 189
Mitchell, J.E., 8, 30, 33, 50, 73, 202, 234-235, 247, 250, 253, 256-258, 279, 283-284, 290-292, 294

Mitchell, M., 644, 678
Mitchell, R., 645, 682
Mitra, G., 641, 644-645, 661, 679, 682, 745
Mitten, L.G., 648, 706
Mitwasi, M.G., 654, 656, 724
Mizrach, M., 646, 689
Mole, R., 648, 706
Mole, R.H., 654, 732
Monroe, G., 643, 675
Monteiro, R.D.C., 209, 212, 292
Moon, D., 654, 716
Moon, I.D., 641, 652-653, 660, 732
Moondra, S.L., 643, 675
Moore, G., 654, 732
Moore, J.M., 654, 715
Moore, R.E., 42, 74
Mora, T., 558, 571
More, J.J., 265, 292
Morin, T.L., 309, 425, 746
Morris, J.G., 643, 654, 675, 729-730
Morrish, A.R., 644, 675
Morrison, I., 558, 570
Moser, P.I., 645, 680
Motwani, R., 116, 122, 127, 140, 145-146
Mukundan, S., 654, 732
Muller, M.R., 645, 681
M/..uller-Merbach, H., 324, 425
Mulvey, J.M., 646, 690
Munro, J.I., 338, 425
Murtagh, B.A., 9, 36, 74
Murty, K., 641, 663
Mutzel, P., 202, 254, 284
Mycielske, J., 654, 732
Myers, D.C., 525, 531

Naccache, P.F., 647, 695
Nair, R.P.K., 652, 654, 714, 732
Nakata, K., 279, 285
Nambiar, J.M., 652, 656, 732-733
Narula, S.C., 653, 658, 733
Natraj, N., 246, 283
Natraj, N.R., 549, 568, 572
Nauss, R.M., 305, 365, 373, 425, 647, 652, 658, 689, 733
Naw, D.S., 646, 690
Nawijn, W.M., 646, 690
Neebe, A., 403, 425, 652, 727
Neebe, A.W., 653, 733
Nelson, J., 648, 703
Nelson, M.D., 648, 706
Nemhauser, G., 649, 706
Nemhauser, G.L., 179, 187, 245, 292, 318, 388, 426, 480, 507, 528-529, 641-642, 645-646, 652, 654, 658, 662-663, 665-667, 685, 688-689, 718-719, 727, 743, 745
Nemirovsky, A.S., 277, 292
Nesterov, Y.E., 277, 292
Neuhauser, G.L., 642, 743
Neumaier, A., 42, 46-47, 50, 72, 74, 180, 187
Neumann, S., 173, 184
Newton, R., 649, 707
Ng, S.M., 641, 659
Nguyen, H.Q., 647-648, 650, 702
Nguyen, T.A., 79, 146
Niccolucci, F., 152, 184
Nicolas, J.M., 79, 143
Niederer, M., 644, 682
Nobert, Y., 648-649, 652, 704, 729
Nobili, P., 85, 146
Noebe, A.W., 652, 658, 723
Noemi, P., 641, 663
Noonan, R., 649, 691

Norback, J.P., 647-648, 650, 698, 707
Norman, R.Z., 642, 667
Northup, W.D., 743
Norton, C.Haibt, 472, 477
Nygard, K.E., 648, 706

O'Connor, A.R., 644, 675
O'Kelly, M., 653, 655, 719
O'Niel, K.K., 645, 684
O'Shea, D., 542, 570
Oakford, R.V., 642, 667
Odier, E., 644, 681
Odlyzko, A.M., 310, 424
Odoni, A.R., 645, 654, 681, 732
Odoni, H., 648-649, 651, 702
Ogbu, U.I., 653, 658, 733
Ogyczak, W., 652, 730
Oley, L.A., 480, 529
Olson, C.A., 648, 707
Ono, T., 116, 141
Oppenheim, R., 641, 662
Orlin, J.B., 643, 669
Orlin, James B., 215, 217, 221, 233, 282
Orloff, C., 648, 705, 707
Orloff, C.S., 653, 733
Orponen, P., 117, 146
Osleeb, J.P., 654, 733
Ostrovsky, G.M., 8, 30, 32, 50, 74
Ostrovsky, M.G., 8, 30, 32, 50, 74
Overton, M., 526, 529
Owen, J., 745
Ozkarahan, I., 643, 675

Padberg, M., 246, 287, 305, 422, 480, 525, 528, 641-642, 652, 663, 717, 745

Padberg, M.W., 245, 247, 253, 292, 302, 313, 365, 372, 419, 423, 426, , 480, 515, 525, 529, 641-642, 659-660
Paessens, H., 648, 707
Pai, R., 201, 292
Paixao, J., 220-223, 293, 643, 675
Paixao, J.P., 645, 682
Paixo, J., 641, 660
Palem, K., 140, 145
Palermo, F.P., 654, 733
Paletta, G., 647, 700
Panconesi, A., 118, 143
Pantelides, C.C., 45, 75
Papadimitriou, C.H., 84, 112, 114, 118, 125-127, 130, 146-147, 318, 426, 465, 477
Pappalardo, M., 180, 183
Pappas, I.A., 643, 675
Pardalos, P., 281, 285
Pardalos, P.M., 130-131, 137, 143, 147, 151, 156, 159-160, 164, 170-171, 176, 179, 182, 184, 187, 189, 196, 198, 213, 217, 235, 237, 239, 272, 277, 285, 287, 290, 293-294, 431, 473-474, 477, 498, 530, 642, 666-667
Parker, M.E., 645, 682
Parker, R.G., 234, 293, 388, 419, 648, 702
Patel, N., 653, 656, 733
Paterson, M.S., 641, 661
Pato, M., 643, 675
Pato, M.V., 645, 682
Patterson, J.H., 645, 686
Patty, B., 645, 677
Patty, B.W., 645, 681
Paules, G.E., 8, 51, 74

Paull, M., 139, 143
Paz, L., 654, 724
Pearce, W., 643, 672
Pecora, D., 79, 146
Pederzoli, G., 654, 721
Peeters, D., 175-176, 186, 652-653, 735
Pekny, J.F., 498, 530
Peled, U.N., 324, 422, 641, 662-663
Pengilly, P.J., 654, 729
Perkins, W.A., 79, 146
Perl, J., 652, 733
Peters, D., 652-653, 726, 729
Peterson, F.R., 649, 707
Pettersson, F., 34, 44, 51, 76
Pferschy, U., 309, 392, 419, 426
Philips, A.B., 119, 146
Phillippe, D., 643, 676
Phillips, A.T., 431, 473-474, 477
Picard, J.C., 654, 734
Piccione, C., 645, 683
Pierce, J., 652, 725
Pierce, J.F., 641, 646, 649, 664, 686, 707
Pierskalla, W.P., 643, 675
Pintelow, L.M., 652, 724
Pirkul, H., 308, 422, 653-654, 734
Pisinger, D., 299, 309, 314, 328, 330, 334-337, 339-340, 343-345, 352, 355, 370, 372, 375-377, 384,386-387, 389-391, 398-399, 403, 406-408, 412, 419, 426-427
Pitsoulis, L.S., 130-131, 147
Plane, D.R., 654, 656, 734
Plateau, G., 308-309, 321, 323, 342-343, 359, 427, 420-421
Platzman, L., 647, 650, 692

Plotkin, S.A., 308, 427, 472, 477
Plyter, N.V., 654, 724
Poljak, S., 151, 179, 187, 277-279, 287
Pollack, M., 649, 707
Pollak, H.O., 642, 666
Polopolus, L., 652, 734
Ponder, R., 644, 678
P:orn, R., 44, 75
Portugal, L., 220-223, 293
Potts, R.B., 643, 648, 670-671, 703
Potvin, J-Y., 648, 708
Powell, W.B., 648, 703
Pratt, V., 452, 474
Prawda, J., 643, 677
Price, E., 644, 675
Price, W.L., 654, 656, 734
Prim, R.C., 216, 293
Pritsker, A.A., 654, 734
Protasi, M., 77, 79, 90, 117, 122, 135-136, 141
Prutzman, P., 647, 650, 693
Pruzan, P.M., 652, 728
Psaraftis, H., 648-649, 651, 702
Psaraftis, H.N., 648-649, 708
Puech, C., 352, 418
Pullen, H.G.M., 648, 708
Purdom, P.W., 83, 137, 144
Puri, R., 79, 144, 147
Putnam, H., 81, 143, 200, 284

Quandt, R., 647, 649, 692
Quesada, I., 30, 32-33, 75

Rabin, M.O., 642, 667
Radzik, T., 429, 430-431, 464-465, 473, 477
Raedels, A., 643, 674
Raft, O.M., 648, 709

Ragavachari, M., 187
Raghavan, P., 127, 146
Ragsdale, C.T., 746
Raimond, J.F., 746
Rajala, J., 653, 727
Ramachandran, B., 498, 530
Ramakrishnan, K.G., 79, 85, 145, 196, 198, 225, 235, 237, 259, 262, 269, 272, 288-289, 293-294, 498, 530
Raman, R., 39, 75
Ramana, M., 277, 294
Ramaswamy, S., 258, 294
Ramirez, R.J., 338, 425
Ramser, J.H., 647, 649, 697
Rand, G.K., 654, 734
Randolph, W.D., 654, 721
Rannou, B., 644, 683
Rao, A., 653, 734
Rao, M.R., 430, 458, 475, 649, 652, 658, 709, 717, 723
Rao, S., 464, 476
Rao, S.P., 201, 292
Rappoport, S.S., 644, 668
Rardin, R., 745
Rardin, R.L., 234, 293
Rath, G.J., 643, 675
Ratick, S., 653, 655, 722
Ratick, S.J., 653-654, 733-734
Ratliff, H.D., 641-643, 647, 649, 653-654, 660, 665, 669, 697, 723, 734
Ratschek, H., 42, 75
Ravindran, R., 30, 32-33, 50, 73
Ray, T., 652, 721-722
Ray, T.L., 652, 720
ReVelle, C., 652-654, 658, 734-735, 737, 740
Read, E.G., 649, 700

Reed, B., 138, 142
Reggia, J.A., 646, 690
Reid, R.A., 643, 652, 657, 676, 738
Reinelt, G., 202, 245-246, 248, 254, 284, 286, 288
Rendl, F., 151, 179, 187, 277-280, 287, 297
Resende, M.G.C., 130-131, 147, 189, 196, 198, 212, 217, 221-222, 225, 235, 237, 239, 259, 262, 269, 272, 277, 282, 288-289, 293-295, 498, 530
Revelle, C., 652-655, 658, 718-719, 726, 732
Revelle, C.S., 652-654, 731
Ribiere, G., 33, 72
Richard, D., 652-653, 735
Richards, D.S., 642, 666
Richardson, R., 649, 709
Rinaldi, G., 202, 245, 247, 253-254, 284, 480, 529
Rinnooy Kan, A., 648-649, 703, 711
Rinnooy Kan, A.H.G., 647-648, 704-705, 745
Rinnooy Kan, H.G., 648, 702
Ritzman, L.P., 643-644, 673, 675-676
Rivest, R.L., 419, 434, 452, 465, 474
Robbiano, L., 558, 571
Roberts, A., 644, 678
Roberts, K.L., 653, 717
Robertson, W.C., 648, 709
Robinson, D.F., 649, 700
Robinson, E.P., 643, 674
Robinson, J.A., 80, 147
Rockafellar, R.T., 161, 187
Roes, A.W., 645, 679

Rogers, J.D., 652, 740
Rojeski, P., 653, 735
Rokne, J., 42, 75
Romig, W., 649, 694
Ronen, D., 648-650, 695
Roodman, G.M., 654, 736
Roos, C., 208, 270, 272, 295-296
Rosen, J.B., 156, 176, 182, 185, 187
Rosenfield, 644, 679
Rosenwein, M.B., 649-650, 699
Rosing, K.E., 654, 736
Ross, G., 654, 730
Ross, G.T., 654, 736
Rossin, D.F., 643, 671
Roth, R., 641, 664
Rothblum, U.G., 186
Rothstein, M., 643, 676
Rouseau, J.-M., 645, 681
Rousseau, J-M., 648, 708
Rousseau, J., 649, 711
Rousseau, J.-M., 644-645, 683
Rousseau, J.M., 643, 673
Roussos, I.M., 164, 185
Roy, B., 641, 664
Roy, S., 646, 688
Roy, T.J.van, 305, 427
Rubin, J., 644, 683
Rudeanu, S., 182, 186
Ruhe, G., 217, 295
Rushton, G., 654, 736
Russell, R., 647, 649, 709
Ryan, D.M., 641, 645, 647, 664, 679, 683, 700
Rydell, P.C., 653, 736
Ryoo, H.S., 41, 75
Ryzhkov, A.P., 641, 664

Sa, G., 652, 736

Saatcioglu, 653, 736
Saedt, A.H.P., 652, 656, 736
Safra, S., 116, 141
Saha, J.L., 649, 709
Sahinidis, N.V., 41, 75
Sahinoglou, H.D., 164, 185
Sahni, S., 313, 333, 337, 376, 423
Salazar, A., 642, 667
Salkin, H.M., 641, 647, 663-664, 691
Saloman, M., 646, 685
Salvelsbergh, M.W.P., 648, 710
Salveson, M.E., 645, 686
Salzborn, F., 648-649, 651, 709-710
Samuelsson, H.M., 653, 658, 733
Sandiford, P.J., 652, 655, 720
Sarkissian, R., 258, 286
Sassano, A., 85, 146, 641, 661
Saunders, M.A., 36, 74
Sauve, M., 648, 697
Savelsbergh, M.M.P, 648, 697
Saydam, C., 653, 736
Scarf, H.E., 550, 571
Schaefer, M.R., 654, 736
Schaeffer, M.K., 654, 727
Schaffer, J., 648, 692
Schaible, S., 430-431, 473-474, 477-478
Scheffi, Y., 645, 684
Schilling, D., 653-654, 734
Schilling, D.A., 653-654, 719, 737
Schmeichel, E., 652, 725
Schneider, J.B., 654, 737
Schneiderjans, M.J., 654, 656, 737
Schoepfle, G.K., 644, 668
Schrage, L., 648, 710
Schreuder, J.A.M., 653, 656, 737
Schrijver, A., 167, 178-179, 183,

185, 187, 246, 286, 483, 526, 529, 535, 537, 550, 553, 566, 570, 571
Schultz, H., 648, 710
Schuster, A.D., 649, 694
Schwartz, J.T., 156, 184
Schwarz, L.B., 654, 736
Schweiger, C.A., 1, 52, 71, 75
Schweitzer, P., 647, 651, 700
Scott, A.J., 652, 737
Scott, D., 645, 683
Scott, K.L., 646, 686-687
Scudder, G.D., 641, 646, 662, 686
Segal, M., 643, 676
Selman, B., 119, 126, 129, 133, 138, 146-147
Serra, D., 654, 734
Sethi, S.P., 310, 424
Sexton, T., 648-649, 694, 710
Shamir, A., 646, 688
Shanker, R.J., 652, 737
Shanno, D.F., 208, 210, 253
Shannon, R.D., 652, 737
Shapiro, J., 746
Shapiro, J.F., 743
Shepardson, F., 643, 645, 676, 682-683
Sheppard, R., 643, 668
Sherali, H.D., 179, 187-188, 217, 283, 479-484, 493, 498-499, 506, 517, 522, 524-527, 530-532
Shetty, C.M., 388, 419
Shi, C.J., 263, 272, 295
Shi, J., 175, 188
Shiloach, Y., 464, 478
Shindoh, S., 277, 290
Shlifer, E., 647, 651, 700
Shmoys, D.B., 308, 427

Shmoys, D.S., 745
Shor, P.W., 647, 695
Showalter, M., 643, 669-670
Showalter, M.J., 643-644, 675-676
Shreve, W.E., 648, 706
Shriver, R.H., 652, 714
Simchi-Levi, D., 647-649, 695, 699, 705
Simeone, B., 85, 144, 308, 421, 641, 662, 663
Simmons, D., 653, 655, 721
Simmons, D.M., 654, 737-738
Simpson, R., 649, 710
Sinclair, G.B., 647, 650, 697
Sinerro, J., 653, 727
Singer, I., 178, 188
Singhal, J., 746
Sinha, A., 305, 365-366, 373, 380, 427
Sinha, L.P., 217, 295
Sivazlian, B.B., 643, 672
Sjouquist, R.J., 480, 529
Skrifvars, H., 44, 75
Slater, P.J., 653, 738
Slorin-Kapov, J., 178, 185
Slutsman, L., 212, 289
Slyke, R. Van, 258, 295
Smith, B., 649, 651, 711
Smith, B.M., 645, 682, 684
Smith, E.M.B., 45, 75
Smith, H.L., 643, 652, 657, 676, 738
Smith, L, 643, 676
Snyder, R.D., 653-654, 738
Soland, R.M., 654, 736
Solomon, M., 647-649, 697, 711
Solomon, M.M., 648, 703
Sorenson, D.C., 265, 292
Sorenson, E.E., 648, 707

Sorger, G., 310, 424
Soumis, F., 648-649, 697-698, 711
Souza, C.de, 305, 421
Spaccamela, M.A., 649, 711
Spears, W.M., 128, 148
Speckenmeyer, E., 83, 142
Spellman, R., 647, 700
Spielberg, K., 641, 652, 663, 725, 738, 742
Spirakis, P., 140, 145
Spitzer, M., 644, 683-684
Stan, M., 133, 145
Stancill, J.M., 652, 738
Stanfell, L.E., 646, 686
Stary, M.A., 652, 716
Steiger, F., 645, 684
Steiger, F.C., 644, 677
Steiglitz, K., 84, 112, 114, 118, 147, 318, 426, 465, 477
Stein, D., 649, 711
Stern, E., 653, 655, 658, 720
Stern, H., 644, 648-651, 684, 695, 711
Stern, H.I., 644-645, 676, 683
Stevenson, K.A., 654, 729
Stewart, W.R., 648-649, 711-712
Stillman, M., 540, 571
Storbeck, J., 653, 716-717
Storbeck, J.E., 653-654, 714, 719, 738
Stouge, L., 648, 702
Stougie, L., 649, 711
Stricker, R., 648, 712
Stulman, A., 653, 731
Sturmfels, B., 536-537, 542, 545, 548-549, 553, 555, 558, 562, 566, 568-569, 570-572
Sudan, M., 116, 122, 140, 145
Suen, S., 140, 143

Suhl, U.H., 480, 529
Sullivan, W.J., 648, 707
Sum, J., 641, 662
Sumichrast, R.T., 648, 712
Sumners, D.L., 643, 669-670
Sussmas, J.E., 652, 716
Sutcliffe, C., 648, 712
Sutter, A., 527
Swain, R., 653, 658, 735, 740
Swain, R.W., 652, 739
Syslo, M.M., 318, 427
Szegedy, M., 116, 140
Szemerédi, E., 139, 142
Szpigel, V., 649, 712

Taillefer, S., 648-649, 704
Talbot, F.B., 645, 686
Tamir, A., 654, 721
Tan, C., 648, 712
Tanga, R., 645, 677
Tanimoto, S.L., 641, 661
Tansel, B.C., 652-654, 739
Tapia, R.A., 226, 235, 285
Tapiero, C.S., 652, 739
Tarasov, S.P., 186
Tardella, F., 149, 163, 169, 171, 173, 178, 184, 188
Tardos, E., 308, 427
Tardos, É., 178, 183-184, 472, 477, 550, 570
Tarjan, R., 464, 476
Tarjan, R.E., 217, 221, 295, 453, 464, 474-476
Taylor, P.E., 644, 676
Taylor, P.J., 654, 739
Tayur, S.R., 549, 568, 572
Teather, W., 644, 677
Tecchiolli, G., 134, 141
Teitz, M., 653-654, 739

Terjung, R.C., 646, 686
Terlaky, T., 208, 220-223, 270, 272, 293, 295-296
Tewari, V.K., 654, 657, 739
Tezuka, S., 474, 476
Thienel, S., 245, 288
Thisse, J.F., 175-176, 186, 652-654, 726
Thomas, R.R., 533, 549-550, 553, 555, 568-569, 570, 572
Thomas, W., 649, 707
Thuve, H., 646, 690
Tibrewala, R., 643, 676
Tibrewala, R.K., 643, 670
Tideman, M., 654, 739
Tien, J.M., 643, 677
Tillman, F., 648, 712
Tind, J., 182, 186
Tjalling, C., 652, 728
Todd, M.J., 247, 277, 292
Toledo, S., 472, 478
Topkis, D.M., 167, 177, 188
Toregas, C., 653, 658, 735, 740
Toth, P., 299, 309, 314, 318, 323, 325-329, 332-334, 336, 340, 342-343, 345, 347-348, 352, 355, 359, 361-362, 382, 384, 387, 391, 394, 396-400, 403-404, 406, 408-410, 415, 418-421, 424-425, 427, 647, 696
Tovey, C.A., 646, 687
Traverso, C., 542, 545-546, 570
Trevisan, L., 116, 148
Trick, M., 136, 145
Trick, M.A., 198, 288
Trienekens, H., 648, 703
Tripathy, A., 646, 690
Trotter, I.E., 646, 688
Trotter, L.E., 642, 667

Trouchon, M., 646, 689
Trubin, V.A., 641, 664
Trudear, P., 647, 698
Trzechiakowske, W., 654, 732
Tsuchiya, T., 225, 294
Tuncbilek, C.H., 480, 493, 517, 525, 531
Tuncel, T., 85, 142
Turcotte, M., 654, 656, 734
T:urkay, M., 37-39, 75
Turner, W., 648, 712
Tuy, H., 156, 162, 176, 182, 186, 188
Tyagi, M., 648, 713
Tykulsker, R.J., 645, 684

Ullman, J.D., 744
Ullmann, Z., 388, 426
Ulular, O., 525-526, 531
Unwin, W., 648, 713
Upfal, E., 140, 142
Urbaniak, R., 569-570, 572
Urbanke, R., 568, 571

Vaidya, P.M., 221, 223, 258, 283, 288, 296
Vaidyanathan, R., 42-44, 75
Valenta, J.R., 646, 690
Valenzuela, T., 654, 656, 724
Valiant, L.G., 454, 478
Valinsky, D., 653, 740
Van Leeuwen P., 648, 713
Van Loan, C.F., 221, 286
Van Roy, T.J., 480, 532, 652, 740
Van Ryzin G., 647, 694
Van Slyke, R., 646, 690
Van Wassenhove, L.N., 646, 652, 656, 685, 724, 732-733

Vandenberghe, L., 151, 179, 188, 277, 296
Vanderbei, R.J., 217, 277, 279, 287, 296
Vanderweide, J.H., 652, 730
Vannelli, A., 263, 272, 295
Vasko, F.J., 641, 646, 664-645, 686, 687
Vazirani, U., 122, 145
Vecchi, M.P., 132, 145
Veiga, G., 196, 212, 217, 221-222, 227, 225, 282, 293-294
Vemuganti, R.R., 573, 642, 646, 665-666, 687
Vergin, R.C., 652, 740
Vial, J.P., 258, 283, 286
Vial, J.Ph., 208, 295
Vijay, J., 652, 740
Vincent, O., 33, 72
Viola, P., 652, 717
Vishkin, U., 464, 478
Viswanathan, J., 9, 20-21, 51, 76
Vlach, J., 263, 272, 295
Vohra, R.K., 653, 722
Vohra, R.V., 643, 677
Vohra, V., 653-654, 738
Volgenant, A., 648, 713
Volz, R.A., 654, 657, 740
Vu'sković, K., 179, 183

Wagner, J.L., 653, 740
Wagner, R.A., 642, 666
Wah, B.W., 83, 137, 144
Walker, J., 372-373, 420
Walker, W., 643, 653, 657, 673, 740
Walker, W.E., 653, 656, 728
Wallacher, C., 464, 478
Walsh, T., 129, 143

Walukiewicz, S., 309, 318, 324, 373-374, 376, 420
Wang, C.C., 642, 667
Wang, J., 221, 225, 291
Wang, P., 212, 289
Wang, P.Y., 646, 690
Wang, Y., 85, 142
Ward, R.E., 645, 684
Warden, W., 647, 650, 692
Warner, D.M., 643, 677
Warners, J.P., 270, 272, 296
Wasil, E., 648, 651, 701
Waters, N.M., 654, 657, 742
Watson-Gandy, C.D.T., 648, 654, 713, 740-741
Watson-Gandy, G., 647, 698
Watts, C.A., 643, 674
Weaver, J.-R., 653-654, 741
Weaver, J.R., 652-654, 717
Webb, M.H.J., 648, 708, 713
Webster, L., 645, 686
Wee, T.S., 645, 685
Weinberger, D.B., 642, 667
Weismantel, R., 277-279, 287, 305, 324, 372, 407, 421, 427, 569-570, 572
Wells, R., 648, 706
Welsh, A., 645, 682
Wendell, R.E., 652, 654, 726-727, 741
Weslowsky, G.O., 654, 729-730
Wesolowsky, G.O., 654, 741
Westerlund, T., 34, 44, 51, 75-76
Weston, Jr., F.C., 654, 657, 741
Westwood, J.B., 646, 687
Wets, R., 258, 295
Whinston, A., 649, 691
White, A.L., 653, 734
White, G.M., 646, 690

White, J.A., 653-654, 722-723, 742
White, W., 649, 713
White, W.W., 645, 687
Whybark, W.E., 652, 728
Widmayer, P., 642, 668
Wierwille, W.W., 654, 722
Wiggins, A., 643, 676
Williams, B., 648, 713
Williams, F.B., 653, 656, 726
Williams, H.P., 480, 532
Williamson, D.P., 85, 144, 276-277, 286
Wilson, G.R., 641, 664-645
Wilson, N., 648-649, 651, 702
Wilson, N.H.M., 645, 681
Winter, P., 642, 667
Wirasinghe, S.C., 654, 657, 742
Witzgal, C., 366, 427
Witzgall, C.J., 653, 724
Woeginger, G.J., 309, 426
Wolf, F.E., 646, 686-687
Wolf, H.B., 643, 674
Wolfe, F.E., 641, 664
Wolfe, H., 644, 677
Wolfe, P., 652, 715, 744
Wolkowicz, H., 151, 179, 187, 276-277, 279-280,, 287, 293, 296-297, 526, 529
Wolsey, L., 305, 421, 569-570, 641, 661, 743
Wolsey, L.A., 179, 187, 245, 292, 305, 313, 318, 324, 426-428, 480, 507, 529, 532, 652, 718, 745
Wolters, J., 649, 713
Wong, C.K., 373, 394, 408, 419, 422, 642, 668
Wong, R., 647, 701
Wong, R.T., 481, 529, 642, 649, 667, 695
Wood, D.C., 642, 646, 668, 690
Wood, D.E., 32, 74
Woodbury, M., 646, 690
Wren, A., 645, 648-649, 651, 684, 711, 713
Wright, J.W., 647, 696
Wright, S., 208, 296
Wu, Y.F., 642, 668

Xu, J., 649, 699
Xu, X., 212, 282

Yannakakis, M., 85, 110, 112, 115, 148
Yano, C.A., 641, 660
Yao, A.C., 453, 478
Yao, E.Y., 641, 662
Ye, Y., 208, 212, 221, 258, 260, 280, 283, 286, 288, 296
Yee, T.F., 60, 76
Yeh, Quey-Jen, 221, 296
Yellow, P., 648, 713
Yesilkokcen, G., 654, 739
Yoshitsugu, Y., 175, 188
Young, D., 649, 713
Young, H.A., 654, 742
Young, J.P., 644, 677

Zamora, J.M., 70, 76
Zangwill, W.I., 170, 188
Zemel, E., 311-313, 321, 324, 338-340, 342, 368, 419, 428, 517-518, 525, 527, 653, 731
Zhang, X., 280, 283
Zhang, Y., 210, 226, 235, 285, 297
Zhao, Q., 210, 276-277, 279-280, 296-297
Zheng, H., 92, 94, 96-97, 142
Ziegler, G., 535, 569, 572

Zionts, S., 649, 709
Zipkin, P., 649, 699
Zoltners, A.A., 305, 342, 365-366, 373, 380, 427-428, 652, 737
Zwick, U., 116, 145

Subject Index

0-1 Knapsack Problem, 302, 306, 318-351
3/4-approximate algorithm for MAX W-SAT, 110
3/4-Approximate SAT algorithm, 108

active literal, 97
adjacency list, 214
adjacent arcs, 586
adjacent edges, 589
adjacent nodes, 586
ADP, 237
affine variety, 542
airline crew assignment problem, 193
ak-approximation algorithm, 106
algebraic theory of Gr:obner bases, 538
allowed, 132
almost strongly correlated instances, 316, 348
annealing schedule, 129
approximation algorithm, 86, 88
aptimization problem, 87
APX, 89
APX-complete, 91
arc list, 214
arcs, 586
assembly line balancing problem, 609-610
asymmetric traveling salesman problem, 520
augmented network, 232
average k-SAT model, 137
avis problems, 361

backtracks, 83
backward greedy solution, 330
balanced filling, 335
balanced insert, 335
balanced remove, 335
balancing, 314
balsub algorithm, 357-358
bal_zem algorithm, 321
base polyhedron, 167
basic moves, 119
basic sets, 37
basic variables, 37
BB, Branch and Bound, 3
beginning node, 586
Bellman recursion, 355
bi-partite planar graph, 5
Bidimensional Knapsack Problem, 308
bilinear programming problem, 176
Bin-packing Problem, 308
binary search method, 436
bipartite graph, 586
blood analysis model, 620-621
Boolean Quadric Polytope, 493
bouknap algorithm, 391
bound, 303
bound-and-bound algorithm, 410
bound-factor products of degree (or order) d, 485
bound-factors, 485
Bounded Knapsack Problem (BKP), 302, 306, 382-394
Bounded Multiple-choice Knapsack Problem, 309
bounded polyhedron, 534
bounding step, 30

bounding, 34
branch and bound methods, 233
branch and cut methods, 242-259
branch-and-bound algorithm, 30-34, 332-335, 373-374, 387-388
branch-and-bound method, 84
branch-and-bound search tree, 237
branch-and-bound tree, 30, 32
branch-and-bound, 303, 397-398
branch-and-reduce algorithm, 41
branching priorities, 32
branching step, 30
branching variable, 43, 47
branching, 34, 84, 239
breadth-first approach, 32
break item, 311, 321
break solution, 321
Buchberger's algorithm, 542

capacitated facility location problems, 234
capacitated transshipment problem, 568
capacity, 463
capital budgeting problem, 626
cardinality bounds, 325
cells, 535
cellular manufacturing problem, 617
central trajectory, 209
chain, 586
Change-making Problem, 307
check clearing, 625-626
Cholesky factors, 211-212
chromatic index, 592, 593
chromatic number, 593
circuit fiber, 551
circuit, 587
class reduction, 373

clause, 198
clausing weighting, 133
closed convex sets, 162
CNF, 80
coefficient matrix, 442
coefficient matrix, 534
coherent, 537
Collapsing Knapsack Problem, 309
combinatorial approach, 202
combinatorial extremum problems, 157
combinatorial optimization problem, 150, 192
Combinatorial Optimization, 303
compact convex sets, 162
completeness in an approximation class, 90
ComputeMaxima algorithm, 457
concave functions, 162
concave quadratic minimization, 180
conceptual interior point cutting plane algorithm, 248
conditional logic, 516
conditonal probabilities, 99
configurations, 133
conjugate gradient method, 221-225
conjunctive normal form, 78
constrained extremum problem, 152
constrained network scheduling problem, 615-617
constraint underestimators, 41
constraint-factors, 501
construction phase, 130
Conti-Traverso algorithm, 543-546
continuous approach, 203
continuous branching variable, 43
continuous embedding, 204
continuous mathematics, 85

Continuous Multiple-choice Knapsack Problem (MCKP), 365
continuous relaxation, 311, 328, 404
continuous trajectories, 206-207
continuously dominated, 366
convex hull extreme point, 161
convex hull representation, 482
convex hull, 161, 244
convex minimization problems, 207
Convex MINLPs, 8
convex underestimators, 41, 47
convex-concave problems, 176-178
core problem, 303, 312, 338-339, 361, 377
core, 339, 344, 361, 377
corrector step, 210
cost vector, 534
cost, 431
cover, 576
coverage, 80
Cramer's rule, 442
crew base, 604
crew scheduling, 603-608
critical item, 311
CTD test, 29
CTDU test, 29
cutting plane algorithm, 245
cutting planes, 242, 244
Cutting problem, 352
cutting stock problem, 613-615
cycle, 587
cyclical scheduling, 599-600

Dantzig bound, 320
days off scheduling, 599-600
dead clauses, 95
deductive reasoning, 129
degree, 558
depth-first approach, 32

derandomization, 99
descent direction, 265-269
diagnostic expert system, 621
diagonal preconditioner, 223
Dinkelbach method, 437
diophantine equation, 351
directed graph, 586
disconnecting set, 594
discrete branching variable, 43
discrete lot sizing and scheduling problem, 610-611
discrete optimization problem, 150, 193
discrete problems, 191
disjunctive linear program, 38
disjunctive programming, 38
distance, 175
distillation sequencing problem, 56-60
distrust-region method, 43
division algorithm, 542
dominance relations, 366, 398
dominance rule, 357
dominated items, 398
DP algorithm, 81
DPLL algorithm, 81
dual affine scaling algorithm, 217-218, 226
dual affine scaling method, 208
dual constraints, 230
dual feasible solution, 230
dual representation, 12
dye_zem algorithm, 369
dynamic programming algorithms, 354-359, 374-376
dynamic programming, 303, 409

e-approximable, 89
ECP algorithm, 34-36, 44

ECP, Extended Cutting Plane, 3
edge, 175, 586
edge-weight function, 463
elementary chain, 587
elementary path, 587
eliminate the unit clauses, 110
elimination term order, 545
elimination, 239
ellipsoid method, 208
ending node, 586
enumerative bounds, 372
equality constraint representations, 495
equivalent vectors, 558
escape, 122
essential, 447, 466
estimation of the nodes, 32
Euclidean distance, 634
evenodd problems, 361
existential fashion, 482
expanding-core algorithm, 343-345, 379
expected performance, 100
expected value, 100
expected weight, 100, 102
exponential-time algorithms, 203
Extended Cutting Plane (ECP), 34-36
extreme point, 161

FA, 3, 9, 36-37
FA, Feasibility Approach, 3
faces, 535
facets, 245
feasibility based range reduction tests, 41
feasible region, 195, 208
feasible solution, 88, 309, 534
fiber, 534

findcore algorithm, 340
finite set, 175
finite subset, 540
fixed charge problem, 627
fixed length clause model, 137
fixed-core algorithm, 342, 379
FIXED-TS algorithm, 134
flight leg, 604
flight segment, 604
flow capacity contraints, 214
flow conservation constraints, 214
flow limitations, 213
flow, 465
formulae equivalent, 112
forward greedy solution, 330
forward star representation, 214
fractional combinatorial optimization problem, 430
fractional optimization, 430
frequency plan, 618
frequency planning problem, 618-619
fully polynomial approximation, 346, 386
fully polynomial-time approximation scheme, 303

gambler's ruin chain, 127
gangster operator, 279
gas pipeline network, 37
GBD algorithm, 13-14
GBD master problem, 11-13
GBD primal problem, 10-11
GBD, Generalized Benders Decomposition, 3, 9-14
GCD algorithm, 27-29
GCD, Generalized Cross Decomposition, 3
Genapprox algorithm, 104

general cutting plane algorithm, 84
general fractional optimization, 430
general set covering (GSC), 578
general set packing (GSP), 578
general set partitioning (GSPT), 578
Generalized Assignment Problem, 309
generalized set covering problem, 84
generalized simplex method, 541
generalized upper bounding (GUB) constrained knapsack polytopes, 493
generalized upper bounding (GUB) constraints, 506-512
generic cost vector, 556, 558
generic, 534
GENRANDOM algorithm, 102
geometric inequality, 105
geometric interpretation, 195
geometric sequence, 441
global approximation, 203, 205
global minimum point, 161, 163-164
global minimum, 161, 208
global optimality conditions, 180
global optimum, 122
GMIN-αBB algorithm, 50-51
GMIN-αBB, General structure Mixed Integer Nonliear αBB, 3
GOA algorith, 23-26
GOA master problem, 23-25
GOA primal problem, 23
Gomory cutting planes, 245
good entries, 471
good models, 480
graded polynomial ideals, 558
gradients, 377

Grasp algorithm, 131
Graver arrangement, 567
Graver basis, 550, 553
greedy algorithm, 330, 367
Greedy Johnson 1 algorithm, 92
Greedy Johnson 2 algorithm, 95
greedy principle, 320
greedy, 91
Gröbner bases to integer programming, 534
Gröbner basis, 542
Gröbner core, 560
Gröbner fans, 558, 559
Gröbner fiber, 553
GSAT, 129-130
GSAT-WITH-WALK algorithm, 130
GTP test, 29
GUC algorithm, 140

H-RTS algorithm, 135-136
half-spaces, 195
heat exchanger network synthesis problem, 60-70
Hessian matrix, 86
Hilbert basis, 553
history, 128, 132
history-sensitive heuristics, 132-136
HSAT, 133
hs_branch algorithm, 333
hyperplane arrangement, 536, 567

ill-conditional problems, 313
ILP feasibility problem, 85
ILP, 83
iminimal test set, 547
incomplete QR decomposition, 223
indicator, 226
inductive inference, 85, 192-193, 198
infeasibility test, 42

infeasible primal problem, 10
infinitesimal version, 205
information retrieval and editing, 623-625
ingot size, 611
initial best known upper bound, 239
initial branch-and-bound search tree, 237
initial form, 546
initial ideal, 542
initial monomial, 542
initial term, 542
initial upper bound, 237
inner normal cone, 535
inner normal fan, 535
Integer Linear Programming, 302
integer programming problem, 242, 246, 270
Integer Programming Problem, 308
integer programming, 193
interior point algorithm, 85, 275
interior point cutting plane algorith, 255
interior point cutting plane methods, 246-249
interior point methods, 191, 247
interval analysis based algorithm, 42-44
inverse image, 205
inverse strongly correlated instances, 316, 348
Inverse-parametric Knapsack Problem, 309
item r dominates item s, 366

k-SAT model, 137
Karmarkar Method, 205

killed clause, 97
Kirchhoff's Law, 213
Knapsack Problem with GUB, 365
Knapsack Problems, 246, 301, 302

Lagrange Interpolation Polynomials (LIP), 523
Lagrangian relaxations, 322, 404, 406
lattice arithmetic, 534
Lawrence lifting, 553
Lawrence type, 567
length, 175
level 1 factors, 505
level 2 factors, 506
level d factors, 506
Lie group embedding, 204
lift-and-project cutting plane algorithm, 494
linear algorithm, 450
linear complementarity problem, 159
linear fractional combinatorial optimization problem, 433
linear integer programming (LIP), 576
linear mixed integer 0-1 programming problem, 484
linear ordering problem, 193, 247
linear parallel algorithm, 453
linear programming (LP) relaxation, 234
linear programming relaxation, 84
linear programming, 208-212
linear relaxations, 482
list size, 134
literal, 198
local minimum point, 181
local minimum, 208, 261
local optimum point, 120

Subject Index

local search phase, 130
local search, 118-127, 202
lock box location problem, 636
Logic-Based Generalized Benders Decomposition, 40
Logic-Based Outer Approximation algorithm, 39-40
lower bound test, 43
lower bound, 272
lower face, 538
lower semicontinuous, 169
LP relaxation, 84, 103, 242
LP, 33
LP-dominated, 366
LPnew, 251
LS-NOB, 135
LS-OB, oblivious local search, 122-123

m-concave, 165
m-quasi-concave, 165
Markov chain, 127
Markov process, 126-128
MarkovSearch randomized algorithm for 2-SAT, 127
Martello and Toth upper bound, 323, 384, 395
mass, 94
master problem formulation, 11
master problem, 9
matched, 80
matching, 590
mathematical modeling, 3, 5-6
mathematical problems, 628
MAX 2-SAT, 122
MAX W-k-SAT, 79
MAX W-SAT, 79
MAX-k-SAT, 79
MAX-SAT problem, 78

MAX-VAR BOUNDED SAT problem, 118
MAX-VAR SAT problem, 117-118
MaxCost algorithm, 435
MaxCost2 algorithm, 456
MaxCostRecursive algorithm 456-457
maxima-computing phase, 455
maximum cardinality constraint, 325
maximum cut problem, 249
maximum flow problem, 215, 462
maximum flow, 465
maximum matching problem, 590
maximum profit-to-time ratio cycle problem, 458
maximum profit-to-time ratio cycles, 458-460
maximum-mean cycle problem, 460
maximum-mean cycles, 460-462
maximum-ratio spanning tree, 453
maximum-ratio spanning-tree problem, 453
maximum-surplus cut, 463
MaxMeanCut problem, 464
MaxRatioCut problem, 462
MaxRatioCycle problem, 458
MaxRatioPath problem, 434
mean-weight cost, 431
mean-weight surplus, 463
Megiddo's parametric search method, 430, 448-457
memory-less, 128
minimal non-face, 540
minimal test set, 540
minimax problem, 170
minimum cardinality constraints, 325
minimum cost network flow problem, 213, 215

minimum covering problem, 592
minimum-cost flow problem, 464
minimum-cost spanning-tree problem, 453
minknap algorithm, 344
Minkowski integrals, 536
Minkowski sum of line segments, 567
Minkowski sum, 535
Minkowski summands, 535
MINLP, mixed-integer nonlinear programming, 3, 6
MINOPT, Mixed-integer Nonlinear OPTimization, 3-4, 51-56
mislead, 126
mixed discrete-continuous optimization, 6
mixed-integer 0-1 constraint region, 487
mixed-integer zero-one polynomial programming problems, 482
model-finding procedure, 129
monomial ideal, 546
monotonicity test, 42
most fractional variable rule, 32
mt2 algorithm, 342
mtm algorithm, 411
mts algorithm, 360
mtu2 algorithm, 400
mulknap algorithm, 412-413
multi-commodity minimum disconnecting problem, 594
Multi-constrained Knapsack Problem, 302, 308
multi-commodity network flow problem, 216
multi-linear mixed-integer zero-one polynomial programming problems, 494
multiple choice constraints, 507
multiple depots and extensions, 633
Multiple Knapsack Problem (MKP), 302, 307, 402-417
multiple runs, 128
Multiple-choice Knapsack Problem (MCKP), 302, 306, 364-381
Multiple-choice Nested Knapsack Problem, 309
Multiple-choice Subset-sum Problem, 309

negative clause, 97
negative literal, 96
neighbors of the origin, 550
Nested Knapsack Problem, 309
network flow problems, 212
network programming, 212-217
network, 175, 463, 586
neural network approaches, 86
Newton method, 430, 437-448
NLP solver, 33
NLP, 33
NOB & OB, two-phase local search algorithm, 124
node with the lowest lower bound, 32
node-arc incidence matrix, 214
node-node adjacency matrix, 214
nodes, 586
non-fractional version, 432
non-negative real line, 538
non-oblivious functions, 122
non-standard set of disjunctions, 482
nonbasic sets, 37
nonbasic variables, 36

nonconvex MINLPs, 40
nonconvex optimization problem, 259
nonconvex potential function minimization algorithm, 264
nonconvex potential function minimization, 259-272
nonconvexity test, 43
nondeterministic machine, 87
Nonlinear Knapsack Problem, 309
nonlinear program, 150
nonlinear pump configuration problem, 44
normal fan, 535
normal form, 542
normally equivalent, 535
NP-complete problem, 79
NPO problem, 87
NPO-complete, 91

OA algorithm, 16-17
OA master problem, 15-16
OA primal problem, 15
OA, Outer Approximation, 3, 14-18
OA/ER algorithm, 18-19
OA/ER/AP algorithm, 20-23
off-line, 132
on-line, 132
optimal face, 226
optimal process flowsheets, 5
optimal slope, 368
optimal solution, 195, 368
optimality based range reduction tests, 41
Original Conti-Traverso, algorithm, 545
original input matrix, 206
original parameters, 206

outer approximation methods, 14

P-center problem, 637
P-median problem, 638
packing problems, 304
packing, 576
parallel algorithm, 453
parameter GAP, 433
parameterized ellipsoid, 273
parametric problem, 431
partial enumeration, 323
partitioning, 577
path, 175, 587
PCP, probabilistic checkable proof, 116
PDPCM, 208
performance ratio, 88
persistency issues, 522
personnel scheduling problem, 598
piecewise concave, 168
piecewise convexity, 168
piecewise quasi-concave, 168
piecewise quasi-convexity, 168
pi_branch algorithm, 334
plant location problem, 634-636
point configuration, 537
pointed Gröbner fan, 560
pointed secondary fan, 564
political districting problem, 622-623
polyhedral complex, 535, 537
polyhedral cone, 534
polyhedral fan, 535, 537
polyhedral geometry, 534
polyhedral properties, 313
polyhedron, 165
polymatroid, 167
polynomial-time algorithms, 203

polynomial-time approximation scheme, 89, 303
Polynomial-time, 386
polytope, 534
positive literal, 96
potential function, 86
preconditioned conjugate gradient algorithm, 218-220
preconditioner, 221-225
predictor step, 210
predictor-corrector method, 212
primal feasible solution, 232
primal formulation, 10
primal problem, 9
primal-dual algorithm, 334
primal-dual predictor-corrector method, 208
primitive edge of the circuit fiber, 552
primitive edge of the Grövner fiber, 555
primitive non-zero vector, 551
primitive, 551
problem entry, 53
problem solution, 53-54
process flowsheets, 3
process superstructure, 5
process synthesis problem, 3-4
prohibited, 132
prohibition parameter, 134
projective transformation, 205
property, 108
pseudo-costs, 33
pseudo-polynomial time, 310
pseudo-polynomially solvable, 310
PTAS, 89
PTAS-reduction, 90
PURE LITERAL algorithm, 140
pure literal, 83

quadratic assignment problem (QAP), 236-242, 497
Quadratic Assignment Problem, 309
quadratic assignment, 37
quadratic integer programming problem, 277
Quadratic Knapsack Problem, 308
quadratic programming problem, 159
quasi-concave functions, 162
quicksort algorithm, 339

RANDOM algorithm, 100
random clause model, 137
randomization, 99
randomized 1/2-approximate algorithm for MAX W-SAT, 99-102
randomized algorithms for MAX W-SAT, 99-110
randomized greedy construction, 131
Randomized Rounding, 105
REACTIVE-TS, 134
reasoning, 129
rectilinear distance, 634
reduced Gröbner basis, 542
reduction algorithms, 398-402, 407-408
reduction procedure, 303
reduction, 90, 331, 386
refinement of two fans, 536
Reformulation Linearization Technique (RLT), 479, 484-485
reformulation, 45
regular triangulation, 537
regular, 537
relative boundary, 163
relative interior, 163
relaxation-penalization, 152

Subject Index

replacement technique, 355
representation of process alternatives, 5
residual capacity, 465
residual demand, 465
residual network, 113
restricted network, 232
reverse star representation, 214
right hand side vector, 442, 534
rounding, 85
row generation scheme, 595

S-factors, 500
SAMD algorithm, 132
Satisfiability (SAT) Problem, 198
satisfiable, 80, 198
satisfied clauses, 97
saturated lattice, 534
scheduling personnel, 598
search trajectory, 119, 133
secondary cone, 563
secondary fan, 563
secondary polytopes, 537, 563
selection, 83, 239
semidefined programming, 116
semidefinite programming relaxations, 276-280
semidefinite programs, 179
separable, 313
separating resolvents, 85
separation problem, 246
separation routine, 246, 253
service facility location problem, 639-640
set covering (SC) problem, 575, 577
set covering model, 577
set packing (SP) problem, 575, 577
set packing model, 577

set partitioning (SPT) problem, 575, 577
set partitioning model, 577
set partitioning polytope, 493
shift scheduling, 601
Simplex Method, 196, 208
simulated annealing algorithm for SAT, 128
sing spin glass problem, 248
single depot vehicle routing, 631-632
sink nodes, 213
sink, 463
slacks, 86
smallest clause, 83
SMIN-αBB, 46-50
SMIN-αBB, Special structure Mixed Integer Nonliear αBB, 3
source nodes, 213
source, 463
spanning tree preconditioner, 221
spare parts allocation problem, 613
sparse constraints, 515
spatial branch-and-bound algorithm, 45
Special Structures Reformulation Linearization Technique (SS-RLT), 502
splitting rule, 81
stability number of a graph, 588
stage, 451
standard assignment problem, 591
standard monomials, 542
state polytopes, 537, 558, 559, 560
states, 310
Steiner points, 596
Steiner problem in graphs, 596
stochastic search trajectory, 128
straightforward merging process, 455

strictly concave function, 162
strictly concave, 156
strictly piecewise concave, 168
strictly quasi-concave function, 162
strongest possible cutting planes, 245
strongly correlated data instances, 379
strongly correlated instances, 316, 348
structured problem models, 137
structures, 430
subdivision, 537
subedge, 175
sublattice, 166
submodular, 166
subset of continuous space, 207
subset-sum data instances, 380
subset-sum instances, 316
Subset-sum Problem (SSP), 351, 307
superstructure, 5
support, 535
surplus, 463
surrogate relaxation, 328
surrogate relaxed problem, 404
symmetric flow, 112
symmetric network, 112

temperature, 128-129
term order, 542
termination test, 239
test sets, 537, 547
testing and diagnosis problem, 620-621
threshold effects, 138-140
threshold value, 138
tighter bound, 323
tighter reduction, 332

timetable scheduling problem, 619-620
todd problems, 361
topology, 207
toric ideal, 542
toric variety, 542
tour scheduling, 602-603
trace minimization problem, 276
tramp-steamer problem, 458
transshipment nodes, 213
traveling salesman problem, 629-631
tree, 587
triangulation, 537
trust region approach, 86, 260
TS, Tabu Search, 132
two-constraint knapsack problem, 326
two-phase local search algorithm, 124

Unbounded Knapsack Problem (UKP), 306, 394-402
unbounded, 162
uncapacitated minimum cost network flow problem, 216
uncapacitated network flow problem, 214
unconstrained, quadratic pseudo-Boolean programming problem, 482
uncorrelated data instances, 315, 348, 379
undirected graph, 586
uniform fractional combinatorial optimization problem, 433
unimodular, 566
UNISAT models, 86
UNIT CLAUSE algorithm, 140

unit clause rule, 83
unit flow cost, 213
universal Gröbner basis, 553
universal test sets, 537, 550
unrestricted location, 633
unsatisfiable, 198
upper bound, 84
upper-bound test, 42

Valiant's parallel algorithm, 454
Value-independent Knapsack Problem, 351
variable upper bounding (VUB) constraints, 506, 512-515
varialbe bounds, 48-49
vertex packing problem, 510, 588
vertex packing, 588
vertices, 175, 586

Weak Minimum Principle (WMP), 163
weaker bounds, 376
weakly correlated data instances, 379
weakly correlated instances, 315, 348
weight, 78, 94, 431, 463
weighted MAX-SAT problem, 78
weighted median problem, 321
weighted Vertex Packing Problem, 589
WeightedMaxCost algorithm, 449
wire routing problem, 200
wounds, 95

zig-zag instances, 380
zonotope, 536, 567